Applications of Photonic Technology

Applications of Photonic Technology

Edited by

George A. Lampropoulos
A.U.G. Signals Ltd.
Toronto, Ontario, Canada

Jacek Chrostowski
National Research Council of Canada
Ottawa, Ontario, Canada

and

Raymond M. Measures
University of Toronto
Institute for Aerospace Studies
Downsview, Ontario, Canada

Springer Science+Business Media, LLC

Library of Congress Cataloging-in-Publication Data

International Conference on Applications of Photonic Technology, Sensing, Signal Processing, and Communications (1994 : Toronto, Ont.)
Applications of photonic technology / edited by George A. Lampropoulos, Jacek Chrostowski, and Raymond M. Measures.
p. cm.
Includes bibliographical references and index.

1. Photonics--Congresses. I. Lampropoulos, George A. II. Chrostowski, Jacek. III. Measures, Raymond M. IV. Title.
TA1520.I56 1995
621.36--dc20 95-32967
CIP

Based on proceedings of an International Conference on Applications of Photonic Technology, Sensing, Signal Processing, and Communications, held June 21–23, 1994, in Toronto, Ontario, Canada

ISBN 978-1-4757-9249-2 ISBN 978-1-4757-9247-8 (eBook)
DOI 10.1007/978-1-4757-9247-8

Originally published by Plenum Press, New York in 1995.
Softcover reprint of the hardcover 1st edition 1995

10 9 8 7 6 5 4 3 2 1

PREFACE

In this book we present a snapshot of the state of the art in photonics in 1994, showing typical applications and emerging new ones; discussing the key technologies behind these applications, their limitations, and prospects. The articles in this book are extended versions of the papers presented at the first International Conference on Applications of Photonic Technology (ICAPT'94), held in Toronto, Canada, on June 21-23, 1994.

Photonics has been recognized as one of the key technologies for the 21st century, as electronics was the technology of the 20th centrury and electrical engineering changed the life of people in the 19th century. According to the recent report of the Organization for Economic Cooperation and Development in Paris (OECD), the market for photonics will grow dramatically in the next 10 years with an expected world-wide expenditure of US $230 billion from some US $30 billion in 1992.

The explosion of information technology was the largest driving force for the deployment of photonic technology. It created insatiable demand for ever-higher data transmission and processing rates, which cannot be sustained by electronics alone. Boosted by the enormous investment of the telecommunications and defense industries, the demand for photonics (or optoelectronics) is steadily increasing. It is solidly established in the long-haul communications, laser printers and CD-ROMs.

Today lasers are used to treat cancer, map and process DNA, explore coastal waters and detect pollution, and directly fabricate complex 3D objects. Fiber optics lies at the heart of the information superhighway, including ultrafast networks discussed in the book, such as the Bellcore Research Network and the Ottawa-Carleton Reseach Infrastructure, OCRInet.

To list just a few other applications, photonics offers exciting prospects for medical diagnosis and treatment, and the new field of smart structures, which encompasses marine, space and aircraft structures, in addition to buildings and bridges, such as the forerunner of a "smart bridge," built in Calgary in 1993. Photonics is being widely used in satellite surveillance for military and civilian applications, such as national security and prevention of drug trafficking. It is also being applied in radar communications and control, satellite communications, and sensing in general.

Part I covers telecommunications. Part II discusses optical routing, computing and processing. Part III is dedicated to optical measurements from space to industrial applications and electro-optical signal processing. Part IV reveals the broad potential for fiber optic sensing ranging from medical to structural monitoring. Part V shows various photonic components: lasers, detectors, integrated optics - including current research and development, which will determine photonic applications in the next century.

G.A. Lampropoulos
J. Chrostowski
R.M. Measures

CONTENTS

Part I: OPTICAL TELECOMMUNICATIONS

Part IV: FIBER OPTIC SENSING

FIBER TO THE DIGITALLY PASSIVE COAX: THE MULTIMEDIA DELIVERY NETWORK OF THE FUTURE?

Jack B. Terry

BNR
Ottawa, Canada

ABSTRACT

An alternative Cable TV architectural approach to that of the presently planned Fiber to the Feeder (FTF) architecture is described for switched video-on-demand applications. This approach, which builds on digital technologies used for non-switched digital services, also enables an economic, smooth and scalable introduction of broadband multimedia services into evolving Cable TV distribution networks. The approach also addresses the present and future issues of interference ingress, privacy and security, power consumption, upstream protocols, testing and maintenance. Both modernization and new-build configurations using this approach are included.

BACKGROUND AND MOTIVATION

Fiber optics transmission pays for itself very quickly when used in existing Amplitude Modulated Vestigial Sideband (AM-VSB) Cable TV distribution serving around 1000 to 2000 homes and provides a platform for future digital services.

Today's introduction of 6 MHz channelized 64-QAM (Quadrature Amplitude Modulation) digital transmission for non-switched near-video-on-demand (NVOD), services will pave the way for future switched television and two-way interactive multimedia services. However the implementation of such systems is strictly paced by the need to control incremental cost as digital penetration and service take expands.

Video compression and efficient digital modulation techniques are about to be used to provide a digital services architecture in fiber/coax distribution. Today this involves digital use of existing, vacant, channels and bandwidth extension using existing NTSC 6 MHz channelization. It is unlikely that two or more programs simultaneously selected by the user (picture-in-picture or VCR recording) will be available within the same 6 MHz channel. Thus for this level of service the set-top converter would require separate tuners, echo cancelers, demodulators, de-interleavers and forward error correcting functions for each program being received. An evolution in transmission strategy which provides greater flexibility at lower cost is a significant challenge. In addition, just as in the past for analog TV services, any new strategy must support and interwork with digital services and structures that already exist.

Outside plant transmission performance for digital signals is relatively good. However, in-home coax, splitters and TV sets create excessive and rapidly changing microreflection echoes which make spectrally efficient digital transmission very difficult. Direct end-to-end transmission

Applications of Photonic Technology
Edited by G.A. Lampropoulos *et al.*, Plenum Press, New York, 1995

through the outside plant and in-home wiring cannot fully exploit their separate potential capabilities. As more services become switched, cost and power consumption of analog fiber systems and Headend digital modulators become more dominant.

The above factors provide motivation to research alternative approaches to digital transmission architecture for switched Cable TV distribution. At the same time this provides an opportunity to re-examine other present and future issues. Issues of upgrade cost, interference ingress, privacy and security, system power consumption, upstream protocols, testing and maintenance continue to create significant challenges for the Cable TV network designer.

DIGITAL TRANSMISSION OF BROADCAST SERVICES

An analog fiber feeding a neighborhood of around 1000 to 2000 homes results in excellent video quality. For both broadcast and near-video-on-demand (NVOD) services there is less economic incentive to bring analog fiber closer to the subscriber. Obtaining more bandwidth on the coax distribution will likely be of greater priority than bringing fiber deeper into the neighborhood.

Today's approach to Cable TV control and billing include use of smart set-top converters and telephone lines. Upstream control via the fiber/coax distribution system for pay-TV or NVOD services is an option. However reverse transmission capability in the coax plant for upstream control is often unsatisfactory due to upstream interfering signal ingress from the drops or home wiring. Such ingress is often intermittent and difficult to locate and isolate.

Asynchronous Transfer Mode (ATM) or similar label addressing formats are currently being considered by the Cable TV industry as a means of providing service bit rate flexibility. However today's standard ATM format (5 byte header plus 48 byte payload) results in an overhead of approximately 10% To counter this, the advantages of uniformity for multimedia and flexibility for video applications are significant. While conservation of spectrum in coax distribution is an issue to be addressed, obtaining 10% greater bit-rate in fiber in this rapidly evolving technology is less of a problem.

DIGITAL TRANSMISSION OF SWITCHED SERVICES

Deployment of QAM modulators and Analog Fiber Feeders

For switched digital services, such as VOD with VCR-like user control, separate QAM modulators are required for much smaller groups of subscribers. A Headend serving 10,000 VOD subscribers at an average service level of one switched 3 Mb/s digital video program, or equivalent, per household would require 1000 QAM modulators, each handling 30 Mb/s. Many analog fiber feeder systems would also be needed to reach closer to the subscribers in order to achieve the necessary coax spectrum re-use. This spectrum re-use results from further sub-division of traditional tree-and-branch Cable distribution networks.

Migration of QAM modulators

Migration of switched services 6 MHz QAM modulators to neighborhood Digital Fiber Terminating units (DFT) would enable digital fiber system components to be used. Electo-optic components for transmission of binary digital signals are today many times lower in cost, and consume somewhat less power, than those needed for multi-spectrum analog applications . Analog fiber system cost decrease has flattened off while the bit-rate of digital fiber systems continue to evolve rapidly without corresponding increases in per-system cost.

To provide a digital service mix of 50% NVOD and 50% VOD, occupying a total of approximately 200 MHz of spectrum above the AM-VSB channels, requires sixteen 30 Mb/s (6 MHz band) QAM modulators at the Headend for NVOD plus sixteen 30 Mb/s QAM modulators for each separately served VOD neighborhood. Using 3 Mb/s video coding this equates to 160 channels of NVOD plus 160 channels of VOD. Use of an ATM format would allow complete flexibility between video quality and bit-rate.

Using modern low power technology, a DFT containing sixteen QAM modulators, plus digital fiber receiver and coax drive amplifier, could fit within a standard Line Extender housing. Such an arrangement would offer 480 Mb/s of switched VOD services. A configuration which can offer this service level to each neighborhood of 150 homes is shown in Figure 1.

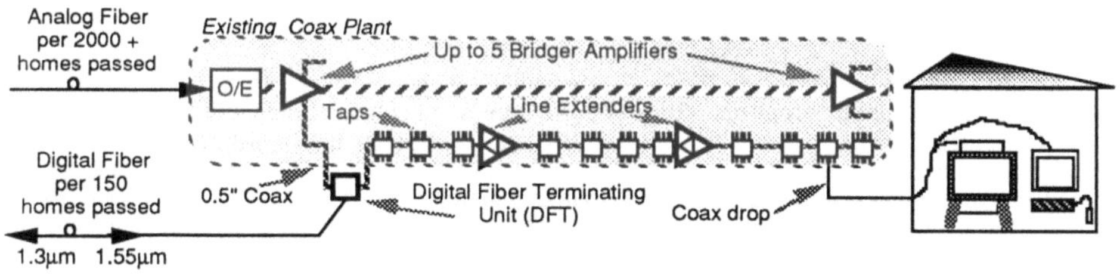

Figure 1. Cable TV distribution using a combination of Analog and Digital Fiber

Early introduction of services could be achieved very economically by initially introducing the VOD spectrum at the input of each Bridger Amplifier thus offering 480 Mb/s of service to perhaps 450 homes.

Scalable Switched Services Capacity

To avoid the need for Line Extender bandwidth upgrades, installation of reverse amplifiers and to provide for future increase in digital capacity, digital fiber could be extended closer to the subscriber. Such a configuration, shown in Figure 2. below, could be called "Digital Fiber to the Digitally Passive Coax"

In this configuration each DFT containing QAM modulators inserts digital signals on individual coax spans, just after each line extender. Passive filters are required at the input of Line Extenders to provide a 75 ohm Digital Termination (DT) for the higher frequency switched VOD services while passing the lower frequencies to the next span in the cascade.

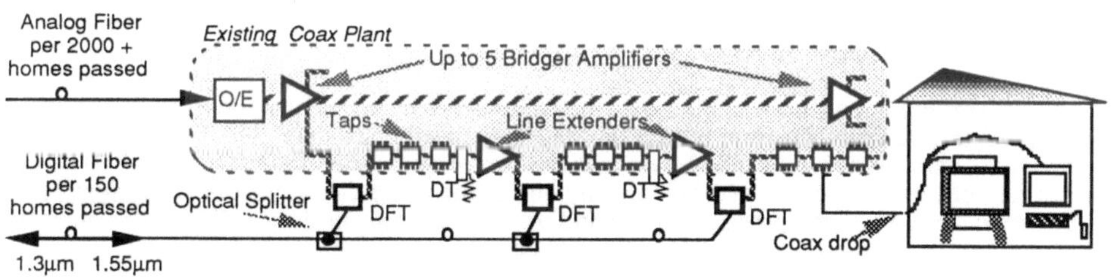

Figure 2. A "Digital Fiber to the Digitally Passive Coax" configuration

In time, as VOD market and digital fiber bit-rate economics evolve, a doubling of bit-rate on the digital fiber, together with a commensurate increase in QAM modulators, is possible. This configuration could potentially offer 500 Mb/s of NVOD via the analog fiber plus at least 1 Gb/s of switched services to each group of 150 homes passed, via the digital fiber.

Each Digital Fiber Terminating (DFT) unit has access to all digital channels on its fiber feeder. QAM modulators could be provisioned, perhaps in multiples of four, according to service take needs, in each DFT, thus keeping cost low during early introduction and service expansion.

At the Headend additional savings may be achieved by connecting a single digital services output port to several digital fibers, either electrically or by using passive optical splitters. If insufficient fiber strands are available, the optical splitting function could be located in the service neighborhood.

The above methods of achieving service scalability are critical to the economic introduction of switched digital video services.

Gaining Additional Spectrum Without Outside Plant Upgrade

In the arrangement shown in Figure 2., the switched channel spectrum is inserted after each Bridger or Line Extender amplifier. Taps are often pre-engineered for higher frequencies than the Line Extenders Thus, without the need to further re-engineer the distribution network, it is possible to move the switched channels higher in the spectrum. This offers the potential to free-up spectrum for more NVOD services provided via the analog fiber. A somewhat higher coax digital driver transmit level would be required for the switched VOD services to accommodate increased coax high frequency loss .

In a system which has not already been upgraded for NVOD, the NVOD services could, more economically, be carried on the digital fiber, modulated, and inserted after each existing amplifier.

Upstream Capabilities

Adopting the "Digital-Fiber-to-the-Digitally-Passive-Coax" configuration brings with it another advantage. The passive coax can easily provide a higher bandwidth, above-band ("top split") upstream path for the VOD command signals and potential data, voice or multimedia revenue generating services.

The choice of modulation technique for upstream in coax is not critical. However, since upstream signal source relative timing and phase are not easy to align, the digital receiver must be capable of fast synchronization at the start of each transmission burst.

A significant proportion of spare fiber strands exist in many installations, thus the upstream path to the Headend could be carried on a separate fiber. Alternatively, and more economically where there are no spare fiber strands, wavelength division multiplexing (WDM) may be adopted. The cost of 1.3 μm / 1.55 μm diplexers, when located inside existing DFT housings and in the arrangement shown in Figure 2, is today around $6 per home passed, a small expense for such a capability.

Upstream Interference Ingress

In the configuration shown in Figure 2., the DFTs also receive and pass on to the headend the upstream signals and traffic from homes on their associated coax span The issue of upstream ingress, although eased significantly by this subdivision of the distribution network into 50 homes passed sections, is still not fully resolved. Any one of the many digital set-top converters could still potentially disable the upstream control and thus service capabilities of all 50 subscribers on their passive coax span -- not a small service complaint issue in a VOD world !

AN ALTERNATIVE SWITCHED DIGITAL SERVICES ARCHITECTURE

As services become more personal and multimedia, degrees of service reliability, security, privacy, etc., become more sensitive, perhaps even safety, issues. Visual medical help is an example of such an issue.

The wired telephone network today offers individual line service using a star distribution for much of the access network and at the subscriber drop. Could a Cable MSO achieve an equivalent to star configuration economically and incrementally ?

QAM Modulators at the Tap

Why not provide QAM modulators in the coax Tap, one per coax drop, thus offering a star configured service ? Such an arrangement is illustrated in Figure 3.

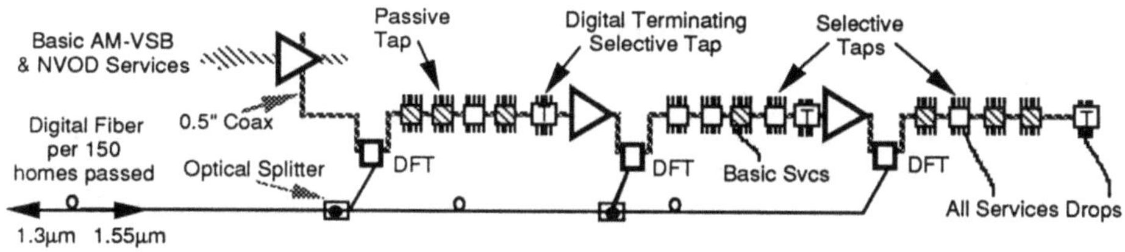

Figure 3. Digital Distribution System Employing QAM Modulators in the Tap

Multiple QAM modulators could be achieved using a pair of silicon chips. The cost and power requirements of such modulators is potentially less than their Headend or coax span driving counterparts. All modulators would operate at the same frequency, thus allowing greater sharing of functions. Filter requirements are also somewhat less As switched services become more prevalent the ratio of QAM modulators to coax drops will ultimately approach unity, thus volume becomes an additional cost lowering factor.

Use of a single 6 MHz band would achieve 30 Mb/s of services to each home. Overall system power consumption could potentially be less than that required for a conventional FTF configuration offering VOD services. The potential service capabilities could be much greater.

Impacts on the Headend and Services

In a system providing a QAM modulator per coax drop, it becomes economic to also transmit all digital NVOD channels on the digital fibers and to select these at the Tap. This also eliminates the need for NVOD QAM modulators at the Headend.

The probability of many of the NVOD channels being required simultaneously by a small group of homes becomes very small. Thus if selection means is provided at the Headend, a much wider variety of NVOD programs could be made available.

In such a Digital-Fiber-to-the-Digitally-Passive-Coax configuration all digital services would thus be delivered as switched services. The difference between NVOD and VOD is in the required form of service storage and the method of combining these services at the digital fiber laser interface. NVOD could be provided by a number of multichannel digital video tape players while true VOD would be provided from local and / or remote random access servers. An example of such a configuration is shown in Figure 4.

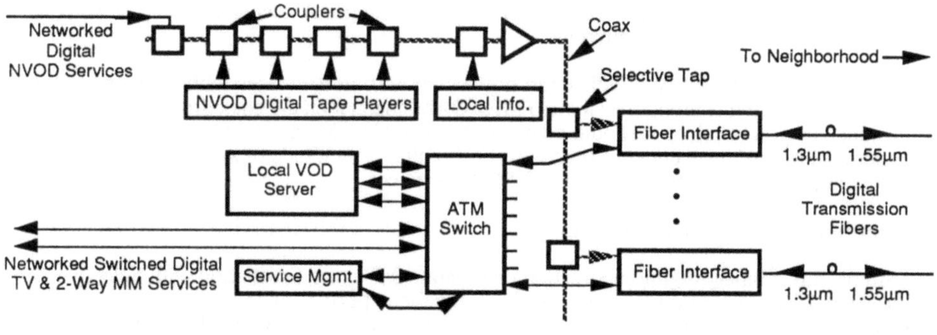

Figure 4. Headend Providing Combination of Digital NVOD and VOD Services

Impact on Digital Set-top Converters

Using this "all services switched" approach, a single QAM receiver function within a digital set top converter may now provide, simultaneously, multiple selected channels. For example, this capability could be used with picture-in-picture TV sets and while recording another program on the VCR. Although a significant cost saving, multiple channel viewing still requires a separate video decoder per channel being viewed. However, future VCRs may be capable of directly recording the compressed digital signals, thus avoiding their need for video decoders.

If conditional access AM-VSB services are also made available on digital channels, lower cost, digital only, set-top converters could be used in conjunction with "Cable ready" TV sets. Users not yet subscribing to the new digital services would continue to use their existing, analog-only, set-top converters. Incentives for "analog" subscribers to become "digital" subscribers could, in addition to VOD options, include the provision of very superior video quality NVOD for existing analog Pay-TV programs.

Local upstream control

Some commands from a remote control to a digital set-top converter are created with the expectation that the required service is already being transmitted on the drop. In such cases the set-top converter would not generate an upstream request signal to the network. Since the Tap has to select the required channel from the distribution coax, it must receive a command signal to do so. Thus, unless the set-top converter can be re-programmed to achieve this using a common standard, some form of external interface function is needed.

An external upstream interface function could take the form of a miniature infra-red (IR) receiver plus very low power FSK modulator. Such a function would cost very little using existing consumer electronics technology. This added local upstream IR interface device could be connected to any in-home coax cable using a low through-loss "T" adapter.

If multiple viewing stations exist in a home, individual upstream control interfaces would need some form of local identification. Simple DIP-switch coding (set by the vendor or consumer) could be used to achieve this. Unlike the situation with cordless phones and garage door openers, there is no requirement to coordinate identification with other nearby consumers. Dry battery powering could be used since all power consuming circuit functions could remain inactive except when IR command signals are received. If the local upstream control function is later integrated into the Set-top Converter or digital-ready TV set, this electronic identification could be its electronic serial number.

Transmission Advantages

Another advantage of locating the QAM modulator in the Tap is that higher signal levels on the drop may be applied, thus making the in-home wiring much more immune to interfering signals or electrical noise. Thus the QAM receivers would be somewhat more tolerant of in-home coax echo effects, a problem which has been underestimated until recently. Re-convergence of QAM receiver echo-canceling functions, which occur when a TV set changes AM channel, may be somewhat faster in the absence of other impairments.

The level and variability of echoes in home coax wiring decreases with frequency, due to cable losses. Thus transmitting the single 6 MHz QAM channel on the coax drop at the higher end of the band is a good choice. It should be noted that a single 6 MHz band QAM modulated signal of higher than normal level contains much less energy than a multitude of lower level QAM bands, thus reducing potential degradation of AM-VSB TV services.

Modulation on Coax Distribution

Simple direct binary digital modulation on separate 0.5" coax distribution with Selective Taps and a good 75 ohm termination can produce excellent performance well beyond 1.5 Gb/s. Transmitter and receiver functions are very simple, low cost and low power. However duplication of the distribution coax and large connectors at the Taps, particularly during the initial period when VOD service take is low, is considered too costly.

A single digital VSB signal could be introduced in each digitally passive span of 0.5" distribution coax already carrying an AM-VSB lineup. A carrier frequency of 1 GHz with its main sideband extending downwards in frequency is a potentially good choice. Using simple binary modulation this could offer over 450 Mb/s of digital service capacity within 250 MHz of lower sideband bandwidth. Pseudo-random scrambling of the digital stream would be necessary to ensure spectral compatibility with the AM-VSB services.

QPSK (or 4-QAM) has been tested extensively with success for inter compatibility with AM-VSB signals on shared coax distribution systems and is thus an alternative to digital VSB. The trade-off here is slightly higher spectral efficiency of QAM vs. the relative cost and power in technology needed to achieve it.

To allow for frequency loss tilt on the coax, some pre-emphasis at higher frequencies is needed. Selective Taps could contain an equalizing function, controlled by signal level, to provide frequency equalization when they are located close to the digital transmitter. Similar functions exist in today's Cable TV coax amplifiers.

Transmission Capacity Scaling

Echo canceling functions in very broadband digital transmission systems are difficult to realize. However the performance of existing 0.5" coax distribution is consistent, with worst case cumulative echoes of less than 5%.[1] Thus 4-level digital modulation (4-VSB or 16-QAM) could be used without the need for echo canceling functions. One gigabit transmission would require only about 300 MHz of bandwidth. Discrete technology for such high speed systems is available today. However for easier receiver integration and lower power, use of two 150 MHz or three 100 MHz transmission bands may be preferable. Sub-equipping of receivers in the Selective Taps enables a further cost saving option

In a Cable plant modernization using Selective Taps only (providing complete isolation of echoes and interference ingress from the drop and home coax wiring), it is possible to use 8-level modulation such as 8-VSB or 64-QAM. Eight-level modulation could offer an additional 50% increase in digital capacity on the distribution coax.

More Digital Capacity on the Drop

Use of, for example, 12 MHz QAM modulation (in place of 6 MHz) on the drop could provide perhaps 50 to 60 Mb/s of service to each home and would not impact, significantly, the cost or power consumption of the Selective Tap. However the complexity of the QAM receiver echo canceller function in digital Set-top Converters would be somewhat greater.

For data or multimedia services there exists the option of using a simple, more robust, modulation on the coax drop and home wiring, thus avoiding issues of high level microreflection echoes in home wiring. Using the Selective Tap approach, bandwidth on the drop is abundant.

Conditional Access, Privacy and Security

In configurations using a mix of Passive and Selective Taps it is necessary to encrypt the digital services to prevent service theft or eavesdropping.

Using Selective Taps for all coax drops would ensure that only the selected digital signals destined for a particular drop exist on that drop. This arrangement may permit all digital services to be transmitted "in the clear", thus simplifying both the Headend and future Digital Set-top Converters. In addition this arrangement offers the advantage of being Set-Top Converter "pirate-proof".

A low-pass filter must be placed in the path feeding the AM-VSB band signal onto the drop. This is needed to remove the digital distribution spectrum, thus providing clear spectrum for digital transmission on the drop and preventing eavesdropping of digital signals. This low-cost passive function can be shared across all outputs of the Selective Tap and requires one or two high attenuation notches in the digital bands.

Power Conservation

The cost of a Watt consumed for a year is between $1 and $2. Thus all efforts must be made to minimize power consumption across an entire system. This includes the Headend and all functions right up to the subscriber drop.

The architecture proposed in this paper emphasizes low power technology, modular provisioning, and sharing of functions as much as possible. In addition there is the possibility of powering down any function not currently in use, for example the QAM modulators and upstream circuits in the Selective Tap Other opportunities for reduced power consumption must continually be sought.

The power needs of Selective Tap common and coax drop circuits, which may use mostly low power integrated circuits, can be made very moderate. Thus power for these circuits can be provided via the 0.5" distribution coax. It is a pre-requisite of the "QAM modulator in the Selective Tap" approach that overall system power is less than in a FTF system providing equivalent NVOD and VOD services.

NEW BUILD CABLE TV DISTRIBUTION

Using Less Fiber

In the above method of upgrading or introducing switched digital services into existing Cable TV distribution structures it was necessary to create a form of "Superdistribution" by lashing the digital fibers to the distribution coax support strands up to the last passively accessible coax span.

In a new build situation it is possible to achieve the same subscriber coverage per digital fiber feeder by running the fiber to a Fiber Terminating Unit located at the center of a span of at least 150 homes passed. Such an arrangement is illustrated in Figure 5.

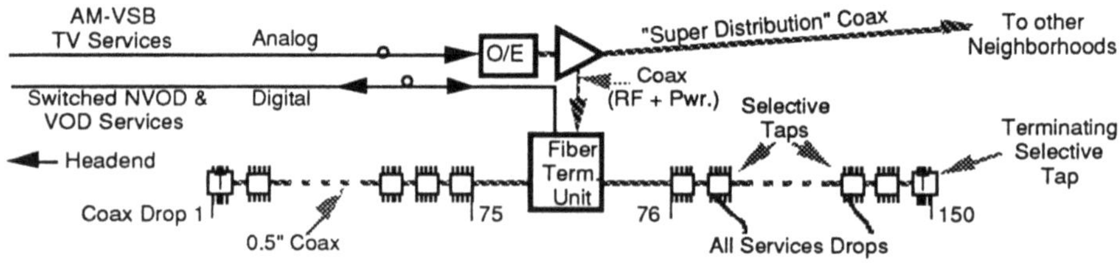

Figure 5. A Fiber to the Passive Coax new-build configuration

Longer Digitally Passive Coax Distribution Spans

A number of factors make it possible to achieve longer passive coax distribution spans than in conventional Cable TV network engineering :

- since the drop is buffered, it no longer contributes to the end-to-end loss.
- today's mostly air (or gas bubble) spaced coax has lower loss than the traditional 87% velocity factor cable used in recent years.
- higher digital signal transmit levels are permissible since they need not share passage through any amplifiers.

- lower noise amplifier technology has potential to provide a few more dBs of margin.

If NVOD services are distributed on the Digital fiber and offered in the same way as VOD, then the analog facilities spectrum may be constrained to 450 MHz. Using amplifiers specified for 110 channel capacity in this 60 channel application would permit higher AM carrier levels per channel. This offers the opportunity to add further to transmission loss margin. It is possible that 10% to 15% more homes could be passed using somewhat longer passive coax distribution spans, with commensurate savings in fiber feeders and Headend transmission and switching equipment.

HIGHLIGHTS

Using the "Fiber to the Digitally Passive Coax" and "Selective Tap" architectural approach described in this paper :

- there is no need to bring analog fiber closer to the subscriber for NVOD or staggercast services, a fiber per 1000 to 2000 homes provides adequate performance.
- scalable transmission capacity can be achieved, using provisionable and bandwidth-modular elements, at low cost and with low power in digitally passive coax distribution.
- there is no need to upgrade bandwidth or add reverse amplifiers in existing Bridger or Line Extender amplifiers.
- evolved systems which use only Selective Taps do not require encryption and have the advantage of being very secure, or "pirate proof", since only the authorized signals exist on the coax drops.
- complete isolation of coax drops and home wiring eliminates effects of interfering signal ingress.
- only Selective Taps can transmit upstream digital signals, thus it is relatively easy to achieve a simple protocol in the local distribution. This eliminates the effect of Set-top Converter babbling
- the numbers of upstream transmitters are fewer. Since they exist only within Selective Taps they are under the specification and control of the Cable Network operator. Thus the network operator remains in control, regardless of the performance of set top converters and, potentially, digital TV sets.
- conservation of spectrum on the coax drop and home wiring provides future opportunities for broadband multimedia telecommuting, telepresence and other futuristic services. Possibilities for home LAN, etc., exist -- a topic well beyond the scope of this present paper.
- from an operations and maintenance perspective the structure of this network evolution provides for margin testing and isolation of sections of the distribution network. This offers potentially lower cost maintenance by anticipation and fixing of potential troubles before they become apparent to the subscriber.
- existing designs of digital set-top converters will accommodate the proposed architectural evolution. However greater functionality at lower cost is possible in new digital-only converters. Analog set-top converter users are unaffected.
- simultaneous reception of multiple digital channels using a single QAM receiver function in a set-top converter can be achieved, thus reducing cost of multichannel reception capability. Scalable service bit-rates can be accommodated, using an ATM format, from basic through to enhanced quality wide-screen entertainment and, eventually, HDTV.

- incorporation of the needed local upstream signaling (to the digital channel Selective Tap function) within future set-tops would be trivial. Providing this function externally for existing digital set-top converters, using a separate non-intrusive remote control IR interface, would also cost very little.

SUMMARY

The digitized cable TV distribution network evolution advocated in this paper for switched services addresses a number of today's problems and tomorrow's potential issues. It provides a smooth path from very economic introduction of broadcast-like digital TV services right through to advanced multimedia applications. At the same time the directions taken reduce and minimize the need to replace or upgrade Cable network components or consumer equipment. Significant additional advantages accrue in new-build or rebuild situations.

ACKNOWLEDGMENTS

In this paper I have highlighted some alternative approaches which go beyond presently planned Cable TV directions. I am, however, indebted to a number of people within the Cable industry for their guidance and encouragement. I would also like to thank my employer, BNR, for it's support of my research activities and permission to publish this paper.

REFERENCE

[1] Waltrich J.B., Glaab J.B., Ryba M. and Muller M.A., Jerrold Communications, Applied Media Lab., "The Impact of Microreflections on Digital Transmission Over Cable and Associated Media", 1992 NCTA Technical Papers, pp. 321-336.

BIOGRAPHY

Jack Terry's professional career started in Television at The Marconi Company, in England. For a number of years he was involved in development, field installation and training in areas of high and low power TV Broadcast transmitters, as well as studio equipment and microwave links. Later he initiated and led much of the PCM development work at Marconi, resulting in the first UK field trial of a PCM Telephony multiplex product This digital telephony work later expanded into PBX and Exchange system design

In 1974 Jack emigrated to Canada to join BNR. He soon assumed managerial and technical responsibility for the hardware design of Northern Telecom's DMS-100 Central Office Switch product, while specializing in its Line Circuit technology and electrical integrity. Over 50 million line circuits based on this work are today in service in North America. Jack's switching work subsequently evolved to ATM switch architecture and technology research. During the past three years he has resumed his career in the field of Television, in this case Digital Television and related multimedia. Today, as an assistant vice-president, he heads a Broadband Systems and Technology research laboratory in BNR, Ottawa, whose activities are focused on "Fiber to-the Coax" transmission technologies and systems.

Jack is a Fellow of IEEE, a Member of SCTE and a Member of Sigma XI

THE ROLE AND CHALLENGES OF FIBER OPTICS TECHNOLOGY IN THE ACCESS TELECOMMUNICATIONS NETWORK

Rolando Oliver, Samih Elhage, and Stéphane É. Brisebois

Stentor Resource Centre Inc.
700 de la Gauchetière West, Office 18W2
Montréal, Québec, Canada H3B 4L1

Telecommunications are quickly evolving as new options appear. The normal evolution of technology renders a full basket of services that users already demand and will continue to demand in the future. Customers require larger aggregates of bandwidth and a larger control over the choice of their information providers / correspondents. While business customers' service requirements are well understood, those related to the residential customers present many issues that need to be resolved. This paper puts forward some fiber technology solutions to accommodate anticipated residential applications.

The role of fiber optics technology in the access network

The telecommunication industry is studying several architectures to support today's and tomorrow's voice, data and video service requirements. These architectures include: fiber architectures such as Fiber-to-the-Home (FTTH), hybrid architectures such as Fiber-to-the-Curb (FTTC) and Hybrid Fiber-Coax (HFC), copper architectures such as Asynchronous Digital Subscriber Line (ADSL), and wireless architectures. The focus of this paper will be on the fiber and hybrid architectures only.

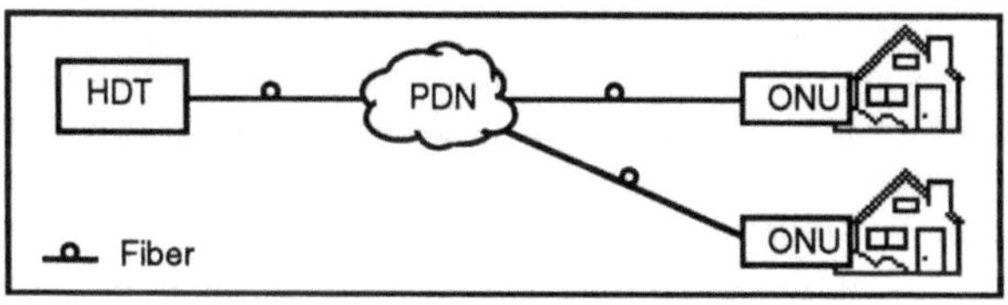

Figure 1 . Generic FTTH Architecture

Of all access architectures, it appears obvious that from the functionality point of view the preferred alternative is a passive fiber all the way from the main switching and distribution centers to the user, or FTTH. In this architecture, a Host Digital Terminal (HDT) located at the central office or in the field (see figure 1) communicates with subtending Optical Network Units (ONUs) via a Passive Distribution

Network (PDN). Each subscriber have a dedicated ONU. With FTTH, all subscribers would then be served with the future broadband capacity through fiber[1]. However in the short term this choice is not cost efficient when carried out on a small scale. Mass deployment is essential to reduce the systems costs[2]. The cost effectiveness of FTTH systems lies within devices that are used. Instead of using costly discrete technology, today's optical modules will rely more and more on hybrid integration techniques, whereby different types of optical components may be mounted on a single substrate. Soon, it will be possible to use waveguides to integrate some of the optical functions. In the long term, further cost improvement will come with monolithic devices that will concentrate all the optical functions. Development in the field of Large Scale Integration (LSI), especially in CMOS technology, will also have an impact on FTTH cost reduction. Presently, the industry has tried to find ways of applying fiber technology to traditional services — switched-voice and data.

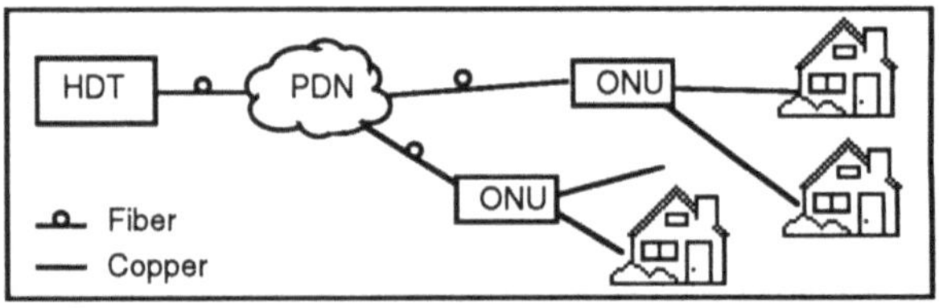

Figure 2 . Generic FTTC

In the recent years, much effort has been put by telcos into FTTC systems (see figure 2). The FTTC architecture is very similar with FTTH, but ONUs are located at the curb side and shared among subscribers to decrease per line costs. While FTTC offers a contending solution for narrowband applications, it allows for the deployment of fiber close to the home. This access architecture can be upgraded to provide broadband services quite easily. FTTC is one of the options retained by telcos especially for "desert start" applications.

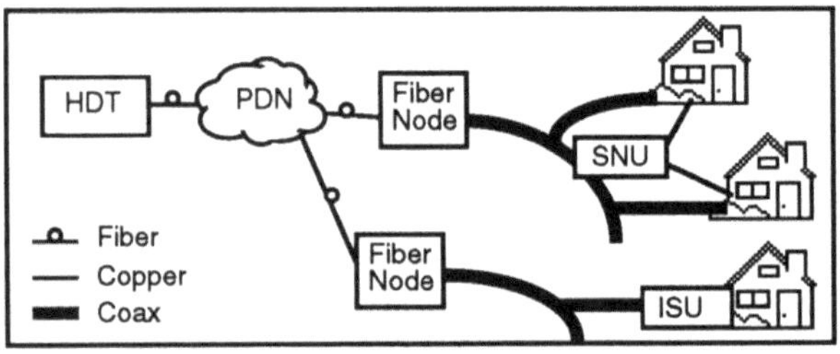

Figure 3 . Generic Hybrid Fiber Coax

Another option is taking fiber within a certain distance from the user, and completing the remaining distance with coaxial cable. This appears to be one of the most promising architectures for delivering tomorrow's services. The industry calls it Hybrid Fiber Coax (figure 3). Here again, an HDT manages multiple optical nodes linked up to a passive network. However, these nodes are called Fiber Nodes and differ from ONUs as signal management is performed in another box located closer to the subscribers. Two options prevail: in the first one a Service Network Unit (SNU) shared by a neighborhood is used to extract telephony signals, while broadband is accommodated by a dedicated coax cable to each home; the second option uses only a coaxial cable to the home, while a dedicated Integrated Service Unit (ISU) extracts telephony signals for the inside copper wiring. HFC appears nowadays economical for bringing multimedia and telephone services to existing and new areas. This has resulted in an avalanche of announcements by most major telephone administrations in North America and around the world, notably PacBell, Ameritech, US

West and Southern New England Telephone. These firms claim to have adopted new network architecture and design based on HFC.

The challenges of fiber optics technology in the access network

It is common to select an access platforms according to three following factors: *capacity*, *cost* and *reliability*. The first means that it should be capable of attaining at least the same transmission rate per user as the transport and switching components of the network. This factor often goes in contradiction with the second — the cost. There is little economic sense in investing now for the capacity that will only be required in the future. The planners should rather assess the ease with which an architecture may evolve to answer future services needs. Also, architectures that require larger investments should not be put aside when the services they bring generate the required revenues. An access network should thus answer four basic criteria: *affordability, scalability, evolvability* and *reliability.*

1)Affordability : The up-front cost of the architecture should be low. The deployment of such a technology should be in line with the low initial revenues that they will generate. Therefore, they should present an initial first cost that is commensurate with the initial applications.

2) Scalability : The incremental investment of such a solution follows as much as possible its service upgrade, and is thus closely followed by extra revenues.

3) Evolvability : As future service requirements are not fully defined, and as technology is not thoroughly mature, an evolution of fiber architectures is not only possible but desirable. However, this evolution should require a minimum construction work at that point in time. Also, long term migration to FTTH is presumed desirable.

4) Reliability: Some elements of the access system are deployed in the outside plant. In most cases they are exposed to very hostile environments. Thus the reliability criterion requires that it be very robust and easy to maintain. Architectures that push the electronics at both ends — in the central office and in the customer premise — should be promoted. In all cases, the final performance will also depend on the system's characteristics related to each manufacturer.

A Qualitative evaluation of access solutions

To see how several technologies match some of these criteria, and for illustration puropes, we will look at three options: FTTH, FTTC, HFC (see Table 1). In this article, focus is put on the application to existing networks.

Table 1 - Technologies comparisons

	FTTH	FTTC	HFC
Affordability	Low	Medium	High
Scalability	High	Low	High
Evolvability	High	Medium	Medium
Reliability	High	Medium	Medium

FTTH : Great option, but not affordable. This is the best option in the long term as it presents high bandwidth capacity for the future, and as it constitutes a reliable platform with a passive outside plant network and sheltered electronics. Its scalability is high as the dedicated equipment will be upgraded only when the subscriber requires complex services, and thus generates extra revenues. However, its affordability is low, as expensive optoelectronics (lasers and receivers) are dedicated to a single subscriber. Today's narrowband services and expected broadband services cannot support the inherent overall cost of this architecture.

FTTC : Limited scalability: FTTC is an economic approximation of FTTH, since it shares the cost of optoelectronics between many subscribers (usually 12 to 16). FTTC is thus more affordable than FTTH. However, this sharing of optoelectronics renders this architecture less scalable: as soon as one subscriber requires broadband services in a given area, the ONU serving the neighborhood has to be upgraded.

HFC : Sound solution for now and tomorrow. HFC presents many of the advantages of FTTH and FTTC, but it is more affordable than FTTH and more scalable than FTTC. By using analog carriers to transport both analog and digital signals, HFC is able to support today's mostly analog broadband signals. And with the help of signal modulation techniques, digital video and telephony can also be carried by the HFC network. Thus, HFC offers a large amount of bandwidth, sufficient to answer the short and medium term requirements. In terms of affordability, HFC allows the sharing of expensive optoelectronics between many subscribers. Similar to FTTC, HFC necessitate the deployment of outside equipment which makes it less reliable than FTTH.

Fiber Optics is the most promising technology to up-grade today's access telecommunications networks and to offer the high transmission rates required for future services. However, the architectures must also satisfy cost and reliability requirements. While an all passive optical access network is ideal, it is not affordable. The hybrid architectures that have been developed as compromises to satisfy the four criteria that define winning deployment strategies: *affordability*, *scalability*, *evolvability* and *reliability*.

REFERENCES

1 Yukou Mochida, *Technologies for Local-Access Fibering*, in **IEEE Communication Magazine**, Feb. 1994, p.64- 72.

2 Tetsuya Miki, Toward the Service-Rich Era, in **IEEE Communication Magazine**, Feb. 1994, p.39-34.

SOME ARCHITECTURAL CONSIDERATIONS FOR LARGE-SCALE FAIL-SAFE ATM-SONET NETWORKS

Marion R. Finley, Jr.[1] and Hawa Sininta

Department of Mathematics and Computer Science
University of Quebec at Montreal
Montreal, Qc. H3C 3P8 Canada

1. Introduction

In this paper we examine the introduction of ATM and SONET technologies in large-scale wide-area networks (WAN's) such as those used in the bulk electric power distribution industry, as this term is used by the Northeast Power Coordinating Council (NPCC)[1]. We take as an example of such a network the internal communications system at Hydro-Québec, the principal supplier of electrical power in the Province of Quebec, although the principles to be discussed could be applied to many other similar networks.

2. The Context

The telecommunications networks associated with large-scale electric power grids typically span large distances, perhaps as much as a thousand kilometers. A network of 735 and 450 kV high-tension wires link remote generating stations to the urban areas and to various industrial and non-local consummers of electric energy. In these networks, the guard wires may be used to install telecommunications cables. Recently there has been a trend towards an all-digital system using single-mode optical fibers, yielding a telecommunications system that parallels the electric grid distribution system. The early applications of such a telecommunications subsystem were confined to certain basic command and control functions necessary to the running of the electric grid and internal telephony. Typically, the corporation also will have implemented an internal packet-switched data communications network and will use local area networks extensively throughout for administrative and other purposes.

The optical fiber cables that now span much, if not all, of the electric grid give rise to the possibility of creating an all-digital broadband integrated services telecommunications network based on a single common switching and transmission infrastructure. This gives,

[1]The authors wish to thank Yvon Depratto, Ing., Direction Télécommunications, Division, Réseau de transmission, Hydro-Québec for the guidance he graciously offered us, and Professor Hafedh Mili for the support from the LARC laboratory at the University of Quebec at Montreal. The ideas expressed in this article do not represent any policy or attitude of the Hydro-Québec Corporation and reflect the authors' views alone.

in effect, a private broadband ISDN (B-ISDN) network capable of offering services such as telephony, command and control functions, security, including video surveillance of remote locations, data communications, and video teleconferencing. In addition, it is possible to envisage limited public offerings of data, voice and video services in order to service remote areas into which the public telephone and the private cable television networks cannot easily penetrate.

3. Broadband ISDN: ATM and SONET

Broadband ISDN is based on the asynchronous transfer mode (ATM) technology for switching and transfer functions and on the synchronous optical network (SONET) (in North America) or the synchronous digital hierarchy (SDH) (just about everywhere else) technologies for transport functions[2,3,4]. These technologies would therefore be the technologies of choice for the internal broadband integrated services networks that are being developped by the electric bulk power industry. We note in passing that these networks imply a mix of wide-area network (WAN), metropolitan area network (MAN) and local area network (LAN) functionalities. Any architectural proposition for these networks must encompass these three classes of subnetwork.

The actual development of these networks has been gradual. The reasons for this include the issue of internal corporate education in, and awareness of, the implications of the emerging technologies as well as purely technical matters relating to them. For example, ATM devices are just now emerging and remain largely untested for the types of applications envisaged, whereas SONET devices are available and currently being installed in a variety of networks. The result is that at this moment SONET technology dominates and perhaps competes with ATM in some applications. Among the advantages offered by SONET technology are architectural simplification due to use of SONET devices such as add-drop multiplexers (ADM's) and digital cross-connect devices (DCS's), standardization *in principle* of E/O interfaces and a network-wide management facility.

We note a trend due to these advantages towards replacing already existing asynchronous optical fiber links with SONET links.

Synchronization of SONET devices is based on a hierarchy of stratum clocks whose quality is determined as a function of their stability and accuracy over time. We note that the location of the clocks and the manner in which their signals are to be distributed must be determined as part of the network architectural specification.

At this writing, a number of ATM test-bed and pilot projects are being carried out, including some for the MAN-LAN side of the WAN-MAN-LAN mix mentioned earlier, in particular for LAN interconnect as such. Typically a large corporation may have several hundred internal LAN's for which most conventional interconnect schemes, such as X.25 and ethernet backbones, are inadequate. The fiber data distributed interface (FDDI), so long the hope of many who desired to have high-speed LAN's and convenient MAN backbones, is now perceived to be inadequate as well. There are, however, some attempts to create the so-called "fast Ethernet," and other LAN interconnect schemes. These compete with ATM and, at this moment, it is difficult to foresee which technology will emerge the victor. Suffice it to say the ATM, with its promise of offering extremely large switching bandwidth and "bandwidth on demand," is a very strong contender.

4. Architectural Considerations

The power system's command and control system requires of the telecommunications network a high degree of robustness. The system is subject to extreme weather conditions, high winds, low temperatures and icing, as well as to fire and other natural phenomena and sabotage. For example, a downed link requires swift reaction in order to avoid damage to the power grid or to to the generating stations. System reaction times of the order of 10 msecs over long distances may be required. The problem is then to find ATM/SONET-based architectures that will be highly robust and rapidly self-restoring so as to preserve the critical QOS parameter values, or, failing this, to allow graceful degradation of critical services. For a given set of QOS parameter values, a viable architectural candidate implies specification of the physical and logical network topologies, the network elements to be used, and the layered protocol functions. Candidate architectures must clearly respect any specified financial contraints as well. For such architectures, the topological candidates are the usual ones: stars, rings, busses (perhaps with CDMA), hybrids (star-ring, ring-star), concatenated (star-star, ring-ring), hierarchical (ring of rings, star of stars) or *layered* topologies, the layers having different functionalities, and meshes. Dual bidirectional counter-rotating ring topologies, using SONET technology, have attracted much attention. Such rings offer several important advantages such as by-pass and loop-back techniques to provide "self-healing," that is, rapid restoration of operation in the case of link or node failure and an architectural simplicity in, for example, routing functions. SONET technology lends itself naturally to such applications and would permit a coherent LAN-MAN-WAN hierarchy and network management facility.

However, dual counter-rotating self-healing rings may pose difficulties, in particular if they span large areas such as those covered by electric power grids. For example, if the link between two adjacent nodes n_i and n_{i+1} fails, then traffic that would normally pass between these nodes is rerouted from node n_i through n_{i-1}, n_{i-2}, ... , n_{i+2}, back to n_{i+1} using the loop-back mechanism offered at each node of such counter-rotating rings. Hence, functionality is preserved, at the expense of an increased propagation time due to having to traverse additional links. This might violate the QOS delay constraint. Therefore, for normal functioning, the ring architecture may be adequate, but under certain fault conditions, it would fail to maintain the QOS needed.

This suggests that a mesh architecture might be best, at the expense of complicating the routing functions. One might consider a mesh overlay, perhaps using alternative technologies, for example micro-wave, to preserve essential network functions. Simulation studies applied to Canadian telephone networks, suggest that mesh networks permit better performance than do rings[5,6,7].

In our current work, we are developping a simulation/knowledge-base model for the study of such systems using an object-oriented approach. Results from this study will form the object of future publications.

5. Conclusions

SONET ring hierarchies are promising candidates for kinds of networks needed by the bulk power industry, However, ring deficiencies may have a serious impact upon the QOS. Careful overlay of a mesh topology may suffice to alleviate this impact. ATM may be introduced at the DCS or higher-lever as it becomes available. At this time, ATM is a strong candidate for LAN-MAN interconnect and SONET links are rapidly replacing traditional asynchronous links in many applications such as CATV trunking.

6. References

1. Northeast Power Council. *Basic Criteria for Design and Operation of Interconnected Power Systems*, revised version, NPCC, New York, 1988.
2. G. L. Byeong, M. Kang, J. Lee. *Broadband Telecommunications Technology.* (Artech House, Boston, 1993).
3. R. Händel, M. N. Huber. *Integrated Broadband Networks - An Introduction to ATM-Based Networks.* (Addison-Wesley, 1991).
4. D. Wright. *Broadband - Business Services, Technologies and Strategic Impact.* (Artech House, Boston, 1993).
5. W. D. Grover. "Case Studies of Survivable Ring. Mesh and Mesh-Arc Hybrid Networks." (1991).
6. W. D. Grover et al: "Performance Studies of a Self-healing Network Protocol in Telecom Canada Long Haul Networks." (1990).
7. Y. Okanoue et al: "Integrated Self-Healing Techniques for SONET Survivability," *NEC Res. & Development, Vol.33. No.4.* (October 1992).

DESIGN, IMPLEMENTATION, AND SYSTEM INTEGRATION FOR A MULTI-GIGABIT WDM NETWORK

Hossein Izadpanah and Chinlon Lin

Bellcore
331 Newman Springs Road
Red Bank, New Jersey 07701, USA

1. Introduction

A three-node, multi-wavelength, multi-Gb/s exploratory optical network is implemented over the interoffice fiber plant of New Jersey Bell. The network, known as Bellcore Research Network (BRN) interconnects the three Bellcore Research Laboratories at Red Bank (Navesink, NVC), Raritan River Center (RRC) and Morristown (MRE) in New Jersey. The network provides a realistic testbed for research in advanced optical network technologies, architectures, and network management such as ATM transport over switched SONET networks, and a distributed operations and management architecture. The network lends itself well to the testing of new services such as remote access from NVC to a MASPAR super computer and HIPPI-LAN located at MRE. SMDS experiments and shared database research can also be conducted between the three locations. Other ongoing research projects include video-based services such as Video Window, Cruiser, video-on-demand, and remote access to the Aurora Gigabit testbed.

In this paper, we describe the design, implementation, and the system integration aspects of the network optical layer and present results on the evaluation of the network performance. Specifically, we demonstrate a flexible upgrading strategy for the existing interoffice network to support multi-wavelength channels operating at multi-Gigabit/s data rates using advanced optical technologies such as wavelength-division multiplexing (WDM), Erbium-Doped Fiber Amplifiers (EDFA), and dispersion-compensating fibers (DCF). In addition, opportunities for using the network for testing new services are discussed.

2. BRN Network Topology

Figure 1 shows the schematic network configuration of the BRN which consists of three network layers: ATM, SONET, and an optical layer. The first layer consists of ATM subnetworks, with several local ATM switches each, at NVC, RRC and MRE, respectively. The subnetworks are interconnected by SONET OC-12 (622 Mb/s) backbone logical rings using 1310/1550 nm wavelength transmission over the same fiber pair. For increased reliability, the OC-12 SONET layer is configured as a unidirectional, path protection switched self-healing ring. Add-Drop Multiplexers (ADM) at the OC-12 rate in each location provide OC-3c and DS3 access to the ATM subnetworks.

Applications of Photonic Technology
Edited by G.A. Lampropoulos *et al.*, Plenum Press, New York, 1995

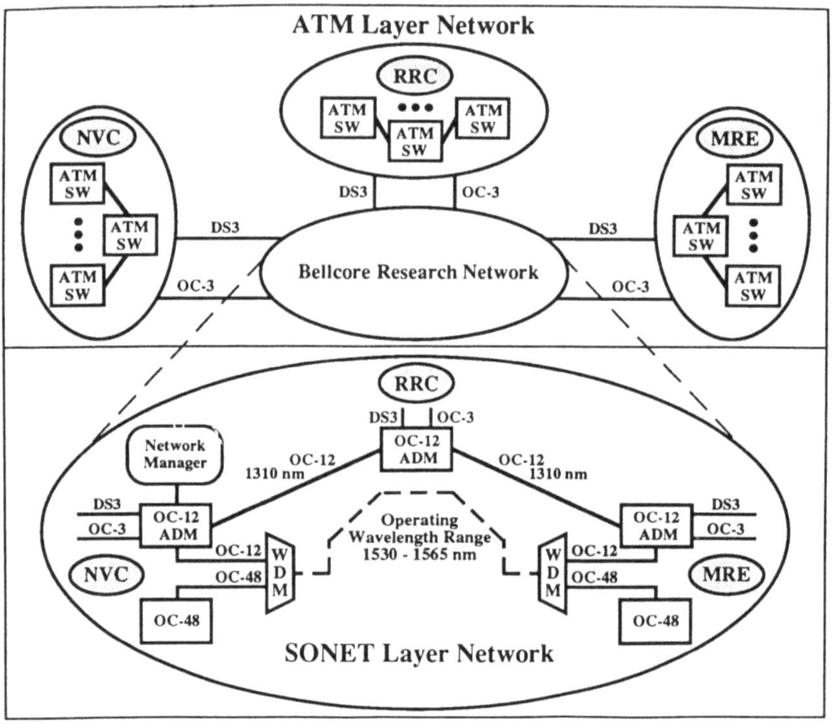

Fig. 1 - The BRN 3-node network configuration

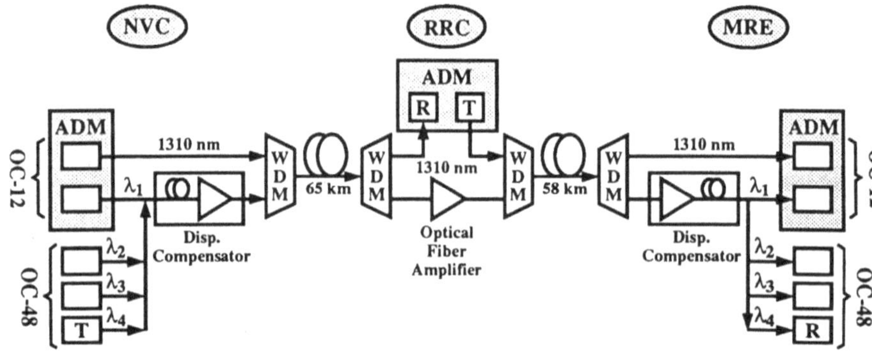

Fig. 2 - Dual-window, multi-wavelength network optical layer (one way only)

3. Multiwavelength Multi-Gb/s System Implementation

In the three-node network described above, there is an additional point-to-point, multi-wavelength, bi-directional direct link between NVC and MRE (123 km), with an EDFA inline optical amplifier at the RRC location. The system consists of one pair of OC-12 transceivers that is required for the logical rings mentioned above, and up to three wavelengths at OC-48. The 1550 nm channels are combined (split) using a 4x1 (1x4) fiber passive coupler at the transmitter (receiver) ends. The desired channel is selected using a tunable optical filter. The 1310 and 1550 nm channels at OC-12 and OC-48 share the same fibers by using 1310/1550 nm WDM fiber couplers. Multiplexing the 1310 and 1550 nm optical channels onto the same fiber facilities represents one of the near-term applications of WDM in interoffice and access networks.

In order to minimize the overall dispersion around 1550 nm, passive DCFs are used to overcome the large dispersion of the deployed conventional single-mode fibers [3,4]. The DCF is a specially designed single-mode fiber having a large negative dispersion (≈ -65 ps/km-nm) at 1550 nm. The DCF is placed in cascade with the transmission SMF, either at the transmitter or at the receiver end, to tune the effective zero dispersion wavelength of the link to ≈ 1550 nm. For broadband WDM dispersion-compensation design, one can take advantage of chirped-pulse-compression at wavelengths experiencing a net negative dispersion (due to over-compensation) to obtain dispersion-penalty-free transmission and even a receiver sensitivity improvement. The use of DCF has proven effective and practical for single and multi-wavelength WDM systems for both 2.5 and 10 Gb/s transmission and up to 246 km in fiber length[5-6].

The overall BRN optical layer is shown in Fig. 2 depicting the EDFAs, DCFs, and WDM optical components. The network accommodates the 1310 and 1550 nm wavelengths at the OC-12 SONET layer, and also multi-wavelength at the OC-48 rate. For clarity, only the NVC-RRC-MRE link is shown.

4. Transmission Test Results

Typical BER characteristics measured in both the 1310 and 1550 nm windows on the individual BRN network segments, together with the baseline of the receiver, are shown in Fig. 3. Maximum system degradations of < 1 dB at 622 Mb/s and < 0.5 dB at 2.5 Gb/s were measured at 1310 and 1550 nm, respectively. The low error performance of 1550 nm signal obtained is attributed to the effectiveness of the DCF. For the MRE-NVC link we have approximately 0.5 dB receiver sensitivity improvement due to overcompensation. The results of the measured BER penalty (or improvement) for transmission of WDM channels over different link lengths of the dispersion-compensated BRN fibers at 2.5 Gb/s are summarized in Table 1. Effective compensation is achieved over 20 nm of spectral range and more than 200 km of SMF length. At 1544 nm wavelength, the NVC-MRE round-trip link (246 km) was tested over a 3-day period without a single error when the received optical power was set to 3 dB above the receiver sensitivity. This indicates a BER-performance better than 2×10^{-14}. The above results show that the technologies for WDM implementations are available now, and upgrading interoffice network with WDM is not only feasible but also practical.

In addition to transmission experiments at OC-48, an initial transmission experiment (eye pattern only) has been performed at 10 Gb/s over NVC-RRC-NVC fiber (130 km in length) indicating the effectiveness of the DCF scheme deployed in the BRN.

5. Network Control, Services, and Applications

Transaction Language 1 (TL1) is used to control and set up the required configurations on the OC-12 ADMs for the three Bellcore locations. The TL1 interface to the OC-12 ADMs supports the following operational functions on the local and remote sub networks: alarm surveillance, performance monitoring, protection/synchronization switching, control, memory administration, and protocol provisioning. While the ADMs were set to a path-protection switched ring, a fiber cut was simulated by disconnecting one of the transmitter fiber. The ADMs then reconfigured themselves automatically to the protection path. Other network experiments performed were changes in low speed channel cross-connect and

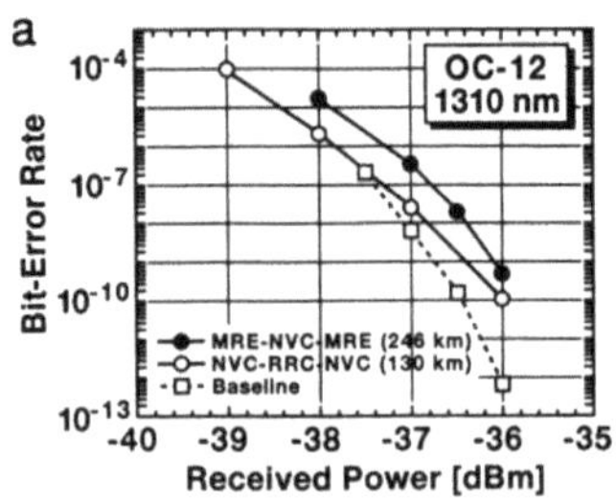

Fig 3a - Measured BER results for regenerated 1310 nm channel

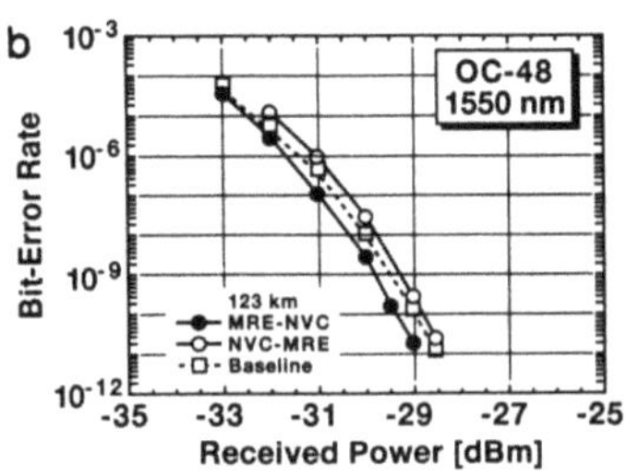

Fig. 3b - Measured BER results for 1550 nm channel

practices on the adding/dropping a signal to/from the high-speed OC-12 at local and remote sites. In the following, we describe a few system experiments and network demonstrations conducted on the BRN which demonstrates the potential and flexibility of the network using SONET technologies and architectures:

1. A HIPPI (High-Performance Parallel Interface) test signal at 800 Mb/s was successfully transported over the 2.5 Gb/s channel on the NVC-MRE-NVC (246 km) path, when it was opto-electronically regenerated at MRE.

2. The BRN was used for broadband Cruiser/Touring teleconferencing sessions between NVC and MRE by transporting the digital video signal from an ATM switch on the ADMs OC-3c channels.

3. Two FDDI rings at MRE and NVC were interconnected using one of the DS3 channels. The FDDI is used to support multimedia traffic over the Internet-wide MBONE. Applications include desktop video/audio teleconferencing, shared white boards, and real-time weather satellite imagery. The research also investigates the interoperability issues in a heterogeneous Touring Machine environment which uses both analog and digital media.

4. Bidirectional broadcast-quality digital video conferencing has been established successfully between MRE and NVC over one of the DS3 channels, at both 1310 nm and 1550 nm, and when the network is configured as a unidirectional ring. Also, a multi-party single screen video conferencing experiment was conducted by combining the video signals from three remote sources and re-broadcasting them to all locations on a single DS3 channel. This option provides the possibility of multi-location, high-quality video conferencing using BRN facilities.

5. A bi-directional link connections at DS3 is planned to access the Switching laboratory in NVC from the Integrated Digital Loop Carrier (TR-303) Laboratory in Morristown. The access will be used to conduct technical analysis of remotes and switch interfaces.

6. Summary

We have described the implementation of a dual-window multi-gigabit WDM research network over the existing interoffice fiber plant using advanced optical technologies. Several SONET/ATM network experiments were performed and service applications were demonstrated using ATM switching and OC-3c and DS3 interfaces.

7. Acknowledgments

We would like to acknowledge the collaboration of many of our Bellcore colleagues and Corning Inc. for providing the dispersion-compensating fibers and the EDFAs.

Table 1. Measured dispersion penalty (or receiver improvement) for different BRN link length

λ (nm)	Link Length 67 km	Link Length 123 km	Link Length 24 6 km
1534	- 0.4 dB	- 2.2 dB	- 1.5 dB
1544	- 0.7 dB	- 0.7 dB	- 0.1 dB
1546	- 0.5 dB	Not Meas.	Not Meas.
1554	< 1.0 dB	< 1.0 dB	+ 1.1 dB

8. References

1. P. Kaiser, "Bellcore Research Network," *LEOS Topical Meeting on Gigabit Networks*, Santa Barbara, CA, July 1993.

2. H. Izadpanah et al, "Bellcore Research Network: Optical SONET/ATM Layer Implementations and Applications," *OFC '94,* TuO3, San Jose, CA, Feb. 1994.

3. H. Izadpanah et al, "Dispersion Compensation in 1310 nm-Optimized SMFs Using an Equalizer Fiber, EDFAs and 1310/1550 nm WDM," *Electron Lett.*, vol. 28, pp 1469-71, July 1992.

4. H. Izadpanah et al., "System Applications of Dispersion-Compensating Techniques in Upgrading Interoffice 1310 nm-Optimized Fiber Networks," *GLOBECOM' 93,* paper # 51.2, Houston, Texas, Nov. 1993.

5. J. M. Dugan et al, "All-Optical, Fiber-Based 1550 nm Dispersion-Compensation in a 10 Gb/s 150 km Transmission Experiment Over 1310 nm-Optimized Fiber," *OFC '92,* Post deadline paper PD14, San Jose, CA, February 1992.

6. H. Izadpanah et al., "Multiwavelength Dispersion-Compensation for 1550 nm Transmission at 2.5 and 10 Gb/s over 1310 nm-Optimized SMF," Paper TuA5.1, *ECOC '92*, Berlin, Germany, 1992.

LASER DIODE FREQUENCY CONTROL FOR MULTIWAVELENGTH (WDM) OPTICAL COMMUNICATIONS

M. Têtu, B. Villeneuve*, C. Gamache, M. Guy, C. Paquet,
C. Latrasse, M. Poulin, M. Breton, P. Tremblay

Centre d'optique, photonique et lasers (COPL)
Département de génie électrique
Université Laval
Ste-Foy, Québec, G1K 7P4, Canada

*Advanced Technology Laboratory
Bell-Northern Research
Ottawa, Ontario, K1Y 4H7, Canada

INTRODUCTION

The simultaneous use of many wavelengths will increase the throughput of point to point optical communication systems and lead to the development of networks where the routing is achieved using the wavelength values as the keys for identification.[1] For relatively dense systems, the laser wavelengths will have to be controlled precisely to avoid crosstalks or wrong identification. In this work, two distinct approaches are considered to set and to maintain the wavelength of an ensemble of laser sources in the 1.5 μm region.

MULTIPLE OPTICAL FREQUENCY GENERATOR

Based on an Optical Resonator

A device offering multiple optical frequency references can be used to frequency-lock the various transmitter lasers. For this approach, we have devised a scheme to calibrate absolutely a Fabry-Perot resonator.[2] A well identified resonator transmission mode is frequency-locked to a laser already frequency-locked to an atomic or a molecular resonance. This sets the free spectral range of the Fabry-Perot resonator to a precisely known value. Afterward, the transmitter lasers are frequency-locked to different transmission modes. The precision to which the absolute value of each laser frequency can be set is limited actually to

35 MHz.[3] This limitation is due to the lack of precision on the knowledge of the frequency of the atomic or molecular frequency references available in the 1.3 μm and 1.5 μm regions.

Figur 1 gives an overall diagram of the multiple wavelength generator based on this approach.[4] ˈı he Fabry-Perot resonator is made of two parallel flat mirrors. Each laser is frequency-modulated at low frequency to achieve synchronous detection and frequency-locking. The transmission data rate is at 2.5 Gb/s. Work is under progress in order to find if the low frequency modulation has an effect on the bit error rate (BER). Also, by selecting different laser frequency spacings (FSR), we will study the effects on the performance of a long-span communication link of induced crosstalks by the fiber nonlinearities.[5]

Based on a Frequency Offset Comparator

In a second approach, the transmitter lasers frequencies are measured directly and tuned to prescribed values. For this method, we have shown the possible use of an integrated optical-sum-frequency generator as a wavemeter giving a resolution of about 2 GHz.[6] The generator is a multilayered nonlinear crystal.[7,8] In this system, the frequency of a transmitter laser is compared to the frequency of a reference laser and the offset frequency is automatically controlled via an electrical feedback loop. When the reference laser is frequency-locked to an atomic or a molecular resonance, the system becomes absolutely calibrated. We have shown that the laser frequencies can be set and maintained to given values with a precision of 1.1 GHz.[9]

Figure 2 shows a block diagram of the generator we are designing using the frequency offset comparator technique. Each laser frequency is compared sequentially to the frequency of a master laser. The measured frequency offset is compared to a selected value, after calibration of the comparator, and precise laser tuning is accomplished through the injection current and temperature. Work is under progress with lasers operating at 1.5 μm and calculations show that a resolution of the order of a few 100 MHz is foreseeable.[10]

MASTER LASERS

To insure reproducibility and absolute calibration, the multiple frequency generators require the use of a master laser, i.e. a laser frequency-locked to a precisely known atomic or molecular resonance. Our work covers three methods suitable for this purpose.

Lasers Frequency-Locked to Acetylene Resonances

We have done spectroscopic studies of the rotation-vibration transitions of $^{13}C_2H_2$ and C_2HD molecules which exhibit optical references in the 1510 nm to 1550 nm band[11] (included in the Erbium doped fiber amplifiers band). As an example, figure 3 shows the absorbance of $^{13}C_2H_2$ centered at 1.535 μm for a 5 cm long cell containing a pressure of 26.2 torr; 114 lines are available for laser frequency-locking purpose.

Lasers Frequency-Locked to Rubidium Vapor Transitions

Figure 4 shows the energy level diagram for the first hyperfine states of ^{87}Rb atoms and the possible optical transitions between them. When the vapor is pumped with a laser operating at 780.2 nm, the transitions between the first excited states can be interrogated

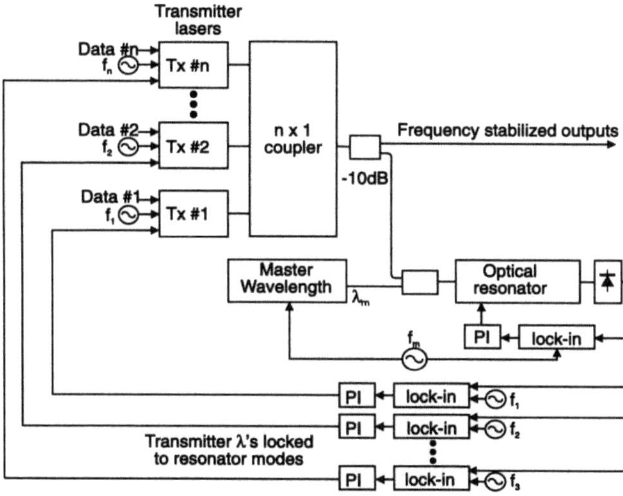

Figure 1. Multifrequency generator based on a calibrated optical resonator.

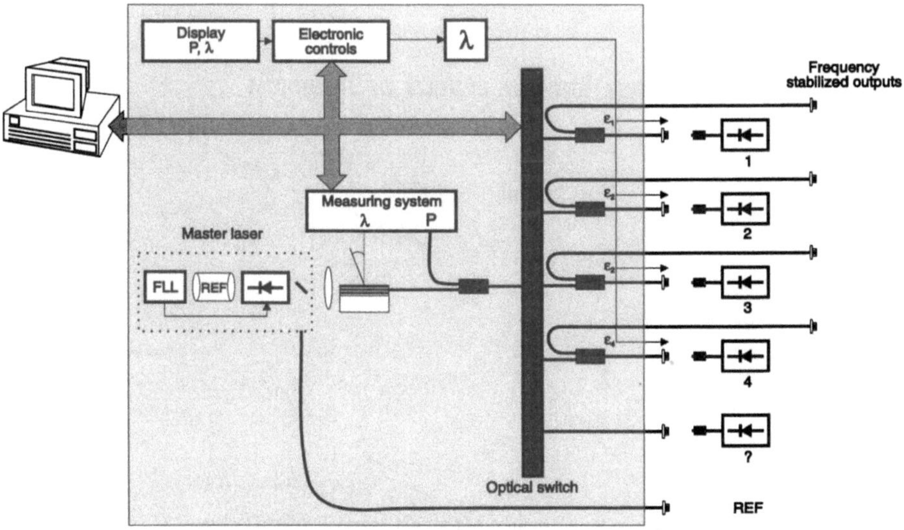

Figure 2. Multifrequency generator based on a frequency offset comparator.

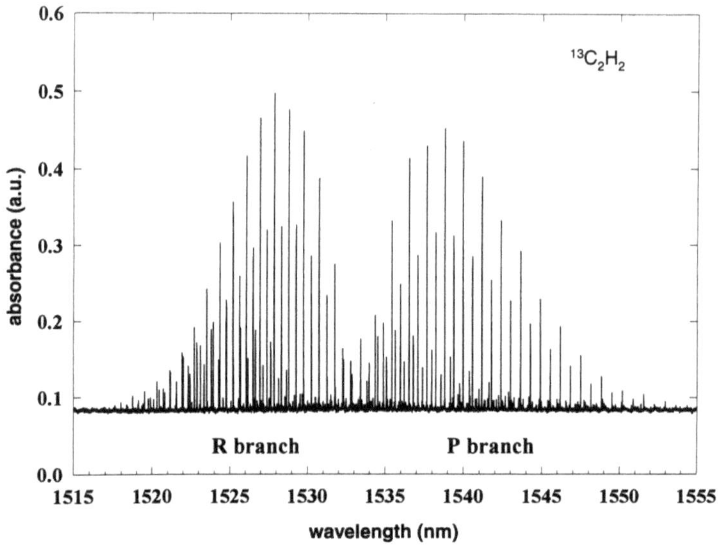

Figure 3. Absorption spectrum of acetylene $^{13}C_2H_2$ in the vicinity of 1535 nm.

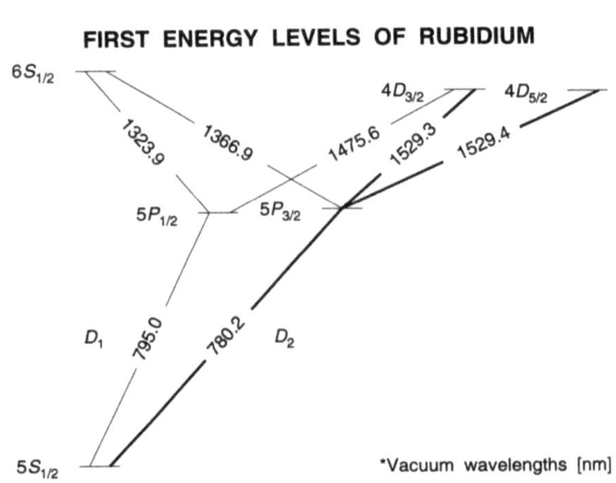

Figure 4. Fine structure of the first energy levels of ^{87}Rb.

with a laser operating around 1.529 μm. We have shown that it is possible to frequency-lock such a laser on any of these transitions.[12] Moreover the locked laser is dither-free.

Frequency Doubled Lasers Frequency-Locked to Rubidium Vapor Transitions

We also studied the use of a second-harmonic generated signal from a 1.56 μm DFB laser through a bulk nonlinear crystal to frequency-lock this laser to a ground based transition of rubidium atoms at 780.2 nm. Using a single pass $KNbO_3$ crystal, 2.2 nW optical power at 780 nm has been produced with 11.3 mW at 1560 nm. The locking loop error signal resulting from this approach indicates that the 1.56 μm laser behavior is similar to the one obtained for a laser operating directly at 780 nm and frequency-locked in the same conditions.[13,14]

FREQUENCY STABILITY COMPARISON

The frequency stability characterization of an oscillator is expressed in terms of the relative frequency fluctuations and is measured, in the time domain, by the two sample Allan variance.[15,16] Usually, with very high precision oscillators, two similar but independant oscillators are compared together and the statistics is done on their frequency difference. With lasers, the heterodyne technique is used to generate an IF signal corresponding to their frequency difference.

Figure 5 gives the square root value of the Allan variance as a function of averaging times between 0.0001 s and 1000 s for a 1.5 μm laser when operated in three different situations. The open circles represent the results obtained when the laser is free running. We see that the relative frequency fluctuations stay almost constant to a level of 2×10^{-8} . This level corresponds to rms frequency fluctuations of about 4 MHz. The dark circles are obtained with a laser frequency-locked to an optically pumped rubidium vapor. We see an improvement over the free running laser of about two orders of magnitude for 100 s averaging time. The dark squares are obtained with a laser frequency-locked to a resonance line of acetylene. We see an improvement of frequency stability of a factor three over the results for a laser frequency-locked to an optically pumped rubidium vapor. Further studies are in progress in order to find the ultimate limit achievable with each method.

CONCLUSION

We have presented two methods to realize a multiple frequency generator. In the first one, an absolutely calibrated optical resonator is used to produce a very large amount of evenly spaced reference lines. These lines are used to frequency-lock transmitter lasers with an absolute precision of about 35 MHz. The second method uses a frequency offset comparator to measure sequentially the frequency of each laser. The reference laser being frequency-locked to a known atomic or molecular resonance, the laser frequency can be set and maintained with an accuracy of about 1 GHz.

Finally, we gave a brief summary of the results obtained recently on the design and study of master lasers. We compared 1.529 μm DFB lasers frequency-locked to acetylene gas or to a rubidium vapor optically pumped. They both show a frequency stability of about 2×10^{-10} for averaging time of about 100 s.

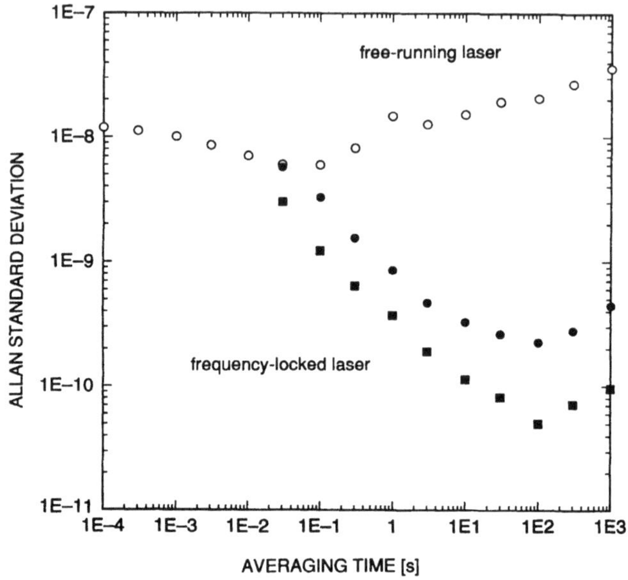

Figure 5. Lasers frequency stability measurements in the time domain: ° free-running, • frequency-locked to rubidium, ♦ frequency-locked to acetylene.

REFERENCES

1. C. A. Brackett, "Dense wavelength division multiplexing networks : principles and applications", IEEE J. Selected Areas Comm., vol 8, n° 6, 1990.

2. R. Boucher, B. Villeneuve, M. Têtu, M. Breton, "Calibrated Fabry-Perot etalon as an absolute frequency reference for OFDM communications", IEEE Photon. Technol. Letters, Vol. 4, No. 7, July 1992.

3. M. Têtu, N. Cyr, C. Gamache, R. Boucher, M. Breton, P. Tremblay, "Ensemble of laser sources with evenly spaced absolute frequencies for optical division multiplexed systems", SPIE, Proc. 1837, paper 29, Boston, 1992.

4. B. Villeneuve, R. Boucher, M. Têtu, M. Breton, "Frequency stabilization of multiple transmitters and receivers in OFDM networks", Photonics'92, pp. 3.17.1 - 3.17.6, Ottawa, Ontario, 1992.

5. F. Matera, M. Settembre, "Exploitation of fiber bandwidth in links with optical amplifiers", paper ThC2, Technical Digest OFC'94, San Jose, CA, 1994.

6. M. Guy, M. Têtu, B. Villeneuve, M. Svilans, "Optical frequency measurement for multichannel network using sum-frequency generation in multilayer waveguides", Electron. Letters, Vol.29, No.11, May 27, 1993. pp.975-976.

7. R. Normandin, S. Létourneau, F. Chatenoud, R. L. Williams, "Monolithic, surface-emitting, semiconductor", IEEE J. Quant. Electron., vol 27, no. 6, 1991.

8. D. Vakshoori, S. Wang, "Integrable semiconductor optical correlator, parametric spectrometer for communication systems", J. Ligthwave Technol., Vol. 9, No. 7, 1991.

9. M. Guy, M. Têtu, B. Villeneuve, M. Svilans, "Optical frequency control for DWDM networks using sum-frequency generation in multilayer waveguides", Photon. Tech. Letters, Vol. 6, No. 3, March 1994.

10 M. Guy, M. Têtu, C. Latrasse, B. Villeneuve, M. Svilans, "Laser optical frequency control at 1.33 and 1.55 μm by using a non linear multilayer semiconductor waveguide", Technical Digests, Conference on Lasers and Electro-Optics, paper CTh A6, Anaheim, Ca. May 1994.

11. C. Latrasse, M. Têtu, M. Breton, N. Cyr, R. Roberge, "C_2HD absorption lines in the vicinity of 1529 nm for frequency-locking", Techn. Digest OSA Annual Meeting, MS2, Toronto, Canada, September 1993,

12. M. Breton, P. Tremblay, N. Cyr, C. Julien, M. Têtu , "Observation and characterization of ^{87}Rb resonances for frequency locking purpose of a 1.53 μm DFB laser", SPIE Proc. on Frequency-stabilized lasers and their applications, 1837, 1992.

13. M. Têtu, B. Villeneuve, N. Cyr, P. Tremblay, S. Thériault, M. Breton. "Multiwavelength sources using laser diodes frequency-locked to atomic resonances", IEEE J. Lightwave Technol., Vol. 7, No. 10, 1989.

14. M. Poulin, C. Latrasse, M. Têtu , M. Breton, "Second harmonic generation of a 1560 nm DFB laser using a $KNbO_3$ crystal for frequency-locking to the ^{87}Rb-D_2 line at 780 nm", to be published, Optics Letters.

15. D. W. Allan, "Statistics of atomic frequency standards", Proc. IEEE, vol 54, n°2, 1966.

16. J. Barnes *et al.*, "Characterization of frequency stability", IEEE Trans. Instrum. Meas., vol IM-20, n°2, 1971.

A SINGLE-HOP WAVELENGTH-ROUTED LOCAL AREA NETWORK[1]

Mark W. Janoska and Terence D. Todd

Department of Electrical and Computer Engineering
McMaster University
Hamilton, Ontario, Canada L8S 4K1

INTRODUCTION

Future LANs and MANs may be based upon multichannel, single-hop network architectures constructed using wavelength-division-multiplexing (WDM) and passive star couplers. Due to device limitations, the number of available channels may be smaller than desired. In this case, wavelength spatial re-use will be required to support an acceptable number of stations and performance. Recently, hierarchical wavelength re-use architectures have been proposed for this purpose [1].

In this paper, a two-level hierarchical extension of our Distributed Channel Controller Network (DCCN) [3] architecture is considered. The network is referred to as the Extended Distributed Channel Controller Network or EDCCN. The physical design is similar to that originally proposed in [1] and used in [2]. A simple scheme is proposed for achieving dynamic scheduling in this type of network. The technique permits single hop, packet-switched media access, using only a single fast tunable optical transmitter and two fixed or slow tunable optical receivers per station.

NETWORK ARCHITECTURE

The physical network architecture of EDCCN is an extension of the Distributed Channel Controller Network (DCCN) as described in [3]. DCCN is a single passive optical star coupler network. Each user station is equipped with a single fast-tunable optical transmitter and a fixed-tuned (or slowly tunable) optical receiver. Each wavelength

[1]This work is supported by the Telecommunications Research Institute of Ontario (TRIO)

channel in DCCN is controlled by a channel controller (CC), which is designed to assist stations with media access by reducing the amount of state information that each must track. The channel controllers are co-located and communicate control information over a low-speed parallel backplane.

The DCCN design had two potential system limitations however. The first is an optical power budget limit and the second is a performance limit due to a limited number of wavelengths. EDCCN addresses these limitations through the use of a two level hierarchy which incorporates wavelength spatial reuse.

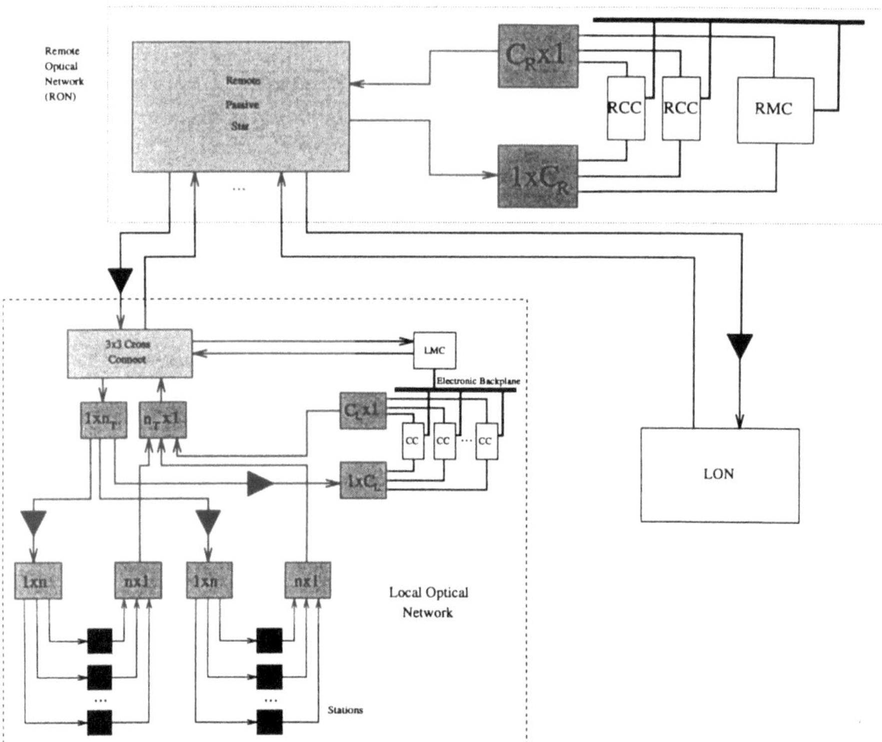

Figure 1. EDCCN Physical Architecture

Figure 1 illustrates the physical topology. The network is divided into M local optical networks or LONs and a single remote optical network or RON. Each user station has access to one LON and the RON. As in [2], the usable optical bandwidth is divided into local and remote wavebands. Each of these is then further divided into a set of local and remote wavelength channels. The channel controllers for the local and remote wavelengths have the same functionality as in DCCN [3]. They provide assistance with media access and synchronization information for their associated stations. A 3×3 wavelength cross-connect links each LON into the RON. It routes the local waveband

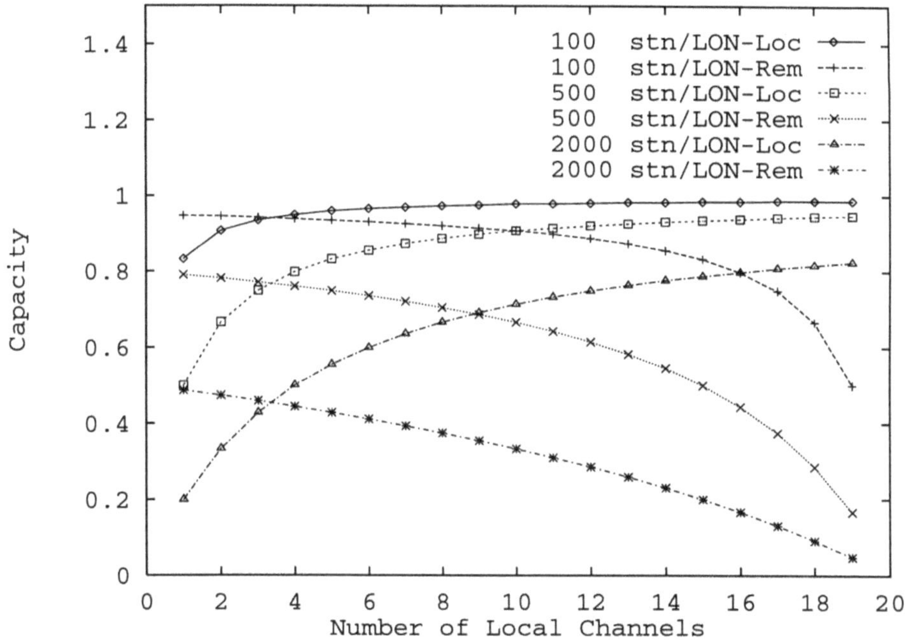

Figure 2. Local and Remote Capacity

from the local input fiber to the local output fiber, creating a passive broadcast-and-select topology within the LON. In addition, it routes the remote waveband from the local input fiber to the remote output fiber and the remote waveband from the remote input fiber to the local output fiber. Finally, it creates a shared signalling network by routing the local waveband from the LMC input fiber to the remote output fiber and the local waveband from the remote input fiber to the LMC output fiber.

Each user station contains two slowly tunable receivers plus a fast tunable transmitter, all capable of tuning to all wavelengths. Each station is pre-assigned a "Local HOME" channel (LHOME) within the local waveband to which it fix-tunes one of its receivers. More than one station can have the same LHOME assignment. Each station is also pre-assigned a "Remote HOME" channel (RHOME) within the remote waveband to which it fix-tunes its other receiver. Stations on different LONs may also share the same RHOME channel assignment. A station uses its LHOME channel to receive data from all other stations on the same LON and its RHOME channel to receive data from stations on other LONs. As in DCCN, each LON has one CC assigned to each local wavelength. The RON consists of a broadcast-and-select topology between all stations on all LONs using the remote set of wavelengths. As in the LON case, a remote channel controller or RCC is associated with each remote wavelength. Using common synchronization signals, each local and remote CC implements an aligned frame structure on each channel.

NETWORK OPERATION AND PERFORMANCE

The local media access protocol for EDCCN is identical to the centralized version of DCCN [3]. Blocks of slots are allocated between each wavelength and all others. A hybrid opto-electronic technique is used where requests are allocated by each channel controller for access to slots on all destination channels for all stations attached to its home channel. In this paper we consider the use of the same protocol in the RON. Although this decouples remote and local operation, the remote channel capacity can become degraded by the large request and allocation overhead if the number of stations is large.

The peak capacity for EDCCN can be easily written as a weighted sum of the normalized local and remote channel capacities, namely $S_T = C_R S_R + C_L \sum_{i=0}^{M} S_{Li}$. Here C_R and C_L are the number of remote and local wavelengths, S_R is the remote capacity, S_{Li} is the capacity of LON i and M is the number of LONs. In Figure 2 the local and remote capacities [3], S_R and $S_{Li} = S_L$, are plotted assuming $C_R + C_L = 20$, $M = 5$, 100 data slots per frame on each channel and a data slot size of $L = 10$ minislots. Different curves are provided for various numbers of stations per LON. It can readily be seen that when the number of stations per LON is large, then the remote capacity can suffer greatly, unless the number of local channels allocated is decreased. Also it is clear that there is a local/remote wavelength allocation that results in the same local and remote channel capacity.

CONCLUSIONS

In this paper we have considered the problem of dynamic scheduling in a 2-level WDM hierarchy with spatial re-use. Both local and remote connectivity is achieved using a straightforward extension of DCCN [3]. This permits dynamic operation with a very small amount of hardware per station. However, the normalized remote capacity may suffer greatly if the number of stations per LON is large or if the remote wavelength allocation is too small.

REFERENCES

[1] K. W. Cheung, "Scalable One-Hop Wavelength-Division Multiplexed Crossconnect", Electronics Letters, 27(4), pp.367-369, 1991.

[2] P. W. Dowd, K. K. Bogineni, K. A. Aly, J. A. Perreault, "Hierarchical Scalable Photonic Architectures for High-Performance Processor Interconnection", IEEE Transactions on Computers, Vol. 42, No. 9, pp.1105-1120, 1993.

[3] M. W. Janoska and T. D. Todd, "A simplified Optical Star Network Using Distributed Channel Controllers", Proc. of GLOBECOM'93, pp.481-487, Houston, Texas, December 1993.

OCRInet: AN ATM BROADBAND RESEARCH NETWORK

Alan Aitken[1], Gavin McLintock[1], Jim Yuan[1], Don Assaff[1], Jim Freeze[2], Robert Kuley[3], Jacek Chrostowski[4], Leo Lax[5], Steve Morrison[6]

[1] OCRInet Inc. 400-340 March Road, Kanata, Ontario K2K 2E4

[2] Bell Ontario, 2 Fieldway Road, Floor 5, Etobicoke, Ontario M8Z 3L2

[3] Communications Research Centre, 3701 Carling Ave., P.O.Box 11490, Station "H", Ottawa, Ontario K2H 8S2

[4] National Research Council, Montreal Road, Ottawa, Ontario K1A 0R6

[5] Newbridge Networks, 600 March Road, P.O. Box 13600, Kanata, Ontario K2K 2E6

[6] Northern Telecom, NT Multimedia Communication Systems, P.O. Box 3511, Station "C", Ottawa, Ontario K1Y 4H7

OCRInet

OCRInet Inc. is a federally incorporated, not-for-profit organization which has developed from an initiative of the Ottawa Carleton Research Institute (OCRI). It was formed on November 17, 1993 and commenced regular network operations on January 23, 1994. It is a partnership between industry, government and post-secondary institutions, and includes service and equipment providers as well as the user community.

OCRInet began with an initial twelve (12) research nodes. Other nodes will be connected as research needs and interests are developed (Figure 1).

The founding partners of OCRInet Inc. are Algonquin College, Bell Canada, Bell Northern Research, Carleton University, Communication Research Centre, Gandalf Technologies, Mitel Corporation, National Research Council, Newbridge Networks, Northern Telecom, Ottawa Carleton Research Institute (OCRI), Stentor Resource Centre, Telecommunications Research Institute of Ontario (TRIO), Telesat Canada and University of Ottawa.

Network Design

OCRInet is a multi-vendor network incorporating ATM equipment from Newbridge Networks and Northern Telecom in the backbone. Equipment from several manufacturers are installed at the research nodes, reflecting both the desire to develop a comprehensive network capability and to optimize the network for different applications.

Applications of Photonic Technology
Edited by G.A. Lampropoulos *et al.*, Plenum Press, New York, 1995

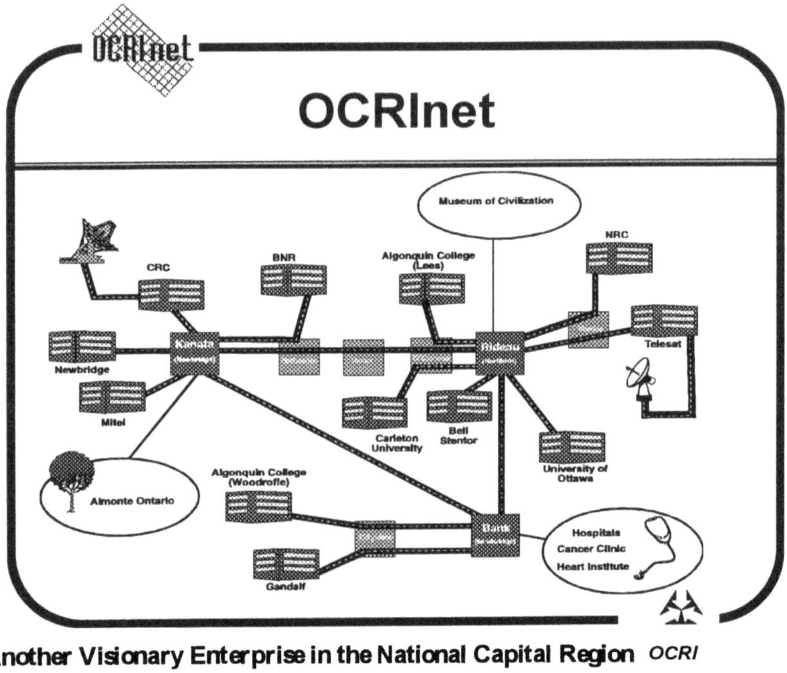

Figure 1

At present 2 DS-3 fibre links connect each of the three central office nodes. Two DS-3 connects are made from a central office location to each of the twelve research nodes. The network uses the User to Network (UNI) interface standards between nodes. The same standards are used to connect the research nodes with the backbone switches. In the absence of standards for Switched Virtual Circuits, OCRInet was established using Permanent Virtual Circuits. The network allows both Constant and Variable Bit rate connections. The circuits can be used for Point to Point communication for interactive/broadcast type video and for one way multicast to allow for broadcast message/data transfers.

Most of the connections between research nodes are between ethernet networks. A number of tests have been undertaken, however, to provide ATM from desktop to desktop, a feature which will be used increasingly in future communication networks. The research network provides facilities for testing different ATM and mixed network architectures.

There is access to satellite services through the CRC and Telesat nodes. Experiments are planned to develop interactive services which can be delivered over the satellite networks to remote locations. There is an interest in up-grading the present capability for delivering medical, educational and public services to remote locations. Major museums have indicated strong interest in delivering their services to distant communities through information retrieval systems accessible over broadband networks.

CRC has installed a Broadband Applications and Demonstration Laboratory (BADLAB) which is designed specifically to provide facilities to access the network, for researchers and companies which do not have a permanent connection to OCRInet.

Objectives of OCRInet

The driving forces for the network are;

- To improve the understanding of the design, management, costing and use of ATM broadband networks

- To develop applications which utilize the potential of broadband communications, in particular the unique capabilities of ATM technology.

The network is designed as a self-contained and operated research network with limitations on the connections which can be made to other networks. Connections are permitted under the following conditions;

1. The connections must be for research purposes;

2. The connection must not be for activities which would normally be delivered over a commercial network;

3. Connections may be made to networks under the control of the OCRInet partner and only for the purpose of research.

Initial Experience

In line with the goal to establish a multi-vendor network, the partners have considerable freedom on the selection of equipment connected to the network, subject only to prior testing of equipment from new suppliers to ensure that it is compatible with the network operation. Testing to-date has indicated that there is good interoperability at the lower layers with limited interoperability of applications.

The error rate for lost packets on a network configuration which twice passed the data through 5 ATM switches from three manufacturers, (four models) was $2*10^{-10}$. The latency through the test network of 10 ATM switches was approximately 2 millisecs. Multi-point video conferencing has successfully demonstrated the feasibility of sending audio, video and data over OCRInet when the network is operating near the maximum data rate. Under certain overload conditions on the network some delays have been observed which are related to the both the ATM network configuration and the application.

At two recent trade shows in Ottawa, OCRInet members were connected to the trade show sites, enabling multi-point communication between the various locations and the trade show. In addition to connecting to the OCRInet Metropolitan Area Network, an additional connection was made through an ATM Wide Area Network to the BCE Shareholder's meeting in Toronto. This complex network, which was demonstrated through the use of desktop video conferencing services resulted in the system being operated at its maximum capacity at some of the switches in the network. In total 12 ATM switches from three manufacturers were involved in the network, together with a variety of other ATM interfaces. LAN to LAN and ATM to the desktop configurations were involved in the demonstration.

Future Plans

The network will expand to accommodate additional users. The emphasis will be to connect users who have clearly defined programs for product and service development leading to new applications that will add value to the future Information Highway. Applications for connecting to the network have been received from hospitals, government services, schools, companies and research groups from across Canada. On July 1, OCRInet and Rnet in B.C. will be connected through DS-3 service as part of the Stentor-Unitel-Canarie agreement to establish a test network for future networking technology and applications. It is envisaged that these connections will be extended to connect other broadband research networks across Canada.

Applications

ATM is still a very new technology which offers many advantages as the switching technology for future communication networks. A summary of the reasons for using ATM technology are shown in Table 1.

Table 1
Advantages of ATM Technology

Feature	Benefit
Fixed 53 Byte Cells	High speed hardware based switching Improved price/performance Low latency Compromize cell size for video, voice, data
Connection oriented	Better control of network Information transmitted over a single path Pay by bandwidth-on-demand (pay by million of cells used) Virtual circuits can be shared with other users improving the use of bandwidth Congestion control
Dedicated, switched bandwidth	Each user has access to the full bandwidth. Use bandwidth as required
Scaleable technology	Since the standards specify only the cell format, the technology is scaleable to a range of transmission speeds.
Virtual LANs	Simplified management Simplified modification of LANs

Applications and R&D Projects which will be run on the network include;

1. Network Management: There is still a limited management capability for ATM networks. Issues such as traffic flow patterns will be studied and compared with traffic models which have been developed.

2. Electronic Newspaper: Building on the publishing, multimedia and distribution skills of the OCRInet partners, a project has been initiated to develop a multimedia news delivery project focussing on the Information Highway. This will act as a test for a full information retrieval system, incorporating advanced multimedia techniques, to be delivered over broadband networks. Many issues arise regarding the configuration of the information, delivery, breadth of information, users and advertising.

3. Medical Services: Major opportunities exist for using advanced networks to reduce the cost of services and to improve the quality of services to centres away from the major urban areas. A project has been initiated to develop and demonstrate applications in this area.

4. Distance training and computer based training: It is essential to make training material readily available to all of society to meet the increasing need for real-time training. This will ensure staff remain current with new technologies and emerging tools which enhance a company's competitive position. This approach to training and education will also have wide applicability in the drive to improve the quality and improve the cost-effectiveness of general education services.

5. Shared CAD Facilities: The concept of the shared workspace will be used to prove-in the capability of remote design teams to work together interactively to jointly develop new products and software using shared resources.

The partnerships which have been formed to develop each of the new applications, include university researchers and industry development staff. These projects build on established research capabilities to develop next generation products which require the bandwidth available on the megabit networks.

OCRInet provides a unique and flexible network to prove-in new applications for broadband networks and to stimulate cooperative research into methods of operation, applications and use of ATM networks, and for the development and testing of ATM equipment.

EFFICIENT CODING AND MAPPING IN PHOTON COMMUNICATION SYSTEMS

G. Cancellieri, F. Chiaraluce, E. Gambi

Dipartimento di Elettronica ed Automatica, Università di Ancona
Via Brecce Bianche, 60131 Ancona (Italy)

INTRODUCTION

Optical communication using photon-counting techniques is very promising for achieving ultimate sensitivity detection limits. In its basic form, an ideal photon counting channel is characterized by two levels of received signals: the one null (0), the other with λ received photons on average (1). No background noise is assumed to affect the transmission and no additional disturbance is present in the receiver. The statistical properties of photon arrivals are those of a pure Poisson process so that, if we set the decision threshold in the receiver between zero and one photon, the probability that a high transmitted signal is received as a low signal results in $p=\exp(-\lambda)$, while the probability that a low transmitted signal is interpreted by the receiver as a high signal is null. Photon counting, therefore, gives a typical example of asymmetric channel. In order to minimize the bit error rate at the output of the receiver, a classical solution is the adoption of an orthogonal pulse position modulation (PPM)[1], to which a suitable external code is applied. PPM is based on the subdivision of the symbol time duration T_s into q adjacent time slots, so that it can be arranged assigning a high level of transmitted signal to only one time slot per symbol. Owing to the properties of the channel, such an alphabet can suffer the presence of erasures only (and not of errors), so that the application of a Reed-Solomon (RS) code, i.e. a code with powerful capabilities of erasures filling, greatly improves the transmission performance[2]. As a counterpart, we have a remarkable bandwidth expansion factor β, which is equal to $q/\log_2 q$ for a q-ary PPM and is increased by the factor n/k (n and k being, respectively, the number of coded and information q-ary symbols for each transmitted codeword) after the application of the RS code.

RS codes, however, were originally conceived for fighting errors in additive white gaussian noise (AWGN) symmetric channels. So, one can think that better performance are achievable using solutions able to exploit, more directly, the ideal photon counting peculiarities. The aim of the present paper is to show that, when small, and then practical, values of β are considered, more efficient schemes can be found, based on suitable choices of the input a-priori probabilities, of the assignation rule of an information bit pattern to the various permitted symbols (i.e. mapping) and, when applied, of the block coding strategy.

Applications of Photonic Technology
Edited by G.A. Lampropoulos *et al.*, Plenum Press, New York, 1995

GENERAL CONSIDERATIONS

The most obvious approach, for improving the bit error rate performance of an ideal photon counting channel, consists in the adoption of asymmetric error correcting codes[3], that is codes correcting only $1 \rightarrow 0$ type transition errors. Because of such specialization, they are characterized by a weight distribution which is better than that of symmetric error correcting codes (for example BCH codes) with the same number of information bits and parity check bits. These codes are usually applied starting from a binary source which emits low and high levels with the same a-priori probability. A more attractive technique consists instead in diversifying such input distribution, assigning to signal 1 a probability x to be transmitted generally different from that, $1 - x$, assigned to signal 0[4]. In this way, maximum values of the transmission efficiency, that is the information rate per photon, can be reached.

The same objective of different a-priori probabilities for the two input signal levels can be pursued introducing transmission alphabets which appear as suitable extensions of the PPM format described above. In particular, the so called multi pulse position modulation (MPPM)[5] and the complete multi pulse position modulation (CMPPM)[6], increasing significantly the number of permitted symbols for a fixed value of q, allow to limit the bandwidth occupancy, so compensating the penalty involved by the appearance of errors at the receiver. Setting r an integer greater than one, in MPPM each symbol has exactly r time slots at high signal level, whereas CMPPM comprehends all the symbols having 1, 2,..., r time slots with high signal level. In principle, efficient puncturations can be conceived, in order to obtain alphabets with reduced number of symbols but with special correction capabilities[7]. For example, we can constrain the asymmetric distance to be equal to two, so realizing an alphabet intrinsically able to correct single erasures. Unfortunately, transmission efficiency is reduced by puncturation to the point that, at a parity of bandwidth occupancy, a complete MPPM alphabet can be always found, with quite comparable global performance. Another crucial point regards the role of mapping. A good assignation of the information bit patterns outgoing the source to a particular symbol of the alphabet greatly reduces the so called assumed bit error rate, that is the probability that the values attributed to unknown bits, consequent to transmission errors which produce unpermitted symbols, are incorrect.

Some of the schemes described above will be compared, in the next paragraph, with reference to a number of explicative cases. For the sake of brevity, we will omit any analytical detail which, however, can be found in the references quoted. Particular emphasis will be given to the results achievable in the region of small β, since it is that of major practical interest. For making our analysis complete, also convolutional codes should be considered[8]. It is known, in fact, that short constraint length convolutional codes with Viterbi decoders compare favourably, as regards performance and complexity, with the system proposed in[2]. Nevertheless, the main justification of their success can be found in the adaptability to use soft-decision demodulation, much harder to be realized using RS codes. This property, in turn, seems significant only in presence of strong background noise which, instead, is neglected in our analysis. For such reason, we will not examine, in the following, convolutional codes, leaving their performance characterization to the next paper.

RESULTS

Let us consider the curves of Fig. 1 where the bit error rate P_{eb} is plotted, as a function of the transmission efficiency ρ, expressed in bit/photon, for a value of $\beta \approx 4.66$. Such a bandwidth expansion factor can be achieved using a PPM transmission with q=8, to which a RS(7,4) code is applied (curve a), but also using a binary transmission with different a-priori probabilities[3] (curve b) or a binary asymmetric code with n=127 and k=27, directly applied to the binary equiprobable source (curve c). The curves have been plotted assuming, along

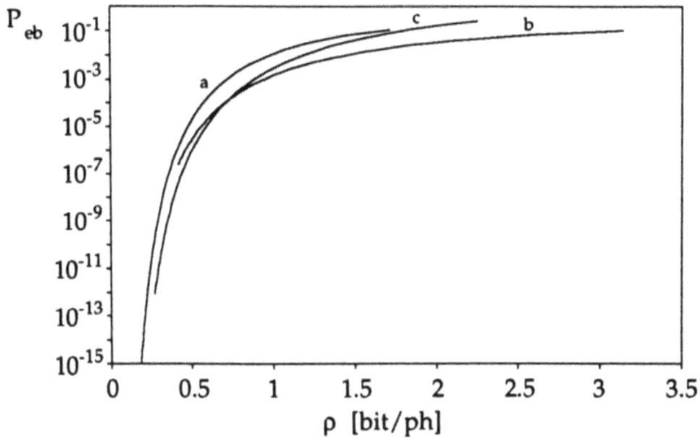

Fig. 1 - Bit error rate as a function of transmission efficiency for some schemes with $\beta \approx 4.66$.

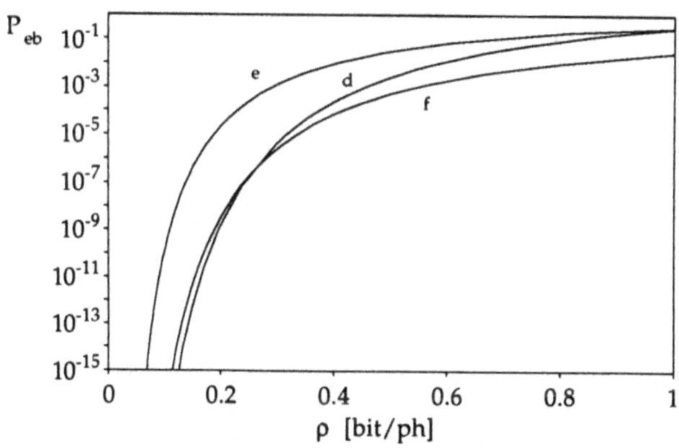

Fig. 2 - Effects of puncturation on an MPPM alphabet with a very large number of symbols.

Table 1 - Optimum mapping for an alphabet able to correct single erasures.

Bit pattern	Symbol pattern
000	0000000
001	1110000
010	1001100
011	0101010
100	0010110
101	0011001
110	0100101
111	1000011

them, different λ. It is clear from the Figure that solutions b and c show performance better than a. In particular, the assumption of different a-priori probabilities of the input binary symbols allows to maximize the transmission efficiency. Unfortunately, very small values of P_{eb} can be reached only increasing significantly the number of photons. Correspondingly, the adoption of the asymmetric code becomes preferable: it gives a bit error rate which is even two orders of magnitude smaller than that provided by the PPM plus RS code solution.

When it is important to realize a significant reduction of the bandwidth expansion factor, transmission alphabets characterized by a very large number of symbols are necessary. Elimination of some of them could be suggested, with the goal of attributing, to the punctured alphabet so obtained, the capacity of correcting a finite number of erasures. An example is shown in Fig. 2, where we have denoted by d the curve relative to a set of 759 symbols obtained by puncturation of an MPPM alphabet with q=24 and r=8. The bandwidth expansion factor of the original alphabet ($\beta \approx 1.23$) is more than doubled by puncturation (up to $\beta \approx 2.51$), remaining, however, within acceptable limits. On the other hand, all the residual symbols have weight 8 and are characterized by an asymmetric distance equal to 4; as a consequence, this alphabet is intrinsically able to correct 3 erasures. Incidentally, it is worth to stress that it is obtained as a subset of the Golay (24, 12) binary code, well known in the theory of AWGN channels. The performance improvement due to puncturation is clearly evident in the Figure, where the result obtained by the original alphabet (curve e) has been plotted for the sake of comparison. On the other hand, several complete MPPM alphabets can be found, characterized by the same bandwidth expansion factor of the punctured solution. For example, we can assume q=30 and r=3, so obtaining the curve denoted by f in Fig. 2, which is only slightly worse than curve d in the region of P_{Eb} sufficiently small. Since a quite similar curve can be achieved with different choices of q and r, we can conclude that if the transmission efficiency (and not, simply, the average number of received photons) is the parameter of interest, no particular advantage is taken from the application of this kind of puncturation. Such a statement was not so obvious before our analysis.

Curves of Fig. 2 have been obtained under the pessimistic hypothesis that the bits relative to an unpermitted received symbol are randomly chosen. This assumption leads to an assumed bit error rate equal to 1/2. In reality, it is just in the nature of the asymmetric channel the possibility of reducing this value through a suitable mapping. An example is reported in Table 1 for a punctured alphabet derived by an MPPM with q=7 and r=3 and able to correct single erasures. The alphabet includes also the sequence everywhere null, which has been added in such a way that each symbol carries an integer number of bits. For values of λ sufficiently high to make negligible the contribution to the bit error rate due to three or more erasures, it is possible to show that this optimized mapping leads to an assumed bit error rate approximately equal to 0.2857, even more favourable than that obtained by a Gray code.

REFERENCES

1. J. R. Pierce, Optical channels: practical limits with photon counting, *IEEE Trans. Commun..*, COM-26: 1819 (1978).
2. R. J. McEliece, Practical codes for photon communication, *IEEE Trans. Inform. Theory*, IT-27: 393 (1981).
3. T. Uyematsu, K. Yamazaki, O. Hirota, M. Nakagawa, K. Sakaniwa, Effect of asymmetric error correcting codes in photon communication systems, *Trans. IEICE*, E71: 850 (1988)
4. G. Cancellieri, Transmission efficiency in photon counting channels, *IEEE Trans. Commun.*, COM-37: 183 (1989).
5. H. Sugiyama, K. Nosu, MPPM: a method for improving the band-utilization efficiency in optical PPM, *J. Lightwave Technol.*, LT-7: 465 (1989).

6. G. Cancellieri, F. Chiaraluce, E. Gambi, A new method for improving the bandwidth utilization efficiency in optical PPM, in: "Lecture Notes in Physics", 378, C. Bendjaballah, O. Hirota, S. Reynaud, eds., Springer-Verlag, Berlin Heidelberg (1991).
7. G. Cancellieri, F. Chiaraluce, E. Gambi, PPM transmission over a photon counting channel: comparison among various transmission formats, in preparation.
8. J. L. Massey, Capacity, cutoff rate and coding for a direct-detection optical channel, *IEEE Trans. Commun.*, COM-29: 1615 (1981).

STIMULATED BRILLOUIN SCATTERING IN OPTICAL FIBERS WITH DISTRIBUTED GAIN

Mário F. Ferreira

Department of Physics, University of Aveiro
3800 Aveiro, Portugal

INTRODUCTION

Recent experimental and theoretical studies have shown the desirability and feasibility of constructing distributed fiber amplifiers (achievable in optical fibers lightly doped with rare-earth ions) of as many as tens of kilometers in length for use in fiber communication links[1]. In this context, it is important to study the stimulated Brillouin scattering (SBS) in optical fibers in the presence of distributed gain. In fact, we can easily admit that even a weak narrow-band signal (pump) will determine the appearance of an intense amplified Stokes wave, which, besides to deplete the pump, will affect strongly the optical source[2]. Furthermore, the appearance of higher Stokes orders will become also readily observable.

Of all nonlinear processes in optical fibers, stimulated Brillouin scattering has the lowest threshold for narrow-band light. An irradiated light wave creates a sound wave, off which light is scattered. The backscattered Stokes wave thus generated is Doppler shifted in frequency by an amount determined by the velocity of sound in the glass. The presence of optical gain will cause growth in both the pump and the Stokes power, thus significantly affecting this nonlinear process.

Bayvel et al.[3] solved numerically the SBS equations, considering only the pump and the first-order Stokes waves, for the case of a lossy fiber. In this paper we report the results of numerical solutions of the SBS equations in optical fibers with distributed gain instead of loss, including an equation for the second-order Stokes wave. We also present an approximate analytical solution of the three nonlinear power coupled equations, which becomes exact for the case of an optical fiber with no net gain or loss.

THEORY

The equations describing the evolution of the signal (P_p), first-order Stokes (P_{s1}) and second-order Stokes (P_{s2}) along an optical fiber with distributed amplification are:

$$dP_p/dz = -g_B P_p(P_{s1} + P_{n1})/A + \gamma P_p \tag{1a}$$

$$dP_{s1}/dz = -g_B P_p(P_{s1} + P_{n1})/A + g_B(P_{s1} + P_{n1})(P_{s2} + P_{n2})/A - \gamma P_{s1} \tag{1b}$$

$$dP_{s2}/dz = g_B(P_{s1} + P_{n1})(P_{s2} + P_{n2})/A + \gamma P_{s2} \tag{1c}$$

Applications of Photonic Technology
Edited by G.A. Lampropoulos *et al.*, Plenum Press, New York, 1995

where g_B is the SBS gain constant, A the effective core area, P_{n1} and P_{n2} the Stokes input noise powers, and $\gamma = g - \alpha$, g being the amplifier gain (assumed independent of z) and α the fiber loss. The Eq.s (1a) - (1c) have no analytical solutions when $\gamma \neq 0$. For $\gamma = 0$, however, we have the following invariant:

$$P = P_p(z) - (P_{s1}(z) + P_{n1}) + P_{s2}(z) + P_{n2} = P_p(0) - (P_{s1}(0) + P_{n1}) + P_{s2}(0) + P_{n2} \quad (2)$$

Inserting Eq. (2) into Eq.(1) gives:

$$dP_{s1}/dz = -2(g_B/A)(P_{s1} + P_{n1})P_p(0).\exp\{-(g_B/A)\int_0^z(P_{s1} + P_{n1})dz'\} + (g_B/A)(P_{s1} + P_{n1})(P + P_{s1} + P_{n1}) \quad (3a)$$

$$P_p(z) = P_p(0).\exp\{-(g_B/A)\int_0^z(P_{s1} + P_{n1})dz'\} \quad (3b)$$

$$P_{s2}(z) = P_{s2}(0).\exp\{(g_B/A)\int_0^z(P_{s1} + P_{n1})dz'\} \quad (3c)$$

From Eq.s (3a) - (3c) we obtain the following solutions:

$$P_p(z) = 2P_p(0)[P_{s2}(0) + P_{n2}][C.\exp\{-Bg_Bz/A\} + 1]/ [(P + B)C.\exp\{-Bg_Bz/A\} + P - B] \quad (4)$$

$$P_{s1}(z) = 2B^2C.\exp(-Bg_Bz/A)/[((P+B)C.\exp\{-Bg_Bz/A\}+P-B) (C.\exp\{-Bg_Bz/A\} + 1)] - P_{n1} \quad (5)$$

$$P_{s2}(z) = [(P+B)C.\exp\{-Bg_Bz/A\}+P-B] /[2(C.\exp\{-Bg_Bz/A\} + 1] - P_{n2} \quad (6)$$

where

$$B = \sqrt{P^2 - 4P_p(0)(P_{s2}(0) + P_{n2})} \quad (7)$$

$$C = [B + 2(P_{s2}(0) + P_{n2}) - P]/[B - 2(P_{s2}(0) + P_{n2}) + P]. \quad (8)$$

The Eq.s (4)-(8) show that the Brillouin Stokes powers build up not only from the injected signals $P_{s1}(L)$ and $P_{s2}(0)$ but also from the spontaneous emission P_{n1} and P_{n2} along the fiber. These noise powers can be viewed as equivalent input Stokes signals provided that spontaneous Brillouin scatterings have developed fully enough to have significant amplification. Thus, the amplified spontaneous scattering could be modeled alternatively by ignoring the spontaneous contribution altogether in (1) with assumption of an equivalent signal of P_{n1} and P_{n2} at fiber input. In the presence of real input signals $P_{s1}(L)$ and $P_{s2}(0)$, P_{n1} and P_{n2} can be omitted since one usually has $P_{s1}(L) \gg P_{n1}$ and $P_{s2}(0) \gg P_{n2}$.

RESULTS AND DISCUSSION

The fourth order Runge-Kutta algorithm was used in solving the Eq.s (1a) - (1c) using a step size $\Delta\zeta = \Delta z/L = 0.002$, L being the fiber length, and $P_{s1}(L) = P_{s2}(0) = 0$, $P_{n1} = P_{n2} = 0.5$ nW [4]. The accuracy of the numerical method was verified by comparison with the exact analytical solutions to the equations given by Eq.s (4) - (8) for the case $\gamma = 0$.

Fig.1 illustrates the spatial dependence of the pump (a), first-order and second-order Stokes powers (b) against the normalized distance $\zeta = z/L$, for an initial pump power $P_p(0) = 1$ mW and for several values of the fiber distributed gain. We assume a fiber length L = 20 km, an effective core area A = 28.26 x 10^{-12} μm^2, and a Brillouin gain coefficient g_B = 2.2 x 10^{-11} m/W, this value corresponding to the case of a nonpolarization maintaining optical fiber.

In the case of a lossy fiber, the pump power decreases monotonically along the fiber (Fig. 1a), while the first-order Stokes power remains five orders of magnitude below the pump power (fig. 1b). For a fiber with no net gain or loss ($\gamma = 0$) the pump power decreases slightly in the first one tenth of the fiber length to 0.81 mW and remains thereafter practically constant, while the first-order Stokes power increases to 0.19 mW and the second-order one remains negligible. When the fiber has a positive distributed gain the pump power presents a non-monotonous evolution and increases after passing through a minimum value (at $z/L \approx 0.2$). The variation of the pump power spatial distribution is more significant for greater values of the distributed gain (Fig. 1a). For the case γ = 2 dB/km, both the first-order and the second-order Stokes powers become of the same magmitude of the pump power.

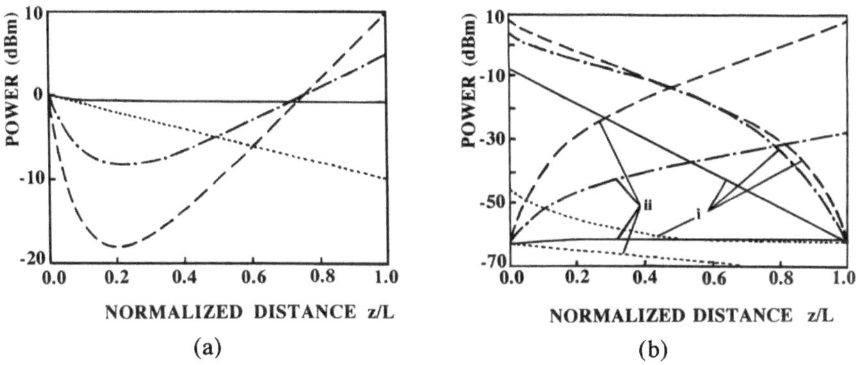

Figure 1. Pump power (a), first-order (i) and second order (ii) Stokes power (b) against the normalized fiber length for γ = -0.5 dB/km (-------), γ = 0 dB/km (———), γ = 1 dB/km (— · —) and γ = 2 dB/km (— — —).

It can be verified that for a fiber with no net gain or loss ($\gamma = 0$) the pump power is reduced by less then 1% for fiber lengths $L \leq 12$ km, which corresponds to the first-Stokes power condition $P_{s1}(0) < 10\ \mu W$. In this case, the second-order Stokes power remains negligible. However, for a fiber with a net gain γ = 1.5 dB/km the second-order Stokes power increases by 7 orders of magnitude and surpasses the first-order power for L > 26 km (Fig. 2). In this case, the first-order Stokes power achieves the value $P_{s1}(0)=10\ \mu W$ ($P_{s1}(0)/P_p(0)$ = 1%) for L = 4.7 km, which corresponds to an effective fiber length $L_{eff}=\{\exp(\gamma L)-1\}/\gamma \approx 12$ km, similar to the value refered above for the case γ=0. We can also observe from Fig. 2 that the output pump power varies non-monotonically with the amplifier length, presenting a local maximum for $L \approx 6$ km.

From Eq.s (1a) - (1c) the following relation between the pump power and the second-order Stokes power can be obtained:

$$\log_{10}[P_{s2}(L)] + \log_{10}[P_p(L)] = \log_{10}[P_{s2}(0)P_p(0)] + 2\gamma L/\ln(10) \qquad (9)$$

This relation can be confirmed from Fig. 2. In particular, when $P_p(L) \approx$ constant we see that $P_{s2}(L)$ (in dBm) increases with L and presents a slope given by $2\gamma/\ln(10)$.

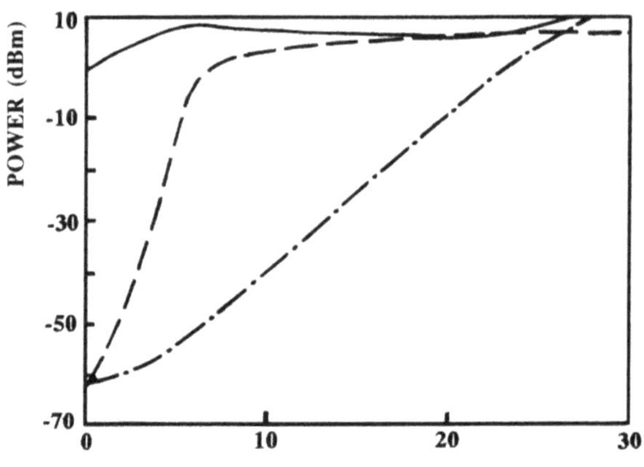

Figure 2. Pump (———), first-order (— — —) and second-order (—— - ——) Stokes power against the amplifier length for the case γ = 1.5 dB/km.

CONCLUSIONS

We have numerically solved the SBS equations for the case of an optical fiber with distributed gain or loss, including the second-order Stokes wave. An analytical solution for the case of no net gain or loss was also obtained. Considering a typical input signal (pump) power of 1 mW, the results show that the presence of distributed gain in an optical fiber leads to ready generation of multiple Stokes orders. It can be easily anticipated that above certain values of the fiber length and of the fiber gain the threshold power will become comparable with the thermal noise power in the fiber, meaning that threshold will always be exceeded. We also observed a significant variation of the signal power along the fiber amplifier. Concerning this aspect, an optimum situation is achieved using an optical fiber with approximately zero net gain or loss and a maximum length of about 12 km. These conclusions are particularly important when lightly doped optical fibers are used as long distributed amplifiers in optical communication links.

REFERENCES

1. K. Rottwitt, A. Bjarklev, J. H. Povlsen, O. Lumholt, and T. P. Rasmussen, "Fundamental design of a distributed erbium-doped fibre amplifier for long-distance transmission", J. Lightwave Technol., 10:1544 (1992).
2. M. F. Ferreira, J. F. Rocha, and J. L. Pinto, "Noise and modulation performance of Fabry-Perot and DFB semiconductor lasers with arbitrary external optical feedback", IEE Proc. - J, Optoelectron., 137:361 (1990) .
3. P. Bayvel, and P. M. Radmore, "Solutions of the SBS equations in single-mode optical fibres and implications for fibre transmission systems", Electron. Lett., 26:434 (1990).
4. M. F. Ferreira, J. F.Rocha, and J. L. Pinto, "Impact of stimulated Brillouin scattering on fibre Raman amplifiers", Electron. Lett., 27:1576 (1991).

LIMITATIONS IMPOSED BY STIMULATED RAMAN SCATTERING ON 1550 nm HIGH-ENERGY PULSE TRANSMISSION

C.W. Barnard,[1] P. Myslinski,[1] J. Chrostowski,[1]
X. Pan,[2] and M. Kavehrad[2]

[1]Institute for Information Technology, National Research Council of Canada, Ottawa, Ontario, Canada K1A 0R6
[2]Department of Electrical Engineering, University of Ottawa, Ottawa, Ontario, Canada K1N 6N5

INTRODUCTION

Several applications have been proposed that transmit high-energy 1550 nm pulses over optical fiber. These include remote optical powering[1] and remote optical switching.[2,3] Pulses with peak power greater than 1 W and nanosecond duration are affected by stimulated Raman scattering (SRS). The high-energy pulses generate Stokes pulses by the process of amplified spontaneous Raman scattering[4]. This paper presents an experimental and theoretical study of the limitations imposed by SRS on pulses with peak power up to 60 W and duration on the order of 10 ns. Experiment shows and theory confirms that there is an optimum input power for a given pulse duration and distance that maximizes the energy transmitted over long fibers. To the best of our knowledge, this is the first demonstration of Stokes pulse generation from a pump pulse within the 1550 nm window.

EXPERIMENT

Pulses from a 1530 nm Q-switched erbium-doped fiber laser[5] with pulse duration (defined as the full-width at half maximum) of 13 ns or 18 ns were launched into standard single-mode fiber of length from 12 km to 50 km. The input peak power was adjustable to over 60 W. The temporal and spectral distributions of the output pulses were monitored at 12 km intervals with a photodiode and an optical spectrum analyzer. Figure 1a shows the measured output powers for 13 ns pulses propagated through 12 km of fiber. For input powers less than about 8 W, the output pulses resemble the input pulses with 0.25 dB/km attenuation. When the input power is greater than the Raman threshold (~8 W), a Stokes pulse trails the pump pulse by 20–30 ns. The output pump pulse is shorter than the input pulse because of depletion of the rear of the pulse by the Stokes pulse. As the input power is increased above the threshold, the peak Stokes power increases to a maximum for a 33 W input and then decreases to a plateau for input powers greater than 43 W. For input powers far above threshold there is also a smearing, in time and frequency, of the Stokes

Applications of Photonic Technology
Edited by G.A. Lampropoulos *et al.*, Plenum Press, New York, 1995

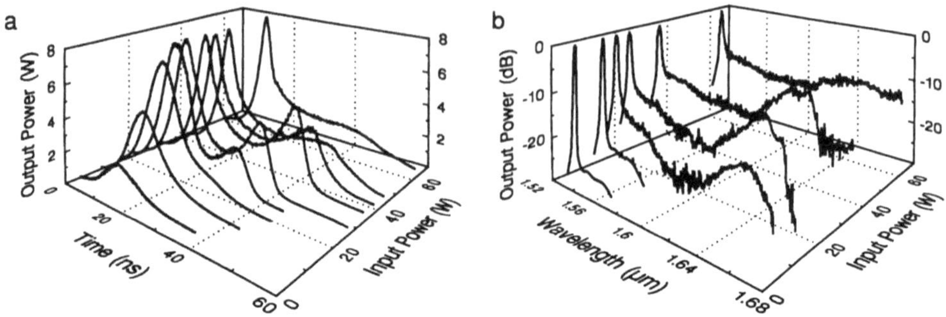

Figure 1. Observed output (a) pulse shape and (b) spectra after 12 km of fiber for 13 ns pulses.

pulse due to the group velocity variation across the Stokes spectrum. As shown by the output spectra in Fig. 1b, the Stokes pulse has a broad spectrum (~30 nm) centered around 1640 nm. For very high input powers a continuum is generated with spectral width larger than 150 nm.

As shown in Fig. 1a, the walkoff of the pump and Stokes pulse, caused by chromatic dispersion, increases with input power since the Stokes pulse is generated at an earlier point in the propagation for greater input powers. The measured walkoff was about 2.4 ns/km, which is expected for 115 nm spectral separation in standard SMF-28 single-mode fiber with 20 ps km^{-1} nm^{-1} dispersion.

In Fig. 2a the measured output pump power and pulse duration are shown as functions of the input peak power. Above threshold, the pump is depleted until its peak power is below the threshold. Thus, above threshold the output pump power is insensitive to the input power and is fixed at about 8 W. Beyond threshold the rear of the pump pulse is depleted by the Stokes pulse, which decreases the output pump pulse duration and hence decreases the output pump energy. Therefore, for a given pulse duration there is an optimum input power that maximizes the transmitted pump energy. This experimental conclusion is confirmed by the theoretical simulations described in the next section.

THEORY

The propagation of the pump and Stokes pulses was modeled with two coupled differential equations that accounted for attenuation, first-order and second-order dispersion, self-phase and cross-phase modulation, and Raman gain.[4] The equations were solved

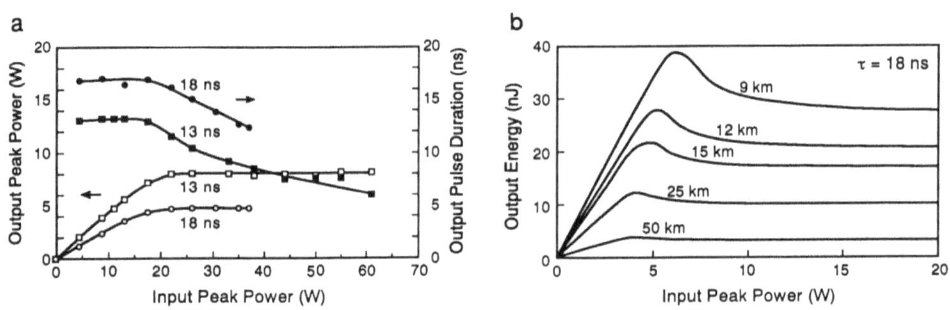

Figure 2. (a) Observed output pump pulse power and duration vs input peak power for 13 ns and 18 ns pulses transmitted through 12 km of fiber and (b) theoretical output pump energy vs input peak power for 18 ns input pulses.

by the beam propagation method.[4] Ideally, at each distance step the spontaneous Raman generation and Raman amplification should be integrated over the entire Raman bandwidth. Reasonably good agreement with experiment was obtained, however, by assuming that the Raman generated light is monochromatic at the peak of the gain spectrum and that the Raman gain coefficient equals the value at the gain peak. These assumptions caused the simulation to predict slightly greater Raman conversion than observed experimentally.

The number of Stokes photons spontaneously generated along the fiber was approximated by an equivalent Stokes c.w. input power:[6]

$$P_S(0, t) = \frac{\sqrt{\pi} hc\Delta f}{2\lambda_S} \left(I_p g_R L_w\right)^{-1/2} \tag{1}$$

where Δf is the Raman bandwidth, I_p is the input pulse peak intensity, g_R is the Raman gain coefficient, and L_w is the walkoff length defined by $L_w = T_0 \left|1/v_{gp} - 1/v_{gS}\right|^{-1}$, where $T_0 = T_W/1.763$ for a sech pulse with duration T_W and v_{gp} and v_{gS} are the pump and Stokes group velocities. This assumption was tested by simulating propagations with different equivalent input Stokes powers. Increasing the equivalent Stokes input power by three orders of magnitude led to only a doubling of the output Stokes power. The model is therefore almost insensitive to the equivalent Stokes input. The calculations used a pump wavelength of 1530 nm, Stokes wavelength of 1630 nm, pump attenuation of 0.25 dB/km, Stokes attenuation of 0.8 dB/km, group index of 1.462, $\Delta g_R = 3.2 \times 10^{-14}$ m/W, $\Delta f = 7$ THz, and a second-order dispersion of -21 ps^2/km at the pump wavelength.

As expected, the theoretical Raman conversion was higher than that observed experimentally. In agreement with experiment, the theory predicts depletion-induced narrowing of the pump pulse and saturation of the output pump power. The distance threshold (minimum length of fiber required for significant Stokes conversion) decreases as the input pump power increases. As a consequence, for a given fiber length there is an input power threshold (minimum input power for significant Stokes conversion). As the input power increases, the depletion rate increases and hence the distance over which the Raman conversion occurs decreases. For example, an 11 W, 18 ns pulse reaches the Raman threshold after 6 km and depletion then occurs from 6 to 10 km with an initial depletion rate of 1.3 dB/km , and a 19 W input pulse reaches the Raman threshold after 3 km and depletion then occurs from 3 to 5 km with an initial depletion rate of 4 dB/km.

Figure 2b shows how the theoretical output energy depends on input power. Increasing the input power yields an increase in the output pump energy until the Raman threshold is reached. Further increasing the input power results in less output energy because of pump depletion. For higher input powers the depletion is complete, resulting in constant output energies. For shorter fibers the peak is more pronounced since the threshold occurs closer to the end of the fiber. Hence, Raman conversion is incomplete for a larger range of input powers. For a fixed propagation distance, the pump depletion is stronger at higher powers, since the propagation distance to reach threshold is shorter. Therefore, as shown in Fig. 2b, for a given pulse duration and propagation distance, there is an optimum input power that maximizes the output pulse energy. This agrees with the experimental results shown in Fig. 2a.

For a given pulse duration and fiber length, the input power that maximizes the transmitted pump energy occurs at the peak of each curve shown in Fig. 2b. The optimum transmitted peak power and energy are plotted in Fig. 3 for the indicated pulse durations. The curves in this figure were derived by finding, for a given length and pulse duration, the input power that maximized the output energy. For pulses longer than 1 ns, the optimum peak power is just below the SRS threshold, which is independent of pulse

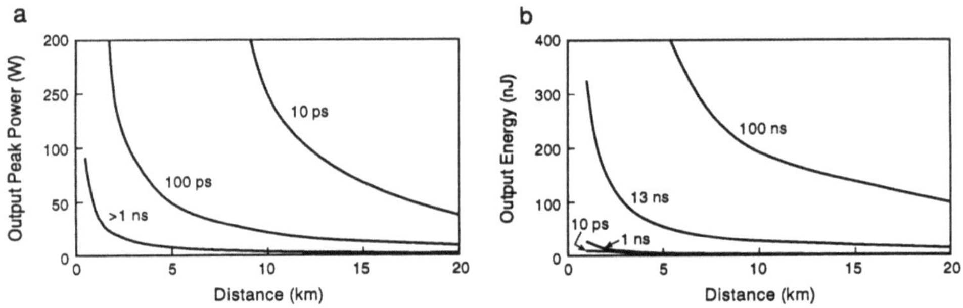

Figure 3. Optimum transmitted (a) pulse power and (b) energy.

duration. Because of walkoff, pulses shorter than 1 ns are not fully depleted by Raman conversion, so that higher peak powers can be transmitted.

For a given output pump energy there is a trade-off between the pulse duration and distance. For example, if a remote optical switch requires 20 nJ of energy from a distance of 10 km, then the pulse duration and hence switching time are limited to being greater than about 10 ns. More pulse energy can be transmitted by using longer pulses with peak powers below the Raman threshold.

CONCLUSIONS

This study demonstrates the SRS of high-energy 1550 nm pulses transmitted over long single-mode fibers. Experiment shows and theory confirms that the amount of energy or power that can be delivered within a given time is limited by SRS, regardless of the input power. There is an optimum input power that maximizes the energy delivered for a given pulse duration and distance. For efficient transmission of a given amount of energy, the input power must be below the SRS threshold. For applications that require a certain peak power, SRS can be avoided by transmitting very short pulses. For applications that require a certain pulse energy, SRS can be avoided by transmitting longer pulses with lower peak power. And, if a certain energy and rise time are required, SRS limits the transmission distance.

REFERENCES

1. T.C. Banwell, R.C. Estes, L. A. Reith, P.W. Shumate, and E.M. Vogel, Powering the fiber loop optically — A cost analysis, *J. Lightwave Technol.* 11:481 (1989).
2. P. Myslinski, C. Barnard, G. Cheney, J. Chrostowski, B. Syrett, and J. Glinski, Nanosecond all-optical gain switching of an erbium-doped fibre amplifier, *Opt. Commun.* 97:340 (1992).
3. C. W. Barnard and J. Chrostowski, Polarization independent transmissive/reflective optical switch, U.S. patent application 08/099453 (1993).
4. G. P. Agrawal, "Nonlinear Fiber Optics", Academic Press, Toronto (1989).
5. P. Myslinski, J. Chrostowski, J.A. Koningstein, and J.R. Simpson, High power Q-switched erbium fiber laser, *IEEE J. Quantum Electron.* 28:371 (1992).
6. D. Schadt, "Contributions to Nonlinear Pulse Propagation in Single Mode Optical Fibers," Ph. D. Thesis, The Royal Institute of Technology, Stockholm (1989).

ON A FORMULATION FOR THE PULSE JITTER INDUCED BY SPONTANEOUS-EMISSION NOISE OF OPTICAL AMPLIFIERS

Adolfo V.T. Cartaxo

DEEC/IT, Instituto Superior Técnico
Av. Rovisco Pais, P-1096 Lisboa, Portugal

1. Introduction

Up to now, research in optical amplifiers has been done mainly concerning noise and gain features[1]. Much less attention has been paid to the pulse jitter, at the optical amplifier output, induced by spontaneous-emission noise. In[2], the average arrival time of the pulse at the amplifier output was used as time reference upon which the pulse jitter was defined. Here, this approach is used to evaluate the pulse jitter at the output of an optical fibre link due to the amplified spontaneous emission (ASE) of optical amplifiers. To achieve this purpose, the pulse power at the output of a system composed by an optical amplifier followed by the optical fibre is derived (section 2). In section 3, the pulse jitter is evaluated and an illustrative example is presented and discussed. Then, assuming statistical independence between the noise due to different optical amplifiers, the output pulse jitter of an optical fibre link (chain of optical amplifiers) is assessed and a comparison with the pulse jitter in soliton propagation is done (section 4). In section 5, conclusions are outlined.

2. Pulse power at the fibre output

We assume that the optical field at the output of the amplifier, $\psi_o(t)$, is the result of the addition of two independent components: the amplification of the input signal and the ASE noise, $n_o(t)$. Denoting by ω_0 the optical carrier angular frequency, the optical field at the amplifier output is expressed as $\psi_o(t) = f(t)e^{j\omega_0 t} + n_o(t)$ where $f(t)$ is the modulation envelope (square root of the power pulse shape) at the amplifier output. The ASE noise field is assumed a zero mean, narrow band Gaussian stationary process and, considering both directions of polarization, its power spectrum is given by[3] $S_o(\omega) = 2n_{sp}(G-1)h\omega/(2\pi)^2$ where n_{sp} is the spontaneous emission noise factor, G is the optical amplifier gain, hf is the photon energy at the optical frequency f ($\omega = 2\pi f$) and the Fourier transform of $\psi_o(t)$ is defined as $\Psi_o(\omega) = \int_{-\infty}^{+\infty} \psi_o(t)e^{-j\omega t}\, dt/(2\pi)$.

Due to the noise stationarity the frequency components of the noise, $N_o(\omega)$, hold the spectral properties[4]: $E\left[N_o(\omega)N_o^*(\omega')\right] = S_o(\omega)\cdot\delta(\omega-\omega')$ and $E\left[N_o(\omega)N_o(\omega')\right] = 0$ for ω and ω' both positive, besides $E\left[N_o(\omega)\right] = 0$.

We express the output noise by $n_o(t) = n(t)e^{j\omega_0 t}$ because it simplifies the expression of the output field which becomes $\psi_o(t) = a(t)e^{j\omega_0 t}$ with $a(t) = f(t) + n(t)$. The power spectrum of $n(t)$ and the power spectrum of $n_o(t)$ are related by $S_o(\omega) = S_n(\omega - \omega_0)$. So, $\Psi_o(\omega) = A(\omega - \omega_0)$ with $A(\omega) = F(\omega) + N(\omega)$.

The propagation of the light at the output of the amplifier along a monomode fibre can be described by the propagation term $e^{-j\beta z}$ with the transmission distance, z, and the propagation constant, β. The amplitude of the lightwave travelling in the fibre can be expressed by means of the Fourier transform of the amplitude spectrum at the input to the fibre[5], $\psi(z,t) = \int_{-\infty}^{+\infty} \Psi_o(\omega)e^{j(\omega t - \beta z)}\, d\omega$, and the optical power is given by $P(z,t) = |\psi(z,t)|^2$. If we approximate the propagation constant β as a function of the difference $\omega - \omega_0$ and terminate this expansion after the third term, the optical power can be written as the Fourier integral

$$P(z,t) = \int_{-\infty}^{+\infty} G(z,x) \exp\left[jx\frac{t - \beta_0' z}{T}\right] dx \tag{1}$$

with $x = \omega T$ where T is the bit period, $G(z,x)$ is defined by the integral

$$G(z,x) = \frac{1}{T} e^{-j\frac{\beta_0'' z}{2}\frac{x^2}{T^2}} \int_{-\infty}^{+\infty} A\left(\frac{x}{T} + \omega\right) A^*(\omega) e^{-j\beta_0'' z \frac{x}{T}\omega}\, d\omega \tag{2}$$

and the prime indicates differentiation with respect to ω and the subscript in β_0' and β_0'' means that the derivative is evaluated at $\omega = \omega_0$.

3. Problem statement: the average arrival time

As in[2], the average arrival time of the pulse at the fibre output is used as the time reference upon which the jitter will be evaluated. The average arrival time of the pulse is defined as[5] $\bar{t}_0 = \int_{-\infty}^{+\infty} tP(z,t)\, dt \,/\, \int_{-\infty}^{+\infty} P(z,t)\, dt$ which, taking into account eq. 1, may be written in terms of $G(z,x)$ as[5]:

$$\bar{t}_0 = jT \left.\frac{\mathrm{d}G(z,x)}{\mathrm{d}x}\right|_{x=0} \cdot \frac{1}{G(z,0)} \tag{3}$$

Differentiating eq. 2 and setting $x = 0$, doing the average arrival time in the absence of noise equal to zero[2] and neglecting small terms of second order, the fluctuating part of the average arrival time (the pulse jitter) is approximated by

$$\bar{t} = \frac{1}{T \cdot G(z,0)} \cdot \left\{ j\left[\int_{-\infty}^{+\infty} \frac{\mathrm{d}N(\omega)}{\mathrm{d}\omega} F^*(\omega)\, d\omega + \int_{-\infty}^{+\infty} \frac{\mathrm{d}F(\omega)}{\mathrm{d}\omega} N^*(\omega)\, d\omega\right] + \beta_0'' z \left[\int_{-\infty}^{+\infty} \omega F(\omega) N^*(\omega)\, d\omega + \int_{-\infty}^{+\infty} \omega F^*(\omega) N(\omega)\, d\omega\right]\right\} \tag{4}$$

where $T \cdot G(z,0)$ is the unnormalized pulse energy. According to eq. 4, $\bar{t}$ is a linear transformation of the noise, $N(\omega)$. As long as the noise $N(\omega)$ is Gaussian distributed, $\bar{t}$ is a Gaussian random variable with zero mean. To characterize the distribution of $\bar{t}$ it is necessary to calculate its variance, $E\left[\bar{t}^2\right]$. Squaring $\bar{t}$ from eq. 4, taking expectation on both sides and according to the spectral properties of the noise we obtain

$$E\left[\bar{t}^2\right] = \frac{1}{T^2 \cdot G^2(z,0)} \cdot \left\{ 2\int_{-\infty}^{+\infty} S_n(\omega) \left|\frac{\mathrm{d}F(\omega)}{\mathrm{d}\omega}\right|^2 d\omega - 4\beta_0'' z \Im\left[\int_{-\infty}^{+\infty} \omega F^*(\omega) S_n(\omega) \frac{\mathrm{d}F(\omega)}{\mathrm{d}\omega}\, d\omega\right] + 2(\beta_0'' z)^2 \int_{-\infty}^{+\infty} \omega^2 |F(\omega)|^2 S_n(\omega)\, d\omega \right\} \tag{5}$$

where $\Im[w]$ means imaginary part of the complex w. The output pulse jitter variance has three terms. The first one is the pulse jitter at the output of the optical amplifier: it is a generalization of[2]. The second one is due to the "interaction" between the skewness of the pulse shape at the amplifier output and the propagation in the fibre: it vanishes when the pulse shape at the amplifier output is even or the fibre length is zero. The last one is also due to the propagation in the fibre: it also vanishes for a zero fibre length.

From eq. 5, it can be concluded that an optical amplifier used as a booster presents an output pulse jitter variance much smaller than as a preamplifier[2].

Figure 1 presents the output pulse jitter root mean square normalized by the bit period versus the distance along the fibre for a bit rate of 10 Gbit/s, an optical bandwidth of 1 nm and a fibre dispersion of 17 ps/(nm·km) (standard monomode fibre at 1550 nm), for three values of pulse duty-cycle: RZ (25%), RZ (50%) and NRZ (100%). It is assumed that the optical bandwidth is only limited by the optical filter, with sharp cut-off and the noise power spectrum is constant over the optical bandwidth with $n_{sp} = 1.3$ and $G = 20\,\text{dB}$ (typical values of an optical preamplifier). The pulse shape (before optical filtering) is considered rectangular with an energy of 0.1 pJ.

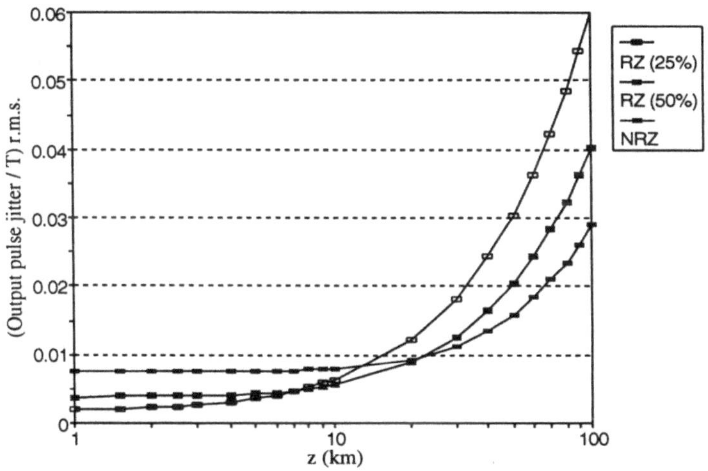

Figure 1. Influence of the pulse propagation in a standard single mode fibre at 1550 nm on the output pulse jitter root mean square at 10 Gbit/s.

We observe that the contribution of the pulse propagation to the output pulse jitter begins to be noticed close to 10 km, at 10 Gbit/s. Anyway, its value is always kept low, except for high distances (around a hundred kilometers). This noticeable contribution at small distances is due to the high dispersion. For dispersion shifted fibres, similar contribution happens above 100 km.

4. Chain of optical amplifiers and the Gordon-Haus effect

To extend the output pulse jitter of one amplifier span to a chain of amplifiers we assume that each amplifier noise is statistically independent of the noise of all the other amplifiers. Thus, the pulse jitter at the output of a chain of amplifiers is Gaussian distributed and its variance is the sum of each amplifier contribution. Note that the pulse shape at the amplifier output is different from amplifier to amplifier due to the fibre dispersion of the backward sections. Hence, the contribution of each amplifier is different from the others and no easily related to even for equal amplifier spans.

It is known that, after each amplifier span, solitons retrieve its original shape except for a small random change in group velocity[6]. This small change produces a random arrival time of solitons (the Gordon-Haus effect[6]). It is very interesting to note that Gordon and Haus' result for soliton jitter variance at the output of one amplifier span corresponds to the jitter variance term (in eq. 5) due to the propagation in the fibre. Indeed, neglecting the first term in eq. 5 and as the soliton pulse shape is even[6], the output pulse jitter variance in soliton propagation along one amplifier span (with length L_a) is reduced to $\left\{E\left[\bar{t}^2\right]\right\}_1 = (\beta_0'')^2 L_a^2 E\left[\bar{\omega}^2\right]$ where $E\left[\bar{\omega}^2\right]$ is the mean square angular frequency deviation from the carrier frequency:

$$E\left[\bar{\omega}^2\right] = \frac{2}{T^2 \cdot G^2(z,0)} \int_{-\infty}^{+\infty} \omega^2 \left|F(\omega)\right|^2 S_n(\omega)\, d\omega \tag{6}$$

5. Conclusions

We used a formulation based on the average arrival time introduced in another paper[2] to derive and characterize the pulse jitter at the output of an optical amplifier span. It was shown how the noise of optical amplifiers is related to the pulse jitter. From this relation, the pulse jitter variance was assessed and its various terms were identified. An illustrative example was presented in which the term due to the pulse propagation begins to be noticed after 10 km of propagation over standard monomode fibre, at 1550 nm and 10 Gbit/s.

The generalization for the pulse jitter at the output of a chain of amplifiers was obtained by considering the noise due to each amplifier is statistically independent of the noise due to other amplifiers. It was shown that the Gordon and Haus' result for timing jitter in soliton propagation can be obtained from the general study presented in this paper. However, more insight is needed in order to provide a thorough explanation of this result.

Acknowledgements

This work is supported by the European Commission within the RACE project R2011 'TRAVEL'.

References

1. *Special issue on optical amplifiers*, *J. Light. Techn.*, 9, Feb. 1991.
2. A. Cartaxo, "A simple formulation for the signal jitter due to the amplified spontaneous emission of optical amplifiers", in *Proc. MELECON'94*, Antalya, Turkey, Apr. 1994.
3. C. Giles, E. Desurvire, "Propagation of signal and noise in concatenated erbium-doped fiber optical amplifiers", *J. Light. Techn.*, 9, pp. 147-154, Feb. 1991.
4. Y. Yamamoto, *Coherence, amplification, and quantum effects in semiconductor lasers*, J. Wiley, 1991, pp. 16-17.
5. D. Marcuse, "Pulse distortion in single-mode fibers", *Appl. Opt.*, 19, pp. 1653-1660, 1980.
6. J. Gordon, A. Haus, "Random walk of coherently amplified solitons in optical fiber transmission", *Opt. Lett.*, 11, pp. 665-667, 1986.

MULTI-DUAL RING CONNECTED SHUFFLE NETWORK -- A BETTER ARCHITECTURE FOR WAN

Feng Liu and Mohsen Kavehrad

Department of Electrical Engineering
University of Ottawa
Ottawa, Ontario
Canada K1N 6N5

Introduction of the problem

Multi-Dual Ring Connected Shuffle Network (Figure 1) was proposed as an optical network architecture for Wide Area Networks. The previous study was carried under a uniform traffic pattern. The performance of the architecture was proved to be better than that of the Perfect Shuffle Network (Figure 2) [1].

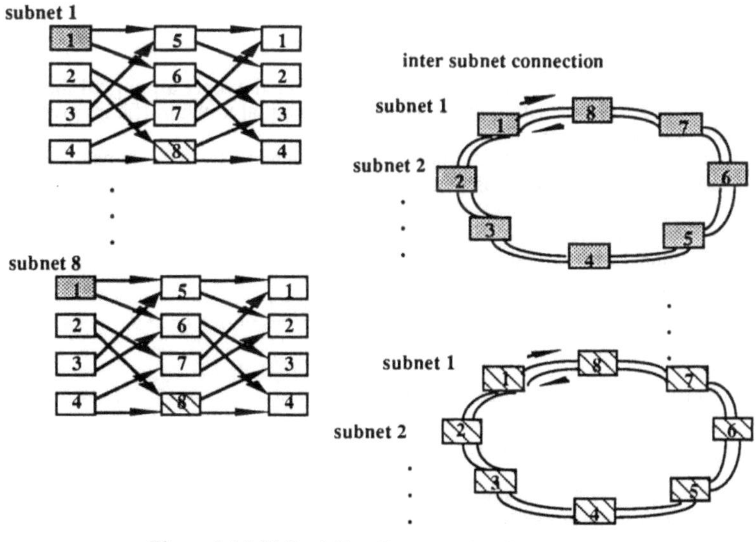

Figure 1 Multi-Dual Ring Connected Shuffle Network

In this paper, the case of nonuniform traffic pattern is considered. Obviously, in practice, the load distribution offered to the system will often be nonuniform, and, to properly assess the capacity of the network, it is necessary to determine how much traffic it can handle under realistic load distributions.

Although unbalanced loads can occur in a multitude of ways, their nature is decided by two factors: traffic intensity and traffic destination. Thus, we use two different approaches to assess the impact of load imbalance. The first one, which we refer to as "extreme value analysis" is an approximate analytical technique that assumes the distribution of the load is uniform, but the intensity of the traffic from individual users has a normal distribution. With these assumptions, it is possible to evaluate the probability that a given channel is overloaded, given the overall average and the variance of the load intensity. In the second approach, in a simulation method we call "random load generation", the intensity of the load generated by each user is assumed to be the same, but the pattern of the load is a variable. In this approach, many load patterns are generated randomly and the utilization of all the channels for each pattern is determined exactly by following a specific traffic routing algorithm.

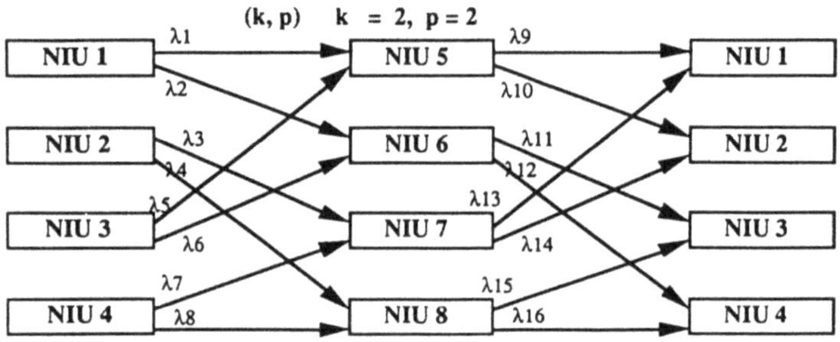

Figure 2 Perfect Shuffle Network

1 Extreme Value Analysis

Balanced mode is the mode in which the traffic distributed evenly among all the channels, i.e., the traffic intensity per channel within the subnetworks be equal to the traffic intensity per channel between the subnetworks. We define:

local traffic intensity as I_1 (node pair), remote traffic intensity as I_2 (node pair),
number of nodes in a subnetwork as n, number of subnetworks as N,
number of average hops in a subnetwork as E[hops].

We assume: $I_1 = f_b I_2$ where f_b *is the load balance factor and* $f_b \geq 1$ <1>

Thus in the "balanced traffic mode", we maintain:

The traffic intensity per channel within the subnetworks

$$= I_1 \times \frac{n \times n \cdot E[hops]}{2n} + I_2 \times \frac{n \times n \cdot (N-1)E[hops]}{2n} = \frac{nE[hops]}{2} \times (f_b - 1 + N)I_2 \quad <2>$$

$$\text{The traffic intensity per ch between subnetworks} = I_2 \times \frac{n \cdot n \cdot \frac{N}{2}(\frac{N}{2}+1) \times 2}{2n} \approx \frac{nN^2}{4} I_2 \quad <3>$$

$$\frac{nE[hops]}{2}(f_b - 1 + N)I_2 = \frac{nN^2}{4} I_2 \quad or \quad f_b = 1 + N(\frac{N}{2E[hops]} - 1)$$

We assume that $\gamma_m (1 \leq m \leq W)$ is the traffic intensity through the channel m and W is the total number of channels in a network. Under the balanced traffic mode assumption, from equation <1> and <2>, we maintain: $\gamma_m = \frac{nN^2}{4} I_2 = \frac{nN^2}{4} * \frac{I_1}{f_b} \quad where \quad 1 \leq m \leq W$

if I_1*'s distribution is* $N(\mu_1, \sigma_1)$; I_2*'s distribution is* $N(\mu_2, \sigma_2)$

the γ_m*'s distribution is* $N(\frac{nN^2}{4}\mu_2, \sqrt{\frac{nN^2}{4}}\sigma_2)$ *or the* γ_m*'s distribution is* $N(\frac{nN^2}{4f_b}\mu_1, \sqrt{\frac{nN^2}{4f_b}}\sigma_1)$

Let $Y_{W*} = Max.\gamma_m (1 \leq m \leq W)$ *where* Y_W *is the traffic intensity on the heaviest channel*

$\therefore$ *The CDF of* Y_W *is the Wth power of the CDF of* γ_m i.e., $F_{Y_W}(y_W) = [F_{\gamma_m}(\gamma_m)]^W$

There is a probability that the traffic intensity on the heaviest channel Y_W could exceed the channel capacity which is assumed as one giga bit per second $P(extreme>1Gbps)$. We calculate the maximum average traffic intensity μ from one node to another under the constraint that only with a very small probability of 10^{-L} (where L is an integer) this case could happen. Then we get the deloading factor: $Deloading\,factor = \frac{Max.\mu\ (non\text{-}uniform)}{\mu_{ideal}(uniform)}$.

The results are shown in Figure 3. In the figure, we define: $Coefficient\ of\ variation = \frac{\sigma}{\mu}$.

The deloading factor shows the sensitivity of a network architecture to the non ideal traffic pattern. The smaller a deloading factor is, the more sensitive the corresponding architecture is. As we can see, the deloading factor decreases with an increase in the coefficient of variation, as expected. Also, the deloading factor of the proposed architecture is higher than that of the Perfect Shuffle Networks which means the proposed architecture is less sensitive to the unbalanced traffic. Another case with a relatively realistic $f_b = 6.820$ and constraints of 10^{-L} ($L = 2, 4$ and 6) is studied, and the result in Figure 4 shows that the Deloading Factor decreases with the increase of L.

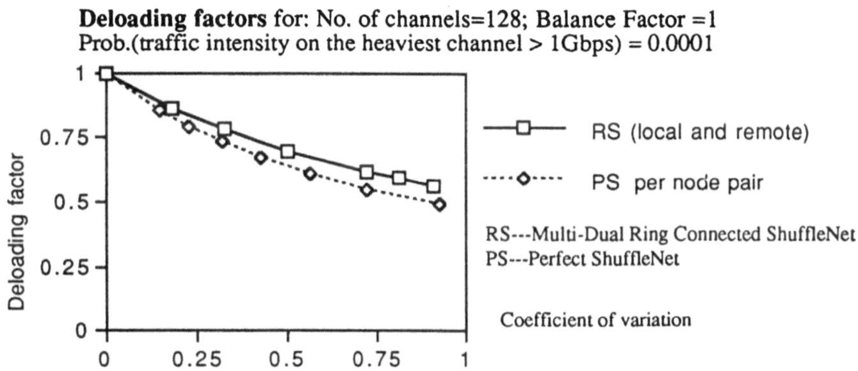

Figure 3 Deloading Factors for W = 128

2 Random Load Pattern

In the previous section, it was assumed that the pattern of load is uniform or balanced, but the intensity of individual traffic items has a random (e.g., normal) distribution. In this section, we take an alternative approach. Namely, we assume that the amount of traffic generated by all the users is the same, but its pattern is random. Specifically, we assume that each user generates traffic at its maximum rate, and that all the traffic generated by a user is destined to one other user, determined randomly[3].

For each randomly generated load pattern, the traffic is distributed over the multihop network using a specific "fix" routing algorithm. The number of traffic items on each channel is calculated for each load pattern, and the number of users on the most used channel is recorded. This process is repeated for very many traffic patterns, producing a histogram of usage of the most used channel. This histogram can be employed to determine the maximum amount of traffic that each user can be permitted to generate. We get the results in Figure 5 and Table 1.

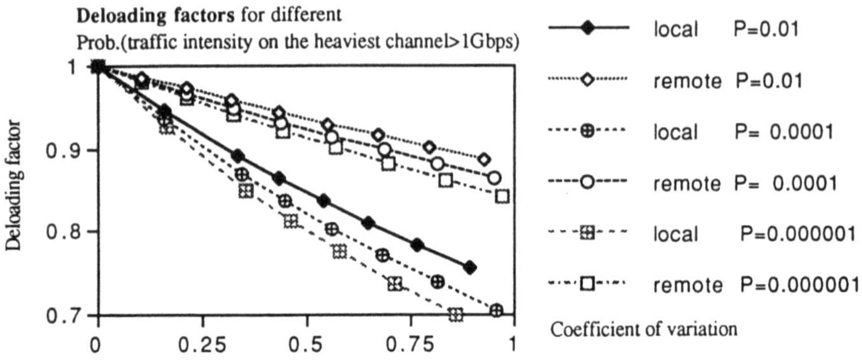

P means Probability of traffic intensity on the heaviest channel >1 Gbps

Figure 4 The effects of different constraints on the Deloading Factors

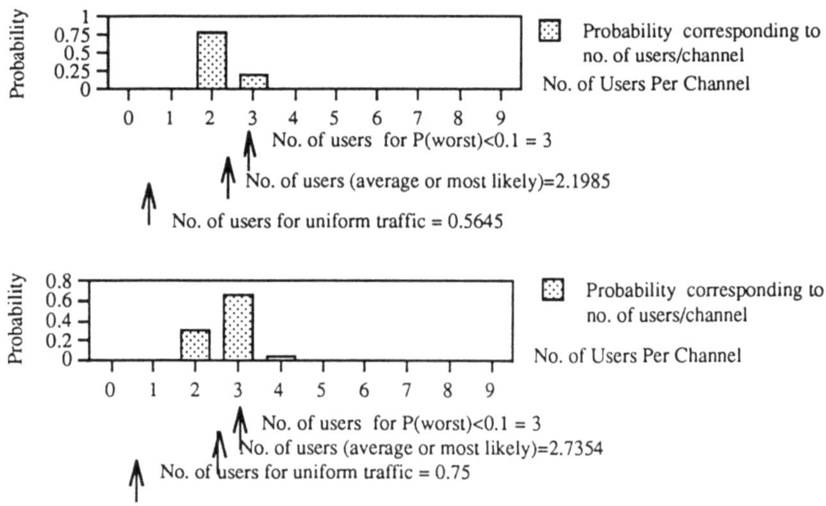

Figure 5 Histogram of usage of most used channel for the Perfect Shuffle Network and the Multi-Dual Ring Connected ShuffleNet (32 NIUs)

Since each NIU or user sends only one traffic item to its destination, we can interchangeably use the terms: "user", "NIU" and "traffic item". Therefore, the average number of users per channel under the uniform traffic pattern is given by:

$$E(user\ /\ channel)_{uniform} = \frac{Total\ No.\ of\ NIUs\ or\ users}{Aggregate\ Throughput\ in\ the\ Unit\ of\ Channel(1\ Gb\ /\ s)} = \frac{Total\ No.\ of\ Users}{\frac{Total\ No.\ of\ Users \times p}{E[hops]}} = \frac{E[hops]}{p}$$

where p is the degree of output of a node.

$$The\ Deloading\ Factor = \frac{Achievable\ Taffic\ Intensity\ per\ User\ (non\ ideal\ traffic)}{Achievable\ Taffic\ Intensity\ per\ User\ \ (ideal\ traffic)}$$

$$= \frac{1/No.\ of\ Users\ per\ Channel\ (non\ ideal\ traffic)}{1/No.\ of\ Users\ per\ Channel\ (ideal\ traffic)} = \frac{No.\ of\ Users\ per\ Channel\ (ideal\ traffic)}{No.\ of\ Users\ per\ Channel\ (non\ ideal\ traffic)}$$

If one must guard against the eventuality of a worst case load pattern, one should use the "maximum number of users per channel" in the non-ideal traffic case. However, it is ordinarily not economical to guard against very low-likelihood, extreme situations. If we are willing to accept the small risk of assuming that an absolute worst case load pattern will not occur, we can assume the probability of worst case is *0.1* and we use the maximum number of users per channel when the worst case does not happen. In this way, we can get the Deloading Factor (P<0.1). Since the results are still too conservative, to obtain a less conservative and more realistic estimate for the network, we choose as the measurement the average number of users per channel which performs to more likely events. Then we get the Deloading Factor (most likely). All the results of Deloading Factors are shown in Table 1.

Table 1 Deloading Factors for the Methods "Random Load Pattern"

		Perfect Shuffle	Multi-Dual Ring Connected ShuffleNet
32 NIUs	P<0.1	0.1882	0.2500
10000 patterns	most likely	0.2568	0.2742
192 NIUs	P<0.1	0.1846	0.2500
20000 patterns	most likely	0.2222	0.2841

In Table 1, we can see that the Deloading Factors of Multi-Dual ring Connected Shuffle Network are higher than the Perfect Shuffle Network under the same conditions showing a significant improvement in the performance for the proposed architecture.

3 Concluding Remarks

In this paper, the effects of load imbalance on the Multi-Dual Ring Connected Shuffle Network, due to either traffic intensity variability or traffic pattern variability, are studied. The results show better performance than those of the Perfect Shuffle Network. The Deloading Factors of the proposed architecture are within the reasonable range (0.3 - 0.9) and insensitive to the network size. Therefore, in the realistic traffic environment, the Multi-Dual Ring Connected Shuffle Network is a very promising choice.

Acknowledgment

This research was supported by a grant from the Canadian Institute for Telecommunications Research under the NCE program of the Government of Canada.

Reference

[1] F. Liu and M. Kavehrad, "Symmetrically Multi-Connected Optical WAN: Multi Dual Ring Connected and Multi Shuffle Connected Shuffle Networks, *IEEE Pacific Rim Conference,* May, 1993.

[2] A.S. Acampora and M.J. Karol, "An Overview of Lightwave Packet Networks" *IEEE Network Magazine* , pp.29-40, January 1989.

[3] M. Eisenberg and N. Mehavari, "Performance of the Multichannel Multihop Lightwave Network Under Nonuniform Traffic" *IEEE Journal on Selected Area in Communications. Vol. 6. No 7, August 1988.*

AN ALL-OPTICAL TUNABLE FILTER FOR LOW-CROSSTALK WAVELENGTH ROUTING IN WDM NETWORKS

F. Khaleghi, M. Kavehrad, and E. Simova

University of Ottawa/ELG, Ottawa, Ont. K1N 6N5

INTRODUCTION

Among various options to exploit the tremendous bandwidth of the optical medium and meet the needs for broadband multimedia telecommunications, multiwavelength optical networks appear to be the most promising. The simple amplification function of EDFA amplifiers, which is independent of the operating data rate and can simultaneously accommodate many wavelength-division-multiplexed (WDM) channels will allow WDM technology and architecture to be utilized in a very cost-effective manner.

The concepts of wavelength multiplexing, demultiplexing and routing form the core of almost all the proposed all-optical WDM networks. In a proposal to DARPA [1], wavelength-routing is done through a network of wavelength cross-connects constructed of acousto-optic tunable filters (AOTF). AOTFs with their broad wavelength tunability, narrow bandwidth and ability to simultaneously route several arbitrarily and separately selected channels are claimed to be the best candidate for performing wavelength-selective functions, i.e., multiplexing, demultiplexing and routing in WDM networks. It was proposed in [2] and experimentally demonstrated in [3] that AOTFs can function as a wavelength router for high-density WDM systems. However, high power consumption and high level of crosstalk are two major problems which make it impossible to use these devices at their present technological state in WDM networks.

In this paper, we introduce an optically tunable filter device with a low crosstalk and a small loss. The filter is basically a waveguide grating with a controlled refractive index modulation. An external light beam controls the refractive index via optical Kerr effect and tunes the grating to a pre-selected wavelength. At a constant intensity of the control beam, we can apply the general results for volume gratings without considering the time response of the material. First, we review the results of coupled wave analysis of volume reflection gratings and apply them to gratings written in a nonlinear material. Next, we present the response of reflection waveguide gratings and apply it to waveguide gratings in a nonlinear medium. A design example is given in a waveguide based on AlGaAs. the design is so that the Kerr material is strongly nonlinear for control beam wavelength, whereas slightly nonlinear [7] in the signal wavelengths. Therefore, typical values of signal power practically do not affect the refractive index of the material.

THEORY OF OPERATION

I. Volume Hologram Gratings:

The formula for the diffraction efficiency of dielectric lossless gratings [4] is:

$$\eta = 1/\{1+(1-\xi^2/\vartheta^2)/\sinh^2\sqrt{(\vartheta^2-\xi^2)}\}, \quad (1)$$

where,

$$\vartheta = j\pi n_1 d/\lambda(c_R c_S)^{\frac{1}{2}} \quad (2)$$

$$\xi = -\vartheta d/2c_S \quad (3)$$

Applications of Photonic Technology
Edited by G.A. Lampropoulos *et al.*, Plenum Press, New York, 1995

and d is the thickness of the grating, n_1 is the refractive index modulation of the grating, λ is the wavelength of the incident light,

$$c_R = \cos\theta, \qquad \text{and} \qquad c_S = \cos\theta - \frac{K}{\beta}\cos\varphi,$$

and ϑ, the detuning parameter is defined by

$$\vartheta = K\cos(\varphi-\theta) - \frac{K^2}{4\pi n}\lambda. \qquad (4)$$

n is the refractive index of the material, $\beta = 2\pi n/\lambda$ and $K = 2\pi/\Lambda$ with Λ the grating period. For unslanted grating, $\varphi = 0$ see Fig. 1, and Bragg incidence, we have $c_R = c_S = \cos\theta$. When θ is small, $c_R = c_S \cong 1$. By substituting (4) into (3), we have

$$\xi = -\frac{1}{2c_S}Kd\cos(\varphi-\theta) + \frac{K^2 d\lambda}{8\pi n c_S}. \qquad (5)$$

By using (5), we can express the parameter ξ directly in terms of wavelength deviation $\Delta\lambda$ and material refractive index deviation Δn to obtain:

$$\xi = \Delta\lambda . K^2 d / 8\pi n c_S \qquad (6)$$

$$\xi = -\Delta n . K^2 d . \lambda / 8\pi n^2 c_S. \qquad (7)$$

In a Kerr medium, the refractive index n in the presence of a relatively high-intensity light changes proportionally to the incident optical intensity. The overall refractive index is therefore a linear function of the optical intensity I,

$$n(I) = n_0 + n_2 I. \qquad (8)$$

Assuming that a control light beam is present, by equating (6) and (7) we have

$$\frac{\Delta\lambda}{\lambda} = \frac{-\Delta n}{n_0 + n_2 I} = \frac{-n_2 I}{n_0 + n_2 I}. \qquad (9)$$

This equation evaluates the required intensity of the control light beam to tune the maximum of the reflected wavelength to another pre-selected wavelength.

The normalized efficiencies from (1) are plotted in Fig. 2 as a function of ξ for different values of ϑ=const.

II. Waveguide Gratings

Waveguide gratings can be made by etching a periodic corrugation on the waveguide surface, as shown in Fig. 3. Accordingly to the scalar coupled-mode theory the waveguide grating transmissivity T and reflectivity R, for the design wavelength at Bragg incidence are given by [4]:

$$T = 1/\cosh^2(\kappa L) \qquad \text{and} \qquad R = \tanh^2(\kappa L) \qquad (10)$$

where L is the length, κL is the normalized grating length, and κ is the coupling coefficient [6]. The normalized lengths can be as high as κL=4. For off-Bragg incidence, i.e., $\Delta\beta \neq 0$ where $\beta = 2\pi n/\lambda$, transmissivity T and reflectivity R are given by:

$$T = \left|1/[\cosh(\Psi) - j(\xi/\Psi)\sinh(\Psi)]\right|^2$$

$$R = \left|1/[((\xi/\Psi) + j(\psi/\mu)\coth(\Psi)]\right|^2 \qquad (11)$$

where

$$\mu = \kappa L \qquad \xi = -\Delta\beta L \qquad \text{and} \qquad \Psi = \sqrt{(\mu^2 - \xi^2)} \qquad (12)$$

By expressing ξ in terms of wavelength and refractive index changes, we can verify that eq. (9) holds as well, for this case. In Fig. 4. reflectivity of the waveguide grating for $\kappa L = 4$ is plotted against ξ (i. e., as a function of the difference in $\beta = 2\pi n/\lambda$ from the design value) for a given normalized length.

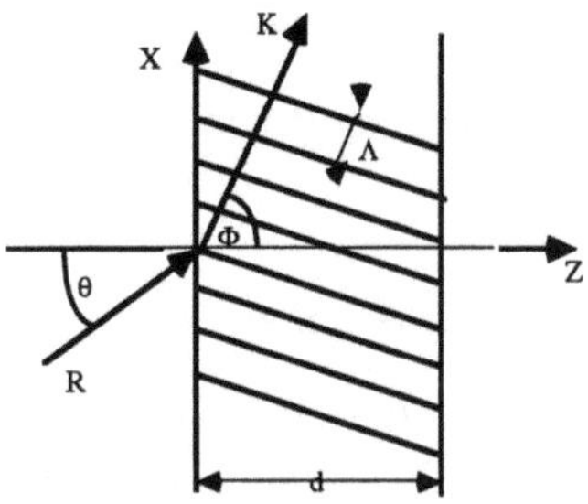

Fig. 1. Model of a thick grating

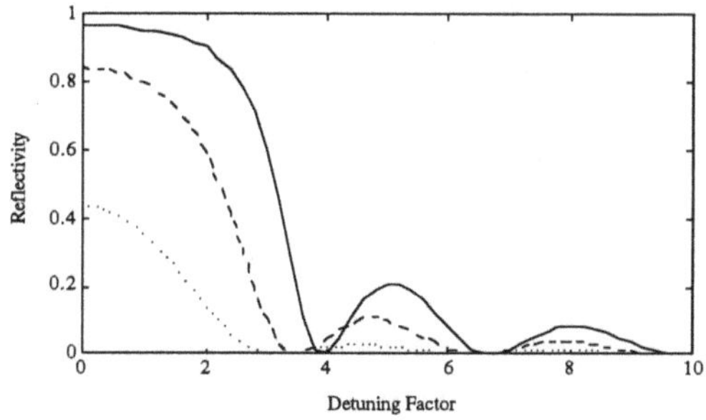

Fig. 2. Reflectivity of a thick grating versus detuning factor, ξ___ $\vartheta=\pi/4$,- - - $\vartheta=2\pi/4$,,. . . $\vartheta=3\pi/4$

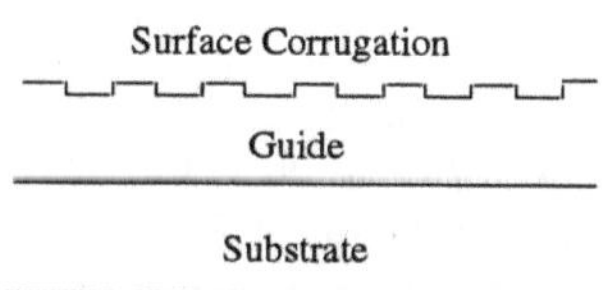

Fig. 3. Schematic of a corrugated reflection grating

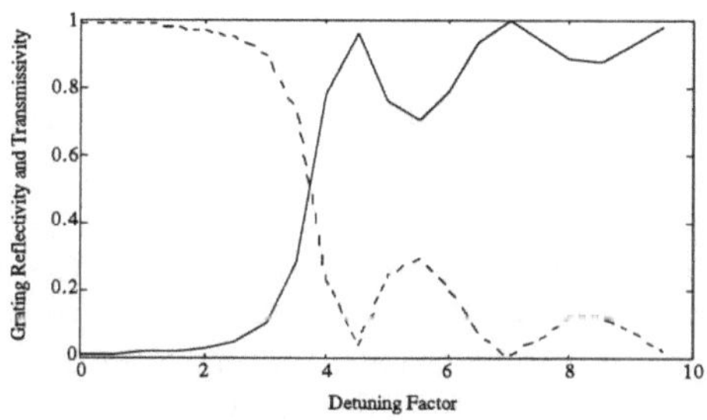

Fig. 4. Waveguide grating reflectivity and transmissivity versus detuning factor, ξ , ___ Ttansmissivity ,- - - Reflectivity

DESIGN EXAMPLE

As a design example, we will consider an AlGaAs based device. The choice of this material is based on its strong nonlinear properties and an already existing technology for the fabrication of gratings and waveguides in these materials.

For AlGaAs the relevant material parameters are estimated as n=3.6, λ=1.55 μm. The material nonlinearity around the resonant frequency is $n_2 = -10^{-8} m^2 / W$; and the absorption is $\alpha = 10^4 cm^{-1}$ [5]. For the resonant frequency, the saturated nonlinear index is $\Delta n_{sat} = 0.1$. This value represents the maximum possible change in the refractive index by exposing the nonlinear material to a high-intensity light beam. The wavelength of the control beam is chosen near the resonant frequency. The strong material absorption in this wavelength can be used to filter out the control beam. Fig. (4) implies that very high peak reflectivities are possible, provided the normalized coupling length κL is large. For $\kappa L = 4$, the reflectivity is 0.999.

I. Power specifications of the control beam and tuning range of the filter:

For changing the maximum reflection set at 1.55μm by $\Delta\lambda$ = 0.5 nm, from eq.(9), we obtain an estimated intensity of the control beam of $1.85 \mu W$ by assuming an effective cross section area of the waveguide of $A = 1.6 \times 10^{-11} m^2$. From eq. (9), one percent change in the refractive index ($\Delta n / n = 0.01$) is equivalent to a wavelength sweep of 15 nm. For the saturated nonlinearity ($\Delta n_{sat} = 0.1$), the tunability range $\Delta\lambda$ is limited to 43 nm which is more than enough for all the existing applications. A control power of about $100 \mu W$ can sweep a spectrum of 30 nm width. In conclusion, the required control power is $3.7 \mu W / 1 nm$ and maximum tunability range is 43 nm.

II. Length and bandwidth specifications of the device:

From Fig. 4, the peak reflectivity for $\kappa L = 4$ is 0.999 resulting in a crosstalk of -30 dB. This effect differs from the crosstalk arising during detection process that is caused by unwanted signals at different frequencies. The degradation by the latter effect is much smaller and is ignored in the present discussion. As seen from Fig. 4, if the device is designed so that the channel spacing corresponds to a detuning factor of 7, $|\xi| = 7$, the interference arising from the sidelobes of the modulated signals would be negligible, as long as the modulating signal bandwidth is much less than the channel spacing. Usually the modulation rate is in the order of 1Gbs which is small enough compared to the channel spacing to ignore the second mechanism of crosstalk generation. For $|\xi| = 7$, from eq. (12), $|\Delta\beta| = 7 / L$, and $|\Delta\beta| = 2\pi|\Delta\lambda| n / \lambda^2$, thus $L = 7\lambda^2 / 2\pi n|\Delta\lambda|$ which leads to a length of 1.4 mm. Longer gratings (up to a few centimeters) are feasible and will result in a very narrow signal spacing, e.g., a channel spacing of $6.24 GHz$ is possible if the length is 1.4 cm. The longer the normalized length of the grating, the higher the peak reflectivity, the lower the crosstalk. The problem with long lengths is due the loss, because of high absorption in semiconductors. However, even 1 cm does not cause any considerable loss for the signal wavelength, with a loss coefficient $\alpha = 1.7 \pm 1.5 cm^{-1}$ [7]. A 1 mm length results in 0.3 dB loss.

CONCLUSIONS

We have introduced an optically tunable grating filter with the following advantages:

(i) broad tuning range (40 nm); (ii) narrow filter bandwidth (FWHA< 1 A°); (iii) fast optical tunability (10 nsec); The wavelength tuning time is limited by the response time of the material in the wavelength of the control beam.; (iv) relatively low insertion loss (< 5 dB) mainly due to the fiber-to-waveguide coupling loss. The intrinsic loss of the device is around 0.3 dB; (v) low crosstalk (<-30 dB); (vi) Multiwavelength selection. Multiple wavelength selection is possible by cascading the filters.

ACKNOWLEDGEMENTS

This work was supported by the Telecommunications Research Institute of Ontario (TRIO), Photonic Networks and System Thrust.

REFERENCES

1. C. Brackett, A. Acampora, et. al, JLT, Vol. 11, No. 5/6, pp. 736-753, 1993
2. M. Irshid, and M. Kavehrad, JLT, Vol. 10, No. 1, pp. 101-108, 1992
3. W. Way, et al, IEEE Photo. Tech. Lett., Vol. 4, No. 4, pp. 402-405, 1992
4. H. Kogelnik, The Bell System Tech. Journal, Vol. 48, No. 9, pp. 2909-2943, 1969
5. W. Nie, Advanced Materials, No. 7/8, pp. 520-545, 1993
6. R. R. A. Syms, ``Optical GruidedWaves and Devices," 1992
7. C. N. Ironside, J. S. Aitchison, and J. M. Arnold, IEEE Journal of Quantum Elect., Vol. 29, No. 10, pp. 2650-2654, 1993

BIT-ERROR-RATE OF AN SCM/WDMA OPTICAL NETWORK EMPLOYING POLARIZATION SCRAMBLING UNDER OPTICAL BEAT INTERFERENCE

Mohammad M. Banat and Mohsen Kavehrad

Department of Electrical Engineering
University of Ottawa
161 Louis Pasteur, Ottawa
Ontario, Canada K1N 6N5

I. INTRODUCTION

A mixed analytical-numerical technique is used to calculate the average bit-error-rate (BER) for a two-user per channel subcarrier multiplexed wavelength division multiple access (SCM/WDMA) optical network. The network is subject to optical beat interference (OBI)[1-5]. The BER is found through elaborate simulation and numerical computation. It is shown that the system performance is severely degraded by OBI, even at very low subcarrier bit rates (e.g., 50 Mbps), when polarization scrambling is not used. However, the network can support much higher subcarrier bit rates (up to 500 Mbps/channel) with sufficiently high polarization scrambling rates (about 100 GHz). Using Dense WDMA, the network is shown to support more than 150 Gbps total throughput.

II. OBI EFFECTS ON SCM/WDMA NETWORKS

A two-user per optical channel SCM/WDMA system was simulated using MATLAB with the subcarrier center frequencies f_1 and f_2 separated by twice the electronic filter bandwidth, under the assumption of f_2 being the larger. The first subcarrier was centered at 4 GHz. Both subcarriers were PSK-modulated at an equal rate of $\boldsymbol{R_b}$ bps. The electronic BPF bandwidth $\boldsymbol{B}$ was equal to the subcarrier modulation bit rate $\boldsymbol{R_b}$. The PSK modulated subcarriers intensity modulated two lasers whose frequencies were assumed to fluctuate uniformly in a range $\boldsymbol{R}$ of 50 GHz. Both users had an equal laser intensity modulation index value of $\sqrt{2}$. With this setup, we evaluated the average carrier-to-interference ratio (CIR) and average probability of error $\boldsymbol{P(e)}$ for the first user.

Applications of Photonic Technology
Edited by G.A. Lampropoulos *et al.*, Plenum Press, New York, 1995

Figure 1 shows the CIR as a function of the simultaneous optical carrier frequency separation. It can be seen from this figure that there is a sharp drop in the CIR when the carrier separation $F_x = |f_1 - f_2|$ satisfies:

$$|F_x - f_1| < B/2 \tag{1}$$

This demonstrates how serious OBI can be when the optical carriers get too close to one another.

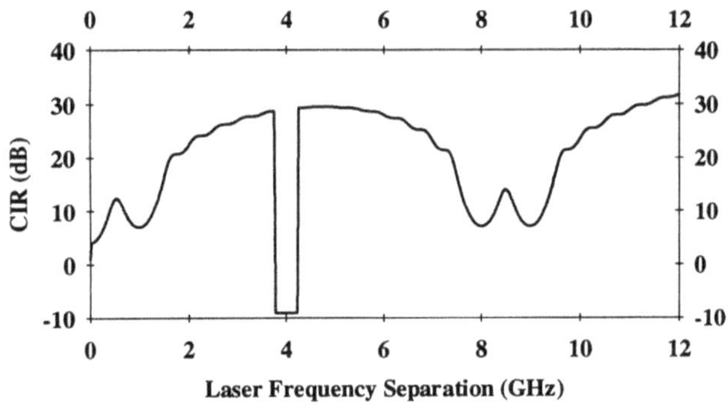

Figure 1. *CIR* as function of optical carrier separation with no polarization scrambling

Figure 2 shows the CIR averaged over all possible optical carrier separations. In this figure, the CIR is plotted versus the subcarrier bit rate to test the achievable network capacity under OBI. As Figure 2 shows, the CIR becomes very poor at subcarrier bit rates higher than about 150 Mb/s.

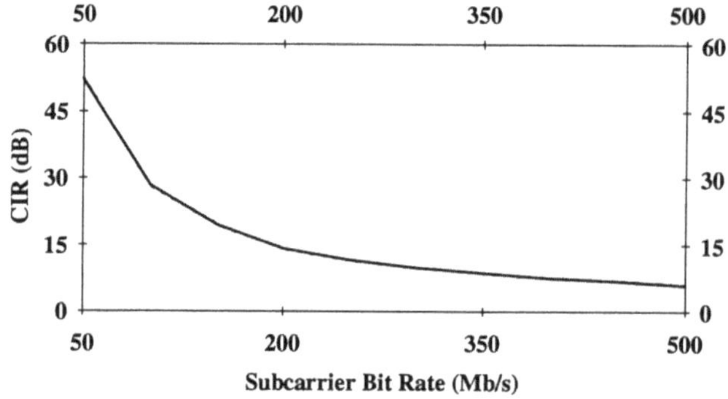

Figure 2. *CIR* averaged over optical carrier separation with no polarization scrambling

Figure 3 shows the simulation result for the system average probability of error $P(e)$. As can be seen from the figure, the probability of error does not fall below 10^{-3} even at bit rates as low as 50 Mb/s. Furthermore, there is almost a linear increase in $P(e)$ at bit rates higher than about 200 Mb/s.

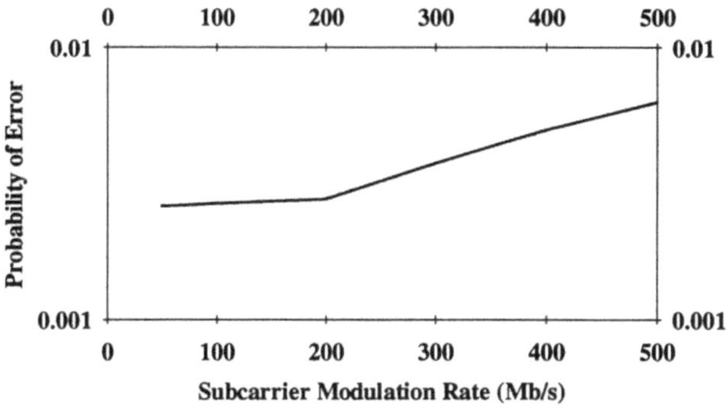

Figure 3. Probability of error with no polarization scrambling

It can be seen from the above figures and the above discussions that the network performance under OBI is not acceptable. In the next section, we demonstrate how polarization scrambling can bring significant improvements to the network performance and throughput. Polarization scrambling will be shown to support network throughputs as high as 150 Gb/s.

III. SCM/WDMA PERFORMANCE WITH POLARIZATION SCRAMBLING

To reduce the OBI level at the output of the electronic BPF, we propose using polarization scrambling. In this process, the laser field is passed through a controllable refractive index material like $LiNbO_3$. The control signal is a high-rate pseudonoise sequence capable of shifting the input field phase by $\pm\pi/2$ radians.

To study the effect of polarization scrambling on the CIR, the system was simulated using MATLAB. The same parameters as in section II were used, with the subcarrier bit rate assuming values of 250 and 500 Mb/s. Figure 4 shows the resulting CIR. Significant improvements in the CIR are achieved at high PN rates. Shown also in the same graph is the threshold CIR value needed in the case of Gaussian OBI to obtain a 10^{-9} error probability. It can be shown that the OBI does become Gaussian for PN rates required to achieve these threshold CIR values.

$P(e)$ is shown in Figure 5 in the case of a 500 Mb/s subcarrier bit rate. For PN rates less than 60 GHz, the OBI was not found to be Gaussian and the probability of error was found by numerically computing the electronic detector decision variable, comparing it to zero, and finding the proportion of times in which it is negative.

When the PN rate exceeds 60 GHz, the OBI becomes Gaussian, and we can integrate over its pdf to find $P(e)$. The OBI pdf is totally determined by its variance estimated by the variance of the equivalent Gaussian random variable. A numerical search algorithm using least squares minimization was used to determine the variance value that best fits the simulated values of the OBI characteristic function.

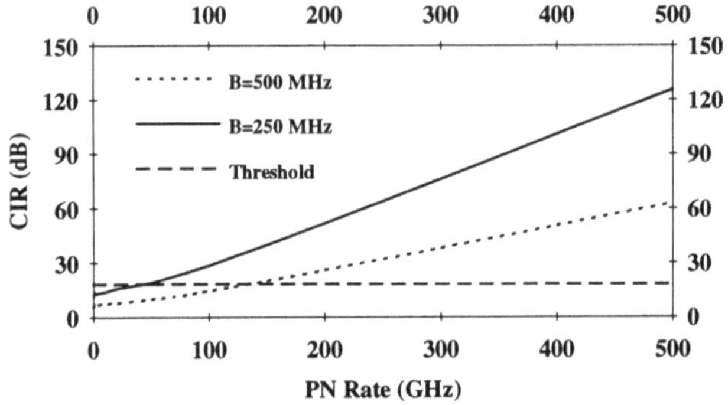

Figure 4. *CIR* with polarization scrambling. Threshold of *CIR*=18 is shown

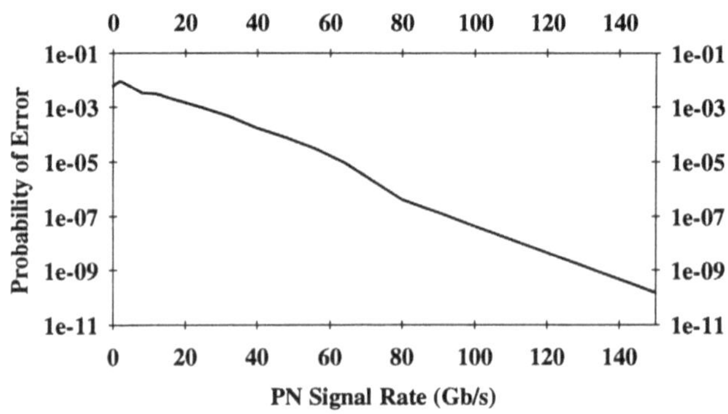

Figure 5. Probability of error with polarization scrambling

With this in mind, $P(e)$ can be calculated from:

$$P(e)=\frac{1}{\sqrt{2\pi\sigma_{obi}^2}}\int_{-\infty}^{-m_1T_b/4} e^{-x^2/2\sigma_{obi}^2}\,dx \tag{2}$$

Note that, T_b is the bit duration, m_1 is the first user intensity modulation index, and $m_1T_b/4$ is the signal term of the PSK matched filter receiver output. The last result can be written in terms of the complementary error function in the form:

$$P(e)=\frac{1}{2}\operatorname{erfc}\left[\frac{m_1T_b}{4\sigma_{obi}\sqrt{2}}\right] \tag{3}$$

As can be seen from Figure 5, the probability of error decreases almost linearly with the increase in the PN polarization scrambling rate.

It is possible at PN rates around 100 GHz to achieve a 10^{-9} error probability with two subcarrier each transmitting at 500 Mb/s. This makes the throughput of each optical channel equal to 1 Gb/s. With a laser frequency uncertainty range of 50 GHz and a polarization scrambling rate of 100 GHz, optical channels can be allocated at about 200 GHz spacing. Assuming a total fiber bandwidth of about 30 THz, 150 optical channels can be supported. This results in a total network throughput of 150 Gb/s.

IV. CONCLUSIONS

Polarization scrambling using PN signals has been demonstrated to significantly improve SCM/WDMA networks performance against OBI. A network throughput as high as 150 Gb/s was shown to be achievable with two subcarriers per optical channel. Although the results indicate that high-rate PN signals are needed, it should be emphasized that this is true only in the worst case of identical field polarizations. When field polarizations are different and varying with time, the required PN signal rates are drastically reduced, resulting in practically implementable polarization scrambling.

The network BER at high PN rates was estimated using a mixed analytical-numerical technique. This technique is based on least squares fitting of the output decision variable characteristic function to a Gaussian characteristic function. Excellent fitting was observed.

REFERENCES

1. N. K. Shankaranarayanan, S. D. Elby, and K. Y. Lau, "WDMA/subcarrier-FDMA lightwave networks: limitations due to optical beat interference,", *J. Lightwave Tech.*, Vol. 9, No. 7, pp. 931-943, Jul. 1991
2. T. H. Wood and N. K. Shankaranarayanan, "Measurements of the effect of optical beat interference (OBI) on the bit-error-rate of a subcarrier-based passive optical network (PON),", *Proc. OFC'93*, pp. 231-233
3. T. H. Wood and N. K. Shankaranarayanan, "Operation of a passive optical network with subcarrier multiplexing in the presence of optical beat interference,", *J. Lightwave Tech.*, Vol. 11, No. 10, pp. 1632-1640, Oct. 1993
4. M. M. Banat and M. Kavehrad, "Reduction of Optical Beat Interference in SCM/WDMA Networks using Pseudorandom Phase Modulation,", *J. Lightwave Tech.*, Accepted for publication.
5. M. M. Banat and M. Kavehrad, "Effects of laser direct intensity modulation index on optical interference in subcarrier multiplexed wavelength division multiple access networks,", Submitted for publication.

INFLUENCE OF THE OPTICAL FIBRE DISPERSION ON THE SQUARE SYNCHRONIZER TIMING JITTER

Adolfo V. T. Cartaxo, Augusto A. de Albuquerque†

DEEC/IT, Instituto Superior Técnico,
Av. Rovisco Pais, P-1096 Lisboa, Portugal
†Temporarily with the European Commission, DGXIII.B

1. Introduction

Very high bit rates are being increasingly used on direct detection optical systems. The influence of the clock recovery circuit (CRC) inaccuracies is of special significance in these systems since a small timing error may represent a high fraction of the bit period. Recently, we evaluated analytically the performance of the square synchronizer (a CRC employing a squarer as nonlinearity) in a direct detection optical receiver with the optical power incident on the photodetector modelled by a binary synchronous signal with equally likely symbol realizations: a general expression for the dynamic timing jitter (timing jitter root mean square) normalized by the bit period, E_{dyn}/T, valid for any type of photodetector (PIN or APD), optical pulse shape, pulse shape at the input to the squarer, prefilter, narrow bandpass filter, noise type dominance and signal to noise ratio (S/N) was presented[1]. In[2] the influence of the various types of noise on the square synchronizer performance in binary direct detection optical communications was analysed for a very narrow optical pulse shape (Dirac pulse shape). The generalization of this analysis to more realistic pulse shapes incident on the photodetector (which take into account the optical fibre dispersion) is the objective of this paper. In section 2, a summary of the dynamic timing jitter computation in the square synchronizer on direct detection optical communications is presented. In section 3, we present numerical results that describe the influence of the optical fibre dispersion on the square synchronizer performance and we provide a thorough physical explanation of these results for two noise dominant situations, the signal shot-noise and the circuit noise. In section 4, main conclusions are drawn.

2. The dynamic timing jitter computation

Theoretical analysis of the conventional CRC with a cyclostationary signal (with period T, the bit period) at its input has already been investigated[1,2,3]. The simpler version of the square synchronizer consists of a squarer followed by a narrow bandpass filter with transfer function, $H_T(\omega)$ tuned close to the bit frequency, $\omega_b = 2\pi/T$, and an

Applications of Photonic Technology
Edited by G.A. Lampropoulos *et al.*, Plenum Press, New York, 1995

impulse generator[1]. The performance of the CRC is based on the timing jitter statistical characterization through its dynamic timing jitter. Assuming that the bandpass filter is band-limited between 1/2 and 3/2 of the bit rate, E_{dyn}/T is given by[1,2,3]

$$\frac{E_{dyn}}{T} = \frac{1}{2\pi} \frac{\sqrt{\frac{1}{2}\int_{-0.5}^{+0.5} [C_0(\eta) - C_2(\eta)]\, d\eta}}{|\alpha_1 \cdot H_T(\omega_b)|} \tag{1}$$

where η is the normalized frequency defined by $\eta = \omega/\omega_b - 1$,

$$C_0(\eta) = B^N_{\tilde{y},\tilde{y},0}\left(2\pi(1+\eta)\right) \cdot \left|H_T\left(\omega_b(1+\eta)\right)\right|^2 \tag{2}$$

$$C_2(\eta) = \Re\left[B^N_{\tilde{y},\tilde{y},2}\left(2\pi(1+\eta)\right) \cdot H_T\left(\omega_b(1+\eta)\right) \cdot H_T\left(\omega_b(1-\eta)\right) \cdot e^{-j\cdot 2\mu_1}\right] \tag{3}$$

where $\Re[\cdot]$ means real part of $[\cdot]$ and $\mu_1 = \arg[\alpha_1 H_T(\omega_b)]$, α_1 corresponds to the index 1 coefficient of the complex Fourier series expansion of the expected value of the signal, $y(t)$, at the output of the nonlinearity and the coefficients $B^N_{\tilde{y},\tilde{y},u}(2\pi v)$ (u, integer) are the normalized index u coefficients of the Fourier transform relative to the time delay of the complex Fourier series expansion of the autocovariance of $y(t)$. Assuming the pulse shape at the input to the square synchronizer is band-limited to the bit rate frequency, the coefficients α_1, $B^N_{\tilde{y},\tilde{y},0}(2\pi v)$ and $B^N_{\tilde{y},\tilde{y},2}(2\pi v)$ were calculated[1].

3. Discussion of numerical results

Numerical results were obtained by assuming that the pulse shape at the input to the squarer is the raised-cosine in the frequency domain[4] with a bandwidth equal to 0.75×bitrate (in Personick's notation[4] corresponds to $\beta = 0.5$); the optical pulse incident on the photodetector is Gaussian shaped[4] with the normalized Fourier transform given by $H'_p(y) = \exp[-(2\pi\alpha y)^2/2]$ where α is the pulse spread, normalized by the bit period, due to the fibre dispersion (typical values of α are $\alpha = 0.2$ for medium spread, $\alpha = 0.5$ for high spread and $\alpha = 0$ for the Dirac pulse shape); the bandpass filter is the tuned tank circuit with a quality factor, Q, of 100. The influence of the fibre dispersion on the dynamic timing jitter is analysed by comparing two noise dominant situations, signal shot-noise and circuit noise, with typical values of −30 dB and of 30 dB for N_t/N_s, respectively[2]. In all the figures S/N (dB) is taken as a parameter, the extinction ratio is considered zero and the circuit noise is assumed white.

Figure 1 shows E_{dyn}/T versus α for $N_t/N_s = -30$ dB: a monotonical increase of E_{dyn}/T with the fibre dispersion is noticed for all S/N. Physically, this behaviour is associated with the photodetection process of these optical pulse shapes: at low dispersion (close to the Dirac pulse shape), the photons arrival of each pulse is located in a narrower time interval leading to a better knowledge of the bit period position than when the photons arrival is spread over the whole bit time and, hence, to a better performance of the synchronizer. Statistically, the increase of the fibre dispersion corresponds to a decrease of the cyclostationarity of the signal shot-noise (so, a decrease of the clock information in this kind of noise) and to a move towards the wide sense stationarity[5] (which arises for very high fibre dispersion) and, consequently, to a situation closer to the one corresponding to the white circuit noise. This can be observed by comparison of figure 1 with figure 2, which refers to $N_t/N_s = 30$ dB: at high fibre dispersion, E_{dyn}/T is slightly greater in shot-noise dominant situation.

Different variation of E_{dyn}/T with S/N can also be noticed: much less variation of E_{dyn}/T at low dispersion than at high dispersion. This behaviour is also explained

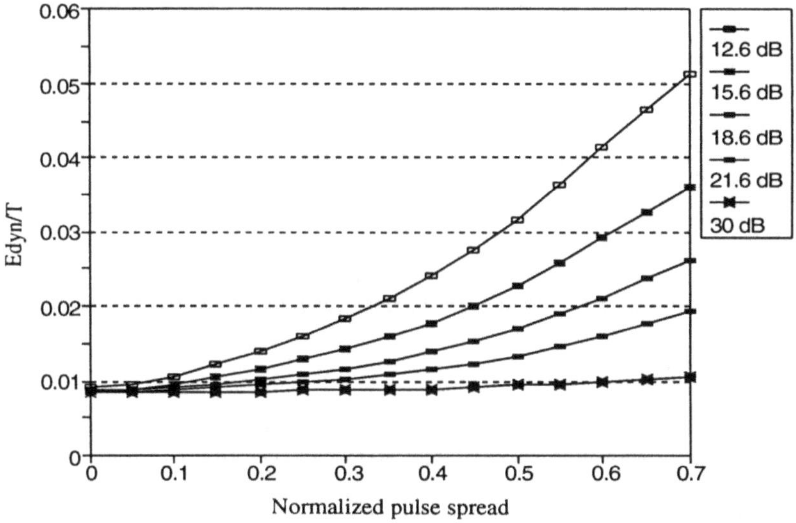

Figure 1. E_{dyn}/T as function of α for $N_t/N_s = -30\,\text{dB}$ taking S/N (dB) as a parameter.

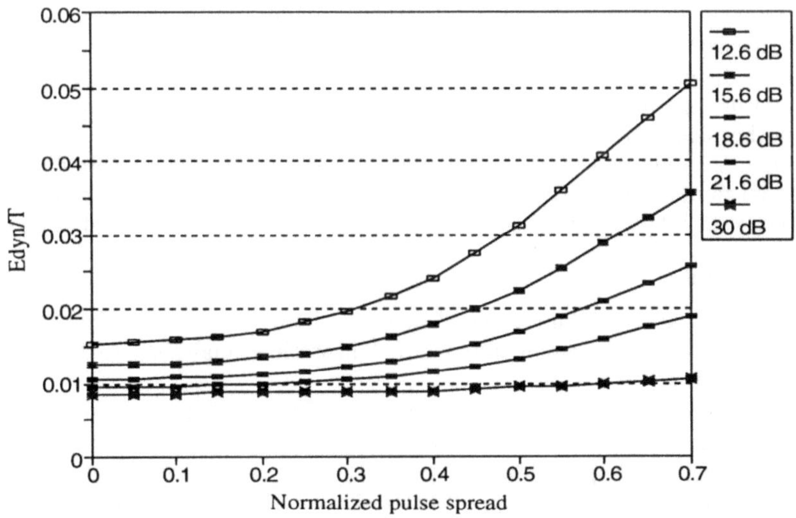

Figure 2. E_{dyn}/T as function of α for $N_t/N_s = 30\,\text{dB}$ taking S/N (dB) as a parameter.

physically by the greater clock information the signal shot-noise possesses at low dispersion for all S/N. This means that, for low dispersion, an increase in the signal shot-noise power (decrease of the S/N) only scarcely affects the clock performance.

Another important remark results from the comparison of E_{dyn}/T for high S/N and for low S/N. With zero dispersion, the excess of E_{dyn}/T due to the noise does not reach E_{dyn}/T due to the signal, corresponding to high S/N (for circuit noise dominance, which corresponds to a higher difference between these values, E_{dyn}/T is about 0.0086 for $S/N = 30\,\text{dB}$ and about 0.0125 for $S/N = 15.6\,\text{dB}$). This observation had already been referred by Franks and Bubrouski[6] for the circuit noise dominance which, following them, leads to not taking into account sometimes the contribution of the noise to E_{dyn}/T. However, even for zero dispersion it does not seem correct to neglect this contribution. Other dispersion situations show greater E_{dyn}/T for low S/N (in both noise type dominances) and, hence, the noise contribution to E_{dyn}/T cannot be ignored.

The main conclusion from these figures is: the Dirac pulse (or very narrow pulse) is the received optical pulse shape which minimizes the dynamic timing jitter when the pulse shape at the input to the squarer is raised cosine in the frequency domain.

4. Conclusions

The influence of the optical fibre dispersion on the square synchronizer performance was discussed and a thorough physical explanation of the obtained results was provided.

The main conclusion of this paper is that, for low signal to noise ratios (but still leading to acceptable error probabilities) and *under quite general conditions*, the performance of the square synchronizer is better in a signal shot-noise dominant situation (typical of long distance communications with APD-receivers), followed by the white circuit noise dominant situation (typical of long distance communications with PIN-receivers). This statement is a generalization of the conclusion presented in[2] wherein a particular pulse shape incident on the photodetector was assumed. However, it must be emphasized that the previous classification depends also on other factors, e. g., at high fibre dispersion, an APD in a signal shot-noise dominant situation reveals a very slightly worse performance than a PIN, with white circuit noise dominance.

Another conclusion is that the noise contribution to dynamic timing jitter can be very high and may be the main contribution particularly at high fibre dispersion and low signal to noise ratio. Hence, the noise contribution cannot be ignored, specially in a system with a few regenerators (most of the direct detection optical systems).

References

1. A. Cartaxo, A. Albuquerque, "Jitter in the square synchronizer on direct detection optical communications. I: Theoretical analysis", to be pub. in *IEE Proc. Pt. J.*
2. A. Cartaxo, A. Albuquerque, "Influence of the various types of noise on jitter performance in binary direct detection optical communications", *IEE Proc. Pt. J*, 137, Dec. 1990, pp. 375-378.
3. A. Cartaxo, A. Albuquerque, "Jitter characterization in optical fibre communication", *Proc. IEEE MELECON'89*, Lisbon, Apr. 1989, pp. 565-568.
4. S. Personick, "Receiver design for digital fiber optic communication systems, I and II", *Bell System Tech. J.*, 52, July-Aug. 1973, pp. 843-887.
5. A. Cartaxo, *Optimização do circuito de sincronismo em receptores ópticos com detecção directa*, Ph. D. Thesis (in Portuguese), Lisbon Tech. Univ., Aug. 1992.
6. L.E. Franks, J.P. Bubrouski, "Statistical properties of timing jitter in a PAM timing recovery scheme", *IEEE Trans. Commun.*, 22, July 1974, pp. 913-920.

EVOLUTION OF PULSE PARAMETERS IN NONLINEAR DISPERSIVE FIBERS

Mirosław Florjańczyk and Réal Tremblay

Centre d'Optique, Photonique et Laser
Département de Physique, Université Laval
Cité Universitaire, Québec, G1K 7P4, Canada

INTRODUCTION

Nonlinear properties of short-pulse propagation in optical fibers are being studied intensely because of their applications in optical communications. The theoretical foundation for the analysis of propagation is the wave equation including dispersion and nonlinear effects. Solution of the wave equation often requires extensive numerical simulations, which have to be repeated independently for different cases.

Approximate analytical methods may overcome these difficulties and give simplified analytical results applicable to many cases. We derived approximate practical formulae describing the evolution of pulse parameters in nonlinear optical fibers. Here, we present some results for the evolution of the pulse width near zero-dispersion wavelength.

POTENTIAL-WELL DESCRIPTION

The propagation of light pulses in nonlinear fibers is described by the following wave equation (nonlinear Schrödinger equation)

$$i\frac{\partial E}{\partial z}=-\frac{\lambda^2 D}{4\pi c}\frac{\partial^2 E}{\partial t^2}-\frac{2\pi n_2}{\lambda A_{eff}}|E|^2 E \tag{1}$$

where $E(z, t)$ is the pulse amplitude, z is the distance of propagation, t is the time variable in a frame moving with the pulse and c is the velocity of light in vacuum. Second order dispersion is determined by the wavelength-dependent dispersion coefficient D, while n_2 is the nonlinear refractive index and A_{eff} is the effective area of the guided mode. Fiber loss is assumed to be compensated by periodically spaced amplifiers. Calculations are performed for a class of different pulses launched at the beginning of the fiber. The initial pulse form can be hyperbolic secant, Gaussian, Lorentzian, parabolic or any other function corresponding to the most frequent models in correlation measuring techniques[1,2].

Applications of Photonic Technology
Edited by G.A. Lampropoulos *et al.*, Plenum Press, New York, 1995

A pulse with amplitude E_p, width t_p, chirp parameter c_p and having the envelope form f is given by

$$E(t,z) = E_p f(t/t_p)\exp(-\frac{i}{2}c_p(t/t_p)^2) \tag{2}$$

A variational approach[3,4] is used in obtaining evolution equations for the pulse parameters. For all considered pulses there exists a simple mechanical analogy representing the pulse width as a coordinate of a mass particle moving in a potential-well

$$\frac{1}{2}\left(\frac{d(t_p/t_0)}{dz}\right)^2 + \Pi(t_p/t_0) = 0 \qquad \Pi(t_p/t_0) = \frac{\mu}{(t_p/t_0)^2} + \frac{\nu}{(t_p/t_0)} + K \tag{3}$$

where t_p is the distance-dependent actual pulse width and t_0 is the width at z=0. The potential-well Π is determined by the initial pulse parameters (wavelength λ, width t_0, peak power $P_0 = E_0^2$) and the fiber experimental parameters (dispersion D, effective area A_{eff} of the guided mode, Kerr coefficient n_2). These parameters are regrouped in constants μ and ν

$$\mu = \frac{\lambda^4 D^2}{2\pi^2 c^2 t_0^4}\frac{a_2}{a_1} \qquad \nu = -\frac{P_0 \lambda\, n_2 D}{c t_0^2 A_{eff}}\frac{a_3}{a_1} \tag{4}$$

The coefficients a_i in constants μ and ν take different values for different initial pulse forms ($x = t/t_0$)

$$a_1 = \int_{-\infty}^{\infty} x^2 f^2(x)\,dx \qquad a_2 = \int_{-\infty}^{\infty} f(x) f''(x)\,dx \qquad a_3 = \int_{-\infty}^{\infty} f^4(x)\,d(x) \tag{5}$$

To this end, the constant K is determined by pulse launching conditions and the amplitude and chirp are related to the pulse width by simple equations analogous to that of Ref.3.

NEAR ZERO-DISPERSION CASE

The potential-well description (3-5) is a generalization of that given for Gaussian and square pulses[3,5] and it is valid for many other pulse forms. An exact solution of the mechanical problem (3-5) is possible, but appears complicated. Instead, additional approximations can be done in order to simplify the resulting formulae. This is the case presented in Fig.1, where we show the evolution of the normalized root-mean-square (rms) value of the pulse width for different pulses and compare the approximate results with exact numerical simulations. There is a very good agreement between the two results. The dispersion parameter is $D = -0.05$ ps/(nm·km) (normal dispersion) and the initial peak power is $P_0 = 1$ mW. Other parameters are $n_2 = 3.2\times10^{-20}$ m^2/W, $A_{eff} = 40$ μm^2 and $\lambda = 1.55$ μm. The initial pulse width t_0 is different in each case. All initial pulses have the same full width at half maximum (fwhm) equal to 200 ps.

In the examples shown in Fig.1, variations of the normalized pulse width t_p/t_0 are small and nonlinear effects dominate over those of dispersion ($\nu >> \mu$). The variational

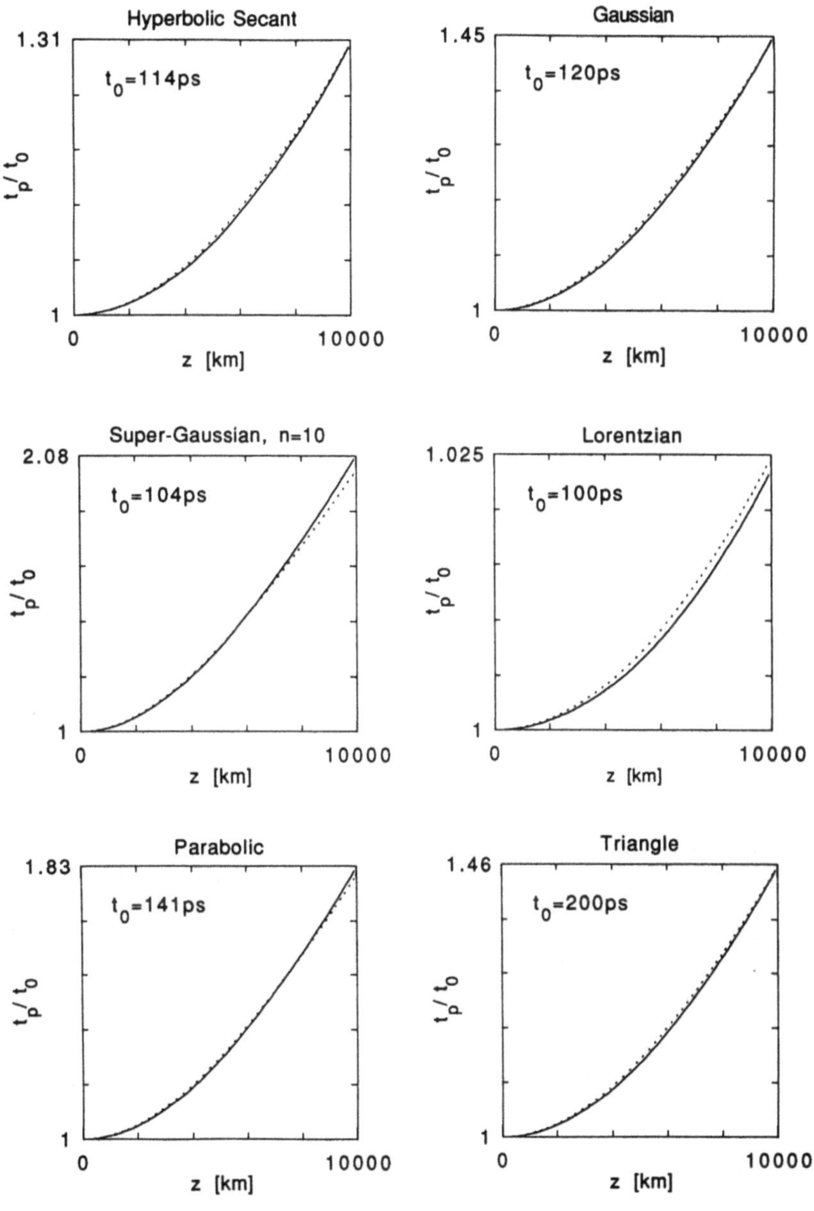

Figure 1. Ratio of rms output pulse width to rms input pulse width as a function of length for different initial pulse forms. Exact numerical results (solid lines) are compared with variational approximate results (dotted lines). See text for details.

evolution of the pulse width is obtained from a formula that comes from a simplification of the exact analytical solution of the problem (3-5). The formula reads

$$t_p / t_0 = \sqrt{1 - a_f \frac{Dn_2}{cA_{eff}} \frac{\lambda P_0}{t_0^2} z^2} \tag{6}$$

where the form factor $a_f = 0.543 a_3 / a_1$ is determined by the function f via Eqs.5 and is independent on other parameters. We list the form factor a_f for different pulses in Table 1.

Table 1. Form factors used in formula (6) ($x = t / t_0$).

Initial Pulse	**Form Function** $f(x)$	**Coefficient** a_f
Hyperbolic Secant	$\mathrm{sech}(x)$	0.440
Gaussian	$\exp(-x^2/2)$	0.768
Super-Gaussian	$\exp(-x^n/2)$	1.611
Lorentzian	$(1+x^2)^{-1/2}$	0.050
Parabolic	$(1-x^2)^{1/2}$ for $\|x\| \le 1$ 0 elsewhere	2.172
Triangle	$(1-\|x\|)^{1/2}$ for $\|x\| \le 1$ 0 elsewhere	2.172
Exponential Cusp	$\exp(-\|x\|/2)$	0.202

It should be stressed that we present the results for the rms width and not for the fwhm of the pulses. One expects a discrepancy between results for the fwhm value obtained from the variational approach and from exact numerical simulations. In the variational method, the dynamics of the pulse is governed by coupled Euler equations for the pulse parameters and not by Eq.1. One uses predetermined pulse forms and there is always a relation between the rms pulse width and its fwhm. On the other hand, in exact simulations, the pulse dynamics is governed by Eq.1, where the pulse form is not preserved and there is no simple relation between the rms pulse width and its fwhm. Nevertheless, the knowledge of rms width provides a useful indication of the pulse deformations[5].

Assuming the variational model, one is also able to predict analytically the evolution of the second-order correlation function[1,2].

REFERENCES

1. K. L. Sala, G. A. Kenney-Wallace & G. E. Hall, IEEE J. Quant. Electron.vol. QE-16 (1980) 990-996.
2. J.-C. M. Diels, J. J. Fontaine, I. C. McMichael & F. Simoni, Appl. Opt., vol. 24 (1985) 1270-1282.
3. D. Anderson, Phys. Rev. A, vol. 27 (1983) 3135-3145.
4. C. Paré, M. Florjańczyk, Phys. Rev. A, vol. 41 (1990) 6287-6295.
5. D. Marcuse, J. Lightwave Technol., vol. 10 (1992) 17-21.

INFLUENCE OF THE NONUNIFORM FM RESPONSE OF DFB LASERS ON CPFSK LIGHTWAVE SYSTEMS WITH NONCOHERENT DETECTION

John Cartledge and Joanne Lennon

Department of Electrical and Computer Engineering
Queen's University
Kingston, Ontario
Canada K7L 3N6

Introduction

Emerging applications of lightwave technology will utilize wavelength division multiplexing (WDM) to transmit a multiple number of optical signals at distinct wavelengths over an optical fibre, with channel selection at the receiver being performed in the optical domain using either an optical filter or optical frequency discriminator (noncoherent detection). Research in this area is based on the fibre bandwidth being more easily accessed in the wavelength domain than the time domain, and the use of multiwavelength technology to perform network functions such as routing, switching and service segregation. The continuous phase frequency shift keyed (CPFSK) modulation format is attractive for these applications because of the ability to generate CPFSK signals by direct modulation of distributed feedback (DFB) lasers, the tolerance of noncoherent receivers to laser phase noise, and the compact spectrum of the modulated signal allows efficient WDM.

The FM response of a DFB laser is typically nonuniform due to carrier and thermal modulation of the active layer refractive index. The thermal effect, which has a time constant on the order of microseconds, is out of phase with the carrier effect thereby creating a dip in the magnitude of the FM response in the kilohertz to low megahertz frequency region. This nonuniformity distorts the CPFSK signal and causes the experimental bit error ratio (BER) performance to depend on the length of the pseudo-random bit sequence (PRBS) used to test the system. In this paper the influence of the nonuniform FM response of DFB lasers on the BER performance of CPFSK systems with noncoherent detection is assessed theoretically and experimentally for receivers based on a tunable fibre Fabry–Perot optical filter.

Modelling

The nonlinear dynamics of DFB lasers are modelled by rate equations which describe the relation between the carrier density $N(t)$, photon density $S(t)$ and optical phase $\phi(t)$[1]

$$\frac{dN(t)}{dt} = \frac{I(t)}{qV_{act}} - \frac{N(t)}{\tau_n} - g_0(N(t) - N_0)\,(1 - \epsilon S(t))\,S(t) \tag{1}$$

$$\frac{dS(t)}{dt} = \Gamma g_0(N(t) - N_0)\,(1 - \epsilon S(t))\,S(t) - \frac{S(t)}{\tau_p} + \frac{\Gamma\beta N(t)}{\tau_n} \tag{2}$$

$$\frac{d\phi(t)}{dt} = \frac{1}{2}\alpha\left[\Gamma g_0(N(t) - N_0) - \frac{1}{\tau_p}\right]. \tag{3}$$

Γ is the mode confinement factor, N_0 is the carrier density at transparency, τ_p is the photon lifetime, β is the fraction of spontaneous emission coupled into the lasing mode, τ_n is the electron lifetime, $I(t)$ is the injected current, q is the electron charge, V_{act} is the active layer volume, g_0 is the gain slope constant, α is the linewidth enhancement factor and ϵ is the gain compression factor. The gain slope constant is given by $g_0 = v_g a_0$ where v_g is the group velocity and a_0 is the gain constant. The laser frequency (chirp) $\Delta v(t)$ is related to the time derivative of the optical phase by

$$\Delta v(t) = \frac{1}{2\pi}\frac{d\phi(t)}{dt}. \tag{4}$$

Using (1)–(4), small–signal intensity modulation and frequency modulation transfer functions can be determined.

Thermal effects are important in the low frequency region of the FM response of a DFB laser. The physical characteristics of the diode chip and submount structure, along with the material thermal conductivity, determine the frequency response of the temperature modulation. A closed form expression for the thermal response has been obtained by considering one-dimensional heat flow in the centre of a laser chip[2]. By taking the Laplace transform of the conduction equation and solving with appropriate boundary and initial conditions, the optical frequency transfer function is given by

$$H_{th}(f) = -K\left(\frac{2}{3\sqrt{\frac{jf}{f_c}}}\left[\tanh\left(\sqrt{\frac{jf}{f_c}}\right) + \tanh\left(\frac{1}{2}\sqrt{\frac{jf}{f_c}}\right)\right]\right) \tag{5}$$

where f_c is the thermal cutoff frequency and K is a constant.

The small–signal transfer function for the total FM response is the vectorial addition of the carrier and thermal responses. The thermal cut-off frequency f_c and the dc FM efficiency ratio K_{dc} (the ratio of the dc thermal and carrier FM efficiencies) are the main factors which determine the low frequency FM response. A characteristic dip in the FM response is observed when $K_{dc} > 1.5$. Fig. 1 shows the magnitude of the FM response for $K_{dc} = 2$ with $f_c = 100$ kHz and 1 MHz.

To allow pattern effects in large–signal pulses to be included in the system model, solutions for the transmitted optical signal are obtained for specific realizations of the data sequence $A_N = \{A_1, A_2, \cdots, A_N\}$. The carrier and photon densities are obtained using a Runge-Kutta algorithm to numerically solve the rate equations. The total time–dependent optical frequency is given by $\Delta v(t) + \Delta v_{th}(t)$ where the carrier contribution is obtained from (4) and the thermal contribution, $\Delta v_{th}(t)$, is obtained from (5) using the injection current waveform.

The lightwave receiver is modelled by a tunable Fabry–Perot optical filter, avalanche photodetector (APD), additive white Gaussian noise process representing the circuit noise, linear filter, sampling device and threshold comparator. The optical filter, which converts the FSK signal to an ASK signal by passing one tone in the CPFSK signal spectrum, has a transfer function which is periodic with transmission peaks separated by the free spectral range (FSR). In WDM systems, all channels are placed within one FSR to avoid ambiguous data reception. Within the vicinity of a transmission peak a Lorentzian lineshape provides an accurate approximation to the filter transfer function

$$H_{filter}(f) = \frac{1}{1 + [\frac{2j(f-f_0)}{f_{FP}}]} \tag{6}$$

where f_0 is the centre frequency of the filter and f_{FP} is the 3 dB bandwidth. The output signal of the receiver is the combination of the filtered output current of the photodetector (a filtered doubly stochastic Poisson process) and the filtered cicuit noise process.

A method of calculating the bit error ratio (or probability of error), P_e, is used which permits consideration of a long PRBS (2^{15}-1 and 2^{23}-1)[3]. This is necessary to allow the effect of the low frequency thermal response to be included. The evaluation of P_e is based on those realizations of the data sequence which consist of consecutive ones and zeros. For example, for a $2^5 - 1$ PRBS, the following sequences and their complements are considered: 00001, 00011, 00111, 01111, and 11111. P_e is calculated based on the number of occurrences of k consecutive ones or zeros appearing in a $2^N - 1$ PRBS, α_k, and the probability of error for the k–bit sequence, $P_e(k)$. $P_e(k)$ is calculated using a truncated pulse train approximation technique. Thus, the probability of error is

$$P_e = \sum_{k=1}^{N} \alpha_k P_e(k) \tag{7}$$

where

$$\alpha_k = \frac{2^{N-k}}{2^N - 1}, \tag{8}$$

and

$$P_e(k) \;=\; \frac{1}{2}\text{erfc}\left(\frac{S - i_0(t_0, A_k)}{\sqrt{2}\sigma_0(t_0, A_k)}\right) + \frac{1}{2}\text{erfc}\left(\frac{i_1(t_0, A_k) - S}{\sqrt{2}\sigma_1(t_0, A_k)}\right). \tag{9}$$

For $m = 0, 1$

$$i_m(t_0, A_k) = \overline{g}\int_{-\infty}^{\infty} [R\, P(t, A_N = m, A_k) + I_d] \cdot h_r(t_0 - t)dt \tag{10}$$

and

$$\sigma_m^2(t_0, A_k) = \sigma_{th}^2 + \overline{g^2}q\int_{-\infty}^{\infty} [R\, P(t, A_N = m, A_k) + I_d] \cdot h_r^2(t_0 - t)dt. \tag{11}$$

Erfc is the complementary error function, S is the threshold level, $i_0(t, A_k)(i_1((t, A_k))$ denotes the average conditioned photocurrent at the output of the receiver filter for $A_N = 0$ ($A_N = 1$, respectively), $\sigma_0^2(t, A_k)\,(\sigma_1^2(t, A_k))$ denotes the sum of the variances of the circuit and shot noise processes at the output of the receiver filter for $A_N = 0$ ($A_N = 1$, respectively), t_0 is the sampling time of the output signal during the Nth bit of the N bit realizations of the data sequence, $\overline{g}$ is the mean avalanche gain, R is the responsivity of the photodetector, $P(t, A_N, A_k)$ is the incident optical signal, I_d is the photodetector dark current, $h_r(t)$ is the combined impulse response of the APD and receiver electronics, σ_{th}^2 is the variance of the filtered circuit noise process, and $\overline{g^2}$ is the mean square avalanche gain. The mean square gain depends on the mean gain and the ionization coefficient ratio. The photocurrents $i_0(t, A_k)$ and $i_1(t, A_k)$ are assumed to be Gaussian random processes since the photocurrents are conditioned on a fixed data sequence. Using this technique the total number of sequences considered is $2N$ for a 2^N-1 PRBS, rather than 2^N as would be the case if all possible realizations of the data sequence were considered.

Results

Of particular interest are the roles of the modulation index and optical filter bandwidth on the receiver sensitivity. The calculated dependence of the receiver sensitivity on the modulation index is illustrated in Figs. 2 and 3 for optical filter bandwidths of 5 GHz and 15 GHz, respectively, f_c = 100 kHz, K_{dc} = 2, and a bit rate of 1 Gb/s. An optimum value of the modulation index m exists due to the tradeoff between a reduced extinction ratio for small values of m and the increased signal distortion arising from the nonuniform FM response for large values of m. The optimum modulation index is more strongly defined for a small optical filter bandwidth. The dependence of the receiver sensitivity on the length of the PRBS decreases as the optical filter bandwidth increases, as a broader filter response

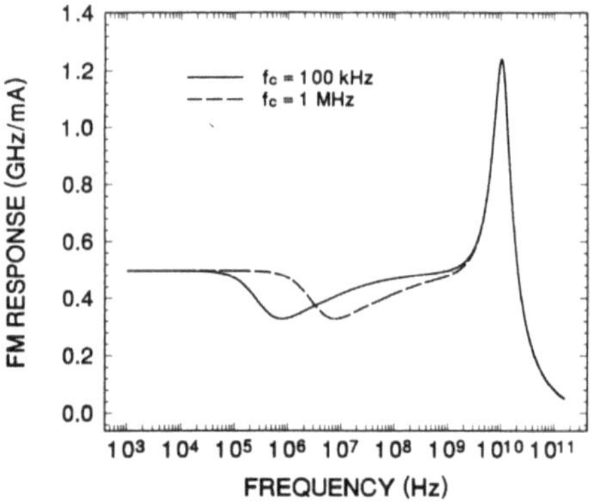

Fig. 1. Magnitude of the FM response.

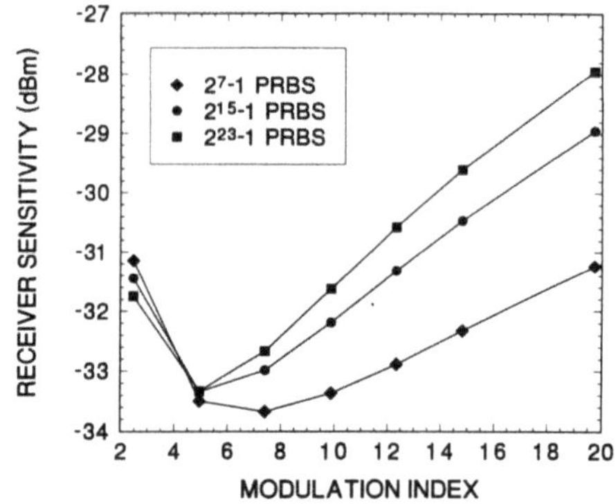

Fig. 2. Optical filter bandwidth is 5 GHz.

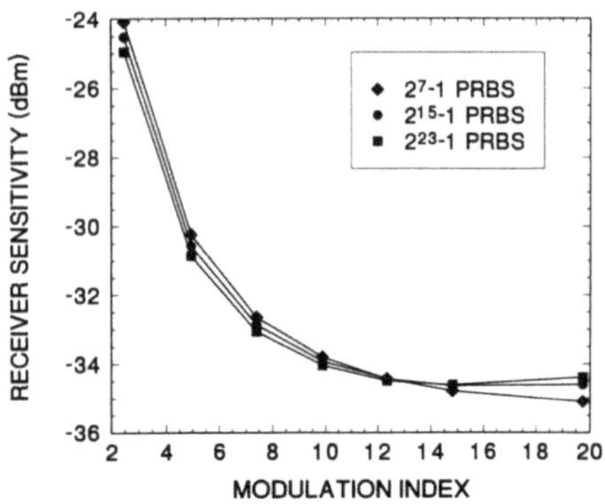

Fig. 3. Optical filter bandwidth is 15 GHz.

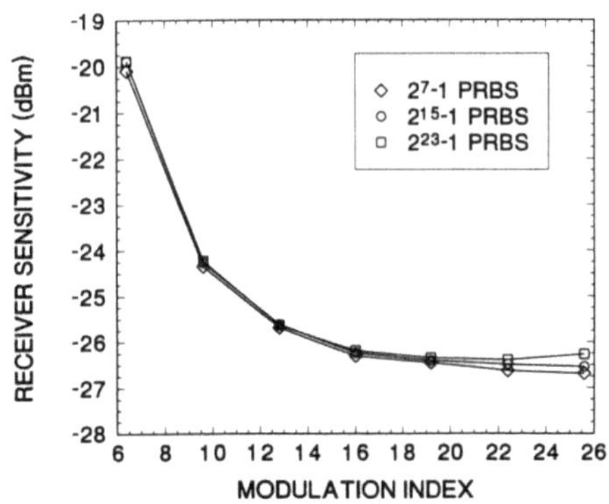

Fig. 4. Optical filter bandwidth is 13.8 GHz.

accommodates the spreading of the signal spectrum due to the nonuniform FM response. Fig. 4 illustrates experimental results for a DFB laser with a dip in the FM response at 5 kHz, an optical filter bandwidth of 13.8 GHz, a *p–i–n* photodetector based receiver, and a bit rate of 1 Gb/s. Qualitative agreement with the theoretical results is obtained, and can only be expected as the actual laser characteristics (FM response, threshold current, light versus current characteristic) differ from those of the numerical model.

REFERENCES

1. R. Tucker, "High speed modulation of semiconductor lasers", *J. Lightwave Technol.*, vol. LT–3, pp. 1180–1192, 1985.
2. G. Pandian and S. Dilwali, "On the thermal FM response of a semiconductor laser diode", *IEEE Photonics Technol. Lett.*, vol. 4, pp. 130–133, 1992.
3. G. Jacobsen, K. Emura, T. Ono, and S. Yamazaki, "Requirements for the LD FM characteristics in an optical CPFSK system", *J. Lightwave Technol.*, vol. 9, pp. 1113–1123, 1993.

A NEW PHOTONIC ARCHITECTURE FOR SIMPLE ROUTING PROCESSORS

Castagnolo Beniamino, Nuovo Michele, Rizzi Maria[1]

[1]Dipartimento di Elettrotecnica ed Elettronica
Politecnico di Bari
Via Re David, 200 - 70125 Bari (ITALY)

INTRODUCTION

Applications such as advanced multimedia workstations, supercomputer interconnections, etc., require a sustained bandwidth of one gigabit per user or even more.

To support such high bandwidth demanding, all-optical networks are extensively employed. In these networks the electro-optical conversion is carried out only by the end users, while the other network functions (switching, multiplexing, amplification, etc.) are carried out in the optical domain.

In the photonic environment, among the huge number of possible network topologies, multihop networks are used because of high throughput, excellent fault-tolerance behaviour and expecially for the existance of simply routing strategies.

The ShuffleNet is the best known example of the multihop architecture[1]. It has represented the basis for different network implementations dedicated to various applications[2].

One of the most pressing problem to be considered in the choice of a topology, is the photonic feasibility of the switching circuits. To this aim, in many cases it is useful to study topologies with easy routing rules, so that the switching decisions, made at every node, are very simple.

In this paper a new multihop network architecture, named Starring, is presented. It is characterized by the property of modularity and by a fair simplicity of routing algorithms. In particular, the network architecture is proposed and one suitable routing strategy is introduced. Based on this algorithm the block diagram of the control logic for the header processing is examined. Finally the conclusions are drawn out.

THE STARRING ARCHITECTURE

The Starring network is a modular structure. In fact it is composed of P modular loops each containing four stations. Communications between loops are performed through two rings. An example of a simple Starring network with two modules is reported in Fig.1. Each station of the network has two input and two output unidirectional links.

In order to increase the number of users, it is sufficient to add r more equal loops. As each standard loop is composed of 4 stations, the number of users that can be added to increase the Starring dimensions is 4r.

The architecture of Fig.1 can also be represented as a space structure (Fig.2). Each modular loop represents one section of a parallelepiped, whose edges are constituted by the rings. Successive sections, inside the structure, are connected each other with

Applications of Photonic Technology
Edited by G.A. Lampropoulos *et al.*, Plenum Press, New York, 1995

unidirectional links. In order to increase the network dimensions, it is easy to add more modular sections.

A simple routing strategy is needed for an all optical implementation, in order to design routing stations including only few all-optical gates. This devised strategy is the following: the informations are composed of a header and of an information field.

h bits of the header, with $h = \lceil \log_2 r \rceil$, contain the address of the loop including the destination station, while k other bits (k = 2) represent the number of the receiving station inside its loop.

Each station performes the following operations on the packet header:

- if the address of the received packet coincides with its own one, the station analyzes the information field of that packet;
- if only the first h bits of the header coincide, the station forwards the packet to the near station on the loop;
- if the h bits of the header of the received packet are different from the first h bits of the header of the processing station, the packet is forwarded through the ring to the near loop.

This routing algorithm is not the optimal one. In fact the connection between two generical stations is not obtained necessarily using the minimal path.

THE HEADER PROCESSING CIRCUIT

The challenging task in the context of optical networks is the photonic implementation of the different apparatuses which compose a station.

In Fig. 3 is presented the optical circuit for the implementation of the described routing strategy.

The XOR gate 1 performes the comparison between the address of the received packet (input A) and the address of the processing station (input B). Following the XOR logic, if the addresses coincide, the result is a sequence of 0 logical values and, in this case, the packet is forwarded to the local receiver; otherwise the packet is forwarded to the second part of the circuit.

The XOR gate 2 operates in the same way but in this case it compares only the first l bits of the header. Infact, they represent only the addresses of the stations inside the same loop. So, if the addresses coincide, at the end of the OR gate there will be a 0 logical value; otherwise there will be a 1 logical value.

In the first case the packet will be forwarded to the next station on the same ring; on the contrary, the packet will be forwarded to the next station on the same loop.

To obtain a single value at the end of the OR gate, it is necessary to splite the signal at the end of the XOR gate 2 into l different delay lines. In this way the OR gate can process these signals simultaneously, obtaining only one logical value: 0 or 1.

In the figure the delay lines L_1 and L_2 are necessary to delay the packet information during the described comparing processes.

COMPARISONS AND CONCLUSIONS

Some important parameters, which are related both the architecture and the routing algorithm, are to be considered in order to compare the Starring network to some other existing structures. They are the network diameter, defined as the maximum distance between two generically selected nodes of the network and the average number of hops needed to connect two generical stations[3].

Denoting with P, the number of the network loops, the diameter is expressed using the following equation:

$$d = 2P + 1$$

The average number of hops varies depending on the transmitting station. If S represents the total number of hops necessary to reach from one station all the others, the corresponding average number of hops is given by: S/(N-1) where N is the total number of the network stations.

It can be also demonstrated that $S_{min} < S < S_{max}$ where:

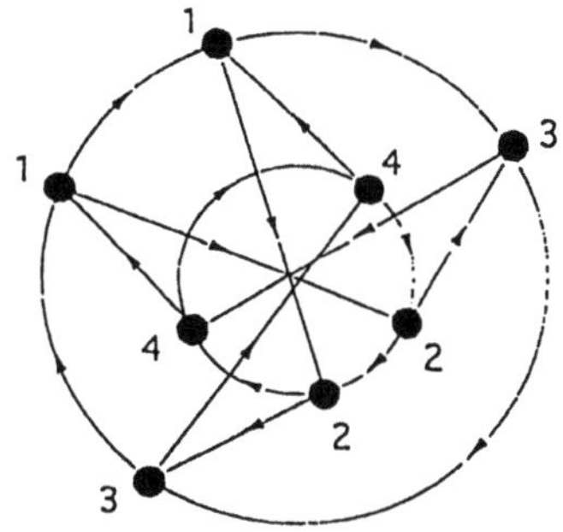

Figure 1. The 8-users Starring architecture

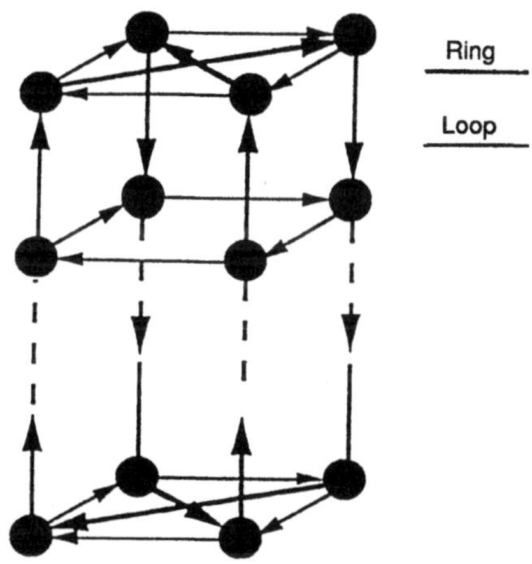

Figure 2. The N-users Starring architecture represented as a space structure

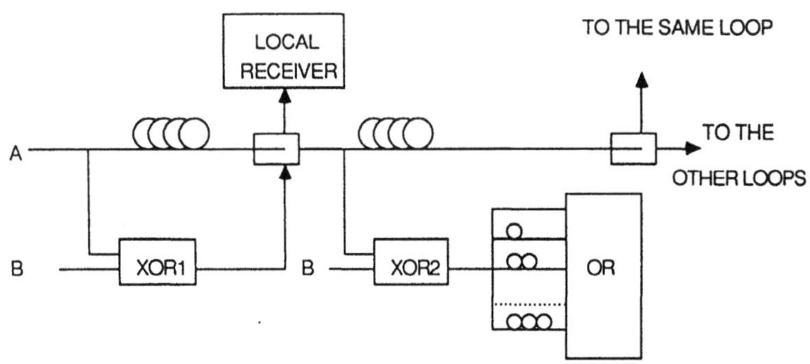

Figure 3. Optical circuit implementing the routing strategy

$$S_{min} = 2P(P-1) \quad \text{and} \quad S_{max} = P(3P+4) - P\,mod2$$

Therefore, the expected number of hops between two randomly selected users is given by:

$$E = [S_{min} + 2S_1 + ...+ (1+ P\,mod2)S_{max}] / P(4P+1)$$

where S_{i+1} is represented by the following iterative expression:

$$S_{i+1} = S_i + [P - (2i+1)] \qquad \text{for } i = 0, 1, ... P/2 - 1$$

Comparisons have been made between the Starring and the ShuffleNet structures regarding the expected number of hops E and the efficiency η which is given by the reciprocal of E. However, owing to the different modularity, the ShuffleNet can be structured only with $N=kp^k$ users. Therefore for N=8, we obtain E=2.6 and η= 0.4 for the Starring and E=3.1 and η=0.32 for the ShuffleNet; for N=24, E=5.2 and η=0.2 and E=3.3 and η=0.3, respectively.

Therefore, adopting the Starring architecture with the given routing algorithm, better results are obtained for low numbers of users.

REFERENCES

1. A.S.Acampora, "A multichannel multihop local lightwave network", *GLOBECOM'87 Conf. Rec.* 1459:1467 (1987).
2. J.H.Patel', "Performance of processor-memory interconnections for multiprocessors", *IEEE Trans. Comp.* 771:780 (1981).
3 T.Yasui, K. Kikuchi, "Overview system network architectural possibilities", *in Proc. Top. Meet. Photon. Switching, Incline Village*, paper WB4 (1987).

ALGORITHMS FOR HIGH-PERFORMANCE COMPUTING WITH SMART PIXELS

Dietmar Fey, Werner Erhard

Friedrich-Schiller-University Jena
Department of Computer Architecture and Communication
Ricarda-Huch-Weg 16
D - 07743 Jena, Germany

1 Introduction

One of the major problems in current parallel high-performance computing systems are limitations in the I/O-bandwidth. This is not only a problem on board level communication but also for chip-to-chip and even gate-to-gate interconnections especially in parallel computers. Optical interconnections between in VLSI technology manufactured electronic circuits as aspired in so-called smart pixels systems may offer in the future a solution[1]. This paper presents well-suited algorithms and architectures for future smart pixels systems consisting of monolithically integrated photodiodes and electronic circuits, which are hybrid mounted to vertical surface emitting laser diodes (VCSEL)[2].

Closely related with the problem of finding suitable architectures and appropriate algorithms is the question for the hardware requirements needed for the implementation. These costs depend for example on the number of necessary optical links and their topology (regular or irregular), the necessary gate density, and the permissible power dissipation in the active components. Hence, we carried out architecture and algorithm studies with simulation tools and derived from the simulation results the required hardware parameters.

Below, we show in section 2 generic declarations concerning the complexity of smart pixels. In section 3 we present a concept for a parallel computer architecture with smart pixels for floating-point operations based on so-called bit algorithms. In section 4 the possibilities for special purpose smart pixels systems are discussed.

2 Complexity of smart pixels systems

Mostly smart pixels systems are considered as SIMD structures in literature. Therefore the complexity of the whole system is mainly determined by the complexity of a single cell. The complexity of a single cell depends on its optoelectronic input/output interface and its data processing capabilities. Assuming monolithically integrated receivers and electronic switching units the complexity is determined by the number of input channels N_{PHD}, and the

Applications of Photonic Technology
Edited by G.A. Lampropoulos *et al.*, Plenum Press, New York, 1995

number of needed transistors N_{TRANS}. In dependence on the required area per receiver inclusive amplifier A_{PHD} and the area per transistor A_{TRANS} the size of a cell can be calculated. Then (1) determines the total number #PE of cells or processing elements (PE) of a smart pixels chip of 1cm² area.

$$\#PE = \frac{A_{CHIP}}{A_{PE}} = \frac{1}{N_{TRANS} \cdot A_{TRANS} + N_{PHD} \cdot A_{PHD}} \left[\frac{1}{cm^2}\right] \tag{1}$$

An important question is, if we should use many small or few large processing elements. To answer this question we defined a new figure, the communication degree of a PE. The communication degree C_D is the relation of the number of photoreceivers to the number of transistors needed in the PE. Multiplying (1) with N_{PHD} we get the number of optical links N_{LINKS} for the whole chip in dependence on the communication degree C_D (2).

$$N_{LINKS} = \frac{1}{1/C_D \cdot A_{TRANS} + A_{PHD}} \quad ; \quad C_D = \frac{N_{PHD}}{N_{TRANS}} \tag{2}$$

The number of optical links N_{LINKS} is an important figure for the demand the optical imaging system has to achieve. Furthermore we assume that the number of the optical senders is not larger than the number of receivers. Then N_{LINKS} can also be used to estimate the power dissipation in the microlaser array. Figure 1 displays N_{LINKS} versus the reciprocal of the communication degree for 150μm² transistor area and 2500μm² photoreceiver area.

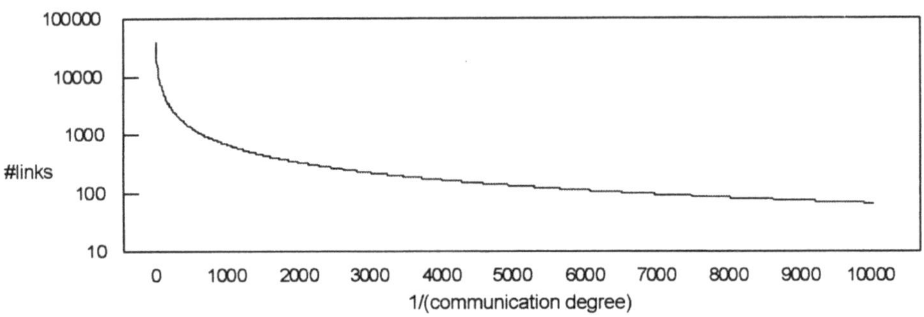

Figure 1. Number of optical links versus communication degree C_D for 1cm² chip size

It can be seen that high communication degrees exploit better the potential of high-dense interconnections optics offers in principle. Furthermore, if we assume two PEs, $PE^{(1)}$ and $PE^{(2)}$, with different transistor and photoreceiver numbers but with the same communication degree, the smaller PE is to prefer, because this allows more parallelism within the smart pixels array (3).

$$C_D^{(1)} = C_D^{(2)} \Rightarrow N_{LINKS}^{(1)} = N_{LINKS}^{(2)} \; ; \quad N_{PHD}^{(1)} < N_{PHD}^{(2)} \Rightarrow \#PE^{(1)} = \frac{N_{LINKS}^{(1)}}{N_{PHD}^{(1)}} > \frac{N_{LINKS}^{(2)}}{N_{PHD}^{(2)}} = \#PE \tag{3}$$

3 Bit algorithms for 3-dimensional smart pixels architectures

For the execution of complex floating-point operations in current microprocessors costly floating-point units are mostly used. Unfortunately such large PEs show a low communication degree. Hence, they profit more by the high time bandwidth than by the high space bandwidth of optics. If appropriate algorithms are available small PEs must not inevitably imply a lower performance. Bit algorithms[3] are advantageous for smart pixels because they need only simple operations like addition, shifting, and memory access to predefined bits. A bit algorithm presents an iterative calculation scheme g for a pair of real numbers $(x_{i+1},y_{i+1})=g(x_i,y_i)$. During this iteration process x_i converges from an initial value to a known value and y_i becomes automatically the desired function value $f(y_0)$. The principle is shown in (4) for the example of a division.

$$\begin{array}{ll} \frac{Y}{X}=\frac{Y\cdot a_1\cdot a_2\cdot\ldots\cdot a_n}{X\cdot a_1\cdot a_2\cdot\ldots\cdot a_n}=Q & \begin{array}{ll} X_{i+1}=X_i\cdot a_i=X_i\cdot(1+2^{-i}) & a_i\cdot X_i\le 1 \\ Y_{i+1}=Y_i\cdot a_i=Y_i\cdot(1+2^{-i}) & \end{array} \\ X\cdot a_1\cdot a_2\cdot\ldots\cdot a_n\rightarrow 1 & \begin{array}{ll} X_{i+1}=X_i & a_i\cdot X_i>1 \end{array} \\ Y\cdot a_1\cdot a_2\cdot\ldots\cdot a_n\rightarrow Q & Y_{i+1}=Y_i \end{array} \tag{4}$$

If nominator and denominator are multiplied with the same factor a_i the quotient in every iteration step remains unchanged. The selection of the factor $a_i=1+2^{-i}$ replaces the required multiplication by the much simpler operation, right shift and addition. Similar algorithms exist for the calculation of frequently needed standard functions like exponential, logarithm, cosine, sine, square root and reciprocal value[3].

For the calculation of each of the mentioned functions with a bit algorithm four operations are fundamental: shifting, addition, selection of the right operand pair (x_{i+1},y_{i+1}) for the next iteration, and testing of the condition. For each of these fundamental operations we defined the functionality for a smart pixel layer. Figure 2 displays a scheme for an optoelectronic 3-D bit-serial and word-parallel architecture based on bit algorithms. Processing layers SEL, ADD and FLAG correspond to smart pixels layers and perform the previous mentioned operations. The layers are attached to a 3-dimensional organized memory DATA to store horizontally every bit position of the mantissas.

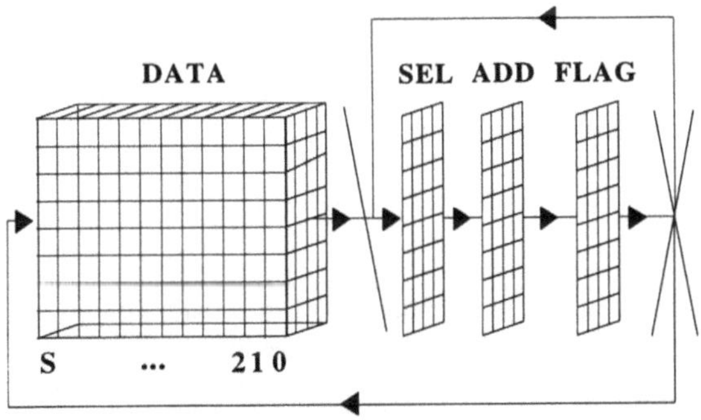

Figure 2. Scheme of a 3D smart pixels architecture for floating-point operations

For the connection scheme between these layers regular connections are sufficient. We verified the functionality with the software tool HADLOP[4], which was especially developed for the simulation of digital optical systems. The maximum parallelism of the whole system is determined by the most complex layer. This is the adder layer with 4 optical inputs and about 100 transistors per smart pixels cell to implement two 1-bit full adders and two memory cells for storing the carry bits. Hence, we get a communication degree of 1/25. With (1) and (2) we can calculate the total number of PEs and links. With the help of the simulation tool we determined the computing performance for one PE.

During our investigations we determined to first order that a smart pixels system, built up on a wafer (15x15inch), and with a gate density of 1400 gates/mm² (5 transistors per gate are sufficient) and an array delay (2 gate delays) of 1-2ns would achieve about 200 GFLOPS (figure 3). Problems would currently arise by the heat dissipation in the microlaser array (~12W/cm²) assuming 1mW dissipated power per laser diode. If a further reduction in the power consumption is possible in the future a massively parallel SIMD system operating near the TFLOP range could be built up on only 5 wafer boards.

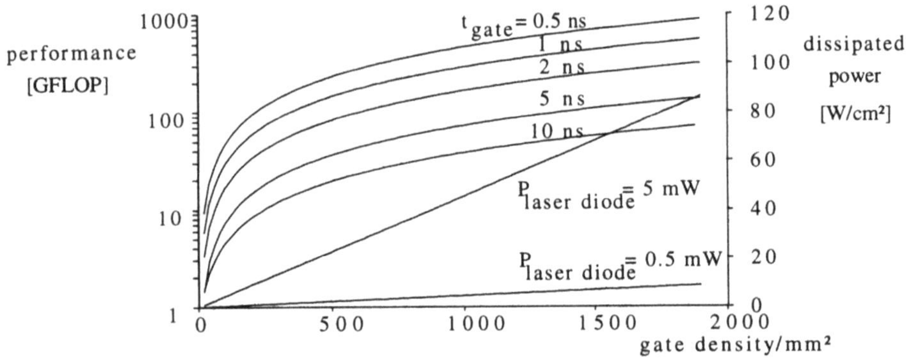

Figure 3. Performance analysis of the massively parallel smart pixels system

4 Special architectures with smart pixels

For special purpose smart pixels architectures systolic arrays or cellular automata supply efficient concepts. Systolic arrays have successfully used in VLSI design. Optoelectronic solutions offer the additional advantage of 3-dimensional implementations. The later increases the performance and simplifies the mapping of real 3D problems. Especially bit-oriented systolic arrays can be well mapped onto digital optical systems as already shown with symbolic substitution architectures[5]. The same technique is convenient for smart pixels systems. Such architectures are well-suited for image and signal processing. Applications are given for example in the field of medical image processing or in real-time decision making during experiments in high-energy physics.

5 Conclusion

We presented a concept for future smart pixel systems based on bit algorithms allowing a space consuming architecture for high-performance scientific computing. If following first order specifications can be fulfilled, this leads to parallel systems with more computing performance than current architectures realized in pure electronics show: an integration density of 7000 transistors/mm², a gate delay of about 1ns, 10000 optical links/cm² and a power dissipation of 0.5 mW per laser diode.

References

1. J.A. Neff, "3D-Computers with smart pixels", IEEE/LEOS Summer Topical Meeting 92 on Integrated Optoelectronics, Santa Barbara, August 1992
2. J.W. Parker,"ESPRIT II - A European Collaborative Programme in Optical Interconnects" in: Digest of Int. Optical Computing Conference, Kobe, Japan 1990
3. T.C. Chen, "Automatic Computation of Exponentials, Logarithms, Ratios and Square Roots," IBM Journal Res. and Dev., July 1972, pp. 380-388
4. D. Fey, "Computer-aided design of digital opto-electronic systems with HADLOP", in: S.D. Smith, R.F. Neal (eds.), "Optical Information Technology", Springer 1993
5. D. Fey, K.H.Brenner, "Digital Optical Arithmetic based on Systolic Arrays and Symbolic Substitution Logic", Int. Journ. of Optical Computing 1, (1990) 153-167

PARALLEL ARITHMETIC ON OPTICAL COMPUTERS BY REDUNDANT BINARY NUMBER REPRESENTATIONS

Giuseppe A. De Biase and Annalisa Massini

Dipartimento di Scienze dell'Informazione
Università di Roma la Sapienza
Via Salaria 113, 00198 Roma, Italy

INTRODUCTION

A parallel arithmetic, suitable for optical computers, can be obtained using two main approaches 1) residue number system[1,2] and 2) redundant number representations[3,4]. In fact using of both approaches it is possible to build totally parallel adders operating by symbolic substitution (SS)[2,5–7] and in constant time (the adding time is independent of the length of the operand digit strings, N). Using a residue number system, the size of the SS Truth Tables required for the carry-free addition heavily increases with numerical range involved[1,2], and these tables depend on the digit position. On the contrary, additions of redundant numbers can be performed in constant time by small SS Truth Tables which are independent of the digit positions.

Recently, the redundant binary (RB) number representation was presented[4]. This representation permits the execution of the main arithmetic operations in an inherently parallel manner and in short time (the algebraic sum requires only two steps), it naturally fits the 2's complement binary number system, it presents the same advantages and efficiency of the MSD representation (which is the one most widely investigated, particularly in the digital optical computing field - see Ref. 8-11), but it requires only two symbols $\{0, 1\}$ (while the radix-2 MSD representation requires three symbols $\{\bar{1}, 0, 1\}$). Using symbolic substitution, the algebraic sum of RB numbers requires small Truth Tables or very simple hardware using other optical technologies.

Here, a family of redundant binary number representations, obtained by generalization of the RB representation, is presented. These representations have properties similar to the those of RB and are suitable for optical computing. In particular the p-RB (packed Redundant Binary) representations present an efficiency greater than RB representation; with p-RB numbers the algebraic sum is always permitted in constant time for any efficiency value. Also p-RB representations fit in a natural way the 2's complement binary number system.

REDUNDANT NUMBER REPRESENTATIONS SUITABLE FOR OPTICAL COMPUTING

A decimal number D can be represented by an MSD radix-2 number by[9–11]:

$$D = \sum_{i=0}^{N-1} a_i r^i \qquad \text{with} \quad r = 2 \tag{1}$$

in this case the digits $a_i \in \{\bar{1}, 0, 1\}$. For $i = \ldots 7, 6, 5, 4, 3, 2, 1, 0$, Eq. 1 generates the following sequences of position weights:

$$\ldots \quad \pm 128 \quad \pm 64 \quad \pm 32 \quad \pm 16 \quad \pm 8 \quad \pm 4 \quad \pm 2 \quad \pm 1$$

The MSD representation is a redundant number representation because more than one representation is possible for any D (see examples in Table 1).

Table 1. Example of some unsigned MSD radix-2 and RB2 numbers.

D	bin.	MSD (radix-2)	RB2
0	000	000	000000
1	001	001 $\quad 01\bar{1} \quad 1\bar{1}\bar{1}$	000001
2	010	010 $\quad 1\bar{1}0$	000100 000010
3	011	011 $\quad 1\bar{1}1 \quad 10\bar{1}$	000101 000011
4	100	100	010000 001000 000110
5	101	101 $\quad 11\bar{1}$	010001 001001 000111
6	110	110	010100 010010 001010 001100
7	111	111	010101 010011 001011 001101

An unsigned integer D can also be given by[4]:

$$D = \sum_{i=0}^{N-1} a_i 2^{i - \lceil \frac{i}{2} \rceil} \quad \text{or} \quad D = \sum_{i=0}^{N-1} a_i 2^{i - \lfloor \frac{i}{2} \rfloor} \quad \text{both with } N \text{ even} \tag{2}$$

where $a_i \in \{0, 1\}$ (the symbols $\lceil\,\rceil$ and $\lfloor\,\rfloor$ represent respectively the rounding to the upper and to the lower integer). D is in redundant binary number representation. For $i = \ldots 7, 6, 5, 4, 3, 2, 1, 0$, Eq. 2 generates the following sequences of position weights:

$$\begin{matrix} \ldots & 8 & 8 & 4 & 4 & 2 & 2 & 1 & 1 & \text{or} & \ldots & 16 & 8 & 8 & 4 & 4 & 2 & 2 & 1 \\ & n & r & n & r & n & r & n & r & & & r & n & r & n & r & n & r & n \end{matrix}$$

Numbers generated by Eq. 2 will be called in RB1 and RB2 representation respectively. In each pair of digits of the same weight nr, the left and right digit are called n (*normal*) and r (*redundant*) digit respectively. From Eq. 2 it follows that RB1 or RB2 representations of a number can be obtained in parallel in one logic step from its binary representation by using the following simple recoding rules:

$$0 \rightarrow 00 \qquad 1 \rightarrow 10 \qquad \text{for RB1}$$
$$0 \rightarrow 00 \qquad 1 \rightarrow 01 \qquad \text{for RB2}$$

Also RB numbers have several representations (some examples of unsigned RB2 numbers are shown in Table 1). This coding operation can be used also for signed numbers if, in analogy with the 2's complement number system, Eq. 2 is changed to:

$$D = -a_{N-1}2^{N-1-\lceil \frac{N-1}{2} \rceil} + \sum_{i=0}^{N-2} a_i 2^{i-\lceil \frac{i}{2} \rceil} \quad \text{or} \quad D = -\sum_{i=N-2}^{N-1} a_i 2^{i-\lfloor \frac{i}{2} \rfloor} + \sum_{i=0}^{N-3} a_i 2^{i-\lfloor \frac{i}{2} \rfloor} \qquad (3)$$

ARITHMETIC BY REDUNDANT NUMBERS

Addition and subtraction of 2-MSD (radix-2) numbers can be performed in parallel by SS in two steps by using suitable Tables.

Table 2. Symbolic Substitution Truth Tables for 2-MSD addition and subtraction. For the columns, the inputs 0 and 1 mean that the input pair is in the sets $\{11, 10, 01, 00\}$ and $\{1\bar{1}, \bar{1}1, 0\bar{1}, \bar{1}0, \overline{11}\}$ respectively.

x_iy_i \ F	1	0
11	1 0	1 0
01 10	1 $\bar{1}$	0 1
00	0 0	0 0
$1\bar{1}$ $\bar{1}1$	0 0	0 0
$0\bar{1}$ $\bar{1}0$	0 $\bar{1}$	$\bar{1}$ 1
$\bar{1}\bar{1}$	$\bar{1}$ 0	$\bar{1}$ 0

x_iy_i \ F	1	0
11 $\bar{1}\bar{1}$	0 0	0 0
00	0 0	0 0
$0\bar{1}$ 10	1 $\bar{1}$	0 1
$\bar{1}0$ 01	0 $\bar{1}$	$\bar{1}$ 1
$1\bar{1}$	1 0	1 0
$\bar{1}1$	$\bar{1}$ 0	$\bar{1}$ 0

y_i \ x_i	1	0	$\bar{1}$
1		1 $\bar{1}$	0 0
0	1 $\bar{1}$	0 0	$\bar{1}$ 1
$\bar{1}$	0 0	$\bar{1}$ 1	

Table 3. Symbolic Substitution Truth Table for RB numbers algebraic sum. This Table acts on *nr* pairs; *u* and *l* indicate the upper and lower row. The lower output pair is shifted left one position respectively.

l \ *u*	00	01	10	11
00	00 00	10 00	00 01	10 01
01	00 01	10 01	00 10	10 10
10	00 01	10 01	00 10	10 10
11	00 10	10 10	00 11	10 11

Table 2 shows the Cherri and Karim Tables[10] (which are obtained minimizing those in Ref. 9) for 2-MSD numbers. To add two numbers, the left and right Tables are, respectively, used in the first and second step. The MSD subtraction is obtained using the center Table in the first step and the right Table in the second one.

The addition between two RB numbers can be simply performed applying twice the rules defined by a Truth Table acting on *rn* digit pairs. Table 3 shows these rules, which applied in parallel on all pairs of two RB numbers, arranged on two superposed and aligned rows, give the sum (or algebraic sum if Eq. 3 is used).

MSD		RB1	
$(7)_{10}$	$001\bar{1}0\bar{1}1$	$(6)_{10}$	00001111
$(27)_{10}$	$00111\bar{1}1$	$(9)_{10}$	00011101
	$00100\bar{1}1\phi$		00001010
	$\phi0000100$		0011110ϕ
$(34)_{10}$	00100010	$(0)_{10}$	00000000
$(-34)_{10}$	$00\bar{1}000\bar{1}0$	$(15)_{10}$	0101101ϕ

Figure 1. Example of addition of two MSD and two RB numbers. The symbol ϕ indicates a padded zero.

In Figure 1 examples of addition of two MSD and two RB numbers are shown. The addition is completed in parallel in constant time (two elemental logic steps in both cases).

EFFICIENCY

To evaluate the efficiency E of a redundant number representation the ratio:

$$E = \frac{R_B}{R_R}$$

(where R_B is the number of bits needed in the binary number system, and R_R is the number of bits needed in a redundant number representation to write the same value) can be used (see Ref. 4). The efficiency, E, is related to the numerical range reached by the different number representations when bit strings of the same length are used. Table 4 shows that RB and MSD radix-2 representations have the same efficiency and both generate bit strings two times as long as with respect to binary numbers.

Table 4. Efficiencies of MSD and RB number representations.

Number representation	digit × pos. weight	bit × digit	R	E
Binary	1	1	1	1
MSD (radix-2)	1	2	2	0.5
RB	2	1	2	0.5

MSD representation can be considered as an extended case of the fixed-radix number system in which both positive and negative weighted digits are allowed in the digit set

$\{\overline{r-1}, \ldots, \overline{1}, 0, 1, \ldots, r-1\}$, where r is the value of the radix used. By using $r > 2$ the efficiency can be increased (see Table 6) but also the SS addition Truth Tables increase in size[11].

p-RB NUMBER REPRESENTATIONS

By generalizing Eq. 2, an integer D can be given by:

$$D = \sum_{i=0}^{N-1} a_i 2^{i - \lceil \frac{i}{p} \rceil} \quad \text{or} \quad D = \sum_{i=0}^{N-1} a_i 2^{i - \lfloor \frac{i}{p} \rfloor} \quad \text{where } p|N \tag{4}$$

where the digits are again $a_i \in \{0, 1\}$ and p is a positive integer. Eq. 4 gives the position weights of p-RB (packed Redundant Binary) numbers. As an example, in the case of 4-RB1 and 4-RB2 numbers, for $i = \ldots 7, 6, 5, 4, 3, 2, 1, 0$, the weights sequences obtained from Eq. 4 are respectively:

$$\begin{array}{ccccccccc} \ldots & 32 & 16 & 8 & 8 & 4 & 2 & 1 & 1 \\ & n & n & n & r & n & n & n & r \end{array} \quad \text{or} \quad \begin{array}{ccccccccc} \ldots & 64 & 32 & 16 & 8 & 8 & 4 & 2 & 1 \\ & r & n & n & n & r & n & n & n \end{array}$$

For $p = 2$, the p-RB1 and p-RB2 representations coincide with the RB1 and RB2 representations respectively, and therefore the p-RB numbers have properties very similar to those of the RB ones. The addition (or algebraic sum) of p-RB numbers can be performed by SS Rule Truth Tables acting on single digits (see Ref. 12 and Table 5).

Table 5. Symbolic Substitution Truth Tables for the algebraic sum of p-RB numbers. These Tables act on single n or r digits; u and l indicate, respectively, the upper and lower row. The lower output digit is shifted left one position.

T_n

l \ u	0	1
0	0 0	1 0
1	1 0	0 1

T_{r1}

l \ u	0	1
0	0 0	0 1
1	0 1	1 1

T_{r2}

l \ u	0	1
0	0 0	1 0
1	1 0	1 1

This operation is completed after $p+4$ elemental logic steps. In the first three steps Table T_{r1} is applied on all r digits and Table T_n is applied on all n digits in parallel. From step 4 to step $p+4$ Table T_{r2} is applied on all r digits and Table T_n is applied on all n digits in parallel. Figure 2 shows an example of addition of 4-RB2 numbers.

The efficiency of p-RB number representations is given by[4]:

$$E = \frac{R_B}{R_R} = \frac{p-1}{p} \tag{5}$$

Table 6. Efficiencies of the r-MSD and p-RB number representations.

MSD	p-RB	E
2-MSD	2-RB	0.500
4-MSD	3-RB	0.666
8-MSD	4-RB	0.750
16-MSD	5-RB	0.800
32-MSD	6-RB	0.833
...	...	...

Eq. 5 shows that efficiency increases as the packing factor, p, increases. In fact, for p-RB numbers one can write:

$$\lim_{p\to\infty} E = \lim_{p\to\infty} \frac{p-1}{p} = 1$$

(Table 6 shows this behaviour). Notice that, for $p \to \infty$, the p-RB2 representation coincides with the natural binary number system:

$$\lim_{p\to\infty} \sum_{i=0}^{N-1} a_i 2^{i-\lfloor \frac{i}{p} \rfloor} = \sum_{i=0}^{N-1} a_i 2^i$$

$(1539)_{10}$	$(00000000011111101111)_{4-RB2}$
$(575)_{10}$	$(00000001111101111011)_{4-RB2}$
	$(00000001000000011100)_{4-RB2}$
	$(00000001111111010110)_{4-RB2}$
	$(00000000011101000010)_{4-RB2}$
	$(00000011000100111000)_{4-RB2}$
	$(00000011011001111010)_{4-RB2}$
	$(00000000001000000000)_{4-RB2}$
	$(00000011010001111010)_{4-RB2}$
	$(00000000010000000000)_{4-RB2}$
	$(00000011000001111010)_{4-RB2}$
	$(00000000100000000000)_{4-RB2}$
	$(00000011100001111010)_{4-RB2}$
	$(00000000000000000000)_{4-RB2}$
	$(00000011100001111010)_{4-RB2}$
	$(00000000000000000000)_{4-RB2}$
$(2114)_{10}$	$(00000011100001111010)_{4-RB2}$
$(0)_{10}$	$(00000000000000000000)_{4-RB2}$

Figure 2. Example of addition of two 4-RB2 numbers.

CONCLUSIONS

Redundant number representations are very suitable for optical computing, but they have efficiencies lower than the binary number system. This fact implies more expensive hardware. By means of p-RB number representations the efficiency can be improved. These representations maintain the same advantages as the RB ones: they naturally fit the 2's complement binary number system, and the constant time algebraic sum permits one to build a fast and inherently parallel arithmetic using symbolic substitution with very small Truth Tables. For these reasons p-RB numbers can be considered as a good trade-off between speed and cost of optical computing devices.

REFERENCES

1. H. L. Garner, "The residue number system", IRE Trans. Electron. Comput. EC-8, 140-147 (1959)
2. C. C. Guest, T. K. Gaylord, "Truth-table look-up optical processing utilizing binary and residue arithmetic", Appl. Opt. 19, 1201-1207 (1980)
3. A. Avizienis, "Signed-digit number representations for fast parallel arithmetic", IRE Trans. Electron. Comput. EC-10, 389-400 (1961)
4. G. A. De Biase, A. Massini, "Redundant binary number representation for an inherently parallel arithmetic on optical computers", Appl. Opt., 32, 659-664 (1993)
5. K.-H. Brenner, "New implementation of symbolic substitution logic", Appl. Opt. 25, 3061-3064 (1986)
6. K.-H. Brenner, A. Huang, N. Streibl, "Digital optical computing with symbolic substitution", Appl. Opt. 25, 3054-3060 (1986)
7. M. M. Mirsalehi, T. K. Gaylord, "Truth-table look-up parallel data processing using an optical content addressable memory", Appl. Opt. 25, 2277-2283 (1986)
8. R. P. Bocker, B. L. Drake, M. E. Lasher, T. B. Henderson, "Modified-signed addition and subtraction using optical symbolic substitution", Appl. Opt. 25, 2456-2457 (1986)
9. Y. Li, G. Eichmann, "Conditional symbolic modified signed-digit arithmetic using optical content-addressable memory logic elements", Appl. Opt. 26, 2328-2333 (1987)
10. A. K. Cherri, M. A. Karim, "Modified-signed digit arithmetic using an efficient symbolic substitution", Appl. Opt. 27, 3824-3827 (1988)
11. A. A. S. Awwal, M. N. Islam, M. A. Karim, "Modified-signed digit trinary arithmetic by using optical symbolic substitution", Appl. Opt. 31, 1687-1694 (1992)
12. G. A. De Biase, A. Massini, "High efficiency redundant binary number representations for a parallel arithmetic on optical computers", Optics & Laser Technology, Special Issue on Optical Computing (in press)

BIT ERROR RATE ANALYSIS OF OPTICAL DATA LINKS FOR COMPUTER COMMUNICATIONS

P. Palacharla,[1] C.W. Barnard,[1] P. Myslinski,[1] J. Chrostowski,[1] R.Neumann,[2] and R. Khalil[2]

[1]Institute for Information Technology, National Research Council of Canada
Ottawa, Ont., Canada K1A 0R6
[2]Canadian Technology Center, Digital Equipment of Canada Ltd.
Kanata, Ont., Canada K2K 2A6

INTRODUCTION

Parallel optical data links have attracted substantial attention in recent years as a potential means for overcoming the electrical interconnections bottleneck in advanced computer systems[1,2]. Optical interconnect technology offers advantages such as high bandwidth-distance product, superior immunity to electromagnetic interference and absence of ground-loop noise. A typical parallel optical data link consists of a transmitter array, a free-space or guided-wave interconnection medium and a receiver array. These links can be used for intra- and inter-computer interconnections. In this paper, we study the bit error rate (BER), one of the main performance criteria of parallel optical data links. In particular, we present details of BER analysis and techniques for experimental measurement of very low BER levels ($< 10^{-15}$) in optical data links. These techniques are valid in the presence of BER floors.

BER REQUIREMENTS AND ANALYSIS

The most important criterion of performance of data links is the percentage of errors the system introduces, the BER. In computers, data is transferred by words of 32 or 64 bits in width, and an error in any one bit causes the loss of the entire word. This can be corrected by the use of error detection and correction (EDC) techniques, which, unfortunately, add latency, cost and reduce computer performance. Typically, high speed computers are operated under the requirement that no bit errors occur over periods of time as long as one month or more. Because of the higher data rates and byte-wide transfers, a fundamentally lower BER is required for computing than for telecom applications. For example, in 32-bit data transfer at 100 Mb/s, a 10^{-9} BER means that more than 1 bit will be dropped every second. Therefore, to ensure data processing reliability, a BER of 10^{-15} or lower is most likely required, which is far below 10^{-9}, the typical BER specification used in telecom systems.

Hence, it is important to evaluate the limitations due to various causes of errors, such as noise and also be able to measure extremely low ($<10^{-15}$) BER levels in optical data links. We first present a brief theoretical framework for the BER analysis of a single-channel optical link whose BER is limited by Gaussian noise in the receiver. A typical receiver consists of a p-i-n photodetector, transimpedance amplifier and decision circuitry (comparator). The signal at the comparator input can be described by $V_j = ZI_j$, where Z is the transimpedance of the amplifier and I_j is the effective input current, $I_j(t) = I_{pj}(t) + n_j(t)$, where j = 0 or 1 corresponds to bit "0" or "1" being transmitted, $I_{pj} = RP_j$ is the photocurrent due to the incident optical power P_j during a single bit, R is the responsivity of the detector, and n_j is the contribution due to noise sources such as thermal noise, shot noise, laser intensity noise etc. The decision circuit compares V_j to a reference voltage V_{th}, which is referred back to the receiver input as the current $I_{th} = V_{th}/Z$. An estimate for the error probability, P_e, or BER can be obtained by approximating the noise density of n_j with a Gaussian distribution, which gives[3]:

$$P_e = \frac{1}{4}\operatorname{erfc}\left(\frac{I_1 - I_{th}}{\sqrt{2}\,\sigma_1}\right) + \frac{1}{4}\operatorname{erfc}\left(\frac{I_{th} - I_0}{\sqrt{2}\,\sigma_0}\right) \tag{1}$$

where $\operatorname{erfc}(x) = (2/\sqrt{\pi})\int_x^\infty e^{-u^2}du$ is the complementary error function, I_0, I_1 are the average currents and σ_0, σ_1 are the rms noise currents for the "0" and "1" bits respectively. The optimum threshold is selected to achieve a minimum BER by equating the number of errors when a "0" or "1" is transmitted. The minimum BER is given by $P_e^{min} = (1/2)\operatorname{erfc}\left(Q/\sqrt{2}\right)$ where $Q = (I_1 - I_0)/(\sigma_1 + \sigma_0)$ is the SNR at the decision circuit. Assuming the various noise sources to be independent, the total variance of the noise current σ_j^2, is the sum of variances of individual noise sources[3]. It can be shown that laser intensity noise degrades the receiver sensitivity and can lead to a BER floor which is unacceptable for computer applications. The presence of a BER floor is due to the dominant contribution of laser intensity noise at high power levels and the noise current being proportional to input power level. This results in the SNR being almost independent of input power and thus, BER attains a floor. The BER performance with respect to modal noise shows similar behaviour as laser intensity noise in optical data links using multimode fiber arrays[4]. In the case of parallel optical links, additional noise sources such as switching noise, crosstalk, timing jitter/skew etc. need to be considered in the performance analysis.

MEASUREMENT OF LOW BER

The most accurate method of BER measurement involves the transmission of a pseudorandom test signal sequence through the data link with the P_e computed by comparing the received bits with a stored replica of the transmitted bit pattern. However, the drawback of this approach is that, for systems with very small probability of error, a large number of bits are needed to achieve a meaningful estimate of the system performance. As an example, for a 95% confidence interval of $[2P_e/5, 8P_e/5]$, about $10/P_e$ bits are required. To evaluate a $P_e = 10^{-15}$ for data that is transmitted at a bit rate of B = 100 Mb/s, the measurement time needed is $t = 10/BP_e \approx 3$ years. Thus, the conventional method of BER measurement is impractical for low BER optical data links.

To reduce the BER measurement time, it is possible to artificially degrade the channel in a controlled manner and thereby raising the error rate to a range (e.g. 10^{-5} to 10^{-10}) that is substantially higher than the BER requirement of the data link. These resultant error rate measurements are then extrapolated to determine the true BER of the data link under test. We present two methods for extrapolated measurement of low BER in optical data links which differ in the channel degradation scheme.

Sinusoidal Interference Method

In the first method, a sinusoidal interference of random phase with respect to the data stream is added at the decision circuitry of the receiver[5]. In practice, a deterministic sine-wave whose frequency is substantially different and is a non-integer multiple or submultiple of the bit rate can be used to simulate a sinusoidal disturbance having (pseudo) random phase. The BER is measured for several levels of sinusoidal interference. These BER data are then mapped by a function which varies linearly with the sinusoid amplitude. The true

BER is obtained by extrapolating the linear fit to the case of zero sinusoid amplitude.

The following analysis gives the mathematical basis for the method. The current signal at the input of the comparator is given as

$$I_j(t) = I_{pj}(t) + n_j(t) + s(t), \quad j = 0, 1 \tag{2}$$

where s(t) is the sinusoidal interference introduced into the system. The reason for adding the interference at the decision circuitry is to avoid any noise contribution from the interference itself. Since $n_j(t)$ and s(t) are mutually independent random variables, the probability density function (PDF) of their sum $n_j(t) + s(t)$ is given by the convolution of the individual PDFs. Using this overall PDF and following the standard derivation for BER, we obtain the error probability,

$$P_e = \frac{1}{4\pi}\int_{-\pi/2}^{\pi/2}\left\{\mathrm{erfc}\left[\frac{Q-\rho_0\sin\theta}{\sqrt{2}}\right] + \mathrm{erfc}\left[\frac{Q-\rho_1\sin\theta}{\sqrt{2}}\right]\right\}d\theta \tag{3}$$

where Q is the SNR defined earlier, $\rho_0 = A_{sin}/\sigma_0$, $\rho_1 = A_{sin}/\sigma_1$ and A_{sin} is the amplitude and θ is the phase of the sinusoidal interference. Notice that with no interference ($A_{sin} = 0$) the error probability is same as in Eq.(1). This integral cannot be evaluated in closed form. It can be seen that the dominant contribution to the integral expression in Eq.(3) occurs in the neighbourhood of $\theta = \pi/2$. So we obtain an upper bound to P_e,

$$P_u = \frac{1}{4}\left\{\mathrm{erfc}\left[\frac{Q-\rho_0}{\sqrt{2}}\right] + \mathrm{erfc}\left[\frac{Q-\rho_1}{\sqrt{2}}\right]\right\} \tag{4}$$

which has a numerical behaviour of $(1/2)\mathrm{erfc}\left((Q-\rho)/\sqrt{2}\right)$ to within an order of magnitude in the range 10^{-4} to 10^{-30}. We use an approximation for the complementary error function[5] $\mathrm{erfc}(x/\sqrt{2}) \approx \exp[-c_1x^2-c_2x-c_3]$, which is within ±0.02 for $10^{-30} \le \mathrm{erfc}(x/\sqrt{2}) \le 10^{-4}$. The coefficients are $c_1 = 0.4926$, $c_2 = 0.2498$ and $c_3 = 0.7912$. Thus, an approximate expression for probability of error is given by

$$P_e \approx \frac{1}{2}\exp\left[-c_1(Q-\rho)^2 - c_2(Q-\rho) - c_3\right]. \tag{5}$$

Solving for (Q-ρ), we obtain

$$Q-\rho = \frac{-c_2+\sqrt{c_2^2-4c_1(c_3+\ln(2P_e))}}{2c_1} = \psi(P_e). \tag{6}$$

Since $\rho = A_{sin}/\sigma$, we have a function Ψ which maps the BER (P_e) to vary linearly with the amplitude of sinusoidal interference A_{sin}. Thus, by fitting a straight line for the BER measurements ($\psi(P_e)$) with respect to A_{sin}, the zero-intercept obtained by extrapolation gives the estimate of Q. This estimate of Q is used to calculate the estimate of the true BER of the optical data link.

We have applied this extrapolation technique to measure the BER of a data link using an LED transmitter and a pin photodetector. The results are shown in Fig.1. We plotted the experimental BER measurements $\psi(P_e)$ ("•" marks) for different sinusoid amplitude levels in Fig. 1 and the solid line is the best linear fit obtained from regression. The zero-intercept ($A_{sin} = 0$) of the straight line is Q = 9.36 and the corresponding error rate is 4.0×10^{-21}. Thus, a BER of 4.0×10^{-21} is expected for the optical data link in the absence of sinusoidal interference.

Threshold Modification Method

In this method, the decision threshold level of the receiver is shifted away from the optimum value which corresponds to the minimum BER[6]. This method is based on the concept that the contributions to BER due to noise in the absence (bit 0) and presence (bit 1) of the signal can be separated in the optical links by operating on either side of the optimum threshold value. Once separated, the BER is a simple expression given by a single (1/4)erfc(.) function (see Eq.(1)). The signal to noise ratio at the decision circuitry is fixed during the measurements. Two sets of BER measurements are taken for different threshold

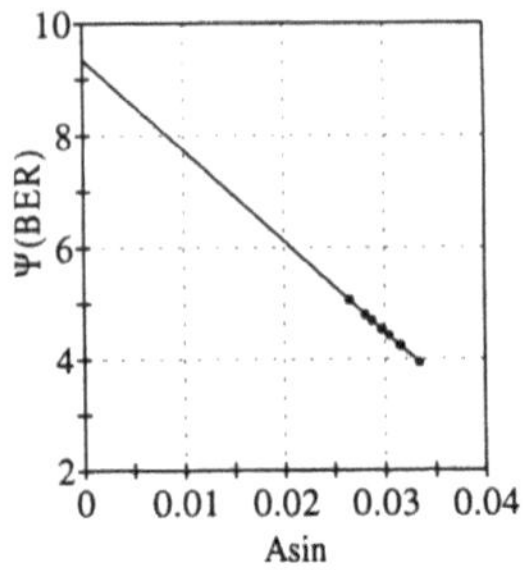

Figure 1. BER measurement using the sinusoidal interference method. (experimental data points : '•', extrapolated linear fit : solid line)

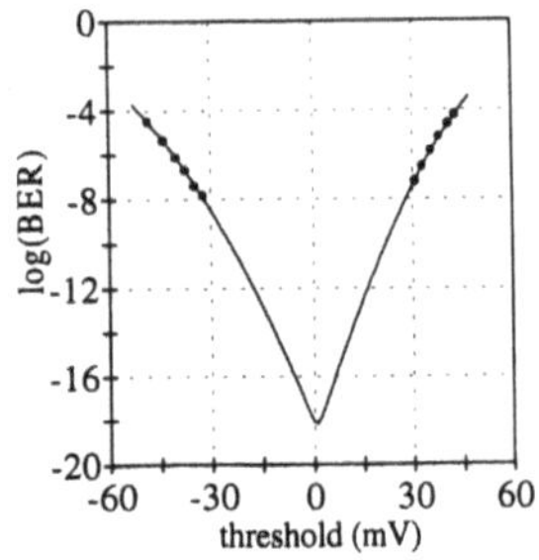

Figure 2. BER versus threshold level. (experimental data points: '•', extrapolated fit: solid line)

values on each side of the optimum threshold. Each set of BER data is transformed by an inverse error function, and then a linear regression is performed with the corresponding set of threshold levels. The equivalent I_j and σ_j , j=0,1 are given by the intercept and slope of the linear fits. These parameter estimates are then substituted into Eq.(1) to obtain the intersection of the two error function curves, which gives the optimum threshold and the actual BER of the data link. The results of this extrapolation scheme are shown in Fig.2. The experimental BER measurements are plotted for different threshold values and the solid line is the best fit. The BER estimate of the optical data link obtained by extrapolation is 6.4×10^{-19}.

CONCLUSIONS

We analyzed and experimentally demonstrated two techniques for measurement of extremely low ($< 10^{-15}$) BER on simple optical data links. These proposed methods can also be used as a design and development tool for optical data links. Work is in progress to extend these techniques to parallel optical data links limited by non-Gaussian noise sources.

REFERENCES

1. A. Takai et. al., "200-Mb/s/ch 100m optical subsystem interconnections using 8-channel 1.3-μm laser diode arrays and single-mode fiber arrays," *IEEE Journal of Lightwave Technology*, 12:260 (1994).
2. R. A. Nordin et. al., "High performance optical datalink array technology," *IEEE Trans. Comp., Hybrids and Manufacturing Technology*, 16:783 (1994).
3. G. P. Agrawal, "Fiber-optic communication systems," John Wiley & Sons, Inc., New York (1992).
4. A. M. J. Koonen, "Bit-Error-Rate degradation in a multimode fiber optic transmission link due to modal noise," *IEEE Journal on Selected Areas in Communications*, SAC-4:1515(1986).
5. D. Chan, "The extrapolation of error rate performances of baseband digital communication systems," Tech. Report TR 1E90-6-73, Bell-Northern Research, October 1973.
6. A. C. Popovici, "Fast measurement of bit error rate in digital links," *IEE Proc.*, Pt. F, 134:439(1987).

THE WAVELENGTH-DEPENDENT REFRACTIVE INDEX CHANGE ASSOCIATED WITH THE BLUE TO PINK MEMBRANE PHOTOCHEMICAL CONVERSION IN BACTERIORHODOPSIN

Richard B. Gross, Angelo T. Todorov and Robert R. Birge

W. M. Keck Center for Molecular Electronics
Syracuse University
Syracuse, NY 13244-4100

INTRODUCTION

Bacteriorhodopsin (BR) is the photosynthetic protein found in the cell membrane of the bacterium *Halobacterium salinarium*.[1, 2] This microorganism thrives in salt brines where the concentration of sodium chloride is roughly four times greater than ocean water. BR's biological function derives from its unique light-transducing properties whereby the absorption of a photon results in the pumping of a proton from the bacterium's cytoplasm to its surrounding environment. This action establishes an electrochemical potential which the bacterium harnesses to synthesize metabolic adenosine triphosphate. During the past several decades much has been learned about bacteriorhodopsin's photochemical and biophysical properties, and the molecular events associated with proton pumping.[3-5] More recently, several groups have demonstrated BR's efficacy as a new type of optical recording material useful in photonic applications that include real-time holography,[6, 7] interferometry,[8] spatial light modulation,[9, 10] and long term optical data storage.[11-13] All of the latter exploit BR's inherent property of photochromism whereby a material under proper illumination reversibly changes its optical transmission properties.[14] Although numerous synthetic photochromes have been reported over the last century, few offer the intrinsic advantages found in BR. These include high light sensitivity in both the forward and reverse directions (large quantum yield), freedom from photodegradation, and the ability to tailor the photochromic properties of the protein through genetic manipulation of the native protein[15, 2], or through the action of external agents such as chemical compounds,[16, 17] humidity, pH,[11] or ionic strength. In regard to the latter, it has been well established that under normal laboratory conditions, BR binds four moles of divalent cations per mole of protein[18] and undergoes the photochemical cycle shown in Figure 1.

Upon the removal of bound cations, or by acidifying a solution of BR, the protein converts to a thermally stable and spectrally-shifted conformer ($\Delta\lambda$~40 nm) called the blue membrane (λ_{max} = 603 nm).[19] Interestingly, the blue membrane is photochemically active

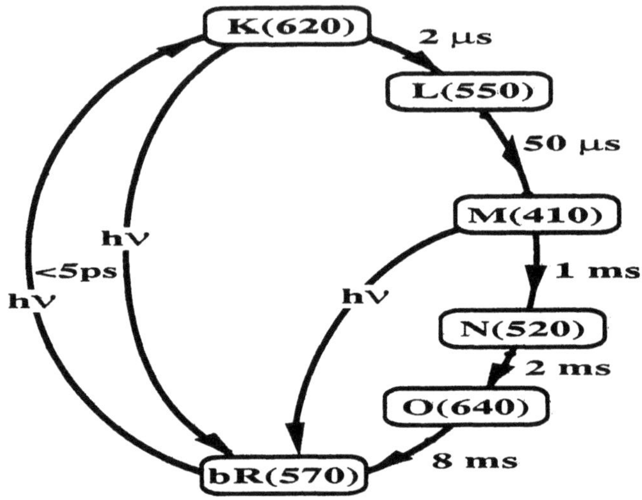

Figure 1. The photocycle of wildtype bacteriorhodopsin. Thermal intermediates are shown as capitalized letters. The inset letters indicate the wavelength of the absorption maximum of each state.

and can be converted to a thermally stable product called the pink membrane (λ_{max} = 503 nm). The blue and pink membrane forms of BR, much like the bR and M forms in the native photocycle, constitute a photochromic pair of states. However, unlike the M to bR thermal reversion which under normal conditions occurs on a millisecond time scale, the pink membrane is thermally stable on a time scale of months or years. Subsequent extended exposure of the pink membrane with blue-green light results in the reformation of the blue membrane. The absorption spectra of the blue and pink membrane forms of bacteriorhodopsin are shown in Figure 2.

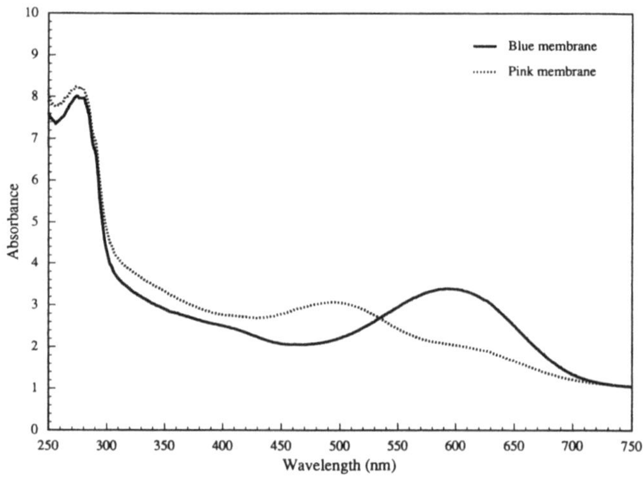

Figure 2. The absorption spectra of the blue and pink membrane forms of bacteriorhodopsin.

The light-induced shift in the absorption band associated with the photochemical conversion of blue membrane to pink membrane results in a concomitant change in refractive index. This change, which follows directly from the Kramers-Kronig relationship, is the fundamental determinant of the diffractive properties of the material, and as such, represents a meaningful property in which to investigate. Similar studies on the native purple membrane form of BR, and genetically derived mutant proteins, have shown that knowledge of the spectral relationship is useful for accurately predicting the holographic properties of bacteriorhodopsin.[20, 21] In this paper, we present preliminary data on the photochemical conversion of the blue to pink form and report results on the measurement of the refractive index change associated with this event. In addition, we present data on the resolution as measured by a USAF 1951 resolution chart.

EXPERIMENTAL

A Mach-Zehnder interferometer[22] was constructed to measure the spectral dependence of the refractive index change associated with blue to pink membrane conversion (See Figure 3). Coherent laser light emitted from a krypton ion laser served as the interferometric probe beam. The probe beam (<1μW) was divided at the entry to the interferometer by means of a beamsplitter (BS1). The two beams were recombined at a diagonally opposed beamsplitter (BS3) via two parallel mirrors M1 and M2. The fringe spacing of the interference pattern for any particular wavelength investigated was kept constant both before and after exposure of the bleaching light. A CCD camera, linked to a TV monitor and a video cassette recorder (VCR), was used to monitor and record the integrity of the interference pattern. A polarizing filter (PA1) placed in one arm of the interferometer was used to equalize beam intensities and optimize the fringe contrast of the interference pattern. A polyacrylamide film containing bacteriorhodopsin (OD_{603nm} = 5.0) was deionized by placing the film in a solution containing a cation ion exchange resin prior to its placement in one arm of the interferometer. A second krypton ion laser, operating at 568 nm, was used as the light source for initiating the blue to pink photoreaction. This beam was made colinear with the probe beam by means of a third beamsplitter (BS2). Interference patterns were recorded by a VCR before, during, and after exposure to bleaching light. In order to obviate contributions to the refractive index change arising from thermal heating of the sample, only frames recorded 5-10 seconds after blue to pink photoconversion were used in the analysis. The recorded images were digitized using an 8 bit video frame grabber and stored on a personal computer.

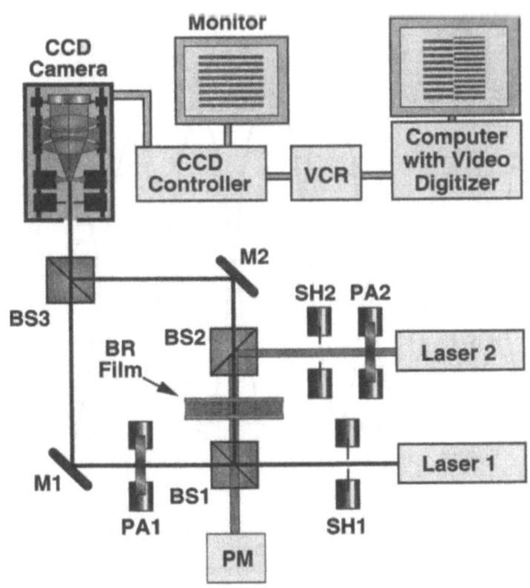

Figure 3. Optical set-up of the Mach-Zehnder interferometer used for measuring the refractive index change associated with the blue to pink photoconversion (See text).

RESULTS

A picture of the processed images of the interference patterns prior and subsequent to the bleaching light is shown in Figure 4. Figure 5 shows the normalized intensity values taken at the same vertical position, perpendicular to the fringes, before and after exposure to yellow light. These values were subjected to a Fast Fourier Transform algorithm to obtain the fundamental grating frequency and the spatial shift between the interference patterns arising from the refractive index change. The magnitude and sign of the change in refractive index originating from the photoconversion can be calculated by noting that:

$$\Delta n = \frac{\lambda_p \, \Delta G}{d \, G} \quad (1),$$

where λ_p is the wavelength of the probe beam (nm), d is the film thickness (nm), G is the fundamental period of the grating pattern (pixels), and ΔG is the spatial shift of the interference patterns maximum (pixels) after bleaching (See Figure 5).

Figure 4. Digitized images of an interference pattern taken before (left), and after (right), the blue to pink photoconversion.

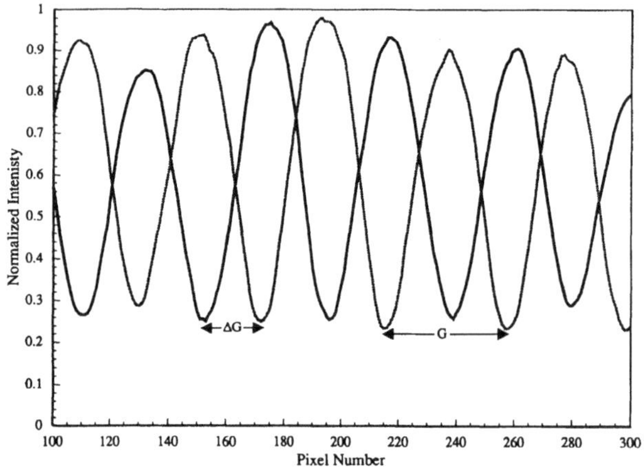

Figure 5. Normalized intensity line profiles of the initial and shifted interference patterns. The fundamental grating period (G) and the spatial shift (ΔG) associated with the phase change are indicated.

The Kramers-Kronig relationship relates the real and imaginary parts of the complex index of refraction, i.e.. the refractive index and the absorption coefficient.[23] The measured values for the change in refractive index were compared with theoretical values numerically derived from the relationship shown in Equation 2. The absorption difference spectrum (ΔA) was obtained by measuring the absorption in the film before and after exposure to the bleaching light. Since the integral in Equation 2 extends over infinite wavelengths and the experimental measurement included wavelengths from the far ultraviolet to the near-infrared, no absorption changes outside these regions were assumed .

$$\Delta n = \frac{2.3026}{2\pi^2 d} \; \text{P.V.} \int_0^\infty \frac{\Delta A \, d\lambda'}{1 - \frac{\lambda'^2}{\lambda^2}} \qquad (2),$$

Figure 6 shows the results of the experimentally and theoretically derived values for the change in refractive index associated with the blue to pink membrane conversion. The accuracy of the experimental data is limited the precision in the measurement of ΔG. We estimate the precision to be ~3 pixels which results in a uncertainty in Δn of $\pm$ 5.6 x 10^{-5}.

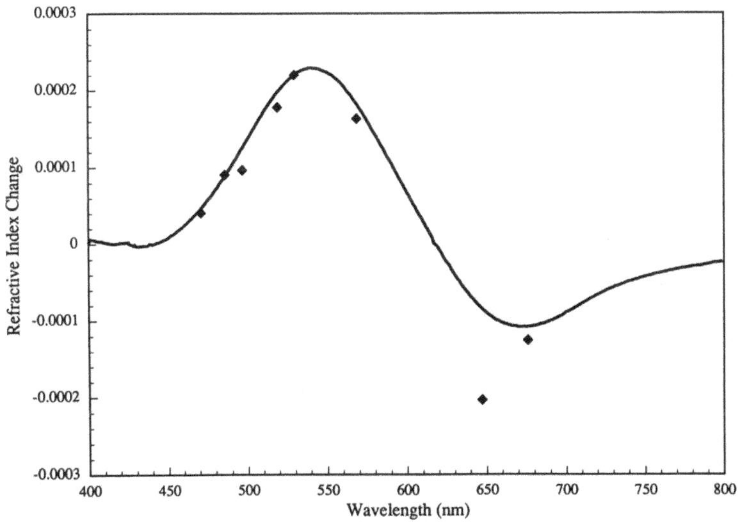

Figure 6. Results of the experimentally (F) and numerically calculated values (solid line), derived by the Kramers-Kronig relation, for the refractive index change associated with the blue to pink membrane photochemical conversion.

Resolution, generally presented in resolvable line pairs per millimeter, describes a material's ability to resolve a standard number of parallel lines of unity space ratio.[24] Although the resolution of a thin film of purple membrane has been measured holographically and is known to exceed 2500 line pairs/ millimeter, we imaged a pattern of a USAF 1951 resolution chart on the blue membrane to obtain a standard measurement of resolution. Figure 7 shows a picture of the resolution chart embossed on the film obtained by photographing a projected image on a nearby screen. The highest resolvable element observed under 40X magnification was found to be Group 6 element 5, which corresponds to 102 line pairs per millimeter. This value is limited by the 700 μm thickness of the film and the depth of field of the microscope objective. These observations nevertheless support the molecular nature of the photochromic process in bR and the virtual non-granularity recording potential of BR.

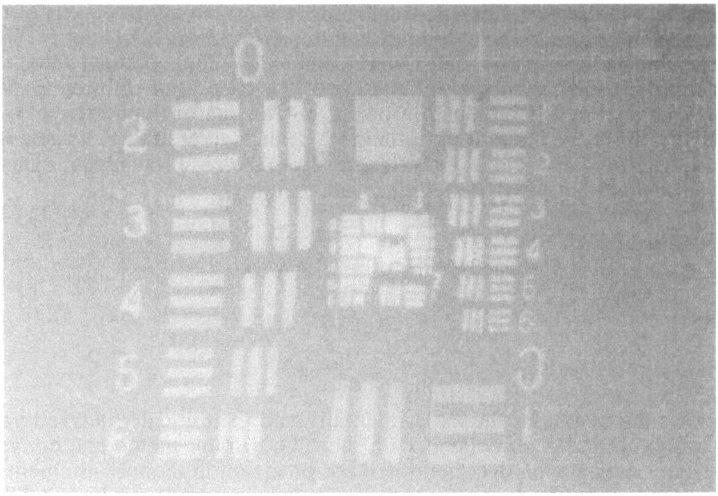

Figure 7. A photograph of the USAF chart embossed on the blue membrane film.

DISCUSSION

It is useful to compare the refractive index change in the blue membrane to that observed experimentally in similar films containing purple membrane (native BR). Previous studies on films of similar optical density show that the refractive index change associated with the blue to pink transition is approximately one numerical order smaller than that associated with the bR to M transition in purple membrane.[21] The reason for the lower value in the blue membrane form of BR results from smaller shift in absorption band associated with the blue to pink transition ($\Delta\varepsilon_{pink->blue}$ ~ 100 nm vs. $\Delta\varepsilon_{bR->M}$ ~ 160 nm). The magnitude of the refractive index change should nevertheless provide adequate diffraction efficiencies which may find use in long term holographic storage applications although no data supporting this supposition can be presented at this time. Other research efforts have shown that the genetic variant D85->N, which is similar in both spectral and photochemical properties to the blue membrane, shows approximately .4% diffraction efficiency per OD sample.[6] Further research focused on the characterization of the photodiffractive properties (sensitivity, photocyclicity, storage time) are needed to access its potential as an optical storage material.

CONCLUSION

In conclusion, we have presented preliminary data on the measurement of the refractive index change associated with the blue to pink membrane photochemistry in bacteriorhodopsin. Further work is in progress to characterize the material's diffractive properties, photocyclicity, and data storage potential.

REFERENCES

1. R. R. Birge. Photophysics and molecular electronic applications of the rhodopsins. *Annu. Rev. Phys. Chem.* 41: 683-733. (1990).
2. D. Oesterhelt, C. Brauchle and N. Hampp. Bacteriorhodopsin: a biological material for information processing. *Quart. Rev. Biophys.* 24: 425-478. (1991).
3. R. A. Mathies, J. Lugtenburg and C. V. Shank. From femtoseconds to biology: mechanism of the light-driven proton pump in bacteriorhodopsin. *SPIE (Biomolecular Spectroscopy)* 1057: 138-145. (1989).

4. R. R. Birge. Nature of the primary photochemical events in rhodopsin and bacteriorhodopsin. *Biochim. Biophys. Acta* 1016: 293-327. (1990).
5. R. Henderson, F. R. S. Schertler and G. R. X. Schertler. The structure of bacteriorhodopsin and its relevance to the visual opsins and other seven-helix G-protein coupled receptors. *Phil. Trans. R. Soc. Lond.* B. 326: 379-389. (1990).
6. N. Hampp, A. Popp, C. Brauchle and D. Oesterhelt. Diffraction efficiency of bacteriorhodopsin films for holography containing bacteriorhodopsin wildtype BRwt and its variants BR_{D85E} and BR_{D96N}. *J. Phys. Chem.* 96: 4679-4685. (1992).
7. R. Thoma and N. Hampp. Real-time holographic correlation of two video signals by using bacteriorhodopsin films. *Opt. Lett.* 17: 1158-1160. (1992).
8. T. Renner and N. Hampp. Bacteriorhodopsin-films for dynamic time average interferometry. *Opt. Comm.* 96: 142-149. (1992).
9. R. Thoma, N. Hampp, C. Bräuchle and D. Oesterhelt. Bacteriorhodopsin films as spatial light modulators for nonlinear-optical filtering. *Opt. Lett.* 16: 651-653. (1991).
10. Q. W. Song, C. Zhang, R. Blumer, R. B. Gross, Z. Chen and R. R. Birge. Chemically enhanced bacteriorhodopsin spatial light modulator. *Opt. Lett.* 18: 1373-1375. (1993).
11. Z. Chen, A. Lewis, H. Takei and I. Nabenzahl. Bacteriorhodopsin oriented in polyvinyl alcohol films as an erasable optical storage medium. *Appl. Opt.* 30: 5188-5196. (1991).
12. R. R. Birge. Protein based optical computing and optical memories. *IEEE Computer* 25: 56-67. (1992).
13. A. Popp, M. Wolperdinger, N. Hampp, C. Brauchle and D. Oesterhelt. Photochemical conversion of the O-intermediate to 9-*cis*-retinal containing products in bacteriorhodopsin films. *Biophys. J.* 65: 1449-1459. (1993).
14. H. Durr and H. Bouas-Laurent. (1990). Photochromism: Molecules and Systems. Studies in Organic Chemistry. 40: 1068.
15. C. Brauchle, N. Hampp and D. Oesterhelt. Optical applications of bacteriorhodopsin and its mutated variants. *Adv. Mater.* 3: 420-428. (1991).
16. M. Yoshida, K. Ohno, Y. Takeuchi and Y. Kagawa. Prolonged lifetime of the 410-nm intermediate of bacteriorhodopsin in the presence of guanidine hydrochloride. *Biochem. Biophys. Res. Comm.* 75: 1111-1116. (1977).
17. N. N. Vsevolodov and V. A. Poltoratsky. Holograms on "Biochrome" a biological photochromic material. *Zh. Tekh. Fiz.* 55: 2093-2094. (1985).
18. C. H. Chang, J. G. Chen, R. Govindjee and T. Ebrey. Cation binding by bacteriorhodopsin. *Proc. Natl. Acad. Sci. USA* 82: 396-400. (1985).
19. C. H. Chang, S. Y. Liu, R. Jonas and R. Govindjee. The Pink Membrane: The stable photoproduct of deionized purple membrane. *Biophys. J.* 52: 617-623. (1987).
20. R. B. Gross, K. C. Izgi and R. R. Birge. Holographic thin films, spatial light modulators and optical associative memories based on bacteriorhodopsin. *SPIE Image Storage and Retrieval Systems* 1662: 186-196. (1992).
21. D. Zeisel and N. Hampp. Spectral relationship of light-induced refractive index and absorption changes in bacteriorhodopsin films containing wildtype BR_{wt} and the variant BRD_{96N}. *J. Phys. Chem.* 96: 7788-7792. (1992).
22. E. Hecht. "Optics," Addison-Wesley, Menlo Park (1987).
23. R. Loudon. "The quantum theory of light," Clarendon, Oxford (1973).
24. C. J. G. Kirkby and I. Bennion. Organic photochromics for spatial light modulation. *IEE Procs.* 133: 98-104. (1986).

CONTINUOUS OPTICAL ASSOCIATIVE MEMORY

Alain Bergeron[1,2], Henri H. Arsenault[2], Denis Gingras[1]

[1]National Optics Institute/Institut national d'optique
369 Franquet, Sainte-Foy,
Québec, Canada, G1P 4N8
[2]COPL, Université Laval,Sainte-Foy
Québec, Canada, G1K 7P4

INTRODUCTION

Associative memories were the first optical neural network architecture to be implemented[1,2]. Previous optical implementations have some drawbacks such as the need for an interruption of the light beam path between the different layers of the neural network due mainly by the lack of suitable optical thresholding units. As a consequence, most of the implementations so far used detectors coupled with light sources[3]. This solution however may cause data bottlenecks while requiring additionnal components.

Paek and Psaltis have also proposed an all-optical associative memory based on a double correlator architecture[4]. However, it lacked translation invariance due to the use of pinholes to suppress sidelobes of the correlation peaks. It also required the use of additional components, such as a powerful laser and expensive optical components, i.e., an image intensifier and a liquid crystal light valve, that makes this approach impractical.

A feed-forward associative memory implementation based on the double correlator architecture is proposed. In order to allow translation invariance, all the pixels of the input plane have to be analyzed. This may be acccomplished by means of an optoelectronic thresholder[5]. This module also allows an uninterrupted propagation path so only one HeNe laser source with a relatively low power of 35 mW is required to operate the associative memory. The holograms required to perform both the correlation and the association are computer-generated holograms with a relatively small number of degrees of freedom. In order to obtain good image quality and sufficient diffraction efficiency, the associative layer's CGHs are computed using the global iterative coding method[6].

Applications of Photonic Technology
Edited by G.A. Lampropoulos *et al.*, Plenum Press, New York, 1995

ASSOCIATIVE MEMORY ARCHITECTUTE

This associative memory is based on the double correlator architecture. Our implementation is shown in Fig. 1. An input image is presented to the first correlator. The choice of filter is crucial as it directly determines the signal that will be transmitted to the second layer of the neural net. The second layer simply performs the recall of the information stored in the second memory. The first filter requires a good discrimination and a sharp peak of correlation to approximate a delta function. The phase-only filter satisfies those two requirements ans is used in the first layer of the neural network.

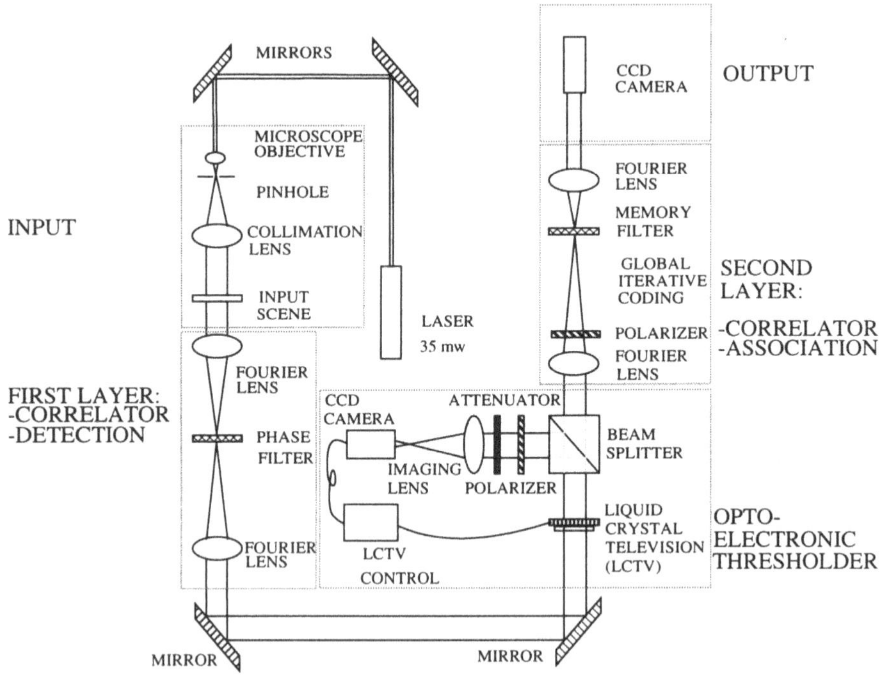

FIGURE 1. Optical memory architecture.

An optoelectronic module was inserted between the two layers to perform the thresholding operation. This module taps the output of the first layer and modulates the output of an electrically addressed spatial light modulator in order to attenuate small signal intensities while keeping higher signal values unchanged. This kind of thresholder allows the light path to be uninterrupted so that the photons arriving at the system output come directly from the input laser without seeing any electronic relay. Part of the photons are detected by the tapping section of the threshold module[5].

It should be noted that the threshold is applied to each point of the output plane of the first layer over a spatial extent given by the resolution of the modulator. Such an operation allows to track the input object even as its location changes. The thresholder maintains the translation invariance property exhibited by the first correlator. The thresholder not only eliminates the cross correlations but also helps to sharpen the detection peak by attenuating the sidelobes with intensities lower than the threshold value. The delta function is more closely approximated after the thresholding operation.

After this process, the recall is carried out by the second layer. In order to obtain a good reconstruction quality, a CGH is calculated with the help of the global iterative coding[6] (GIC). This coding allows the fast computation of the CGH and a good reconstruction quality. Furthermore, the use of the GIC allows to detect, in the first diffraction order, more than eight times the energy detected with conventional coding.

EXPERIMENTAL RESULTS

An optical associative memory corresponding to the set-up of Fig. 1 was built. The filters were recorded on high-resolution photographic film with the help of a laser writer. The input images were binary images of 256X256 pixels representing standard security

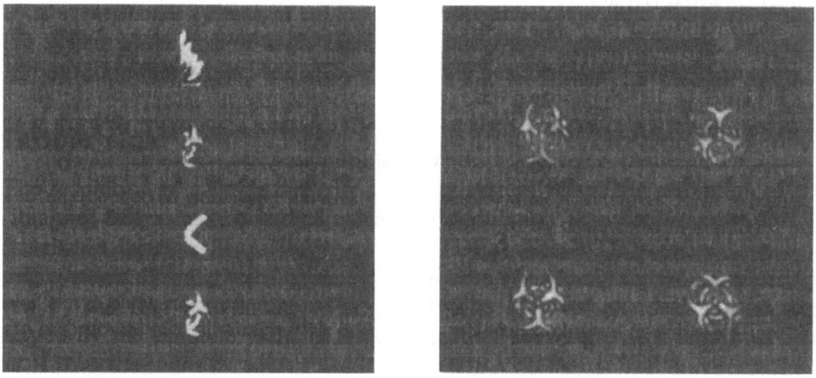

FIGURE 2. Occluded input by 50 % (left) and corresponding output (right).

symbols. The first filter had dimensions of 256X256 pixels, the dimensions of the second filter were only of 128X128 pixels. This small number of degree of freedom required the use of a coding scheme scheme with good performance on small supports, such as the GIC. A single 35 mW He-Ne laser was used to perform the experiments.

The system behavior when a degraded object is presented at the input is illustrated in the Fig. 2. The objects to be detected were the two biological hazard symbols, whereas the fire and wheat symbols have to be discarded. In this experiment, 50% of the input image was occluded. The system recognized the object, and the association was correct. The amount of degradation allowed by the system strictly depends on the choice of the filter in the first layer. A filter showing good discrimination capabilities will allow the neural network to be robust. Note that only one cycle had to be performed to obtain the output.

CONCLUSION

A continuous optical associative memory was implemented. Experimental results show a good behavior of the system in the presence of unwanted objects and of degraded targets. The neural network can be the basis of a system allowing an input scene to be cleaned from unwanted noise or objects while the desired occluded objects could be reconstructed. The architecture is versatil because it can easily be modified by changing the filters in the first and in the second layers of the neural network.

ACKNOWLEDGEMENT

This research was supported by grants from the Natural Sciences and Engineering Research Council of Canada (NSERC), the Fonds pour la formation des chercheurs et l'aide à la recherche (FCAR) program of Québec and the JSTF program of the Canadian Ministry of External Affairs.

REFERENCES

1-N. H. Farhat, D. Psaltis, A. Prata and E. Paek, "Optical Implementation of the Hopfiel Model", Applied Optics, vol. 24, no. 10, pp. 1469-1475, 1985.

2-B. Kosko, "Adaptive Bidirectional Associative Memories", Applied Optics, vol. 26 no. 23, pp. 4947-4960, 1987.

3-S. Jin-Yuan, Z. Yan-Xin, "Optical Pattern Recognition System Based on a Winner-Take-All Model of a Neural Network", Optical Engineering, vol. 32, No. 5, pp. 1053-1056, May 1993.

4-E. Paek and D. Psaltis, "Optical Associative Memory Using Fourier Transform Holograms", Optical Engineering, vol. 26 No. 5, pp. 428-433, May 1987.

5-A. Bergeron, H. H. Arsenault, E. Eustache, D. Gingras, "Optoelectronic Module for Winner-Take-All Operations in Optical Neural Network", to be published in Applied Optics.

6-A. Bergeron, H. H. Arsenault, J. Gauvin, D. Gingras, "Computer-Generated Holograms Improved by a Global Iterative Coding", Optical Engineering, vol. 32, no. 9, pp. 2216-2226.

ILLUMINATION AND DATA COLLECTION TECHNIQUES FOR HIGH PERFORMANCE TIME-INTEGRATING CORRELATORS

N. Brousseau and J.W.A. Salt

Defence Research Establishment Ottawa
3701 Carling Avenue, Ottawa
Ontario, Canada, K1A 0Z4

INTRODUCTION

Time-Integrating Correlators (TIC) are analog optical computers designed to perform the correlation of two signals. They are characterized by the ability to correlate large bandwidth signals at very high speed. The many possible ways to build TICs are well documented in the literature[1]. Numerous applications of these processors have been identified in areas such as the analysis of DNA sequences[2], spread spectrum signal processing[1], finger print identification and high speed searches of large unstructured databases.

This paper presents two issues that are particularly important to the reliability and performance of TICs. The first one is the construction of a line illumination across the detector array having a uniform intensity , low aberration and making efficient use of the laser light. This is achieved by using a lens designed to transform the gaussian light distribution of a laser beam into a uniform line illumination. The second critical issue is the collection of the data by the detector array. The technique used up to now was based on an aberration free wavefront at the detector of the TIC . With this technique, the presence of accidental aberration causes "blind" areas where the correlation peak is not formed. A new technique for formatting the correlation peak is presented in this paper to avoid the creation of "blind" areas, even in presence of a moderate amount of aberration.

A TIC is illustrated in Fig. 1. The two input signals are applied to Bragg cells A and B and their correlation is formed on the detector array. The presence of a peak indicates that the two input signals A and B are identical and conversely the absence of the correlation peak indicates that the two inputs are different.

The peak formed by the correlator appears on a pedestal whose amplitude depends on the profile of the illumination and on the attenuation of the Bragg cells. The peak can be described by the product of a triangular envelope with a sinusoidal spatial modulation of the light distribution (thereafter referred to as fringe) whose period can vary from infinity to a very small value depending on the configuration of the TIC. Figure 2 a) illustrates a peak with an infinite fringe period and a pedestal produced by an illumination having a gaussian profile. Figure 2 b) illustrates an output, with the same pedestal, when a fringe pattern with a finite period is present . The individual outputs from each element of the detector array are visible on the oscilloscope trace. These results are based on contiguous detector elements.

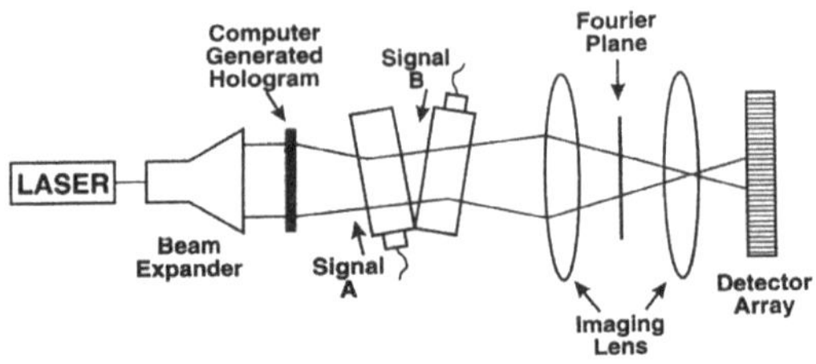

Figure 1 Time-integrating correlator: Mach Zehnder architecture.

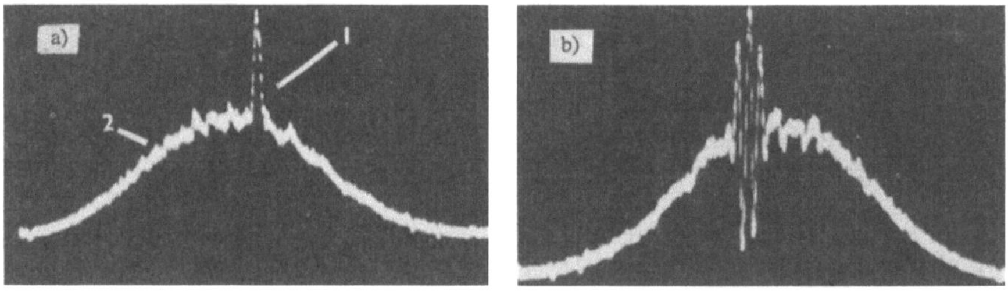

Figure 2 a) Peak without fringes (1) over a gaussian pedestal (2), b) Peak with fringes over a gaussian pedestal.

BEAM EXPANSION FOR UNIFORM ILLUMINATION

The maximum height of the peak above the pedestal is limited to the height of the pedestal itself. Consequently, only short and difficult to detect correlation peaks can be produced where the pedestal is low. This problem can be alleviated if the pedestal has a uniform intensity. Two approaches are possible. In the first approach, the gaussian laser beam is expanded to dimensions much larger than the aperture of the Bragg cells and only the central portion is used: most of the laser light is thus wasted. A better approach consists of using an optical element designed to transform a gaussian laser beam into a diverging line of light with a uniform intensity[3]. The diverging beam is then collimated with a lens. This technique allows the production uniform light distributions (see Fig.3) with little phase error that contain over 80% of the light of the laser beam.

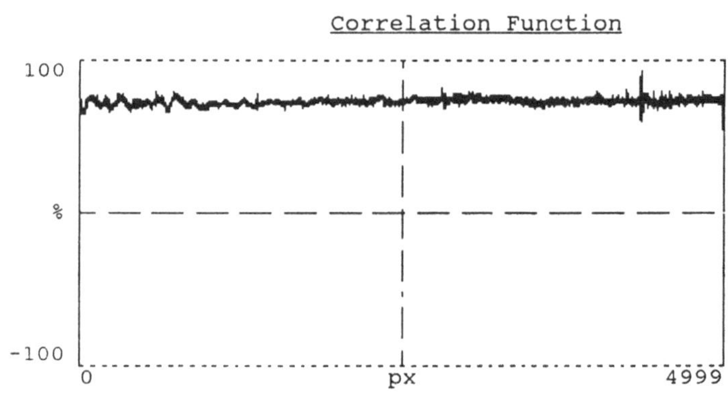

Figure 3 Illumination intensity measured over a 35 mm aperture with a 5000-element detector array.

OPTIMAL FORMAT OF THE PEAK

Let us consider a few examples to illustrate the salient features of the peak detection of a TIC. Two factors are considered: the fringe period and the effect of the phase difference between the input signals. If the fringe pattern has an infinite period, the triangular envelope is multiplied by a uniform background whose amplitude depends on the relative phase of the input signals. For a phase difference of 0° and 90°, the contributions to the peaks are, respectively, a maximum and a null (see Figures 4a and 4b). The formation of a detectable peak thus requires a tight control of the phase difference between the input signals.

Ideally, the output of a TIC should be free of optical phase error and the phase difference of the input signals should be zero to produce a maximum amplitude. This is the only mode of operation that has been proposed for TICs up to now. However, these conditions are quite difficult to achieve and maintain. They require the utilization of diffraction limited optical components, the capability to achieve and maintain near perfect alignment of the whole optical system and the precise adjustment of the phase difference of the signals applied to the Bragg cells. A precise control of the temperature is also required[4].

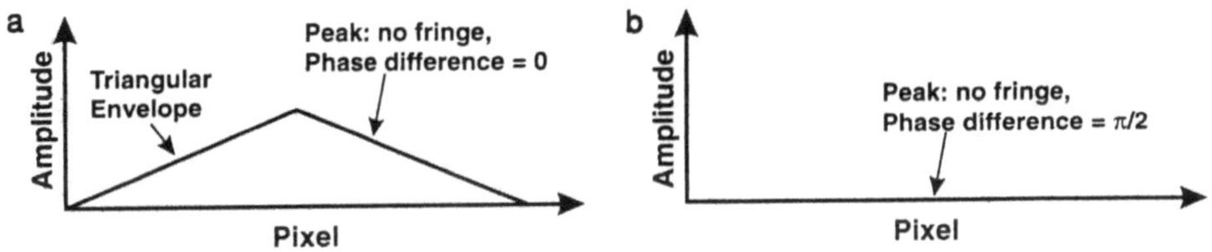

Figure 4 a) Best adjustment of the relative phase of the input signals, b) worst adjustment of the relative phase of the input signals for fringe with an infinite period.

Let us assume that a low frequency fringe pattern is produced by one or a combination of the factors mentionned in the previous paragraph: the detection process may become unreliable. Two extremes results of the detection process are illustrated in Fig. 5. Fig.5a corresponds to a combination of parameters that leads to the detection of a peak that has a large amplitude. Fig. 5b differs only by the location of the apex of the triangular envelope relative to the fringe pattern . The amplitude of the peak produced is, at best, difficult to detect and a "blind" spot in the field of view of the correlator is generated. This is obviously an unacceptable feature for a system that is expected to produce reliable peak detection.

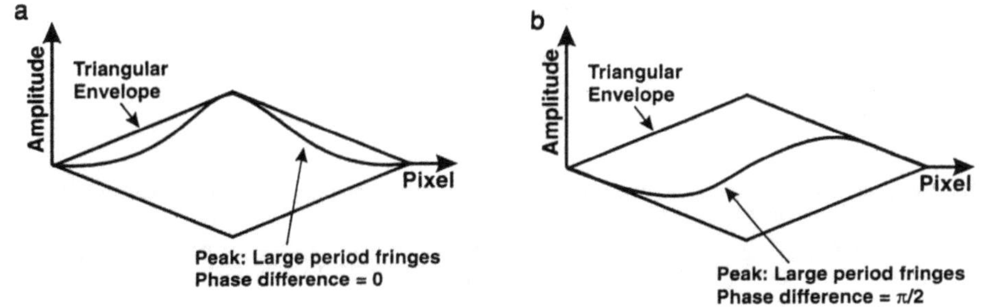

Figure 5 a) Best signal with large period fringes, b) worst signal with large period fringes.

Studies of the detection process, using a computer simulation, have established that it is possible to reach a compromise where stable performance of a fair quality is traded against unpredictable performances that are either excellent or very poor. Stability in the detection process is achieved by purposefully producing a fringe pattern with an optimal period at the output of the TIC. The period is calculated to produce peaks with a minimum amplitude variation for all positions of the detecting elements relative to the fringe system and the triangular envelope. Fig.6 presents the amplitude of the correlation peak for the optimal fringe period, width and positions of the apex of the triangular envelope.

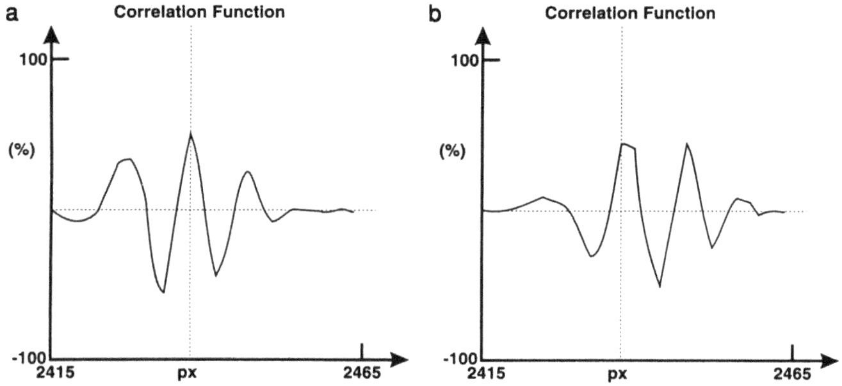

Figure 6 a) Best case with an optimal fringe period, b) worst case with an optimal fringe period.

One particularly important feature of this new detection technique is the minimized deterioration of the amplitude of the peak when the width of the triangular envelope and the fringe period drift from their optimal values. The parameter that is the most likely to drift from its set value is the fringe period p. This is due to the presence of phase error produced by aberration or misalingment of the optical components of the system which causes alteration of the fringe period of an optimal output. It has also been demonstrated that the amplitude of the peak produced by the detection of a fringe system, contrarily to the case with no fringes, is very resistant to temperature change[3].

CONCLUSION

A beam expander designed to produce uniform line illumination for the Bragg cells of a TIC and a data collection technique where the format of the peak is altered to permit reliable detection when a moderate amount of aberration is present have been presented. Experimental results are included in both case.

REFERENCES

1. N.J. Berg and J.N. Lee, 'Acousto-Optic Signal Processing: Theory and Implementation', Marcel Dekker Inc. New York and Basel, 1983, p.291.

2. N. Brousseau, R. Brousseau, J.W.A. Salt, L.Gutz and M.D.B. Tucker, 'Analysis of DNA sequences by an optical time-integrating correlator', Applied Optics, vol.31, no.23, 10 Aug.1992, p.4802-4815.

3. I. Powell, 'Design of a laser beam line expander', Applied Optics, vol.26, Sept 1,1987, p.3705-3709.

4. N. Brousseau, 'Effects of temperature changes on the output of time-integrating correlators', DREO TN 93-30, Nov. 1993.

OPTICAL PATTERN RECOGNITION USING THE WAVELET TRANSFORM

Danny Roberge and Yunlong Sheng

Centre d'Optique, Photonique et Laser,
Département de Physique, Université Laval
Québec, Canada G1K 7P4

1. INTRODUCTION

The wavelet transform (WT) is a new mathematic tool for multiresolution local analysis of non-stationary and fast transient signals. A 2-D image may be considered as a non-stationary signal with edges, textures and deterministic objects located at different locations and orientations. The WT is useful for image coding and compression [1].

We propose to combine the WT with the optical matched filter for automatic pattern recognition. The optical shift invariant WT with the wavelet in the preselected frequency band and orientation enhances local edge features of the image. The optical matched filter performs the correlation between two 2-D WT coefficients for pattern recognition. The wavelet matched filter (WMF) performs the WT and the matched filtering in a single step. The composite wavelet matched filter (CWMF) is a non-linear combination of the WMF's that produces desired outputs for a given set of objects. Both the WMF and CMWF are optically implemented with a couple-mode modulation phase mostly real-time spatial light modulator based on the commercial liquid crystal television. (LCTV).

2. OPTICAL WAVELET TRANSFORM

The wavelet transform (WT) is a correlation between the signal f(x,y) and a family of the wavelets $h_s(x,y)$. The wavelets are generated from a mother wavelet h(x,y) by dilations

$$h_s(x,y) = \frac{1}{s_x s_y} h\left(\frac{x}{s_x}, \frac{y}{s_y}\right) \tag{1}$$

where s_x, s_y are the scale factors. The wavelets must satisfy the admissible condition such that h(x,y) oscillates to have a zero mean. The wavelets should also have some regularity so that the WT is also local in frequency. In the Fourier domain the WT is expressed as:

Applications of Photonic Technology
Edited by G.A. Lampropoulos *et al.*, Plenum Press, New York, 1995

$$W_f(s_x, s_y, x, y) = s_x s_y \iint_{-\infty}^{\infty} F(u, v) H^*(s_x u, s_y v) \exp[i2\pi(xu+yv)] du dv \quad (2)$$

where F(u,v) is the Fourier transform of f(x,y) and $H(s_x u, s_y v)$, referred to as the WT filters, is the Fourier transform of the wavelets $h_s(x,y)$. The output of the optical correlator is the 2-D WT coefficients, $W_f(s_x,s_y,x,y)$, as a function of the continuous translation factors of the WT: x and y. The scale factors s_x, s_y are predetermined by the WT filter [2].

The Gaussian function, $g_s(x,y)=\exp[-(x^2+y^2)/2s^2]$, is local in both space and frequency domains. The derivative of any order n of the Gaussian function may be a wavelet. Its Fourier transform is a band-pass WT filter in the frequency plane and has zero of order n about the zero frequency. Both the wavelet admissible condition and the regularity condition are satisfied. The Mexican-hat wavelet is the second order derivative of the isotropic 2-D Gaussian function with $s_x = s_y = s$:

$$h(x,y) = \nabla^2 g_s(x,y) = \frac{1}{s^2}\left[\frac{x^2+y^2}{s^2} - 2\right] \exp\left(-\frac{x^2+y^2}{2s^2}\right) \quad (3)$$

With the Mexican-hat wavelet the WT may be written as

$$\begin{aligned} W_f(s,x,y) &= f(x,y) * [\nabla^2 g_s(x,y)] \\ &= \nabla^2_{x,y}\left\{\iint_{\infty}^{\infty} f(x',y')\, g_s(x'-x, y'-y)\, dx'dy'\right\} \end{aligned} \quad (4)$$

where * denotes the correlation, The WT is then the second order derivative of the smoothed image $f*g_s$ with the dilated Gaussian function gs(x,y) as the smoothing function. The correlation $f*g_s$ wipes out all structure in the image f(x,y) at scales smaller than the scale s of $g_s(x,y)$. The WT coefficient, $W_f(s,x,y)$, is zero when $f*g_s$ is constant

Fig. 1 From left to right: Original image, Optical wavelet transform of the image with Mexican-hat wavelets of s = 1 and s = 2.

or changes linearly, but exhibits a positive and a negative peak in two sides of the edges of $f*g_s$. A zero crossing occurs at the point of inflection of $f*g_s$. Thus, the WT enhances edges of the image. Figure 1 shows the optical WT of four images of aircraft. The isotropic Mexican-hat wavelet transform enhances the edges of all directions.

3. WAVELET MATCHED FILTER

Edges are the most significant features of images. Pattern recognition using correlation between the 2-D WT coefficients of images can have good discrimination capability. From the Fourier plane expression of the WT, Eq.(2), the correlation $W_f * W_t$ may be expressed as

$$\iint_{-\infty}^{\infty} W_t(s_x, s_y, x', y') W_f^*(s_x, s_y, x'-x, y'-y)\, dx'dy' =$$
$$\iint_{-\infty}^{\infty} T(u,v) H^*(s_x u, s_y v) F^*(u,v) H(s_x u, s_y v) \exp[i2\pi(xu+yv)]\, du dv \quad (5)$$

where T(u,v) is the Fourier transform of an input t(x,y) and the constants on the right hand side of the equation $s_x^2 s_y^2$ are omitted. Thus, the correlation $W_f * W_t$ may be directly obtained in an optical correlator with a filter:

$$G(u,v) = F^*(u,v) |H(s_x u, s_y v)|^2 \quad (6)$$

We call G(u,v) the wavelet matched filters (WMF). It is a product of the conventional matched spatial filter $F^*(u,v)$, and the square modulus $|H(s_x u, s_y v)|^2$ of the WT filter. Figure 2 shows the correlation output of a WMF compared with that of the conventional matched filter. The WMF shows a good discrimination capability [3].

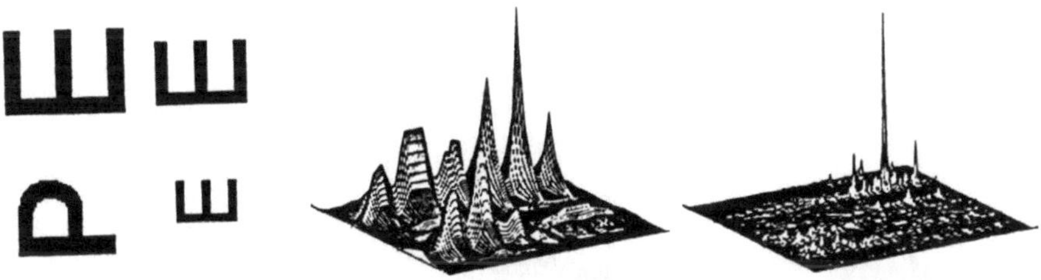

Fig.2 From left to right: Input scene i which the letter E in the right-top corner is to be recognized; Correlation output of the conventional matched filter; That of the WMF

4. COMPOSITE WAVELET MATCHED FILTER

For a given set of training images $\{t_n(x,y)\}$ with n=1, 2, ..., N, the composite filter is able to yield desired outputs for each of the images $t_n(x,y)$. We synthesize the composite wavelet matched filter (CWMF) as

$$G(u,v) = \frac{\sum_{n=1}^{N} \alpha_n \Phi_n(u,v)}{\left|\sum_{n=1}^{N} \alpha_n \Phi_n(u,v)\right|} |H(s_x u, s_y v)|^2 \quad (7)$$

where $\phi_n(u,v)$ is the phase function of the Fourier transform of $t_n(x,y)$. The CWMF would be able to generate a desired output c_k for every input images $t_k(x,y)$ in the training set with k=1, 2, ..., N. We have N non-linear equations in the correlation plane:

$$\left\langle t_k(x,y) \left| FT^{-1}\left\{ \frac{\sum_{n=1}^{N} \alpha_n \phi_n(u,v)}{\left|\sum_{n=1}^{N} \alpha_n \phi_n(u,v)\right|} |H(s_x u, s_y v)|^2 \right\} \right. \right\rangle = c_k \qquad (8)$$

To ensure the desired outputs c_k, the coefficients α_n of the CWMF in Eq.(8) are solved from the system of N equations using an iteration algorithm with the delta learning rule.

The phase-only filters (POF's) $\phi_n(u,v)$ have good discrimination, so that the inner products $\langle\phi_n, \phi_k\rangle$ with $n \neq k$ are approximately equal to zero. The information about the objects stored in the CWMF, $FT^{-1}\{\phi_n(u,v)\,|H(s_x u,s_y v)|^2/|\Sigma a_k\phi_k(u,v)|\}$ are approximately orthogonal. As a result, the iteration algorithm converges very fast with few iterations and the CWMF yields shape correlation peaks without sidelobe.

5. OPTICAL IMPLEMENTATION

For implementation of the CWMFs we built an on-axis optical correlator with two Fourier planes. According to Eq.(7) the CWMF consists of two terms: one is a POF another is the band-pass wavelet filter. A liquid crystal television (LCTV) is in the first Fourier plane for encoding the phase-only term. A photographic mask in the second Fourier plane encodes the wavelet filter $|H(u,v)|^2$. Because of the high light efficiency of the on-axis phase mostly filter on the LCTV, we need in fact only less than 1 μW laser power to operated this optical correlator. Figure 3 shows the optical correlation outputs of the CWMFs designed to recognize 2, 3 and 4 objects and reject the remained objects respectively. The input image is that shown in Fig.1.

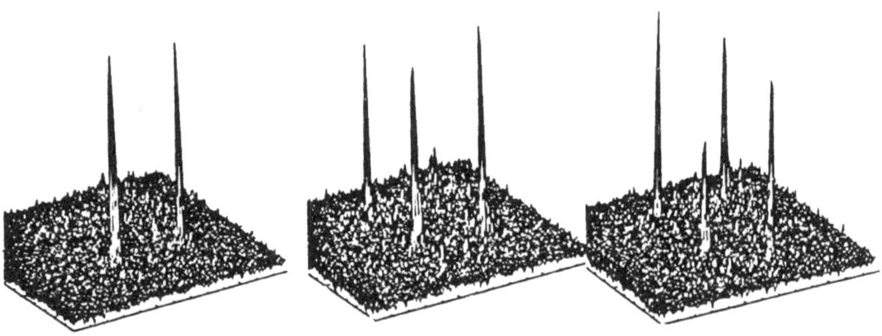

Fig. 3 Optical correlation output of the CWMF for recognize 2, 3 and 4 object and reject the remained objects. The input image is that shown in Fig.1

REFERENCES

1. S. G. Mallat, "A theory for multiresolution signal decomposition: the wavelet representation", Trans. IEEE, **PAMI-11**, 7, 674-693 (1989).
2. Y. Sheng, D. Roberge and H. Szu, "Optical wavelet transform", Opt. Eng., **31**, 1840-1845 (1992).
3. Y. Sheng, D. Roberge, H. Szu and T. Lu, "Optical wavelet matched filters for shift-invariant pattern recognition", Optics Letters, **18**, 299-301 (1993).

OPTICAL REAL-TIME PHASE-ONLY CORRELATOR USING LIQUID CRYSTAL TELEVISION

Yunlong Sheng, Gille Paul-Hus, Lixin Shen
Luiz Gonçalves and Danny Roberge

Centre d'Optique, Photonique et Laser
Département de Physique, Université Laval
Québec, Canada G1K 7P4

1. INTRODUCTION

The spatial light modulator (SLM) is a key element in an optical processor. The principle limitations of the currently available SLMs are the limited phase and amplitude modulation capacity, the limited space bandwidth product (SBWP) and the limited speed. We use the commercial liquid crystal television (LCTV) as the SLM and build a real-time on-axis phase-only optical correlator. This approach allows the efficient use of the SBWP of the LCTV (200 x 200 and 440 x 480 for new type of LCTV) and provides high light efficiency.

We measured the modulation characteristics of the commercial LCTV and determined the optimal condition to obtain the continuous phase modulation in the LCTV. Then, we implemented the programmable kinoforms, the continuous phase-only matched spatial filter, the rotation invariant circular harmonic filters and the composite filters in a real-time optical correlator with the LCTV.

2. MEASUREMENT OF THE OPERATING CURVE

The phase modulation of a SLM is usually measured with the Mach-Zehnder interferometer. We built a wedged shear plate interferometer to measure the phase modulation of the LCTV. This interferometer is very simple, requires no critical alignment and can be used for on-cite testing of the LCTV.

The twisted nematic LCTV provides continuous phase modulation. But the undesired amplitude modulation is cross coupled with the phase modulation. The operating curve describes the coupled phase and amplitude modulation in the complex plane. The operating curve of a LCTV is a function of the angles of the polarizer and analyzer, the brightness and contrast control voltages. We rotated the polarizer and analyzer to obtain the optimal operating curve which is a perfect circle, but slightly shifted from the origin in the complex plane.

Applications of Photonic Technology
Edited by G.A. Lampropoulos *et al.*, Plenum Press, New York, 1995

3. PROGRAMMABLE KINOFORMS

We designed the phase holograms with the modified Gerchberg-Saxton iterative method. The operating curve of the coupled-mode modulation LCTV was adjusted to optimal and was used as the Fourier plane constraint in the iteration. The programmable kinoforms displayed on the LCTV have good quality with the central spot of zero order diffraction reduced to minimum. Figure 1 shows images reconstructed from the kinoforms on the LCTV which is a panel of the Epson Crystal Image video projector with 1 in. x 1 in. aperture size and 200 x 200 pixels [1].

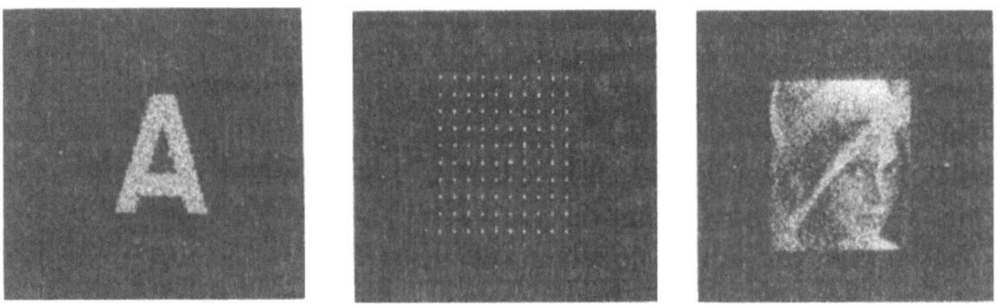

Fig. 1 Reconstructed images from the kinoforms on the programmable LCTV

4. DIFFRACTION OF THE COUPLED MODE LCTV

The main imperfection of the commercial LCTV used as a SLM is the phase mismatching and the coupled amplitude modulation. Juday proposed optimal coding for the coupled mode SLM [2]. We extended the theory of diffraction for multilevel phase holograms of Goodman and Silvestri to the coupled mode modulation LCTV. The on-axis phase mostly hologram displayed on the coupled mode LCTV gives still the true response of the phase-only filter (POF). The coupled amplitude modulation and phase mismatching of the LCTV contribute to other diffraction orders. The zero order yields an image of the input in the output correlation plane. Other diffraction orders representing noise in the correlation output are usually weak and negligible. The theoretical analysis, computer simulation and optical experiments give the percentage of energy of the diffraction orders, as shown in Table 1 [3].

5. REAL-TIME OPTICAL CORRELATOR

In the optical correlator we let polarization of the input beam parallel to the director of the liquid crystal molecules in the front of the LCTV panel. The analyzer was removed. This configuration yields the maximum phase modulation. Figure 2 show the correlation output of an optical correlator with the coupled mode LCTV encoding the continuous POF. The on-axis phase mostly filter yields a very bright autocorrelation peak, so that only about 1 μW laser source power was used to avoid saturation of the CCD camera [3].

Fig.2 From left to right: Input scene; Impulse response of the phase mostly filter on the LCTV; Correlation output

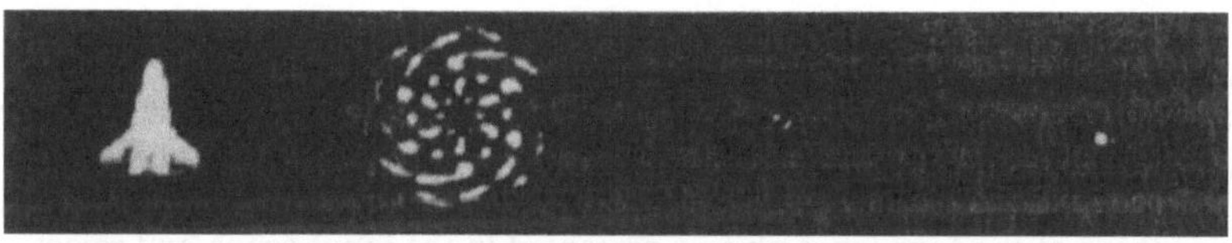

Fig.3 From left to right: Input image; CHF of order m = 6 on the LCTV; Impulse response of the filter; Correlation output

6. OPTICAL CIRCULAR HARMONIC FILTER

The full rotation invariant correlation was obtained with the continuous phase-only circular harmonic filter (CHF) encoded on the LCTV. In the Fourier plane, the amplitude of the CHF's is equal to zero about the origin. We used a small central disc on the optical axis for the amplitude modulation about the origin. Figure 3 shows the optical correlation of the CHF that agrees with the computer simulation. We implement a sequence of CHF's of different orders of the same object in real-time with the LCTV. From each correlation output we take the position of the correlation peaks. The distances between the correlation peaks of the CHF's of different orders are used as the rotation and shift invariant object features for pattern recognition. This geometrical approach of multiple CHF's has the good discrimination capability and the ability against the cross-correlation false alarms.

7. OPTICAL COMPOSITE FILTER

The composite filter or the synthetic discriminant function (SDF) is able to generate any desired outputs for a given set of objects. We implement the SDF with the real-time coupled mode phase mostly LCTV. The phase-only SDF is a non-linear combination of the component POF's of the training objects.

$$S(u,v) = \frac{\sum_{n=0}^{N} \alpha_n \phi_n(u,v)}{\left|\sum_{n=0}^{N} \alpha_n \phi_n(u,v)\right|} \tag{1}$$

where $\phi_n(u,v)$ is the POF of the training set image $f_n(x,y)$. The coefficients in the combination are computed from a system of N equations

$$\langle f_k(x,y) | FT^{-1}[S(u,v)] \rangle = c_k \tag{2}$$

where c_k is the desired output when the input is $f_k(x,y)$. This system of non-linear equations is solved with an iterative method with the delta learning rule proposed by Jared and Ennis. Figure 4 shows the optical impulse response of a phase-only SDF encoded on the coupled mode LCTV and the correlation output. The filter was designed to recognize three of the four input objects presented in the input scene and reject one of the objects.

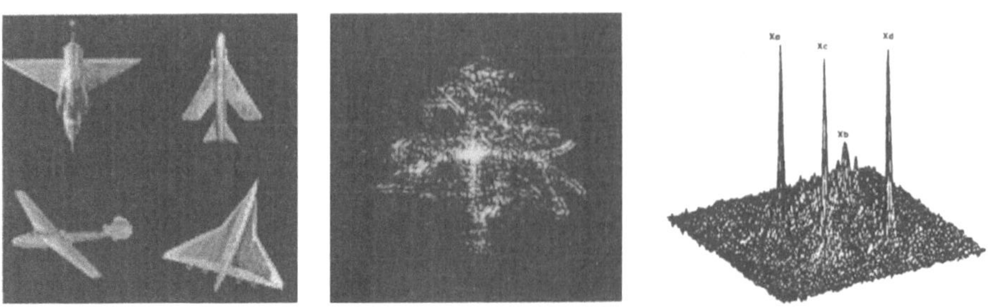

Fig.4 From left to right: Input scene; Optical impulse response of the SDF; Correlation output

REFERENCES

1. L. Gonçalves Neto and Y. Sheng, "Optical real-time kinoform using the liquid crystal television and iterative design", Proc. SPIE, **vol. 2237**, 190-195 (1994).
2. R. Juday, "Correlation with a spatial light modulator having phase and amplitude cross coupling", Appl. Opt. **28**, 4865-4869 (1989).
3. Y. Sheng and G. Paul-Hus, "Optical on-axis imperfect phase-only correlator using liquid-crystal television" , Appl. Opt. **32**, 5782-5785 (1993).

AN INTELLIGENT OPTICAL SENSOR

B. Lam, G. A. Jullien, W. C. Miller

Department of Electrical Engineering
University of Windsor
Windsor, Ontario, N9B 3P4
CANADA

ABSTRACT

An analog VLSI realization and testing of an intelligent optical sensor is described in this paper. The sensor is designed for the use in process control applications requiring image capture or non-contact measurement. The architecture is based on an analog VLSI realization of a multi-layer feedforward artificial neural network with an integrated photosensitive array. The design is implemented in 3μ single-polysilicon CMOS technology.

INTRODUCTION

The design and testing of VLSI realization of an intelligent optical sensor based on multi-layer feedforward neural network with integrated photosensitive array is presented in this paper. The sensor determines the position or surface geometry of an object using non-contact measurement techniques employing either reflected LASER beam-steering methods or structured illumination. The resultant pattern on the photosensitive array is used to generate control vectors for the process at the output neurons. The neural network has 3-layers and has been trained to recognize eight different patterns that represent possible modes of operation in the process being controlled. The photosensitive array consists of 10X10 photosensitive elements that act as the input devices to the neural network and are divided into four sub-arrays. Each of the sub-array consists of twenty-five photosensors that are fully connected to four neurons in the hidden layer. The 16 neurons in the hidden layer are fully connected to five output neurons of the output layer.

TRAINING

The learning strategy is carried out off-board using a modified back-propagation algorithm. The algorithm used a hard-limiting function to generate the synaptic weights. The weights are implemented by transistor sizes which are suitable for VLSI implementation. A recall program using hard-limiting function has been developed to ensure the weights are suitable and proper functionality of the network.

Applications of Photonic Technology
Edited by G.A. Lampropoulos *et al.*, Plenum Press, New York, 1995

NEURON AND WEIGHTING CIRCUITRY

The artificial neuron is comprised of a differential amplifier and an output inverter to obtain a hard-limiting function. The configuration of the neuron is shown in Figure 1a. The neuron compares voltages of the differential pair which are based on the sum of positively and negatively weighted currents. If the sum of the positively weighted current is greater than the negatively weighted current, the voltage of the negatively weighted current of the differential pair will increase. As a result, the current flow through that PMOS transistor will decrease and the output voltage of the differential amplifier will decrease. This voltage drives the inverter amplifier to 5v.

The neuron output changes state when both the differential voltages are equal. The synaptic weights have been implemented by using different size transistors since the channel current of a transistor is directly proportional to the width of the transistor. The width of each transistor is calculated by the training program. The positive and negative weights are separately connected in parallel to form the inputs of the neuron. Each transistor's gate is driven by the output of the neuron from the previous layer or from the photosensitive elements.

PHOTOSENSITIVE ELEMENTS

The photosensor converts light into electrical signals. The photosensor consists of a phototransistor, load circuit and output inverter amplifier. The phototransistor is a vertical parasitic NPN bipolar transistor with a floating base configuration. The substrate acts as a collector, the p-well as the base and the n+ plug as the emitter. Two photodiodes are formed in the pn junction when light enters the silicon and positive charge builds up in the p-well. To compensate the charge difference and to maintain charge neutrality, electrons are injected into the base by the n+ plug. Furthermore, an increase in the overall transistor gain can be performed by making the emitter-base junction small in comparison to the collector-base junction. A net photocurrent is generated and is directly proportional to the incident light intensity. A load circuit with a logarithmic response is used to map the wide range of photocurrents into the desired range between 5v and ground. The load circuit consists of a chain of 3 diode tied MOSFETs which due to their logarithmic response in the subthreshold region[1,3], compress and convert the wide range of photocurrents into the desired voltage signal. A bias voltage controls the sensitivity of the photosensor whose digital level output is attained by using an inverter amplifier. A schematic diagram of the photosensor is shown in Figure 1b.

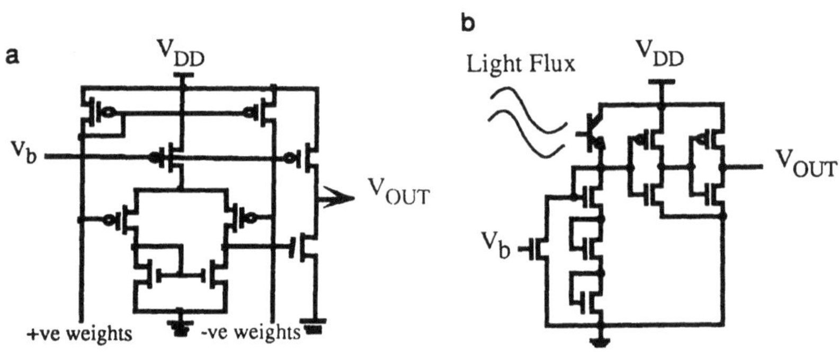

Figure 1. a) The artificial neuron b) The photosensitive element.

VLSI IMPLEMENTATION

The photosensitive array is divided into four groups. Each group consists of 25 photosensors and each photosensor is placed as close to alignment as possible. Furthermore, each group is placed in a symmetric even configuration. The 16 hidden neurons are also divided into four groups. Four hidden neurons are fully connected to one of the sub-group of the photosensitive array. These four groups of neurons are placed around the photosensitive array to ease the routing. The output of the hidden neurons is fully connected to the output layer which is shown in Figure 2.

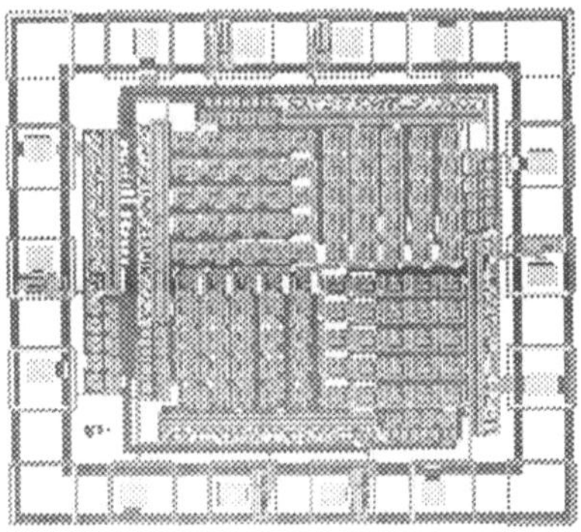

Figure 2. The layout of the whole optically coupled neural network.

TEST RESULTS

The network is trained to recognize eight input patterns for control purposes. The testing of the optically coupled neural network is divided into two main parts. The first part is concerned with the operational performance of components of the neural network. The second part of the testing is focused on the network's ability to act as an intelligent controller.

A test program has been written to determine how the network responds to corrupt input patterns. The program generates patterns with varying noise levels that corresponds to Hamming distances from one to fifty. The test results show that the network is able to recognize almost all the noisy patterns (90% to 97%) with Hamming distance equal to one. From Hamming distance of two to fifty, the program generates 500 different patterns for each training pattern. The results show that the network can distinguish patterns but the performance drops very rapidly, since only eight patterns are trained. An average recognition rate under this scheme is shown in Figure 3.

Since patterns with equal Hamming distance can be totally different, one can conclude that Hamming distance is not an adequate measurement for noisy patterns in this kind of application. A further testing of the network is in process to find a practically useful boundary or envelop whereby one noisy input pattern maps into the output state associated with another input pattern. Each architecture and training strategy that corresponds to a specific application must be assessed experimentally as there is no applicable general nonlinear theory.

The second part of the testing involves electrically examining the fabricated chip. Two integrated circuits have been fabricated in 3μ CMOS technology. The first one is used to evaluate the weighting circuitry, neurons and four photosensor elements. Light patterns are imaged on the photosensitive elements and an output voltage, proportional to light intensity, with a value of 0 or 5 volts, is obtained. Also, the output from the hidden layer of neurons connect to the weighting circuitry driving the output is verified to have the correct neurons states. The tests indicates that the photosensor, neuron and weighting circuitry elements function properly. The complete intelligent sensor architecture is implemented on the second integrated circuit. This network is tested using trained and noisy patterns.

The test patterns are made into a number of 35mm slides. These are then projected onto the photosensitive array using a microscope in conjunction with other lenses. The various patterns are successfully projected onto the photosensitive array. The network shows itself capable of mapping 80% of all input patterns to their correct output patterns. A significant portion of the error is caused by the vibration of the test fixture and the alignment of the lenses.

In summary, the test results of the optically coupled neural network show that the neural network is able to map the application specific patterns corrupted by noise in over 80% of the experiments.

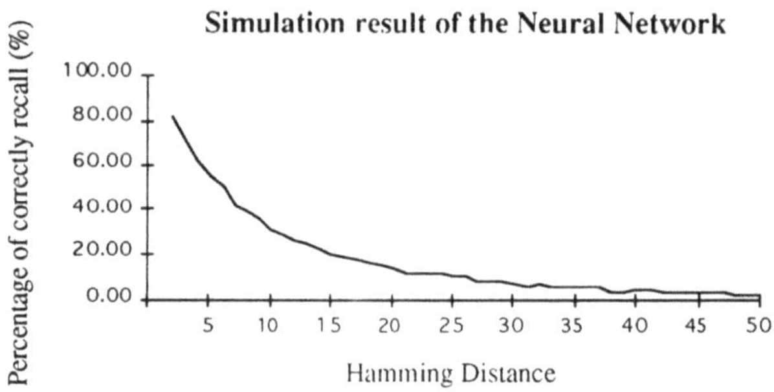

Figure 3. The Simulation Results with varying Hamming Distance.

CONCLUSIONS

An intelligent optical sensor is presented in this paper. Its architecture is based on an analog VLSI realization of multi-layer neural network with an integrated photosensitive array and has been implemented in 3μ single-polysilicon, double-metal, CMOS technology. The network consists of 100 photosensors as its input, 16 neurons in the hidden layer and 5 output neurons. A modified back-propagation algorithm is used and testing is performed by both computer simulations and physical verification of chip operation. The overall size is approximately 4500mm X 4820mm design micron scale. Further improvement will be implemented in 1.2μ CMOS with programmable capability.

REFERENCES

1. C.A.Mead, "Analog VLSI and Neural Systems", pp.67-82, pp.260-261, 1989.
2. P.E.Allen, D.R.Holberg, "CMOS Analog Circuit Design", pp.257-273, 1987.
3. M.A.Sivilotti, M.A.Mahowald, C.A.Mead, "Real-Time Visual Computations Using Analog CMOS Processing Arrays", Proceedings of the 1987 Stanford Conference on Advanced Research in VLSI, pp.295-311, 1987.

DESIGN OF AN OPTO-ELECTRONIC BEAM NULLING NETWORK FOR A MULTIPLE-BEAM ANTENNA

C. Bélisle, L. Mason, and B. Felstead

Communications Research Centre
Industry Canada
Ottawa, Ontario, K2H 8S2

INTRODUCTION

Multiple-beam antennas (MBAs) are attractive for satellite EHF frequency-hopped signal applications because of their wide bandwidth and light weight. Adaptive nulling techniques are often used to suppress jammers and other interference. Nulling entails weighting the received signal from each beam. All weighted signals are linearly combined to provide an output. From estimates of the correlation between the various received beam signals and by minimizing the interference in the output, the weights are determined. Detection of the desired signals in the output is then carried out.

It has been shown[1] that the use of optics allows wideband nulling to be achieved without difficulty. In this paper a network is considered which uses optical technology to carry out the correlation and for implementing the weighting required in nulling.

NULLING ALGORITHM

In a typical nulling application with N beams, the total received band-limited signal, including interference and noise, on the ith beam may be written as

$$y_i(t) = Im\{z_i(t)exp[j2\pi f_{RF}t]\} \tag{1}$$

where $z_i(t) = A_i(t)exp[j\theta_i(t)]$ is the complex envelope and f_{RF} is the RF carrier frequency of the received signal. Let $\mathbf{Z}(t)$ and $\mathbf{W}$ denote the N-vectors of complex envelopes and complex weights respectively. Let $\mathbf{U_S}$ be the steering vector, that is, the vector of beam gains for the desired direction of reception. The optimum weights are determined by minimizing the mean squared value of the output but constrained so that the signal from the direction with steering vector U_S is protected[2]. Thus, if the output after the combiner is

$$e(t) = \mathbf{W^H Z(t)}, \tag{2}$$

then the function to be minimized is

$$F = \mathbf{W^H E\{ZZ^H\} W} + \eta(\mathbf{W^H U_S} - \mathbf{1}) \tag{3}$$

where the superscript H denotes the complex conjugate transpose operation, $E\{\cdot\}$ is the expectation operator, and η is a Lagrange multiplier. Minimizing (3) gives

$$\mathbf{W} = \frac{\mathbf{\Phi}^{-1}\mathbf{U}_S}{\mathbf{U}_S^H \mathbf{\Phi}^{-1}\mathbf{U}_S} \tag{4}$$

where $\mathbf{\Phi}$ is the covariance matrix of $\mathbf{Z(t)}$ with elements

$$\phi_{ik} = E\{z_i z_k^*\} = E\{A_i(t)A_k(t)exp\{j[\theta_i(t) - \theta_k(t)]\}\}. \tag{5}$$

SYSTEM DESIGN

Fig. 1 shows a schematic diagram of an opto-electronic implementation of the weighting and combining as well as a means of evaluating $\mathbf{\Phi}$. The design is based on integrated-optics components, such as Mach-Zehnder[3] (M-Z) modulators and phase shifters. The design uses guided optics, rather than free space signal distribution, to give better temperature stability and lower sensitivity to vibration. The system may be divided into three sections: RF-to-optics conversion, nulling processor, and signal processor.

RF-to-Optics Conversion

The conversion from the electrical domain to optics is done directly at each antenna element using conventional integrated-optics M-Z modulators. Rather than using a CW optical signal as the input to the M-Z, the optical signal is modulated by a local oscillator in the payload. The complex optical E-field at the output of the M-Z can be written as[3]

$$E(t) = \sqrt{2}E_0 exp\{j2\pi\nu_0 t\}cos[\psi_0 + \beta sin2\pi f_{LO}t] \tag{6}$$

where ν_0 is the light wave frequency, E_0 is the amplitude of the unmodulated optical signal at the M-Z input, β is the modulation index, ψ_0 is a constant phase shift introduced by a dc bias, and f_{LO} is the frequency of the local oscillator. By modulating this optical output with the signal from the ith antenna element in a second M-Z device, it may be shown that the optical signal power can be approximated by[1]

$$P_{0i} = (1 + K_1 V_{LO})(1 + K_2 V_i) \tag{7}$$

where K_1 and K_2 are constants, V_{LO} is the first harmonic at the LO frequency and V_i is the first harmonic of the signal received from the ith antenna element. The optical output from the last M-Z in the path from each antenna element then contains a term at the difference frequency (IF) between the incoming RF and the LO. This optical mixing in the M-Z allows much slower and consequently more sensitive optical detectors. After RF-to-optics conversion, delay lines may be included in the optical path to compensate for the antenna dispersion. These delay lines may be required for large antenna apertures or for broadband signals. They may be implemented in integrated optics[4].

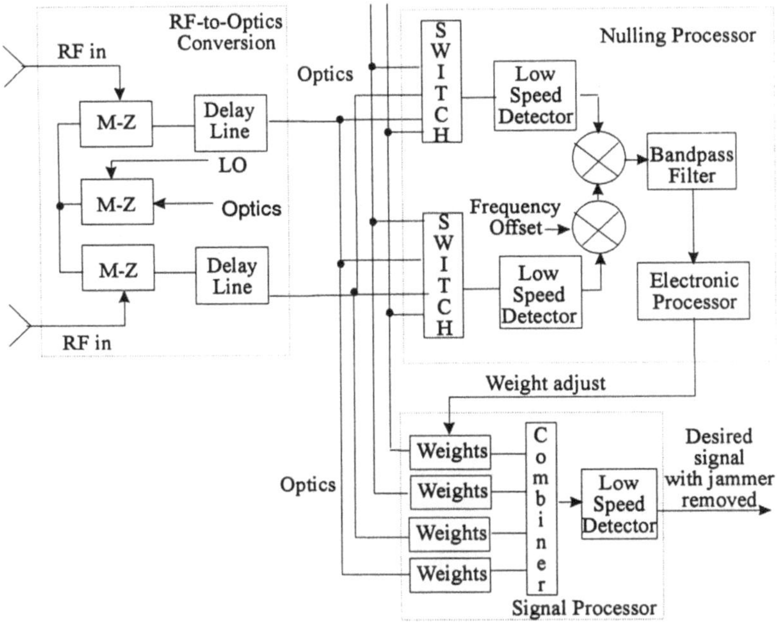

Figure 1. Schematic diagram of opto-electronic beam nulling network.

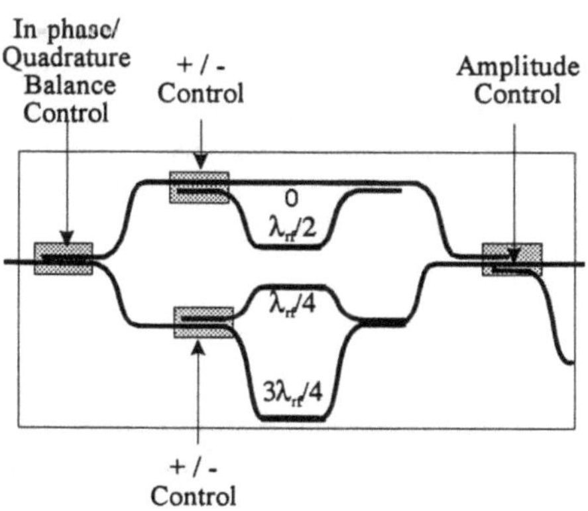

Figure 2. Integrated optics module to realize complex weights.

Nulling Processor

The function of the nulling processor is to evaluate the covariance matrix $\mathbf{\Phi}$ and to compute the weights. The former is done optically while the latter is done by a digital processor. For correlation the signal received by each beam is connected to two optical switches. The appropriate outputs from the switches are then selected to allow evaluation of each element in $\mathbf{\Phi}$. One of the two detected signals must be up-converted by some nominal frequency using a conventional RF mixer. This frequency offset allows the correlation term to be set away from baseband where it could mix with other signals present. The correlation term is evaluated by using a photodetector with bandwidth equal to the frequency offset and integrating the resulting power. Since $\mathbf{\Phi}$ is hermitian, the function only needs to be performed for terms on and above the diagonal. For the above-diagonal terms, both inphase and quadrature components are needed. The weights are computed from (4) by a digital processor which sends the results to the signal processor section.

Signal Processor

The function of this section is to apply the weights to the optical signal and linearly combine the optical powers from the resulting outputs to give an IF signal in which the effect of jammers and noise is minimized. The phase of each weight is realized by an interferometric process similar to an inphase-quadrature modulator at RF. The optical signal is split into two paths as shown in Fig. 2. In one path a time delay of either zero or $\lambda_{if}/2$ may be selected. In the other path a time delay of either $\lambda_{if}/4$ or $3\lambda_{if}/4$ may be selected. This allows the phase of the weight lying in any quadrant to be realized. Recombining the two paths gives an optical signal in which the phase of the modulating signal is shifted by the amount required. The amplitude of the weight is implemented by means of a directional coupler acting on this phase shifted signal.

CONCLUSION

In this paper a beam-nulling algorithm and its implementation by opto-electronic means is described. The network may be used in any of the RF bands and uses Mach-Zehnder modulators to provide RF-to-optics conversion. A unique way to realize the weighting of the various received beam signals, as required for nulling, is given. Since this network can be realized with standard components, an actual fabrication is expected to perform without much difficulty.

REFERENCES

1. A. Sonnenschein and W. K. Hutchison, A design for an electro-optic implementation of a wideband nulling algorithm, *IEEE Military Communications Conference Record*, 0742 (1990).
2. S. Haykin. "Adaptive Filter Theory," Prentice-Hall, Englewood Cliffs, NJ (1991).
3. R. Simons. "Optical Control of Microwave Devices", Artech House, Boston, MA (1990).
4. S. Paquet, M. Tetu, and C. Bélisle, Integrated 5-bit time-delay unit in high silica, *International Conference On Applications of Photonic Technology*, this issue (1994).

SEMICONDUCTOR LASER PHASE-LOCKING FOR OPTICAL BEAM FORMING NETWORKS

Rogerio T. Ramos

Schlumberger Cambridge Research
P.O. Box 153, Cambridge CB3 0HG, UK

This paper presents a comparison between Optical Phase-Lock Loops (OPLL) and Optical Injection Locking (OIL) systems and proposes that optical beam forming networks (OBFN) for phased array antennas can be built using OIL.

A phased array with a high number of elements would benefit from optical fibre distribution, due to its small size and light weight. Optical generation would overcome limitations imposed by the response bandwidth of the laser or external modulator in previously proposed optical fibre beam forming networks [1]. Phase-locked semiconductor lasers can also be applied in coherent detection, optical carrier generation and phase demodulation for optical instrumentation and communication systems [2].

Figure 1 shows a diagram of an OPLL systems [3]. The slave laser is controlled by the loop to oscillate at an optical frequency different from that of the master laser, with a precise offset given by the offset generator. The microwave signal is obtained by heterodyning the two optical signals after transmission through the fibre network. The action of the feedback loop phase modulates the slave laser in order to minimise the phase noise around the beat signal.

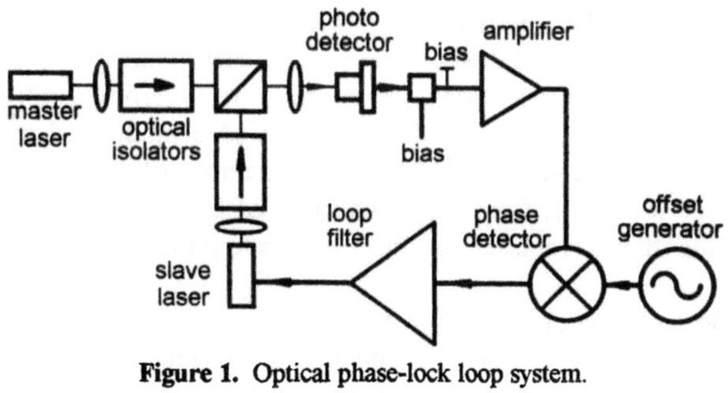

Figure 1. Optical phase-lock loop system.

Applications of Photonic Technology
Edited by G.A. Lampropoulos *et al.*, Plenum Press, New York, 1995

OPLLs can be made using narrow linewidth lasers, such as Nd:YAG lasers or gas lasers. However this kind of system becomes complicated when semiconductor lasers are used, due to their relatively wide linewidth. External cavities can be added to reduce the lasers linewidth [4], but this technique is not very attractive due to mechanical stability problems. Alternatively, the OPLL loop bandwidth can be increased in order to compensate for the wider phase noise spectrum of the beat signal. To achieve wide loop bandwidth, it is necessary to use low loop propagation delay time and wideband phase response lasers [5,6,7]. Reduced size systems were built to diminish the total loop delay time to allow the use of semiconductor lasers without any linewidth narrowing devices, such as external cavities [3,8]. Nevertheless, the commercial implementation of these reduced size systems can only be expected with future system integration.

Figure 2 shows the microwave signal generated by an OPLL as described in [3], using two temperature stabilised double quantum well GaAs/AlGaAs laser diodes, with 2.5MHz linewidth, in a 3ns delay time loop. A locking range of up to 20 MHz was achieved. The signal linewidth is narrow than the resolution bandwidth of the spectrum analyser (3kHz in Figure 2b).

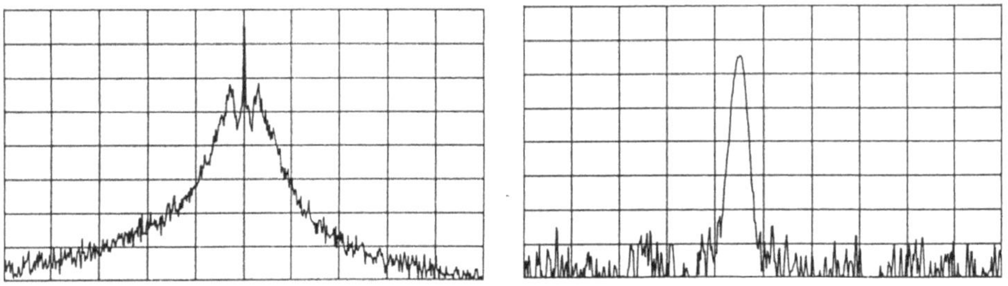

(a) *(b)*

Figure 2. Spectrum of the microwave signal generated by the OPLL: *(a)*: -35dBm reference level, 5dB/division, 6GHz centre frequency, 200MHz span, 300kHz resolution bandwidth. *(b)*: -30dBm reference level, 5dB/division, 6GHz centre frequency, 200kHz span, 3kHz resolution bandwidth.

Another method of achieving optical synchronisation between two semiconductor lasers is through direct optical injection from one laser to the other [7]. OIL systems can provide phase-lock with wide locking range and without the complexity involved in building an OPLL. OIL systems do not require wide frequency modulation bandwidth lasers, allowing locking ranges of the order of GHz to be obtained [9]. The locking range can be easily adjusted by changing the injection rate of the master laser power coupled to the slave laser mode [10]. However, this method presents non-zero steady state phase error and, therefore, can not follow long term drifts of the master laser, contrary to the OPLL systems.

A microwave signal distribution system using an heterodyne OIL system can be obtained by locking two slave lasers to the sidebands of a modulated master laser signal using injection locking techniques [11,12]. The block diagram of such a system is given in Figure 3. Microwave signals at frequencies multiple of the modulating frequency can be generated by beating the slave laser signals on a fast photodetector after optical fibre transmission.

A 3dB optical fibre coupler is used to inject the full spectrum of the master signal into both slave lasers. The master laser is modulated to produce two well defined sidebands and the injection rate for each slave laser is regulated by adjusting the alignment of the fibre/laser coupling. The locking range of each slave laser is controlled by the injection rate in order to make them lock to each side band and not to the master signal carrier. The locked optical signals from both slave lasers are then combined and a microwave signal at a frequency twice of that of modulating signal is generated at the photodetector.

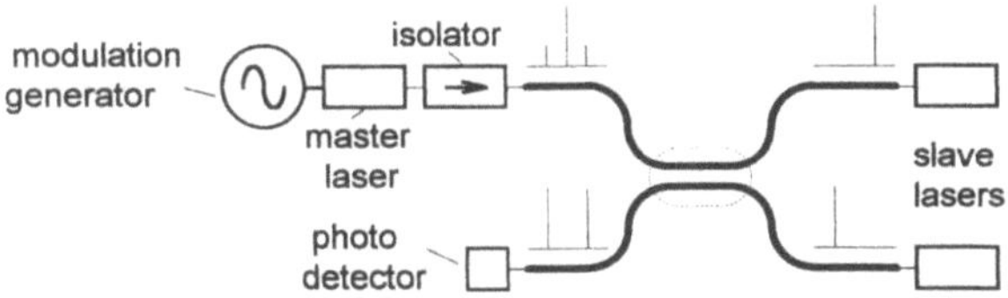

Figure 3. Optical injection locking system.

Figure 4 shows the beat signal between the two slave lasers with two span values. For this system, two DFB (distributed feedback) semiconductor lasers operating at 1.55μm were used as slave lasers, with free running linewidths of 20MHz. The master laser was a commercially available tunable semiconductor laser with a linewidth better than 100kHz [13]. Temperature controllers were used for long term stability of the slave lasers frequencies. The modulation frequency was 3.1GHz, at which the modulation response was strong. The locking range was adjusted to be about 200MHz. Other peaks at 3.1 and 9.3GHz were also present at levels 8dB below the main one at 6.2GHz, due to the presence of residual master signal carrier at the photodetector. Those peaks could be easily removed with electrical filters. The signal was seen to be stable and to present narrow linewidth (better than the resolution bandwidth of the spectrum analyser), confirming that phase locking was achieved.

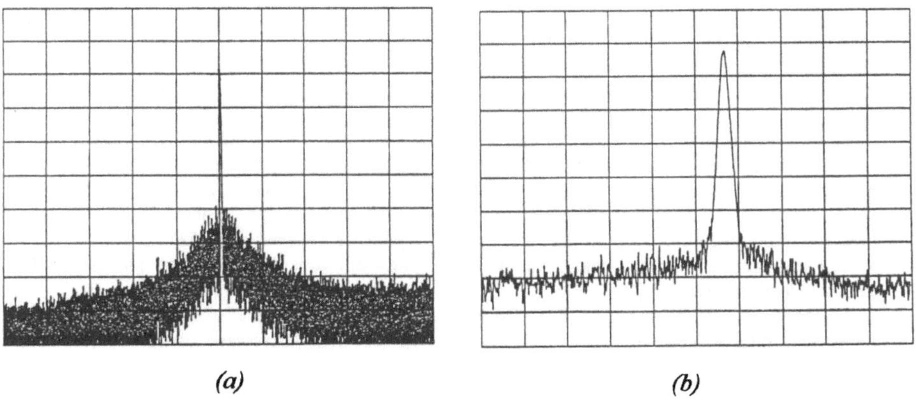

Figure 4. Spectrum of the microwave signal generated by the OIL: *(a)*: -1.5dBm reference level, 6dB/division, 6.2GHz centre frequency, 4GHz span, 3MHz resolution bandwidth. *(b)*: -1.5dBm reference level, 8dB/division, 6.2GHz centre frequency, 100kHz span, 1kHz resolution bandwidth.

Microwave signals at several frequencies multiple of the original modulating signal can be generated by locking the slave lasers to different lines of a comb of optical signals provided by a mode locked master laser. This method can be used for generation and fibre optic transmission of high frequency signals, as wide bandwidth photodetectors can are commonly available.

A combination of Optical Injection and Phase-Lock Loop (OIPLL) [14], given in Figure 5, could be an option to the use of OPLL and OIL. This system can present both a wide locking range, given by the OIL action, and a low phase error for low frequencies, given by the OPLL action [14]. This can improve the system tracking capability and decrease the final phase error variance, compared to the isolated systems.

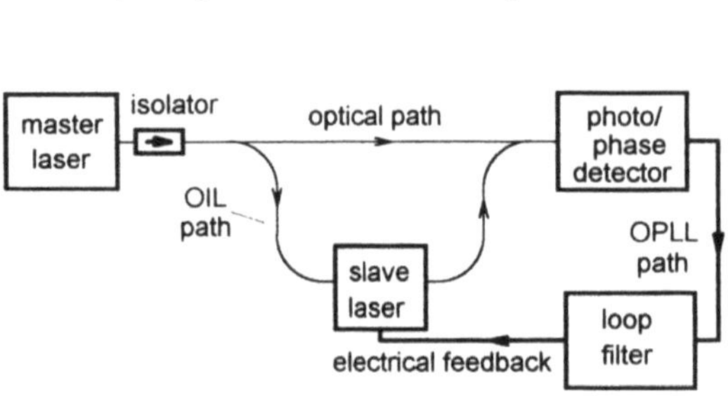

Figure 5. Combined optical injection and phase-lock loop system.

To conclude, OIL presents an alternative way of microwave generation and transmission for OBFN. It normally presents wider locking range and it is simpler to assemble than OPLL systems. However, it is more susceptible to long term fluctuations. Combined OIPLL may be used to overcome this problem.

ACKNOWLEDGEMENTS

To my colleagues at École Nationale Supérieure des Télécommunications, Paris, France, and University College London, UK.

REFERENCES

1. R. Benjamin, C. D. Zaglanikis, and A. J. Seeds, Optical beam former for phased arrays with independent control of radiated frequency and phase , *Electron. Lett.*, 26:1853 (1990).
2. M. Ohtsu, "Highly Coherent Semiconductoir Lasers, Artech House", (1992).
3. R. T. Ramos and A. J. Seeds, Fast heterodyne optical phase-lock loop using double quantum well laser diodes, *Electron. Lett.*, 28:82 (1992).
4. Steele, R.C., Optical phase locked loop using semiconductor laser diodes, *Elect. Lett.*, 19:69 (1983).
5. U. Gliese, E. L. Christensen, and K. E. Stubkjær, Laser linewidth requirements and improvements for coherent optical beam forming networks in satellites, *J. Lightwave Technol.*, LT9:779 (1991).
6. M. A. Grant, W. C. Michie and M. J. Fletcher, The performance of optical phase locked loops in the presence of nonnegligible loop propagation delay, *J. Lightwave Technol.*, LT5:592 (1987).
7. R. T. Ramos and A. J. Seeds, Delay, linewidth and bandwidth limitations in optical phase-locked loop design, *Electron. Lett.*, 26:389 (1990).
8. U. Gliese, T. N. Nilsen, M. Bruun, E. L. Christensen, K. E. Stubkjær, S. Lindgren, B Broberg, A wide heterodyne optical phase-lock loop for generation of 3-18GHz Microwave carriers, *IEEE Photon. Technol. Lett.*, 4:936 (1992).
9. H. L. Stover and W. H. Steier, Locking of laser oscillators by light injection, *Appl. Phys. Lett.*, 8:91 (1966).
10. F. Morgensen, H. Olesen, and G. Jacobsen, Locking conditions and stability properties for a semiconductor laser with external light injection, *IEEE J. Quantum Electron.*, QE21:784 (1985).
11. L. Goldberg, H. F. Taylor, J. F. Weller and D. M. Bloom, Microvave signal generation with injection-locked laser diodes, *Electron. Lett.*, 19:491 (1983).
12. R.T. Ramos, C. Chabran, Optical fibre microwave distribution using injection locked DFB semiconductor lasers, *IEE Colloquium on Microwave Opto-electronics*, digest n. 1994/022:5/1-4, London 27 January (1994).
13. F. Favre and D. Le Guen, 82nm of continuous tunability for an external cavity semiconductor laser, *Electron. Lett.*, 27:183 (1991).
14. R.T. Ramos, P. Gallion, D. Erasme, A.J. Seeds, A. Bordonalli, Optical injection locking and phase-lock loop combined systems, *Optics Letters*, 19:4 (1994).

PROTOTYPE MODEL OF AN OPTICALLY CONTROLLED ARRAY ANTENNA

Claude Bélisle, John Oldham

Communications Research Centre
3701 Carling Ave. Ottawa, Ont.

INTRODUCTION

Active phased array (PA) antennas are being proposed for the next generation of remote sensing synthetic aperture radars (ASAR) because of their electronic beam steering and shaping capabilities, graceful degradation, and out-of-theatre interference rejection capability. However, those antennas are difficult to build with conventional microwave components. Optical technology could play a significant role in providing solutions to the problems encountered with an all-microwave implementation. For example, fibre optics could replace the conventional waveguides or coaxial cables for signal distribution. This would not only reduce considerably the weight and size of the system, but would also allow lower losses in the signal distribution, provide better interference immunity, and give higher bandwidth.

In this paper, an architecture, using opto-electronic technology, is proposed for the fabrication of a synthetic aperture radar for remote sensing application and the prototype model being built at CRC is described.

SYSTEM DESIGN

A vast amount of architectures has been proposed for the optical control of array antennas[1-3]. However, the majority of them were for generic systems where independent phase and amplitude control for every element of the antenna was desired. In ASAR, the phase and amplitude control requirements are not so stringent. It can be shown that the phase control may be done by rows and columns and dynamic amplitude control is required only at the row level on the receive signal. The overall beam profile, $E(\theta,\phi,t)$, may be expressed as the sum of the signal emitted, or received, from every antenna element.

$$E(\theta,\varphi,t) = \sum_{n=0}^{N} \sum_{m=0}^{M} A_{n,m} \exp\left[j\left(wt - \frac{2\pi}{\lambda_{rf}} \sin\varphi \left[x_n \cos\theta + y_m \sin\theta\right] + \phi(x_n,y_m)\right)\right] \quad (1)$$

Applications of Photonic Technology
Edited by G.A. Lampropoulos *et al.*, Plenum Press, New York, 1995

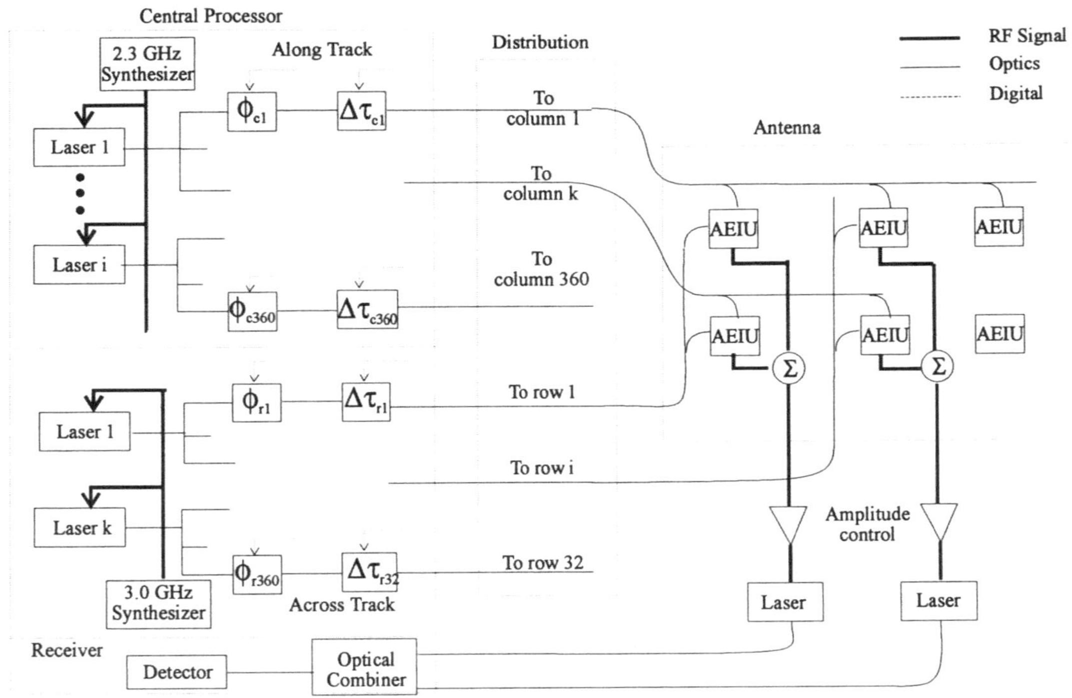

Figure 1 Architecture of an optically controlled synthetic aperture radar. A four element prototype is being tested at CRC.

where $A_{n,m}$ is the amplitude weight of the signal of the element located at position x_n, y_m and $\phi(x_n, y_m)$ the phase weight. From (1), it is seen that in order to obtain a beam in a direction (α, β), the phase weights $\phi(x_n, y_m)$ must equal:

$$\phi(x_n, y_m) = \phi(x_n) + \phi(y_m) \tag{2}$$

$$\phi(x_n) = \frac{2\pi}{\lambda} x_n \sin\beta \cos\alpha \text{ (along track)} \quad \phi(y_m) = \frac{2\pi}{\lambda} y_m \sin\beta \sin\alpha \text{ (across track)}$$

Fig. 1 shows the diagram of the system being built and tested at CRC for this project. Only four elements are being implemented at this point but the complete system is to have over 10000 elements radiating in the C-band (5.3 GHz) and output power at each element, of 450 mW.

The central processor is responsible for the RF-to-optics and optics-to-RF conversion. Phase weighting is also done in the controlled environment of the payload, rather than at the antenna element. As mentioned above, the phase control section is divided into along track (rows) and across track (columns). Two RF signal generators, 2.3 and 3.0 GHz, are used to directly modulate laser diodes. External modulators could be used and would offer better stability but at the expense of higher RF power requirements, higher optical insertion loss and cost. The intensity modulated signal is split into N branches each one corresponding to a row and a column of the antenna. In each branch, a phase shifter and an optical time delay are included. At present, the phase shifts are done on the RF signal, before conversion to optics, but a non-coherent optics phase shifter is being designed to operate on the RF modulation of the optical signal[4]. The time delay is needed to adjust the phase of the optical modulation in order to compensate for the propagation delay through the antenna. The delays are fixed and correspond to the row and column positions. The distribution network carries the phase weighted signals to every element in the associated row or column. At the antenna element (Fig. 2), the two signals must be mixed to obtain the 5.3 GHz centre frequency with the phase equal to the sum of the along track and across track signal phase. Conventional RF mixers, following two photodetectors, are being used at present but research is being done in-house in the area of optical mixing which suggests that the photodetectors could act as mixers[5] (this would eliminate the RF mixers).

On receive, the incoming signal is mixed with a properly phase weighted signal generated in the central processor in the same way as for the transmit signal, and downconverted to an IF frequency. This time sharing of the distribution network can be done since in radar applications, the transmit and receive pulses are not simultaneous. The

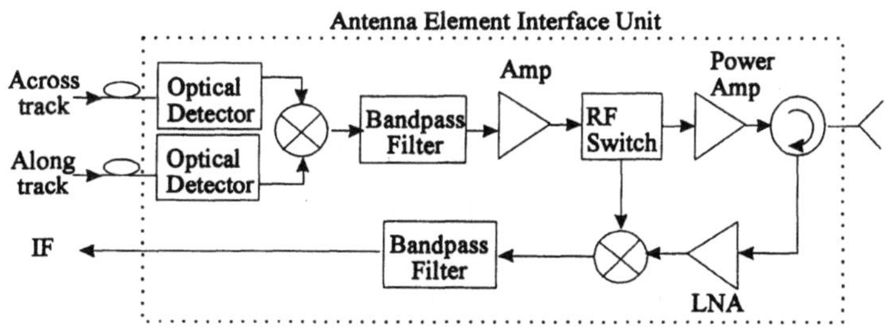

Figure 2 Schematic diagram of the antenna element interface unit.

signals from every antenna element then become in phase and may be coherently added. Those signals are initially added at the IF level in order to increase the power before being converted to optics. Amplitude control can be done in the electrical domain on the IF signal or on the optical signal using optical amplifiers. Once converted to optics the signals are brought back to the central processor and detected.

LINK BUDGET

A link budget was established for the RF and optical power requirements. It was evaluated that 36 lasers of 40 mW each would be required to drive every element of the antenna. The distribution losses, including splitting losses (each laser signal is split in 640 branches) and RF-to-optics conversion loss was calculated to be 37 dB. Optical amplification of 30 dB and RF amplification of 45 dB in the antenna element was assumed. The RF power required to drive those 36 lasers for 100% modulation is equal to 38 dBm.

CONCLUSION

This paper presented a novel architecture for the optical control of synthetic aperture array antennas for application in remote sensing. RF signal phase weighting, which allows beam steering and beam shaping, is done in the payload rather than in the antenna element, away from environmental stress. The number of phase shifters required is limited to the number of rows and columns of the antenna. The phase weight of a particular element is the sum of the phase weights associated with its row and column number. A novel antenna element architecture is proposed which involves opto-electronic mixers rather than conventional RF mixers. An optical link budget was made and it was estimated that 36 laser diodes of 40 mW each would be required to power the antenna. 30 dB of optical amplification and 45 dB of RF amplification at the antenna element would also be needed to increase the RF transmitted power to 450 mW.

ACKNOWLEDGEMENT

This work has been undertaken under the support of the Department of National Defence. The authors would also like to thank Dr. L. Mason and Dr. B. Felstead for their helpful discussion.

REFERENCES

1. J.B. Wood, Real-Time Acousto-Optic Processor for Synthetic Aperture Radar, *SPIE* **1296** (1990) 222.
2. N.A. Riza and D.Psaltis, Acousto-Optic Signal Processors for Transmission and Reception of Phased-Array Antenna Signals, *Applied Optics* **30** (10 Aug 1991), 3294
3. J.H. Hong, Broadband Phased Array Beamforming, *SPIE* **1202** (1989) 134
4. C. Bélisle, L. Mason, B. Felstead, Design of an Opto-Electronic Beam Nulling Network for a multiple beam antenna, *ICAPT* (1994), in this issue.
5. D.K.W. Lam and R.I. MacDonald, GaAs Optoelectronic Mixer Operation at 4.5 GHz, *IEEE Trans. Electron Dev.*, **ED-31**,(1984), 1766

PHOTONICS APPLIED TO PHASED ARRAY ANTENNAS: WORK DONE AT UNIVERSITÉ LAVAL

M. Têtu, M. Chamberland, P. Tremblay, C. Beaulieu, S. Paquet
A. Fekecs, G. Lessard, M.-L. Charès, C. Laperle

Centre d'optique, photonique et laser
Département de génie électrique, Université Laval
Québec (Québec) G1K 7P4, Canada

INTRODUCTION

Space-borne phased array antennas may benefit from the use of optical communications technology and photonic devices to reduce their weight and their power consumption, and also from signal handling by innovative means.[1-3] At Université Laval, we have undertaken a research program to study some promising avenues. We present here a summary of our latest results concerning the microwave signal transport, distribution, processing and generation by optical means.

MICROWAVE SIGNAL TRANSPORT

We have generated pure microwave signals by heterodyning two synchronized semiconductor lasers.[4] The two lasers are injection-locked by different sidebands of an optical signal emitted by a frequency-modulated master laser.[5] A noticeable advantage of the present technique is that the nominal frequency of the heterodyne signal is at a multiple of the modulation frequency of the master laser.

In our experiment, the beat note signal was observed at 6.8 GHz which corresponds to 4 times the modulation frequency, which is 1.7 GHz. As stated in the previous paragraph, the output signals from the two slave lasers are strongly correlated. The power spectrum of the generated microwave signal exhibits a very narrow peak [$<$ 25 mHz (FWHM)] sitting on top of a broad Lorentzian-shaped pedestal (Figure 1). This pedestal has a level of -89 dBc in a 1 Hz bandwidth at 10 MHz offset from the carrier. Figure 2 illustrates the portion of the power spectrum close to the carrier, this corresponds to the power spectral density of the single sideband noise.

Temporal measurements of the phase of the generated microwave signal have been used to estimate the power spectral density of the phase fluctuations (Figure 3). This

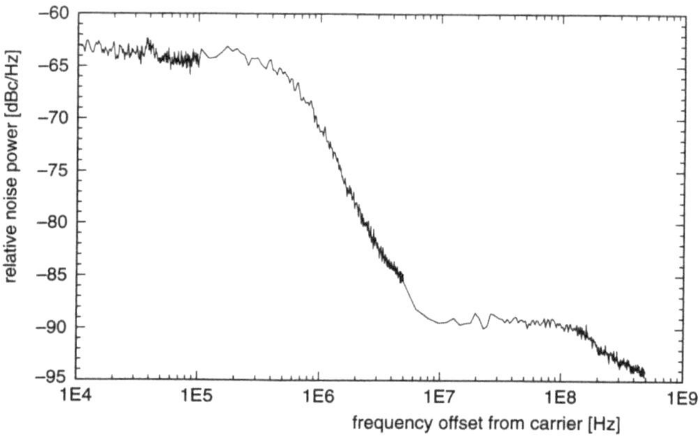

Figure 1. Experimental spectrum of the microwave signal generated by heterodyne between the two slave semiconductor lasers: injection-locked (solid line) and free-running (dotted line).

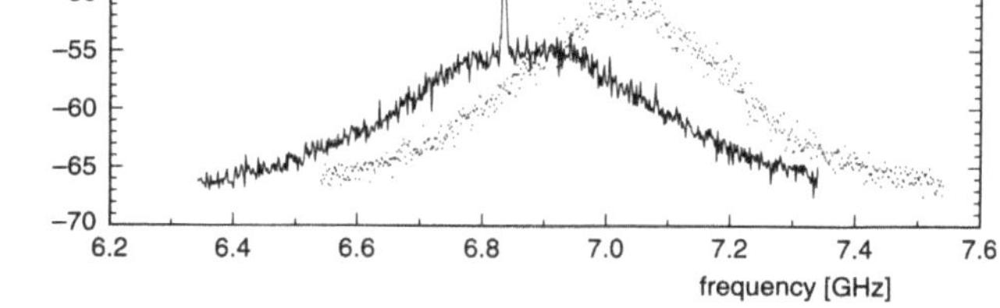

Figure 2. Experimental relative single sideband noise power spectral density [the so-called "$\mathcal{L}(f)$"] for the signal generated by heterodyne between injection-locked slave semiconductor lasers.

last result brings more information about the close-to-the-carrier noise. From the time domain measurements, the relative frequency stability, expressed as the square root of the Allan variance, is observed (Figure 4). It reaches a level of 1×10^{-11} for an averaging time of 1 s.

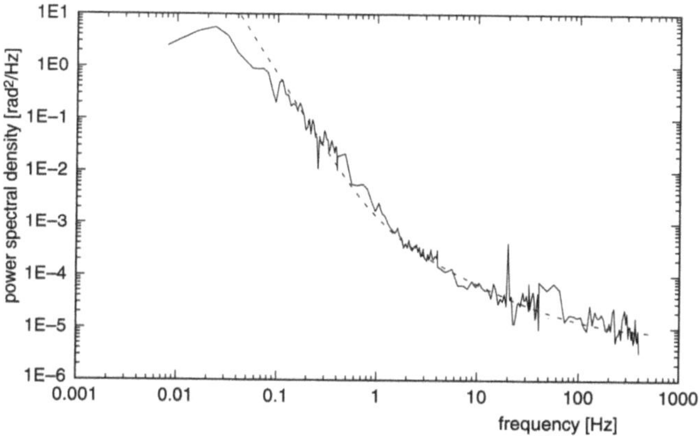

Figure 3. Power spectral density of the phase fluctuations of the 6.83 GHz microwave signal generated by heterodyne between injection-locked semiconductor lasers (solid line). The dashed line represents the truncated polynomial approximation of the spectrum.

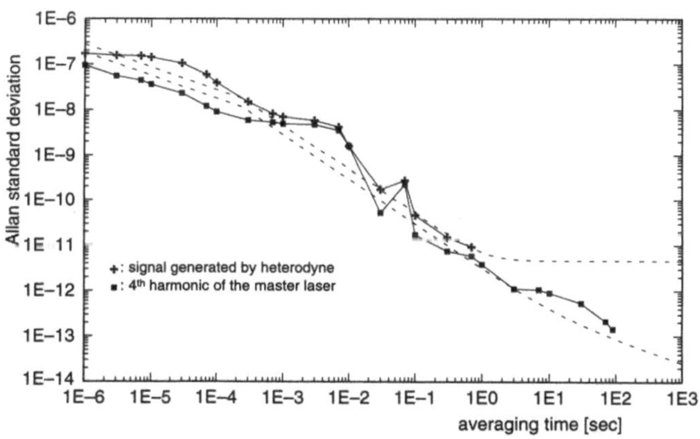

Figure 4. Square root of the Allan variance measured with a coherent receiver tuned at 6.83 GHz. Experimental results (solid lines) and predictions from the phase fluctuations spectrum (dotted lines) are shown for the signal generated by heterodyne and for the 4th harmonic of the modulated master semiconductor laser.

MICROWAVE SIGNAL GENERATION

We have produced microwave signals of several milliwatts by the direct injection-locking of a microwave oscillator with the optical signal emerging from a current-modulated semiconductor laser.[6] The locking range depends directly on the injected power and inversely on the quality factor and the power of the oscillator.[7] We measured the variation of the locking range with injected power when injecting at the fundamental as well as at subharmonics of the microwave oscillator frequency (Figure 5). We note that a locking range of 5 MHz is achievable.

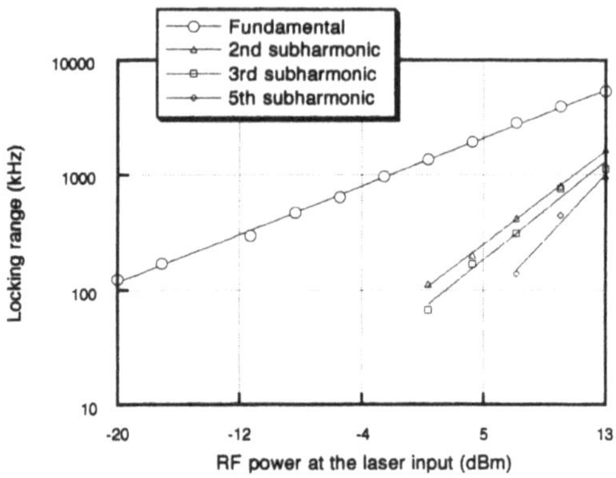

Figure 5. Variation of the locking range as a function of the microwave power at the semiconductor laser input when modulating the laser at the fundamental or at a subharmonic of the oscillator frequency.

Frequency stability measurements done in both the time and frequency domains show a full transfer of stability between the oscillator used to modulate the laser and the injection-locked microwave oscillator. These results are presented in Figure 6.

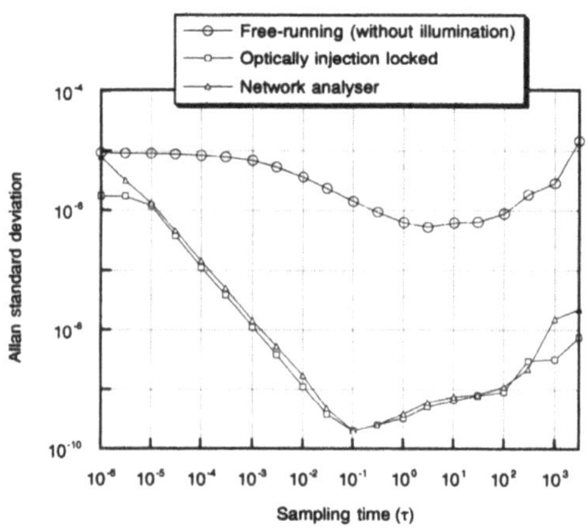

Figure 6. Comparison of the oscillator's Allan standard deviation in the free-running and in the frequency-locked regimes. The frequency stability of the reference used is also shown (triangle).

We are now studying indirect injection-locking of microwave oscillators for an improved locking bandwidth and ease of fabrication.[8] A newly designed circuit will allow an automatic control of the phase of the oscillator relative to the impinging signal.

MICROWAVE SIGNAL PROCESSING

A switchable 5-bit true time delay with selectable optical waveguides made of doped silica on silicon has been designed (Figure 7). The switching is achieved with interferometric couplers controlled through thermo-optic effect on silica.[9] The dissipated power is 500 mW per bit and the total optical insertion loss is 20 dB.[10]

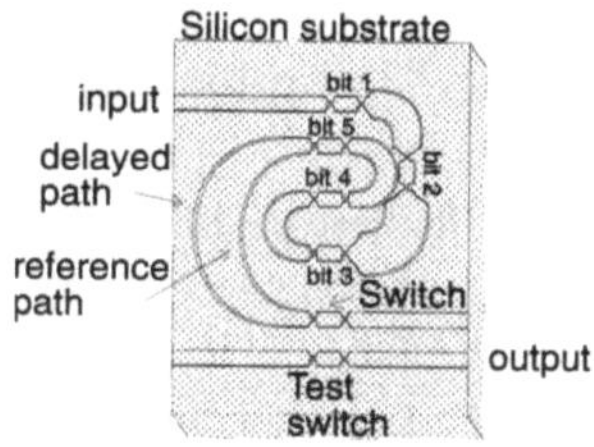

Figure 7. Schematic diagram of the 5-bit true time delay.

MICROWAVE SIGNAL DISTRIBUTION

We are presently setting up an experiment to realize and study semiconductor optical amplifiers[11] used in conjunction with high splitting ratio optical couplers. Such amplifiers are highly suitable for integration but they operate properly only when the residual reflection coefficients of the facets are very low. Up to now we have been able to achieve, on a commercially available laser, a reflection coefficient of 0.008, which is sufficient for the purpose of our test experiment. This coefficient is estimated by comparing the emission spectra of the device under various coating conditions, but under identical temperature and current conditions (Figure 8).

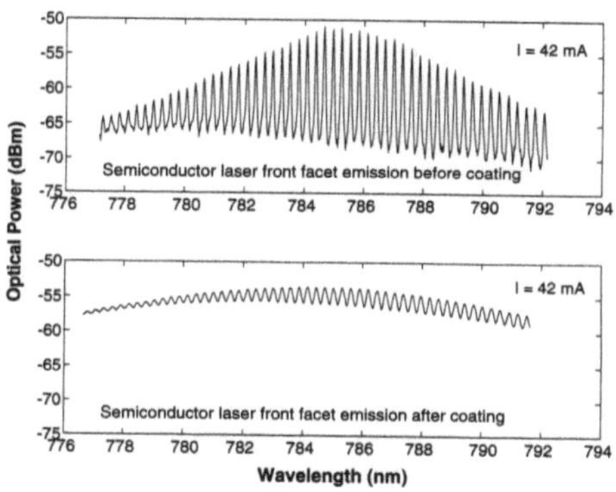

Figure 8. Emission spectra of the semiconductor laser: operated slightly below threshold before coating (upper graph), operated at the same current after coating (lower graph).

ACKNOWLEDGEMENTS

The authors thank *SPAR Aérospatiale*, the *Conseil de recherches en sciences naturelles et en génie*, and the *Centre de recherches sur les communications* for financial support.

REFERENCES

1. L. Goldberg, R. D. Esman, and K. J. Williams, Optical techniques for microwave generation, transmission, and control, *in*: "IEEE Microwave Theory Techn. Symp. Dig.," IEEE, Piscataway (1990).
2. A. J. Seeds, I. D. Blanchflower, and N. J. Gomes, New developments in optical control techniques for phased array radar, *in*: "IEEE Microwave Theory Techn. Symp. Dig.," IEEE, Piscataway (1988).
3. A. J. Seeds, W. I. McMillan, C. R. Pescod, M. J. Wale, and W. S. Birkmayer, Optical control of phased array antennas: a European perspective, *in*: "IEEE Microwave Theory Techn. Symp. Dig.," IEEE, Piscataway (1990).
4. M. Chamberland, M. Têtu, and P. Tremblay, Spectral characterization of microwave signals generated by the heterodyne of injection-locked semiconductor lasers, *in*: "Optoelectronic Signal Processing for Phased Array Antennas IV, Proc. 2155," , B. M. Hendrickson, ed., SPIE, Bellinghan (1994).
5. L. Goldberg, H. F. Taylor, and J. F. Weller, Microwave signal generation with injection-locked laser diodes, *Electron. Lett.* 19:491 (1983).
6. C. Beaulieu, M. Chamberland, M. Têtu, and P. Tremblay, Stability measurements of an optically injection-locked microwave oscillator, *in*: "Frequency-Stabilized Lasers and their Applications, Proc. 1837," Y. C. Chung, ed., SPIE, Bellinghan (1993).
7. R. Adler, A study of locking phenomena in oscillators, *Proc. IRE* 34:351 (1946).
8. D. J. Sturzebecher, X. Zhou, X. Zhang, and A. S. Daryoush, Design of oscillators for optically fed MMW phased array, *in*: "IEEE Microwave Theory Techn. Symp. Dig.," IEEE, Piscataway (1992).
9. M. Kawachi, Silica waveguides on silicon and their application to integrated-optic components, *Opt. Quantum Electron.* 22:391 (1990).
10. S. Paquet, F. Chénard, Z. Jakubczyk, M Bélanger, M. Têtu, C. Bélisle, Optical delay lines in high-silica (SiO_2/Si) waveguides, *in*: "International Conference on Applications of Photonic Technology," G. A. Lampropoulos, ed., Plenum, New York (1994).
11. M. J. O'Mahony, Semiconductor laser optical amplifiers for use in future fiber systems, *J. Lightwave Technol.* 6:531 (1988).

OPTICAL SIGNAL PROCESSING IN INTEGRAL TRANSFORM SOLUTION APPLICATIONS IN PARTICLE SIZING

Janusz Mroczka

Institute of Electrical Metrology,
Technical University of Wrocław,
ul. B. Prusa 53/55. 50–317 Wrocław, POLAND

INTRODUCTION

Light scattering theory in dispersed systems illuminated by a beam of parallel rays can be formulated similarly to the electromagnetic diffraction theory. If it is assumed that the source of flat waves and a radiation detector are placed far from the examined dispersed system sample and that the dispersed phase particles are spherical, the description of light scattering phenomenon corresponds to the Fraunhofer diffraction theory[1].

Light intensity distribution of Fraunhofer diffraction, scattered on a circular obstacle for small angle ranges appears as:

$$I(\Theta) = I(0)\left[\frac{2J_1(x\,\Theta)}{x\,\Theta}\right]^2 \tag{1}$$

where: $I(0)$ — scattered light intensity for $\Theta = 0$, dimension [W/sr],
x — Mie parameter.

In the case of a finite volume of a dispersed system with the dispersed phase particles of $f(a)$ distribution, Eq. (1) is:

$$I(\Theta) = \frac{c}{\Theta^2} I_0 \int_0^\infty x^2 J_1^2(x\,\Theta)\, f(x)\, dx \tag{2}$$

where: c — constant,
$J_1(x\Theta)$ — Bessel function of the first kind of the first order,
λ — wavelength.

Eq. (2) is a linear integral Fredholm equation of the first kind.

SIGNAL PROCESSING OF THE MATHEMATICAL MODEL OF THE MEASURING METHOD WITH THE USE OF SCATTERING AT A SMALL ANGLE

Eq. (2) is a mathematical model of the particle size distribution measuring method. It describes the distribution of the scattered light intensity for a polydispersed system. To estimate the particle size distribution, one has to measure the light scattered intensity (scattering volume function) and to transform Eq. (2). The solution of the Eq. (2) is:

$$f(x) = \frac{1}{c\,x^2} \int_0^\infty D(\Theta)\, h_D(x\,\Theta)\, d\Theta \tag{3}$$

where:

$$h_D(x\,\Theta) = 2\,\pi\, x\,\Theta\, J_1(x\,\Theta) Y_1(x\,\Theta) \tag{4}$$

$Y_1(x\,\Theta)$ — Neuman function

$$D(\Theta) = \frac{d}{d\,\Theta}\left[\frac{I(\Theta)}{I_0}\Theta^3\right] \tag{5}$$

Considering (2) in the expression (5), it has got the following form:

$$D(\Theta) = \frac{\lambda^3}{8\pi^3} \int_0^\infty f(x)\, x^2\, k_D(x\,\Theta)\, dx \tag{6}$$

where:

$$k_D(x\,\Theta) = \frac{d}{d\,(x\,\Theta)} \left[x\,\Theta\, J_1^2(x\,\Theta) - \frac{1}{\pi} \right]. \tag{7}$$

From the metrological point of view two factors can be the reason of errors when determining particle size distribution, using (3).

The first is impossibility of obtaining full information from an experiment, which is formally required in (3). The integrating limits in this expression, regarding an angle Θ at which a scattered light intensity measurement is taken, should be from 0 to ∞. In reality the range of scattering angles Θ is limited by technical possibilities of the experiment. The lower limit Θ_{min} is limited by the width of the used radiation beam as well as a radiation detector system design. The upper limit Θ_{max} results from the fact that for the particles bigger than $x \sim 1$, the scattered light intensity decreases very quickly with the increase of the angle Θ, and at the angle Θ_{max} can be comparable with the noise of the light source as well as the detector itself[2].

Another reason of errors when using Eq. (3) is the necessity of making differenciations upon experimentally measured function of a finite number of arguments. It complicates data processing and sometimes may result in completely mistaken results.

To avoid the problem of differenciation in the form presented in Eq. (2) some transformations have been done. They are the introduction of the function $K(\Theta)$ described by the expression[3]:

$$K(\Theta) = \frac{I(\Theta)}{I_0}\, \Theta^3 . \tag{8}$$

Introducing function:

$$T(\Theta) = K(\Theta) - K(\infty) = c \int_0^\infty x f(x) \left[x\,\Theta\, J_1^2(x\,\Theta) - \frac{1}{\pi} \right] dx \tag{9}$$

for which

$$D(\Theta) = \frac{d\,T(\Theta)}{d\,\Theta}, \tag{10}$$

the expression (3), after integration by parts, is:

$$f(x) = -\frac{c}{x} \int_0^\infty h_T(x\,\Theta)\, T(\Theta)\, d\,\Theta . \tag{11}$$

The kernel of this equation, is:

$$h_T(x\,\Theta) = -\,2\,\pi\, Y_1(x\,\Theta)\,[\,2\,x\,J_0(x\,\Theta) - J_1(x\,\Theta)] - 4\,, \tag{12}$$

where : $x\,\Theta$ — independent variable,

$$h_T(x\,\Theta) = \begin{cases} 2.0 & \text{for } x\,\Theta \ll 1 \\ 4 \sin 2x\,\Theta & \text{for } x\,\Theta \gg 1 . \end{cases} \tag{13}$$

The obtained expressions (3) and (11) are the solution of the Fredholm equation (3) being the mathematical model of the measuring method, using light scattering at a small angle. In this method the mathematical model of the physical phenomenon is the Fraunhofer diffraction.

The visible difference in Eq. (3) and (11) results from different application of the measuring information in the experiment conducted. In the case of expression (3) the differenciated measured value was used, and in the expression (11) the value after initial processing (8). This initial processing applies analogue signals and at the same time in practice can be done without additional errors. The introduction of the initial information processing(expression (8)), frees the measuring method mathematical model (making it a new method, to some extent) from the differenciation (5).

To compare the presented particle size distribution estimation methods, the theoretical values of their errors were determined. In the case of expression (3) this error is[4]:

$$\delta = \frac{1}{f(x)}\, \frac{6}{2\pi x_m} \left(\frac{\cos 8\,\pi\, a\, \Theta_{max}}{(8\,\pi\, a\, \Theta_{max})^2} - \frac{\sin 8\,\pi\, a\, \Theta_{max}}{8\,\pi\, a\, \Theta_{max}} + C\,i\, 8\,\pi\, a\, \Theta_{max} \right) \tag{14}$$

where: $f(x)$ — a priori assumed Gamma distribution,

x_m — expected value,

Θ_{max} — upper integration limit,

$$Ci\,(x) = -\int_x^\infty \frac{\cos y}{y}\, dy,$$

a — particle radius.

In the case of Eq. (11) this error is:

$$\delta = -\frac{1}{f(x)}\frac{3\beta}{\pi}\frac{\Gamma(\mu)}{\Gamma(\mu+1)\Gamma(\mu+2)}\left[\frac{\sin 2x\,\Theta_{max}}{2x\,\Theta_{max}} + Ci\,2x\,\Theta_{max}\right] \tag{15}$$

where: $\Gamma(\mu)$ — Gamma function,

$x = \frac{2\pi a}{\lambda}$ — Mie parameter.

To conduct the signal analysis after integration, which practically can be done by means of the supplementation of the measuring system by an integrator system, a mathematical model of this problem has to be done. Introducing $Z(\Theta)$ function determined by the expression:

$$Z(\Theta) = \int_0^{\infty} T(z)\,dz\,, \tag{16}$$

and integrating by a part of the Eq. (11) analogically to Eq. (3) we obtain in the final form new expression of the particle size distribution function:

$$f(x) = -c\int_0^{\infty} h_s(x\Theta)\,Z(\Theta)\,d\Theta\,. \tag{17}$$

In this expression the function (kernel):

$$h_s(x\Theta) = 4\pi\left\{\left(\frac{1}{x\Theta} - x\Theta\right)J_1(x\Theta)\,Y_1(x\Theta) + J_0(x\Theta)\,[Y_0(x\Theta)\,x\Theta - Y(x\Theta)]\right\} - \frac{4}{x\Theta} \tag{18}$$

has got asymptotes

$$h_s(x\Theta) = \begin{cases} 0 & \text{for } x\Theta \ll 1 \\ -8\cos 2x\Theta & \text{for } x\Theta \gg 1\,. \end{cases} \tag{19}$$

Eq. (17) is a new form of the Fredholm equation solution (2). To compare the method obtained in this way with the methods discussed before, the error has been evaluated. In the case of Gamma distribution, one obtains:

$$\delta = -\frac{1}{f(x)}\frac{3\beta}{\pi}\frac{\Gamma(\mu)}{\Gamma(\mu+1)\Gamma(\mu+2)}\,Ci\,2x\,\Theta_{max}\,. \tag{20}$$

The described mathematical model constitutes a fundamental stage of a real experiment realization in signal processing. Analysing presented three methods of the particle sizing (3;11;17), being the result of the integral equation solution (2) at three different experimental data, we obtain different errors. As (14) does not consider errors resulting from differenciation in the process of the particle sizing on the basis of the assumed physical and mathematical models, the comparison has been done in the case of (15;20). Making a product of these expressions and necessary transformations one obtains:

$$\frac{\delta_i}{\delta_c} = \frac{\dfrac{\sin 2x\,\Theta_{max}}{2x\,\Theta_{max}}}{Ci\,2x\,\Theta_{max}} + 1\,, \tag{21}$$

where: δ_i — error determined by (15),

δ_c — error determined by (20).

Considering that the smallest particle size possible for determination in the physical model of the Fraunhofer diffraction is of 1.6 μm (for non absorbing particles) and that angle Θ_{max} is of 0.17 rad, the value of independent variable ($2x\,\Theta_{max}$) in the Eq. (21) is not less than 5. At this value of the independent variable, the value of the quotient of the special functions in (21) is the biggest and is ~ 2. With the increase of the independent variable, the value of the expression quickly decreases asymptotically towards unity. It means that for small particles the methods in which integrated volume scattering function appers is twice as precise as the method without integration. It has also got its physical explanation. In the case of small particles it is necessary to know the experimentally measured volume scattering function for bigger angles at which the intensity signal from a detector increases, and the appearing turbulences (noise) can be very important. The integration of this signal is the elimination of unnecessary turbulences and at the same time the decrease of the measurement error.

The analysis done here considered the mathematical model (error evaluation) for the practical possibilities of the experiment realization — limited integration range in (3;11;17) and initial measuring signal processing.

Taking in (15;20) 10% as an error value, the maximum angles of the volume scattering function in the function of parameter Mie ($x = 2\pi\alpha/\lambda$), at different Gamma distribution parameters, have been evaluated[5]. The numerical data have been presented in Table 1.

The results presented in Table 1 show the ranges of parameter changes appearing in mathematical models (11;17) in the case of which the measurement error does not exceed 10%. The underlined in the table results regard the range of parameter changes, in the case of which the assumptions limiting the validity of the physical model applied, are fulfilled.

The measuring signal integration being the representation of the volume scattering function enlarges the range of the variable x and the width of the a priori assumed particle size distribution. The measuring averag-

Table 1. The maximum angle value of the light scattering measurement depending on the properties of the a priori assumed Gamma distribution of the signal $K(\Theta)$, $Z(\Theta)$ initial processing way

	μ/x	25	50	75	100	150	200	300
Signal $K(\Theta)$	2	0.29	0.15	0.09	0.06	0.04	0.03	0.02
	4	0.29	0.18	0.13	0.09	0.06	0.05	0.03
	6	0.33	0.21	0.17	0.16	0.10	0.07	0.05
	8	0.40	0.27	0.21	0.17	0.15	0.14	0.08
Signal $Z(\Theta)$	2	0.01	0.01	0.004	0.003	0.002	0.002	0.001
	4	0.01	0.01	0.008	0.008	0.007	0.007	0.007
	6	0.19	0.10	0.08	0.06	0.04	0.03	0.02
	8	0.31	0.17	0.12	0.09	0.06	0.05	0.03

ing signal being the result of its integration is not only an elimination of unnecesary signal turbulences but also the loss of information about its complexity. The result of this is a considerably small admissible value of the maximum measuring angle ($\Theta_{max} \sim 0.1$ rd) of the volume scattering function, which makes this method practically unrealizable.

In the case of linear operation in a measuring signal (8) some limitations occur, concerning the value of parameter x and the width of the a priori assumed distribution; but the minimum value of the maximum angle of the volume scattering function is of 0.02 rd, which in practice is possible to realize.

CONCLUSIONS

In particle sizing, a group of problems dealing with physical model is developed very fast and considerably leaves behind problems that should be solved so that newly developed mathematical models could be applied in practice.

The integral equation (2) solution was done by means of analytical transformations (using integral transforms) without any mistakes. It was possible due to convenient form of the function expressing mathematical model of the physical phenomenon, which is Fraunhofer diffraction. Because the Fraunhofer diffraction is an approximate phenomenon the question appears if the solution of the integral equation (2) is possible and with what error, in the case of more perfect models, for example considering the dispersed phase shape, etc.

In the author's opinion, the problem whose importance has been shown by the above signal processing, constitutes an important turning point in the particle sizing method improvement, when more and more eleborate technologies have been used without any verification works, metrological analysis among others. The measuring method mathematical model should be developed by the use of approximate methods of the integral equation solutions[6;7], considering the specific character of particle sizing.

The alternative solution is regarded by the author in a partial resigning from the quantity of information resulting from the knowledge of the particle size distribution function ($f(a)$). This way has been presented in the author's works[8;9], where the moments, particle size distribution integral parameters have been used for the description of the dispersed system dispersed phase physical properties.

BIBLIOGRAPHY

1. A. Sommerfeld, "Optics," *Lectures on theoretical physics.*, Vol. 4, Acad. Press, New York, 1954.
2. E. L. Dereniak, D. G. Crowe, "Optical Radiation Detectors", J. Wiley and Sone Inc., New York, 1984.
3. K. S. Šifrin, V. I. Golikov, "Izmerenija Mikrostruktury Metodom Malych Uglov," *Trudy GGO*, 1964, vyp. 152, s. 3–15.
4. K. S. Šifrin, "Suscestvennaja Oblast Uglov Rassejania pri Izmerenii Raspredelenia Castic po Razmeram Metodom Malych Uglov," *Fiz. Atmos. Okeana*, 1966, T. 2, No. 9, s. 928–932.
5. J. Mroczka, "Metrological Problems of Scattered Light Application for Particle Size Distribution Investigation in Dispersed Solutions," *Metrology and Measuring Systems*; Monograph No. 4, Warsaw 1990 (in Polish).
6. A. N. Tichonov. V. J. A. Arsenin, "Metody Resenija Nekorrektnych Zadac," IZD. "Nauka", Gl. Red. Fiz.-Mat. Lit., Moskva, 1974.
7. E. C. Titchmarsii, Extensions of Fouriers' integral formula to formulae involving Bessel functions, *Proc. London Math. Soc.*, 1925, Vol. 23, ser. 2, pp. XXII–XXIV.
8. J. Mroczka, Method of moments in light scattering data inversion in the particle size distribution function, *Optics Communications*, 99:147 (1993).
9. J. Mroczka, Metrological problems of scattered light application in particle sizing, The Third International Congress on Optical Particle Sizing, August 23–26, 1993 Yokohama, Japan.

ANALYSIS OF MODE SWITCHING OPERATION OF LASER DIODE BY PHASE CONJUGATE LIGHT INJECTION

Tsutomu Shimura, Masayuki Tamura, and Kazuo Kuroda

Institute of Industrial Science
University of Tokyo
Minato-ku, Tokyo 106 Japan

INTRODUCTION

Injection locking of lasers with mutually pumped phase conjugate mirror has various advantages compared with that using conventional optics such as mirrors and lenses.[1] In this configuration, we need no alignment to inject the light into the cavity of the slave laser, and high coupling efficiency is easily achieved. We have demonstrated injection locking and mode switching operation of 780 nm diode lasers with double phase conjugate mirror[2]. When we tuned the frequency of the injected light and it is close enough to one of the cavity modes of the slave laser, the injected mode started oscillating and the free running mode was suppressed. Under some condition, the free running mode was completely disappeared and only the injected mode was oscillating. In this paper we analyze the mode switching characteristics of this system with the van der Pol equations.

INJECTION LOCKING WITH DOUBLE CONJUGATE MIRROR

We have demonstrated injection locking of diode lasers using photorefractive double phase conjugate mirror. When we tuned the frequency of the master laser and it was close enough to one of the slave laser modes which was not oscillating without light injection, this mode started oscillating. A frequency diagram of the slave laser modes and injection light is shown in Fig.1. The power of the free run mode, which we will call it mode 1, decreased according to the growth of the injected mode, which we will call it mode 2. Oscillating powers of the modes 1 and 2 are determined by the detuning between mode 2 and the injection light, and the injection power. The power of mode 2 increases according to the increase of the power of the injection power and the decrease of the detuning between injection light and the mode 2. We confirmed the mutual coherence between the output beam of mode 2 and the injected light by observing the interference fringe made by these two beams.

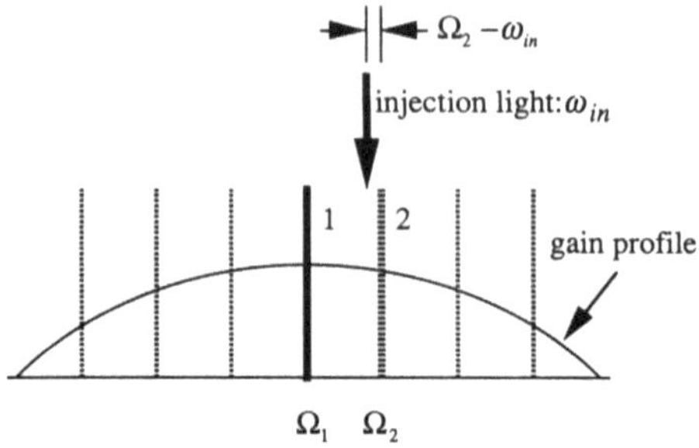

Figure 1. Frequency diagram of the slave laser modes and injection light. Frequencies of free run and the injected mode are Ω_1 and Ω_2, which are not equal to the laser frequencies for these modes, ω_1 and ω_2 in general. The frequency of the injection light ω_{in} is tuned to Ω_2.

MODEL FOR ANALYSIS

We start the calculation with simultaneous van der Pol equations for two modes of the slave laser as[3],

$$\begin{aligned}\frac{d|E_1|}{dt}&=\left(\alpha_1-\beta_1|E_1|^2-\gamma_1|E_2|^2\right)E_1\\ \frac{d|E_2|}{dt}&=\left(\alpha_2-\beta_2|E_2|^2-\gamma_2|E_1|^2\right)E_2+\kappa_2|E_{in}|\cos\Theta\end{aligned} \tag{1}$$

where,

$$\Theta\equiv\sin^{-1}\left[\frac{2\tau_p|E_2|(\Omega_2-\omega_{in})}{\kappa_2|E_{in}|}\right].$$

Here, E_1 and E_2 are amplitudes of the optical field for the free run and injected mode of the slave laser, and E_{in} is that of the injected light from the master laser. We add a driving term, which represents the effect of the injected light to the slave laser, to the second equations. Here α is gain coefficient, β is gain saturation coefficient, γ is the cross saturation coefficient that causes the mode competition, κ_2 is the coupling coefficient of the injected light to the mode 2, τ_p is the photon life time of the cavity, and Ω_2 and ω_{in} are the frequencies of the cavity mode 2 and the injected light. Subscript 1 and 2 denotes the free run mode and the injected mode. We can see the value of the driving term in the second equation determines the oscillation intensities of the two modes. By solving the simultaneous equation (1) for steady state condition by setting,

$$\frac{d|E_1|}{dt}=\frac{d|E_2|}{dt}=0,$$

we can obtain the amplitude for both modes 1 and 2.

NUMERICAL RESULTS

The equations are solved numerically with changing two parameters, injection light intensity $\kappa_2^2|E_{in}|^2$ and detuning parameter Θ. We consider the case that only the mode 2 oscillates when no light is injected to the slave laser. It means the gain for the mode 2 is slightly larger than that for the mode 1.

Results of calculation for fixed detuning is shown in Fig. 2. We set the parameters as $\alpha_1 = 1.2$, $\alpha_2 = 0.9$, $\beta_1 = \beta_2 = 0.9$, $\gamma_1 = \gamma_2 = 1.0$. The crossing point of the two lines is the solution. When the two lines do not cross, we analyze the stability of the solution and find an operating point that a black dot indicates. For large detuning, only the mode 1 is oscillating (Fig.2a). Then the mode 2 starts oscillating as the detuning is getting smaller and exceeds a threshold (Fig.2b). When injection intensity becomes smaller and exceeds next threshold, the mode 1 win the competition and the mode 2 stop oscillating (Fig.2c).

We also examine the oscillation power for each mode when the detuning parameter Θ is fixed with changing the injection power. For small input power only the mode 2 oscillates (Fig.3a), and as the injection power is decreasing, mode 1 starts oscillating (Fig.3b), and finally only the mode 1 survives(Fig3c). These numerical results agree with our experimental results qualitatively.

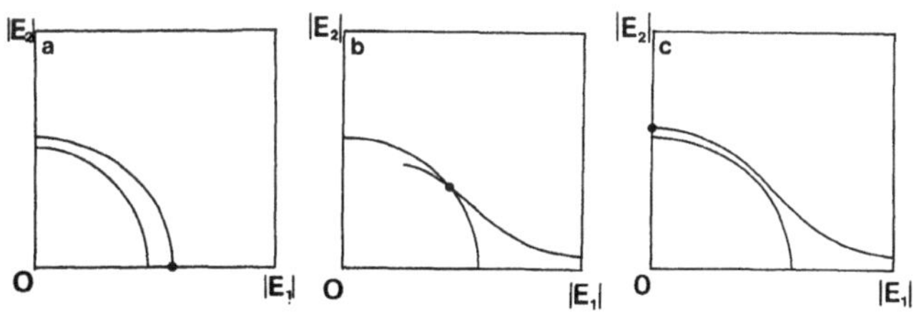

Figure 2. Numerical results of the amplitudes of two modes with light injection. Detuning parameter for each calculation is (a) $\Theta = 1.57$,.(b) $\Theta = 0.5$, (c) $\Theta = 0$, respectively. Injection amplitude parameter is $\kappa_2|E_{in}| = 0.35$.

SUMMARY

We examined injection locking operation accompanying with mode switching. The experimental results are well explained by the simultaneous Van der Pol equations with an injection term. Measuring the detuning and the injection power which gives same oscillation condition, we can compare the coupling coefficient of the injection light to the slave laser κ_2 among different light injection systems.

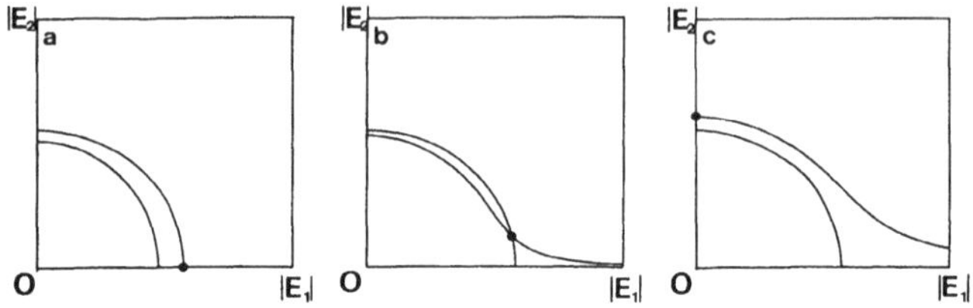

Figure 3. Numerical results of the amplitudes of two modes with light injection for fixed Θ. Injection amplitude is (a) $\kappa_2|E_{in}| = 0$, (b) $\kappa_2|E_{in}| = 0.1$, (c) $\kappa_2|E_{in}| = 0.5$. Detuning parameter is $\Theta = 0.01$.

References

1. M. Segev, S. Weiss, and B. Fischer, Coupling of diode laser arrays with photorefractive passive phase conjugate mirrors, *Appl. Phys. Lett.* 50:1397 (1987) .
2. T. Shimura, M. Tamura, and K. Kuroda, Injection locking and mode switching of a diode laser with a double phase conjugate mirror, *Opt. Lett.* 18:1645 (1993).
3. W. E. Lamb, Jr., Theory of an optical maser, *Phys. Rev.* 134: A1429 (1064).

DESIGN AND PROJECTED PERFORMANCE OF MODIS - A MODERATE-RESOLUTION IMAGING SPECTRORADIOMETER FOR THE EARTH OBSERVING SYSTEM (EOS)

William L. Barnes[1] and Vincent V. Salomonson[2]

[1]Laboratory for Hydrospheric Processes
[2]Earth Sciences Directorate
NASA/Goddard Space Flight Center
Greenbelt, MD 20771

ABSTRACT

MODIS, a 36 band imaging spectroradiometer, is a key component of EOS, NASA's 15-year mission to measure global climate change. MODIS is being developed by Hughes/Santa Barbara Research Center for an initial launch on the EOS AM-1 spacecraft in June 1998. Five subsequent launches are planned on the AM and PM series of EOS spacecraft in order to generate the required 15 years of global coverage. The MODIS design was completed in January 1994, and fabrication of an engineering model is underway with delivery scheduled for February 1995.

The basic requirements for MODIS include 36 spectral bands that range from 0.4 to 14.4 micrometers with bandwidths of 10-50 nanometers in the 20 reflective bands and 50-500 nanometers in the 16 emissive bands. Spatial resolutions of 250, 500, and 1,000 meters at nadir will be scanned across a +\-55 degree swath from a 705 kilometer sun-synchronous orbit to give near two-day global coverage. Requirements for SNR's of several thousand near the equator and NEDT's of < 0.05K at 300K together with dynamic ranges sufficient to enable ocean, land and atmosphere science algorithms result in a twelve bit system with a 24-hour average data rate of 6.1 megabits per second. The MODIS design is a crosstrack scanning radiometer that is 1.11 x 1.61 x 1.17 meters, weighs 234 kilograms and requires an average of 180 watts of power.

The MODIS design has several unique subsystems including a Spectral Radiometric Calibration Assembly (SRCA), a solar diffuser, a Solar Diffuser Stability Monitor (SDSM), and a blackbody calibration target which will allow inflight calibration and characterization of the sensor, thereby enabling the required accuracy of the planned geophysical data products.

INTRODUCTION AND SYSTEM REQUIREMENTS

The National Aeronautics and Space Administration (NASA) has contracted with Hughes\Santa Barbara Research Center (SBRC) for the development of a 36 band Moderate Resolution Imaging Spectroradiometer (MODIS) to be launched aboard the Earth Observing System (EOS)-AM platform in 1998 with up to five subsequent launches on two and one-half year centers aboard the remainder of the EOS AM and PM series of spacecraft.

The MODIS requirements[1] generated by NASA and the scientific community are outlined in Table 1. The 36 spectral bands which were selected and revised by several scientific panels starting[2] in 1984 are listed in Table 2 together with their spatial resolution, spectral bandwidth and primary use. Not listed in Table 1 are the requirements for five years of on-orbit life and an ability to perform spectral, spatial and radiometric characterization of the system post launch. The total set of requirements are designed to enable the at-launch generation of 39 scientific data products ranging from ocean chlorophyll fluorescence and total ozone burden to vegetation indices and, via the EOS series of spacecraft, to generate a 15-year global data set for studying global climate change.

Table 1. MODIS system requirements

Orbit	705 km, 10:30 am descending
Swath	±55°, 2330 km
Spectral coverage	0.407 μm - 14.385 μm, 36 Bands
Spectral resolution	10 nm ≤ $\Delta\lambda$ ≤ 500 nm
Spectral stability	λ & $\Delta\lambda$ Stable to ≤ 2 nm (VIS/NIR Bands) λ & $\Delta\lambda$ Stable to ≤ 1% (All other bands)
Spatial resolution	250 m; 500 m; 1000 m at nadir (502 m; 1003 m; 2006 m; at ±55°)
Spatial registration	≤ 0.1 IFOV
IFOV	0.354 mr, 0.709 mr, 1.418 mr (all ±6%)
Radiometric range	0.002% ≤ρ≤ 100%, 3K ≤T≤ 500K
Dynamic range	NEΔL to Lmax (requires 12 bits)
Radiometric performance	57 ≤ SNR ≤ 1087, @ 70° Solar Zenith, Typical L 0.05K NEDT @ 300K to 2.0K NEDT @ 500K
Polarization insensitivity	≤ 2%, 0.43 μm ≤λ≤ 2.2 μm
Calibration accuracy	Absolute Calibration: ±1%(λ >3 μm); ± 5%(λ<3 μm); ±2% reflectance (λ <3 μm)

MODIS DESIGN

Since June 1988, SBRC[3,4,5] has been developing a MODIS design to address all of the requirements discussed above. This effort culminated in a final design review in January 1994, that demonstrated a system design capable of meeting or exceeding all but a few of NASA's demands. Key parameters from this design are listed in Table 3. The sensor is a crosstrack scanning spectroradiometer that views a swath 2330 kilometers wide by 10 kilometers along-track at nadir in all 36 spectral bands every 1.477 seconds. The incoming radiance is divided via dichroic beam splitters onto four focal planes, two of which are cooled to 85K, and then imaged through bandpass filters onto linear detector arrays of 10 to 40 elements each depending on the required spatial resolution. Dynamic ranges and required signal-to-noises (SNR's) are such that outputs from each of the spectral bands must be digitized to 12 bits. Requirements that have driven the design are a need for all 36 spectral bands to be registered to 0.2 IFOV (instantaneous field-of-view) and a capability for the sensor to setttle within 0.5 percent of a typical land or ocean radiance two kilometers after scanning over the edge of a cloud.

Table 2. MODIS spectral bands

Band	λ	IFOV	Δλ	Purpose (Examples)
	Land and cloud boundaries			
1	659 nm	250 m	50 nm	Veg. Chlorophyll Absorption
2	865 nm	250 m	40 nm	Cloud and Veg. Land Cover Transformation
	Land and cloud properties			
3	470 nm	500 m	20 nm	Soil, Vegetation Differences
4	555 nm	500 m	20 nm	Green Vegetation
5	1240 nm	500 m	20 nm	Leaf/Canopy Differences
6	1640 nm	500 m	20 nm	Snow/Cloud Differences
7	2130 nm	500 m	50 nm	Land and Cloud Properties
	Ocean color			
8	415 nm	1000 m	15 nm	Chlorophyll
9	443 nm	1000 m	10 nm	Chlorophyll
10	490 nm	1000 m	10 nm	Chlorophyll
11	531 nm	1000 m	10 nm	Chlorophyll
12	555 nm	1000 m	10 nm	Sediments
13	667 nm	1000 m	10 nm	Sediments, Atmosphere
14	681 nm	1000 m	10 nm	Chlorophyll Flourescence
15	750 nm	1000 m	10 nm	Aerosol Properties
16	865 nm	1000 m	15 nm	Aerosol/Atmospheric Properties
	Atmosphere/Clouds			
17	905 nm	1000 m	30 nm	Cloud/Atmospheric Properties
18	936 nm	1000 m	10 nm	Cloud/Atmospheric Properties
19	940 nm	1000 m	50 nm	Cloud/Atmospheric Properties
26	1375 nm	1000 m	30 nm	Cirrus Clouds
	Thermal			
20	3.75 μm	1000 m	0.18 μm	Sea Surface Temperature
21	3.75 μm	1000 m	0.05 μm	Forest Fires/Volcanoes
22	3.96 μm	1000 m	0.05 μm	Cloud/Surface Temperature
23	4.05 μm	1000 m	0.05 μm	Cloud/Surface Temperature
24	4.47 μm	1000 m	0.05 μm	Tropospheric Temp./Cloud Fraction
25	4.52 μm	1000 m	0.05 μm	Tropospheric Temp./Cloud Fraction
27	6.72 μm	1000 m	0.36 μm	Mid-Tropospheric Humidity
28	7.33 μm	1000 m	0.30 μm	Upper-Tropospheric Humidity
29	8.55 μm	1000 m	0.30 μm	Surface Temperature
30	9.73 μm	1000 m	0.30 μm	Total Ozone
	Surface temperature/Clouds			
31	11.03 μm	1000 m	0.50 μm	Cloud/Surface Temperature
32	12.02 μm	1000 m	0.50 μm	Cloud Height & Surface Temperature
33	13.34 μm	1000 m	0.30 μm	Cloud Height & Fraction
34	13.64 μm	1000 m	0.30 μm	Cloud Height & Fraction
35	13.94 μm	1000 m	0.30 μm	Cloud Height & Fraction
36	14.24 μm	1000 m	0.30 μm	Cloud Height & Fraction

Table 3. MODIS baseline parameters

Scanning	360° scan, double sided mirror, 20.3 rpm, 2.954 sec period
IFOV	0.354 mr (0.25 km), 0.709 mr (0.50 km), 1.418 mr (1.0 km)
Dwell time	83.3 µs (0.25 km), 166.7 µs (0.50 km), 333.3 µs (1.0 km)
Telescope	2-mirror off axis afocal Gregorian, 4X magnification, EPD 17.8 cm
Reimaging optics	Refractive, EFL 380.8 mm, 282.1 mm (LWIR only)
Spectral	36 bands from 0.41 µm to 14.3 µm, dichroic beamsplitters, bandpass filters
Focal planes	Silicon (0.4 µm $\leq\lambda\leq$ 1 µm), PV HgCdTe (1 µm $\leq\lambda\leq$ 10 µm) PC HgCdTe ($\lambda>$ 10 µm), CTIA readouts for PV, bipolar preamps for PC
Detector sizes	Nominally square: 135 µm (0.25 km), 270 µm (0.5 km), 540 µm (400 µm LWIR) (1.0 km)
Detector cooling	Thematic Mapper-type radiative, T~85K
Calibration	Laboratory, ground truth, on-board blackbody, solar diffuser, SRCA, and SDSM
Data rates	10.6 MBPS (day mode), 3.2 MBPS (night mode) with cal and overhead
Size, weight & power	1.11 m x 1.61 m x 1.17 m, 234 kg, 180 w

Although the sensor as described above is a challenging endeavor, the proposed system's true uniqueness derives from the scientific community's need to generate a 15-year data set with sufficient calibration integrity to allow the detection and quantification of subtle changes in global climate[6]. Consequently, the MODIS design includes four calibration subsystems and a view of space for determining a zero radiance level.

Radiometric calibration of the 16 thermal emission bands is performed by the standard two point method used by AVHRR and other thermal imaging systems. This technique is based on transferring an accurate temperature scale from a ground calibration blackbody to the onboard blackbody, using the space view to establish a zero radiance value, measuring any system nonlinearity and temperature dependence prior to launch and assuming that these functions remain constant for the life of the mission. After launch, the onboard blackbody and space are viewed each scan line, the temperature of the system is measured and the scene spectral radiance is calculated for each band using the functional relationships established pre-launch.

Post-launch radiometric calibration of the 20 solar reflective bands is performed by viewing the sun via a solar diffuser while passing over the North pole. The diffuser is protected from contamination and long-term solar exposure by a deployable door. A second level of solar radiance is generated by deploying a transmission screen between the solar aperture and the diffuser. In order to separate any degradation of the diffuse surface from changes in sensor performance, plans are being generated to allow MODIS to view the moon periodically. The moon serves as a stable solar diffuser, thereby allowing comparisons with the onboard diffuser. The Solar Diffuser Stability Monitor (SDSM) will be used to characterize any spectral change in the solar diffuser. The SDSM alternately views the sun through a two percent transmission screen, the solar diffuser and a black target as illustrated in Figure 1. The incoming radiance is collected in an integrating sphere and measured at 10 of the MODIS visible and near-infrared wavelengths via sets of filters and detectors on the wall of the sphere.

The most complex of the onboard calibration subsystems is the Spectral Radiometric Calibration Assembly (SRCA). Functions performed by the SRCA include spectral, spatial and radiometric calibration/characterization. Major components of the SRCA include the source/filter wheel assembly, the monochrometer assembly and the output collimator. A functional diagram of the SRCA is shown in Figure 2. The system includes three motors. The first of these drives a wheel that includes order-sorting filters for use with the spectrometer, a neutral density filter and a beam combiner that enables measurement of the registration of all 36 bands. A second motor drives reticles and slits that are positioned at the entrance and exit plane of the spectrometer. The third motor steps the diffraction grating during the system-level spectral bandpass measurements and replaces the grating with a mirror during the radiometric and registration measurements. Also included are several Si diodes that are used to monitor and control SRCA output.

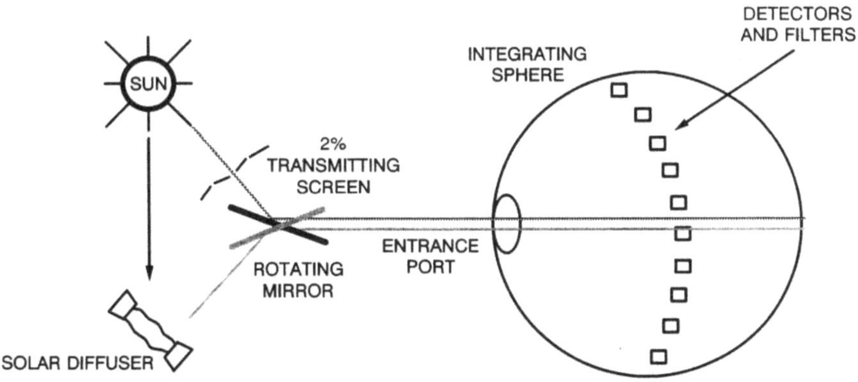

Figure 1. The solar diffuser stability monitor (SDSM)

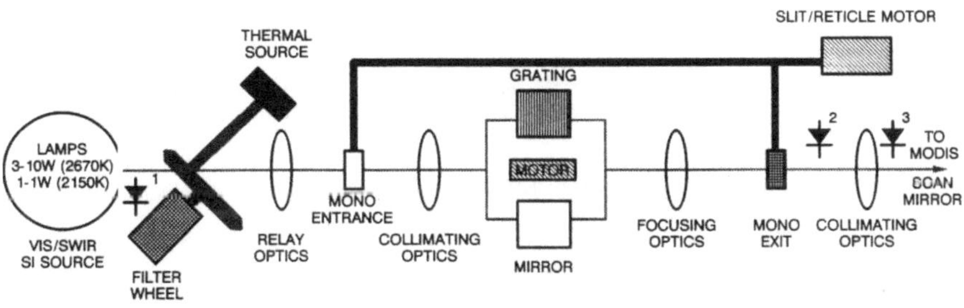

DETECTORS: 1 = SIS FEEDBACK CONTROL, 2 = DIDYIUM SPECTRAL SELF CAL, 3 = SPECTRAL REFERENCE

CALIBRATION MODE	SOURCE	FILTER	MONO ENTRANCE	GRATING/ MIRROR	MONO EXIT	ESTIMATED TIME REQRD
RADIOMETRIC	VIS/SWIR	CLEAR & ND	OPEN	MIRROR	OPEN	20 min
SPECTRAL	VIS/SWIR	ORDER (3)	SLIT	GRATING	SLIT/SiPDs	70 min
SPATIAL	VIS/SWIR & THERM	ITO DICHROIC	OPEN	MIRROR	RETICLES	40 min

* DOES NOT INCLUDE TIME REQUIRED FOR TEMPERATURE STABILIZATION OF HEATED COMPONENTS: THERMAL SOURCE AND SIS FEEDBACK CONTROL DETECTOR

Figure 2. The spectral radiometric calibration assembly (SRCA)

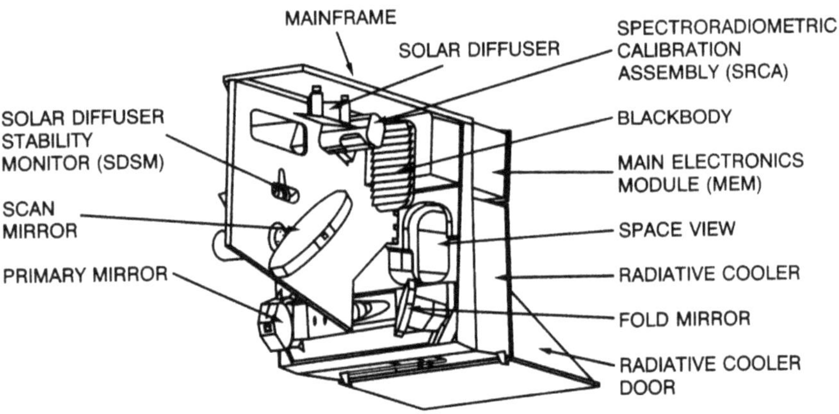

Figure 3. MODIS scan cavity

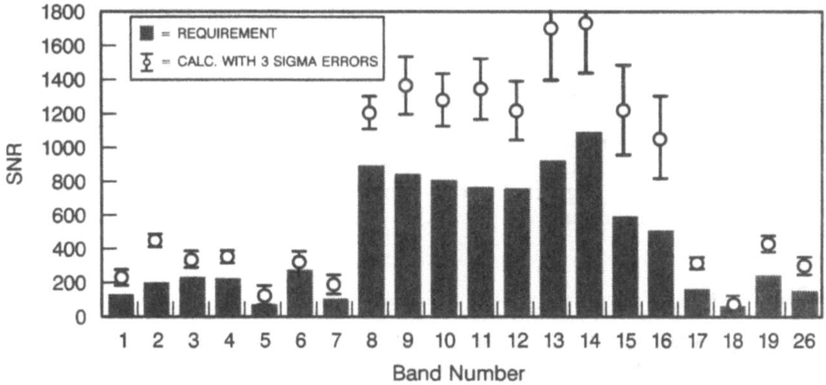

Figure 4. Calculated SNR's for bands 1-19 and 26

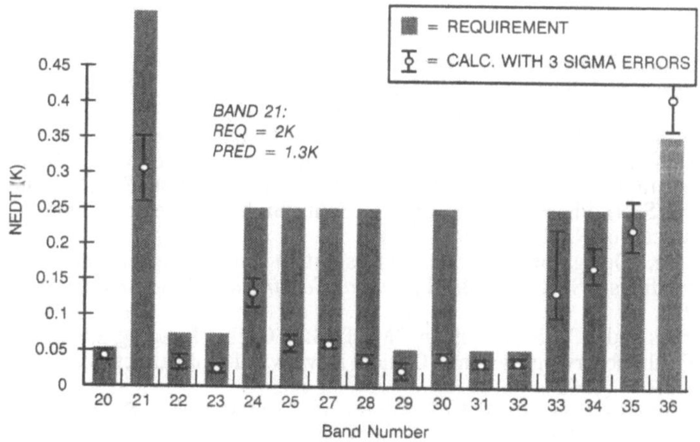

Figure 5. Calculated NEDT's for bands 20-36

The double-sided scan mirror views each of the onboard calibrators and the 110 degree Earth-view swath twice per revolution. A cut-away view of the MODIS scan cavity is shown in Figure 3. Flight is in the direction parallel to the scan mirror's axis of rotation.

MODELED PERFORMANCE

All aspects of the proposed MODIS design have been extensively modeled[7,8,9,10] to establish compliance with NASA's requirements. In addition to the typical signal-to-noise and modulation-transfer models, considerable effort has gone into STOP analysis which is a combination of Structural (NASTRAN),Thermal (NEVADA) and Optical (CODE V) results. STOP analysis is used to insure system stability for the life of the mission.

An example of modeled MODIS performance is shown in Figures 4 and 5 which illustrate expected signal-to-noises (SNR's) for the 20 reflective bands and noise-equivalent-temperature-differences (NEDT's) for the 16 emissive bands. Both figures include 3-sigma error bars. The SNR's of Figure 4 are calculated for a typical radiance at a solar zenith angle near 70 degrees. For the higher solar zenith angles near the equator, the SNR's will be 2-3 times larger.

CONCLUSION

The MODIS design including flight and test hardware and flight and test software is essentially complete. All major subcontracts have been established, piece-parts for both flight and non-flight subsystems are being delivered and assembly of the MODIS engineering model (EM) is underway. The EM will be completed in February 1995 and will undergo extensive testing. Plans are under development to use the mainframe, radiative cooler and scan mirror from the EM on the protoflight model (PM) with delivery of the PM slated for May 1996.

REFERENCES

1. Goddard Space Flight Center, "Specification for the Moderate-Resolution Imaging Spectroradiometer-Nadir (MODIS-N)," GSFC Doc. No. 422-20-02, Goddard Space Flight Center, Greenbelt, Maryland, (1990).
2. W. Esaias and W. Barnes, "Moderate Resolution Imaging Spectrometer (MODIS) Instrument Panel Report," Earth Observing System Reports, Vol. IIb (1986).
3. Santa Barbara Research Center, "Moderate Resolution Imaging Spectroradiometer-Nadir (MODIS-N), Phase-B Final Report," Santa Barbara Research Center, Goleta, CA, Contract No. NAS5-30149 (1989).
4. T.S. Pagano, "Design of the Moderate Resolution Imaging Spectrometer-Nadir," Proceedings ASPRS/ACM/RT, *Remtoe Sensing and Data Acquisition*, 4:374-385 (1992).
5. W.L. Barnes and V.V. Salomonson, "MODIS: A Global Imaging Spectroradiometer for the Earth Observing System," Critical Reviews of Optical Science and Technology, Vol. CR47 (1992).
6. P.E. Ardanuy, D. Han, and V.V. Salomonson, "The Moderate Resolution Imaging Spectrometer (MODIS) Science and Data System Requirements," *IEEE Trans. on Geoscience and Rem. Sensing*, 29 (1991).
7. H.E. Montgomery, H. Ostrow, and G.M. Ressler, "Sensor Performance Analysis," NASA Reference Publication 1241, (1990).
8. K. Forrest, "GSFC Radiometric Model," GSFC Internal Communication, (1992).
9. T. Pagano, "Radiometric Math Model," Santa Barbara Research Center, Goleta, CA, Contract No. NAS5-30800, CDRL No. 101, (1992).
10. T.S. Pagano and J.B. Young, "MODIS radiometric accuracy and sensitivity modeling using the Radiometric Math Model (RMM)," W.L. Barnes, ed., Proceedings SPIE, Vol. 1939 (1993).

MIPAS: A FOURIER TRANSFORM SPECTROMETER FOR LIMB SOUNDING OF THE ATMOSPHERE

André Villemaire, Martin Chamberland, and Jean Giroux

Bomem inc.
450 av. St-Jean-Baptiste,
Québec (Québec) G2E 5S5, Canada

INTRODUCTION

Throughout the years, Bomem get involved in a number of state-of-the-art remote sensing projects from balloons, planes and satellites. MIPAS, a Michelson Interferometer for Passive Atmospheric Sounding is among the recent activities of Bomem. This paper will focus on several aspects of this device designed to measure concentration profiles of stratospheric constituents on a global scale. After this introduction, we present the purpose of MIPAS and its different subsystems. We list the specifications of MIPAS and discuss some details of its different parts. We finally look at the signal processing and the calibration of the instrument.

DESCRIPTION

MIPAS is a limb sounding Fourier Transform spectrometer for the measurement of the thermal emission of the earth atmosphere from the upper troposphere to the lower thermosphere (5–150 km) in the 4.15–14.6 μm spectral range.[1] MIPAS will deliver high resolution (0.035 cm^{-1}) radiance spectra to allow the retrieval of the concentration profiles of a number of important trace gases. These measurements will help a better understanding of the atmosphere in the following research areas:

- stratospheric chemistry: global ozone problem, polar stratospheric chemistry,
- global climatology: global distribution of climate relevant constituents,
- atmospheric dynamics: stratospheric transport phenomena, exchange between troposphere and atmosphere, and
- upper troposheric chemistry: correlation of gas distribution with human activities.

Applications of Photonic Technology
Edited by G.A. Lampropoulos *et al.*, Plenum Press, New York, 1995

MIPAS will make simultaneous measurements of about 20 trace gases including the complete NO_x family as well as several CFC's. MIPAS will fly on ENVISAT–1, the first Polar Orbiting Earth-observation Mission of the European Space Agency. The launch is planned for 1998.

MIPAS can be broken into two main parts, the space segment and the ground segment as illustrated in figure 1. The space segment comprises the front end optics,

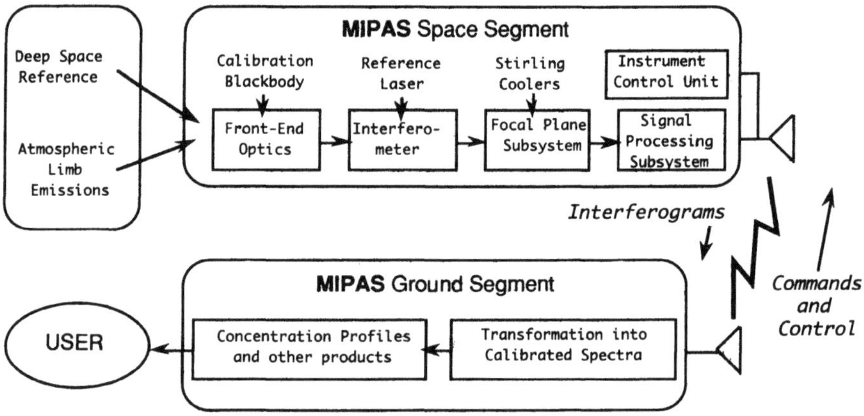

Figure 1. MIPAS functional block diagram

the interferometer, the focal plane subsystem and the signal processing subsystem.[2] The space segment produces interferograms from atmosphere measurements and also from references for calibration purpose. The calibration references are the deep space and an on-board blackbody. The interferograms are sent to the ground segment on earth by an RF link. The ground segment performs the calculations to transform the interferograms into calibrated spectra and to compute the concentration profiles of different constituents. The ground segment also computes the calibration factors from dedicated measurements of the deep space and from an on-board blackbody.

SPECIFICATIONS

MIPAS has the following specifications:

- Instantaneous Field-of-View: 3×30 km^2 (height $\times$ width),
- Elevation pointing: 5 ... 250 km above earth horizon,
- Azimuth pointing: 35° rearward, 30° sideways,
- spectral range: 685 ... 2410 cm^{-1} (14.6 ... 4.15 μm),
- spectral resolution (unapodized): 0.035 cm^{-1} (1 GHz),
- radiometric sensitivity NESR: 70 ... 2 nW/cm^2 sr cm^{-1},
- absolute radiometric accuracy: 1% (14.6 μm) ... 3% (4.15 μm) of input radiance,

The spectral resolution can be selected in the range from 0.035 cm^{-1} to 0.35 cm^{-1}. 4.6 seconds are allowed for the measurement of a spectrum at full resolution while at low resolution (1/10 of full resolution) it is 1 second. These times include the turn-around time.

FRONT END OPTICS

The front end optics comprises the azimuth and elevation scan units for line-of-sight selection, an anamorphic telescope and a calibration blackbody.

The azimuth scan unit selects the azimuth line-of-sight, the field-of-regard (rearward or sideways) and gives access to the internal blackbody. The elevation scan unit is used to select a particular limb for a measurement. The angle covered by this mirror is less than 3° which is sufficient to reach limb heights between 5 km and 250 km. The offset calibration of the instrument is done with deep space measurements corresponding to the 250 km height as pointed with the elevation scan unit. The anamorphic telescope has a magnification of 6 in elevation and 1 in azimuth with an input aperture of 55 × 165 mm^2. The entrance aperture of the interferometer is 55 × 27 mm^2.

Finally, mounted with the azimuth scan unit, there is a calibration blackboby designed to completely fill the instantaneous field-of-view. The blackbody is used for gain calibration of the instrument.

INTERFEROMETER

From the front end optics, the radiant energy is directed to the interferometer. It is a symmetrical dual slide Michelson interferometer with dual input and output ports. Only one of the input ports is used for the measurements; the other is closed with a cold blackbody. The two retroreflectors are moved mechanically on ±5 cm yielding the optical path difference of ±20 cm needed to achieve the required resolution.

The optical path difference measurement system (ODS) is based on a homodyne laser interference. The interference produces fringes and the mechanical motion of the retroreflectors makes the interference pattern to oscillate. The laser used in the ODS is a DFB semiconductor laser operating at 1.3 μm. The speed of the retroreflectors is 2.5 cm/sec giving an oscillation frequency of the interference pattern around 77 kHz. The injection current and the temperature of the DFB laser are stabilized in order to provide a relative frequency stability for the laser of $\Delta f/f = 2.1 \times 10^{-7}$ for a time period of 75 seconds.

FOCAL PLANE

The two outputs of the interferometer are sent to the focal plane. Each beam is split into four spectral channels and distributed to a set of four HgCdTe detectors. The spectral bands are given in table 1. The separation between the channels is done with dichroic

Table 1. Optical range of spectral bands in the focal plane.

Spectral band	wavelength range port 1 [μm]	wavelength range port 2 [μm]
A	10.0 – 14.6	8.4 – 14.6
B	6.5 – 10.0	6.5 – 8.4
C	5.5 – 6.5	5.5 – 6.5
D	4.15 – 5.5	4.15 – 5.5

beamsplitters. All optics and detectors are cooled with Stirling-coolers to reduce their thermal emission.

SIGNAL PROCESSING

On-board signal processing electronics is responsible for the amplification and analog filtering of the detector outputs, the analog-to-digital conversion of each signal (16-bit ADC) and the digital filtering and decimation to reduce the bandwidth of each channel. The signal processing electronics also combines some channels into predefined bands and performs a wordlength reduction as data compression. Finally, it also formats all these data into packets and sends them to the Data Handling & Transmission interface of the platform.

CALIBRATION

The calibration of the instrument consists of several parameters. Radiometric calibration is performed starting with two type of measurements: the deep-space observation for the offset correction and the blackbody measurement for the gain calibration. Offset correction is intended to be performed every 72–144 seconds while gain calibration will be done once a week. Another calibration is required for the exact line-of-sight determination. The measurement for this calibration is performed by looking at predetermined stars entering and leaving the field-of-view of the instrument. Finally, it is also planned to make periodic measurements to determine the instrument line shape.

CONCLUSIONS

MIPAS for Envisat-1 mission presents a new type of spaceborne instrumentation that promises to deliver important data about the composition of the middle atmosphere and upper troposphere. The processing of the data acquired by the instrument is partially done on-board and the rest of the calculations is performed at the ground segment. We briefly described the parts of this Michelson interferometer. To achieve the required precision in the spectra measured by this instrument, periodic calibration measurements are needed.

REFERENCES

1. M. Endemann, G. Lange and B. Fladt, "Michelson interferometer for passive atmospheric sounding - a high-resolution limb sounder for the European Polar Platform ," *Passive Infrared Remote Sensing of Clouds and the Atmosphere*, D. Lynch, SPIE Proceedings vol. 1934, Orlando, April 1993.
2. P. de Zoeten, R. Maurer and R. Birkl, "Optical design of the Michelson interferometer for passive atmospheric sounding," *Passive Infrared Remote Sensing of Clouds and the Atmosphere*, D. Lynch, SPIE Proceedings vol. 1934, Orlando, April 1993.

A POLARIMETRIC HYPERSPECTRAL IMAGING SENSOR

Li-Jen Cheng

Jet Propulsion Laboratory
California Institute of Technology
Pasadena, California 91109

ABSTRACT

This paper reports recent field test results of a polarimetric hyperspectral imaging prototype sensor using a noncollinear acousto-optic tunable filter as a wavelength sorter and a polarizing beam splitter. The objective of this work is to evaluate the AOTF-PHI technology for a variety of remote sensing applications.

INTRODUCTION

Acousto-optic tunable filter (AOTF) is a high resolution, rapidly tunable spectral bandpass filter that uses the diffraction of an incident light beam at a moving grating produced by an acoustic wave in a birefringent crystal.[1] This device has a large aperture and is suitable for imaging applications. The selection of operating wavelength is done by tuning the frequency of a RF power supply connected to a transducer mounted on the crystal. Therefore, the filter can be operated in a sequential, random (wavelength hopping), or multiple wavelength mode with a wavelength switching time in the order of 10 μsec, providing an operational flexibility. Furthermore, AOTFs can provide high spectral resolution ($\lambda/\Delta\lambda$) of 10^2-10^4. This high resolution capability gives opportunities to characterize materials through the remote sensing of reflection, absorption, or emission spectrum. With proper optics and focal plane arrays, one can develops a polarimetric hyperspectral imaging (PHI) sensor capable of providing two spectral image cubes of polarization orthogonally to each other.

In the past, we developed two AOTF-PHI breadboard systems using TeO_2 AOTFs, one operating in the 0.48-0.76 μm range and the other in the range of 1.2 to 2.5 μm. Their performance was investigated in the laboratory environment.[2,3] A visible/near-infrared PHI prototype sensor using a TeO_2 AOTF was recently developed[4] and tested in outdoor field experiments.[5] The sensor was put together with the out-of-the-shelf items. Therefore, its performance is far from the ideal. However, it is capable to provide needed information for illustrating the potential of the AOTF technology for remote sensing applications.

OBSERVATIONS

Figure 1 gives a grayscale image of the test scene in Ft. Huachuca, AZ that was taken with an ordinary 35 mm camera. The scene contained a variety of typical southern Arizona plants and a small number of Army exercise facilities. There were mesquites at the bottom of the image. Above mesquites, there were groups of trees with open areas among them. Some trees were oaks. Beyond the tree-rich region, there was a wide open area with reddish soil, dry grass, a dirt road, bushes, yuccas, and two trees. In the top part of the scene, there were cascaded ridges with a mixture of green trees and yellow dry grass. Some Army facilities and tents were located in the area at the boundary of the tree-rich area and the open area. They included a disk antenna and a pole antenna. The boundary was about 3.1 km away from the instrument. In addition, there was a reflection reference plate of $BaSO_4$ observable as a small bright rectangular object in the upper left part of the picture, that was used for intensity normalization. The distance between the plate and the instrument was about 3.6 km. The observation was carried at the noon time of a sunny day in the spring of 1993. The atmosphere was clear with observable haze only when one looked at distant objects. The instrument directed at the north with 12 degrees off to east.

RESULTS

Vegetation Signatures

Figure 2 gives a spectral image of the scene at 0.67 μm at which the chlorophyll absorption is at maximum. Consequently, oak trees appeared to be dark, whereas mesquites were in gray. In general, all dark and gray areas were related to vegetation. The bright areas contained mainly dry grass. Figure 3 gives reflected intensity spectra of oak and mesquite located in the lower left part of the scene, illustrating that the reflectance of oak is considerably lower than that of mesquite in the 0.62-0.69 μm wavelength range, consistent with the observed image.

The observed spectra also show a sharp rise in reflectance between 0.69 and 0.73 μm, known as the "red edge". The slope and position of the red edge have been directly correlated with leaf chlorophyll concentration.[6] The sharp rise provides an effective way to map vegetation using spectral derivatives. Figure 4 gives a spectral derivative image of the scene at 0.71 μm. Bare ground and dry grass are often bright objects in the intensity image. However, they often do not have chlorophyll. So, their reflection does not have a strong spectral dependence in this wavelength range. Consequently, they are dark in the spectral derivative image. The brightness in the image relates to leaf chlorophyll concentration. Therefore, the spectral derivative image is an effective approach for monitoring health state and stress response in vegetation.

Aerosol Scattering Effects

The existence of atmospheric boundary-layer aerosols affects the magnitude, shape, and polarization of the spectra. The effects become substantial in the observation of distant objects. The measured differences in intensity, spectrum, and polarization due to the distance can provide important information about the aerosol. Using spectral and polarization data of two oak trees located at different distances from the instrument, we obtained the wavelength dependence of the scattering in the air at the observation site.[5]

In general, the scattered light spectrum due to the molecules and aerosol in the atmosphere is expected to a negative wavelength power dependence.[7] Therefore, we obtained a smooth inverted wavelength power function from the observed spectra using an iterative approach. Figure 5 gives the observation data with fitting descending curves of $200.7\lambda^{-2.043}$ for vertical polarization and $56.4\ \lambda^{-2.48}$ for horizontal polarization, where λ is the wavelength. The obtained λ dependence is consistent with the clear atmosphere of the day.[7]

Polarization Signatures

Polarization is an important parameter for analyzing signatures of objects obtained from a remote sensing instrument. Each AOTF observation creates two spectral image cubes with polarization orthogonal to each other. The data reported here are differential polarization defined as the difference in intensity at one corresponding pixel of the two polarization images divided by the sum of the intensities.

Figure 6 gives measured polarization spectra of oak, mesquite, and a unknown tree for illustrating some interesting spectral polarization features of tree leaves. It should be noted that the measured spectrum contains two components. One is due to the polarization created by various scattering mechanisms in the atmosphere. This scattering component is a monotonically decreasing smooth function of wavelength as illustrated in Figure 5. Therefore any significant structure observed in the polarization spectra shall relates to the object property.

Canopies have polarization signatures.[8,9] The spectral dependence of the signatures has not been studies extensively. The data in Figure 6 have revealed that leaves of these trees have a similar polarization spectral signature, namely a sharp descending step at 0.69 μm beyond which the signal is low. The observation suggests that the sharp decreasing feature could be related to the red edge.

Figure 7 gives a spectral polarization image at 0.56 μm. The image illustrates two important observations. All the vegetation have polarization signatures and the measured polarization signal increases with distance. The latter is due to aerosol scattering . This component increases with the observation distance.

Object Detection

There were several military facilities in the boundary area between the tree and open areas. Here we reported only the detection of a camouflaged target located near the center of the scene. We located the target as a dark pattern near the middle of the polarization images at wavelengths around 0.56 μm as shown in Figure 7. Figure 8 gives enlarged positive and negative images. The target pattern also appeared in the secondary spectral derivative images in the same wavelength range. The target was composed of several different objects. Each object has its own spectral dependence. Around 0.56 μm, they appeared together as the observed pattern.

In the color picture, the top part of the target is a small diamond-shape open area surrounded by brushes. However, in the enlarged images of Figure 8, only the brushes at

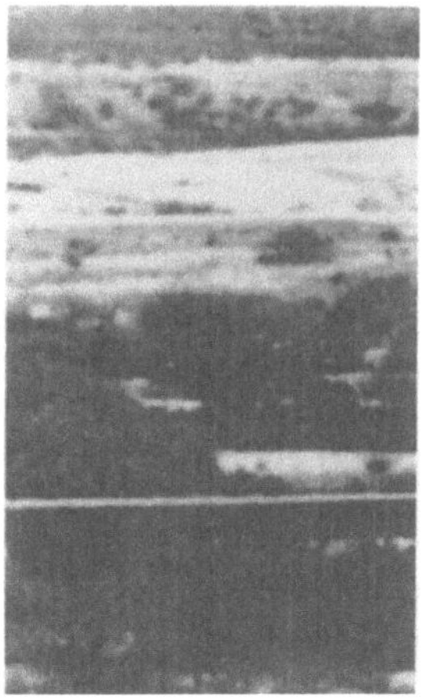

Figure 1. Grayscale image of the scene.

Figure 2. spectral image at 0.64 μm.

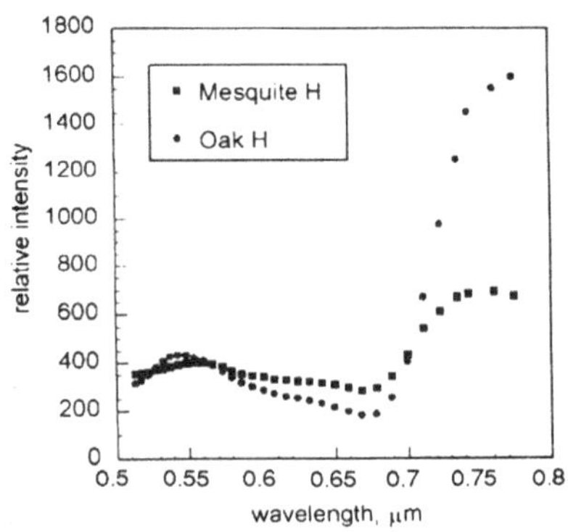

Figure 3. Reflectance spectra of oak and mesquite with horizontal polarization.

Figure 4. Spectral derivative image of the scene at 0.71 μm.

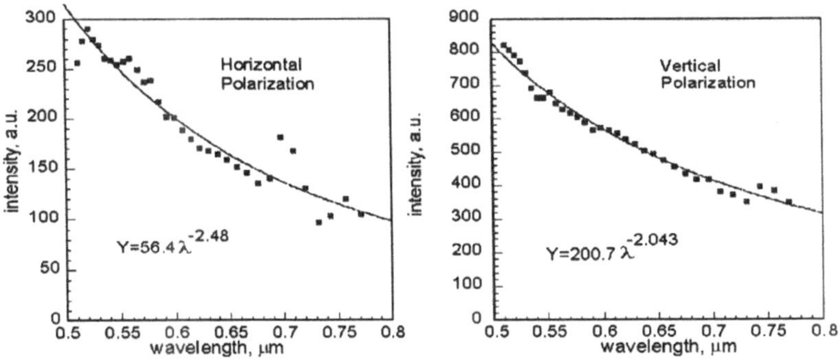

Figure 5. Estimated aerosol-related components of the observed spectra for horizontal and vertical polarization.

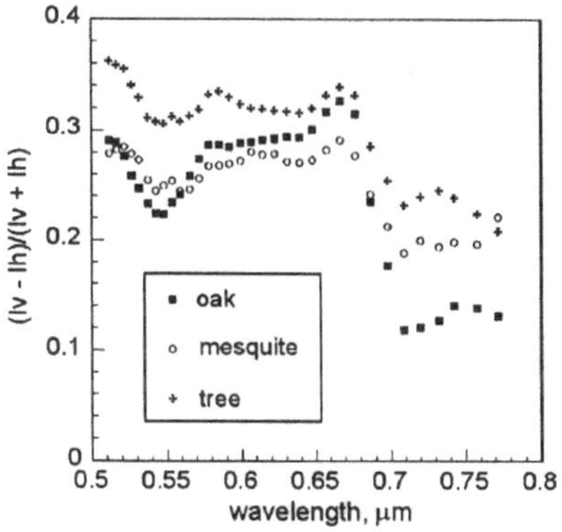

Figure 6. Polarization spectra of oak , mesquite, and unknown tree.

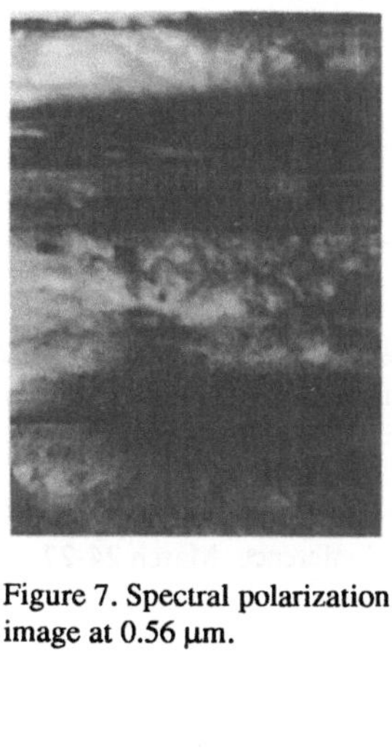

Figure 7. Spectral polarization image at 0.56 μm.

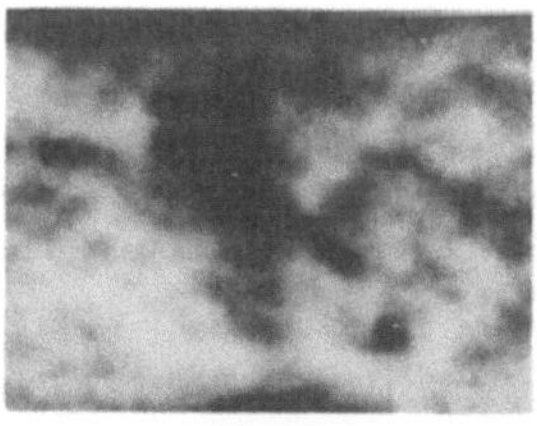
original

inverted

Figure 8. Enlarged positive (upper) and negative (lower) images of the target area.

the left had normal vegetation polarization signal. The rest was two short dark line segments. The darkness of the segments indicated strongly that their origin must be some materials different from those of vegetation. The neighboring area was also darker in comparison with the brush signal. These observations suggest that the line segments and some of their surrounding were man-made objects.

The lower part of the target appeared to be bright green trees with two separated green spherical objects. In the polarization picture, the spherical objects were very dark dots with a dark line between them forming a dumbbell pattern. The dumbbell pattern is likely to be a man-made object. In addition, in the polarization picture, the image of the 'green tree' area was composed with several broad horizontal gray line segments arranged in the shape of an inverted acute triangle.

These contradicting observations between the color and spectral polarization images, we concluded that we detected a camouflaged target area. A detailed observation at the enlarged picture provided a verification that the 'green tree' was green camouflage nets covering unknown objects and neighboring tree tops. The detection was mainly due to the spectral polarization imaging capability of the AOTF-PHI system. It ought to note that the lower part of the pattern was also observable in spectral images of green wavelengths. However, the spectral polarization data provided the evidence for the determination of the target being camouflaged.

CONCLUSION

Our field test results have illustrated the powerful capability of the AOTF-PHI technology for a variety of remote sensing applications, including object detection in the different surroundings. The results also show the complexity of the outdoor environment. In order to take the advantages of PHI, there is a need to understand the fundamental physical phenomena related to the observation and to establish data bases of object signatures and environmental influences.

ACKNOWLEDGMENT

The research described in this paper was performed by the Center for Space Microelectronics Technology, Jet Propulsion Laboratory, California Institute of Technology, and was jointly sponsored by the National Aeronautics and Space Administration, the Planetary Instrument Definition and Development Program, and the Army Space Technology and Research Office.

REFERENCES

1. I.C. Change, "Noncollinear Acousto-Optic Filter with Large Angle Aperture", Appl. Phys. Lett., Vol. 25, p.370 (1974).

2. Tien-Hsin Chao, Jeffrey Yu, George Reyes, David Rider and Li-Jen Cheng, "Acousto-Optic Tunable Imaging Spectrometers", Proceedings of 1991 International Geoscience and Remote Sensing Symposium, Helsinki, Espoo, FINLAND, June, 1991 (IEEE 91CH2971-0), p. 585.

3. Li-Jen Cheng, Tien-Hsin Chao, and George Reyes, "Acousto-Optic Tunable Filter Multispectral Imaging System", AIAA Space Programs and Technologies Conference, March 24-27, 1992, paper no. 92-1439.

4. Li-Jen Cheng, Tien-Hsin Chao, Mack Dowdy, Clayton LaBar, Colin Mahoney, George Reyes, and Ken Bergman, " Multispectral Imaging Systems Using Acousto-Optic Tunable Filter" , in "Infrared and Millimeter Wave Engineering" , SPIE Proceedings Vol. 1874, p. 224 (1993).

5. Li-Jen Cheng, Mike Hamilton, Colin Mahoney, and Geogrge Reyes, "Analysis of AOTF hyperspectral Imagery", To be published in "Algorithms for Multispectral and Hyperspectral Imagery", SPIE Proceedings, Vol. 2231, 1994.

6. For example, D.N.H. Horler, M. Dockray, J. Barber, and A.R. Baringer, "Red Edge Measurements for Remotely Sensing Plant Chlorophyll Contents," Adv. Space Res. 3(2), 273 (1983).

7. E. Schanda, "Physical Fundamentals of Remote Sensing", (Sringer-Verlog, New York, 1986), p.117.

8. P. J. Curran, "The Relationship Between Polarized Visible Light and Vegetation Amount", Remote Sens. Environ., Vol. 11, p. 87 (1981).

9. D.A. Talmage and P.J. Curran, "Remote Sensing Using Partially Polarized Light", Int. J. Remote Sensing, Vol.7, p.47, (1986).

CCDs FOR THE LYMAN FUSE MISSION

R. Murowinski[1], M.J. Deen[2]

[1]Dominion Astrophysical Observatory, National Research Council
Victoria, British Columbia V8X 3X1
[2]School of Engineering Science, Simon Fraser University
Burnaby, British Columbia V5A 1S6

ABSTRACT

Charge Coupled Devices (CCDs) flown in spacecraft are subject to radiation damage from the energetic protons they encounter. This radiation causes degradation in a number of electro-optical characteristics. It is important to understand these effects on a CCD so that performance degradation during a mission lifetime can be anticipated, and so that methods for preventing unacceptable performance degradation can be found. We look at the expected read noise, dark current and charge transfer efficiency performance of the Lyman FUSE CCDs.

1. INTRODUCTION

The Lyman FUSE spacecraft is one which will be subjected to an unusually severe radiation environment. The orbit has a perigee of 600 km and an apogee of 71,000 km. Each orbital period of 23 hours and 56 minutes will see the spacecraft traverse the trapped proton belt twice.

The Fine Error Sensor (FES) on this spacecraft is required to deliver information to the fine pointing system every two seconds so that the instrument pointing can be maintained to within 0.2 arcseconds RMS. The FES CCD needs adequate sensitivity to be able to provide this information with an 85% probability of finding guide objects at any orientation in the sky. In addition, the FES will provide users on the ground with full images of the field for acquisition information.

Proton radiation is known to have many deleterious effects on CCDs: the dark current increases; a number of dark current spikes appear in the image; charge transfer efficiency goes down; and the flat band voltage changes causing the required bias and clocking voltages to change. It is important to understand how these changes will affect the ability of the FES to deliver pointing information to the spacecraft, so that the scientific program schedule can make best use of the spacecraft's abilities.

2. CAMERA DESCRIPTION

The primary mirrors of the telescope are two silicon carbide mirrors and two lithium fluoride (LiF) mirrors, each of which is a normal incidence paraboloid of area ~ 1300 cm^2. Each of the LiF mirrors will feed one of the two FES cameras through off-axis re-imaging optics, resulting in a field of 24 arcminutes square.

The CCD is housed in a dewar and shrouded by 1 inch of aluminum. The CCD attains its operating temperature, in the range of -40°C to -80°C, by means of passive radiators outside the spacecraft and heaters next to the CCD. Electronics in close proximity to the CCD provide clocking and bias voltages, and perform analog signal processing and digitization of the video signal. Guiding data are reduced to the resultant image centroid for each subraster at the collocated FES computer, and then passed to the main computer.

The CCD detective element has a square imaging area with 512 x 512 pixels, each of which have an area of $(20\ \mu m)^2$. The CCD will also incorporate a frame transfer storage section the same size as the imaging area, so that the image integration duty cycle can be kept high while allowing slowscan low noise readout. To meet the sensitivity requirement, the CCD must be a high quantum efficiency and wide optical bandwidth device. A likely candidate is the thinned and backside-illuminated CCD, although transparent gate and thin gate CCDs remain options under study.

In addition to taking entire images, the CCD camera must be able to read up to 6 arbitrary subrasters on the array. It needs to be able to vary the readout rate from 10 kpix/s to 50 kpix/s, and vary the image integration time from 0.1 to 100 s. The pointing requirement of 0.2 arcseconds RMS will be met if the camera can clearly image stars with an integrated signal of a few thousand electrons (e^-) or greater with a noise floor of 10 e^-. Stars of this brightness (about 15th magnitude visually) will occur with adequate probability in the field of the detector.

Software incorporated into the camera system will reduce image noise by means of dark image subtraction and by cosmic ray detection and cancellation, as well as performing the centroiding functions on each of the subrasters.

3. RADIATION EFFECTS

The proton energy spectrum for this mission has been calculated and described elsewhere[1]. From the spectrum, and using TRIM-91 software[2], we numerically integrated the amount of displacement damage expected in the CCD. With this information, and again using the TRIM-91 model, we are able to predict that the same amount of displacement damage in the CCD would occur with a fluence of 1.0×10^9 protons/cm^2 if the protons were mono-energetic with an energy of 1 MeV.

The CCDs which were damaged are model FA1024 CCDs made by Ford Aerospace[3]. They are frontside-illuminated CCDs with 1024 x 1024 pixels each of area $(15 \mu m)^2$, with an inversion barrier implant but without notch technology. The CCDs were irradiated at the University of Western Ontario Tandetron accelerator, with 5.0×10^8 protons/cm^2 on one half of the imaging area and serial register, and with 2.7×10^9 protons/cm^2 on the other half. In addition to the CCDs, a number of discret MOSFETs of similar construction to that of a slowscan CCD sense amplifier were damaged for read noise experiments.

3.1 Read Noise

The noise power spectra of the transistors which had been subjected to proton damage were carefully measured. From the noise spectrum, one can determine what the best noise performance would be if the system noise is dominated by the CCD output MOSFET.

Using these spectra, we were able to model the expected noise as a function of dose and temperature[4]. We could also examine the effect of changing the readout rate by examining the noise as a function of the correlated double sample time constant τ_{CDS}.

From these measurements we find an increase in CCD read noise as a result of proton radiation damage. The noise is a result of induced trapping centres creating excess generation-recombination noise. In general, the noise excess is greater at slower read rates and warmer temperatures. We also find that the bias range over which linear behaviour is expected of an LDD MOSFET is reduced in a damaged transistor. This effect would manifest itself in the CCD as increasingly non-linear behaviour of the CCD output as radiation damage progresses, or as a more restrictive range at which the CCD output MOSFET must be biased to retain some linear performance.

Typical results show that the noise versus τ_{CDS} relationship at the coldest measured temperature of -65°C behaves as expected: the noise is proportional to $\tau_{CDS}^{0.5}$. At warmer temperatures the excess noise of a generation-recombination centre becomes evident as a change in this relationship. We find that for transistors with as much displacement damage as they would receive in 2.5 orbital years, as temperatures approach 0°C or warmer the noise is greater than that of the $\tau_{CDS}^{0.5}$ dependance. Rather than decreasing, in some cases the noise will actually increase as τ_{CDS} increases and the CCD is read out more slowly. A more complete description of these measurements can be found in the literature[4].

3.2 Dark current

Previous measurements[5] of the dark current I_{dark} have found a linear relationship between mean I_{dark} and dosage. They have also shown that the known relationship between I_{dark} and temperature remains valid as damage increases. We can therefore multiply the the known expression for I_{dark} with the proton damage factor to get I_{dark} (in electrons per second per pixel) during the mission

$$I_{dark}=2.5X10^{15}(K+K_{exp}\cdot t)A\cdot T^{1.5}\exp\left[-\frac{E_g}{2kT}\right] \tag{1}$$

where T is the absolute temperature, t is elapsed mission time in years, E_g is the energy gap of silicon, A is the area of a pixel in cm^2, K is the virgin CCD I_{dark} rate in nA/cm^2 and K_{exp} is the rate of dark current increase as a result of exposure to protons. The values we measure are $K = 0.13$ nA/cm^2 and $K_{exp} = 3.62$ $nA/cm^2/year$ in normal mode, while $K = 0.026$ nA/cm^2 and $K_{exp} = 1.81$ $nA/cm^2/year$ in inverted mode. Note that these values are from experiments based on predictions for the FUSE orbit and detector shielding. They would certainly change with other orbits or shielding.

The dominant noise source in a heavily damaged CCD is often the I_{dark} noise. One expects that much of this spatial noise is a result of dark current spikes with fixed locations on the CCD, and could be reduced by means of dark current subtraction.

We measured the total I_{dark} spatial noise in a damaged CCD at fixed temperature and a range of dark exposure times. Two frames of identical dark exposure times were taken for each integration time, as well as a zero second "bias" exposure. The experiment was performed with the CCD operating in inverted and in normal (non-inverted) modes. In Figure 1 we see the noise in these frames as a function of mean dark current charge Q_{dark} in the exposed frame. The solid lines represent noise measured in the [dark1 - bias] frames, and represent the spatial variation of Q_{dark} in the image. The dashed lines are the resultant noise of the [dark1 - dark2] frames. The reduced levels of these lines confirms the spatially-fixed nature of I_{dark} noise. The lower line shows the noise expected from the [dark1 - dark2] frames if the noise were only subject to Poisson statistics in both the temporal and spatial dimensions.

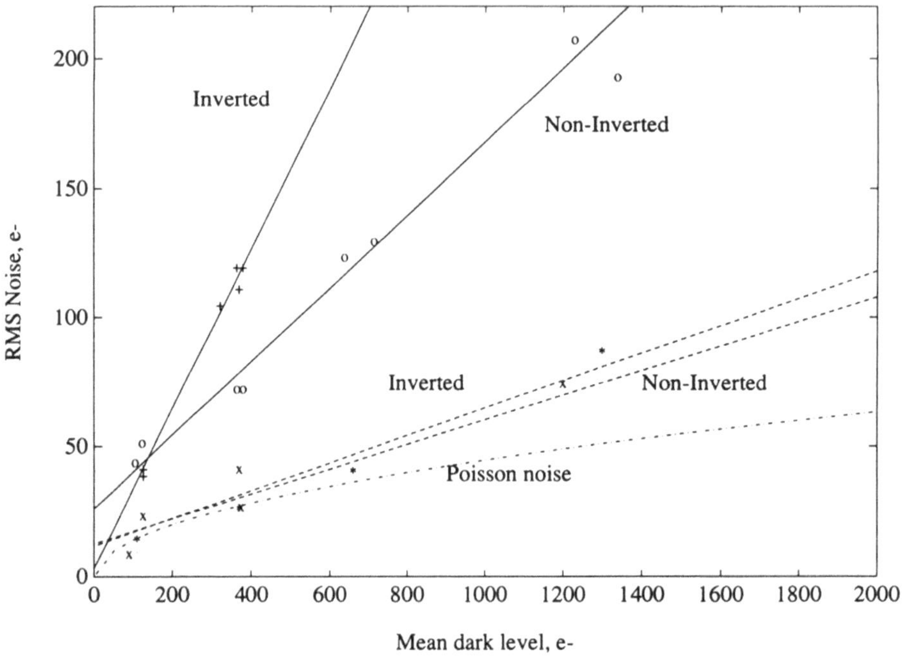

Figure 1. Dark current noise as a function of total dark signal. The solid lines have had bias exposures subtracted, the dashed lines have had 2nd dark frames subtracted. The lower line indicates Poisson noise.

The noise resulting from dark current appears to be proportional to mean dark current, and could be expressed as

$$\sigma(Q_{dark}) = C \cdot \langle Q_{dark} \rangle \tag{2}$$

where the constant C changes with mode of operation and signal processing. When the images are bias-subtracted, C is found to be 0.30 for inverted CCDs and 0.14 for non-inverted CCDs. When the images are dark-subtracted, C is found to be 0.04 in both modes. Clearly, dark frame subtraction is an effective technique for reducing image noise.

It should be noted that the dark frame subtraction is most effective when the dark current reference frame is of the same length as the exposed frame. Attempts to improve the noise cancellation by taking a longer reference frame, which was then scaled to the exposed frame's time, were counterproductive: they did not cancel noise as effectively. This is due to the nonlinear processes of charge transfer efficiency (CTE) in the damaged CCD, as discussed below. A further effect we noted was that the dark current image spatial pattern was not stable. When the dark current image is recorded over periods of many hours, one can see the image 'mutating': new hot pixels occurring and others previously hot becoming less so. It is not clear what proportion of the pixels are changing; it appears that the majority are stable but some few percent are still transitory even a year after the CCD was damaged.

3.3 Charge Transfer Efficiency

Loss of charge in a proton-damaged CCD, or in fact any CCD, is a more complex phenomenon than the simple proportional process implied by oft-quoted CTE figures. In addition to the losses which may occur in charge diffusion as packets are clocked between gates, charge is trapped in a number of trap energy levels within the silicon bandgap. Each trap level has its own characteristic emission time constant and capture time constant, which are different functions of temperature. The amount of charge lost from a pixel is therefore a function of number of traps, energy level of the traps, temperature, clock period, clock overlap, size of packet and the history of the pixels a packet must traverse to arrive at the output. A number of recent papers[6,7] attempt to find order among these variables.

We chose to focus our CTE measurements on experiments which we thought would give us the information most relevant to this mission. One experiment examined the parallel charge loss per row in the case of bright stellar objects on a dark field, very much like the situation the spacecraft FES will most often encounter. The experiment was conducted at -80°C, the lower extreme of the operating temperature range and one likely to present the worst trapping losses.

Figure 2 shows how the charge loss in the parallel direction changes with signal size. In Figure 2(a) the field remains dark through the measurement while the stellar brightness is gradually increased. In Figure 2(b) the stellar brightness remains constant and the field is gradually brightened. This latter measurement simulates the effect of using a pre-flash in the system, increased dark current levels or finding a guide star within an extended object. From the data in Figure 2(a) and in the dark field case, we find that at the 2.5 year damage level suffered by this CCD, the number of electrons lost per row L is

$$L=54\left(1-\exp\left[-\frac{S}{16000}\right]\right) \qquad (3)$$

where S is the average signal in the stellar image, in electrons.

The stellar image is also degraded by the smearing and apparent position change caused by CTE. This effect results in a digitized image which is different from the incident light distribution. Since the image changes are repeatable frame after frame, they should not to first order affect the ability to guide on an object.

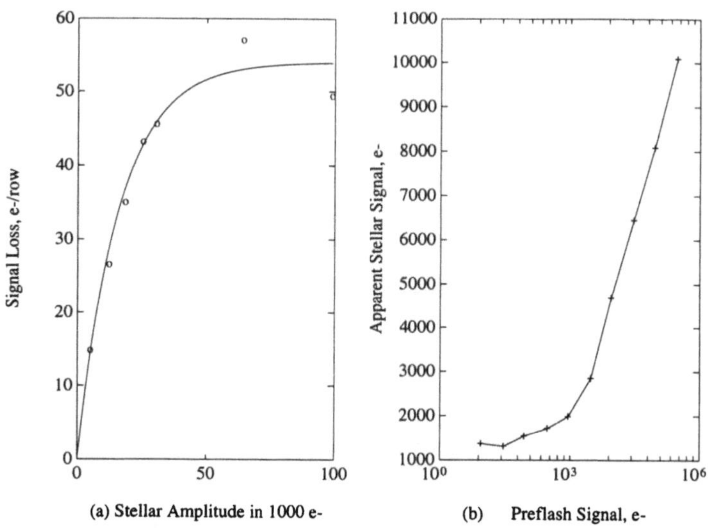

Figure 2. Charge transfer losses when imaging compact stellar-like objects in a dark field, at -80°C and with 2.5 years damage. (a) The amount of signal lost per row in the parallel direction as a function of stellar brightness. The solid line is Equation (3) above. (b) The effect on transfer efficiency by increasing the background field brightness by means of a preflash signal.

4. PERFORMANCE MODEL

A simple empirical model can be created with the above information, which can be used to predict how the CCD signal and noise will change as radiation damage progresses in the FUSE orbit and shielding, and operating at -80°C. Other orbits or shielding would require appropriate scaling. At that temperature, one expects the read noise to remain uncorrelated and the image noise N in electrons for exposure of time τ seconds should be

$$N=\sqrt{N^2_{read}\frac{R_{read}}{5x10^4} \quad + \quad 2.5x10^{15}(\ K+K_{\exp}\cdot t\ )C\cdot\tau\cdot A\cdot T^{1.5}\ \exp\left[-\frac{E_g}{2kT}\right]} \qquad \textbf{(4)}$$

where N_{read} is the read noise at 50 kpix/s and R_{read} is the read rate of the system in pix/s.

The signal expected at -80°C, when imaging compact objects in a dark field would be

$$S=I_0\cdot QE\cdot\tau\left[1-N_{row}\cdot A\cdot t\left(1-\exp\left(\frac{-I_0\cdot QE\cdot\tau}{t\cdot B}\right)\right)\right] \qquad \textbf{(5)}$$

where I_0 is the incident flux in photons/s and QE is the quantum efficiency. A and B are empirically found above to be A = 21.6 $(\text{row}\cdot\text{year})^{-1}$ and B = 6400 electron·year·seconds. Confidence in the results of these models is not better than a factor of two.

5. ACKNOWLEDGMENTS

We would like to thank Dick Bredthauer of Loral Aerospace for the test CCDs, and Rob Lowles at ComDev for system design details. We would also like to thank John Hutchings and Chris Morbey at the Dominion Astrophysical Observatory for useful discussions and encouragement.

6. REFERENCES

1. G. Stassinopoulos and J.M.Barth, "FUSE Elliptical Orbit Radiation Environment," NASA/Goddard publication X-900-91-09.

2. J. Ziegler, J. Biersack, and U. Littmark, "The Stopping and Range of Ions in Solids," Pergamon Press (1985).

3. J. Janesick, T. Elliot, R. Bredthauer, J. Cover, R. Schaefer, R. Varian, Recent developments in large area scientific CCD image sensors, *SPIE* Vol 1071:11 (1989).

4. R. Murowinski, G. Linzhuang, M. Deen, Effects of space radiation damage and temperature on the noise in CCDs and LDD MOS transistors, *IEEE Trans. on Nuclear Science* Vol 40 No 3 (1993).

5. R. Murowinski, G. Linzhuang, M. Deen, Effects of space radiation damage and temperature on CCD noise for the Lyman FUSE mission," *SPIE* Vol 1953:71 (1993).

6. J. Janesick, G. Soli, T. Elliot and S. Collins, The effects of proton damage on charge-coupled devices, *SPIE* Vol 1147:87 (1991).

7. I. Zayer, I. Chapman, D. Duncan, G. Kelly, K. Mitchell, Results from proton damage tests on the Michelson Doppler Imager CCD for SOHO, *SPIE* Vol 1900:97 (1993).

THE MEASUREMENTS OF POLLUTION IN THE TROPOSPHERE (MOPITT) INSTRUMENT

James R. Drummond, G.V. Bailak, G. Mand

Department of Physics, University of Toronto,
Ontario, CANADA, M5S 1A7

INTRODUCTION

The earth's environment is changing through natural and artificial mechanisms. Monitoring these changes on a global basis is becoming an international priority. The troposphere, being the area of the atmosphere which we inhabit, is of great importance although it is extremely difficult to make measurements of this region from space due to the interfering effects of clouds and the nearby surface. However the state of the troposphere, and particularly its chemistry is of considerable and immediate importance. The MOPITT instrument will use the principle of correlation spectroscopy to measure carbon monoxide (CO) amounts at three levels in the troposphere utilising thermal radiation at 4.7μm, and the total column amount of CO and methane (CH_4) using reflected sunlight around 2.3μm. The required accuracy is 10% for the CO measurements and 1% for the CH_4 channels. These figures pose a considerable challenge to the instrument engineering.

INSTRUMENTATION

Correlation spectroscopy uses a sample cell containing the gas to be measured in the path of a simple radiometer and modulates the amount of gas in the cell. The resulting signal can be processed to show enhanced sensitivity to radiance at optical frequencies close to the spectral lines of the gas in question. The MOPITT instrument uses two methods of gas modulation: pressure modulation (Taylor, 1983) and length modulation (Drummond, 1988) to achieve its objectives.

In addition to the use of correlation spectroscopy, MOPITT also uses cooled (100K) detectors and filters to achieve nearly background limited performance. The detectors in both spectral regions are InSb photovoltaic detectors. The use of InSb at short wavelengths requires careful attention to the long-wave blocking of the filters to avoid excessive photon noise from the long wavelength end of the spectrum. A novel feature of this cooled system is the use of a mechanical cooler (a Stirling Cycle Cooler from British Aerospace Plc) rather than a stored cryogen or a radiation cooler. The use of a mechanical system provides a large cooling capacity (>1W) with a long life expectancy enabling us to cool the detectors, filters and rear optics of the instrument.

Applications of Photonic Technology
Edited by G.A. Lampropoulos *et al.*, Plenum Press, New York, 1995

The on-orbit configuration of MOPITT is shown in Figure 1. There are four optical channels, each of which is spectrally split into a 4.7μm channel and a 2.3μm channel. Each of the eight channels has four pixels giving a total of 32 simultaneous independent measurements. Each pixel measures 22km x 22km on the earth's surface and the pixels are scanned across-track by ±14views at a rate to give an almost complete measurement set over a swath of ±320km about the sub-satellite track.

An instrument to detect infra-red radiation is very sensitive to its thermal environment and great care has to be taken to ensure that the environment is stable. MOPITT uses a Capillary Pumped Heat Transfer System (CPHTS) to remove the heat from the instrument to a remote radiator plate. The temperature of the CPHTS plate is constant with time and this enables the instrument to set up a stable temperature distribution which results in stable calibrations.

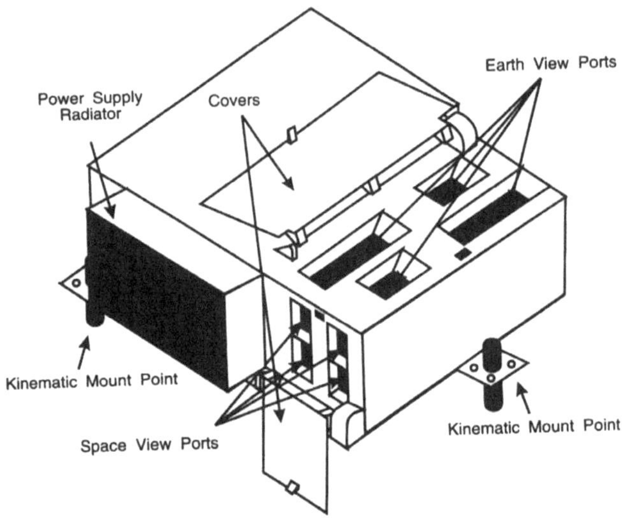

Figure 1 MOPITT on-orbit configuration.

The signal processing of MOPITT uses almost entirely digital technologies to recover two signals corresponding to the total incoming radiation and the difference radiance between the two states of the correlation cell systems. The signal from the detectors is passed through a transimpedance amplifier (TIA) using CMOS technology in the cold section. The resulting high-level signal is transferred to the warm section of the instrument where it is minimally conditioned before being digitised in a delta-sigma analog-to-digital converter at a fairly high rate. The bandwidth of the analog system is determined by the RC combination of the TIA. Further processing of the signal, including the overall bandwidth definition, is then performed in the digital domain. The major advantage of this approach is that the thirty-two signal channels of MOPITT can be easily reproduced with a low part count and high reliability.

CALIBRATION AND CHARACTERISATION

In-flight calibration of the MOPITT instrument is essential and this is provided by blackbody sources which are known to be near-perfect emitters of known temperature. These form one known radiance and a view of "space" provides a second known point for the radiance calibration. In addition to radiance calibration, sensors throughout the instrument monitor temperatures and gas cell pressures to aid in the instrument calibration.

Before launch the instrument will go through a primary calibration and characterisation cycle at the University of Toronto in a purpose-built clean room facility with full vacuum test capabilities. Virtually all of the MOPITT characterisation tests will be performed in a vacuum since this is the operational environment for the instrument and will give a representative thermal environment. A variety of cooled systems surrounding the instrument will mimic the space environment for thermal control through the radiator plates as well as the "space" side view of the calibration.

As well as providing the correct thermal environment the ICF system will be able to support MOPITT for power, commanding, telemetry, cooling and mounting as well as providing a comprehensive measurement capability for the ancillary equipment in the system.

Since the MOPITT instrument will not be accessible to the science team after launch, it is essential that the instrument be fully characterised before launch. Since it is also true that testing often occurs at a very critical point in the program, it is also necessary that the tests be performed quickly and accurately. To this end a test system and a test program is being devised which will fully meet these objectives.

The characterisation tests are divided into a number of categories which determine the equipment required for each test. From these categories a limited number of equipment sets are designed which can perform a number of tests simultaneously or nearly simultaneously. The two major sets of equipment are the Radiometric Test System (RTS) and the Field of View (FOV) Test System (FTS).

The RTS consists primarily of a set of calibrated blackbodies which are used to calibrate the MOPITT internal blackbodies, to verify the noise performance of the system and the system performance as a function of input signal level. In addition, by inserting a test gas cell between the instrument and a calibrated blackbody, the instrument gas response can be evaluated. An adjunct set of equipment for the RTS is a uniform source at 2800K which mimics reflected sunlight from the surface of the planet. Although not an absolute source, the shortwave channels of MOPITT can be calibrated at a spectral input which is much closer to the operational condition than is the case for the calibrated blackbodies which are restricted to temperatures <500K by materials considerations.

The FTS consists primarily of a collimator system large enough to produce a plane beam which covers all the MOPITT FOVs. The collimator is essentially a 2m focal length f/3 reverse telescope with a variety of sources. The most used source will be a 2800K quartz-iodine lamp. The FTS is used to evaluate the FOV of MOPITT and establish the relationship between the FOVs for the various pixels in the various channels. Both the instrument and the collimator can be rotated to map out the responses as a function of input angle and instrument scan angle.

The FTS can also be configured with a monochromator source which enables an end-to-end spectral test to be performed for both the "in-band" (required) signal and the "out-of-band" signal. Because the MOPITT spectral filters are cooled and also slightly temperature sensitive, this test must be repeated at several temperatures.

More details of the MOPITT test equipment and schedule are given in Drummond and Mand (1994).

FLIGHT PROGRAM

The MOPITT instrument will fly on the AM-1 platform of NASA's EOS program in mid-1998. The other instruments on this platform are primarily surface imagers in the visible and infrared (MODIS, MISR and ASTER) and a radiation budget instrument (CERES). The satellite will be launched in a sun-synchronous orbit with a 10:30am equator crossing time. The satellite design life is five years.

ACKNOWLEDGEMENTS

The MOPITT instrument is the result of collaboration between a large number of individuals and agencies. The MOPITT science team consists of: Dr. G.P. Brasseur, National Center for Atmospheric Research (NCAR), Dr. G. Davies, University of Saskatoon, Dr. J.C. Gille, NCAR, Dr. J.C. McConnell, York University, Dr. G.D. Peskett, Oxford University, Dr. H.G. Reichle, NASA Langley and Dr. N. Roulet, McGill University. The MOPITT instrument is funded by the Canadian Space Agency, the US members of the science team by NASA, and the for modelling and simulation studies in Canada by the Natural Sciences and Engineering Council. The prime contractor for the MOPITT instrument is COM DEV Atlantic.

REFERENCES

J.R. Drummond, 1988, A novel correlation radiometer - the length-modulated radiometer, *Applied Optics*, **28**, 2451-2452.

Taylor, F.W., 1983, Pressure Modulator Radiometry, Spectrometric Techniques, Vol. III, Academic Press, 137-197.

J.R. Drummond and G.S. Mand, 1994, The measurements of pollution in the troposphere, MOPITT instrument: overall performance and calibration requirements, submitted to *Journal of Atmospheric and Oceanic Technology*, February, 1994.

AOTF-BASED FOREST FIRE SENSOR: OPTICAL DESIGN

A.B. Hollinger

Canadian Space Agency
6767 rue de l'Aeroport
St. Hubert, P.Q. J3Y 8Y9

P.J. Thomas, R.H. Wiens

Institute for Space and Terrestrial Science
4850 Keele Street
North York, Ont., M3J 3K1

and

E.H. Richardson

EHR Optical Systems
1871 Elmhurst Place
Victoria, B.C, V8N 1R1

INTRODUCTION

The detection of small objects against a highly cluttered background is a scenario with a wide range of practical applications. The focus here is the detection of small incipient forest fires that may be hidden by the forest canopy[1,2]. Multispectral analysis with adaptive signal processing can facilitate the reduction of false alarm rate and enhance the probability of detection for a small object detection system.[3,4]

In the infrared, the spectral features of interest are often broad and well separated. Hence, information from only a few spectral channels is generally required for a specific application such as forest fire detection. Incipient forest fires can have temperatures in the range 600 - 1000 K which will produce substantial infrared radiation at wavelengths near $4\mu m$ provided one avoids the regions of atmospheric absorption.

We examine the suitability of an acousto-optic tunable filter to the problem of detecting incipient forest fires. An Acousto Optical Tunable Filter (AOTF)[5,6,7] is well suited to the acquisition of two dimensional imagery in a narrow wavelength band whose centre wavelength can be changed over an octave or more in less than a millisecond. Because the objective was to evaluate the impact of the AOTF and the IR detector array parameters on the performance of various optical designs, a conservative scenario was used. Such factors as the nature of the forest canopy and alternatives for texture-based signal processing are not addressed here.

Applications of Photonic Technology
Edited by G.A. Lampropoulos *et al.*, Plenum Press, New York, 1995

SYSTEM CONCEPT AND MODEL

Detection of small fires is difficult because they are often hidden by the forest canopy. In our scenario, an airborne optical sensor would nominally operate at an altitude of 3 km above ground level and at an aircraft speed of 50 m/sec. Normally the sensor would view a forest scene and the fire signal must be observed against the background radiation (both diffusely reflected sunlight and infrared emission from the canopy). The instantaneous field of view (IFOV) of each pixel is a major determinant of system performance since fires are assumed to be well below the resolving power of the electro-optical system. Generally one should attempt to decrease the ground footprint so that the fire occupies a larger fraction of the instantaneous field of view. We assume that the total field of view of the sensor is 30 degrees and we have baselined a 256 x 256 InSb array. The IFOV would be approximately 6 m square while our fire target is assumed to occupy an area of 0.3m square.

By using an area array detector and an auxiliary scan mirror, each ground location can be revisited many times from different directions which could enhance the ability to see fire targets which might otherwise be obscured by the canopy. The AOTF allows multiple spectral channels to be acquired for any spatial location and this information can be used to discriminate against false alarms. The system performance will be affected by the number of spectral channels and by the use of the scan mirror as these can affect the integration time of the array which is nominally 1/30 sec .

A spreadsheet model of the system performance has been developed; it characterizes different configurations by the etendue of the optical system. The model estimates the contribution of four sources to the detected signal: narrow band radiation from the fire target , broad band radiation from within the optical system, and both narrow band and broad band radiation from the scene adjacent to the fire target. To simplify the analysis, we have further assumed that the background signal is entirely due to solar radiation that is reflected from a diffuse target with an albedo of 1. This is clearly a worst case, since the infrared radiation from a 300 K scene is substantially smaller than the solar signal. Signal strengths have been estimated from a Lowtran model of the atmosphere. When the target to background ratio is greater than one, only two spectral measurements are required to distinguish a fire target from the background. The performance of different optical systems is characterized by signal to noise ratio for detection and noise equivalent irradiance (NEI).

Optical layouts were analyzed using a ray trace program to estimate the system blur.

OPTICAL DESCRIPTION AND PERFORMANCE

A TeO_2 AOTF was chosen because of the commercial availability of large crystals at reasonable cost; its large birefringence and a slow shear mode give excellent diffraction efficiency as an AOTF . While TeO_2 has high transmittance in the spectral range of interest from 2- 4 μm, there is significant absorption at longer wavelengths; in particular, the external transmittance drops from about 70% in the region of high transmittance to 55% at 4.5 μm and 20% at 5 μm. This has significant implications because the spectral regions of low transparency are regions of high emissivity which lie in the passband of the InSb detector. (Consequently we have assumed that a 4.75 μm cutoff filter would be placed within the detector dewar. This background signal can be measured periodically by tuning the AOTF to a region of high atmospheric absorption.)

TeO_2 operates in a noncollinear mode which allows some spatial separation of the broad band (undiffracted) light from the desired narrow band signal which is linearly polarized. However the cone angle which the material can accept (about 11 degrees) exceeds the angle of deviation (5.9 degrees) between these two beams. This leads to the consideration of two optical configurations: one in which the cone of light is restricted by the subsequent optics to lie within a 5.9 degree cone and a second configuration in which the overlapping beams are separated by using crossed polarizers[5].

Beam Expander

Figure 1 shows three optical groups - a 3-element wide angle foreoptic in front of a 20 mm clear aperture AOTF which is followed by an airspaced doublet camera lens consisting of two, off-the-shelf ZnSe plano-convex lenses. The dewar window precedes

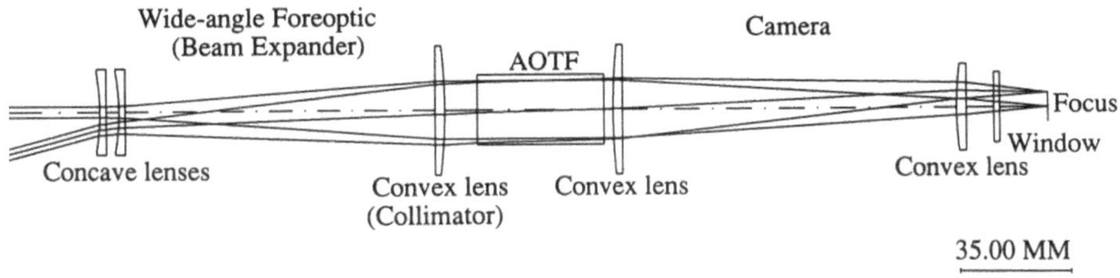

Figure 1. 30 degree beam expander.

the detector (9.4 mm on each side) which lies at the focus of the system. The field of view of the doublet camera has been set to 5.5 degrees and light passing through the AOTF is pseudo-collimated. so that the diffracted and undiffracted light fields are separated by about .65 degrees for rays near the edge of the pupil. Together this allows us to achieve a 30 degree field of view. To achieve a 50% diffraction efficiency for polarized light, about 6 W of acoustical power is required.

The signal reaching the detector is dominated by the emission from the AOTF which fills each photosite to approximately 7% of capacity. By comparison the narrow band signal from the scene is about 2% of the charge capacity of the detector. Figure 2 shows the relative strength of the target and the narrow band signals from the scene at wavelengths of high atmospheric transmission. The ratio increases with wavelength but the target signal is attenuated by the transmission of the TeO_2 and the cutoff filter. The fire signal near the peak of the spectral distribution increases by about 50 times as the target temperature varies from 500 - 1200K; for the four longer wavelength bands, the fire and narrow band scene signals are nearly equal for target temperatures of 750 K. Observation of equal amplitude signals at the short wavelength band is indicative of a target which is very hot. We note that hot fires are always brighter than the diffuse solar signal at the long wavelengths so that we can discriminate against specular reflections which would be characteristically brighter at the shorter wavelengths.

Figure 3 shows the signal to noise ratio when optical blur is not included. At the long wavelengths the signal to noise ratio can approach 100 for targets at 750K. The high signal to noise ratio means that we can confidently conclude that a fire is present in the background signal.

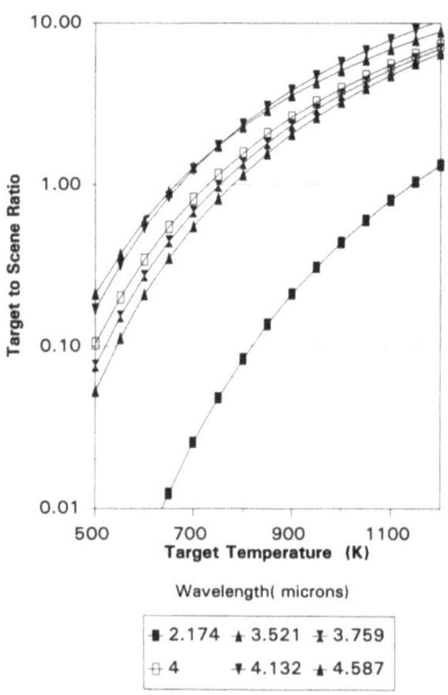

Figure 2. Target to scene ratio versus target temperature at selected wavelengths. AOTF and scene temperature are 300K.

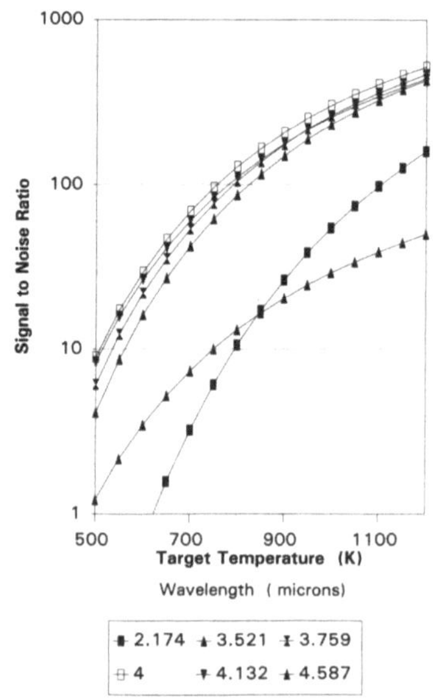

Figure 3. Signal to noise ratio versus target temperature at selected wavelengths. AOTF and scene temperature are 300K.

The optical blur has also been evaluated. For spots 15 degrees (4.5 mm) from the axis, 90% of the rays lie within a 47 μm diameter while 50% lie within a 25 μm diameter. At this angle, the effective focal length is 16.8 mm, compared with 17.6 mm on axis, thus there is barrel distortion and the effective entrance aperture has a diameter of 3.4 mm. The beam increases to 18 mm diameter at the AOTF. When diffraction for the 3.5-5 μm region is taken into account, the image is broadened and 50% of the encircled energy lies within a 30 μm diameter, 70% within a 46 μm diameter and 90% within 107 μm diameter.

The blur diameter exceed the size of an individual pixel; when this is taken into account, the signal to noise ratio drops by about 25% and we can typically detect a noise equivalent irradiance <1.5E-10 W/cm^2 μm through the regions of high atmospheric transparency.

Reimager

We have also examined the radiometric properties of a reimager configuration. In this configuration (not shown) a five element fore-optic images the far-field light at the centre of the AOTF. An additional five element group reimages the far field onto the detector. As we have noted above, we can increase the angular acceptance of the AOTF by using polarizers to separate the narrow band and broadband light before they reach the detector. Acoustic power requirements could be reduced by 50% if the clear aperture of the AOTF could be reduced to 10 mm. Because the angle accepted by the AOTF has doubled, the etendue of this configuration, is almost identical to the beam expander configuration.

We have analyzed this system when the overall transmission of the two lens groups (of five elements each) is assumed to be comparable to the two groups in the beam expander configuration (three lens elements). The model accounts for the difference in throughput due to the polarization states of the different sources. Figure 4 shows the signal to noise ratio of this system as a function of the peak transmission of the polarizer (i.e., aligned parallel to the desired component of the diffracted beam). For comparison purposes, the condition without a polarizer is shown at the extreme right. We note that the signal to noise ratio is degraded unless the polarizer has a peak transmittance >80%. The signal to noise ratio is further degraded if the increased number of elements further decreases the transmission. (For example, if each of the four extra elements had a transmittance of 95%, a further 20% loss is introduced.)

There is a further difficulty with the polarizer -- namely, the extinction ratio has to be adequate to attenuate the broadband scene. In fact, the broad band scene will give signals which exceed the narrow band component unless the rejection ratio is greater than 100. This combination of high transmission and extinction ratio is generally not available.

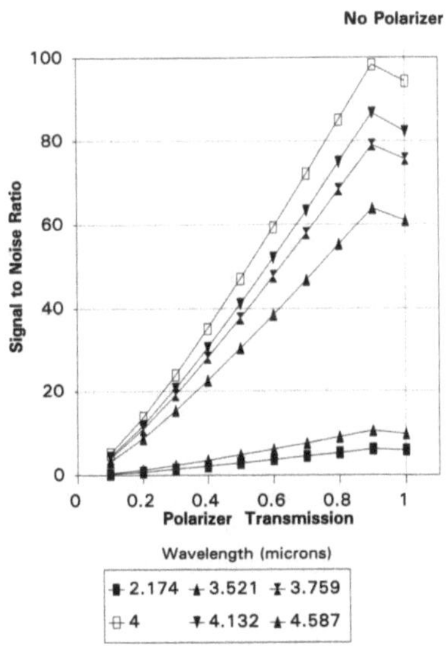

Figure 4. Signal to noise ratio versus transmission of the polarizer at selected Wavelengths. Target temperature is 750K; AOTF and scene temperature are 300K.

CONCLUSIONS

We have shown that the beam expander configuration for detecting forest fires provides adequate signal to noise ratio for detecting unobscured fires $\approx 0.25\%$ of the ground foot print. The analysis has indicated that the perceived benefit of a substantially smaller AOTF in a reimager configuration cannot be realized unless polarizers of very high quality are obtained.

ACKNOWLEDGMENTS

When most of this work was carried out, A.B. Hollinger was with the Institute for Space and Terrestrial Science. The authors wish to acknowledge financial support from the Ontario Ministry of Economic Development and Trade through the Institute for Space and Terrestrial Science and (for A.B. Hollinger) the Canadian Space Agency. Funding for portions of the work from the Defence Research Establishment Valcartier was much appreciated, as was the cooperation of Benoit Montminy of that organization.

REFERENCES

1. R.A. Wilson, "The Remote Surveillance of Forest Fires", Appl. Opt. 5 (6), 899-904 (1966).

2. P.J. Thomas and Nixon O, "Near Infrared Forest Fire Detection Concept", Appl. Opt. 32 (27), 5348-5355 (1993).

3. P.J. Thomas, E. Lithopoulos, S. Pelletier and S. Law, "Image Processing for Forest Fire Detection", Canadian Conference on Electrical and Computer Engineering, Toronto, 13-16 September 1992, paper WA3.31.1.

4. P.J. Thomas, A.B. Hollinger and R.H. Wiens, "Adaptive Infrared Forest Fire Sensor", SPIE Vol. 1969, "Infrared Imaging Systems: Design, Analysis, Modelling and Testing IV", pp 370-381 (1993).

5. I.C. Chang, "Tunable Acousto-Optic Filters: an Overview", SPIE Vol 90, "Acousto-Optics", pp 12-22, 1976.

6. J. Yu, Tien Hsin Chao and Li-Jen Cheng, "Acousto-Optic Tunable Filter (AOTF) Imaging Spectrometer for NASA Applications: System Issues", SPIE Vol. 1347, "Optical Information Processing Systems and Architectures II" (1990)

7. Cui Yan and Zhang Minghui, "Study on spectrally agile staring sensor using acoustooptic tunable filter", SPIE Vol. 1693 "Surveillance Technologies II", pp 238-243 (1992).

THE HISTORY OF LASER BATHYMETRY

Sebastian Sizgoric, John Banic and Paul LaRocque

Optech Inc.
100 Wildcat Road
Toronto, Ont. Canada

ABSTRACT

This paper gives a coarse-grained historical overview of activities in airborne Lidar hydrography (ALH), starting with the general operating principles of the technique, then followed by a synopsis of the principal, worldwide ALH efforts over the past two and a half decades. This will lead to a discussion of the design trade-offs and performance of the SHOALS system, Optech Inc's recently concluded ALH program for the US Army Corps of Engineers. Finally, a comparison is given of the current ALH operational capabilities (depth, area coverage, accuracy, reliability, cost-effectiveness) to the currently used acoustic methods.

INTRODUCTION

Airborne lidar hydrography is a relatively new technique aimed at augmenting nautical chart production and bathymetric mapping capability in shallow coastal waters. The potential of water-penetrating airborne laser radar to provide cost effective characterization of underwater topography to depths as great as 50 m (depending on water clarity) has triggered a number of R & D efforts worldwide over the past two and a half decades. Three recently concluded programs which were aimed at providing operational ALH tools, are Optech Inc's "SHOALS" program for the US Army Corps of Engineers, the Royal Australian Navy's "LADS" program, and Saab/Optech's "HAWKEYE" program.

The motivation to develop ALH technique to operational status is primarily two fold: the speeding up of the surveying tasks that exist under the current mandate of the various hydrographic agencies, and the anticipated, significant savings in the cost-per-unit-area-surveyed. Millions of km^2 of uncharted waters with depths of less than 50 m, that still exist world-wide, present a backlog of hundreds of ship-years to conventional acoustic, surveying. Additionally, much of surveyed, navigation routes have bottom types that is sufficiently dynamic, in the short term, to require frequent resurveying. The much greater speeds, than that of surface vessels, of ALH platforms have already well demonstrated, that the needed surveys can be done faster and more economically. Furthermore, because airborne, the use of ALH to carry out whatever surveys may be suitable to its capability provides a benefit of flexible deployment in distant areas

Applications of Photonic Technology
Edited by G.A. Lampropoulos *et al.*, Plenum Press, New York, 1995

with small operational windows, or shallow areas unsafe for survey boats. And, though perhaps not yet adequately appreciated throughout all of the hydrographic community, ALH offers, as standard, the benefit of virtually uniform x-y sounding distribution.

PRINCIPLES OF OPERATION

Figure 1 illustrates the operating principle of an ALH system. Laser-generated optical pulses, of wavelength suitable for propagation through water, are transmitted from an aircraft towards the water. An optical receiver, co-located with the transmitter, detects the pulse reflections from both the water surface and the bottom. The water depth is determined from the elapsed time between these two reflection/scattering events, after accounting for the operating geometry and corrections for propagation-induced biases and waveheight and tide effects. In the operational scenario shown, the laser pulses are scanned sequentially across the water surface to produce, when combined with the aircraft's forward velocity, a swath of soundings of uniform spatial distribution. The horizontal co-ordinates of the soundings are determined from the knowledge of the aircraft position, laser-beam exit angles, the aircraft altitude and attitude, and the measured water depth.

SUMMARY OF THE MAIN ALH EFFORTS

In the pre-1970 era most ALH-type work was sponsored by the US Navy, related to submarine detection (Ott, 1965 & Sorenson, 1966) and, at that time, classified. The first work, reported in the open literature came from the University of Syracuse (Hickman, 1969) in 1969. During the early-to-mid 1970's experimental, profiling ALH-type systems were deployed by NASA (Kim, 1974) and the US Navy (Cunningham, 1972) in the US, by Canada Centre for Remote Sensing (CCRS)/Optech in Canada (Ryan, 1980), by the Weapons Research Establishment (WRE) of The Australian Royal Navy (RAN) (Clegg, 1978), and by the Defense Research Institute (FOA) in Sweden. These efforts gave rise to follow-on work in the mid-late 1970's: in the US a joint effort by NASA, NOAA and US Navy, with AVCO as the contractor, produced a scanning Airborne Oceanographic Lidar (AOL) (Avco, 1975). In Australia, WRELADS-2 followed, another scanning system, built upon the experience of WRELADS-1, the initial profiling prototype. In the early 1980's in the US, an attempt to take the AOL experience into the operational domain resulted in the HALS program (Hydrographic Airborne Laser Sounder) (Houck, 1980), sponsored by DMA, NASA and the US Navy, with AVCO as contractor. In Australia, during the same period, WRELADS-2 carried on with an extensive set of field trials and data evaluation (Abbot, 1980). In Canada a similar process was taking place with the profiling Mark-2 ALH system, while in Sweden, FOA carried out again similar trials with Optech's assistance, using Canada's system augmented with a scanning capability (Steinvall, 1981).

In the mid-1980's Optech developed the LARSEN-500 ALH system, an operational prototype, with support from the Canadian Hydrographic Service (CHS) and CCRS (Banic, 1986). In the latter-1980's Sweden's FOA sponsored the development by Optech of a prototype scanning ALH system. This FLASH-1 system was delivered to FOA in 1988. In the US HALS, after prolonged programmatic difficulties, was resurrected within the confines of a wider effort on the part of the US Navy, the Airborne Bathymetric System (ABS) (Harris, 1986). The ABS flew until 1988. During this period work also began in China on a Blue-green Oceanographic Lidar (BLOL) and in Russia (the GOI system). In the late 1980's Optech Inc. built the ALARMS (Airborne Laser Radar Mine Sensor) scanning system for the US DARPA which, though essentially an ALH system, had as its primary objective the detection of water-column-suspended mines. During 1988/89 SHOALS, the US Army Corps of Engineers sponsored, operational ALH system, was being built by Optech Inc. (Banic, 1990), and LADS, the RAN sponsored operational ALH system, was being built by BHP Industries (Compton, 1988). During the late 80's and early 90's, Thomson-Sintra-ASM of France developed a profling prototype systems which was tested sometime in 92/93 with unknown results. In 1991 the Swedish Department of Defense ordered two ALH systems with Saab Instruments as prime contractor and Optech as the major subcontractor.

The current (spring 1994) status of ALH is as follows. There have been no known recent results from the Chinese or Russian systems. In France, Thomson-Sintra-ASM has decided not to pursue any further ALH activity. The FLASH-1 system was recently retired in Sweden to make way for the new HAWKEYE systems. Among the systems currently flying the AOL is still active although it is used primarily as a multipurpose platform and most of the recent work has been in fluorosensing. Another active system is the Oceanographic Water Lidar (OWL). Not much is known about the OWL system due to its classified nature but it must have some hydrographic capability as it has recently been discussed in this context at Hydrography '94 (Norfolk, Virginia).

If we restrict ourselves to only the active systems used primarily for ALH we have the following:

(a) In Canada, the LARSEN-500 is still being operated by Terra Surveys of Sidney, British Columbia. It has been used in Malaysia, the northern US and more Arctic work is planned.

(b) In Australia, the LADS program has completed field trials and is surveying waters around Australia.

(c) In Canada/Sweden, based on the SHOALS design, Optech Inc. has recently delivered to SAAB the second of the two HAWKEYE systems destined for the Swedish Royal Navy and the Swedish Hydrographic Department. These systems are being flown over Swedish waters in a Boeing Vertol helicopter.

(d) In Canada/US, Optech Inc. has successfully completed the field acceptance trials of SHOALS and recently delivered the system to the US Army Corps of Engineers. SHOALS was taken almost immediately to Florida Bay to do an operational survey off the southern tip of Florida.

SHOALS OBJECTIVES

The primary objective of the SHOALS program was to provide an airborne bathymetric survey capability concomitant with the US Army Corps of Engineers' "reconnaissance" and "condition" type surveys on a variety of navigational channels throughout, and around, the United States. To this end SHOALS was made compatible with a Bell 212 helicopter platform, has a nominal operating altitude of 200 m; a swath width variable from zero to 40 degrees; and a daytime, maximum depth penetration capability (designated by the product of the water-diffuse-attenuation-coefficient, k, and the water depth, d) of $(kd)_{max} > 3$.

A second objective of the SHOALS program was to make the SHOALS system simultaneously suitable to the needs of the broader and international hydrographic community. This has been achieved, by (a) showing that the SHOALS performance capabilities meet the International Hydrographic Organization (IHO) standards, (b) making SHOALS compact enough to be compatible with a variety of both fixed-wing and rotary platforms, and (c) adopting design/manufacturing approaches that, for subsequent units, would make it affordable to a greater variety of users.

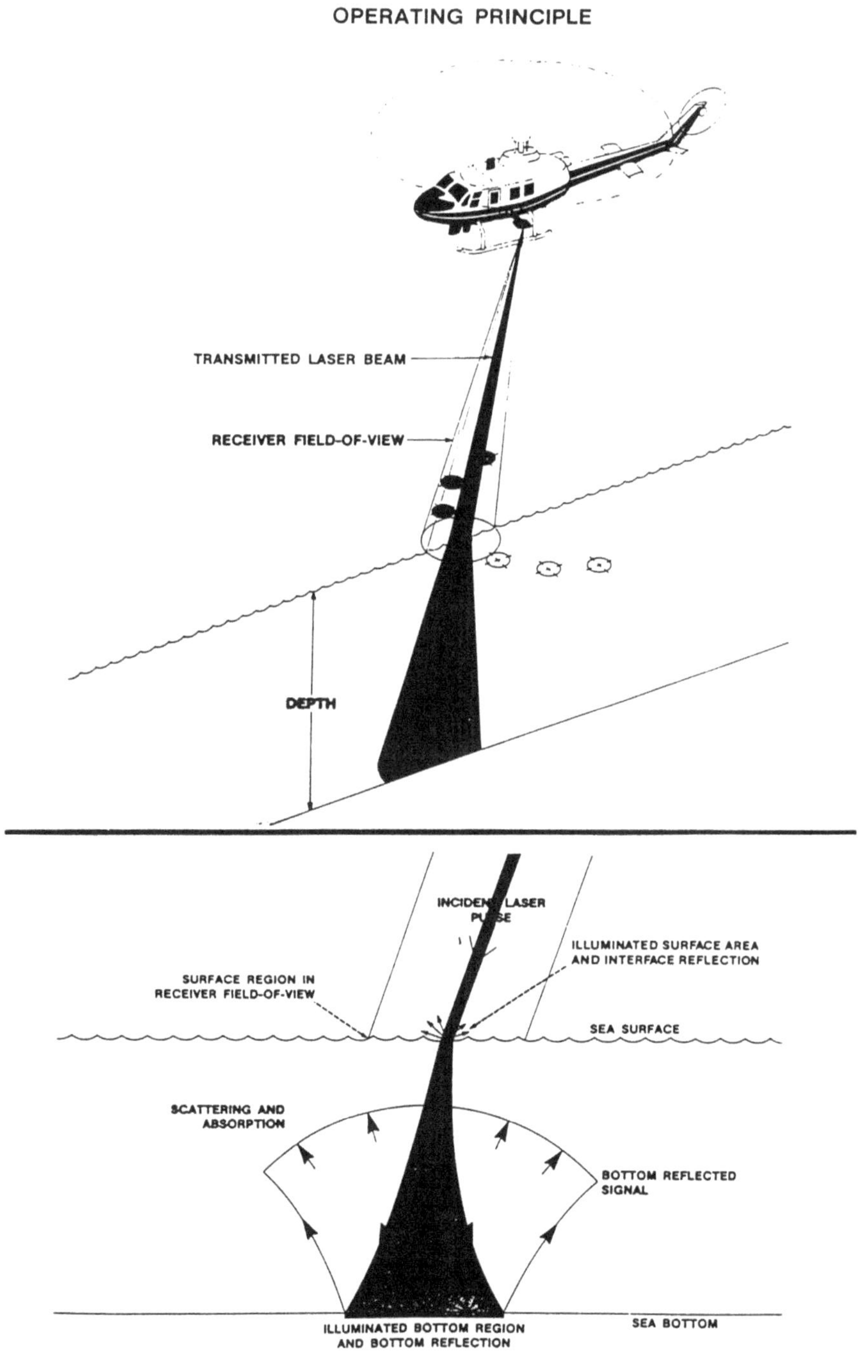

Fig. 1. Airborne Laser Hydrography Schematic of Principle of Operation.

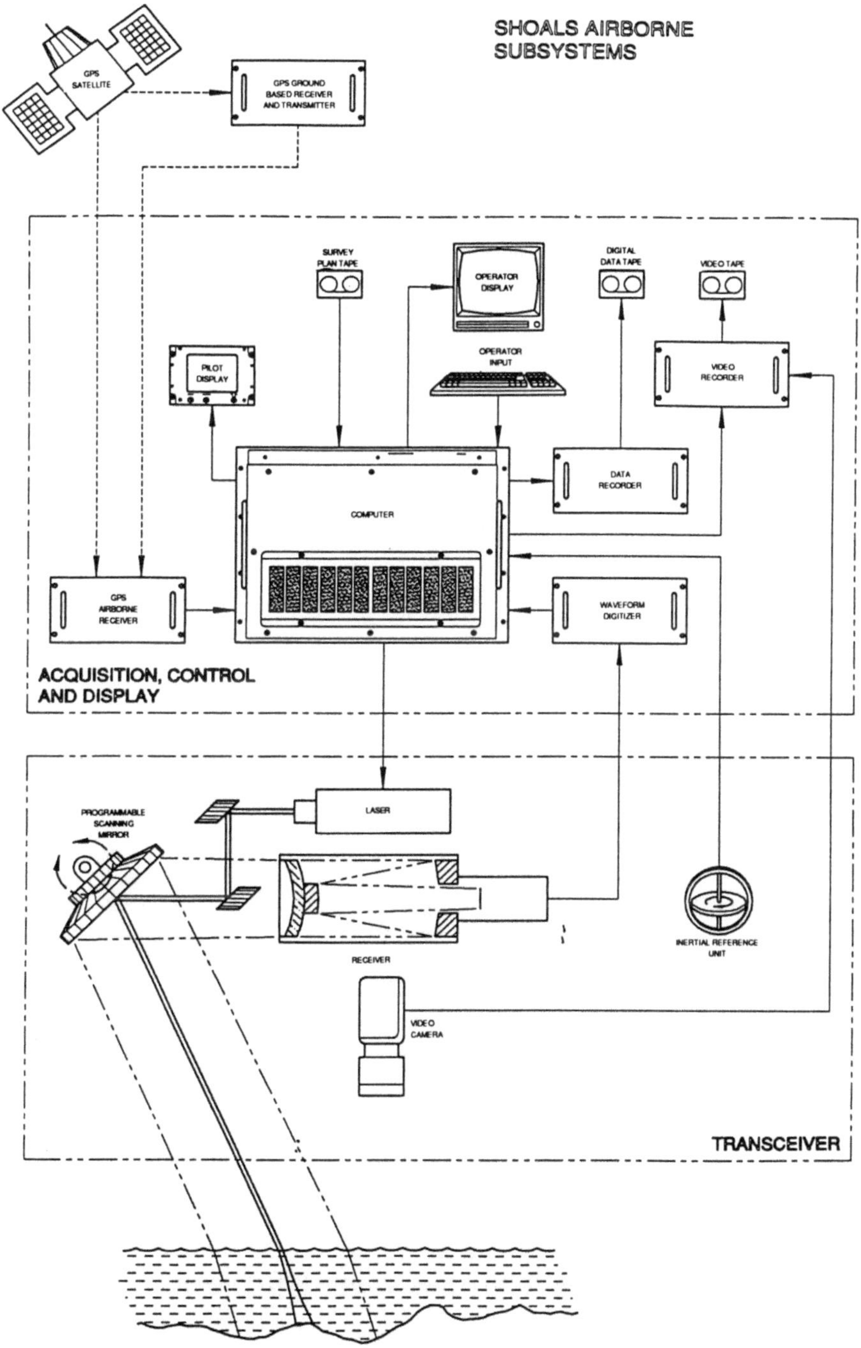

Fig. 2. SHOALS Airborne System Block Diagram.

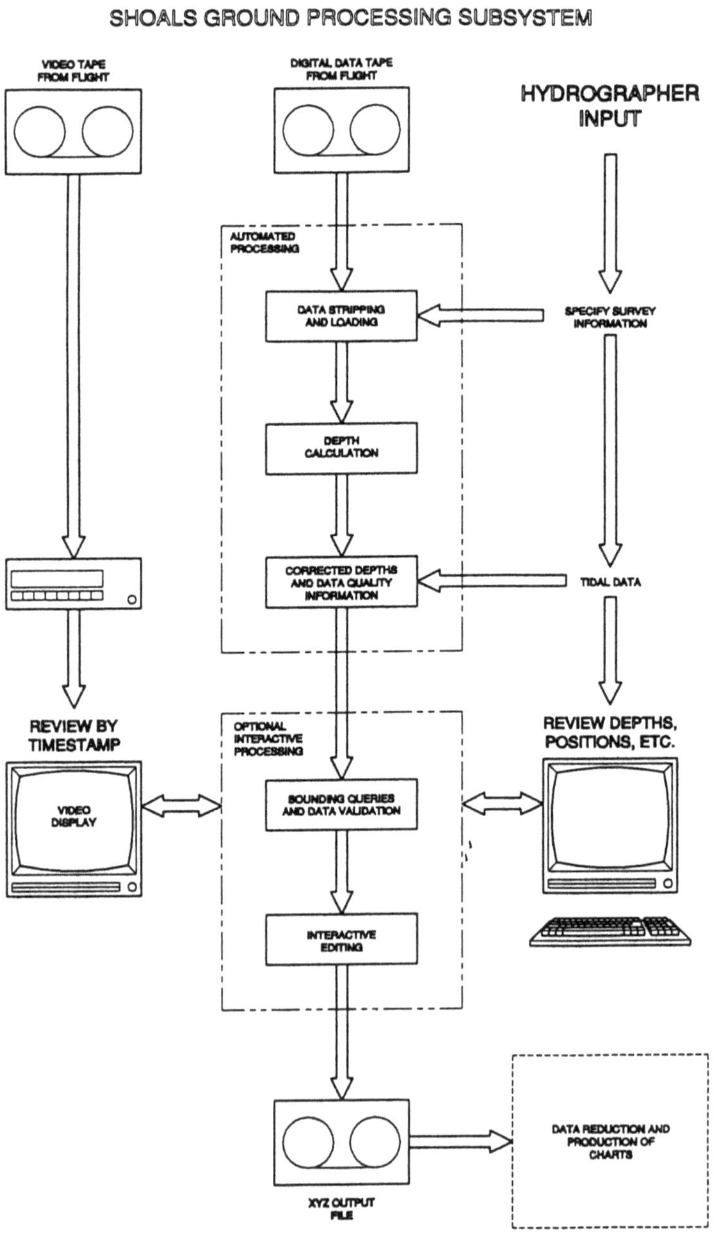

Fig. 3. SHOALS Post Processing System Block Diagram.

SYSTEM OVERVIEW

The SHOALS system is shown in two block diagrams. It consists of three major airborne subsystems shown in Figure 2, and a ground-based data processing subsystem illustrated in Figure 3.

The transceiver subsystem generates laser pulses, scans the laser beam, performs the detection and preprocessing of the optical return signals and provides a high resolution video image of the sounded area. The positioning subsystem receives aircraft position data from GPS system and, with an inertial reference system provides the angular orientation of the transceiver. The above, together with the scan-angle data, are used to determine the horizontal position of the soundings. The data acquisition, control and display subsystem provides the operator interface, captures and stores all data and displays the system status and quality control parameters. Additionally, it provides information and guidance displays for flight-line management by the pilot. The ground-based data processing subsystem analyses the acquired data to produce corrected depths, corresponding horizontal positions and assigns confidence values to each sounding. It enables data examination and editing by the operator, generates the desired XYZ output and provides hard-copy (sounding plot) data output.

SUMMARY OF ALH CHALLENGES

An operational ALH system needs to provide an optimized solution to a number of diverse, and difficult technical problems. Areas that have historically been challenging in most ALH design-and-build efforts have generally been among the following:

- reliable, power-efficient, high PRF laser of adequate pulse energy and suitable emission wavelength
- accurate and reliable determination of the instantaneous water-surface position at the laser-beam entry point
- compression of the large dynamic range in the received-signal intensity to a level suitable for transient recorders
- accurate algorithm for depth extraction
- reliable and accurate algorithm for instantaneous mean-water-level determination (i.e. waveheight corrections)
- efficient handling/processing of very large data sets
- high automation level in the data processing to "validated XYZ" stage
- acceptable data-processing-time/data-acquisition-time ratio
- compatibility of processed-data format with current hydrographic usage

DISCUSSION OF SOME SHOALS TRADE-OFFS

As it is not considered within the scope of this paper to provide an exhaustive trade-off analysis of an ALH system design, this section discusses only a few of the technical challenges identified in the previous section, this with a view of providing some visibility into the trade-offs that were adopted in the construction of the SHOALS system.

(a) Laser - optimal transmission in most coastal waters dictates that the operating wavelength of an ALH system, if not tuneable, be somewhere in the 490-550 nm spectral range. Historically, the most frequently selected source has been the frequency-doubled Nd-YAG laser, operating at 532 nm. Most workers have concluded that for readily manageable receiver/scanner apertures (typically 20-30 cm diameter) a laser pulse energy of 5-10 mJ, at 532 nm, was a good trade-off with the depth penetration capability, desired sounding density and area-coverage rate (limited by the concomitant maximum laser available PRF) and the laser weight/power-consumption/reliability. The SHOALS design also settled on a choice of a Nd-YAG laser, frequency doubled, with a PRF of 200 pps and with a pulse of 5 mJ energy and 5-6 ns width. As there was hardly one such item available off-the-shelf from laser manufacturers, the SHOALS laser was developed specifically to its purpose, relying heavily on previously proven subsystems, many of these from military systems. The newly-advancing diode-pumped Nd-YAG technology had not yet progressed adequately to make a laser with SHOALS requirements easily affordable, so initially a hybrid approach was pursued which consisted of a diode-pumped oscillator and a flashlamp-pumped amplifier. The delivery of this laser became problematic so the SHOALS laser became a compact flashlamp design. The laser's 200 Hz PRF allows uniform sounding density of 6 m nominal grid spacing, for an aircraft at an altitude of 200 m and ground speed of 40 m/s (80 knots), while permitting survey-data collection at 20 km^2/hr. Furthermore, SHOALS' ability to vary the altitude and scan angle under computer control allows a ready operational trade-off between the sounding grid-spacing and survey area rate, thus permitting a finer, or coarser, sounding density than the nominal, depending on the requirement of any given mission.

(b) Water-surface location - the SHOALS system has adopted the approach of sending a simultaneous, infra-red beam collinear, with the green one for each laser pulse transmitted. The beams are transmitted at a constant angle of incidence of 20° with respect to the water surface for reasons of minimizing raw depth bias errors due to beam propagation in the water. Since the raw water depth is first determined relative to the instantaneous local water surface, a precise knowledge of the time - location of the air/water interface event on the backscattered intensity-vs-time signature is required. However, since the water surface is highly variable in regard to its reflective/scattering properties, driven primarily by the wind/ wave conditions, the return signal intensity of the air/water interface event of both the infra-red and the green beams can vary by many orders of magnitude over a range of environmental conditions that would be operationally acceptable.

To ensure reliable availability of an air/water interface event in the detected backscattering signature the SHOALS system utilizes three separate signal channels: the infra-red channel that yields a backscattering signature up to the water surface, the green channel that yields one through, and below, the water surface, and a red channel (Raman return stimulated by the excitation of the water molecules by the green beam) that yields the backscattering signature only from a relatively thin surface water layer. None of the three channels, by itself, can provide reliable, and/or accurate, air/water interface returns over the entire range of operating altitudes and environmental conditions and, hence, a prioritized combination of all three is used.

(c) Data handling - the SHOALS system could, during an operational day's surveying collect up to several gigabytes of airborne data. The ground-based data handling/processing system had, as a target, the objective of processing one day's acquired data overnight. It had to move the airborne data from the airborne storage medium, through processing steps, and into files available for editing. Datafile format must allow for a variety of searches as part of the processing editing task and for the ability to provide easily new user searches at later times.

An analysis of the approaches to meet optimally the above requirements narrowed to a relational database. It allowed quicker access to data organized into smaller files, with search criteria as keys. It provides efficient use of disk space. Although most searches are predefined, new searches can be easily added without programming by using the end user query system to look at any of the recorded data. The tremendous flexibility of a database has been very useful in characterizing system performance. This flexibility has a cost however, in processing time. The demonstrated post processing ratio was 5.5:1. Each hour of collected data took 5.5 hours to be stripped and automatically processed from the airborne mission tape. It should be mentioned however that the somewhat less accurate realtime depths was able to be stripped and displayed in less than 1:1.

LIMITATIONS AND PERFORMANCE

The main limitations of an ALH system are its maximum and minimum depth capability, weather-related phenomena, and bottom structure. Water clarity will typically limit the system's maximum depth penetration capability to less than 50 meters in very clear water, and 20-30 meters in moderately clear waters. (Areas of high turbidity significantly degrade ALH performance, and need to be identified as part of the survey planning process.) The SHOALS system was able to detect depths greater than 37 meters more than 50 kilometers offshore. Closer to shore the minimum depth measurement capability of the SHOALS system was demonstrated to be approximately 1 meter. Weather-related phenomena such as high surface waves, heavy fog and precipitation, and surface sun glint, also degrade the system performance by decreasing depth penetration and/or depth measurement accuracy. Similarly, operation in areas where the bottom is 5heavily vegetated or covered with "fluid mud" will, as with acoustic techniques, present serious challenges to the system's ability to recognize the bottom signal. The performance of the SHOALS system in the presence of these limitations is still being characterized.

COMPARISON WITH ACOUSTIC SYSTEMS

It is difficult to make a widely applicable comparison between the SHOALS system and a variety of acoustic approaches being used today, as most such comparisons are meaningful only if done for specific scenarios and, better yet, after-the-fact. However some comments of a general nature can be made:

- depths accessible to sounding by the ALH approach with a SHOALS-type system, even under the best of conditions, are at maximum 50 m, and more typically 20-30 m. It is therefore only in this depth range than any comparisons can be meaningful.

- area coverage rate will, for the most part, be greater with an ALH system by from 10 to 100 times, depending on the type of area to be surveyed, the sounding density required, the survey resources deployed, and the distance of the area to be surveyed from the base where the survey resources are normally located.

- the accuracy of the two different approaches is comparable within the 50 m depths considered here, although ambiguities exist with both techniques when soft/extended bottom-type/structure is present. The horizontal accuracy is governed by the same parameters for both approaches and is, therefore, largely equivalent.

- operational reliability for the ALH systems is still being established. Early systems have been of prototype nature and, consequently, not overly reliable. However, operational systems such as SHOALS, HAWKEYE and LADS have been designed for reliable operation. For example, the SHOALS system logged over 120 flight hours during the field trials from January to March of 1994, with only light routine maintenance required.

- cost effectiveness analyses (Golaszewski, 1989 & Enabnit, 1978) have indicated an anticipated performance from 5-10 times to the advantage of a SHOALS-type system, but actual cost effectiveness will be highly dependent on the specific survey mission. Canadian experience, over the range of Canadian coastal waters, with a historically relatively low-reliability, relatively-low sounding-frequency system has demonstrated cost savings in the range of 2-4 times over previously used acoustic approaches.

CONCLUSIONS

Now that ALH is coming of age there remains little doubt of its effectiveness to enhance the ability of the hydrographic agencies to carry out their mandate. Systems like SHOALS, HAWKEYE and LADS have begun to confirm, in a statistically more meaningful manner, the actual extent of the operational enhancement that ALH offers. As a new tool in the hydrographic arena, complimentary to the acoustic one, ALH is likely to stay.

REFERENCES

ABBOT, R.H., "WRELADS II Trials", Proceedings Laser Hydrography Symposium, Australian Defence Research Centre, Salisbury, South Australia, p.188 (1980).

AVCO EVERETT RESEARCH LABORATORY, "Airborne Oceanographic Lidar System Final Report", NASA Contract No. NAS6-2653 (1975).

BANIC, J., SIZGORIC, S., and O'NEIL, R., "Scanning Lidar Bathymeter for Water Depth Measurement", Proceedings of SPIE Conference on Laser Technology and Applications, Quebec City, Quebec, SPIE Vol. 663, p187 (1986).

BANIC, J., SIZGORIC, S. and LILLYCROP, W.J., "Second-Generation Airborne Lidar System for Hydrographic Application", Oceanology International 90 Conference, Brighton, U.K. (1990).

COMPTON, J.S. and HUDSON, M.A., "New Charting Technology in Australia: The Laser Airborne Depth Sounder", International Hydrographic Review, Monaco, p. 145 (1988).

CUNNINGHAM, L.L., "Test Report on Pulsed Light Airborne Depth Sounder (PLADS)", Naval Oceanographic Office Technical Note 6620-102-72, U.S. Navy (1972).

CLEGG, J.E. and PENNY, M.F., "Depth Sounding from the Air by Laser Beam", Journal of Navigation 31, p. 52 (1978).

ENABNIT, D.B., GOODMAN, L.R., YOUNG, G.K., and SHAUGHNESSY, W.J., "The Cost Effectiveness of Airborne Laser Hydrography", NOAA Technical Memorandum NOS 26 (1978).

GOLASZEWSKI, R. and BAROL, D., "An Economic Analysis of the Optech Helicopter Lidar Bathymeter System", Gellman Research Associates Inc. Report (1989).

HARRIS, M.M., HICKMAN, G.D. and BOOKER, R., "Development of the Airborne Bathymetric Survey System", Proceedings of the Second Biennial National Ocean Service International Hydrographic Conference, Norfolk, Virginia, p.50 (1986).

HICKMAN, G.D., and HOGG, J.E., "Application of An Airborne Pulsed Laser for Near-Shore Bathymetric Measurements", Remote Sensing of Environment 1, p.47 (1969).

HOUCK, M.W., "The Hydrographic Airborne Laser Sounder (HALS) Development Program", Proceedings Laser Hydrography Symposium, Australian Defence Research Centre, Salisbury, South Australia, p.35 (1980).

KIM, H.H., and RYAN, P., Proceedings of "The Use of Lasers for Hydrographic Studies", NASA Wallops Station, National Aeronautics and Space Administration, Wallops Island, Va. (1974).

OTT, L.M., "Underwater Ranging Measurements Using Blue-Green Laser", NAVAIRDEVCEN Report No. NADC-AE-6519, Naval Air Development Center, Warminster, Pa., 1965 (Confidential).

RYAN, J.S., and O'NEIL, R., "Field Trials of an Airborne Lidar Bathymeter", Proceedings of the 19'th Annual Canadian Hydrographic Conference, March 20, 1980, Halifax, Nova Scotia, J. Canadian Hydrographers Assoc. (1980).

SORENSON, G.P., HONEY, R.C., and PAYNE, J.R., "Analysis of the Use of Airborne Laser Radar for Submarine Detection and Ranging", SRI Report No. 5583, Stanford Research Institute, 1966 (Confidential).

STEINVALL, O., KLEVEBRANT, H., LEXANDER, J., and WIDEN, A., "Laser Depth Sounding in the Baltic Sea", Applied Optics 20, p. 3284 (1981).

ARCTIC LIDAR MEASUREMENTS OF THE MIDDLE ATMOSPHERE

D.P. Donovan, J.A. Whiteway, W. Steinbrecht and A.I. Carswell

Institute for Space and Terrestrial Science and
Department of Physics and Astronomy, York University
4700 Keele Street, North York, Ont., Canada, M3J 1P3

INTRODUCTION

In January of 1993 the Atmospheric Environment Service (AES) of Canada opened the Arctic Stratospheric Observatory (AStrO). The observatory is located close to Eureka (80° N, 86° W) in the Canadian high arctic which is also the site of a long running weather station. AStrO is one of the primary arctic stations of the NDSC (Network for the Detection of Stratospheric Change). The NDSC is a network of remote-sounding research stations for observing and understanding the physical and chemical state of the stratosphere. At these stations the ozone and key ozone related parameters are measured by a variety of sensor systems. The group from the Institute for Space and Terrestrial Science (ISTS), with headquarters at York University in Toronto, has the responsibility of undertaking the lidar measurements at AStrO for AES. Working with the ISTS lidar group are Japanese lidar researchers from the Meteorological Research Institute, Communications Research Laboratories and Nagoya University.

There are at AStrO two lidar systems which have been developed and installed in collaboration with Optech Incorporated of Toronto [1]. One is a Differential Absorption Lidar (DIAL) system designed to measure ozone in the stratosphere at altitudes between about 10 and 50 km. The second is an elastic backscattering, Rayleigh/Mie lidar designed for the dual role of measuring the Polar Stratospheric Clouds (PSCs) and the tropospheric arctic haze. The PSC/Haze lidar has been operated by the Japanese groups and the DIAL system by the ISTS researchers. This paper presents an overview of the data obtained by the ISTS group during the first two seasons of operation at AStrO.

MEASUREMENTS

The DIAL system is based upon a Lumonics XeCl excimer laser which transmits short (13 nsec) pulses of 308 nm radiation at a repetition rate of 300 Hz with an average output power of about 60 W. These pulses are transmitted through a cell containing pressurized hydrogen gas to generate 353 nm radiation via stimulated Raman scattering.

Applications of Photonic Technology
Edited by G.A. Lampropoulos *et al.*, Plenum Press, New York, 1995

The 308 nm radiation is absorbed by ozone whereas the 353 nm light is not. As the dual wavelength beam propagates upward the differential attenuation of the beam as a function of range can be used to provide a quantitative profile of the ozone concentration by the well-known DIAL measurement method [2] .

The AStrO DIAL system has been configured to operate only during the nighttime with the intention being to have it operational throughout the arctic winter period from early November to late March. In the winter of 1993 data were obtained for the months of February and March. For the last winter, data were obtained from the beginning of November 1993 to the middle of March 1994. In addition to the DIAL ozone measurements, the 353 nm radiation was also used to preform measurements of atmospheric density, temperature and stratospheric aerosol levels.

Since the eruption of Mount Pinatubo in June of 1991 stratospheric aerosols have remained at elevated levels. This increased aerosol burden has had a significant effect on the stratosphere and upon the sensors making a variety of remote atmospheric measurements. In particular, this aerosol has introduced an increased uncertainty in the quantitative accuracy of DIAL measurements of ozone [3]. As a result at AStrO there has been considerable effort devoted to the observation of the aerosol characteristics and the effects of the aerosol on the ozone DIAL measurements.

The DIAL system accuracy in the presence of aerosols may be improved by detecting the inelastic (Raman) backscattering from atmospheric nitrogen rather than the elastic (Rayleigh) scattering[4]. During the first winter of operation the DIAL system was operated only in the Rayleigh mode. It was found, however that there was a very substantial aerosol loading present and during the recent winter the AStrO DIAL system was modified to incorporate the Raman mode as well as the Rayleigh. Typically the Raman mode is used for ozone measurements in the lower altitude region (below about 20 km) where the aerosols reside. The Rayleigh mode is used for altitudes above about 20 km.

OBSERVATIONS

A summary of ozone measurements made during February and March 1993 is shown in Fig. 1. At the time of these measurements a Raman capability had not yet been installed and significant levels of stratospheric aerosol were present, usually limiting the lower altitude of reliable DIAL ozone retrievals. The figure is a composite of DIAL ozone measurements and electrochemical sonde data. Above about 17 km the data shown have been derived by DIAL measurements while below this altitude the sonde measurements are used. At altitudes above the aerosol layer, when ozone sonde data was available, it agreed very well with the DIAL measurements.

In addition to the ozone measurements the DIAL reference wavelength (353 nm) was used to perform measurements of the thermal structure of the stratosphere and mesosphere. In the absence of aerosols the amount of light backscattered from a given range is proportional to the atmospheric density at that range. From this information and the assumption of hydrostatic equilibrium one can measure temperature up to 70 km within a few degrees Kelvin [5].

A summary of temperature measurements made during February and March of 1993 is shown in Fig. 2. As was the case with the DIAL ozone measurements, the presence of stratospheric aerosol limited the lower altitude of the lidar temperature measurements. In this case the temperatures shown below about 20 km have been provided by radiosondes launched twice daily from the nearby (less than 15 km) weather station. When sonde temperature measurements were available above about 20 km they

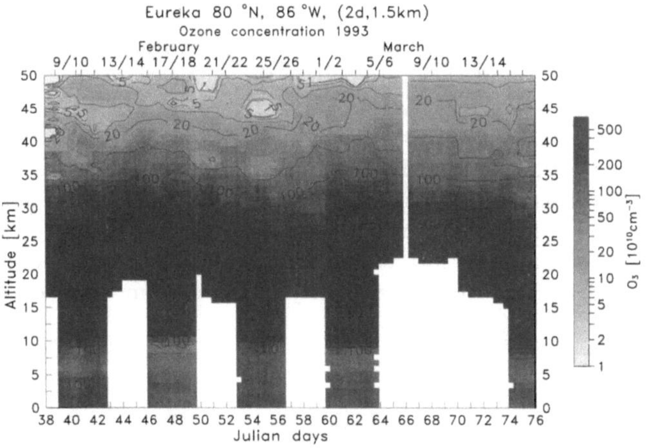

Figure 1. Ozone concentràtion over Eureka in February and March 1993. Below about 17 km the data shown are from ozone-sondes launched once a week. Above 17 km the lidar measurements have been used.

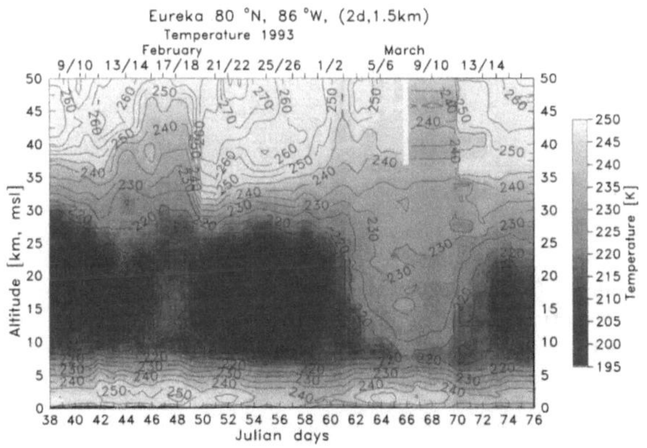

Figure 2. Temperature over Eureka in February and March 1993. The temperature below about 20 km is from radiosondes launched twice daily. Above the upper end of the radiosonde it is derived from the lidar measurement at 353 nm. The dates of the lidar measurement are indicated by the tick marks at the top.

agreed well with the lidar derived temperatures, usually within a few degrees K even though the measurements may be separated in time by a few hours.

Both the temperature and ozone measurements show a large degree of correlation and variability associated with the dynamics of the atmosphere around the polar region. Of particular note is the sudden stratospheric warming and the corresponding change in observed ozone levels. This event began about the twenty–first of February and is accompanied by a cooling in the mesosphere [6].

In Fig. 3 we show two measured ozone profiles along with the measured scattering ratio ($R = (\beta_{molecular} + \beta_{aerosol})/\beta_{molecular}$) taken on Feb. 23, 1994. The "Raman" ozone profile shown by the thick line was derived using the 332 and 385 nm Raman N_2 backscatter returns which contain no aerosol backscatter contribution while the other profile shows the ozone derived using the conventional DIAL method with the 353 and

308 nm returns which do contain an aerosol contribution. The "dip" in the ozone profile at 15 km is present in both the Raman and conventional measurement and is a real feature while the dip that shows up only in the conventional profile at 12 km is an artifact due to a change in the aerosol properties.

With regard to Fig. 3, the scattering ratio shown at 353 nm was found from the ratio of the 353 nm return to the Raman return at 385 nm. R is seen to become very noisy above 25 km, this is due to the fact that the Raman backscatter cross section is some three orders of magnitude lower than the corresponding Rayleigh cross section and the signal to noise ratio is low at higher altitudes. Further modifications to the system are planned to improve the sensitivity of the system with regards to Raman measurements.

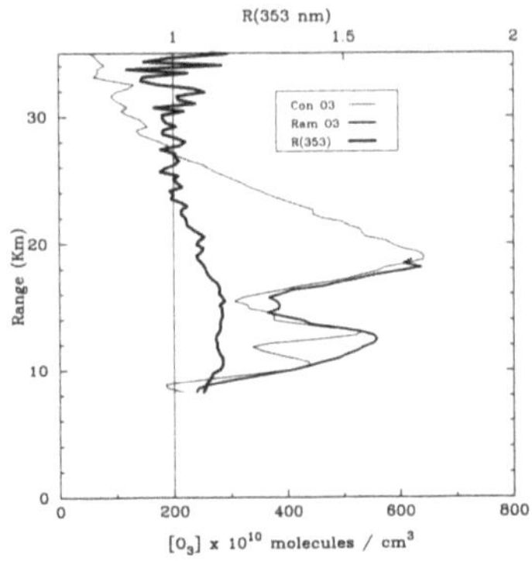

Figure 3. Ozone profile and scattering ratio (ratio of the sum of molecular and aerosol scattering to the molecular scattering alone) for Feb. 23, 1994. The trace labeled "Con O3" shows the ozone profile obtained by using the conventional DIAL method with the 353 and 308 nm returns while the line labeled "Ram O3" shows the ozone derived using the Raman N_2 backscatter returns which contain no aerosol scattering contribution.

REFERENCES

1. A.I. Carswell, A. Ulitsky and D.I. Wardle,"Lidar Measurements of the Arctic Stratosphere", *SPIE Atmospheric Radiation*, **2049** (1993) 9-23

2. G. Megie and R.T. Menzies,"Complementarity of UV and IR Differential Absorption Lidar for Global Measurements of Atmospheric Species", *Appl. Opt.* **19** (1980) 1173-1183

3. W. Steinbrecht and A.I. Carswell, "Evaluation of the Effects of Mt. Pinatubo Aerosol on Differential Absorption Lidar (DIAL) Measurements of Stratospheric Ozone", *J. Geophys. Res.* (1994)

4. T.J. McGee, M. Gross, R, Ferrare, W.S. Heaps and U.N. Singh, "Raman DIAL Measurements of Stratospheric Ozone in the Presence of Volcanic Aerosols",*Geophys. Res. Lett.* **20** (1993) 955-958

5. T. Shibata, M. Kobuchi and M. Maeda, "Measurements of Density and Temperature Profiles in the Middle Atmosphere with a XeF Lidar", *Appl. Opt.* **25** (1986) 685–688

6. J.A. Whiteway and A.I. Carswell, "Lidar Observations of Thermal Structure and Gravity Wave Activity in the High Arctic During a Stratospheric Warming". *J. Atmos. Sci.* In press (1994).

AN OVERVIEW OF THERMAL IMAGERS AND THEIR BASIC COMPONENTS

Paul Chevrette

Electro-Optics Division
Defence Research Establishment Valcartier
2459, Pie XI Blvd., North (P.O.Box 8800)
Courcelette, Québec, Can., G0A 1R0

INTRODUCTION

The primary interest of thermal imagers arises from their waveband of operation, which is their fundamental difference from visible (0.4-0.7 µm) and near infrared (0.7-2.5 µm) counterparts; it is normally located within either the 3-5 µm band or the 8-12 µm band, corresponding to atmospheric transmission windows. Within those thermal wavebands, called short wave infrared (SWIR) and long wave infrared (LWIR) respectively, the blackbody emission of the objects and background in a scene overcomes the reflected energy from the sun or any other external illuminators. Hence, in the infrared (IR) wavebands, objects are self-emitting which makes the thermal imager a day/night viewing system.

The first generation of thermal imagers, using various detector configurations with scanning mechanisms, have been on the market for many years now. The second generation is rapidly growing and taking over the market with the maturing cooled staring focal plane array (FPA) technology. Meanwhile, a third generation characterized by its low cost and compactness is already emerging, taking advantage of the new uncooled FPA technology.

This paper is an overview of the basic principles and technologies used in the design of thermal imagers and their components.

GENERIC THERMAL IMAGING SYSTEM

A generic thermal imager normally contains the following components: an objective lens system, a scanning mechanism, a detector lens, a detector with its cooling system, a detector biasing circuit and preamplifier, a video amplifier, an optional signal processing circuit, an optional post-processing circuit, and a display system. Figure 1 illustrates such a system. This arrangement may vary in simplicity up to the complete elimination of some of

Applications of Photonic Technology
Edited by G.A. Lampropoulos *et al.*, Plenum Press, New York, 1995

the components depending mainly on the detector technology used, as will become apparent in the next paragraphs, where the characteristics and particularities of the main components are described. The electronic circuitry, amplifiers and processors will be left out of the description as they pertain to a technology which can be found in any standard video system.

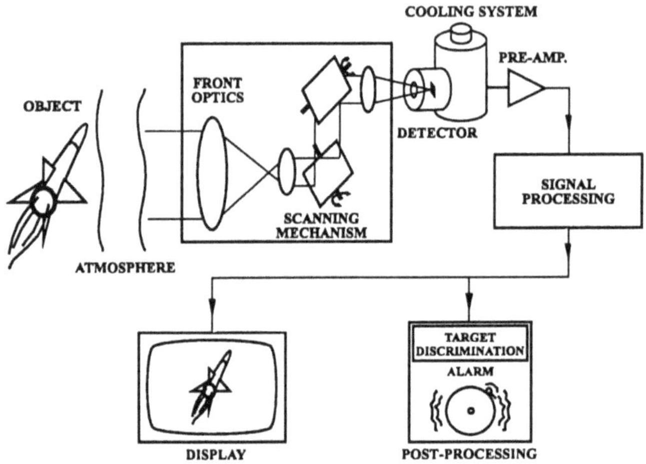

Figure 1. Components of a generic thermal imager.

FRONT OPTICS

The front optics, or objective, can be any combination of catoptric (all mirrors), catadioptric (mirrors and refractive elements) or dioptric (all refractive) elements[1,2]. The advantage of a catoptric system is that it has no chromatic aberration and hence can be used in any waveband. The most commonly used materials for refractive elements in the IR are mainly Ge, ZnSe, ZnS or Si. Other exotic materials may be selected depending on the specific application and the environment constraints. The front optics can be classified in two groups: afocal optics and focussing objectives. Afocal optics or telescopes are found principally in scanning systems. Their exit beam parallelism leaves all the required space for the scanning mechanism while minimizing the aberrations that may be introduced by scanning in a converging beam. Since the scanning is effectuated in a parallel beam, there is no axial displacements of the image in the detector plane. The system field of view (FOV) may be changed simply by changing the telescope magnification without modifying the scanning mechanism and the detector lens behind it. The effective F# of the overall system is the same as the F# of the detector lens. The scanning mechanism should be positioned in the exit pupil of the telescope in order to reduce the size of the scanning mirrors. Focussing objectives are used more commonly in imagers with FPA detectors, where the scanning mechanism has been removed. In all cases, the narcissus effect, which creates a cold spot in the center of the image as the cold detector is reflected on itself by some optical surface, should be eliminated.

DETECTOR TYPES

There are basically two types of detectors currently used in thermal imagers: the thermal detectors and the quantum or photon detectors.

In the thermal detectors, the radiant flux is transformed into heat that changes the temperature of the detector[3]. The detector materials are selected to possess a property that varies significantly with temperature, such as the resistance, the pyroelectric polarization, the junction voltage or the dielectric constant. Since the inherent detection principle is based on a temperature change, the sensitive part of the detector must be thermally insulated from its environment and does not need to be cooled. Tremendous advance has been realized in the development of these uncooled detectors over the past few years, in an effort to increase their sensitivity and arrange them in high density FPAs. Pyroelectric ceramic (barium strontium titanate) or vanadium oxide materials micromachined or etched in a suspended bridge configuration to make an insulating air gap are currently used.

The photon detector types used in the IR wavebands are mostly made of semiconductor materials[4]. Their inherent principle of operation is based on the capture of the energy from an incident photon to bring an electron from the valence band in the crystal to the conduction band. To reduce the thermal noise and the generation-recombination noise in order to increase their sensitivity, they must be cooled[5]. Photon detectors are subdivided into two categories: photoconductor (HgCdTe is the most common material for both SWIR and LWIR bands) and photovoltaic detectors (HgCdTe and InSb mainly in the SWIR band). Photovoltaic detectors are obtained by making a junction between two semiconductors of the n and p type respectively. A junction between a metal and a semiconductor is called a Schottky barrier detector. The first large IR FPA cameras to be available on the market were of this type and are based on PtSi technology (SWIR band).

More recently, a new type of photon detector has appeared, called quantum well intersubband photon detector (QWIP)[6,7]. Those detectors are fabricated by depositing several layers of GaAs and AlGaAs materials on a substrate to create wells in the conduction band of the crystal (Figure 2). Subbands of energy levels are created within those wells, the energy gap between the subbands being strictly dependant on the thickness of the AlGaAs layer. When a bias voltage is applied across the layers, electrons can be kicked out of the wells or forced to tunnel through by photons with the appropriate energy. The most obvious advantage of this type of detector is that their waveband sensitivity can be easily tuned from the near to the far infrared simply by adjusting the thickness of the layers. Multilayers of various thicknesses may be stacked in order to obtain multiwaveband detectors on the same substrate, individual waveband being selectable through the bias voltage. Moreover, QWIPs can be made into large focal plane arrays. Further development in this domain are being pursued because of the high interest into spatially co-located multiwaveband detectors for applications where the synergism of different bands may be used profitably.

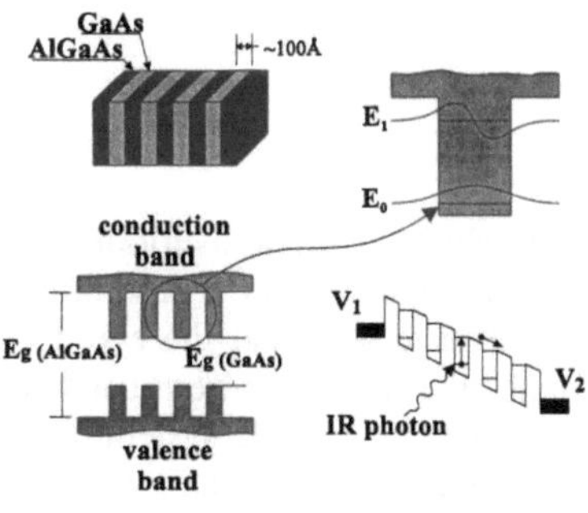

Figure 2. Multi-quantum well detector.

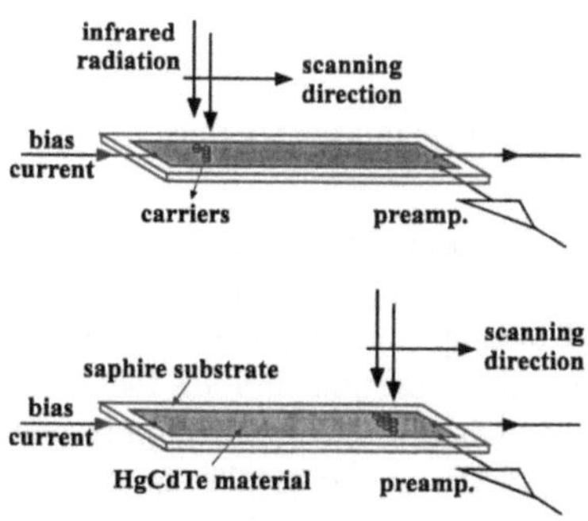

Figure 3. SPRITE detector.

DETECTOR ARRANGEMENTS

Since the advent of thermal imagers, the detector arrangement has evolved from a single detector mounted on a cold probe in a dewar to today's FPA configuration. Between those extremes, linear arrays parallel to the direction of scan, serial arrays colinear with the direction of scan wih time delay and integration (TDI) of the signal from each element, and a parallel and serial combination making a two-dimensional arrangement were produced. The signal-to-noise ratio increases as the square root of the total number of detectors. Parallel configurations reduce the required amplifier bandwidth and eliminate part of the scanning mechanism in the case of full field parallel arrays. A variant of the serial TDI array called the SPRITE detector (Signal Processing Into The Element) was developed in the UK. This is a detector which is significantly longer in the scanning direction. The scanning speed of the image is optically adjusted to match the electron drift speed within the detector material (Figure 3). This causes an integration of the charges as an image point is scanned across the detector element, giving a signal equivalent to the signal obtained in a TDI serial arrangement. The advantage is that only two wires are connected to the detector, using a single amplifier, which reduces the heat load on the cooling system. In multidetector systems, each individual detector element must be corrected for gain and offset in order to yield an overall field of reasonable uniformity.

SCANNING MECHANISMS

The complexity of the scanning mechanism depends on the detector arrangement. For a single detector, scanning needs to be performed along both X and Y axis in order to cover the complete object field[8,9]. Mirrors are usually the most effective way of scanning a field, being free of chromatic aberrations. More exotic approaches use refractive polygons or counter-rotating pair of prisms, but in addition to their non-linearity, they introduce strong chromatic aberrations and hence are not suitable for high resolution systems.

The advent of staring FPAs has eliminated the bulk of the scanning mechanisms in thermal imagers, but because of their discrete sampling of the object field at fixed positions, their resolution is now limited by the sampling theorem and is a factor of two lower than their scanned counterpart having the same size detector. To overcome this problem, a technique called microscanning or dithering is used where the image is resampled after it has been displaced a fraction of the pixel pitch over the detector array. This microscanning is performed in two dimensions in 2x2, 3x3, or 4x4 steps. An oversampling of the image is thus introduced, which tends to eliminate the aliasing problem and to approach the frequency limit of a system with continuous scanning[10]. The resolution gain here is in both X and Y axis.

VISUAL PSYCHOPHYSICS

An important part of an imaging system design which is often overlooked is the interface to the human visual system, that is the display. Particularly in the case of displays viewed through an ocular lens, care should be taken to make sure that the eye does not become the resolution limiting factor. It has been found that the eye-system bandwidth is optimized when the magnification is such that the subtended angle of the system linespread function (LSF) at the display is about 0.5 mrad, which is a little over the eye resolution limit[11]. Over- and under-magnification may result in severe degradation of system performances. Other factors to consider are the presence of flicker in the display and the ratio of surround luminance to display luminance, which should remain between 0.1 and 1 for comfortable usage.

NEW CONCEPT

Thermal imagers are currently used for general surveillance applications, search and rescue operations and, in many military scenarios, for target detection and recognition. For those purposes, they are often mounted on an aircraft and man operated for a systematic search of a given area. Current systems are scanning systems which possess two FOVs: a wide field (WFOV) for detection purpose and a narrow field (NFOV) for recognition and identification. Normally, the WFOV is used during the search process and when a target is detected, the system is switched to the NFOV for more details in order to recognize the target.

There is a fundamental problem with those systems: the scanning mechanism being the same for both FOVs, the sensitivity is much less in the WFOV than in the NFOV, leading to better detection performances in the NFOV provided the operator is pointing the imager in the right direction.

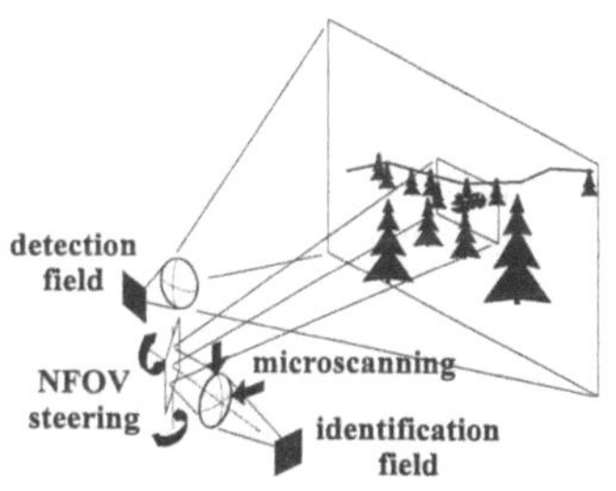

Figure 4. Dual FOV system with microscanning in the NFOV for increased resolution.

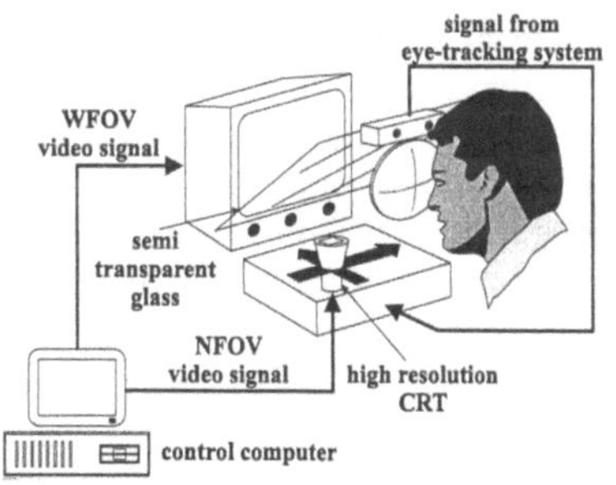

Figure 5. Adapted display with eye tracking system to position the NFOV in the central vision.

To overcome this problem, a new concept called the IR Eye is being investigated at DREV, based on the FPA technology and inspired by the principle of operation of the human vision. The human eye has multiple FOVs of different resolution and sensitivity, the central vision having the highest resolution and lowest sensitivity while the peripheral vision possesses a very high sensitivity but a low resolving power. The IR Eye concept is an attempt to simulate the human eye behavior using two FPAs covering simultaneously two FOVs: a low resolution but high sensitivity WFOV of about 60° for search and detection, and an NFOV of the order of 6°, mobile within the WFOV, with high resolution but lower sensitivity, for recognition and identification. The increased sensitivity of the WFOV is obtained by adjusting the integration time and frame rate of the FPA. The high resolution of the NFOV is the result of a microscanning (up to 4x4 steps) of the image on the FPA. The concept will be demonstrated in the 3-5 µm band since the FPA in this band are readily available.

Figures 4 and 5 schematically illustrates the system and its adapted display. The display will present a fusion of images from both FOVs on a 1:1 magnification ratio, with sufficient resolution for the individual pixels to be resolved by the central vision of the observer. To surmount the resolution problem brought by the display of an oversampled image (NFOV) of 1024x1024 pixels, covering about 1/10 of the overall field (WFOV), an optical projection is used whereas two CRTs (or LCDs) are viewed by the observer superposed through a half-reflecting glass. The projection is seen by mean of a biocular lens which takes care of the 1:1 magnification ratio between the sensor and display FOVs. The high resolution CRT (LCD) presenting the NFOV is mounted on an X-Y translation table in order to match the position of the corresponding optics within the WFOV. The area of the WFOV corresponding to the NFOV is blacked out to avoid interference between the two images. Ultimately, the NFOV optics and display will be slaved to an eye tracking system, allowing to systematically present the high resolution image to the central vision of the observer and to take advantage of the high detection sensitivity of his peripheral vision over the WFOV area, thus making a perfectly adapted display. A variety of image projection schemes are being investigated.

CONCLUSION

The basic principles of operation of thermal imaging systems have been described as well as the main technologies used in the fabrication of their components. Based on the new FPA detector technology, a new concept for improving the performances and displaying the image information of surveillance systems for search operation has been put forward. It is obvious that the future of thermal imagers lies on the rapid growth of FPA technology and scanning systems are bound to disappear as the requirement for higher resolution isotropic imagery becomes more stringent. On the other hand, the advent of uncooled highly sensitive FPAs is a promise for low cost, light weight and small volume systems which could bring an oubreak of the use thermal imagers in the industry and day to day life.

REFERENCES

1. G. Gaussorgues, Les optiques utilisées en infrarouge, *in*: "La Thermographie Infrarouge," Technique et Documentation, Paris (1981).
2. J.M. Lloyd, Optics, *in*: "Thermal Imaging Systems," Plenum Press, New York (1975.
3. R.E Flannery and J.E. Miller, Status of uncooled infrared imagers, *Proc. SPIE Infrared Imaging Systems* 1689:395 (1992)
4. G. Gaussorgues, Les divers types de détecteurs quantiques, *in*: "La Thermographie Infrarouge," Technique et Documentation, Paris (1981).
5. K. Seyrafi, Noise in optical systems, *in*: "Electro-optical system analysis", Electro-Optical Research Company, Los Angeles (1973).
6. B.F. Levine, Quantum-well infrared photodetectors, *Journal of Applied Physics* 74:1 (1993).
7. H.C. Liu,J. Li, J.R. Thompson, Z.R. Wasilewski, M. Buchanan, and J.G. Simmons, Multicolor voltage-tunable quantum-well infrared photodetector, *IEEE Electron Device Letters* 14: (1993).
8. G. Gaussorgues, Les systèmes d'analyse spatiale, *in*: "La Thermographie Infrarouge," Technique et Documentation, Paris (1981).
9. J.M. Lloyd, Scanning mechanisms, *in*: "Thermal Imaging Systems," Plenum Press, New York (1975.
10. F.P. Blommel, D.J. Bradley, P.N.J Dennis, The effect of microscan operation on staring infrared sensor imagery, *SPIE Infrared Technology XVII* 1540:653 (1991).
11. J.M. Lloyd, Visual psychophysics, *in*: "Thermal Imaging Systems," Plenum Press, New York (1975.

REFRACTION AND LONG-RANGE FLIR IMAGES

A.J. Beaulieu

BCI Co.
Ste-Foy, Québec
Canada, G1W 1L9

INTRODUCTION

With the advent of focal plane arrays, IR sensors are increasingly being used for long-range applications with increased resolution to allow object identification. For sea applications the detection of near surface objects is often limited by the horizon and substantial image distortions due to atmospheric refraction are encountered. In some cases, targets can be detected well beyond the geometrical horizon limit while under different conditions, the horizon limit can be much smaller than the geometrical value and mirages can occur. For the prediction of the performance of the IR sensors, it is essential to take these effects into account. Ship-borne IR sensors to detect near sea-level targets, such as ships or low-altitude aircraft, are particularly sensitive to the refraction phenomena. This is why accurate models have been developed to cover the Marine Boundary Layer (MBL) which is the atmospheric layer immediately above the water surface. The predictions of these models are so significant that a sea-shore measurement program was undertaken to verify the validity of the models and to give a higher degree of confidence in its application to a variety of weather scenarios. The present report summarizes the model development, its experimental validation and the system implications of the effects of refraction.

MBL MODEL

In the boundary region extending from the sea surface to approximately 50 meters altitude, the most significant refraction effects are caused by the large gradient of the temperature and the water vapor density in this region. The effects of refraction in this region were first noticed with radar where the large difference in the refractive index of the water vapor is the most significant factor leading to the "ducting" effect for radars. The origin of the phenomena is the rapid variation of the temperature and the water vapor near the surface of the sea. Numerous models have been developed which are all based on the steady state conditions that exists in a "well-mixed" gas. The uniform mixing is due to the turbulence caused by the wind and the vertical behavior of meteorological parameters can be described by a logarithmic expression.

In the development of a meteorological profile, the model which has been retained by DREV as the most detailed is that of Walmsley[1], which was extended by Kel Research and DREV to include the water vapor effects. The type of expression obtained for the air temperature is of the form:

$$T_a(z) = T_{sea} + A \ln(z/z_o) + \Psi(z/L)$$

where z is the height, T_{sea} is the sea temperature, A and z_o are constants, L is the Monin-Obukhov length and $\Psi(z/L)$ is a fine correction term which is dependent on the stability of the atmospheric conditions. This was called the WKDMBL model.

The gradient of the temperature, the water vapor and the atmospheric pressure give rise to a variation of the refractivity profile, which is dependent on the wavelength and which is particularly strong at radar wavelengths. For IR propagation, it is the temperature and the pressure gradients which dominate since the refractive index of the water vapor is comparable to that of dry air at these wavelengths. There exists two refraction categories depending on whether the light rays are bent towards the sea surface or away from it. The first category, which produces super refraction, is representative of conditions where the air temperature is higher than that of the sea and corresponds, in general, to what is called "stable" conditions in meteorological terms. The "unstable" conditions exist when the air temperature decreases with height and gives rise to sub refraction. These are the more frequent conditions.

Stable conditions correspond to conditions where the ray bending allows an IR sensor to see beyond the geometrical horizon limits. When strong enough, this effect makes it possible to see objects on the surface at an infinite range. Sub refraction conditions have the opposite effect in that it shortens the sea-horizon limit. However, the more evident characteristic of sub refraction is the appearance of mirages. When observing objects beyond the sea horizon range, rays which travel nearest the water surface are more strongly bent upwards than higher rays. A "critical ray" can be defined as the lowest ray that does not touch the water surface and the depression angle of that ray is referred to as the horizon depression angle. At ranges beyond the sea horizon range, the critical ray being the most strongly refracted one will be higher than that corresponding to slightly higher angles. At some given range this critical ray will cross another ray corresponding to a substantially higher look angle. An object at that point of intersection will therefore be observed at two simultaneous different elevation angles, giving rise to two separate images. This is a form of mirage which can often be observed although this is often mistakenly regarded as a water reflection.

To analyze the refraction effects, it is necessary to compute the path of light rays for a large number of observation angles. If one assumes that the water surface is perfectly flat, the critical ray could be considered to pass within a fraction of a millimeter of the surface. It would then experience a very large refraction since the gradients are extremely large very close to the surface. This is where the wave height comes in. If the waves were perfectly sinusoidal, with a peak to peak amplitude of H_w, the critical ray would have its lowest height above the mean sea level of $\frac{1}{2} H_w$. This ray would only come very close to the water surface when it is near the top of a wave and experience a much reduced average angular refraction. A model taking this phenomenon into account has been developed at DREV and is known as the Wavy WKDMBL model or the WWKD model.

EXPERIMENTAL METHOD

To evaluate experimentally the refraction conditions, it is necessary to observe images of targets at long ranges and measure the image distortions in the vertical direction, and/or to determine the horizon limit. Well-calibrated objects can be observed with visible cameras or with FLIRs, the advantage of the visible region being a higher resolution for a given optics size and should give results very similar to IR sensors. Some IR measurements must also be done to validate the near equivalence of visible and IR propagation predicted by the model. A number of trials were carried out over the last 3 years. The use of high power lamps proved to be quite adequate point sources for the visible and mid-IR wavebands while propane heaters were used for far-IR imagers. The experimentally simplest method consists in using two such sources, one being set at a given height and used as a reference while the height of the second one is varied to provide a mapping of the angular separation as a function of the height difference. This method however was found to lead to a data reduction problem, the number of images to be analyzed being large and the accuracy of the measurements of the physical separation was not as high as would be desired under many experimental conditions. For an extended trial, the best method is to use an array of lights or small targets that only requires the processing of a single image for the measurements of the angular separations and the physical measurement of the sources' separation has to be carried out only once, allowing for a greater precision and repeatability of the data.

The most complete trial was held last October as part of an international exercise called MAPTIP, which was hosted by Holland and took place over the North Sea near the Dutch coast. For this trial, Holland provided a meteorological platform that was some 10 Kilometers off shore and a research ship which was dedicated to this exercise and would run along predetermined tracks in an area covering from 1 Km off shore to over 30 Km in a 100 degree sector. An array of 8 lights was mounted on the platform. The ship was set-up with an array of 6 lights distributed from the lower deck to the top of the mast which were used as reference targets as well as the chimney for the IR sensor. The ship's position was recorded with a GPS system with an accuracy of better than 50 m and a one second time resolution.

Three different imaging systems were used by the DREV team. A visible and an IR camera were located side by side in a beach house with their observation axis 7 meters above the mean water level while another visible camera was mobile could be located anywhere on the beach or near-by dunes so that its observation heights could be changed from 3 to 12 meters above the sea level. These cameras were equipped with high resolution telescopes with focal lengths of 1.4 and 2.0 meters. In all cases the pixel resolution was better than 10 microradians or 2 seconds of arc. Simultaneous observations were made by different participants such as the German team who was located some 13 meters above sea level and accurate ranges were provided by the Norwegian group using laser range finders.

At least once a day, the ship would follow a radial course starting from a point near the beach where the observation equipment was set-up and sailed at a constant speed toward a predetermined point some 30 Km away. During those runs, the aspect of the ship was nearly constant except when it had to avoid the local fishing traffic. The images of the ship were continuously recorded on video tapes for the entire linear run. Since the ship had to meet the requirements of other participants, most of the day was spent sailing along trajectories that were not suitable for our type of analysis. To follow the evolution of the refraction conditions during the entire day, the cameras were oriented to observe the lights on the meteorological platform. The apparent position of the lights on the platform could be determined with an accuracy of the order of approximately 10 cm providing sufficient precision to evaluate the refraction profile.

Another advantage of the participation in this joint-trial was that the meteorological conditions were observed by different groups at different points such as the meteorological platform, from a buoy some 3 Km off shore and from the ship as it traveled throughout the operation zone. The associated difficulty was in resolving the ambiguities between the different and sometimes conflicting weather reports.

METHOD OF ANALYSIS

The ship's results provided the simplest method of analysis since one only needs to observe the time of occurrence of certain events such as the disappearance of a light beyond the horizon or the appearance of the mirage of a light at a known height above the sea level. The ship's position is obtained from the ship's log. The inaccuracies of this technique are dependent on the sea state which affects the pitching and the heave of the ship. To account for ship's motion, the time of certain events must therefore be based on the frequency of appearance of a phenomenon. For example, the horizon masking of a given light is taken as the time at which the light disappears 50% of the time over a period longer than the natural ship's pitch or roll frequency (10 seconds). Since this depends on the observer's estimate, it contains a certain degree of error. These uncertainties are overweighed by the ability to make observations at very long ranges (up to 25 Km) where the refraction effects are maximized.

The analysis of platform lights is fundamentally more accurate because of the absence of target motion but more limited because they are at a fixed range and require verification by correlating two different sets of images taken from different heights. They also represent a much greater data reduction effort, the video images having to be selected, digitized and digitally analyzed for accurate information. The results of these analyses are then compared with the model prediction to detect divergences between measurements and predictions.

It has been found that the meteorological conditions, particularly the air and sea temperatures, the wind velocity as well as the sea level and the wave height all play a significant role in the accurate prediction of the refraction effects. The uncertainties with the provided weather data require that the various parameters be slightly readjusted to obtain a very good match. For example, an adjustment of 0.1°C of the air temperature was required to obtain a good match in a given condition, which is well within the margin of uncertainty of data provided. This adjustment is however very time consuming since for every change of meteorological conditions, the profile of the refractivity has to be calculated and corresponding ray tracing computed for a large number of observation angles to determine the image distortion pattern at the range of the observation. Each of the main meteorological parameters mentioned above affects the image distortion in a different way and to determine the matching conditions requires the evaluation of a fairly large set of meteorological data. In general, it has been possible to establish that the model used is accurate although in some cases the validity region can be seen to decrease to altitudes well below the 50 meters that is generally assumed to a value as low as 10 meters.

Another effect of refraction is to change the effective contrast of the target with respect to the background if the target's image is smaller than the resolution of the sensor, including the atmospheric transmission effect due to the turbulence and forward scattering by the atmosphere. The point spread function in the visible wavelength can be larger than 100 μr under certain conditions which means that the image of a point target is distributed over such an angle. As images become smaller than this point spread angle of the angular resolution of the imager, the target contrast decreases. Thus for super refraction conditions, the over the horizon targets that can be detected are compressed and become smaller than the sensor resolution, which means that the contrast decreases with range. On the other hand, under sub refraction conditions, the image of objects very near the horizon masking limit will appear very extended, which increases the contrast especially if the sensor has a large instantaneous field of view.

IMPACT ON IR SYSTEMS

The effects of refraction are very significant on the detection range of ship-borne IR sensors which are used to detect low altitude objects at long ranges. Under sub refraction conditions, which occur over 70% of the time, the horizon masking range can be as little as one half the geometrical value. Under these conditions, an approaching object will first appear as a very high-intensity object some 0.5 mr above the edge of the water at the horizon. As in gets closer, this image will split into two, the lower image being an inverted image and gradually approaching the sea edge. The apparent intensity of the objects changes as the images become more compressed. Then as the range is further decreased, the lower image suddenly disappears. Low resolution imagers might well detect the appearance the of object when it is near the horizon masking limit because of the image expansion, and after a short period looses the target that has shrunk to its natural size, and only reacquires it when much closer. Most common false alarm rejection systems would ignore the first detection as being too short to be a real target. Signal growth algorithms which are sometimes used to estimate the range of approaching targets and false target rejection, are completely baffled by such a behavior since the apparent signal decreases as the target approaches.

Under super refraction conditions, a similar approaching target would first be detected at the water line but its image would be compressed vertically. Thus the maximum detection range is strictly limited by the sensitivity of the sensor and not the horizon masking. As such a target approaches, its apparent angular position would remain almost constant but the contrast would grow more rapidly than normal because the image compression decreases with range. This could be interpreted as a much closer target by signal growth signal processors.

CONCLUDING REMARKS

The WWKD model for predicting atmospheric refraction effects has proven to be quite accurate in the majority of some 100 meteorological conditions during which accurate image distortion effects have been measured. Unfortunately, no strong super refraction conditions were experienced during the most recent trials and earlier measurements of these phenomena were not refined enough to provide an accurate validation. In a few cases, there was some evidence that the marine boundary layer conditions did not extend above 10 meters and that a definite layering of the vertical profile of the atmosphere occurs, especially in the afternoons and when the wind was coming from the land. This is probably a coastal effect which would not be met in high seas conditions. However, since naval operations often take place near shores, this phenomenon needs to be explored further.

The model verifications have been done primarily at visible and mid-IR waveband and not at radar wavelengths where the effects of water vapor are dominant. The validation of the model at radar radiations is much more complex because the inherently lower resolution of radar receivers and the multipath problem associated with the specular reflection of the radar radiation by the sea surface. Long range radar measurements will be needed to further validate the model.

REFERENCE

1. Walmsley, J. L., "On theoretical wind speed and temperature profiles over the sea with applications to data from Sable Island, Nova Scotia", Atmosphere-Ocean 26 (2), pp. 203-233. (1988)

MIRAGE FORMATION IN THE INFRARED REGION AND ITS EFFECTS ON THERMAL IMAGING APPLICATIONS

R.N. Singh, S.S. Negi, A.K. Sahay, and V. Rajeshwar

Instruments Research and Development Establishment
Raipur, Dehradun - 248008 (India)

Abstract

During one of our recent trials to evaluate the performance of an advanced target acquisition system comprising thermal imager, CCD camera and Laser range finder, mirage effect in 8-12 μm band has been prominently observed.These observations have been validated with some of the existing theories for mirage formation in terms of maximum and minimum ranges and the height of the target. Salient design features of thermal imager has been outlined. Apart from mirages, the effect of temperature on image quality and range performance of thermal imager has also been studied. A near Real Time Image Processing algorithm has been proposed to suppress mirages and improve the image quality.

Introduction

Over the centuries, scientists have been fascinated by "Optical Illusions". One such phenomenon is appearance of "Mirages" which are caused by curvilinear propagation of light in an optically stratified medium having abnormal vertical distribution of temperature and pressure.

Basic physics of mirage can be well understood by Snell's law of refraction. However, a number of theories have been proposed to explain the phenomenon and their utility[1-4]. The phenomenon of mirage has been extensively studied in the visible region but little information exist for IR region. Feinberg et al[5] have investigated the magnitude of refractivity using CO_2 Laser transmission close to ocean surface. However, for the first time, Singh et al[6] have recorded thermal picture of mirage formation in 8-12 μm.

The present paper is in continuation of our earlier observation which has been validated on the basis of conditions established by Sodha et al[4]. Effects of temperature on image quality and range performance have also been given. Near real time image processing algorithms have been suggested to improve the image quality in presence of mirages.

Thermal Imager

Thermal Imager, {Fig (1) & optical scheme Fig (2)}, operates on the principle of parallel scan with 2:1 interlace utilizing a linear MCT array cooled to 77 K. The front end of the thermal camera is an afocal telescope . A high precision glass polygon scanner, having 12 facets, operates in between the afocal telescope and the final image forming lens system. An optical encoder linked to the shaft of the scanner drive assembly produces blanking and synchronisation pulses for the CRT. The objective of the NFOV channel has an RC system and that of WFOV is a single germanium lens . The resolutions in the two FOVs are 72 μrad and 250 μrad respectively. A digital scan converter has been incorporated to display images in a CCIR-B format. The MRTD plot of the system is given in Fig (3) which correlates very well with the field performance. Thermal photographs of 'A' & 'B' vehicles are given in Fig (4).

A high resolution visible CCD camera has also been used to compare mirage formation in the two wavebands.

Experimental Conditions and Observations

This investigation was carried out in a flat barren desert land during day time from 0900 to 1600 hrs (IST). Atmospheric temperature varied from 26° to 35° C and RH from 78% to 25%. Sighting devices were tripod mounted at a height of 3.5m above the ground . Target under observations were vehicles of height 2.3m. The distance between target and devices was varied from 1 Km to 5 Km. The inferior mirage of the targets located beyond 2 Km could only be observed in both visible and IR cameras.

Thermal & CCD images of the target with mirage at 4 Km are reproduced in Fig (5) & (6) respectively. As per Sodha et al the maximum and minimum ranges at which mirage could be seen are

$$R_{max} = \Pi - \sin^{-1}\{\exp(-Y_D)\} - \sin^{-1}\{\exp(-Y_T)\} \qquad (1)$$
$$R_{min} = \Pi/2 - \sin^{-1}\{\exp -(|Y_D|) - (|Y_T|)\} \qquad (2)$$

Where Y_D and Y_T are normalised device and target heights expressed as : $Y_D = \alpha(Y_d)$ and $Y_T = \alpha(Y_t)$, here 'α' being refractive index profile parameter, Y_d and Y_t being absolute heights of the device and target in meters respectively.

During the course of experiment, temperature and pressure were recorded {Fig (7)} at different heights from the ground to observation locations. According to Edlin[7], refractivity 'N' is given by,

$$N = (\mu - 1) \times 10^6 = \{77.6 + 0.584/\lambda^2\} P/T \qquad (3)$$

Where 'μ' is refractive index of air, 'P' the barometric pressure in mbar and 'T' the temperature in Kelvin. Refractivity data over measured temperature and pressure range have been generated for both visible (0.5 μm) and IR (10 μm) region {Fig (8)}.

Further to calculate R_{max} & R_{min} 'α' may be considered as exponential function:

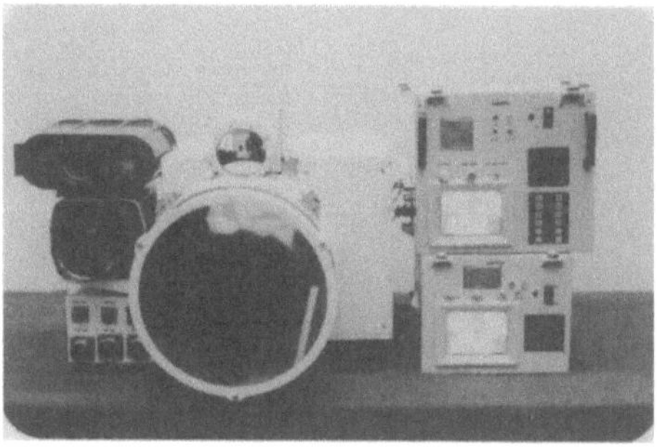

- SPECTRAL RANGE : 8-12μm
- SCANNING : PARALLEL
- FIELD OF VIEW
  - NARROW : 1.5^{0}X0.8^{0}
  - WIDE : 5.2^{0}X2.8^{0}
- OPTICS : 240MM, f2 RC SYSTEM
- DETECTOR : MCT ARRAY
- VIDEO OUTPUT : CCIR-B
- IFOV (WFOV) : 250μrad
  - (NFOV) : 72μrad
- RANGE
  - DETECTION : ≥ 5Km
  - RECOGNITION : ≥ 4Km

FIG. 1. THERMAL SIGHTING SYSTEM

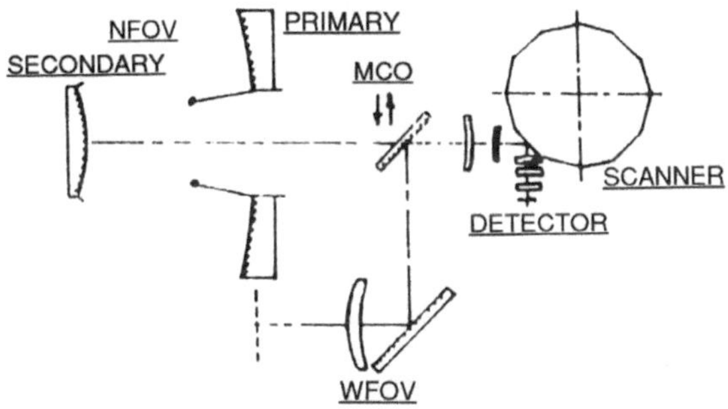

FIG. 2. OPTICAL LAYOUT OF THERMAL IMAGER

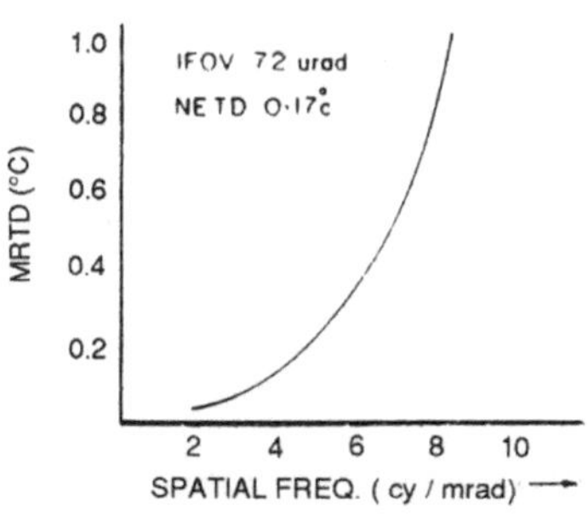

FIG. 3. MRTD PLOT

FIG. 4. NORMAL THERMAL IMAGE (4 Km)

FIG. 5. THERMAL MIRAGE (4 Km)

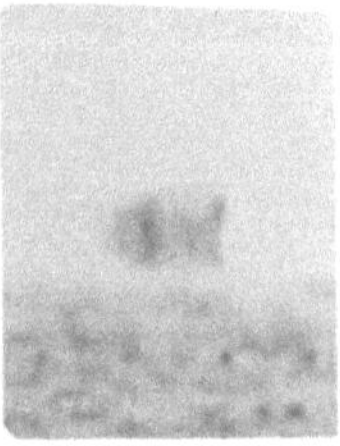

FIG. 6. VISIBLE MIRAGE (4 Km)

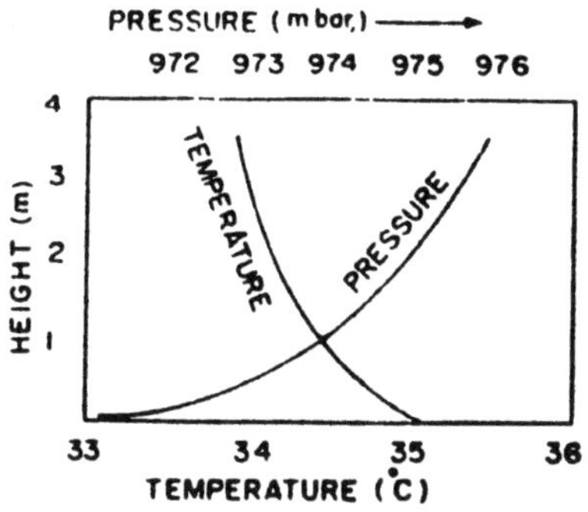

FIG. 7. TEMPERATURE AND PRESSURE PROFILE

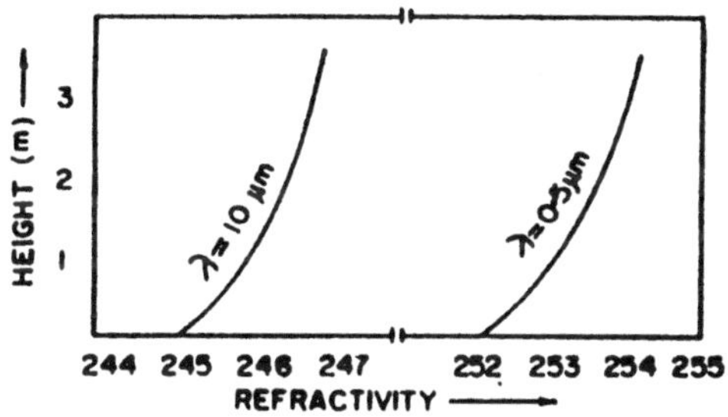

FIG. 8. REFRACTIVITY PROFILE

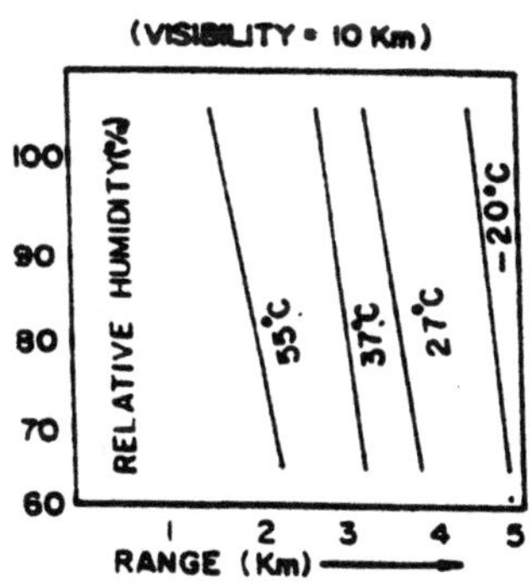

FIG. 9. EFFECT OF TEMPERATURE

$$\mu_h = \mu_o \exp(\alpha h) \tag{4}$$

Where μ_h & μ_o are the refractive indices of air at h meters height & at the ground. Using equations (3) & (4), α is found to be 6.12 x 10^{-7}/m, and 6.30 X 10^{-7}/m for IR and visible wavelengths. This shows that quality of image in presence of mirage is degraded to the same extent in both wave bands. From equation (1) & (2) the maximum and minimum range for IR is 6.12 Km & 1.99 Km and for visible it is 6.08 Km & 1.95 Km. These theoretical results are in close agreement with our experimental observations for the given target & observer system.

Image quality of thermal imager is also dependant on the prevalent temperature and humidity conditions. MRTD equation embodying these factors is used to evaluate recognition ranges for a given set of system parameters {Fig(9)}. Our analysis indicates that ambient temperature is more vital.

Correction For Mirage

To suppress effect of mirages, real time image processing technique need to be resorted to. However, in present case, a simpler image folding technique has been proposed. Suppression of mirages could be achieved by first isolating the mirage affected images from the total scene (to be identified by contrast) just above the ground followed by finding the lines of symmetry in the isolated parts of the scene. In the final display the portions below the lines of symmetry are omitted in each frame.

Acknowledgement

Authors express their gratitude to Dr. APJ Abdul Kalam, Director General, Defence Research & Development Organisation, Govt. of India and Dr. OP Nijhawan, Director, Instruments Research & Development Establishment, for their constant encouragement and interest in the work.

References

1. Raman, C.V. and Pancharatnam, S., 1959, Proc. Ind. Acad. Sci. A, 49, 251-61.
2. William, H.Mach and Alistair, B. Fraser, 1979. Applied Optics, Vol. 18, No.11.
3. Lehn. W.H,1983, Jr. Opt. Soc. Am. Vol.73, No.12.
4. Sodha. M.S. et al . Brit J. APPL. Phys., 1967, Vol.18.
5. Feinberg, R., Hughes, H.G, 1979, Appl. Optics, Vol.18, 1915.
6. Singh, R.N, et al., 1994, Applied Optics, Vol.33.
7. Hand book of Geophysics (Macmillan, New York, 1960), Chap.13

PULSED LASER PHOTOACOUSTIC SPECTROSCOPY OF GASES: TRACE GAS DETECTION AND MULTIPHOTON ABSORPTION STUDIES

D.D. Markusev[1)], M. Terzic[2)], P. Vujkovic Cvijin[1)] and J. Jovanovic-Kurepa[1)]

[1)] Institute of Physics, P.O. Box 57, 11001 Belgrade, Yugoslavia
[2)]University of Novi Sad, Institute of Physics, P.O. Box 224, 21001 Novi Sad Yugoslavia

INTRODUCTION

Laser photoacoustic spectroscopy in the midd-infrared spectral region is a powerful spectroscopic technique usable both for highly sensitive trace gas detection (linear regime) and for the study of multiphoton absorption (nonlinear regime). For the latter, high peak power pulsed laser excitation is typically used. The experimental technique is applicable both for atmospheric trace gas monitoring (linear absorption), and for the study of processes leading to the generation of advanced materials (multiphoton spectroscopy).

The technique of linear pulsed laser photoacoustic spectroscopy provides the temporal resolution of the photoacoustic signal, which makes it possible to separate the sample gas absorption from the background window and cell wall absorption signals, resulting in an improved signal/noise ratio. The rejection of a large portion of background signals by temporal resolution leads to a high signal/noise ratio[1,2]. The experimental technique and high detection sensitivity was demonstrated previously by investigating the detection of SO_2 in the ultraviolet spectral region[1]. It is of great interest to extend the technique into the middle infrared, where an abundance of fundamental absorption transitions of many gases of interest exists.

EXPERIMENTAL APPARATUS

The excitation radiation is provided by a line tunable pulsed TEA CO_2 laser (up to 300mJ TEM_{00} mode, 1 J higher order mode, spectral range 9.2μm-10.8μm). The photoacoustic cell is of cylindrical shape, with an extension placed opposite to the microphone in order to avoid the interference of the acoustic wave reflected from the cell wall[1]. The cell is closed at each end with Brewster angle windows and it incorporates a condenser microphone to monitor the acoustic signal. The acoustic waveform and the laser excitation energy level are digitized by a waveform recorder and analyzed by a computer.

Applications of Photonic Technology
Edited by G.A. Lampropoulos *et al.*, Plenum Press, New York, 1995

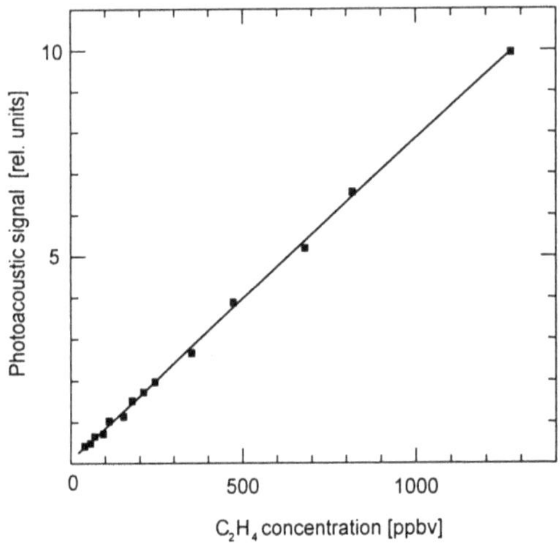

Fig. 1. Photoacoustic signal vs. ethylene (C_2H_4) concentration.

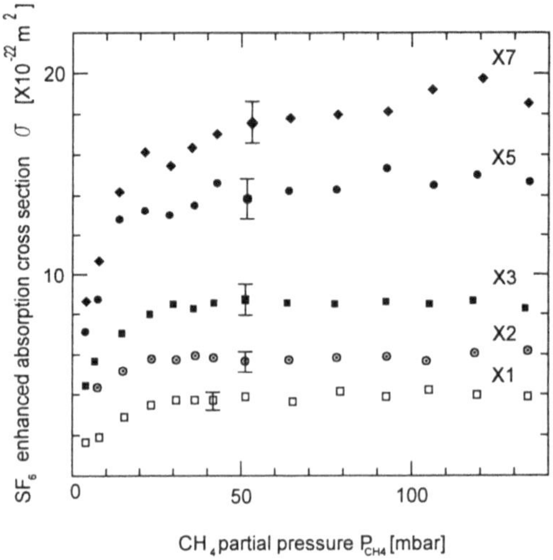

Fig. 2. The enhanced absorption cross sections σ for SF_6-CH_4 mixtures irradiated by 10P(16) CO_2 laser line. The partial pressure of SF_6 is 0.47 mbar. □ - fluence ϕ =0.18 J/cm^2; ⊙ - fluence ϕ =0.23 J/cm^2 (multiplied by 2); ■ - fluence ϕ =0.26 J/cm^2 (multiplied by 3); ● - fluence ϕ =0.29 J/cm^2 (multiplied by 5) and ◆ - fluence ϕ=0.42 J/cm^2 (multiplied by 7).

EXPERIMENTAL RESULTS

The sensitivity of the pulsed laser photoacoustic apparatus operating in the linear regime was investigated by using ethylene (C_2H_4), whose absorption coefficients are well known[3]. For the detection sensitivity studies, the 10P(14) CO_2 laser line with the highest ethylene absorption coefficient was used[3]. Excitation energy of 100mJ per pulse was found to be appropriate, still avoiding optical saturation. The pulse repetition frequency was 1 Hz. The experimental results of trace gas detection of ethylene with our pulsed photoacoustic spectrometer are presented on Fig.1. The detection limit was found to correspond to 1.5 ppbv of ethylene, which results in the minimum detectable absorption coefficient of 4.9×10^{-8} cm^{-1}. This detection sensitivity compares favorably with the sensitivity of modulated cw technique[4]. It is achieved here with an experimental apparatus which is simpler, less susceptible to acoustical interferences, and more robust for field use than amplitude modulated cw photoacoustic instruments.

High peak power associated with the TEA CO_2 laser excitation suggests the use of the photoacoustic spectrometer described above for the study of multiphoton excitation phenomena[5]. The apparatus for multiphoton absorption spectroscopy differs from the one described above mainly in the amount of the excitation energy (up to 1J), the use of focusing lens to increase the energy density, and in the photoacoustic cell construction (cell adapted for high vacuum). Experimental apparatus was developed for simultaneous absorption measurements, both by transmission and by the photoacoustic spectroscopy. The effects of collisions of SF_6 gas with nonabsorbing partner molecules (CH_4) were investigated. The enhanced absorption cross sections of SF_6 in collision with nonabsorbing CH_4 molecules were measured. The results are presented on Fig.2, for the fluence range of 0.18-0.42 J/cm^2. The partial pressure of SF_6 was 0.47 mbar. The accuracy of the measured cross sections, given by error bars indicated on Fig.2, is 7-16%. The experimental results obtained were analyzed based on the predictions of the generalized two-level model proposed for multiple photon processes. The analysis showed that the rotational relaxation process is the dominant process responsible for the enhanced absorption by the SF_6 molecule, undergoing collisions with CH_4 (nonabsorbing buffer gas) at $p_{CH4} \leq 25$ mbar. At $p_{CH4} > 25$ mbar, both vibrational and rotational relaxation processes contribute with similar efficiency.

CONCLUSIONS

The experimental results, together with the inherent simplicity of design, suggest that the linear pulsed photoacoustic technique is potentially advantageous for trace gas detection, relative to the classical modulated cw approach. To obtain the same sensitivity, modulated cw excitation is usually enhanced by either multipassing the excitation beam through the sample cell, or by using acoustically resonant operation, or both. The technique presented here is also demonstrated to be a valuable tool for the study of multiphoton absorption, which we plan to apply further to the study of laser-induced chemical processes leading to the generation of new materials, e.g. advanced ceramics[6].

REFERENCES

1. P. Vujkovic Cvijin, D.A. Gilmore, M.A. Leugers and G.H. Atkinson, *Anal. Chem.* **59**, 300 (1987)
2. P. Vujkovic Cvijin, D.A. Gilmore and G.H. Atkinson, *Appl. Spectrosc.* **42**, 770 (1988).
3. J. Hinderling, M.W. Sigrist and F.K. Kneubuhl, *Infrared Phys.* 27, **63** (1987).
4. M.W. Sigrist, *J. Appl. Phys.* **60**, R83 (1986).

5. J.L. Lyman, G.P. Quigley and O.P. Judd, in: *"Multiple-Photon Excitation and Dissociation of Polyatomic Molecules"*, ed. by C. D. Cantrell (Springer-Verlag, Berlin, 1986).
6. E. Borsella, L. Caneve, R. Fantoni, S. Piccirillo, *Il Nuovo Cim., Note Brevi*, **11D**, (N.3), 487-493 (1989).

3-D BLUR ESTIMATION AND RESTORATION OF SEQUENTIAL SPACE-BASED INFRARED IMAGES

George A. Lampropoulos[1] and James F. Boulter[2]

[1]A.U.G. Signals Ltd., 560 Lauder Ave.
Toronto, Ontario, M6E 3J6
[2]Defence Research Establishment Valcartier
2559 PIE XI North (P.O. Box 8800), Courcelette
Quebec, Canada G0A 1R0

1. INTRODUCTION

Infrared satellite images are often of low resolution and degraded by additive noise, linear blurs, and nonlinear effects due to scintillation and jitter noise. By using sequential frames, the presence of nonlinear effects may be suppressed. In fact, nonlinear effects might contribute to improvements of such image resolution by applying multi-frame processing where sub-pixel resolution approaches are used.

Objects hardly seen in infrared images shall be restored prior to any other image processing operation. An infrared image processing system could be briefly described by the following three blocks. In the first block, image calibration and conditioning is taking place. The second block is for image restoration. In an image restoration process often a number of input image parameters are required. We call them restoration parameters and they are estimated on line. These restoration parameters differ from one restoration algorithm to other. Some restoration algorithms are more sensitive to the accuracy of these parameters than others. Finally, in the third block, feature extraction algorithms are applied for applications such as: target detection, edge detection, discrimination, identification, and recognition. The selection of the appropriate features or algorithms for a specific application is a very difficult task. It should be emphasized that with a feature selection process one often rejects information which might be extremely important to the intended application of a user.

In this paper we address ways to estimate the parameters of the blurring function. For restoration of multi-frame images we apply three-dimensional (3-D) restoration algorithms developed by A.U.G. Signals Ltd. in cooperation with the Defence Research Establishment Valcartier. In these algorithms, the well known parametric Wiener and inverse filters are special cases. This work was supported by the Defence Research Establishment Valcartier under the contract W7701-1-2591/01-XSK.

2. LINEAR IMAGE RESTORATION

Linear image restoration is applied in order to reduce additive noise and the effects of linear blurring functions. When linear blurring functions are singular, the restoration process becomes complicated. Often, as in the Wiener filtering, the additive noise is used as a regularization parameter for reducing the artifacts of the inversion process (deblurring). In other cases, such as the Constrained Least Squares (CLS) and Resolution to Noise Tradeoff (RNT) algorithms, additional regularizing functions and parameters are used to reduce the inversion artifacts in the deblurring process at the zero points of a blurring function spectrum without introducing additional artifacts at the other regions of the spectrum. In this paper the 3-D CLS and RNT are used for testing the blur estimates on real images. It should be noted that the inverse and Wiener filters are special cases of the above two algorithms.

In the following the CLS algorithm is outlined for completeness of the 3-D restoration section. Its closed form solution is given by

$$\hat{\mathbf{f}} = [H^t \Lambda H + \alpha C^t C]^{-1} H \Lambda \mathbf{g} \tag{1}$$

where:

$$\mathbf{g} = \begin{bmatrix} \mathbf{g}_1 \\ \cdot \\ \cdot \\ \cdot \\ \mathbf{g}_N \end{bmatrix} = \begin{bmatrix} H_{11} & \cdot & \cdot & \cdot & H_{1N} \\ \cdot & \cdot & \cdot & \cdot & \cdot \\ \cdot & \cdot & \cdot & \cdot & \cdot \\ \cdot & \cdot & \cdot & \cdot & \cdot \\ H_{N1} & \cdot & \cdot & \cdot & H_{NN} \end{bmatrix} \begin{bmatrix} \mathbf{f}_1 \\ \cdot \\ \cdot \\ \cdot \\ \mathbf{f}_N \end{bmatrix} + \begin{bmatrix} \mathbf{n}_1 \\ \cdot \\ \cdot \\ \cdot \\ \mathbf{n}_N \end{bmatrix} \tag{2}$$

or,

$$\mathbf{g} \;=\; H\mathbf{f} \;+\; \mathbf{n} \tag{3}$$

where $\mathbf{f}_1, \ldots \mathbf{f}_N$ are vectors of the N channels (frames). Each vector $\mathbf{f}_i$, $i = 1, ..., N$ is created from the lexicographic ordering of the ith frame pixel values; Λ is a diagonal matrix given by

$$\Lambda \;=\; \begin{bmatrix} \lambda_1 I_1 & 0 & \cdot & \cdot & \cdot & 0 \\ 0 & \cdot & \cdot & \cdot & \cdot & \cdot \\ \cdot & \cdot & \cdot & \cdot & \cdot & \cdot \\ \cdot & \cdot & \cdot & \cdot & \cdot & \cdot \\ \cdot & \cdot & \cdot & \cdot & \cdot & \cdot \\ \cdot & \cdot & \cdot & \cdot & \cdot & 0 \\ 0 & \cdot & \cdot & \cdot & 0 & \lambda_N I_N \end{bmatrix}; \tag{4}$$

α is a regularization parameter and C is an image dependent matrix. The Laplacian is a popular choice for the matrix C and it is used in our work.

The closed form solution of the CLS algorithm, given in equation (1), is the solution of the following constrained optimization problem:

$$\begin{aligned} &minimize \;\; ||C\mathbf{f}||^2 \\ &subject\ to \;\; ||H\mathbf{f} - \mathbf{g}||^2 \;=\; ||\mathbf{n}||^2 \end{aligned} \tag{5}$$

For simplicity all values of λ_i (Lagrange multipliers) are taken equal to one. It should be noted that the CLS algorithm could also take the forms:

- For $\lambda_i = 1, \ i = 1, ..., N$ and $\alpha = 0$ it becomes an inverse filter.
- For $\lambda_i = 1, \ i = 1, ..., N$ and $C = R_{ff}^{-\frac{1}{2}} R_{nn}^{\frac{1}{2}}$ it is a parametric Wiener filter, and for $\alpha = 1$, the familiar Wiener filter, where R_{ff} and R_{nn} are the signal and noise autocorrelation matrices.

3. BLUR ESTIMATION

Blur estimation is the process of identifying the blur parameters on a linearly degraded image. In most image blur estimation algorithms, the image is assumed to be modelled by an Auto-Regressive (AR) model, while the blurring function is an Mean-Average (MA) model. Using this assumption, a number of algorithms have been proposed for the AR and MA parameter identification, including the recent expectation-maximization (em) and bispectrum related methods. It has been shown through simulated and experimental results[1] that these methods are not accurate enough and do not offer significant improvements in real infrared images due to large estimation errors. This is true even in the case where the order of the MA blur (i.e. the size of the blur) is known.

This paper presents two techniques for blur estimation. The first method is based on a general approach that is model independent. Blur parameters are estimated using the singular points of the degraded image spectrum. The second is a model based approach. In this method only out-of-focus, motion and detector sampling blurs are considered. Their parameters are estimated at the singular points of the degraded image spectrum, which coincide with the singular points of the blurring functions.

3.1 A New Image FIR Blur Estimation Method with Arbitrary Phase

In this section a new method is presented for estimating a class of blurring functions. The assumptions here are as follows: a) the original image does not have zeros in the Fourier spectrum, and b) The blurring function has zeros in the Fourier spectrum which are greater or equal to the number of parameters of the blurring psf. These assumptions are valid in most cases and agree with the assumptions made in all previous techniques (i.e. The blurring psf is approximated by an Finite Impulse Response (FIR) system). Here, the blurring psf could be quite general. The mask of the blurring psf is assumed known, but it could also be estimated at an additional computation cost.

Algorithm description

Let a blurred image be represented as a 2-D rational z-domain polynomial, i.e.

$$G(z_1, z_2, z_3) = \frac{B(z_1, z_2, z_3)H(z_1, z_2, z_3)}{A(z_1, z_2, z_3)} \tag{6}$$

with

$$
\begin{aligned}
A(z_1, z_2, z_3) &= \sum_{(m,n,q)\in S_A} a(m,n,q) z_1^{-m} z_2^{-n}, z_3^{-q} \\
H(z_1, z_2, z_3) &= \sum_{(m,n,q)\in S_H} h(m,n,q) z_1^{-m} z_2^{-n}, z_3^{-q} \\
B(z_1, z_2, z_3) &= \sum_{(m,n,q)\in S_B} b(m,n,q) z_1^{-m} z_2^{-n} z_3^{-q}.
\end{aligned} \tag{7}
$$

where S_A, S_H, S_B are the regions of support for $A(z_1, z_2, z_3)$, $H(z_1, z_2, z_3)$ and $B(z_1, z_2, z_3)$ respectively. For $z_1 = e^{j\omega_1}$, $z_2 = e^{-j\omega_2}$ and $z_3 = e^{-j\omega_3}$ it is assumed that

$$A(\omega_1, \omega_2, \omega_3) \neq 0, \quad \forall[\omega_1, \omega_2, \omega_3] \tag{8}$$

and

$$H(\omega_{1k}, \omega_{2l}, \omega_{3r} = 0, \;\Rightarrow\; I_{H0}^2 = \{(k, l, r)\} \tag{9}$$

Statement 1: (Necessary Condition)
It is necessary that the number of elements of I_{H0}^2 be greater than the elements of S_H for the identification of the elements of S_H (i.e. $\{h(m, n, q\})$.

Proof:
By normalizing $G(\omega_1, \omega_2, \omega_3)$ such that $G(0,0,0) = 1.0$ the following set of equations is formed:

$$
\begin{aligned}
&\tfrac{B(0,0,0)}{A(0,0,0)} \sum_{(m,n,q)\in S_H} h(m,n,q) = 1.0 \\
&\tfrac{B(\omega_{1k},\omega_{2l},\omega_{3r})}{A(\omega_{1k},\omega_{2l},\omega_{3r})} \sum_{(m,n,q)\in S_H} h(m,n,q) e^{-j\omega_{1k}m} e^{-j\omega_{2k}n} e^{-j\omega_{3r}q} = 0.0, \; for \; (k,l,r) \in I_{H0}^2
\end{aligned} \tag{10}
$$

The above equations are normalized. Hence,

$$
\begin{aligned}
&\sum_{(m,n,q)\in S_H} h(m,n,q) = 1.0 \\
&\sum_{(m,n,q)\in S_H} h(m,n,q) e^{-j\omega_{1k}m} e^{-j\omega_{2k}n} e^{-j\omega_{3r}q} = 0.0, \; for \; (k,l,r) \in I_{H0}^2
\end{aligned} \tag{11}
$$

and they are expressed as

$$D\mathbf{h} = \mathbf{e} \tag{12}$$

where

$$
D = \begin{bmatrix}
1 & \cdots & 1 \\
e^{-j\omega_{1k_1} m_1} e^{j\omega_{2l_1} n_1} e^{j\omega_{3r_1} q_1} & \cdots & e^{-j\omega_{1k_1} m_M} e^{j\omega_{2l_1} n_M} e^{j\omega_{3r_1} q_M} \\
\cdot & \cdots\cdot & \cdot \\
\cdot & \cdots\cdot & \cdot \\
\cdot & \cdots\cdot & \cdot \\
e^{-j\omega_{1k_Q} m_1} e^{j\omega_{2l_Q} n_1} e^{j\omega_{3r_Q} q_1} & \cdots & e^{-j\omega_{1k_Q} m_M} e^{j\omega_{2l_Q} n_M} e^{j\omega_{3r_Q} q_M}
\end{bmatrix} \tag{13}
$$

and

$$
\mathbf{h} = \begin{bmatrix}
h(m_1, n_1, q_1) \\
\cdot \\
\cdot \\
\cdot \\
h(m_M, n_M, q_M)
\end{bmatrix}, \tag{14}
$$

and $\mathbf{e} = [1, 0, ...0]^t$. It can be seen that eqn (12) requires that the leading dimension of D be greater or equal to M, i.e. $Q + 1 \geq M \;\Rightarrow\; Q > M$, where Q is the number of elements of I_{H0}^2. □

Statement 2: (Sufficient condition)

If $Q > M$ it is sufficient that $dim\{D\} = M$ for a unique least squares solution of eqn (12), where $dim\{D\}$ denotes the dimension of D.

Proof:
If $dim\{D\} = M \Rightarrow dim\{D^tD\} = M \Rightarrow \mathbf{h} = [D^tD]^{-1}D^t\mathbf{e}$ which is the least squares solution of eqn (12). □

In the case where additive noise affects the blurred image, we have

$$G(\omega_1, \omega_2, \omega_3) = \frac{A(\omega_1, \omega_2, \omega_3)H(\omega_1, \omega_2, \omega_3)}{B(\omega_1, \omega_2, \omega_3)} + N(\omega_1, \omega_2, \omega_3) \tag{15}$$

where $N(\omega_1, \omega_2, \omega_3)$ is the additive noise spectrum. If the noise spectrum is constant across the frequency range, then assign

$$|\hat{G}(\omega_1, \omega_2, \omega_3)| = |G(\omega_1, \omega_2, \omega_3)| - min_{(\omega_1, \omega_2, \omega_3)}|G(\omega_1, \omega_2, \omega_3)| \tag{16}$$

Finally, we determine the set I_{HO}^2 as the set of $(\omega_{1k}, \omega_{2l}, \omega_{3r})$ that satisfy

$$|\hat{G}(\omega_{1k}, \omega_{2l}, \omega_{3r})| \leq \epsilon_1 \tag{17}$$

where ϵ_1 is a small real number which is proportional to the noise spectrum standard deviation.

In the next section a model based method is proposed which can be viewed as a special case of the method described in this section as it deals with a priori known models of blurring functions. Although the basis of the method is the same, the solution is different and it is adjusted to the used blur models.

3.2 A Model Based Approach for Image Blur Estimation in Two Dimensions

Let's start this method by considering a blurring function (say out-of-focus blur). Over the discrete spectrum we identify the points where

$$H_f(\omega_{1k}, \omega_{2l}) = 0 \tag{18}$$

and therefore we can find I_{HO}^2 and evaluate the performance index:

$$J_f(R_f) = \sum_{(i,j)\in I_{HO}^2} \big|\, |G(\omega_{1k}, \omega_{2l}| - |H_f(\omega_{1k}, \omega_{2l}; R_f)|\, \big|. \tag{19}$$

The above performance index becomes minimum at $R_f = R_f^*$, where R_f^* is the correct out of focus blur of the blurred image $G(\omega_1, \omega_2)$.

To identify the value of R_f where $J_f(R_f)$ is minimum the following test is used

$$\begin{array}{ll} R_f = R_f^* & if\ 1 - \frac{2\ J_f(R_f)}{J_f(R_f+\Delta_2)+J_f(R_f-\Delta_1)} \geq \epsilon_2 \\ R_f \neq R_f^* & otherwise \end{array} \tag{20}$$

where Δ_1 and Δ_2 are small real numbers such that:

$$J_f(R_f - \Delta_1) \geq J_f(R_f); \quad and \quad J_f(R_f + \Delta_2) \geq J_f(R_f) \tag{21}$$

and ϵ_2 is a small real number.

In the case of uniform motion blur we find the corresponding I_{HO}^2 where

$$H_M(\omega_{1k}, \omega_{2l}) = 0 \tag{22}$$

and the corresponding performance index

$$J_M(X,Y) = \sum_{(i,j)\in I_{H0}^2} \mid |G(\omega_{1k},\omega_{2l}| - |H_M(\omega_{1k},\omega_{2l};X,Y)| \mid \tag{23}$$

Since, the parameters X and Y are from separable functions the above performance index can be further simplified as

$$J_M(X) = \sum_Y J_M(X,Y) \tag{24}$$

and

$$J_M(Y) = \sum_X J_M(X,Y) \tag{25}$$

The values of X^* and Y^* are half of the uniform blur extent in the image spatial directions and can be found in a similar way as R^* using the corresponding equations to (20)-(21) by replacing $J_f(R_f)$ with $J_M(X)$, or $J_M(Y)$, respectively.

4. 3-D HYBRID RESTORATION USING 2-D BLUR IDENTIFICATION

In this section a method is presented that compensates blur estimation errors of a 2-D blur estimation algorithm. It is shown that if 2-D blur estimation errors in a sequence of frames of the same object are mutually uncorrelated and zero mean, then they could be reduced by the method presented in herein.

Consider a 3-D Restoration algorithm (i.e. the 3-D Constrained Least Squares (CLS)) that operates on a limited number of sequential frames (e.g. 3 frames). By shifting the 3-D restoration algorithm over the sequential frames and using the 2-D identified blurs for every frame, a sequence of restored frames are obtained. Namely,

$$\hat{\mathbf{s}}_i(m,n) = \mathcal{R}\{\mathbf{g_i(m,n)}\} \tag{26}$$

where i denotes the central frame in the window of the frames at which the restoration is performed, and the vector arrays $\mathbf{s}$ and $\mathbf{g}$ represent the set of frames at which the 3-D restoration is performed. The 3-D restoration does not completely restore the frames due to blur estimation errors, regularization artifacts and residual additive noise. Hence,

$$\hat{\mathbf{s}}_i(m,n) = \Delta H_i(m,n) ** \mathbf{s}_i(m,n) + \psi_i(m,n) \tag{27}$$

where $\Delta H(.)$ is the residual blur error after 3-D restoration and $\psi_i(m,n)$ is the residual additive noise at frame i. Assuming that the target is located at the same location at every frame (through registration) the image

$$\bar{s}(m,n) = \sum_i \sum_l \hat{s}_{i,l}(m,n) = s(m,n) ** \sum_i \sum_l \Delta H_{i,l} + \sum_{i,l} \psi_{i,l}(m,n) \tag{28}$$

is the result of the proposed 3-D hybrid restoration algorithm. From eqn (28) it can be seen that if the restoration errors have zero mean blur psf errors and zero mean noise over the sequence of frames, then the above hybrid 3-D restoration technique is asymptotically efficient.

On the other hand direct 3-D restoration over all available frames shall be better since multiframe correlation over a larger number of frames shall result in a better estimation of the restored target. However, as the number of frames increase, the computational and memory requirements increase proportionally.

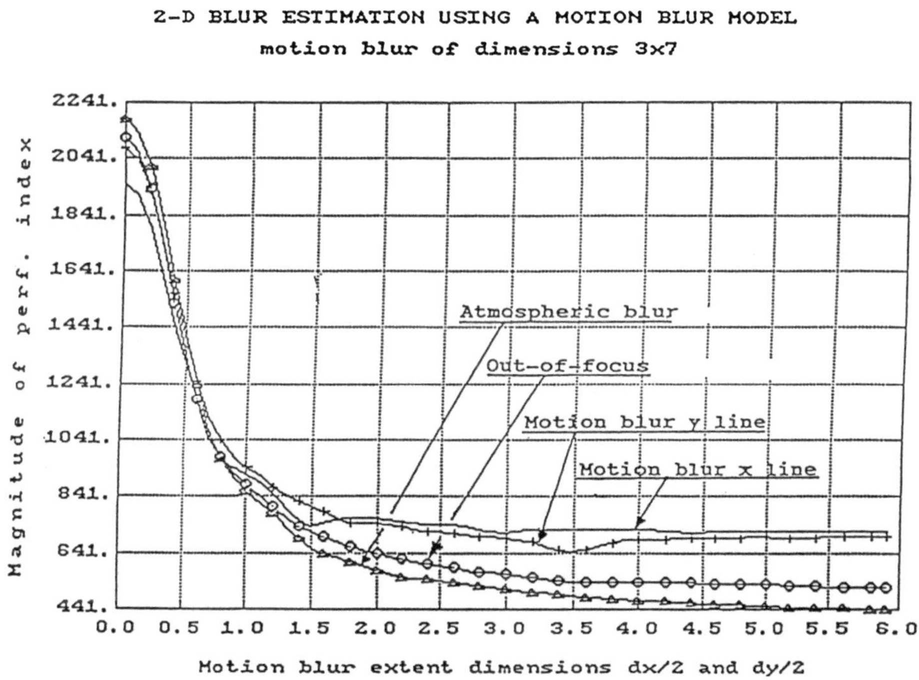
2-D BLUR ESTIMATION USING A MOTION BLUR MODEL
motion blur of dimensions 3x7
2241.
2041.
1841.
1641.
1441.
1241.
1041.
841.
641.
441.
Magnitude of perf. index
Atmospheric blur
Out-of-focus
Motion blur y line
Motion blur x line
0.0 0.5 1.0 1.5 2.0 2.5 3.0 3.5 4.0 4.5 5.0 5.5 6.0
Motion blur extent dimensions dx/2 and dy/2

Figure 1

As a final remark, we can relate the proposed hybrid 3-D restoration to the overlap-add approach used in power spectral estimation.

5. EXAMPLE

Due to limited space, in this example we present only the motion blur identification in the model based approach. In Figure 1 it is shown that the size of the motion blur can be identified from the local minima of the motion penalty functions in the x and y dimensions. In the example, the image has motion blur of size 3×7. Hence, local minima appear at $x = 1.5$ and $y = 3.5$, which are half the size of the blur dimensions in the x and y dimensions, respectively, as expected (see the frequency domain sinc function of the blurring function).

6. CONCLUSION

In this paper we have presented blur estimation algorithms and their incorporation to 3-D image restoration. This is part of ongoing work in Space Based Infrared Processing.

7. REFERENCES

1. A.U.G. Signals Ltd., "Space Based Infrared (SBIR) Multi-Dimensional Image Processing", Report No. R&D-0002/93, September 1993.

MICROSCANNING SETUP FOR FOCAL PLANE ARRAYS

Jean Fortin, Paul Chevrette and Robert Plante

Electro-optics Division
Defence Research Establishment Valcartier
2459, Pie XI Blvd., North (P.O. Box 8800)
Courcelette, Quebec, CAN., G0A 1R0

INTRODUCTION

Infrared focal plane arrays (FPA) are becoming widely used in the field of thermal imaging. As compared to older 1-detector scanning systems, these sensors showed tremendous improvement. However, FPAs have a major drawback over scanning systems as, for a given instantaneous field of view (IFOV), they can only reproduce half the frequencies scanning systems do. This is due to the Nyquist theorem which stipulates that at least two samples must be taken for each cycle[1] in the image.

In a surveillance system, it might be worthwhile to increase the resolution of the imaging system at the cost of some extra mechanical parts and processing. Microscanning was introduced in that respect and allows to get the best from each pixel in an array.

SINGLE DETECTOR SCANNING SYSTEMS VS STARING ARRAY SYSTEMS

The resolution of a single detector scanning system versus the resolution of a staring array imaging system can be studied using the Modulation Transfer Function (MTF)[2] and the Modulation Detector Transfer Function (MTDF)[3]. The MTF is defined for 1-detector scanning systems. The highest frequency that such a system can reproduce (aliasing limit) is given by the first zero in the MTF curve. The MTF of an ideal system would be represented by the absolute value of a sinc function showing a first zero at a frequency equal to 1 over the IFOV.

In the case of detector matrices, the MTF is no longer defined. The MDTF is used instead. In order to measure the MDTF, a point source (smaller than the IFOV of 1 detector) is imaged onto the FPA. The array is then read out and the modulus of the Fourier transform of the output signal is computed. The aliasing frequency is again given by the first zero of the MDTF function.

It is noticed that the width of the MDTF depends on the position of the image of the point source over the detector matrix. In one case, it may fall just in between two pixels leading to an MDTF half of the width it would have if the image had fallen in the center of one pixel. These extreme cases represent respectively the worst and the best MDTF. The real aliasing limit of an imaging system is given by the worst MDTF curve. In an ideal case, this first zero appears at 1 over 2 times the IFOV (as predicted by the Nyquist theorem).

By comparing the MTF and the MDTF curves, it is observed that for a given IFOV, an FPA reproduces only half the frequencies a scanning system does. In that respect, 1-detector scanning systems are more efficient. We will now see how microscanning can remove that limitation.

MICROSCANNING

The microscanning technique can be seen as an oversampling process[4]. A series of images representing the same scene is taken while displacing each time the image over the array by a fraction of the pixel pitch. The oversampled image is built by interlacing all the pixels from all the images in both directions. The left hand side of figure 1 represents the 4 images of a 2x2 microscan operation. The image on the right hand side of figure 1 represents the resulting oversampled image.

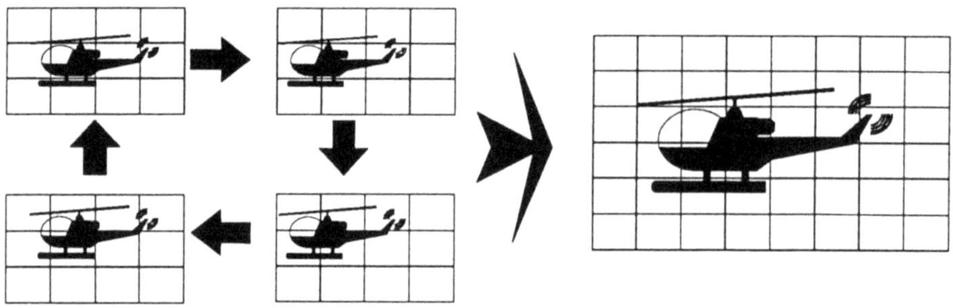

Figure 1. Representation of the microscanning process.

In order to determine the resolution improvement due to microscanning, one may compare the MDTF obtained with and without microscanning. From the definition of the MDTF previously mentioned, it is possible to develop a modified expression for the MDTF in the case of a microscanned array of detectors. For the sake of simplicity, the case of a linear array was considered hereafter. The results can be directly transposed to the 2-dimensional case.

The modified expression for the MDTF of a microscanned array is represented by eq. 1,

$$\mathrm{MDTF} = \mathrm{FFT}\left[\left(\left(\delta(u) \otimes \sqcap_{Ps}\right) \times \text{Ш}_{\frac{Pp}{N}}\right) \otimes \sqcap_{\frac{Pp}{N}}\right] \qquad 1$$

where FFT stands for the Fourier transform, $\delta(u)$ a dirac impulse in the spatial domain (representing a point source), $\sqcap_w$ a window of width w, Ш_d a comb of unitary amplitude impulses and period equal to d, Ps the sensitive area of a pixel, Pp the pixel pitch and N the number of steps in the microscanning process. $\otimes$ and $\times$ represent respectively convolution and multiplication operations. It should be kept in mind that in a real situation, the point source is degraded by the optics placed in front of the FPA. This degradation is not included in eq. 1.

The worst MDTF was calculated in 4 different scenarios: no microscanning, 2x2, 3x3 and 4x4 microscanning in the case where $Ps=Pp$ (representing a fill factor of 100%). The results shown in figure 2 present the normalized (with respect to the IFOV) aliasing limit obtained in each case. One can see that by increasing the number of steps in the microscanning process, the aliasing limit increases, reaching 1 for an infinite number of steps. The gain thus obtained by increasing the number of steps decreases gradually. Results are slightly different when $Ps<Pp$. In that case, for a large number of microscan steps, the aliasing limit is higher than before as the IFOV must now be calculated on the basis of the pixel size instead of the pixel pitch.

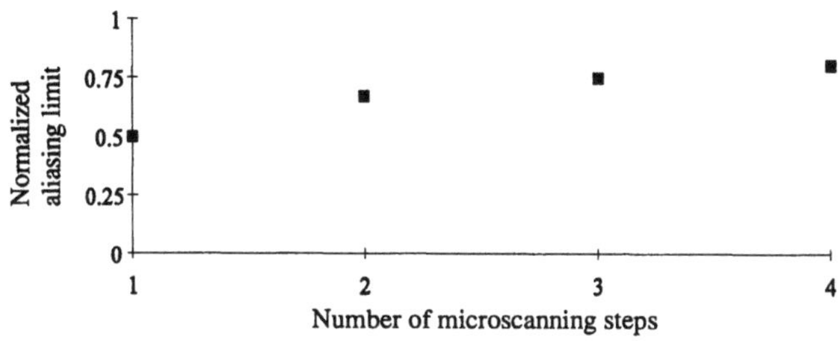

Figure 2. Aliasing limit obtained with a staring array imaging system with and without microscanning. The vertical axis represents the aliasing limit normalized with respect to the IFOV.

MICROSCANNING SETUP

The performance of the microscanning process was tested using a real FPA device. Figure 3 gives an overview of the setup used. Microdisplacements of the image over the array are obtained by the use of 2 micropositionners mounted on each side of a 2-axis translation table in such a way that horizontal and vertical motions are allowed. The position is controlled by the use of a microcomputer and a piezo controller device. A 100mm focal length lens is attached to the translation table and is aligned in the optical path such as to image a target on a 256x256 detector FPA. The target is seen at infinity through a 64" collimator. The selected array is from Amber Engineering (AE 4256). It is made of InSb and is sensitive in the 3-5µm band. Each detector is square and has a fill factor of 67%. The detector pitch is specified to be 38µm. Images from the camera are acquired using an Oculus-500 frame grabber. This board also stores the oversampled images and displays them on a high-resolution video monitor.

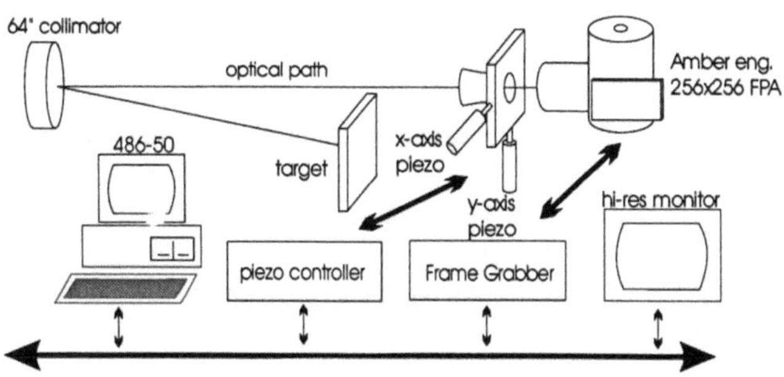

Figure 3. Overview of the setup used to evaluate the performance of the microscanning process.

RESULTS

From the value of the detector pitch previously given, one can calculate the maximum observable frequency in a single image to be 1.3 cy/mrad. By doing a large number of microscan steps, this limit should be brought to 2.6 cy/mrad or better because of the smaller size of a detector with respect to the detector pitch. Figure 4 shows the results of a 4x4 microscanning experiment. The target used in that case presents frequencies increasing from 0.434 cy/mrad to 2.2 cy/mrad. The image of figure 4a shows the direct output of the camera. It is seen that frequencies in the right part of the image are indistinguishable. Figure 4b shows the oversampled image. A clear improvement is observed as the highest frequency is clearly resolved.

Figure 4. Image produced a) without microscanning b) with a 4x4 step microscanning operation.

CONCLUSION

The microscanning process improves the resolution of a given FPA device and brings it near to the same level it is with a scanning system. However, microscanning presents two major drawbacks. First, for a given frame rate, it reduces the integration time available to build a complete oversampled image. One must make sure that the integration time required for each image multiplied by the number of steps in the microscanning process does not exceed the available frame period. Second, microscanning requires the processing of a large amount of data which can become overloading. A compromise between the number of steps, the sensitivity required and the possibility of processing data in real time must thus be reached.

Microscanning also opens the way to a technique called superresolution[5]. Using that technique, one can increase the resolution of a microscanned image and bring it to the level of the microdisplacements done. This will be investigated by future work performed with the system.

REFERENCES

1. D.J. Bradley and P.N.J. Dennis, Sampling effects in CdHgTe focal plane arrays, *SPIE on Infrared Technology and Applications.* 590:53 (1991).
2. W.L. Woolfe and G.J. Zissis, "The Infrared Handbook", Infrared information analysis center, Washington DC, (Revised Edition 1985).
3. W. Schuberth, Characterization of the spatial resolution of focal plane arrays, *in:* "Proposal for the RSG.16 subgroup on focal plane arrays", Forschungsinstitut für optik, Tübingen (1991).
4. F.P. Bloomel and P.N.J. Dennis, The effects of microscan operation on staring infrared sensor imagery, *SPIE on Infrared Technology XVII.* 1540:653 (1991).
5. G. Jacquemod, C. Odet and R. Goutte, "Superresolution technique applied to CCD sensors", *Proc. EUSIPCO-88*, 2:847 (1988).

APPLICATION OF MULTIRESOLUTION TRANSFORMS TO IMAGE ANALYSIS AND ENHANCEMENT

Simon Yu and Dennis Gregoris

SPAR Aerospace Ltd., Brampton, Ontario, Canada L6S 4J3

1. INTRODUCTION

In this paper we will illustrate the application of a specific type of multiresolution transform, the Wavelet Transform, for image analysis and enhancement. The focus will be on the Discrete Wavelet Transform (DWT) and its complementary transform, the Discrete Scaling Transform (DST) [1].

The DWT and DST are multiresolution transforms defined as the inner products of the signal and a basis function. The high pass wavelet function forms the basis for the DWT and the associated low pass scaling function forms the DST basis. The DWT of a discrete sequence x[n] can be evaluated iteratively using a Quadrature Mirror Filter (QMF) pair H and G, representing the low pass and high pass filters respectively together, followed by downsampling [1]. The resulting output coefficients from the G filter of each iteration or scale level, j, are the DWT coefficients (d_1, d_2, ... d_j) and the corresponding coefficients from the H filter are the DST coefficients (a_1, a_2, ... a_j). Higher subscript values correspond to coarser scales. Figure 1 displays the difference between the Discrete Fourier Transform (DFT), the DWT and the DST in terms of their frequency and spatial characteristics. The spatially localized basis functions make the DWT and DST well suited to analyze non-stationary signals such as *1/f* noise. Section 2 and 3 describe example applications of the DWT and DST for fractal dimension estimation and dim target detection.

2. IMAGE ANALYSIS: FRACTAL DIMENSION ESTIMATION

A *1/f* -type process is defined as process whose power spectrum is of the form $S(f) \sim \sigma_x^2 / |f|^\gamma$ where the corresponding fractal dimension D is defined as $D = (5-\gamma)/2$. Due to the octave frequency band characteristics and uncorrelated nature of the wavelet basis functions, it is shown [2] that the DWT of the *1/f* process has a variance progression property, $var\ d_j = \sigma^2\ 2^{-\gamma(J-j)}$, where J denotes the coarsest scale. Hence D can be estimated from the slope of the logarithm of the variance calculated using linear regression.

Applications of Photonic Technology
Edited by G.A. Lampropoulos *et al.*, Plenum Press, New York, 1995

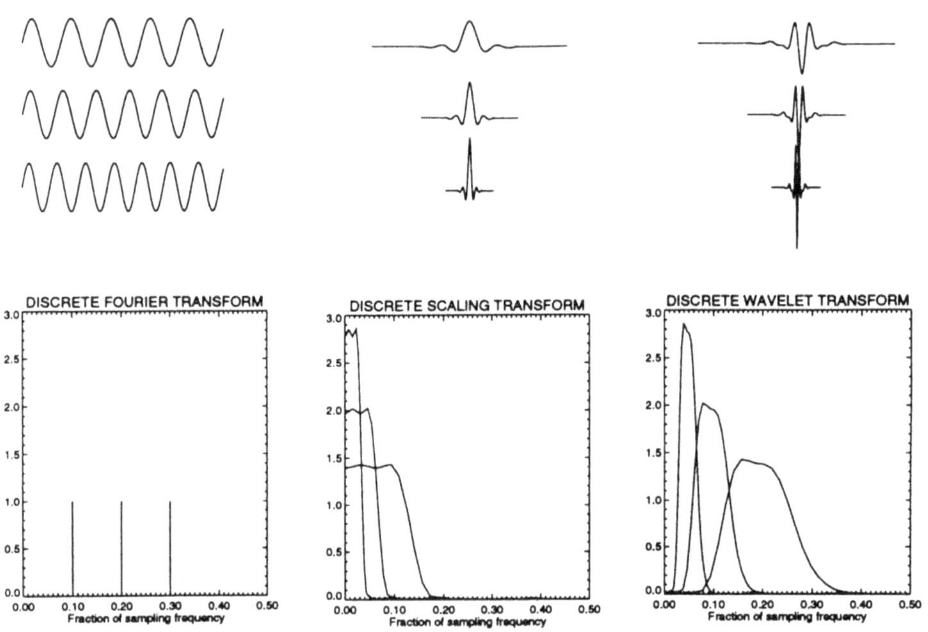

Figure 1 Frequency and Spatial Characteristics of the DFT, DST, DWT.

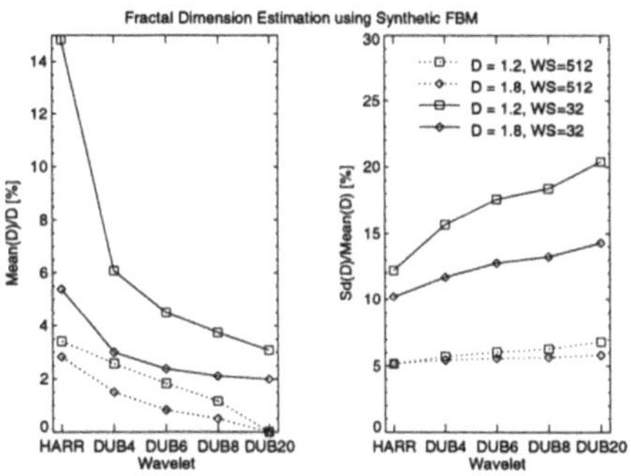

Figure 2 Estimating fractal dimensions using wavelets

Alternatively, the estimation problem can be reformulated into a Maximum Likelihood (ML) problem [5]. The ML method is more accurate than regression methods for a small number of data samples. We applied the ML algorithm to a set of synthetically generated *1/f* noise sequences and used different wavelet types and sample window sizes to compute D. The result is shown in Figure 2 which shows estimation errors as functions of wavelet type (i.e. Haar and Daubechies wavelets).

As noted in [4], the interscale wavelet coefficient correlation can be made small by using wavelets with a large number of vanishing moments which manifests as a flatter frequency response near dc. The smaller interscale correlation more closely resembles the *1/f* noise model described in [5]. Hence it is expected that such wavelets will reduce the dimension estimation error and Figure 2 corroborates this. However, the mean value of the estimation is reasonably accurate even for wavelets with moderate number of vanishing moments (e.g. 4 or 6 coefficient Daubechies wavelets: DUB4 or DUB6). Moreover, the standard deviation of the estimation actually increases slightly for wavelets with larger number of vanishing moments (20 coefficient Daubechies wavelet: DUB20). Overall, low order Daubechies wavelets represent good choices because of their accuracy and small size.

3. DIM TARGET DETECTION

The application of Wavelet Transforms to enhance and detect small dim targets in Forward Looking Infrared (FLIR) imgery led to the development of the Scale Subtraction Filter (SSF). The SSF is a non-linear spatial filter [3] which operates on the DST of an image. The SSF has the form:

$$\begin{aligned} A &= a_1 - k_1 \min[a_1, f_1(a_3)] \\ B &= a_4 - k_2 \min[a_4, f_2(a_5)] \\ \textit{Enhanced Signal} &= A - k_3 \min[A, f_3(B)] \end{aligned}$$

$k_{1,2,3}$ are scale weightings and $f_{1,2,3}$ are local order statistic operators. The filter enhances objects by performing non-linear background subtraction. It has demonstrated excellent detection performance of low contrast targets in clutter backgrounds. Figure 3 illustrates the detection performance of the SSF relative to the high performance Holmes Double-gated spatial filter method. A notable feature of the SSF over conventional filters is that its multiresolution nature enables it to enhance a wide range of different sized targets without modification.

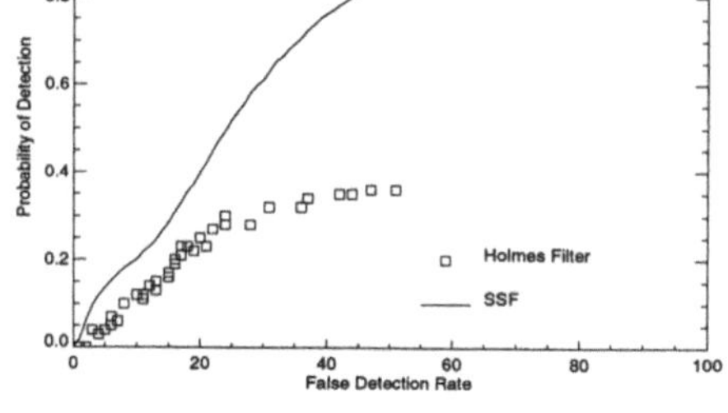

Figure 3 Detection performance of SSF and Holmes Filter on FLIR image sequence.

4. SUMMARY

Multiresolution Wavelet Transforms were successfully applied to image analysis and enhancement. In fractal image analysis applications, low order Daubechies wavelets were found to be the most suitable. In image enhancement applications, a new multiresolution spatial filter based upon the DST was developed to effectively enhance and detect dim targets in FLIR imagery.

5. ACKNOWLEDGEMENTS

The work was performed in part under the Defence Industrial Research Contribution Agreement 2207EV01 and in part for the Defence Research Establishment Valcartier under Contract W7701-2-1476/01-XSK. The studies resulted from on-going research and development in signal and image processing at SPAR Aerospace Ltd..

6. REFERENCES

1. S.Mallat, *A Theory for Multi-resolution Signal Decomposition: The Wavelet Representation*, IEEE Trans. P.A.M.I., **11**, **7** (1989).
2. J.Ramanathan and O.Zeithouni, *On the Wavelet Transform of Fractional Brownian Motion*, IEEE Trans. Info. Theory, **37**, **4,** p.1156-1158 (1991).
3. D.Gregoris, S.Yu, S.Tritchew, and L.Sevigny, *Wavelet transform-based filtering for the enhancement of dim targets in FLIR images*, Proc. SPIE, **2242** p.573-583 (1994).
4. A.Tewfik and K.Kim, *Correlation Structure of the Discrete Wavelet Coefficients of Fractional Brownian Motion*, IEEE Trans. Info Theory, **38**, **2**, p.904-909 (1992).
5. G.Wornell and A.Oppenheim, *Estimation of Fractal Signals from Noisy Measurements using Wavelets*, IEEE Trans. Signal Processing, **40**, **3**, p.611-623 (1992).

A NEW CLUTTER MODEL FOR SAR IMAGES

V. Anastassopoulos[1], G.A. Lampropoulos[1] and M. Rey[2]

[1]A.U.G. Signals Ltd., TRDL, 560 Lauder Ave.
Toronto, Ontario, M6E 3J6
[2]Defence Research Establishment Ottawa
3701 Carling Ave., Ottawa, Ontario, K1A 0Z4

INTRODUCTION

A new generalized compound pdf (GC-pdf) is proposed to model the amplitude statistics of SAR images. This model is introduced because the K-pdf which has been used in the past[1] for this purpose, occasionally fails to do so. Such cases are met when the speckle of the clutter deviates from Rayleigh pdf due to the small area of the resolution cell. The speckle and the modulation components of the new compound pdf are described using the generalized gamma pdf ($G\Gamma$-pdf). A thermal noise component is incorporated in the statistical model. Although it is not possible to have a closed form expression for the model itself, its moments can be obtained in relatively simple expressions and consequently be used for clutter characterization. Using the equivalence of theoretical and experimental moments, real data show very good agreement with the proposed model. The K-pdf is proved to be a special case of the GC-pdf.

THE MODEL

The proposed compound model results when the speckle and its underlying mean obey to the Generalized Gamma distribution ($G\Gamma - pdf$):

$$f(x) = \frac{b}{s\Gamma(\nu)}(\frac{x}{s})^{b\nu-1}exp[-(\frac{x}{s})^b] \quad (1)$$

where s is the scale parameter, ν is the shape parameter and b is the power parameter. The Generalized Gamma pdf can describe satisfactorily the speckle when deviation from Gaussian statistics is observed. This is because most of the probability density functions that have been used to model radar clutter, such as the Gaussian, Rayleigh, Exponential, Gamma, Weibull and Lognormal are special cases of the $G\Gamma$-pdf. Deviations from Gaussian statistics are observed when the resolution cell becomes small

Applications of Photonic Technology
Edited by G.A. Lampropoulos *et al.*, Plenum Press, New York, 1995

enough [2]. This happens when the azimuth (cross-range) resolution is high. This can be achieved if electrically large antennas (SAR sensors) are employed.

The incorporation of the thermal noise into the statistics of the radar echo can be carried out if the speckle of the radar clutter and the thermal noise are added so that the new speckle component x is again described by the $G\Gamma$-pdf:

$$p(x/s) = \frac{b_1}{\Gamma(\nu_1)} \frac{x^{b_1\nu_1-1}}{(\sqrt{s^2+2\sigma^2})^{b_1\nu_1}} exp[-(\frac{x}{\sqrt{s^2+2\sigma^2}})^{b_1}] \tag{2}$$

where s is the mean of the speckle (modulation component) which has been increased by $2\sigma^2$ (the power of the thermal noise).
The mean s of the speckle is assumed $G\Gamma$ distributed:

$$p(s) = \frac{b_2}{\Gamma(\nu_2)} \frac{s^{b_2\nu_2-1}}{a^{b_2\nu_2}} exp[-(\frac{s}{a})^{b_2}] \tag{3}$$

The density $p_x(x)$ is as follows:

$$p_x(x) = \int_0^\infty p(x/s)p(s)ds \tag{4}$$

The moments are derived using the previous expressions and the conventional definition:

$$\begin{aligned} E[x^k] &= \int_0^\infty x^k p_x(x)dx = \\ &= \frac{b_1 b_2}{\Gamma(\nu_1)\Gamma(\nu_2)} \frac{1}{a^{b_2\nu_2}} \int_0^\infty \frac{s^{b_2\nu_2-1}}{(\sqrt{s^2+2\sigma^2})^{b_1\nu_1}} exp(-\frac{s}{a})^{b_2} \times \\ &\times \int_0^\infty x^{k+b_1\nu_1-1} exp(-\frac{x}{\sqrt{s^2+2\sigma^2}})^{b_1} dx = \\ &= \frac{\Gamma(\nu_1+\frac{k}{b_1})}{\Gamma(\nu_1)\Gamma(\nu_2)} \frac{b_2}{a^{b_2\nu_2}} \int_0^\infty (\sqrt{s^2+2\sigma^2})^k s^{b_2\nu_2-1} exp[-(\frac{s}{a})^{b_2}]ds \end{aligned} \tag{5}$$

substituting $s^{b_2} = w$ we have:

$$E[x^k] = \frac{\Gamma(\nu_1+\frac{k}{b_1})}{\Gamma(\nu_1)\Gamma(\nu_2)} \frac{1}{a^{b_2\nu_2}} \int_0^\infty w^{\nu_2-1}(w^{\frac{2}{b_2}}+2\sigma^2)^{\frac{k}{2}} exp(-\frac{1}{a^{b_2}}w)dw \tag{6}$$

the form (6) for k=0,2,4,6,8... gives closed form expressions for the moments of the GC clutter plus thermal noise:

$$E[x^0] = 1 \tag{7}$$

$$E[x^2] = \frac{\Gamma(\nu_1+\frac{2}{b_1})}{\Gamma(\nu_1)\Gamma(\nu_2)}[a^2\Gamma(\nu_2+\frac{2}{b_2})+2\sigma^2\Gamma(\nu_2)] \tag{8}$$

$$E[x^4] = \frac{\Gamma(\nu_1+\frac{4}{b_1})}{\Gamma(\nu_1)\Gamma(\nu_2)}[a^4\Gamma(\nu_2+\frac{4}{b_2})+4\sigma^2a^2\Gamma(\nu_2+\frac{2}{b_2})+4\sigma^4\Gamma(\nu_2)] \tag{9}$$

In case that the K-pdf is assumed ($b_1 = b_2 = 2,\ \nu_1 = 1$), the previous expressions give the moment expressions provided by Watts[2].

On the other hand eqn (6) gives for k=1,3,5... infinite series expressions for the moments:

$$E[x^k] = \frac{\Gamma(\nu_1+\frac{k}{b_1})}{\Gamma(\nu_1)\Gamma(\nu_2)} \frac{1}{a^{b_2\nu_2}} \int_0^\infty w^{\nu_2-1} w^{\frac{k}{b_2}}(1+\frac{2\sigma^2}{w^{\frac{2}{b_2}}})^{\frac{k}{2}} exp(-\frac{1}{a^{b_2}}w)dw \tag{10}$$

If we assume that $\frac{2\sigma^2}{w^{\frac{2}{b_2}}} < 1$ (the power of the mean of the clutter $w^{\frac{2}{b_2}} = s^2$ is expected to be greater than the power of the thermal noise $2\sigma^2$) then the parentheses in the integral of (10) can be expressed in an infinite power series according to the following mathematical formula:

$$(1+x)^p = 1 + px + \frac{p(p-1)}{2!}x^2 + ... + \frac{p(p-1)...(p-n+1)}{n!}x^n + ... \quad 0 \leq x < 1 \quad (11)$$

From this expansion the first six terms give very good approximation to the quantity $(1+x)^p$, while the rest of them can be ignored. In case that x is almost equal to one the error using the above approximation is smaller than the .01 of the quantity. For smaller values of x the approximation error becomes negligible. Using the above expansion for the term in parentheses in the integral of (10) we obtain for the odd moments:

$$E[x^k] = \frac{\Gamma(\nu_1 + \frac{k}{b_1})}{\Gamma(\nu_1)\Gamma(\nu_2)} a^k \sum_{n=0}^{5} \lambda_n (\frac{2\sigma^2}{a^2})^n \Gamma(\nu_2 + \frac{k}{b_2} - \frac{2n}{b_2}) \quad (12)$$

The coefficients λ_n can be easily evaluated as being the coefficients of the polynomial in (11) when $p = 0.5$.

SPECIAL CASES

The GC-pdf is a generalized statistical model which can be expressed in integral or infinite series form. However, this model can give as special cases a large number of compound probability density functions, in closed form, some of which are examined below.

In case that the power parameter b, is the same for both the speckle and the modulation components in (2) and (3), and if the thermal noise power is zero, we have from (4) after some manipulations:

$$p(x) = \frac{2b}{a\Gamma(\nu_1)\Gamma(\nu_2)} (\frac{x}{a})^{\frac{b}{2}(\nu_2+\nu_1)-1} K_{(\nu_2-\nu_1)}[2(\frac{x}{a})^{\frac{b}{2}}] \quad (13)$$

We call the pdf given by (13), Generalized K-pdf (GK-pdf). The well known form of the K-distribution is obtained when $b = 2$ and $\nu_1 = 1$ i.e. it is a special case of the Generalized K-pdf . Similarly (13) results to Weibull-pdf if $\nu_2 = 0.5$ and $\nu_1 = 1$. Therefore, the Weibull and the K-pdf are two different distributions which result from the same general form, the GK-pdf, and different values in its parameters. Other special cases of the proposed model can be obtained when the $b_i s$ take integer values, using the *generalized hypergeometric functions.* The *generalized hypergeometric functions* possess interesting mathematical properties so that they are suitable for further mathematical manipulations.

EXPERIMENTAL RESULTS

The estimation of the clutter parameters can be based on the equivalence of theoretical and experimental moments. Using the GC-pdf plus thermal noise model, six parameters must be estimated from the data. These are, $\nu_1, \nu_2, b_1, b_2, a$ and σ^2. The first six experimental moments can be used to estimate these parameters. The procedure followed here to estimate the unknown parameters is to find the global minimum of the quantity:

$$A(\nu_i, b_i, a, \sigma) = \sum_{\mu=1}^{6} \mid (E[x^\mu] - m_\mu)/m_\mu \mid \tag{14}$$

where $E[x^\mu]$ are the theoretical and m_μ the experimental moments. Results taken using simple optimization techniques gave values for the parameters different from those obtained using Watts[2] equations, although the global minimum was not quite reached. Real SAR data, acquired with the CCRS (Canada Center for Remote Sensing) C/X-BAND SAR have been used to validate the proposed pdf. The results are provided in Table I. Theoretical moments are given in normalized form with respect to experimental values. The proposed model gives the better results compared with the existing ones.

Table I. Moments at a specific pixel of the SAR image, using various models.

moment	Experimental	K-pdf	Watts[2]	GC-pdf+thermal noise
1st	32.34	0.77	0.34	1.00
2nd	1860	1.00	1.00	1.00
3rd	200322	1.13	0.88	1.00
4th	$3.83\ 10^{7}$	1.00	1.00	0.98
5th	$1.05\ 10^{10}$	0.78	0.97	0.97
6th	$3.45\ 10^{12}$	0.62	1.00	1.00
7th	$1.22\ 10^{15}$	0.54	1.13	1.08
8th	$4.54\ 10^{17}$	0.51	1.41	1.16

CONCLUSIONS

A generalized compound model, namely the GC-pdf, is proposed to represent the statistical characteristics of SAR clutter. The Generalized K-pdf is also defined, while the well known K-distribution is a special case of the Generalized K-pdf. Based on the equivalence of experimental and theoretical moments, the proposed GC-pdf is proved to be superior to the K-pdf.

ACKNOWLEDGEMENT

The authors gratefully acknowledge the assistance of Tom Lukowski, Data Acquisition Division of CCRS for providing us with the data.

REFERENCES

1. L.M. Novak and M.C. Burl, "Optimal Speckle Reduction in Polarimetric SAR Imagery",*IEEE Trans. on Aerospace and Electronic Systems*, Vol. AES-26, No. 2, pp. 293-305, March 1990.

2. S. Watts, "Radar Detection Prediction in K-Distributed Sea Clutter and Thermal Noise", *IEEE Trans. on Aerospace and Electronic Systems*, Vol. AES-23, No. 1, pp. 40-45, January 1987.

DESIGN AND REALIZATION OF FLUORESCENT IMAGES ANALYSIS ALGORITHMS FOR NONINVASIVE TUMOR DIAGNOSIS

L. Biancardi,[1] R. Cubeddu,[2] F. Docchio,[1] G. Sansoni,[1] P. Taroni,[2] L.Valentini[2]

[1]Dipartimento di Elettronica per l'Automazione, Università degli Studi di Brescia, Italy I-20133
[2]Centro di Elettronica Quantistica e Strumentazione Elettronica, Politecnico di Milano, Italy I-20133

INTRODUCTION

The diagnosis of malignant tissues with fluorescent probes excited by visible or even infrared radiation has been fostered by the concurrent use of fluorescent substances and laser sources in photodynamic therapy of tumors. Various techniques have so far been proposed to implement diagnostic systems for malignant tissues, especially for endoscopy.[1,2] One of the most promising among them in terms of selectivity is the so-called Time Gated fluorescence Imaging (TGI) technique which has been recently proposed by the authors.[3,4] The approach followed relies on the detection of an exogenous fluorophore that localizes in malignant tissues selectively with respect to surrounding healthy tissues. The fluorophore of election for the experiments is Hematoporphyrin Derivative (HpD), a photosensitizer extensively used in the clinical practice. For diagnostic purposes, HpD can be profitably excited at a wavelength around 400 nm. It emits a red fluorescence (615-630 nm), having a lifetime of about 15 ns. The tissue auto fluorescence, which is also excited by the blue light, gives a background fluorescence that mainly extinguishes within 3-5 ns. The difference in the decay time of the two components can be used to suppress the auto fluorescence background. To this purpose the fluorescence of the suspected portion of tissue is acquired after a delay greater than 15 ns with respect to the excitation. Assuming a selective localization of HpD in the tumor, the image of the long living fluorescence maps the extension of the neoplasia.

A set of preliminary experiments have shown promising results. However, there are some limitations to the application of the technique in the clinical diagnosis, mainly due to the fact that, even making use of TGI, still the fluorescence signal may be very weak. This may result in dark images, presenting a compressed gray scale, leading to low contrast pictures. These limitations should be overcome, especially in view of the use of the TGI instrumentation in conjunction with an endoscope.

The aim of the work presented here is the design and the implementation of the algorithms able to increase the signal-to-noise ratio of the images, to analyze the characteristics of the acquired images in terms of contrast and brightness and to on-line perform the enhancement of the fluorescence with respect to the background. Particular care

Applications of Photonic Technology
Edited by G.A. Lampropoulos *et al.*, Plenum Press, New York, 1995

has been taken to maximize the elaboration speed: the refresh time of the enhanced images is less than 2 sec. Several categories of fluorescence images have been observed and analyzed, and, for each of them, the enhancement strategy which best matches the diagnosis requirements has been implemented.

THE TGI SYSTEM

In the TGI technique,[3,4] HpD is exited by a dye laser, pumped by a Nitrogen laser operating at atmospheric pressure, which generates sub-nanosecond pulses. The dye used for the experiments is DPS, lasing around 400 nm. A commercial intensified CCD video-camera (SIM Security & Electronic System GmbH - Mod. ICCD-05), equipped with a Microchannel Plate image intensifier, acquires only the fluorescence light that falls within a pre-selected time window, properly delayed with respect to the excitation pulses. Use of the intensifier allows the acquisition window to be as short as 5 ns, and this results in a good discrimination between fluorophores whose lifetimes differ for at least few nanoseconds. A delay/pulse generator (Stanford Research Mod. DG535) allows the selection of both width and delay of the acquisition window with respect to the excitation pulse. For the experiments, the tumors have been implanted in mice intradermally with HpD (10 mg/kg body weight) 12 hours before the experiments. The dye laser output is coupled to an optical fiber, whose distal end faces the mouse at a distance of about 15 cm. The resulting spot is rather uniform, with an excitation intensity on the sample around 75 $\mu W/cm^2$. The fluorescence images are acquired using a 50 mm Nikon lens when the diameter of the tumors is about 6-8 mm.

The TGI system has been equipped with two dedicated boards (Advanced Frame Grabber AFG, from Imaging Technology), plugged into a personal computer. Images acquired by the video-camera are digitally converted into a pixel matrix of 768 (columns) x 572 (rows). The AFG memory is 1024 x 1024 pixels x 16 bits. A 16-bit arithmetic unit and an 8 x 8 multiplexer perform arithmetic and logic functions. The hardware architecture allows all process involving input data to be performed in real time. The AFG provides a separate auxiliary memory which allows images to move from one memory to the other to gain additional processing space or to continue processing in parallel with acquisition. Input and output look-up tables allow transformations of the 256 intensity levels avoiding any calculation or processing delay. Images are visualized by a monitor connected to the AFG boards.

PROCEDURES FOR IMAGE ACQUISITION AND ANALYSIS

A suitably designed control unit integrates the various tasks and allows the user to interact with the application. The first task is the continuous acquisition of the images, which allows the regulation of the instrumentation parameters. The images are then averaged: from the examination of an extensive set of fluorescent images it results that a good improvement of the signal to noise ratio is obtained by averaging upon 8 frames, which requires about 0.3 sec. While the average task is running, no images are displayed, to avoid flickering.

The output image is then analyzed. The information contents of the image are examined on the basis of the histogram.

Two fluorescence images obtained by the TGI instrumentation are shown in Figures 1.a and 1.b. Picture 1.a presents an image of good quality, where the HpD fluorescence spot is well separated from the background. Picture 1.b shows a poorly contrasted, dark image.

a)

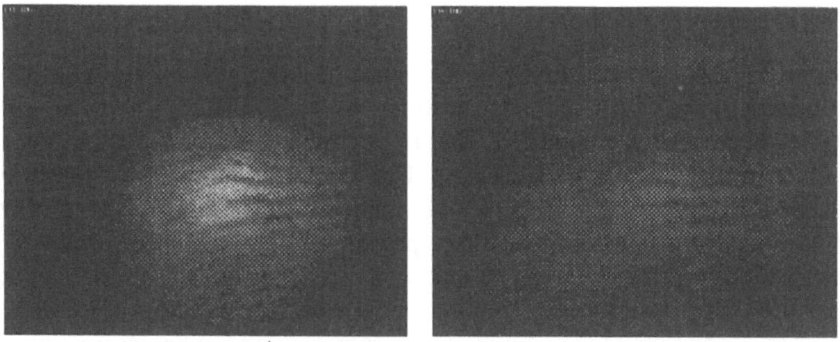

b)

Figure 1. Fluorescence images obtained by the TGI instrumentation. *1.a*: Fluorescence image of good quality; *1.b*: fluorescence image of poor quality.

The histograms of these images, plotted on a logarithmic scale, are presented in Figure 2. The histogram of image 1.a (filled area) shows that the gray levels extend over the full dynamic range: the maximum gray level is 230. The lower peak which corresponds to the image background, is well separated from the levels occupied by the fluorescence.

In the histogram of image 1.b (solid line), the gray levels equal or very close to zero are absent, due to the noise introduced by the video-camera intensifier, which presents a non zero mean value. The peak at low values of intensity, due to the uniform, dark background of the image is still well observable, but the low brightness of the fluorescence spot result in the marked decrease of the histogram curve in correspondence of increasing values of the gray levels. Finally, the dynamic range is only partially occupied (the maximum gray level is equal to 106).

From our experiments, it results that fluorescent images presenting histograms similar to that of the image of Figure 1.b are rather common: for these images contrast enhancement is required.

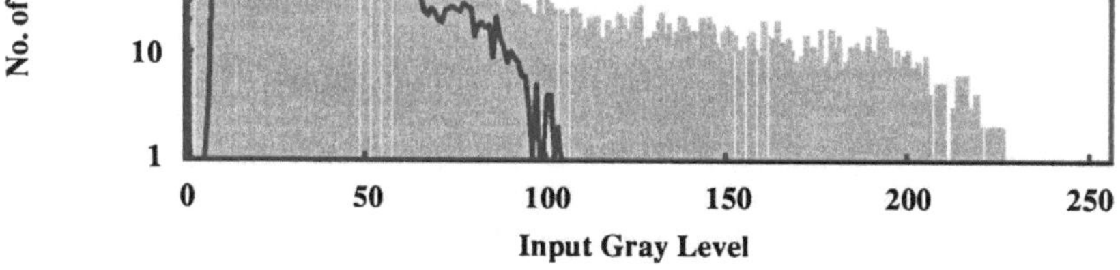

Figure 2. Histograms of the images in Figure 1.a (filled area), and 1.b (solid line).

Image contrast enhancement

In order to enhance the fluorescence while leaving the background unchanged, a suitable nonlinear stretching of the histogram has been developed. The basic scheme of the transformation is presented in Figure 3. The input gray levels are mapped to the output gray levels by a segmented transformation curve. This input-output curve is shown in the figure (right vertical scale), superimposed to the log plot of the gray level histogram of the image of Figure 1.b (left scale). A purposely designed procedure has been implemented to define five intervals on the input gray levels, and to modify the gray levels in each interval according to the slope of the corresponding transformation segment. The intervals are defined by the x-coordinate of the endpoints of the corresponding segments on the transformation line.

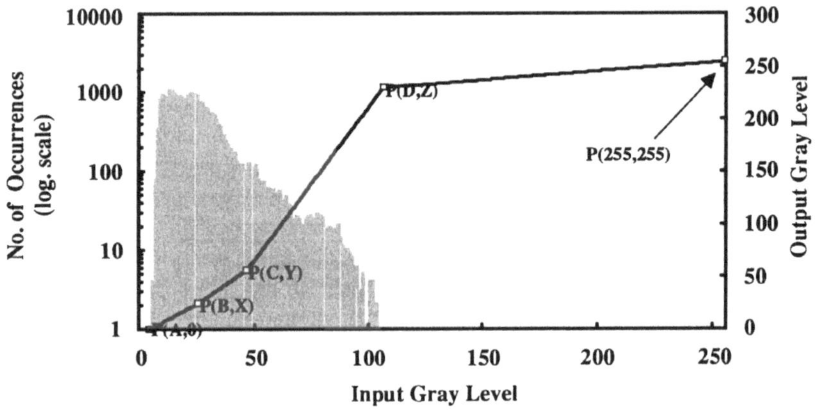

Figure 3. Description of the non-linear transformation.

All gray levels belonging to the first interval 0-A, are mapped to 0 by sliding the whole histogram to the left. In this way, the noise of the image intensifier is eliminated. The second interval, A-B, corresponds to the histogram background peak, and is left unchanged (segment slope=1). The gray levels of the fourth interval C-D, which correspond to the fluorescence, are markedly stretched. The intermediate interval B-C is also stretched although to a lesser extent, because it corresponds to the gray levels of the intermediate region between the background and the fluorescence. The gray levels of the last interval D-255 are transformed to ensure that all 256 output gray levels are filled.

General criteria for the detection of the transformation points

The evaluation of points A, B, C and D is adaptive, i.e. based on the characteristic pattern of the histogram. Point A is set in correspondence of the peak initial level, while point B is chosen in correspondence of the gray level at the end of the peak. Points A and B are obtained by differentiating the histogram sequence and by detecting both the first zero crossing point and the first local minimum of the derivative. Particular care has been devoted to guarantee flexibility of the detection: in fact, a smoother kernel is used to eliminate the irregularities in the histogram curve, thus avoiding that each of them is detected as a peak.[5] The examination of a number of images showed that a good criterion to eliminate the contouring effect, is to set D to be equal to the image maximum gray level. Finally, the position of point C is determined by the following weighted average:

$$C = \frac{K \cdot B + D}{K + 1} \tag{1}$$

In Eq. (1), K represents the average weight. The higher K is, the higher is the enhancement of the intermediate zone between background and fluorescence. Point X is set equal to B-A, while evaluation of point Y is performed by the following equation:

$$Y = X + \frac{(Z - X) \cdot (C - B)}{(D - B)} \cdot \alpha \tag{2}$$

where $\alpha < 1$. Eq. (2) guarantees that, if Z is suitably chosen, the slope of the stretching line from point P(B,X) to point P(C,Y) is smaller than that of the line from point P(C,Y) to point P(D,Z). From our experiments it result that Z=230 is a good choice.

EXAMPLES OF THE PERFORMANCES OF THE ALGORITHM

Particular care has been taken to make the elaboration as flexible as possible, in order to obtain the optimal stretching for each image under elaboration. In Figure 4 the effect of the contrast enhancement on the image of Figure 1.b is presented. The algorithm determines the following coordinates of the transformation points (see Figure 3): A=4, B=X=25, C=46, Y=56, D=106, Z=230. It can be noted that the fluorescence is well enhanced with respect to the background. The intermediate zone between the background and the fluorescence is more evident with respect to the original image. However, it is well separated from the fluorescence.

Figure 4. Effect of the enhancement on the image of Figure 3.

The procedure described so far operates efficiently on a great number of different histograms. To perform equally well with multiple-peak histograms, it has been integrated by a user-activated peak merging procedure. Figure 5.a shows a typical example of a bimodal-histogram image, combining the initial peak of the dark background to the peak corresponding to the area surrounding the target. Both peaks have to be treated as background in the enhancement procedure. The additional procedure operates a peak-search routine, and combines the two peaks into a single histogram region, which undergoes discrimination with respect to the fluorescence tail to be enhanced. Whereas the slide-stretching procedure alone would have resulted in poor enhancement, the combined procedure (Figure 5.b) results in a marked enhancement of the fluorescence of the tumor tissue with respect to the surrounding area of the image.

a)  b)

Figure 5. Example of the elaboration of a fluorescent image characterized by a marked visibility of the background. *5.a*: Original image; *5.b*: enhanced image.

Further procedures implemented in the diagnostic system include histogram windowing, whereby the operator can select a portion of the image to be elaborated, in order to limit the enhancement only to the potentially tumor-bearing areas, and to avoid enhancement of strongly fluorescent areas as bones, surgical tools, etc.

CONCLUSIONS

The aim of this work has been the development of flexible image-enhancement procedures, originally intended to be combined to the Time-Gated Imaging system, but which can be successfully applied to any type of imaging technique for diagnostic purposes. Far from being thought as an alternative to the decisional ability of the surgeon, the ensemble of procedures described can effectively integrate the diagnostic process. A typical implementation of the diagnostic system provides the on-line presentation of the original image combined (side-by-side or superimposed) with a quasi-on-line replica of it after the enhancement, with or without pseudo-coloring. All procedures may be push-button activated, in order to keep the surgeon free from complex procedures during diagnosis.

REFERENCES

1. S. Andersson-Engels, J. Johansson, K. Svanberg, and S. Svanberg, Fluorescence imaging and point measurement of tissue: applications to the demarcation of malignant tumors and arteriosclerotic lesions from normal tissue, *Photochem. Photobiol.* 53, 887 (1991).
2. R. Baumgartner, H. Fisslinger, D. Jocham, H. Lenz, L. Ruprecht, H. Stepp, and E. Unsöld, A fluorescence imaging device for endoscopic detection of early stage cancer-Instrumental and experimental studies, *Photochem. Photobiol.* 46, 759 (1987).
3. R. Cubeddu, P. Taroni, G. Valentini, G. Canti, Use of time-gated fluorescence imaging for diagnosis in biomedicine, *J. Photochem. Photobiol.* 12, 109 (1992).
4. R. Cubeddu, P. Taroni, G. Valentini, Time-gated imaging system for tumor diagnosis, *Opt. Eng.* 32, 320 (1993).
5. M. I. Sezan, A peak detection algorithm and its application to histogram-based image data reduction, *Computer vision, Graphics and Image Processing*, 49, 36 (1990).

AN OPTICAL PRESSURE MEASUREMENT SYSTEM FOR WIND TUNNEL TESTING

X.J.Gu,[1] J. Coll,[2] D. Lye,[2] F.A. Ellis,[3] V.D. Nguyen,[3] J. Bureau [3]

[1]Ontario Laser and Lightwave Research Center
60 St. George St., Suite 331, Toronto, Ont., Canada M5S 1A7
[2]de Havilland Inc., Garratt Boulevard, Downsview, Ont., Canada M3K 1Y5
[3]Institute for Aerospace Research, Montreal Rd.
Ottawa, Ont., Canada K1A 0R6

ABSTRACT

In this paper, a new optical pressure measurement system will be described. Pressure distributions measured with this technique are presented and compared with those measured by conventional pressure taps. The tests showed that the OPMS measurements of surface pressure distribution were comparable to those measured by conventional methods. However the OPMS offers a complete static pressure mapping and pressure visualization on the surface of the model.

1. INTRODUCTION

A new optical pressure measurement technique has recently received a wide attention. Some major aircraft manufacturers and research institutes are currently involved in the research and development of new measurement techniques based on pressure sensitive paint.[1] The success of this technology could save a considerable amount of cost in model fabrication and instrumentation.

Ontario Laser and Lightwave Research Center (OLLRC) together with de Havilland Inc., who initiated this project, and Institute for Aerospace Research (IAR) at NRC has set up and tested an Optical Pressure Measurement System (OPMS) in a small high speed wind tunnel. The pressure sensitive paint was developed by two research groups (M. Winnick and I. Manners) at Department of Chemistry, University of Toronto. The primary objective of the test was to compare the pressure distribution over a two dimensional model measured by OPMS with that of conventional pressure taps. The second objective was to determine various factors that might affect the OPMS measurements such as oil contamination, ambient light, temperature changes etc..

The OPMS measurement is based on a phenomenon called oxygen quenching of luminescence. A surface coating is formed by mixing photoluminescence molecules in an oxygen permeable polymer. The luminescence excited by an appropriate light source will decrease as

Applications of Photonic Technology
Edited by G.A. Lampropoulos *et al.*, Plenum Press, New York, 1995

oxygen concentration rises. As a results, the luminescence from the coating varies as a function of the partial pressure of oxygen. The intensity of the luminescence can be related to the static pressure of the air at the coated surface by the Stern-Volmer relationship,[2]

$$I_0/I = A + B\, p/\, p_0, \qquad (1)$$

where I is the luminescence intensity measured at pressure p and I_0 is intensity measured at reference pressure p_0, A and B are the constants. A digital camera measures the intensity distribution over a model and the pressure distribution can be computed.

2. INSTRUMENTATION

An accurate Optical Pressure Measurement System based on pressure sensitive paint consists of several subsystems. These include an illumination system, a detection system and a data processing system. The first two subsystems have to be customized to meet the requirements of the wind tunnel configuration.

The 2-D model was illuminated with a krypton ion laser operating at visible. Using visible light as a excitation source reduced significantly the photo-degradation of the paint.[3] A trifurcated optical fiber bundle was used to deliver the light over the model surface. Care was taken to arrange three fiber ends to generate an uniform illumination. A color filter was used to separate the illumination light from the luminescence of the model.

A high resolution CCD camera (model TM-745E, Pulnix America Inc.) was used as a detector. The camera has 768 x 494 pixels with a cell size of 8.4 μm x 9.8 μm. The luminescence light from the model was collected with a television lens (Cosmicar f/1.4) and imaged onto the CCD camera. A variable aperture was used to reduce the light intensity measured by detector. The intensity resolution of this detection system is 8 bits. The CCD camera records images at video speed i.e. 33 ms per frame. The integration of image is allowed but not the average of the frame.

A commercial image measurement and analysis software (OPTIMAS, Infrascan Inc.) was used for post processing the images. The software can perform arithmetic operations of images.

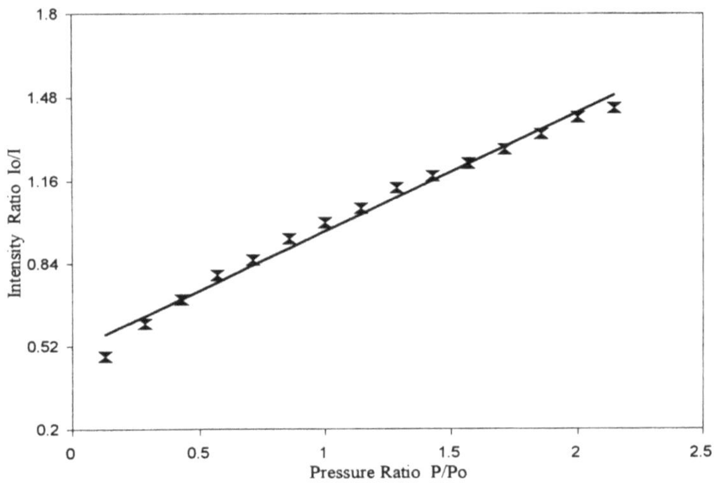

Figure 1. Calibration curve showing I_0/I versus P/P_o for one of pressure sensitive paint measured at room temperature, the solid line is a linear least square fit for the data.

The wind tunnel tests were conducted at the High Speed Aerodynamics Laboratory of IAR at NRC. The wind tunnel is a blow-down type. The Mach number varied from 0.50 to 0.58 and the Reynolds number was approximately 5 million per foot during the tests. . The top side of the test section is made of clear glass so the model can be illuminated and imaged during blowdown. The pressure sensitive paint was applied to a semi-circular aerofoil model of thickness-to-chord ratio of 0.2. The model had 9 conventional pressure taps located on the model centerline and was mounted on the floor of the tunnel's test section.

3. RESULTS AND DISCUSSIONS

3.1 Calibration of Pressure Sensitive Paint

A calibration curve was determined by taking the readings of luminescence intensity over a pressure range of a 0.05 - 2.0 atm in a vessel (see Fig.1). The resulting intensity reading at 1.0 atm, or ambient pressure p_0 is taken to be I_0 . The data indicates a nearly linear relationship between I_0/I versus p/p_0, as predicted by the Stern-Volmer relation. The intercept A and the slope B are determined by least-squares fitting to be 0.51±0.04 and 0.46±0.012, respectively at an ambient temperature of 23 °C. Notice that the sum of A and B is unity within experimental error as expected.

3.2 Pressure Distribution Measured with OPMS

To apply the Stern-Volmer model, two images of the aerofoil model were taken from the top: one was at wind-on case where the pressure distribution was unknown; and the other at wind-off case in which the pressure distribution was a constant on the surface. The wind-off image was divided by the wind-on image. The resulting image is shown in Fig.2 for one of the tests. The darken color corresponds to lower pressure and higher Mach number, while the lighter

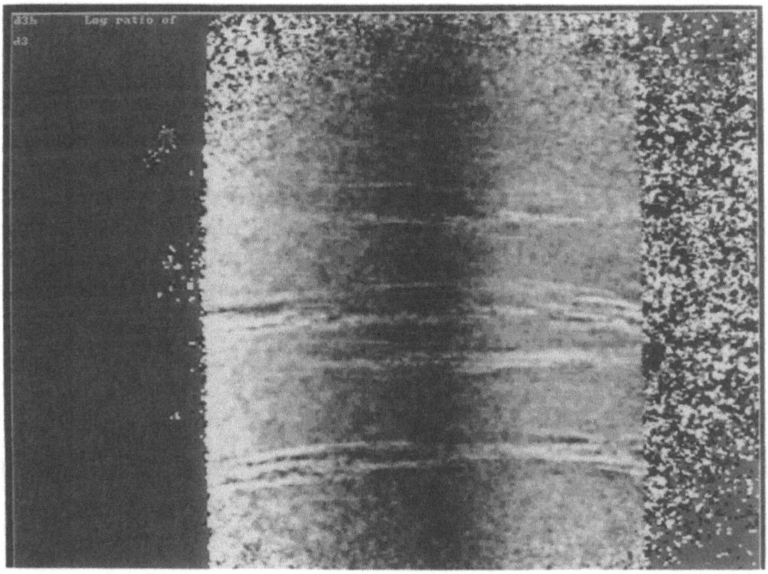

Figure 2. The image of pressure distribution over a aerofoil model at M=0.507 measured with the OPMS. The flow is from the left to right.

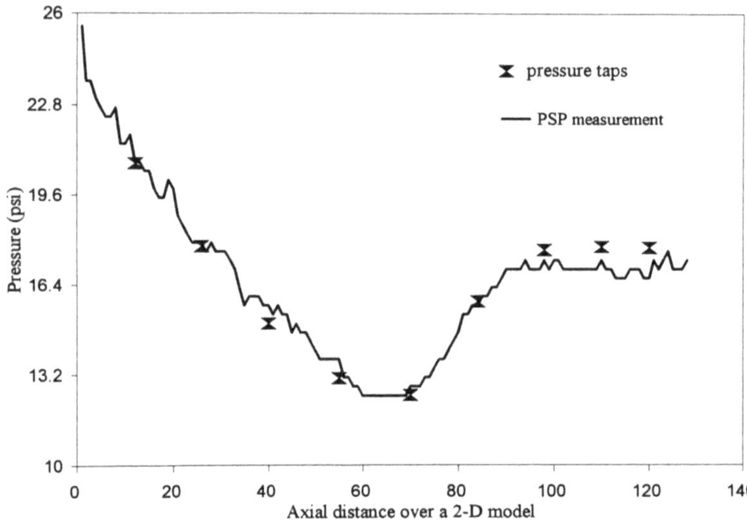

Figure 3. Comparison of pressure measurements using the OPMS and conventional pressure taps at M=0.507.

color corresponds to higher pressure and lower Mach number. The streaks seen in the photograph were from residual oil drops blown onto the glass window of the wind tunnel caused by a leak from the control valve upstream that has been repaired later.

A spatial resolution was less than 1 mm along the centerline of the model in this test. Even it is higher than that of pressure taps, further improvement can be achieved by frame averaging of the images and by using large CCD arrays.

A line of the intensity ratio data I_0/I in Fig.2 was selected along the centerline row of the pressure taps and these data were used to calculate pressure distrubution according to Eq. (1). The line was selected to avoid the oil streaks. Fig.3 shows a comparison of the pressures determined by the OPMS and those measured by pressure taps. It can been seen that the agreement between the two measurement techniques is very good. The peak Mach number is the same and the location and strength of the weak shock wave downstream of this peak is accurately observed. Additionally the area of essentially constant pressure after the shock wave, due to the separation of the model boundary layer is also well correlated for both measurement techniques.

In conclusion, the OPMS measurements agree well with those of conventional pressure tap measurements over a 2-D aerofoil model in a small wind tunnel. It has demonstrated that the OPMS offers a complete static pressure mapping. The images of OPMS also provide a means for flow visualization on boundary layer transition, separation and reattachment lines as well as vortex and shock wave locations.

REFERENCES

1. J. Kavandi, J. Callis, M. Gouterman, G. Khalil, D. Wright, E. Green, D. Burns, B. Mclachlan, *Rev. Sci. Instrum.*, **61**(11), Nov. 1990, pp. 3340-3347.
2. O. Stern and M. Volmer, *Phys. Z.*, **20**, 189 (1919).
3. X.J. Gu, *Internal report.*

HIGH SPEED WEB INSPECTION USING INTELLIGENT TDI CAMERAS

J.W. Roberts, S.D. Rose, S.G. Chamberlain, G. Maroscher[1]

DALSA INC.
605 McMurray Road
Waterloo, Ontario., N2V 2E9
[1] E.I. DU PONT DE NEMOURS & COMPANY (INC.)

ABSTRACT

Inspection systems for wide web materials have been unable to effectively image fine defects as they are detected. The amount of data produced by highly parallel video inspection cameras can exceed 400 MBytes/sec. The system described in this paper is capable of analyzing and displaying a detected image within seconds of the event using a single frame grabber and a 386 computer. The system can operate at processing speeds of greater than 400 MBytes/sec since it makes use of a novel post processing algorithm within the camera itself. The video cameras are based on TDI (Time Delay and Integration) technology to provide high grey scale resolution at high data rates and low light levels. The system has an adjustable resolution ranging from 256 to 24,000 pixels per line scanned. The scanning rate is adjustable to a maximum of 20,000 line scans per second.

1. INTRODUCTION

In this paper we describe a new real time inspection system that is designed to scan continuous web media at high speed. The system has proven itself capable of reducing product wastage and in eliminating some product handling steps.

The ability of the system to detect and display images on line and in real time allows a line operator or process engineer to correlate the various product defect types with the physical process. The purpose of the work was to demonstrate the feasibility of TDI based inspection for film and to develop better components such as advanced TDI cameras and electronics for use by integrators in the development of advanced high performance inspection systems.

2. ARCHITECTURE OF THE SYSTEM

A basic system diagram is shown in Figure 1. The inspection system consists of a bank of cameras arranged in parallel across the length of the web to be scanned, a computer console containing a single computer and frame grabber, an illuminator and supporting mechanical and electrical interfaces to the web machinery. The inspection system employs a massively parallel approach to the image capture with a front end algorithm to reduce the data flow to the items of interest only. Therefore, the actual bandwidth of the system for post processing was limited to the size of objects that were imaged.

The image input method is based on a high performance Time Delay and Integration (TDI) camera. This method has been well understood as a means of achieving excellent image detection performance on objects in motion by synchronizing the "shutter" of the TDI to the motion of the object that is imaged. [1] The cameras operate in parallel with from 1 to 8 channels per camera. For a 12 camera system, this can result in a maximum of 96 parallel processing channels. TDI technology allows for high sensitivity under low light conditions with excellent stop action [2-4].

All processing electronics are housed within a single console unit. The vast majority of the data processing is performed within the system cameras. As a result the data processing requirements of the central computer are minimal. This allows a single pc computer interfaced with a frame grabber to be used despite the fact that the video data is captured at more than 400 MBytes per second. The cost and complexity of the system are thus considerably less than with other possible system approaches with currently available equipment.

The cameras are housed within a self contained environmentally isolated module. The cameras are self aligned within the box and can be removed and replaced while the production line is in operation. The inspection system can continue to operate during camera replacement. The system can be configured to include from 1 to 12 cameras.

The illumination design is transmissive. Other illumination schemes can be used with this system approach. All components are mechanically and electrically isolated from the production line.

3. CAMERA TECHNOLOGY

The system employs DALSA modular cameras [5]. In particular, the base technology is that of a DALSA standard 8 tap TDI camera. Integrated into the base camera is an abnormal pixel screening circuit. This algorithm is key to the reduction of data reaching the console. With this algorithm only useful data is passed from the camera to the console buffer memory. The screening algorithm is contained within a removable, re-programmable logic array. This approach permits future modifications to the screening algorithm. The camera requires external loading of the parameters that define the screening algorithm. These parameters are provided externally from the host computer. Parameter loading is required at power-up but can be repeated at any time as required. The modified TDI camera has dimension 3.5"x3.5"x14.5" and weights approximately 4 lbs and is compatible with the DALSA modular camera family [5].

Cameras are housed within a environmentally isolated beam. The beam components use accurate machining to permit field replacement of cameras without the necessity of re-alignment. At the web surface the inter-camera overlap tolerance is 0.040".

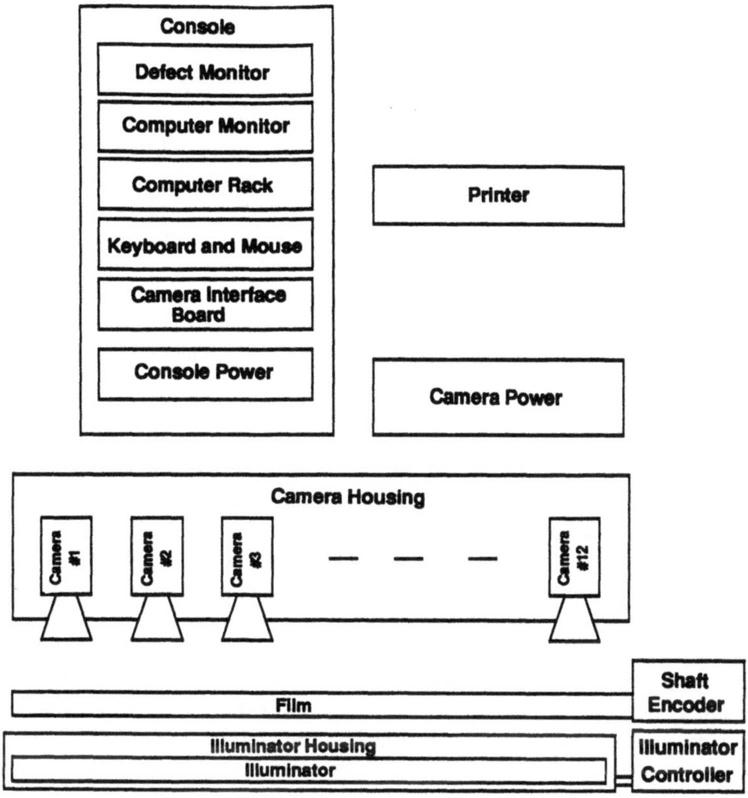

Figure 1. System Architecture Block Diagram

4. SYSTEM ARCHITECTURE

In discussing the basic architecture of the system we begin with the camera and work back to the host computer.

In terms of functionality the camera can be divided into two parts.

i) Raw Data: An eight tap, 2048X96 element TDI CCD sensor is used. Analog video signals from each tap of the CCD are digitized into eight bits and communicated to the abnormal pixel screening portion of the camera electronics.

ii) Abnormal Pixel Screening: The bulk of the data processing required by the system is performed within the camera itself. An abnormal pixel screening algorithm has been realized as a electronic circuit board and integrated as part of an otherwise standard TDI camera. There are separate screening circuits for each of the eight channels within a camera. As there are 12 cameras in the system we have a total of 96 screening circuits running in parallel. This massive parallel processing is really the heart of the system and is the reason that the system can process so much data with only a single frame grabber and 386 computer. The screening algorithm functions by comparing each pixel with a reference window. If the pixel lies outside the window (ie above the upper grey level bound or below the lower grey levelbound) the pixel is tagged as abnormal and retained. The definition of the reference window is set by an algorithm run by the host computer.

The act of throwing away "normal" pixels results in a loss of pixel contiguity. In order to be able to reassemble abnormal pixels into defects at some later time we must attach a physical position address to each abnormal pixels. A two dimensional (x,y) pixel address is incorporated into the pixel description. This address is used later, within the frame grabber, to re-assemble the individual abnormal pixels into the defect image.

Image data is transmitted from the cameras to both a standard frame grabber and the host computer via a single interface circuit board. The interface is described below.

i) Frame Grabber/Cameras Interface: As with most frame grabbers, it can accommodate only a single channel of video. Since the inspection system is comprised of up to twelve cameras there must be a twelve to one MUX. This MUX is performed on the camera interface board. Each pixel from a camera is comprised of 20 bits. As the pixel data is passed from the camera to the frame grabber, 4 additional address bits are added.

Pixel data from each camera is stored within a FIFO located on the interface board. These FIFOs are polled in sequence. A single line of data is read out during each polling cycle. That is, data from the current camera FIFO is read continuously until an end of line (EOL) is detected. Polling then switches to the next camera. Data from this camera FIFO is read until an EOL pixel is detected and so on. In this way the data from the cameras always remain synchronized. Data from all the camera FIFOs are read onto a common 24 bit data bus. In principle this data could then be transferred immediately into the frame grabber. However most frame grabbers can handle no more than 16 bits. To get around this problem we break the 24 bit pixel into a packet comprised of three 8 bit bytes in series. Three synchronization bits are added for a total of 11 bits. The sync bits are used by the frame grabber to uniquely define each byte in the packet.

ii) Host Computer/Cameras Interface: The interface board also is used for communication between each of the cameras and the host computer. This communication is used to down-load the pixel screening parameters from the computer into the cameras. The computer can also access the pixel data from the cameras if necessary.

All of the raw pixel data from the cameras is processed by a single image processing board (frame grabber) with on board VRAM and a TMS34010 micro-processor. The VRAM can hold up to 16,000 pixels. The TMS34010 assembles individual pixels into separate defect runs. Individual runs are stored within RAM. A terminated run is defined as a defect entity. The frame grabber can process pixels into defects at a maximum rate of about 22,000 pixels/sec. Once a defect entity is assembled it must be passed on to the host computer via a software interrupt.

The host computer has several functions mainly related to house keeping. These include data storage, defect mapping, system monitoring and the GUI.

5. REAL TIME DEFECT CAPTURE

The system is currently installed and operating within an industrial setting. A typical example of defect image is shown in figure 2. This image was captured in real time while the production line was operating. The image quality is sufficient to enable the line operator to identify the source of the defect and take corrective action.

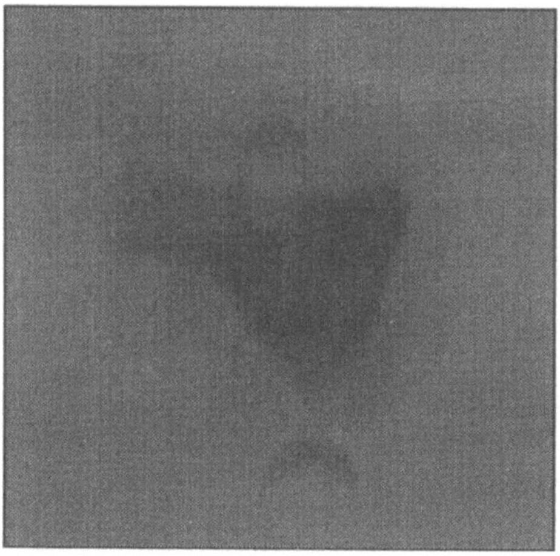

Figure 2. This is an image of a gel pattern with a diameter of approximately 0.2".

6. CONCLUSIONS

A system for real-time, on-line continuous web inspection has been described. We discussed the general design philosophy as well as details of its component parts. The heart of the system is the camera itself which contains pixel screening based on a comparison of the pixel grey level with a abnormal pixel threshold window. Screening of pixels reduces the amount of data to a level that is compatible with a single frame grabber and 386 pc. Images of some actual defects acquired during on-line inspection are included.

7. ACKNOWLEDGEMENTS

The development of this system was conducted by DALSA under a joint research and development agreement between DALSA INC. and E.I. DU PONT.

8. REFERENCES

1. Gudmundsen, R.A. Solid State Integrating Image Motion Compensating Imager, United State Patent Appl. No. 366,828, June 4, 1973.

2. Chamberlain, S.G. (1989). A New 2048 X 96 High Photosensitivity Wide Dynamic Range High Speed Imager, Proc. SPIE Technical Symposium, November 1989, Philadelphia, PA.

3. Washkurak, W.D., S.G. Chamberlain and P.T. Jenkins (1989). A Time Delay And Integration CCD Image Sensor For High Speed And Low Light Level Applications. Proc. Electronic Imaging 89, April 1989, Pasadenia, CA. 2-2562-567.

4. Barbe, D.F. (1989). Time Delay and Integration Image Sensor. Solid State Imaging, 1st Edition, Jespers, Van der Weile, White, Noordhoff, Leyden, The Netherlands. pp. 659-671.

5. Doody, B.C., Washkurak, W.D., Jenkins, P.T., Miethig,M. Hood, S.J., Prince, N.D. High Speed Variable Scan CCD Camera For Machine Vision and Direct Computer Interface. Proc. Electronic Imaging 91,Boston,Mass., pp. 63-71.22

DISTORTION OF BAR CODE SIGNALS IN LASER SCANNING

Stephen J. Shellhammer

Symbol Technologies, Inc.
116 Wilbur Place
Bohemia, NY 11716

1 Introduction

One of the most common methods of reading bar codes is to scan them with a laser beam and process the received optical signal. The received light is proportional to the integral of the beam profile multiplied by the region of the bar code covered by the beam. The finite spot size of the laser beam acts as a filter on the original bilevel bar code signal. This filtering can cause intersymbol interference between adjacent edges of the bar code signal. An edge detector attempts to reconstruct the original bilevel signal from this filtered signal. The edges detected in the filtered waveform are in different locations than those in the original bilevel bar code signal resulting in a distortion of the bar code signal. If this distortion is too large, it can prevent the bar code from being properly decoded. Thus, there is a limit on the spot size of the laser beam that can be used to scan a given density bar code.

The laser beam profile that is most often used in bar code scanners can be approximated by a Gaussian distribution. The width of the beam changes depending on the distance from the scanner to the bar code.[1] As the size of the laser beam becomes too large, relative to the bars and spaces of the bar code, it becomes impossible to read the bar code. Thus, the working range (i.e. the range of distances between the laser scanner and the bar code) over which the bar code is readable is limited by this growth in the spot size and the maximum spot size that can be tolerated.

In this paper a detailed analysis of the distortion of bar code signals caused by the finite spot size of the laser beam is performed. In Section 2 a general method of analysis is proposed for determining the edge shifts in the bar code signal due to the filtering of the optical beam. In Section 3 the technique is applied to a comprehensive study of one of the most popular bar code symbologies, Universal Product Code (UPC). A measure of distortion is proposed which relates directly to the readability of the bar code. A study of which UPC bar code characters are most sensitive to the edge shift is performed. The distortion of the most sensitive UPC characters is shown as a function of the optical spot size.

Applications of Photonic Technology
Edited by G.A. Lampropoulos *et al.*, Plenum Press, New York, 1995

2 Edge Shifts of Gaussian Filtered Bilevel Signal

A bar code is a sequence of vertical bars and spaces where the information stored in the bar code is encoded in the relative widths of the bars and spaces.[2] The specific pattern of the bars and spaces depends on the coding method used, often referred to as the symbology, as well as the actual data that is encoded in the bar code. In this paper we will deal with the most common type of bar code the Universal Product Code (UPC).[3]

The bar code is scanned by a flying spot laser scanner. The scanner receives the laser signal reflecting off the bar code. The laser beam is assumed to scan at a fixed velocity and have a Gaussian beam profile. The received signal is a time domain function that results from scanning the bar code. Mathematically the bar code can be represented, in the time domain, as a sequence of step functions of alternating polarity,

$$x(t) = \sum_{i=1}^{N}(-1)^{(i-1)}u(t-T_i) \tag{1}$$

where $u(t)$ is the unit step function and the T_i represents the time when the signal transitions from either black to white or white to black (i.e. the edges). Here we assume that the edges are ordered: $T_1 < T_2 < \ldots < T_N$. The width of the bars and spaces are given by the difference between adjacent edges (e.g. $T_{i+1} - T_i$). This bar code signal, $x(t)$, is filtered by the optical spot which has a Gaussian impulse response,

$$h(t) = \frac{1}{\sqrt{2\pi\sigma^2}}e^{-(t^2/2\sigma^2)} \tag{2}$$

where the standard deviation (in the time domain), σ, is just the standard deviation in the spatial domain divided by the velocity, $\sigma = \sigma_x/v$. It is often common to refer to the *spot size* of the optical beam which is given by 4σ, in the time domain (or $4\sigma_x$ is the spatial domain). The received signal is the convolution of the bar code signal and the optical beam,[4]

$$y(t) = x(t) * h(t) = \int_{-\infty}^{\infty} x(\tau)h(t-\tau)d\tau. \tag{3}$$

Given the received signal, $y(t)$, there are numerous methods of detecting the edge locations and reconstructing an estimate of the original bar code signal. One of the more common methods is to detect the inflection-point of the signal which is given by the zero crossing of the second derivative. The edge locations are given by the solutions to the equation: $\ddot{y}(t) = 0$. Due to the linearity of the convolution equation we can write,

$$\ddot{y}(t) = \dot{x}(t) * \dot{h}(t). \tag{4}$$

Since $x(t)$ is a sum of unit step functions of alternating polarity then $\dot{x}(t)$ is a sum of impulses of alternating polarity,

$$\dot{x}(t) = \sum_{i=1}^{N}(-1)^{(i-1)}\delta(t-T_i). \tag{5}$$

Substituting this into Eq. (4), simplifying, and equating to zero gives,

$$\ddot{y}(t) = \sum_{i=1}^{N}(-1)^{(i-1)}\dot{h}(t-T_i) = 0. \tag{6}$$

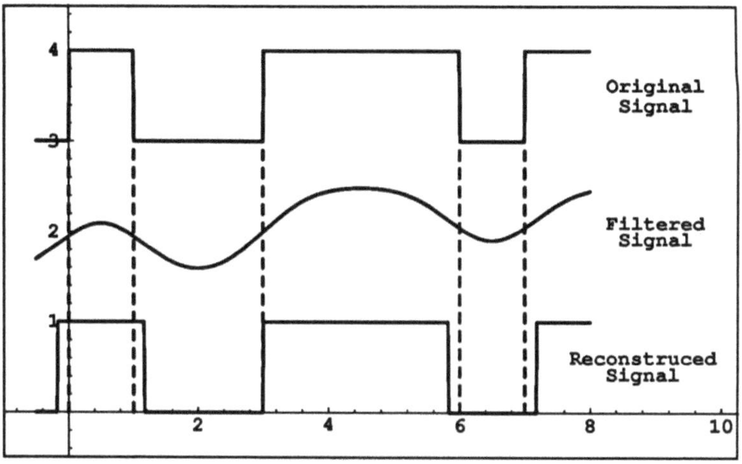

Figure 1. Original Bar Code Signal, Filtered Signal, and the Reconstructed Signal, with $\sigma = 0.6$

Substituting the derivative of $h(t - T_i)$ into Eq. (6) gives the following equation for the edge locations,

$$\boxed{\sum_{i=1}^{N}(-1)^{(i-1)}(t - T_i)e^{-((t-T_i)^2/2\sigma^2)} = 0.} \tag{7}$$

Since this equation is nonlinear it is best solved using a numerical root finding routine like that available in Mathematica.[5] In finding the root of a nonlinear equation it is always necessary to have an initial estimate to initiate the search. Here, the search is initialized at the original edge locations, T_i. In solving Eq. (7) we obtain the inflection-points of $y(t)$ which are estimates of the edges in $x(t)$. If we call these inflection-points $\widehat{T}_i$ then the reconstructed bar code signal is given by,

$$\hat{x}(t) = \sum_{i=1}^{N}(-1)^{(i-1)}u(t - \widehat{T}_i) \tag{8}$$

Figure 1 shows a segment of a UPC bar code scanned by a Gaussian beam with $\sigma = 0.6$. Shown are the original bilevel bar code signal, the Gaussian filtered signal, and the reconstructed bilevel signal.

3 Distortion of UPC Characters

In this section the method of determining the shift in the edges due to Gaussian filtering is applied to the UPC bar code symbology. A UPC bar code consists of a sequence of characters surrounded by guard bars. Each UPC character is composed of two bars and two spaces. The width of the bars and spaces are multiples of a unit width called a *module*, and they sum to seven modules.[2,3] Each UPC character can encode a single digit from 0 to 9. The objective in scanning the bar code is to process the received signal and determine what character combination was originally printed.

Once there is a method of determining the edge locations of the Gaussian filtered signal, several questions can be posed which relate to bar code scanning. First, which character is most distorted by the optical beam? Second, for what value of σ is the

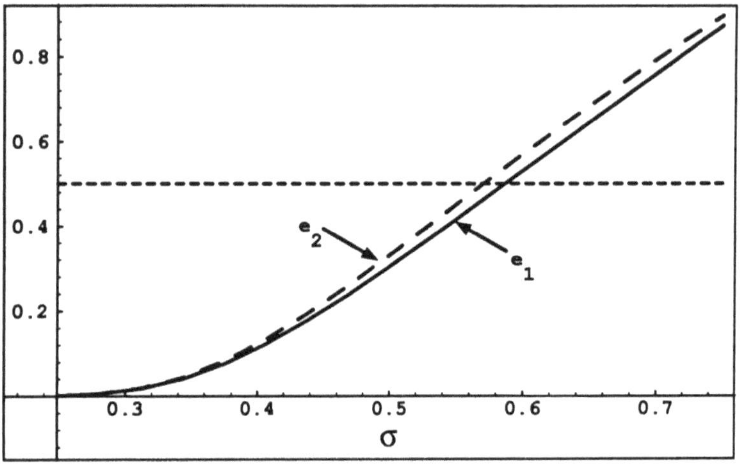

Figure 2. The Solid Line is e_1 for Character "3", and the Dashed Line is e_2 of Character "5", both as a Function of σ for a Unit Module Size

distortion sufficiently large so that the character is no longer decodable? In order to answer both of these questions, it is necessary to introduce some measure of distortion. The proposed distortion measure is directly related to the decodability of the UPC characters.

It is convenient to group the widths of the two bars and two spaces representing a UPC character into a vector. For example, the bar space pattern for the UPC character "7" can be written in vector notation as $\mathbf{u} = (1\ 3\ 1\ 2)^T$, where the components of $\mathbf{u}$ are the widths of the bars and spaces. In the decoding procedure it is useful to deal with the vector represented by the sum of adjacent elements. The reason for this is that the sum of adjacent elements (one bar and one space) is not effected by uniform ink spread.[3] The vector representing the sums of adjacent pairs, $\mathbf{v}$, is given by $v_i = u_i + u_{i+1}$ for $i = 1, 2, 3$. For example, the UPC character "7" represented this way is $\mathbf{v} = (4\ 4\ 3)^T$. In some of the literature the first vector, $\mathbf{u}$, consisting of the widths of the bars and spaces is referred to as the x-sequence and the second vector, $\mathbf{v}$, consisting of sums of adjacent pairs is referred to as the t-sequence.

The standard decode procedure for a UPC character begins by normalizing the measured widths of the bars and spaces so they sum to seven. The widths of the bars and spaces are obtained from the locations of the detected edges, $\widehat{T}_i$. We define this as the normalized vector $\hat{\mathbf{u}}$. This is an estimate of the original vector $\mathbf{u}$ representing the UPC character. An estimate of $\mathbf{v}$ can easily be obtained adding adjacent pairs of elements of $\hat{\mathbf{u}}$, namely $\hat{v}_i = \hat{u}_i + \hat{u}_{i+1}$ for $i = 1, 2, 3$. Next, each component of estimate vector, either $\hat{\mathbf{u}}$ or $\hat{\mathbf{v}}$, is rounded off to the nearest integer and that vector is compared to the patterns for the different characters. For most characters the $\mathbf{v}$ vector is used since it is insensitive to ink spread; but, for some characters it is necessary to use the $\mathbf{u}$ vector.

Therefore, if any of the components of $\hat{\mathbf{u}}$ or $\hat{\mathbf{v}}$ differ from the component of actual character, $\mathbf{u}$ or $\mathbf{v}$, by more than 0.5 then the character will not be decoded. This suggests the use of an ∞–norm[6] in defining a distortion measure for UPC characters,

$$e_1 \triangleq ||\hat{\mathbf{u}} - \mathbf{u}||_\infty = \max_i |\hat{u}_i - u_i| \tag{9}$$

$$e_2 \triangleq ||\hat{\mathbf{v}} - \mathbf{v}||_\infty = \max_i |\hat{v}_i - v_i|. \tag{10}$$

The first type of error apples to decoding when using the vector of bar and space widths; while the second applies when using the vector composed of the sum of adjacent element widths. So, in general, both are necessary.

An exhaustive study was performed to find the UPC characters that were most sensitive to the Gaussian filtering. In doing this study, it was necessary to not only consider the current UPC character but also the adjacent UPC characters since the Gaussian filtering causes the edges in those characters to interfere with the current character. So for each character it was necessary to consider all possible adjacent characters. Actually, only those edges that are within one to two modules of the current character have any effect. In comparing different UPC characters, it was determined that character "3" has the highest e_1 error, and character "5" has the highest e_2 error, for a given spot size. It turns out that some characters have large errors of one type but a relatively small error of the other type, so it is necessary to consider both types of errors. To summarize the results Figure 2 shows a plot of the errors (with the worst case adjacent characters) for these two worst case characters as a function of σ, where the module size is assumed to be unity (i.e. σ has been normalized by the module size). The dotted line at 0.5 show the threshold at which the characters become undecodable. This occurs at about $\sigma = 0.58$.

4 Conclusions

A method of determining the edge location of a Gaussian filtered bilevel signal is proposed. The method is applied to a systematic study of distortion of UPC bar code characters. A measure of distortion of a UPC bar code is introduced which is derived from the standard decode procedure. The characters that are most sensitive to this type of systematic distortion were identified and the distortion of those characters was plotted as a function of the optical spot size.

This method could be used to study other bar code symbologies to determine which bar code symbologies are most sensitive to this type of distortion. Also, other distortion measures could be introduced to study the merits of different decoding procedures.

References

[1] A. Yariv, *Optical Electronics.* Holt, Rinehart, and Winston, 4th ed., 1991.

[2] T. Pavlidis, J. Swartz, and Y. P. Wang, "Fundamentals of bar code information theory," *IEEE Computer*, pp. 74–86, April 1990.

[3] Uniform Code Council, Inc., *UPC Symbol Specification Manual*, January 1986.

[4] E. Barkan and J. Swartz, "System design considerations in bar code laser scanning," *Optical Engineering*, vol. 23, pp. 413–420, July/Aug. 1984.

[5] S. Wolfram, *Mathematica: A System for Doing Mathematics by Computer.* Addison-Wesley, 2nd ed., 1991.

[6] G. H. Golub and C. F. Van Loan, *Matrix Computations.* Johns Hopkins University Press, 1983.

APPLICATION OF CCD DETECTOR IN THE MULTIFREEDOM DYNAMIC MEASUREMENT

Qixiang Zhu

Institute of Optics and Electronics
Chinese Academy of Sciences
P.O.Box 350,Shuangliu Chengdu
610209,P.R.China

1. INTRODUCTION

The real-time measurement for the motional parameters of the flexible nozzle is a significant technical problem in the development of the rocket vector control system. Since the nozzle is flexibly connected, so when the nozzle is operating,it has no stationary swing center. The multifreedom motion of the nozzle in operating brings great difficulty in accurately measuring the nozzle motion.

By using multi-CCDs as detectors,we have measured the parameters of six degrees of freedom motion of the nozzle in real-time through measuring the linear displacement in the different parts of the nozzle with photoelectronic noncontact measurement. The method has the advantages of high accuracy, real time processing and not influenced by the deformation of the nozzle.The measuring principle,mathematical model ,the measuring system based on this principle and its technical specifications are presented in the paper.

2. THE MEASURING PRINCIPLE AND THE MATHEMATICAL MODEL

When the nozzle is operating,it is an object with swing center ,moving in 3-D , twisting (α_z) and swinging in double (α_x, α_y) directions along the swing axis (Fig.1).Its motion is a complicated space motion occured simultaneously in six degrees of freedom.

Using the rectangular coordinate formed by the crossing point of the nozzle axis line in the initial state with the plumb line through the point as measuring reference. We measure the nozzle motional paramenters in six degrecs of freedom according to the displacement of the marking line outside the nozzle wall.

The system uses the photoelectronic measuring heads to measure the marking line on the nozzle in the three rectangular directions by noncontact method.Its measuring principle and constitution are showing in Fig.2.

The various measuring light path involved in each measuring head is an imaging measuring path within the limited range, so during the measurement,the nozzle swing and the swing center displacement will cause the variation of objective distance of the nozzle marking line,as will as the variation of the image height. During the six degrees of freedom of nozzle no matter the swing angle or the swing center displacemcut, they will effect each other. So the realtime correction for object distance variation and the mutual effects of six varying parameters must be considered during the operating of the measuring system.According to the geometric optics principle considering the corrections of the object distance variations and the mutual effects of six parameters,the six motional parameters expressions can be given to the multiple measuring light path with the three measuring heads:

$$
\begin{aligned}
\alpha_x &= \mathrm{arctg}\{\{X_1'[L_1 + X + R(1-\cos\alpha_z) + (H-z)\mathrm{tg}\alpha_y - R\mathrm{tg}(\alpha_y/2)\mathrm{tg}\alpha_y] \\
&\quad /F1 - X_2'[L_2 + X + R(1-\cos\alpha_z) + (h-z)\mathrm{tg}\alpha_y - R\mathrm{tg}(\alpha_y/2)\mathrm{tg}\alpha_y]/F_2\} \\
&\quad /(H-h)\} \\
\alpha_y &= \mathrm{arctg}\{\{Y_1'[L_5 + Y + R(1-\cos\alpha_z) + (H-z)\mathrm{tg}\alpha_x - R\mathrm{tg}(\alpha_x/2)\mathrm{tg}\alpha_x] \\
&\quad /F_5 - Y_2'[L_6 + X + R(1-\cos\alpha_z) + (h-z)\mathrm{tg}\alpha_x - R\mathrm{tg}(\alpha_x/2)\mathrm{tg}\alpha_x]/F_6\} \\
&\quad /(H-h)\} \\
Z &= Z'\{L_7 + X + R(1-\cos y) + R_z\sec\alpha_x\sin\alpha_y - R(\sec\alpha_x - 1) - R_z\sin\alpha_x\mathrm{tg}\alpha_z\} \\
&\quad /F_7 - R_z(\sec\alpha_x - 1) - R_z\sin x(\cos\alpha_y - 1) + R\sin\alpha_y - Y\mathrm{tg}\alpha_z \\
X &= Y_1'[L_5 + Y + R(1-\cos\alpha_z) + (H-h)\mathrm{tg}\alpha_x - R\mathrm{tg}(\alpha_x/2)\mathrm{tg}\alpha_x]/F_5 - R\sin\alpha_z \\
&\quad \sec\alpha_y - (H-h)\mathrm{tg}\alpha_y \\
Y &= X_4'[L_4 + X + R(1-\cos\alpha_z) + (H-h)\mathrm{tg}\alpha_x - R\mathrm{tg}(\alpha_x/2)\mathrm{tg}\alpha_x/F_4 - R\sin\alpha_z \\
&\quad \sec\alpha_x - (H-h)\mathrm{tg}\alpha_x \\
\alpha_z &= \arcsin\{\{Y + (H-h)\mathrm{tg}\alpha_x - X_3'[L_3 - X - (H-z)\mathrm{tg}\alpha_y + R\mathrm{tg}(\alpha_y/2)\mathrm{tg}\alpha_y]/ \\
&\quad F_3\}/(R\sec\alpha_x)\}
\end{aligned}
$$

where

$X_1' - X_4'$:the measured value of the marking line displacement measured in the image plane by the measuring sysem in the direction rectangular to the X axis;

$Y_1' - Y_2'$:the measured value of the marking line displacement measured in the image plane by the measuring sysem in the direction rectangular to the Y axis;

Z':the measured value of the marking line displacement along the Z axis direction measured by the maeasuring system in the image plane;

$F_1 - F_7$:the focal length of the corresponding measuring light path;

R: the distance between the marking line and the swing axis;

R_z: the vertical distance from Z marking line to the swing center;

$L_1 - L_7$: the initial object distance of the corresponding measuring light path;

H, h : the distance form the measuring head center in the vertical direction to the swing center;

$\alpha_x.\alpha_y.\alpha_z$: the swing angle of nozzle along X,Y,Z axis;

X, Y, Z : the displacement of the swing center.

3. THE MEASURING SYSTEM AND THE CONSTITUTION

Based on the above mentioned principle a measuring system was developed,the specifications of the measuring system are as following:

(1)Measuring parameters:the motional parameters $X, Y, Z, \alpha_x, \alpha_y, \alpha_z$ in six degrees of freedom are measured at the same time;

(2)Measuring range: $X, Y, Z : 0 - \pm 10mm; \alpha_x, \alpha_y : 0 - \pm 8°; \alpha_z 0 - - \pm 4°$

(3)Measuring accuraay: $\Delta\alpha_x; \Delta\alpha_y, \Delta\alpha_z \leq 50" \Delta X, \Delta Y, \Delta Z \leq \pm 0.3mm$

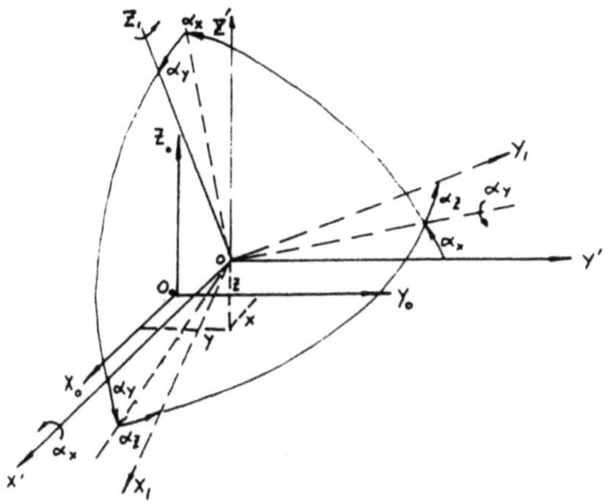

Fig.1

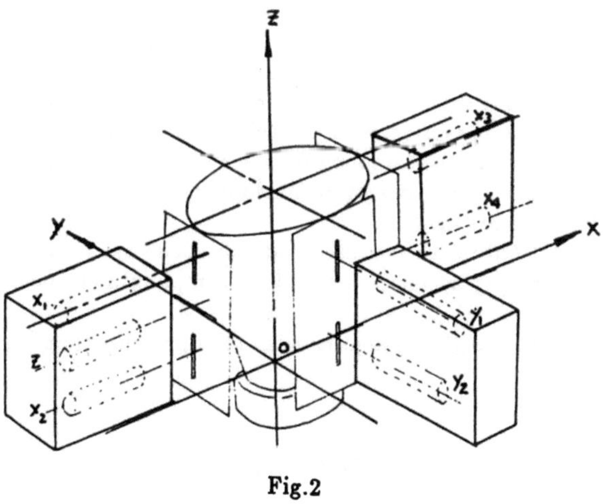

Fig.2

(4)Measuring speed and delay Sampling and processing rate: $\geq$ 500Hz; Signal extracting,calculating and transmitting delay: $\leq$ 3.5ms

(5)Output mode digital output: 16 bit serial output; analog output: 12 bit D/A converted output.

(6)Display six analog outputs are displayed by 4 bit digits at the same time.

The measuring system constitution and the block diagram.

Constitution:

(1)Auto luminescent target marking board;

(2)Photoelectronic measuring head;

(3)Signal extracting and processing electric cabinet;

(4)Operating control and display devies.

The block diagram is showing in Fig. 3.

4. THE ASSURENCE OF THE TECHNICAL SPECIFICATIONS OF THE MEASURING SYSTEM

The main specifications of the measuring system are as following:

(1)Measuring accuracy;

(2)Sampling accuracy;

(3)Information delay.

In order to fulfil the above specifications we adopt the following measures:

(1)The measures of insuring the measuring accuracy.

(A) The optical distortion and asymmetric aberration of the various measuring light path reach the smaller value.Optical distortion is less than 0.08%, chromatic aberration is up to 0.05% and spot symmetry is good.

(B)The CCD linear array sensors with small pixel size are adopted .The pixel size is $7\mu m \times 7\mu m$, thus the measuring error caused by the CCD pixel size will be minimum.

(C) Saturated amplifier and binary float threshold voltage processing are adopted in the signal extracting circuits. So the signal leading edge and following edge are steep. In this way the measuring error caused by the fluctuation of the threshold level can reduced.

(D) The position of the measuring points are determined by positioning of spot edges symmetric center, thus the measuring error caused by spot diffusion can be avoided.

(2)The measures insuring the sampling frequency

(A)The suitable CCD clock driving rate is adopted, thus the signal amplitude value and the required sampling rate are insured.

(B)A COMPAQ386/40 microcomputer and the 80387 assisting processor are used as data sampling and operating system, so the data sampling and opearting time is within 1.8ms.

(C)Simplify the data processing algorithm,so the operating time is reduced.

5. THE TESTING RESULTS OF THE MEASURING SYSTEM

When the system is finished,the testing results are as following:

(1)The measuring errors of the varous motions:

α_x is within $\pm 8°$: $\Delta\alpha_x \leq 24.4"$

α_y is within $\pm 8°$: $\Delta\alpha_y \leq 19.2"$

α_z is within $\pm 4°$: $\Delta\alpha_y \leq 25"$

X is within 10mm $\Delta X \leq$ 0.1mm

Y is within 10mm $\Delta Y \leq$ 0.1mm

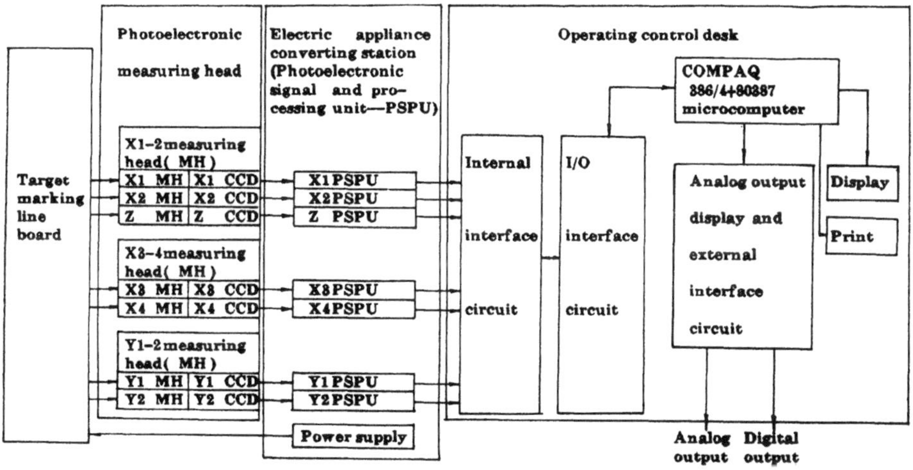

Fig. 3. Block diagram of real-time measuring system for nozzle in six degrees of freedom.

Z is within 10mm $\Delta Z \leq 0.1$mm .

(2)Sampling and processing rate: 526 Hz

(3)Signal delay: 3ms

(4)Display accuray of analog output: 3%

From the above resules we see that the angular error is less than 0.8%, displacement error is less than 3%, sampling and processing rate is larger than 500Hz.

ACKNOWLEDGEMENT

The author gratefully acknowledges the support of K.C.Wang Education Foundation, Hong Kong.

EVALUATION OF SILICON WAFER SURFACE QUALITY BY COHERENT RADIATION SCATTERING

N. Miron and D. G. Sporea

Institute of Atomic Physics, Department of Lasers
P.O. Box MG-6, Magurele, Romania, R-76900

INTRODUCTION

The inspection of surface quality of opaque materials by optical methods is now a very attractive topic with many people involved and many papers covering this subject. The theories describing phenomena produced by interaction of a laser beam with a conductive surface are studying two main aspects: specular reflection and scattering of the incident beam. The spatial distribution of scattered and specularly reflected flux is strongly dependent of the surface quality. Reciprocally, the analysis of this distribution will give an information about surface quality. The analysis of the interaction between a light beam and a rough surface started long time ago. The theoretical analysis and measurement of the scattered beam and of the specularly reflected beam are reported in many papers. The specular reflective component R_s produced by a light beam incident normally to a perfect conductive surface can be expressed by:

$$R_s = R_0 \exp[-(4\pi\sigma)^2 / \lambda^2], \tag{1}$$

where R_0 is the reflectance of a perfect surface of the same material, σ is the *RMS* value of the roughness, defined as the *RMS* deviation of the surface from the mean surface level, λ is the wavelength of the incident light beam. In formula (1) is considered the total specularly reflected light flux.

If the light wavelength is long enough compared with the surface roughness, the total diffuse reflectance (specular component) R_d can be expressed by:

$$R_d = R_0 (4\pi\sigma)^2 / \lambda^2. \tag{2}$$

Similar to the meaning of the formula (1), the total diffusely reflected light flux is considered in formula (2).

The theoretic assumptions encountered either into an old paper[1] or in more recent papers[2-18] are more or less close to the real case, because the interaction of a light beam with a rough surface is difficult to put into rigorous mathematic equations. For practical evaluation of the surface roughness, some experimental considerations must be taken into account. If the surface is perfectly flat (no roughness at all), the reflected light flux is given only by the well-known reflection law. If the surface is not perfect, its irregularities (roughness) will produce a specular reflection with an angular distribution dependent on roughness and a scattered flux with a spatial distribution dependent also of surface roughness. The sum of the specular light flux and of the scattered light flux is equal to the incident light flux (however, the absorption is considered negligible). The scattered flux is strongly dependent on the surface roughness and this dependence is used in this application for detection of surface defects. For the very well polished surface of the silicon wafer, the defects can be considered as a local roughness. The measurement of the total light flux scattered by a defect -- total integral scatter *TIS* - has an information about the defect size, though this dependence is not linear. For defects detection, the non-linearity of the dependence is not important. The most important feature for this purpose is its sensitivity.

THE OPERATING PRINCIPLE

On the surface of the silicon wafer under test is directed at normal incidence a very sharply focused laser beam, having a waist diameter < 0.1 mm. The only scattered light flux *TIS* is collected and detected. The specular light beam is rejected by the optical setup. The photoelectric signal generated by each defect is proportional with its cross-section scattering area, not with its size. This means that a defect with a small geometric area can produce a photodetected signal greater than a larger size defect. The photodetected defect signal is encoded by amplitude and the defect code value is loaded into a computer *RAM* (random access memory). For the inspection of the whole wafer surface, the laser beam must explore it fully. In this implementation, the beam explores the wafer surface under an Archimede's spiral, from the wafer center to its edge (Figure 1).

Each detected defect must be associated with its position on the wafer surface. For this purpose, the surface is theoretically divided in circular tracks (3) and angular sectors (4). The track width is 75 % of the laser spot diameter on the wafer surface, so any encountered defect is detected twice to avoid false defects detection. The area inside a contour defined by four adjacent circular tracks and four adjacent angular sectors is defined as a surface element (5). The area of a surface element (5) is almost constant over the wafer surface. This is done by defining a different number of angular sectors per surface area, form a wafer center to its boundary, even during the inspection process. This dynamic definition is done via a software programmable counter, which is set for a division ratio from few units to few hundreds, when the laser beam moves from the wafer center to its boundary. Each surface element has an associated code: a number, increasing from 1 (the wafer center), to few ten thousand (to the wafer boundary). The defect position on the wafer surface is reported as a defect located inside a surface element (5). If a surface element has many defects of different sizes, just the greatest defect of this element is reported. The associated amplitude code of each defect is successively stored in *RAM*, from the first surface element until the last one, so there is no need to store also the code of each surface element.

The spiral scanning of the wafer surface is done by a simultaneous rotation of the wafer around its center (*O*) and a translation along its radius. Putting the wafer into a such movement, the fixed-position laser beam scans the entire surface over an Archimede's spiral from center to boundary. When the laser beam reaches the contour of the exclusion zone (2), data defect code acquisition from the photodetector is disabled. The wafer movement still continues until the inspection beam reaches the wafer contour: the flat (7) first and the circular contour (1) last. At this moment, the wafer surface inspection is completed. The scattered signal generated by the intersection of the laser spot (6) with the wafer boundary - the flat and the circular contour - is also detected with the defect photodetector. A discrimination of this signal with a voltage comparator for a value much higher then the defects level, will give a code for the wafer boundary points.

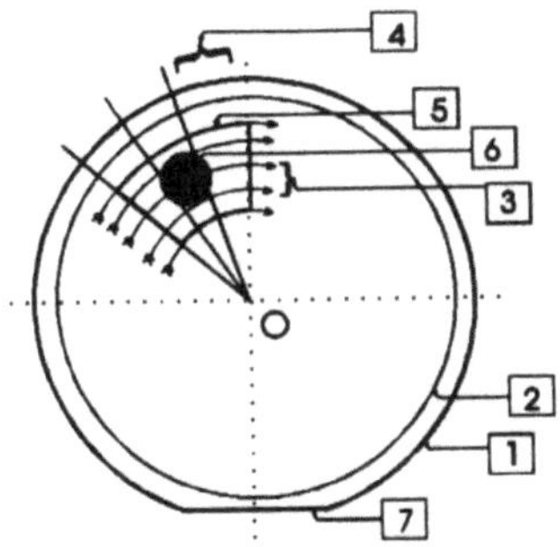

Figure 1. The scanning of the wafer surface by a laser beam; the surface element has a bolded contour and is shown enlarged: 1-wafer circular contour; 2-exclusion zone contour; 3-track; 4-angular sector; 5-surface element; 6- laser spot on wafer surface; 7- wafer flat; *O*- wafer center.

During the wafer surface scanning by the laser beam, in the computer *RAM* is created an image of the surface defects, including wafer boundary. This image is displayed on the monitor screen, even when the data acquisition is performed. Thus, the monitor continuously shows the defects map of the wafer under test. Moreover, defects' statistic is also displayed.

THE EQUIPMENT DESCRIPTION AND OPERATION

The description and operation of the laser based equipment for detection of surface defects of silicon wafers by laser beam scattering will be done in conjunction with Figure 2.

The silicon wafer under test (1) is put on the surface of a vacuum chuck (2) and held on this surface until the completion of the inspection. The vacuum chuck (2) is put simultaneously into a rotation around its center and into a translation along its radius, by a driving mechanism (3). An incremental angular transducer (4) generates angular electric signals, from which end-of-sector (*EOS*) signals are obtained. Every *EOS* signal means the end of the current sector and the beginning of the next sector. The *EOS* are counted by a programmable modulo *N* sector counter (*SC*). The division ratio *N* of the *SC* is software programmable with continuously increasing values, when the inspection laser beam moves from the wafer center to its boundary, for generating surface elements of almost equal area, over the wafer surface. The output of *SC* is the end-of-track (*EOT*) signal, which marks the end of every track (turn) of the wafer and gives a command for one-step linear movement of the wafer along the radius.

The wafer under test (1) continuously turns around its center (*O*) under the action of a *DC* motor and moves linearly along its radius under the action of a stepper motor. Every *EOT* signal will produce a linear movement of the wafer with a 0.075 mm pitch.

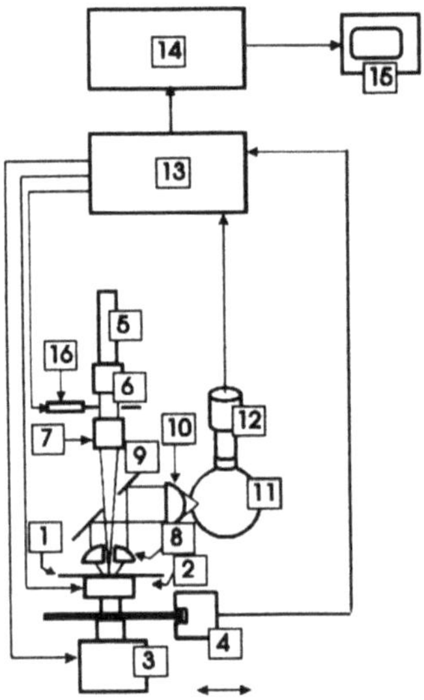

Figure 2. Block-diagram of the equipment for detection of surface defects of silicon wafers by laser beam scattering: 1-wafer under test; 2-vacuum chuck; 3-wafer driving mechanism; 4-angular incremental transducer; 5-He-Ne laser; 6-beam expander; 7-collimating objective; 8-parabolic collecting lens; 9-totally reflective mirror; 10-focusing lens; 11-integrating sphere; 12-photomultiplier; 13-computer interface; 14-IBM-AT compatible computer; 15-computer monitor; 16-shutter.

The beam emerging from a He-Ne laser (5) is collimated and expanded by a beam expander (6) and is focused to the surface of the wafer under test (1), by a collimating objective (7). The objective (7) has a focal length f = 250 mm and is specially corrected for λ = 633 nm. It produces on the wafer surface a spot with < 0.1 mm diameter at $1/e^2$ intensity. The only scattered beam produced by the surface defects (and haze) is collected by the parabolic lens (8) and is sent to an integrating sphere (11). The light beam is directed from the output of the collecting lens (8) to the integrating sphere (11), by the totally reflective mirror (9) and the focusing lens (10). One surface of the collecting lens is parabolic, the other surface is spherical. The lens aperture is 48 mm and its front focal length is 35 mm, so the light collecting angle is 68°. This lens has also a 4 mm circular hole concentric with the lens aperture. Through this central hole passes the inspection beam focused by the objective (7) on the wafer surface. Thus, the influence of the collecting lens on the inspection beam focused by the objective (7) is practically avoided. For the same reason, the mirror (9) has 6 mm circular hole concentric with the inspection beam passing through it.

The detection of the scattered light flux entered into the integrating sphere (11) is made with a photomultiplier (12). This detector type was chosen due to the very high dynamic range of the scattered light flux and to its very weak low level. The integrating sphere (11) produces an uniform specific irradiance on its inner surface, independent of the angle of incidence of the light scattered beam and of its spatial distribution. Thus, amplitude variations of the photodetected signal with non-uniformity of the photocatode response over its area are avoided. The signal generated by the photomultiplier (12) is sent to the computer interface (13), for a binary 3-bit amplitude coding. This coded value will be stored to computer *RAM*, by *DMA* (direct memory access) input port operations.

The interface (13) connects the equipment to an *IBM-AT* compatible computer (14). Via this interface, the computer (14) sends commands to:

• switch-on the vacuum valve (not shown on the Figure 2), to lock the wafer on the chuck;

• start and stop the *DC* motor for turning the wafer around its center;

• move forward and backward the wafer along its radius;

• shut the laser beam for background light level measurements;

Via the same interface, the computer receives signals from:

• the vacuum relay (not shown on the Figure 2), to confirm the wafer locking;

• the opto-coupler detecting the beginning of the linear travel (inspection laser beam in the wafer center);

• the opto-coupler detecting the end of the linear travel (inspection laser beam on the wafer contour);

• the incremental angular transducer (4);

• the photo-multiplier (12).

The checking-up procedure for detection of surface defects of silicon wafers is performed in the following steps:

1. A robotic arm puts on the vacuum chuck the wafer under test (to be inspected).

2. From an operator-selectable menu, the operator types the following parameters:

2.1. wafer diameter;

2.2. width of the exclusion zone;

2.3. wafer code;

2.4. user-defined levels for amplitude encoding of photodetected defect signals: haze, level #*1*, level #*2*, level #*3*, level #*4*;

2.5. user-defined colors, associated with each defect level, selectable from a 16-color palette;

2.6. number of accepted defects on the whole wafer surface, for each defect level.

2.7. the start of measurement command.

3. If the wafer is properly locked on the chuck, the wafer test begins. This is signaled by the monitor, which displays the pattern shown in Figure 3.

The screen area is divided in wafer map area and the command/comment area. The displayed numbers of accepted defects in "Accepted No. of defects" column are those typed in step 2.6.

4. The computer sends a command for opening the vacuum valve to lock the valve on the chuck and then tests the wafer locking, by reading the status of a vacuum relay. If there is positive answer, the next step will be performed.

5. The *DC* motor supply is turned-on. The wafer turns and defects data acquisition begins from now. The amplitude of the photoelectric signal generated by the photomultiplier

is compared with four voltage levels, according to the defects levels programmed in the step 2.4. The result of this comparison is the binary defect code associated with a surface element, as shown in Figure 3.

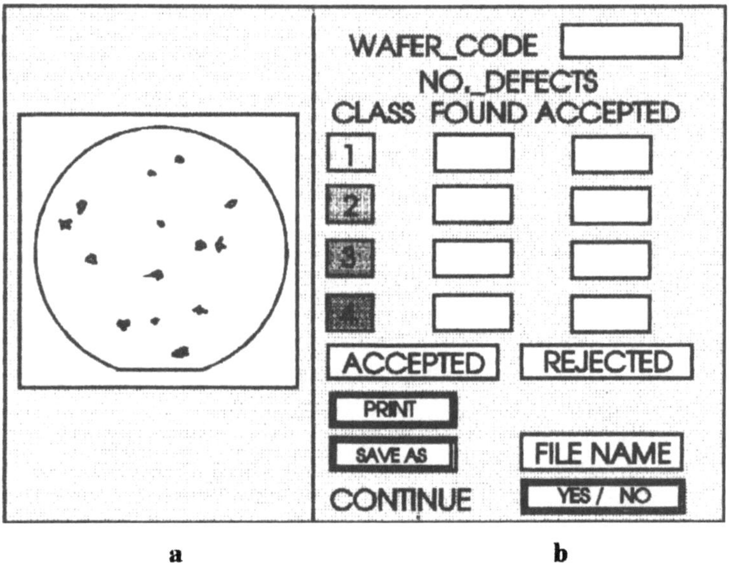

Figure 3. The pattern shown by the monitor during the wafer test: *a*-wafer map area; *b*-command and comment area.

The binary defect code is stored to computer *RAM* by *DMA* operations, after a complete scanning of a surface element. The monitor image is updated with this defect:

- a dot is lit on the wafer map into a position according to the physical position of the defect on the wafer and with the color defined in the step 2.5 and,
- the number in the appropriate case of the column "No. of found defects" is incremented by one.

Every time a wafer turn is completed, a signal for linear displacement along the radius with a 0.075 mm step is generated.

6. The operation as described in step 5 is continuing until the exclusion zone contour is encountered. The detection of this contour is made by software, counting the pulses generated at every complete turn of the wafer. Data acquisition stops in this moment, but the wafer is still moving until the flat and the circular wafer contour passes through the exploring laser beam. The very strong defect signal generated by now is used to detect the wafer edge and to draw the wafer contour on the monitor screen.

7. The wafer inspection is completely performed now. The statistics of the defects found during inspection is compared with the accepted number of defects for each defect level and the Accepted / Rejected decision is displayed.

8. The operator can choose now from three options:

8.1. Check pass: Yes / No;

8.2. Save the defects map and their statistics into a file and specify the file name;

8.3. Print the screen image.

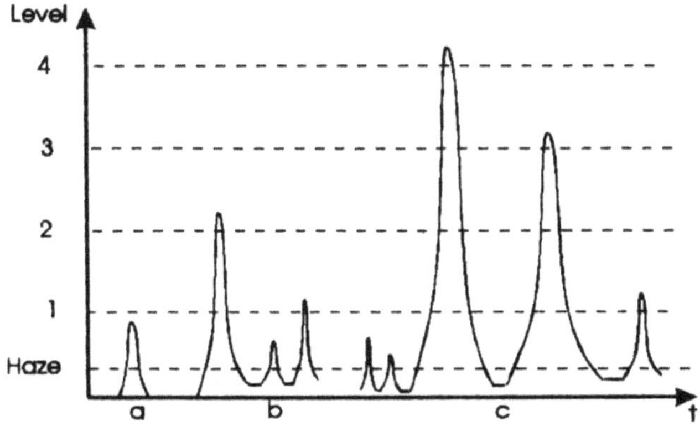

Figure 4. Level encoding of the photoelectric signal generated by the photomultiplier, considered over a surface element: *a*-no defects; *b*-level *#2* defect; *c*-level *#4* defect.

The optical signal scattered by the surface defects has a very wide dynamic range. For an appropriate photodetection of this signal, two measurements are made before the beginning of every checking-up procedure: dark level signal measurement and no-defect signal measurement.

At the dark level measurement, the laser beam is switched-off by the mechanical shutter (16) (Figure 2). The offset of the photomultiplier pre-amplifier is adjusted until its output voltage is slightly above zero. The zero defect measurement is made with no wafer on the chuck and with the chuck in its initial position. There is a piece of high-quality wafer (reference surface) in the chuck center, just below its surface. The optical flux scattered by this reference piece of wafer is considered emerging from a "perfect surface" and from the corresponding photodetected signal is generating the discrimination level for "haze".

No measurements with a wafer having reference defects were performed, due to the lack of an appropriate clean room. Instead, evaluations of the defects location (defect map) and their statistics, were made on the same wafer sets, with this equipment and with *SURFSCAN* 164 equipment, made by *TENCOR Instruments USA*, were done. The evaluations were made into a clean room class 1000 and about 85% agreement of results was obtained.

ACKNOWLEDGMENTS

The authors wish to express their gratitude to *TEHNOFINA S.A.* Company, and especially to Dr. N. Blaga for sponsoring the research and development of this equipment. We are expressing also many thanks to Mr. Eugen Frumosu from the *Research Institute for Fine Mechanics*, who developed all data processing software of this equipment.

REFERENCES

1. H.E. Bennett, J.O. Porteus, Relation between surface roughness and specular reflectance at normal incidence, *J. Opt. Soc. Am.* 51:123 (1961).

2. W.J. Alford, Automatic high speed inspection of semiconductor wafers, *in*: "Proceedings of ICALEO'82: Inspection, Measurement and Control," Laser Institute of America, Toledo (1982).
3. D. Gilsinn, Th. Vorburger, Lin-xiang Cao, Ch. Giauque, F. Scire, E. Clayton Teague, Optical roughness measurements of industrial surfaces, *in*: "Optical Techniques for Industrial Inspection," P.G. Cielo, ed., The International Society for Optical Engineering, Bellingham (1986);
4. T. Matsubara, T. Toyota, A. Fujiyama, Hot slab surface inspection by laser scanning method, *in*: "Optical Techniques for Industrial Inspection," P.G. Cielo, ed., The International Society for Optical Engineering, Bellingham (1986).
5. Y. Shibata, Y. Asano, K. Kurita, M. Shiozumi, On-line glossmeter for stainless steel sheets, *in*: "Optical Techniques for Industrial Inspection," P.G. Cielo, ed., The International Society for Optical Engineering, Bellingham (1986).
6. M. Fairlie, J.G. Akkerman, R.S. Timsit, Optical techniques for the surface evaluation of bright aluminum sheets, *in*: "Optical Techniques for Industrial Inspection," P.G. Cielo, ed., The International Society for Optical Engineering, Bellingham (1986).
7. P. Wägli, R. Huber, E. Morger, Noncontact surface flow detection by fiber optic reflection sensors, *in*: "Optical Techniques for Industrial Inspection," P.G. Cielo, ed., The International Society for Optical Engineering, Bellingham (1986).
8. M. S. Virdee, High accuracy form measurement of specularly reflecting surfaces by laser autocollimation, *in*: "Ultraprecision Machining and Automated Fabrication of Optics," D.L. Decker and R.A. Jones, eds., The International Society for Optical Engineering, Bellingham (1986).
9. H. H. Hurt, P.A. Temple, J.M. Bennett, D.L. Decker, V.A. Hodgkin, Near-angle scatter measurement on diamond-turned surfaces, *in*: "Ultraprecision Machining and Automated Fabrication of Optics," D.L. Decker and R.A. Jones, eds., The International Society for Optical Engineering, Bellingham (1986).
10. J. C. Stover, Overview of current scatterometer measurements and the impact on optical systems, *in*: "Metrology of Optoelectronic Systems," E.M. Granger, ed., The International Society for Optical Engineering, Bellingham (1987).
11. M. Hercher, Two-frequency laser techniques for precision measurement of surface profiles and retardation, *in*: "Metrology of Optoelectronic Systems," E.M. Granger, ed., The International Society for Optical Engineering, Bellingham (1987).
12. B. S. Yilbas, K. Danisman, M. Yilmaz, Z. Yilbas, Development of the computer controlled electro-fiber optic system for surface roughness measurement of metals, *in*: "Metrology of Optoelectronic Systems," E.M. Granger, ed., The International Society for Optical Engineering, Bellingham (1987);
13. M Saillard, D. Maystre, Scattering from metallic and dielectric rough surfaces, *J. Opt. Soc. Am. A* 7: 982 (1990).
14. E. Max, T. V. Vorburger, Direct and inverse problems for light scattered by rough surfaces, *Appl. Optics* 29:3613 (1990).
15. P. Tran, A. A. Maradudin, V. Celli, Backscattering enhancement from a dielectric surface, *J. Opt. Soc. Am. B* 8:1526 (1991).
16. J. A. Sanchez-Gil, N. Nieto-Vesperinas, Light scattering from random rough dielectric surfaces, *J. Opt. Soc. Am. A* 8:1270 (1991).
17. L Cao, T. V. Vorburger, A. G. Liebermann, T. R. Lettieri, Light scattering measurements of the rms slopes of rough surfaces, *Appl. Opt.* 30:3221 (1991).
18. H. Wang, Light-scattering method for characterizing optical surfaces, *Optical Engineering* 32:2904 (1993).

POLYMER OPTICAL FIBER SENSORS FOR STRUCTURAL SENSING APPLICATIONS

Peter L. Fuhr , Dryver R. Huston , Timothy P. Ambrose

Intelligent Structural Systems Research Institute
College of Engineering
University of Vermont
Burlington, VT 05405 USA

INTRODUCTION

The research area of "Smart Materials and Structures" represents one of the blossoming stars in multi/cross-disciplinary applied and fundamental research. Activities abound both in the fundamental area of investigation into new "smart or intelligent" materials as well as the application of existing materials into structural elements and systems.

The recent advent of advanced materials with internal sensory and reaction capabilities, i.e. 'intelligence,' has opened the door for many useful applications in civil structures, e.g. bridges, buildings, dams, in contrast with aerospace structures. Structures built with intelligent materials will be able to sense external loads, assess the internal stresses and damage caused by the loads, and, if necessary, respond appropriately. Although relatively few structures are presently built with such materials, the potential market for the application of these materials can be quite large[1]. The most probable candidates for the initial application of these materials will be high-performance structures such as skyscrapers, including intelligent buildings, bridges, and those facilities where failure presents large safety concerns, such as nuclear waste containment vessels, bridge decks, dams, and structures under construction.

Large scale structures that have actually had fiber optic sensors embedded into them are still relatively few. Holst and Lessing[2] have installed fiber optic displacement gages in dams to measure shifting between segments. Fuhr et al.[2] report the installation of fiber optic sensors into the Stafford Building at the University of Vermont. Similar installations have been completed at the Winooski One dam in Winooski, VT[2], in the newly constructed physics building on the campus of the Dublin City University in Ireland[2], and in a railway overpass bridge in Middlebury, VT[2]. Researchers at the University of Toronto, in conjunction with City of Calgary (Alberta, Canada) officials, have embedded fiber optic strain gauges into the prestressed concrete support girders on a highway overpass[3].

Polymer Based Fiber Optic Sensors

Almost all of the above mentioned applications of fiber optic sensors in civil structures have used silica as the base material for the transmission of light. However, silica may not be the ideal material for use in fiber optic sensing of civil structures. Silica, while having

Applications of Photonic Technology
Edited by G.A. Lampropoulos *et al.*, Plenum Press, New York, 1995

excellent optical properties, has rather fragile mechanical properties. Silica glass is brittle, breaks rather easily, and cannot tolerate alkalai environments, such as occur in concrete. Polymer-based fibers offer an attractive alternative to silica-based fibers for civil structural sensing applications. Polymer fiber is much more rugged than silica fiber. It can undergo large deformations by yielding plastically without cracking. This plastic deformation capability opens up some possibilities for use in crack detection schemes using Optical Time Domain Reflectometry where the key reflections occur from the plastically deformed cross sections.

It has been predicted that the market for plastic optical fiber (POF), used mainly in display, signage, industrial control, medical and automotive lighting applications, will increase from a current level of $50 million worldwide to over $250 million by the year 2001[4]. In spite of the inherent attenuation and bandwidth problems associated with POF, this represents an annual growth rate of approximately 20%. About 90% of the estimated 300,000 km of POF now placed worldwide was produced by Mitsubishi Rayon Co., Toray Industries and Asahi Chemical Co. This dominance is partly because Japanese producers have been much more aggressive than others in developing new high-end uses of POF, most notably in automobiles, industrial process control systems, data transmission and consumer electronics.

POF Data Specifics

While we have investigated and compared various POFs from various manufacturers, Toray Polymer Optical Fiber Cord, Type PG was used in this study. This fiber exhibits the characteristic spectral transmission associated with PMMA-based fibers[5]. The PG Series POF had a core refractive index, $n_1 = 1.492$; cladding refractive index, $n_2 = 1.405$; a numerical aperture, N.A. = 0.50; and associated acceptance angle of 60°. POF attenuation characteristics were measured using a simple cut back method at $\lambda = 660$ nm and 25° C.

It is the physical properties of the POF that make their use quite valuable for embedded civil structural applications. Tensile stress and elongation properties of a polyethylene jacketed PG series POF yield allowable maximum loads of: Tentative Tensile Load: less than 95% of yield strength or 7% elongation; Continuous Tensile Load: less than 10% of yield strength or 0.9% elongation. By contrast, silica based similar diameter optical fiber has an allowable maximum elongation of 0.02%. Intensity based fiber optic sensors have inherent limitations unless some form of self-referencing scheme is used due to the ambiguities caused by component transmission fluctuation or sensing parameter variation. The variation in transmission when a static compressive load deforms the fiber yields allowable maximum compressive loads as well as sensing information for an intensity-based fiber deformation sensor. Light transmission versus compressive load using a contact tip of 70 kg / 5 cm resulted in maximum allowable compressive loads of: Tentative Compressive Load: 50 kg / 5 cm; Continuous Compressive Load: 15 kg / 5 cm. The transmission rate versus static bending radius yields the classic bending fiber optic sensor. Measurement of the POF fiber transmission versus bending radius yields bend radius sensing information as well as the maximum allowable bending radius before fiber fracture. Light transmission versus bending radius (mm) yielded maximum allowable bending radius of: Tentative Bending (no breakage, no damage): 1 mm; Continuous Bending (transmission rate = 80%): 9 mm.

In instances where cyclic POF bending is to occur (e.g., a vehicle counting fiber optic sensor), the longer term transmission level versus number of bending cycles should be measured. Cyclic bending with parameters: bending radius = 50 mm, bending angle = 90° was performed on the PG series fiber resulting in the transmission level versus number of bend cycles. The overall ruggeddness of the POF was loosely measured using cyclic drop impacts. This test roughly simulates the loose concrete dropping onto exposed fiber conditions present at many civil structures' construction sites. In these measurements a drop tip (1 kg / 2 cm) was dropped from a height of 5 cm onto the POF while light transmission has been studied.

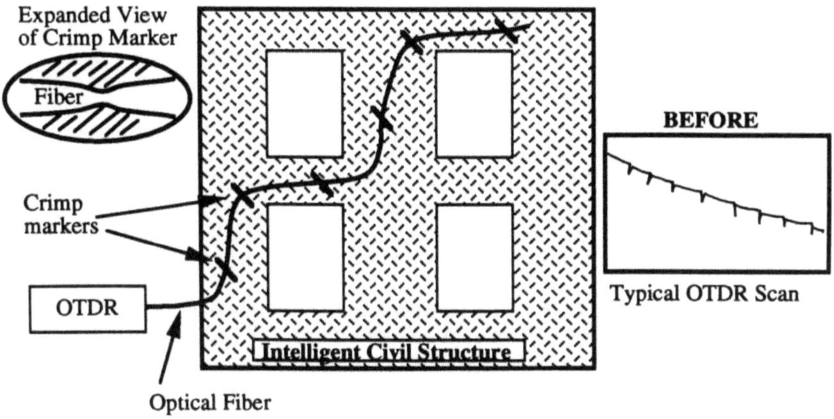

Figure 1. OTDR using POF which has been embedded into a civil structure.

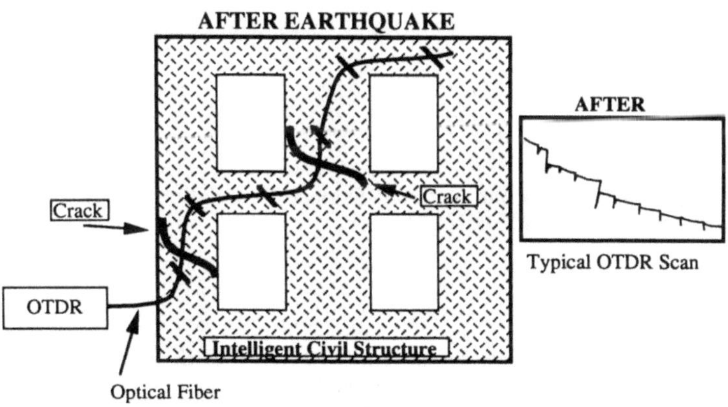

Figure 2. OTDR scanning of the POF instrumented structure may reveal internal damage status.

OTDR Structural Damage Detection Using POF

The resiliency of POF indicates that in certain structural health monitoring sensing applications, the current method using glass optical fibers may be significantly improved upon through the use of POF. The ability of POF to sustain significantly larger bending, drop impacts and elongation than glass optical fiber allows its embedding into structures with a subsequent utilization as an improved structural health sensor. The use of OTDR techniques for damage detection in structures has been previously reported by Huston et al. Such OTDR sensing of structural health may be performed using crimped "markers" located at distinct lengths along the POF. The crimps locally change the boundary conditions along the POF cause slight decreases in the received optical intensity thereby serving as key indicators or location markers along the fiber. When embedded into a structure, as shown in Figure 1, the associated OTDR scan of the POF shows the markers. After an "event", such as an earthquake, has mechanically stressed the structure, if a crack occurs, as shown in Figure 2, it may alter the OTDR scan signature. Such OTDR integrity verification of embedded optical fibers is routinely performed at the structrues which the authors have instrumented.

The benefits of using POF lie in the fiber's inherent ruggedness and ability to survive large cracks (where glass fiber will catastrophically break). The detriment of POF usage in OTDR scanning is the relatively high signal attenuation at those near-IR wavelengths where most commercially available OTDR units operate. Significant system improvement wil arise when OTDR's use visible light sources.

Summary

The inherent robustness of plastic optical fiber makes it a potential candidate for certain sensing applications associated with civil structures. While the attenutation and spectral response of POF limits its use in long lead length fiber optic sensors, the sensor lengths found in the civil structures in which the authors have been involved, typically < 90 m, due not preclude POF usage. The mechanical and thermal limitations of certain types of POF have been studied with values associated with PG Series POF manufactured by Toray Industries reported in this paper.

Integration of POF with OTDR interrogation of embedded fiber will result in a system capable of significant improvement, in terms of robustness, of providing health monitoring of civil structures.

Acknowledgement

The authors wish to thank the National Science Foundation for their support of this project under Grant Number BCS-9024390.

References

1. D.R. Huston, "Smart Civil Structures - An Overview", SPIE Paper No. 1588-21, Proc. SPIE Fibers '91 Symposium, Boston, MA, Sept. 1991.
2. D.R. Huston, P.L. Fuhr, T.P. Ambrose, and D.A. Barker, "Intelligent Civil Structures - Activities in Vermont", to be published in the Journal of Smart Materials and Structures, Dec. 1994.
3. R. M. Measures et al, "Bragg Grating Laser Sensing System Integrated with a Two-Span Concrete Road Bridge for Monitoring Graphite Prestressing Tendons", SPIE Paper No. 2191-51, Proc. of SPIE Smart Structures and Materials '94, Orlando, FL, Feb. 1994.
4. KMI Corp., "Plastic Optical Fiber - Market Study", as reported in Photonics Spectra, April 1994, p. 68.
5. Toray Corp., Technical Bulletin, 1994.

FLIGHT TESTS OF FIBER-OPTIC INTERFEROMETRIC STRAIN GAUGES FOR LOAD MONITORING OF AIRCRAFT STRUCTURES

Norbert Fürstenau, Douglas. D. Janzen and Walter Schmidt

Institute for Flight Guidance
German Aerospace Research Establishment (DLR)
Lilienthalplatz 7, D-38108 Braunschweig, Germany

INTRODUCTION

A large number of fiber optic interferometric strain gauge concepts has been investigated during recent years (e.g. [1,2,3]), however only little experience exists up to now with respect to aerospace applications under real world conditions[4,5]. In this paper we report in-flight dynamic load monitoring of plates made from carbon fiber composites, which are screwed to the main wing spar of a Cessna C207A. Several flight tests have been performed with four different sensors since June 92. The fringe counting readout of passive quadrature demodulated double-polarization Michelson interferometers with balanced interferometer arms (OSG)[3] is compared to the strain indicated by conventional resistive strain gauges (ESG). Of particular interest is the response of the interferometric sensor to the quasistatic maneuver-dependent load variations given strong simultaneous structural vibrations due to engine and aerodynamic effects.

EXPERIMENTAL SETUP AND PRINCIPLE OF OPERATION

Fig. 1 shows a schematic of the latest version of the interferometric sensor with opto-electronic passive quadrature demodulation unit. The polarized light from a HeNe laser is

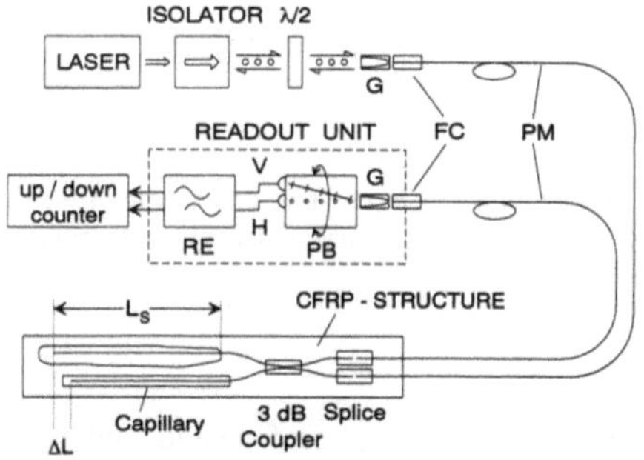

Figure 1. Schematic of interferometric strain gauge (latest version) with CFRP plate, readout electronics and data acquisition. G = GRIN-lens fiber coupler, FC = connectors, PM = polarization maintaining fibers, path imbalance $\Delta L < 3$ mm.

Applications of Photonic Technology
Edited by G.A. Lampropoulos *et al.*, Plenum Press, New York, 1995

guided to the interferometer via 3 m of cabled polarization maintaining fiber, after passing a 30 dB-optical isolator. The downlead is spliced to the input arm of a 3 dB directional coupler. Two equal length coupler arms with silvered end faces serve as reference (R) and sensing (S) arms of the interferometer. They are adhered to the surface of a thin carbon-fiber composite plate (CFRP) of rectangular cross section (thickness 2h =1mm, width =10 mm, length of strained section l=94.5 mm) which in turn is screwed to the main wing spar of the test aircraft as sketched in fig.2. A conventional resistive strain gauge beside the OSG serves as a reference. The length of the fixed section of the bare S arm is L_S = 50 to 70 mm. A section $L_{SA} \leq L_S$ of the S arm is covered by the adhesive (Fig.2). The bare R arm is isolated from the structural strain by feeding it through a capillary fiber of 150 µm inner diameter, which in turn is glued to the surface. Alternatively a bend of the R-arm may serve for this purpose[4] as indicated in fig.2. The interference signal is guided to the demodulation unit via a PM-output fiber. The collimated output beam is split into two orthogonally polarized (horizontal H, vertical V) and phase shifted components using a beam displacing prism (PB), with the polarizer axes aligned parallel to the birefringent axes of the PM-fiber. The H and V interference signals are described by[3]

$$P_{H,V} = \frac{P_0 T_{H,V}}{4(1+\mu_{H,V})} \left(1 + \mu_{H,V} \cos(\Delta\Phi + \Phi_{H,V}) \right) \tag{1}$$

P_0 = input power, T = interferometer transmissivity, µ=fringe contrast, and $\Phi_{H,V}$= phase offset due to path imbalance and fiber birefringence. Initial adjustment of the offset phase difference $\Delta\Phi_{HV} = \Phi_H - \Phi_V \approx \pi/2$, which serves for eliminating the fringe counting ambiguity, is performed by rotating the input polarization via orientation of the PM fiber in the splice (fig. 1), or by means of photoelastic polarization adjusters[3,4] inside the housing of fig.2. The gauge factor $G = \Delta\Phi / \Delta L_s = \Delta\Phi/\varepsilon/L_s$ (ε = relative strain $\Delta L_s/L_S$, L_S = length of sensitive fiber section) is obtained as[4] G = 1.318 deg/µε/mm = 14.6 ± 0.1 counts/µm. Counting half fringes of both quadrature signals (π/2 - phase shift increments) this corresponds to a strain resolution of G^{-1} = 0.0683 µm/count (±0.0001). The H and V signals are fed into an up-down counter after passing through the readout unit (RE) which serves also for symmetrizing and automatic intensity-offset control.

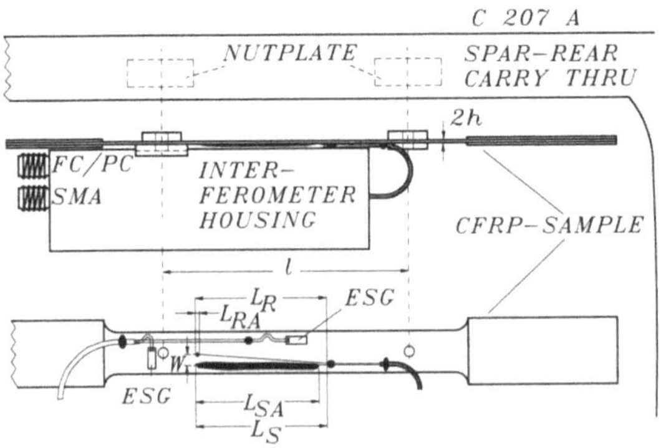

Figure 2. CFRP sample with surface adhered fiber optic and electrical strain gauges. Fixing positions at main spar are indicated. Early version of interferometric sensor[4] with polarizing beam splitter in the housing and two multimode output fibers.

Flight profiles are characterized by aircraft speed and vertical acceleration. Data is acquired using a single board unit containing a Z80 processor controlling a 16 channel A/D-converter with 8 µs conversion time and 10 MHz up-down counters. The data acquisition board is controlled by a 286 Laptop which is also used for storing the data for off line evaluation. The OSG readout electronics with He-Ne laser, the strain gauge bridge amplifier, and the data acquisition unit are mounted in a flight test rack which is fixed to the aircraft seat rails via shock mounts.

EXPERIMENTAL RESULTS AND DISCUSSION

Figure 3 shows the results of a ground test in the form OSG versus ESG data. R and S arm of the sensor are surface adhered as shown in fig.2, with pre-stress applied to the S-arm (L_S = 50 mm). The wing tip is manually lifted and allowed to return slowly to its rest position, repeatedly. The sensitivity of the OSG is seen from the last squares fit to be 0.683±0.004 counts/με (=13.7 counts/μm), which is about 6% lower than the predicted value, probably due to incomplete isolation of the reference arm from the strain.

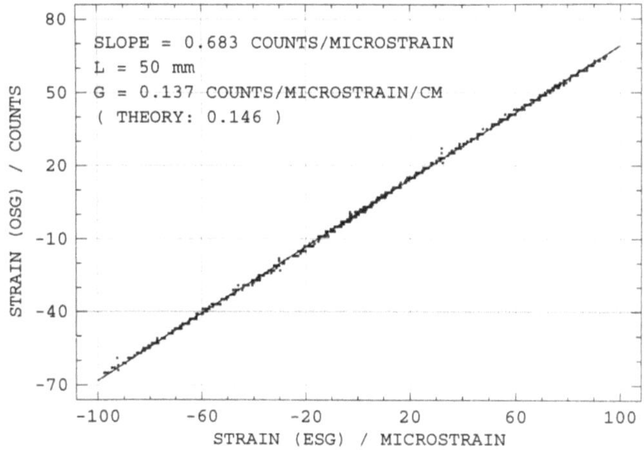

Figure 3. Ground test with manual wing loading. OSG vs ESG for data. Solid line shows least-squares fit.

Flight tests are performed with four slightly different sensor configurations (different adhesive, different method of R arm isolation) and under varying flight conditions, such as alternating speed and acceleration (parabolic flight paths).

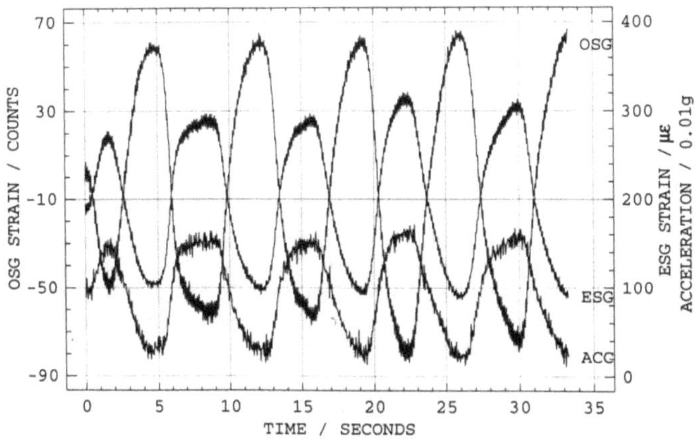

Figure 4. OSG (left ordinate), ESG, and accelerometer data (both right ordinate) from alternating acceleration flight. Sampling rate 30 Hz.

As an example of the flight test data, fig. 4 shows the responses of ESG, OSG and accelerometer to a series of 5 acceleration cycles between roughly 0g and +1.6g (normal to the wing) over a period of 35 seconds. The slope of the least squares fit to the OSG vs. ESG data of fig. 4 yields G = 14.0±0.1 counts/μm (=0.64 counts/με with L_S=46 mm) which differs by 4 % from the theoretically predicted value. During the experiments significant variations of offset phase $\Delta\Phi_{HV}$ and fringe contrast are observed, in particular when using the hard two component (HBM X60) adhesive. Polarization instabilities of the bulk optics light source system and birefringence variations at the adhesive covered fiber sections[4] account for these effects. If the $\Delta\Phi_{HV}$ variation approaches π/2, the counter looses track and an arbitrary offset is introduced. The peak-to-peak amplitude of the vibrational noise on the time series plots is between 5 and 10 με.

Figure 5 shows the power spectral densities of the data from a 2 second 500 Hz acquisition during normal steady flight. The majority of the prominent spectral peaks corresponds to subharmonics and harmonics of the engine rotation rate. The spectra show good agreement of ESG and OSG response up to at least 250 Hz.

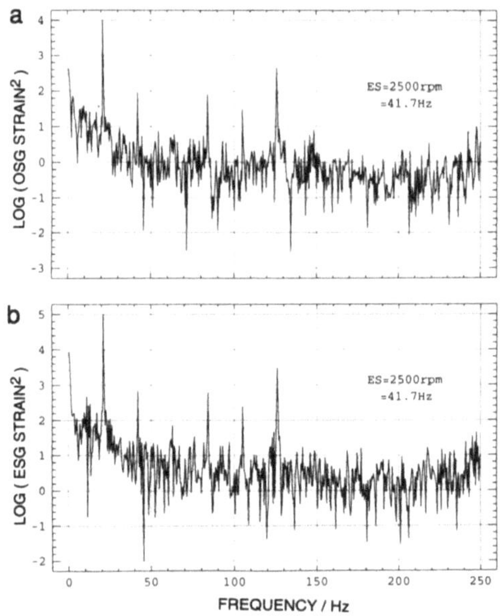

Figure 5. Power spectral densities of 2 sec 500 Hz acquisition during steady normal flight (ES = engine speed). a) OSG output, vertical axis: log10(counts^2); b) ESG output: log10(ADC-digits^2), 2 digits = 1 με.

CONCLUSION

In flight strain measurements on an aircraft wing structure, have been performed using passive quadrature demodulated fiber optic interferometric strain gauges with fringe counting readout. The data taken during single runs exhibit reasonable agreement between theoretical and experimentally determined sensitivity. The stability of the present optical setup proves sufficient for continuous readout over time periods of the order of minutes. It is expected to be improved by using compact temperature stabilized laser diode light sources and reducing the adhesive induced birefringence effects.

ACKNOWLEDGEMENT

This work was partially funded by the German ministery of research under the German-Canadian science and technology agreement.

REFERENCES

1. T.Valis, D. Hogg, R.M. Measures, Fiber-optic Fabry-Perot strain rosettes, *Smart Mater. Struc.* 1:227 (1992)
2. S.R.Waite , R.P.Tatam and A.Jackson, Use of optical fibre for damage and strain detection in composite materials, *Composites* 19:435 (1988)
3. N.Fürstenau, W. Schmidt, H.-C.Goetting, Simultaneous interferometric and polarimetric strain measurements on composites using a fiber-optic strain gauge, *Appl. Opt.* 31:2987 (1992)
4. N. Fürstenau, D.D. Janzen, W. Schmidt, In-flight strain measurements on structurally integrated composite plates using fiber-optic interferometric strain gauges, *Smart Mater. Struc.* 2:2987 (1993)
5. K.A.Murphy, M.F.Gunther, A.M.Vengsarkar, R.O.Claus, Fabry-Perot fiber-optic sensors in full scale fatigue testing on an F-15 aircraft, *Appl. Opt.* 31:431 (1992)

SELECTED APPLICATIONS OF FIBER-OPTIC SENSORS BASED ON HIGHLY BIREFRINGENT FIBERS IN ENGINEERING MECHANICS

Wojtek J. Bock,[1] Waclaw Urbanczyk,[2] Marc R.H. Voet[3]

[1]Laboratoire d'optoélectronique industrielle, Université du Québec à Hull
P.O.Box 1250, St. B, Hull, Qué., J8X 3X7 Canada
[2]Institute of Physics, Wroclaw University of Technology
50-370 Wroclaw, Poland
[3]Glötzl GmbH, Forlenveg 11, 76287 Rheinstetten, Germany

INTRODUCTION

The need for fiber-optic sensor (FOS) technology in the broad area of engineering mechanics is now well documented. Civil engineering structures such as highway and railway bridges, dams, tunnels, large buildings and underground mines must resist environmental and in-service loads due to winds, earthquakes, traffic, thermal effects, construction or environmental damage. For safety and maintenance purposes their response must be carefully monitored. This information is needed to assess their overall "health" and to increase engineering knowledge for future projects. Pressure and strain gauges, accelerometers and tiltmeters are among the variety of electrical and non-electrical transducers which can be directly embedded in concrete, earth, rock, borehole or tunnel lining to sense and monitor structural loads and responses. None are well suited to more demanding civil engineering environments. Non-electrical hydraulic devices are costly in terms of labour-intensive inspection, maintenance and repairs; they are difficult to multiplex and impossible to use for dynamic measurements. Electrical measuring devices are out of the question in open-air structures (risk of lightning), and in mining environments (risk of explosion). For these applications fiber-optic sensors offer significant metrological improvement: electrical passivity, high bandwidth, safety in corrosive or explosive environments, immunity to EMI, high sensitivity, miniature dimensions, possibility of remote operation and direct compatibility with increasingly present fiber-optic data transmission networks.

As great deal of research has already been done on every imaginable sensing mechanism[1], the implementation issues such as packaging, interconnecting and above all reliability and multiplexing are now the focus of attention. Research on FOS technology for specific applications in civil engineering (CE) is still, however, in its infancy mainly because three important aspects have to be addressed simultaneously. These aspects are: developing an accurate, reliable and cost-effective sensor system; safe embedding and interfacing of sensors in structural elements; and reliable and cost-effective optical multiplexing and signal processing for embedded sensor arrays.

Various effects of mechanical perturbation on the transmission of light in optical fibers[2] have been researched and some may have application in CE, but the focus has been on qualitative rather than quantitative measurements[3]. In highly birefringent (HB) optical fibers, the phase effects induced by twist, bend, elongation, stress and pressure have practical relevance for the construction of polarization controllers, switches, modulators and sensors. We will concentrate our attention specifically on pressure measurements, since quasistatic and dynamic pressure is one of the most important parameters in geotechnical and structural measurements: a universal pressure sensor can conveniently be configured to detect uni- and multidirectional stresses, loads or forces. With the exception of our work[4,5,6] the research on radial stress and hydrostatic pressure effects on optical fibers has been mainly theoretical[7]. Other sensing mechanisms have also been studied but very few papers mention the accuracy of measurement achievable with a particular sensor (e.g., -/+5%[8] or -/+10%[9]), because of significant problems still associated with leading fibers, temperature drift, apparent readings and low repeatibility of the sensor signal. Fabry Perot and Bragg grating sensors have been evaluated recently for strain sensing. If embedded in a real environment, however, both fail due to false readings induced by ambient temperature gradients (a 40% error have been reported[10]), uncontrollable hysteresis and high cost.

SAFE EMBEDDING - FIBER-OPTIC STRESS CELL CONCEPT

Direct embedding of FOS in concrete using a variety of metallic, epoxy or plastic coatings has been heavily researched in the last 6-7 years. So far this approach has not been successful[11]. The concept of "smart" structures and skins with fiber-optic sensors embedded in composite structural materials was initially tempting in aircraft technology; despite immense effort, however, just one in-flight test has been reported[12] and practical applications have never materialized. The extension of this concept to structural monitoring would be even more difficult: a) embedding fiber into composite material is still an unsolved issue: a 65% reduction of the composite compression strength due to the presence of embedded fibers have been reported[10]; b) still unsolved are issues of distinguishing a desired measurand from others that equally influence the sensor output and interconnecting the embedded fibers with external networks; c) composite materials cost much more than concrete; d) several sensing systems proposed to date are unstable, costly and unlikely to be applicable.

In this paper we propose a safe and reliable method for embedding and interfacing the FOS (of pressure, load or stress) in concrete, rock or soil materials under harsh environmental conditions. This simple and practical method does not impair the mechanical properties of evaluated structures and effectively combines modern fiber-optics with elements of traditional hydraulic measurement technology used in CE for over 40 years. We believe it will be easily accepted by the CE community, where reliability and proven record of operation are the most important requirements for any measuring instrumentation. Unlike the smart skin concept, our approach can be immediately applied to any kind of structure and for all kinds of geotechnical instrumentation.

The standard technique of stress analysis in large-scale structures with pressure cells involves various configurations of hydraulic measuring systems based on the compensation method, in which stress in the material surrounding the pressure cell is compensated by automatically adjusted pneumatic or hydraulic pressure within the device. This simple principle is illustrated in Fig.1 using the example of a Glötzl standard pressure cell.

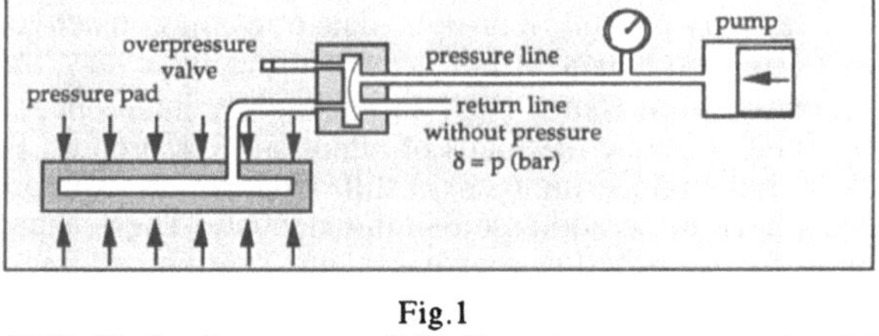

Fig.1

Standard Glötzl hydraulic pressure cell for direct stress measurement up to 600 bar

A thin flat cell filled with a hydrostatic pressure medium (oil or mercury) and equipped with a sensitive diaphragm valve is cast into the concrete or placed in some other solid medium depending on the particular application. If a stress is acting in the direction perpendicular to the plane of the cell, the pressure of the finite volume of hydraulic fluid enclosed inside the cell will increase. This pressure can be calibrated using an external pressure pump, also shown in Fig.1. When both pressures are equal, the diaphragm valve allows hydraulic fluid to flow into the return pipe; the rate of this flow can then be associated with the cell pressure and thus with the stress exerted on the cell. This type of system requires installation of great lengths of pressure tubing and creates serious maintenance problems when the number of discrete sensors increases. To avoid such problems and to capitalize on the potential advantages of fiber-optic technology, we replaced the hydraulic measurement system by a direct read-out of the interior pressure in the cell using a polarimetric fiber-optic sensor presented later in this paper, with the pressure cell already optimized for stress measurements being basically unchanged (such optimization involves adapting the plate surface/thickness factor to minimize both the measurement error and the influence of the different modulus of elasticity between the cell and the surrounding material).

The assembly of the fiber-optic pressure cell (FOPC) is composed of two metal plates which are joined together using a welded sealing to form a pressure pad as shown in Fig.2. The dimensions of the flat jack cell are 70mm x 140mm x 6mm, with slight longitudinal deformation in the form of a V-groove to house a polarimetric pressure sensor in a temperature-compensated configuration and with two specially designed pressure leadthroughs to input and output the leading fibers. The FOPC shown in the photograph is connected to the cover of a pressure chamber, where it was placed for calibration purposes.

Fig.2
Fiber-optic pressure cell (FOPC) embeddable in concrete, rock or soil materials

PRINCIPLE OF THE POLARIMETRIC PRESSURE SENSOR

We have already patented[13] and published elsewhere[6] the principle of the polarimetric pressure sensor. Briefly, it employs a highly birefringent optical fiber in a temperature-compensated configuration. Such fibers are characterized by two perpendicular polarization eigenmodes having different propagation constants β_x and β_y. If a linearly polarized light is launched into the fiber at 45° to the two principal polarization axes, both eigenmodes are equally excited and the resulting polarization state will rotate from linear through elliptical to circular and back again with a repetition distance equal to beat-length parameter of the fiber.

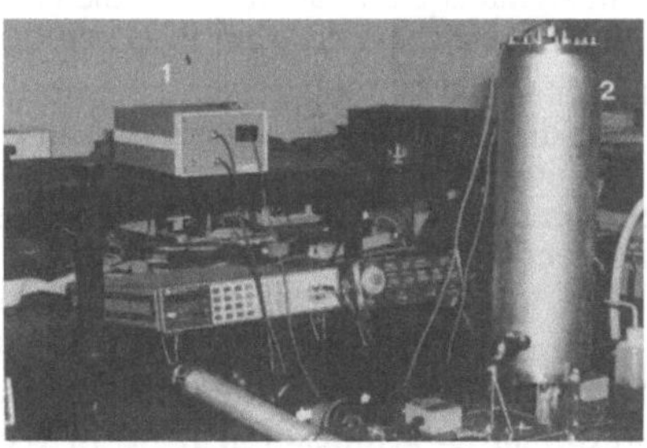

Fig. 3
FOPC undergoing calibration tests up to 70 bar in a pressure chamber

We have shown experimentally[5] and it has since been confirmed by recent theoretical considerations[7] that, contrary to long-standing belief, the birefringence of HB PM fibers depends significantly on external hydrostatic pressure. This birefringence sensitivity to pressure occurs because of the difference in compressibility coefficients between the stress-applying regions and the surrounding cladding, or due to geometrical effects in a HB fiber. For instance, the resulting pressure sensitivity of birefringence in the range up to 100 MPa was found to be about 9.5 rad/MPam for York HB 800 fiber and about 1.4 rad/MPam for the elliptical core Corning fiber with 800 nm cut-off wavelength.

CALIBRATION OF THE FOPC

The instrumentation system of the FOPC, shown in Fig.3 in a configuration designed to calibrate the sensor against a primary pressure standard, involved a semiconductor laser diode emitting at 837 nm and directly pigtailed to a 3M polarizing fiber serving to guide a linearly polarized light parallel to one of the two eigenaxes of the fiber with a polarization extinction ratio of as high as 1000/1. To allow the incoming light to equally excite both polarization modes in the sensing element, the leading fiber has to be fusion-spliced at 45° relative to its birefringence axes. High accuracy of azimuthal alignment is required in this case in order to avoid instabilities and diminution of device sensitivity. In this case the system is "all-fiber" and does not require any additional polarizers or phase retarders. The analog conversion of the pressure-induced phase shift into intensity-pressure characteristics was provided by an electronic unit (1) driven by specially designed data-acquisition software and connectorized to the leading fibers. For calibration purposes, the FOPC was placed inside a large pressure chamber (2) filled with a liquid medium and fed directly from a Harwood DWT-35 deadweight tester with an accuracy of at least 0.1% traceable to NIST.

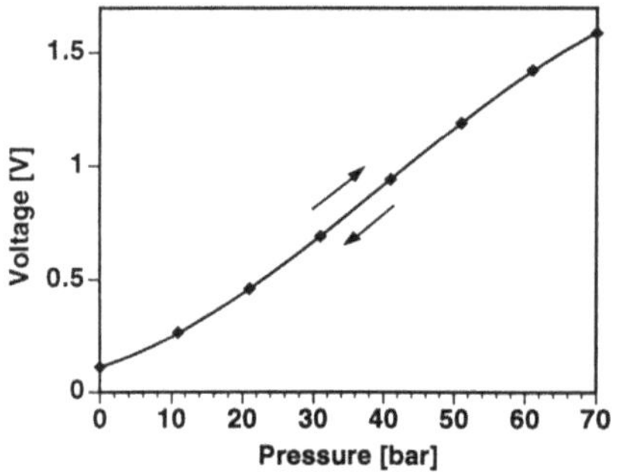

P (bar)		0	11	21	31	41	51	61	70
T = 10°C	→	0.111	0.271	0.470	0.702	0.950	1.195	1.421	1.578
	←	0.109	0.267	0.464	0.696	0.944	1.190	1.418	1.577
T = 23°C	→	0.112	0.264	0.459	0.690	0.941	1.190	1.422	1.587
	←	0.110	0.261	0.454	0.684	0.935	1.188	1.421	1.586

Fig. 4

Calibration data for increasing and decreasing pressure at 10°C and 23 °C for the polarimetric FOPC

In Fig.4 we present some calibration data for the temperature-compensated fiber-optic pressure cell in polarimetric configuration for pressures up to 70bar and for two temperatures 10°C and 23 °C. To show the performance of this device more accurately we also include a table containing all the measurement results. As the full pressure range of 70 bar corresponds to the increase of measurement signal equal to about 1.5V, the targeted accuracy of 1% of FS will mean a total error of 15mV. It can be seen from the table that the maximum difference occurs for the readings of sensor signal at 70bar, taken at temperatures of 10°C and 23 °C, and does not exceed 10mV. The differences of about 5 mV observed at the same pressure and temperature values while increasing and decreasing the pressure are due to the local fluctuations of chamber temperature and disappear at the noise level after about 10-15 minutes. Neither hysteresis nor cumulative drift of the zero signal after multiple pressure cycles was observed. Results of the stability tests up to 12 hours at ambient temperatures (21°C +/-3°C without stabilization) are shown in Fig.5. It is worth noting, that the stability test included repeatedly bending and squeezing the leading fibers and heating them to about 60°C. The resulting total noise can be estimated at 0.2% of FS. Long term stability test results for this system will soon be available. As for now, its accuracy is better than 1% and sensitivity better than 0.1% of FS.

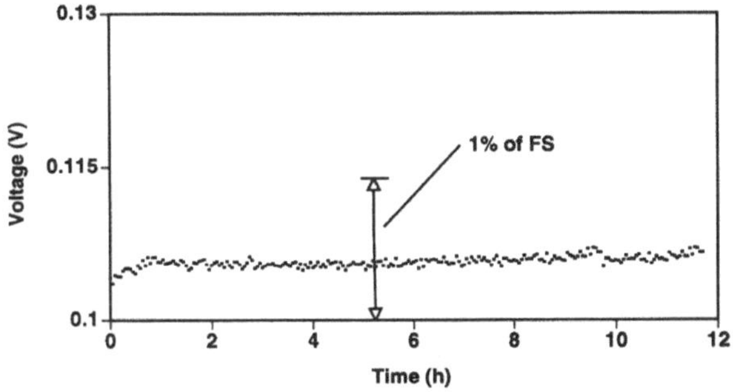

Fig. 5
Stability test of the FOPC

WHITE-LIGHT INTERFEROMETRIC PRESSURE SENSOR

The second configuration of a pressure sensor based on HB fibers exploits the principle of white-light interferometry. It uses a broad band light source and a CCD detection unit to recover and process the white-light interference output pattern from the sensor. The transducer consists of HB leading and sensing fibers once again arranged in temperature-compensating configuration, with electronic scanning and with an internal step-delay line making the receiving interferometer very stable, reliable and inexpensive[14]. Due to the dependence of propagation constants and polarization mode dispersion in HB fibers on pressure, the resulting WLIP registered by the CCD will be transversally shifted with pressure. The instrumentation of this system is shown in Fig.6.

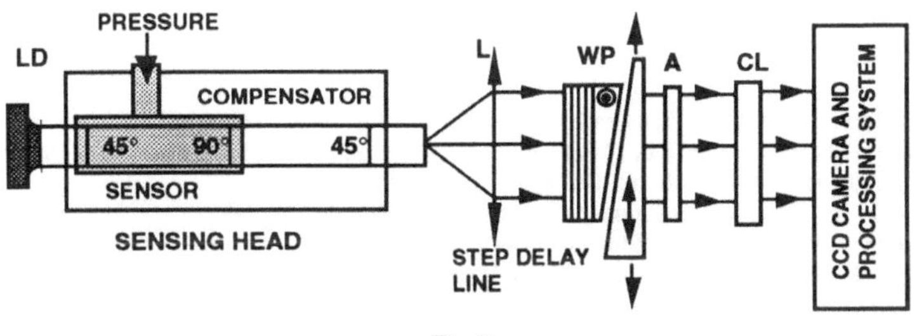

Fig.6
White light interferometric pressure sensor

Two stages of signal processing and the calibration data for this sensor are shown in Fig.7a,b, respectively. The measurement data indicate a very good repeatibility of the sensor

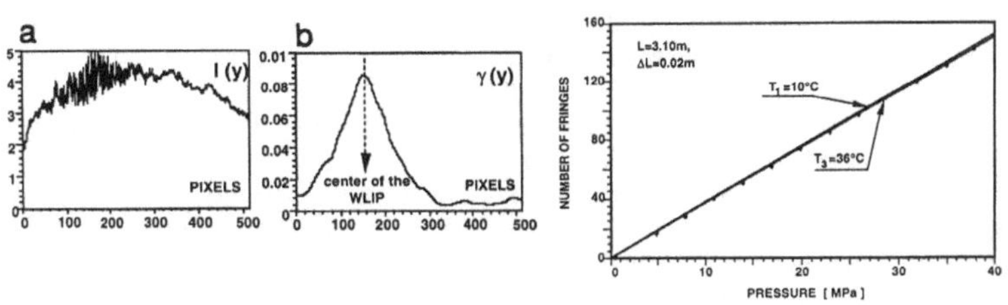

Fig. 7
Signal processing stages (a) and calibration data (b) for WLIPS at two different temperatures

signal and only a residual response to temperature changes. This response depends on the noncompensated length of the sensor as well as on the temperature-pressure cross-sensitivity effect, which is different in different types of HB fibers. An overall accuracy of 0.5% of FS can be achieved with this sensor, which in addition will be completely immune to light intensity fluctuations induced by source instability. An industrial prototype of this device is under development.

CONCLUSIONS

In this paper we present two reliable fiber-optic sensor systems developed at UQAH for pressure, stress or load evaluation, directly and safely applicable to any required concrete structure such as railway or highway bridges and hydroelectric dams or to geotechnical measurements as pore water, soil or rock pressure measurement systems at mining or waste-disposal sites. We developed a simple and practical method of embedding the FOS in concrete or other materials. This method, based on the concept of a fiber-optic concrete-embeddable pressure cell, effectively combines modern fiber-optic with elements of traditional hydraulic measurement technology used in civil engineering for over 40 years.

Performance of the first industrial prototype fiber-optic pressure sensor based on the concept of polarimetric measurement and such concrete-embeddable stress cell is discussed in the paper: we have successfully developed, built and initially tested a complete, reliable and inexpensive fiber-optic prototype sensor system for stress and load evaluation for pressures up to 10 MPa with accuracy better than 1% (including stability, repeatability and temperature drift in the range of about 20 deg.), completely immune to any kind of perturbation in leading fibers and with an excellent power budget allowing for remote operation at up to several km. No similar performance has ever been reported in the literature in the context of any FO pressure sensor. The second configuration of a pressure sensor based on HB fibers exploits the principle of white-light interferometry. Overall accuracy of 0.5% of FS can be achieved for this sensor, which in addition is immune to light intensity fluctuations induced by source instability. An industrial prototype of this device is under development.

REFERENCES

1. E.Udd, Editor, "Fiber Optic Sensors", John Wiley & Sons, New York, 1991.
2. S.Y.Huang, J.N.Blake, B.Y.Kim, "Perturbation effects on mode propagation in highly elliptical core two-mode fibers", J.Lightwave Technol., vol. 8, 1, 23-33, 1990.
3. M.Sherif et al., "Optical response of sapphire multimode optical sensor for ceramic composite applications", *Proc. SPIE,* vol. 2072, 244-51, (1994).
4. W.J.Bock, W.Urbanczyk, "Measurement of polarization mode dispersion and Modal Birefringence in HB fibers by means of electronically scanned shearing type interferometry", *Appl.Opt.,* vol. 32, 30, 5841-8 (1993).
5. W.J.Bock, A.W.Domanski, T.R.Wolinski, "Influence of high hydrostatic pressure on beat length in HB optical fibers", *Appl.Opt.,* vol. 29, 24, 3484-8 (1990).
6. W.J.Bock, T.R.Wolinski, A.Barwicz, "Development of a polarimetric optical fiber sensor for electronic measurement of high pressure", *IEEE Trans.Instrum.Meas.,* vol. 39, 715-21 (1990).
7. K.S.Chiang, "Pressure induced birefringence in a coated highly birefringent fiber", *J.Lightwave Technol.,* vol. 8, 1850-5 (1990).
8. B. Grossman et al., "Fiber-optic pore water pressure sensor for civil engineering applications", *Proc. SPIE,* vol. 2072, 342-351, (1994).
9. J.Kollé et al., "Embedded polarimetric differential stress sensor", *Proc. SPIE,* vol. 2070, 55-64, (1994).
10. G.P.Carman et al.,"Implementation and evaluation of an embedded extrinsic Fabry-Perot fiber-optic strain rosette sensor", *Proc.SPIE,* vol. 1798, 237-46 (1993).
11. A.Mendez, T.F.Morse, L.J.Reinhart, "Measurement of the internal strain in concrete elements using embedded fiber sensors", *Proc.SPIE,* vol. 2072, 362-69 (1994).
12. N.Fürstenau, D.D.Janzen, W.Schmidt, In-flight strain measurements on structurally integrated composite plates using fiber-optic interferometric strain gauges", *Smart Mater.Struct.* vol. 2, 147-56, (1993).
13. W.J.Bock, A.W.Domanski, US Patent 4 920 261 (1990), Canadian Patent 1 322 109 (1993).
14. W.J.Bock, W.Urbanczyk, M.Beaulieu, J.Wojcik "White-light interferometric fiber-optic pressure sensor", *Proc. IEEE Instr.Meas.Tech.Conf.,* Hamamatsu, Japan, May 10-12, 406-11, (1994).

BRAGG INTRA-GRATING STRUCTURAL SENSING

S. Huang, M. LeBlanc, M. M. Ohn, and R. M. Measures

University of Toronto, Institute for Aerospace Studies
4925 Dufferin St., Downsview, Ontario, Canada, M3H 5T6

INTRODUCTION

Fiber-optic intracore Bragg gratings have been of interest for a variety of sensing applications,[1-3] especially for smart structures.[4] In most smart structures applications, the strain along a fiber grating sensor is uniform and the strain is measured by the shift of the reflective wavelength of the grating. However, in some situations such as near a fiber debonding zone or near a structural damage point, strong local strain gradients are produced. Under such strain fields, the grating spectrum will not only be shifted but also distorted due to the chirp induced in the grating period by the imposed strain gradient. Based on this spectrum distortion, it is possible to use Bragg gratings as quasi-distributed sensors to sense strain gradients over a small region (sub-centimeter). We have calculated the reflection spectrum of nonuniform Bragg gratings by using the T-matrix method. This has enabled us to determine the effects of strain gradients on the spectrum. Experiments have verified these theoretical results.

THEORETICAL MODEL

Nonuniform gratings, including nonuniform grating period, nonuniform index modulation depth, and even nonuniform average refractive index, have been analyzed in different ways.[5-7] We present here a method utilizing the T-matrix formalism[8] which offers a straightforward way to analyze any kind of grating structure.

In a uniform grating with length l, the electric fields of the backward and forward modes can be expressed as $a(z) = A(z)\exp(i\beta z)$ and $b(z) = B(z)\exp(-i\beta z)$ respectively where β is the wave propagation constant. Their complex amplitudes $A(z)$ and $B(z)$ obey the coupled-mode equation:[9]

$$\begin{cases} \dfrac{dA}{dz} = i\kappa B\exp[-i2(\Delta\beta)z] \\ \dfrac{dB}{dz} = -i\kappa A\exp[i2(\Delta\beta)z] \end{cases} \qquad (0 \le z \le l) \qquad (1)$$

Applications of Photonic Technology
Edited by G.A. Lampropoulos *et al.*, Plenum Press, New York, 1995

where κ is the coupling coefficient which is related to the index modulation depth Δn, Λ is the grating period length, and $\Delta\beta = \beta - \pi / \Lambda = 2n\pi(1/\lambda - 1/2n\Lambda)$ is the differential propagation constant associated with detuning from the Bragg condition, n is the average refractive index. From Eq. (1), the close form solution of $a(z)$ and $b(z)$ can be obtained under boundary conditions $a(l)$ and $b(0)$. Therefore, the T-matrix of this grating can be readily formed by the solutions $a(0)$ and $b(l)$ plus the boundary conditions (see Figure 1(a)).

For a nonuniform grating, we divide it into many segments. Each segment can be assumed to be uniform, but have slightly different κ, Λ, and n from its neighbors. Each segment is treated as a uniform grating with its own T-matrix. The final transfer function of the segment series can be expressed by the product of all T-matrices (see Figure 1(b)). Since there is no backward input, we begin with the right-side vector elements for the first segment (at far end) as $a(l) = 0$, $b(l) = b_l$. After calculating the product of all T-matrices, we obtain the final left-side vector elements, $a(0)$ and $b(0)$, and the subsequent spectral reflectivity of the grating: $R(\lambda) = |a(0,\lambda) / b(0,\lambda)|^2$. It can be seen that we can actually assume any non-zero value for b_l as it will not affect the value of $R(\lambda)$. The selection of the number of pitches in every segment depends on the nonuniformity of grating. For grating regions having very high nonuniformity, a segment can be limited to only one pitch, which shows the high flexibility of this T-matrix method.

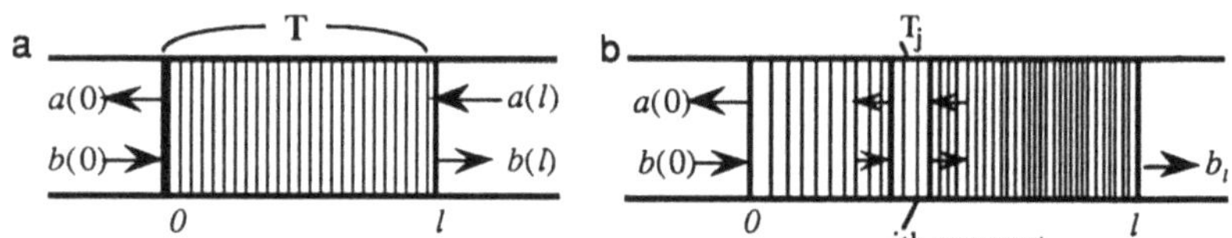

$$\begin{bmatrix} a(0) \\ b(0) \end{bmatrix} = [T] \begin{bmatrix} a(l) \\ b(l) \end{bmatrix} \qquad \begin{bmatrix} a(0) \\ b(0) \end{bmatrix} = [T_M][T_{M-1}]\cdots[T_j]\cdots[T_1] \begin{bmatrix} 0 \\ b_l \end{bmatrix}$$

Figure 1. Illustrations of T-matrix model for (a) uniform grating and (b) nonuniform grating.

CALCULATION RESULTS

By using the T-matrix formalism, we have determined the reflective spectra for different nonuniform distributions of grating period length $\Lambda(z)$ and the refractive-index modulation depth $\Delta n(z)$. The calculation results show a strong dependence of the reflective spectrum on the chirp of period lengths. With an increase of the grating-period chirp, the spectrum width increases accompanied with a decrease of the reflectivity. For linearly chirped gratings, the spectral responses for different chirping grades are shown in Figure 2. Fig. 2(a) is for a grating with uniform index-modulation depth and Fig. 2(b) is for a grating with Gaussian distribution of index modulation depth. It can be seen from Fig. 2 that, compared with the uniform Δn case, a Gaussian-distributed $\Delta n(z)$ makes the reflection spectrum much smoother, but less sensitive to the period chirp. Current gratings generally have Gaussian-distributed $\Delta n(z)$ because of the method used to write them. Writing uniform gratings is also possible, for example using a point-by-point technique.

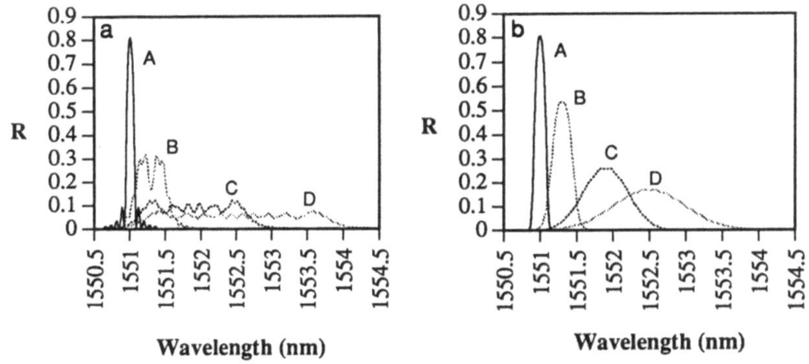

Figure 2. Spectral response of Bragg grating subjected to a strain which is linearly changing over the grating length, from zero strain at one end to (A) 0 $\mu\varepsilon$, (B) 387 $\mu\varepsilon$, (C) 1161 $\mu\varepsilon$, and (D) 1934 $\mu\varepsilon$, respectively, at the other end. Figure 2(a) is for a grating with uniform Δn $(= 7\times10^{-5})$, and figure 2(b) is for a grating with Gaussian-distributed $\Delta n(z)$ $(= 1.28\times10^{-4}\exp\{-10[(z-l/2)/l]^2\})$. Both gratings have a length $l = 1$ cm, and an initial period length (without strain) of 517 nm.

EXPERIMENT WITH EMBEDDED BRAGG INTRA-GRATING SENSOR

A fiber-optic grating (made by United Technologies Research Center) was embedded in a specially designed cantilever beam whose shape induced a high strain gradient in the gauge region upon application of an end deflection. The grating length was measured as 13.7 mm based on the red fluorescence emitted over the grating region when launching 457.9 nm Argon-ion laser light into the fiber. Fig. 3(a) shows strain profiles along the grating length for four different end-deflection conditions of the cantilever beam. Curve A corresponds to the original (unloaded) beam. With increasing the end-deflection (curves B, C, and D), both the average strain and the strain gradient increase. Note that the shape of this strain profile is slightly nonlinear.

The measured grating reflection spectra under those strain conditions are shown in Fig. 3 (b), as the solid lines. Also shown in Fig. 3 (b) are the theoretical curve fits based on the T-matrix method assuming a Gaussian distribution of the index modulation depth $\Delta n(z)$ selected to best match the experimental broadening. The expected tendencies of shift towards longer wavelength (with increasing average strain) and broadening (with increasing gradient) are clearly illustrated in both the experimental and theoretical sets of curves. We can also observe, in both sets, a progressive asymmetry of the reflection curves with increasing beam deflection. This is due to the nonlinearity of the strain profile. It can be seen in Fig. 3 (a) that the strain gradient (the slope of the curve) is greater in the regions of larger strain compared with the regions at lower strain. In the spectrum curves, this corresponds to more flattening and broadening in the higher wavelengths part of the spectrum. Both the experimental and theoretical curves display this expected feature.

Significant discrepancy of the theoretical curves from the experimental ones can also be seen in Fig. 3(b). In particular, the reduction in peak reflectivity and the spectral broadening seem to evolve at a faster rate in the theoretical curves compared with the experimental ones. Unfortunately, it is difficult to perfectly fit the experimental results without knowing the detail of the grating structure, especially its exact index-modulation profile.

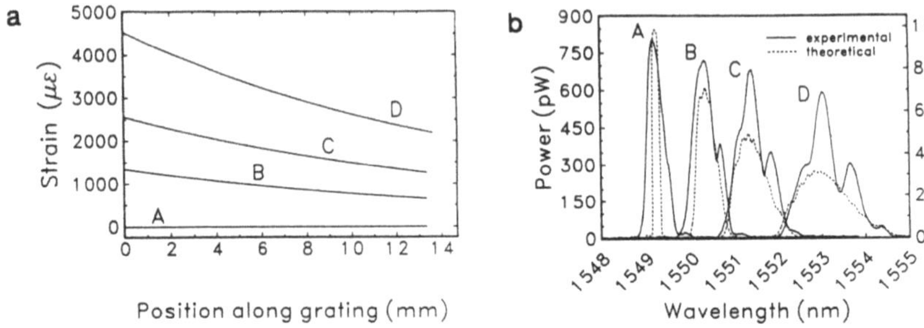

Figure 3. Experimental results to illustrate the effect of strain gradient on the reflection spectrum of an embedded Bragg grating. (a) Strain profile along the grating produced by deflecting the specially designed loading beam for four different deflection conditions. (b) The measured reflection spectra under these strain conditions and their theoretical curve fits. The fitting curves were obtained using the T-matrix formulation assuming the following index-modulation depth profile: $\Delta n(z) = 1.5\times10^{-4}\exp\{-5.66[(z-l/2)/l]^2\}$.

CONCLUSIONS

We have shown both theoretically and experimentally that important information on the strain distribution within a material can be extracted from the details of the reflection spectrum of a fiberoptic grating. This is based on the fact that each sub-length of a grating contributes to the overall spectrum according to its local state of strain, thus enabling it to be used as a quasi-distributed sensor. This unique Bragg intra-grating (BIG) sensor shows a great potential in the field of smart structures for monitoring high strain gradients.

ACKNOWLEDGEMENTS

This work was supported by the Ontario Laser and Lightwave Research Center, the Ontario Centre for Material Research, and the Natural Science and Engineering Research Council of Canada.

REFERENCES

1. W.W. Morey, G. Meltz, and W.H. Glenn, "Fibre optic Bragg grating sensors," *Proc. SPIE*, vol. 1169, pp. 98-107 (1989).
2. S.M. Melle, A.T. Alavie, S. Karr, T. Coroy, K. Liu, and R.M. Measures, "A Bragg grating-tuned fibre laser strain sensor system," *IEEE Photonics Technol. Lett.*, vol. 5, pp. 263-266 (1993).
3. A.D. Kersey and T.A. Berkoff, "Fiber-optic Bragg-grating differential-temperature sensors," *IEEE Photonics Technol. Lett.*, vol. 4, pp. 1183-1185 (1992).
4. R.M. Measures, "Smart composite structures with embedded sensors," *Composites Engineering*, vol. 2, pp. 597-618 (1992).
5. K.O. Hill, "Aperiodic distributed-parameter waveguides for integrated optics," *Appl. Opt.*, vol. 13, pp. 1853-1856 (1974).
6. H. Kogelnik, "Filter response of nonuniform almost-periodic structures," *Bell Syst. Tech. J.*, vol. 55, pp. 109-126 (1975).
7. L. Poladian, "Graphical and WKB analysis of nonuniform Bragg gratings," *Physical Review E*, vol. 48 (1993).
8. H.A. Haus, p. 78 in "Waves and Fields in Optoelectronics," N. Holonyak, ed., Prentice-Hall, Inc., Englewood Cliffs, New Jersey (1984).
9. A. Yariv and M. Nakamura, "Periodic structures for integrated optics," *IEEE Quantum Electro.*, vol. 4, pp. 233-253 (1977).

RADIOFREQUENCY SUBCARRIER ENCODED FIBRE OPTIC STRAIN GAUGE FOR GROUND ANCHORS

D. Walsh[†], W.C. Michie[†], B. Culshaw[†], B. Noharet[‡], J. Chazelas[‡], M. Turpin[‡] and T. Herbst[*]

[†] Dept. E.&E.E., University of Strathclyde, 204 George St., Glasgow G1 1XW, Scotland
[‡] Thomson CSF, L.C.R., Domaine de Corbeville, 91404 Orsay Cedex, France
[*] Dyckerhoff & Widmann, P.O.Box 810280, D-81902 München, Germany

INTRODUCTION

Optical sensing schemes have shown particular promise for a wide variety of applications in civil engineering. Some advantages of fibre optic sensors over their rivals are their immunity to electromagnetic interference, distributed sensing capability and the fact that they can be embedded in and are compatible with various composite materials. An area of considerable interest has concerned the measurement of elongation/strain experienced by various in situ structural elements using optical fibres. Obviously, to operate effectively in such an environment the sensing fibre has to be suitably protected to not only withstand the rigours of the application, but also the rough treatment envisaged during transportation and handling. The approach proposed by THOMSON CSF was to form a fibre optic strain sensor by pultruding a protective coating around the sensing fibre.

A specific civil engineering application where it is desirable to monitor strain changes, is along the bond length of a loaded ground anchor. Fig.1 outlines the basic design of a DYWIDAG permanent bar anchor. The three principal sections of the ground anchor are; the bond length, the free stressing length and the anchorage and its main purpose is to transfer the load from the anchorage at the surface, along the free stressing length to the ground at some point along the bond length. Its chief constituent is a high strength steel tendon whose total length varies between 15m to 100m with an outer sheathed diameter of 65mm to 200mm. On site the prefabricated ground anchor is installed into a borehole and locked into position with a cementitious grout. Up to now the efficiency of the load transfer capacity of the anchor when embedded can only be estimated by stress tests using hydraulic jacks placed at the anchor head. Therefore there is an acute need to integrate a strain monitoring system within the ground anchor at the assembly stage to examine the exact strain distribution over the anchor length and its subsequent variation over the service life of the anchor.

Standard optical fibre interferometric strain sensors are extremely sensitive[1] - greater than 10^6 rad/m.ϵ. However, their high sensitivity to small optical phase changes also has a drawback because their cosinusoidal output transfer function only has an unambiguous measuring range extending over one optical wavelength (i.e. of the order of a micrometre). Of course there are other techniques employed to increase this measuring range, most notably white light interferometry, but in situations where large fibre length changes are expected, such as structural monitoring, these methods become impractical. In such cases it may be more advantageous to sacrifice some of the sensitivity of optical fibre interferometers for an extension of the useful measuring range. This can be implemented with a system based upon a microwave subcarrier technique which has been traditionally employed in rangefinding and geodetic surveying applications[2].

This paper will firstly examine the operation of the RF modulated system with a standard acrylate coated optical fibre as the sensor and then investigate the response of a rugged pultruded rod optical fibre sensor attached to a high strength steel tendon (the principal component of a permanent bar anchor) as it undergoes tensile strain.

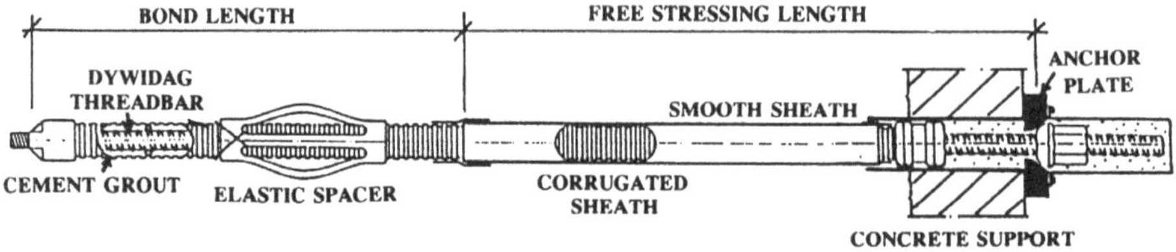

Figure 1. DYWIDAG permanent bar anchor for soil and rock.

PRINCIPLE OF OPERATION

As a single mode of light travels down a glass fibre its phase changes according to $\phi=2\pi nL/\lambda_{op}$ where n is the effective index of the mode, L is the fibre length and λ_{op} is the free space optical wavelength. If the fibre is now subject to an axial strain ϵ then the phase of the light changes because both the physical length and the refractive index of the fibre core varies with strain. The strain sensitivity can be written in terms of the physical properties of the glass[1] as

$$\frac{1}{L}\frac{\partial\phi}{\partial\epsilon} = \frac{2\pi n}{\lambda}\left[1-\frac{1}{2}n^2\left[(1-\mu)P_{12}-\mu P_{11}\right]\right] \tag{1}$$

where P_{11} and P_{12} are the strain optic coefficients of the glass and μ is its Poisson's ratio. For standard silica we find that $P_{11}=0.121$, $P_{12}=0.27$ and $\mu=0.17$. Using an optical wavelength of $1.3\mu m$ and a refractive index for the fibre core of 1.452 we can determine that the strain sensitivity of an optical fibre at this wavelength is $\Delta\phi_\epsilon=5.51 rad/m.\mu\epsilon$.

In a basic radiofrequency (RF) modulated fibre optic system a laser source is modulated at a frequency in the range of 100MHz to 1GHz (i.e. RF wavelengths, λ_{RF}, of 3m to 0.3m). The modulated light is transmitted down an optical fibre and detected at its output end. A double balanced mixer combines the amplified electrical signal from the receiver with another directly from the signal generator. Since the mixer has two input signals at the same frequency it acts as a phase detector and its low pass filtered output is a d.c. voltage proportional to the cosine of the phase difference between its two input signals. Thus if the fibre is strained the RF subcarrier experiences a proportional phase change and hence the d.c. output of the phase detector changes accordingly. Comparing the response of the RF modulated system to that of the optical interferometer above, we find that at a modulation frequency of 1GHz, $\lambda_{RF}=0.3m$ and therefore, substituting for λ in Eq.1 we find, $\Delta\phi_{RF\epsilon}=23.9\mu rad/m.\mu\epsilon$. Clearly the electrical subcarrier strain system is over five orders of magnitude less sensitive than is the case with the optical interferometric strain sensor. Nevertheless, there is a correspondingly larger unambiguous range over which the fibre can be extended before the subcarrier signals experience a relative optical path length change of one RF wavelength. This greater measuring range can be of particular benefit for monitoring strain in large civil engineering structures such as ground anchors.

EXPERIMENTS AND DISCUSSION

(i) Acrylate Coated Fibre

Initially the optical strain sensitivity of an acrylate coated $1.3\mu m$ singlemode fibre was determined to be $\Delta\phi_\epsilon=5.5\pm0.04$ rad/m.$\mu\epsilon$, using a fibre optic Mach-Zehnder interferometer. If this is compared to the value calculated previously using the standard parameters for silica based fibres it is found to be in close agreement.

The response of the RF subcarrier encoded system, depicted in Fig.2, with the acrylate coated fibre was then investigated. A $1.3\mu m$ wavelength laser diode was intensity modulated at a frequency of 1GHz using a signal generator and the resultant modulated output light was coupled into a 10m length of standard $1.3\mu m$ singlemode telecommunications fibre. 2.8m of this fibre was clamped between two programmable translation stages to strain the fibre. The output of the receiver was then mixed with a reference signal directly from the signal generator and the resultant d.c. output from the mixer, which was proportional to the phase difference between the two input signals, was monitored with a digital voltmeter.

Fig.3 displays the output voltage of the mixer and the equivalent microwave phase change at a frequency of 1GHz as the optical fibre is extended up to 10mm. The resultant plot shows a linear relation between axial strain and the induced phase change. The resolution of the microwave system in the short term is limited by the uncertainty in the output voltage reading from the phase detector. For the previous test the voltage error was ± 0.05mV, which was equivalent to a phase error ± 0.12mrad and hence yielded a length resolution of $\pm 5\mu$m. Obviously for long fibre lengths this could yield sub-microstrain resolution. However, long term drift in the output signal, due to; temperature, fluctuations in the amplitude of the input signals to the mixer and variations in the oscillator frequency, can lead to far more significant phase errors. Nevertheless all of these effects can be suitably compensated using reasonably straightforward optical or electronic techniques, so the $\pm 5\mu$m length resolution for the system gives an idea of the measurement possibilities with this microwave strain sensing scheme.

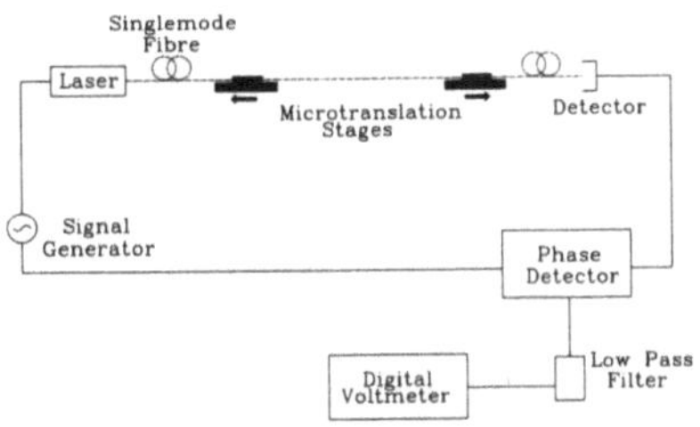

Figure 2. RF subcarrier encoded strain sensing system.

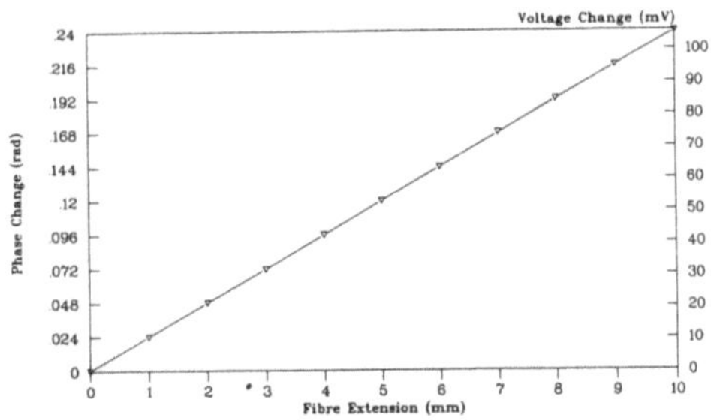

Figure 3. Strain response of an acrylate coated singlemode fibre.

(ii) Pultruded Fibre Rod Sensor

Following preliminary experiments at THOMSON CSF the pultruded fibre optic sensor was supplied for these tests. The sensor consisted of a standard 1.3μm acrylate coated singlemode optical fibre with a 2mm outer sheath of 67% E glass fibres and 33% epoxy pultruded around the optical fibre. The optical fibre was positioned approximately in the centre of this rod. The Young's modulus of the complete sensor was determined to be 49GPa which reflects the high content of silica in the rod (the Young's modulus of the optical fibre is 70GPa). The strain sensitivity of the pultruded rod was determined using a tensile testing machine to be $23.5 \pm 0.4\mu$rad/m.$\mu\epsilon$ at a frequency of 1GHz. Evidently this value agrees with that determined above for the acrylate coated optical fibre alone, indicating that the strain is fully transferred from the rod through to the optical fibre and that the pultrusion process does not alter the strain sensitivity of the sensing fibre.

An 87.5cm sensing length of the pultruded rod was bonded to a DYWIDAG prestressing bar (1.7m long and 32mm diameter) using a silica loaded epoxy. The steel bar was then locked into position in a tensile testing machine and the input and output fibre pigtails from the sensor were connected into the microwave subcarrier strain system described earlier. Electrical strain gauges were mounted on the steel surface and also on the surface of the pultruded rod to assess the strain transfer across the interfaces and these were electrically connected to a data logging and readout system. For comparative purposes an extensometer was also fitted to the steel tendon. During the tests the load on the steel was increased manually up to 750kN (equivalent to 4.5‰), which was close to the test load of such a ground anchor.

The voltage output from the mixer of the microwave sensing system and also the extensometer output as the load on the tendon is increased and decreased is depicted in Fig.4. The pultruded rod sensor exhibits a linear response with load which correlates well with the extensometer output, though the extensometer displayed a greater incidence of hysteresis. The load-strain curves, plotted in Fig.5, compare the operation of the pultruded rod sensor with that of the electrical strain gauge mounted on the prestressing steel bar and also on the sensor surface. The fibre response is observed to be linear with strain within experimental error (no compensation was employed for temperature, amplitude drift etc.) and also agrees extremely well with the strain gauge mounted on the sensor surface. This corroborates our previous assertion that there is full transfer of strain between the pultruded rod and the sensing fibre. However a 5% discrepancy is found between the strain experienced by the gauges mounted on the steel and on the sensor surface. This was attributed to a poor transfer of strain through the epoxy to the pultruded rod due to shear effects. This is not surprising since the epoxy matrix has a Young's modulus of about 4GPa while the sensor (i.e. pultruded rod plus optical fibre) is listed as 49GPa.

Therefore the glass/epoxy pultruded rod appears to fulfil two principal functions; it has a linear strain response and also provides a robust, protective sheathing for the sensing optical fibre. To minimise the interface problems between the sensor and the ground anchor, the outside of the pultruded rod could be coated with a metal and braised to one of the steel tendons. This would indeed form a rugged sensing unit which would also be compatible with standard optical fibre connectors.

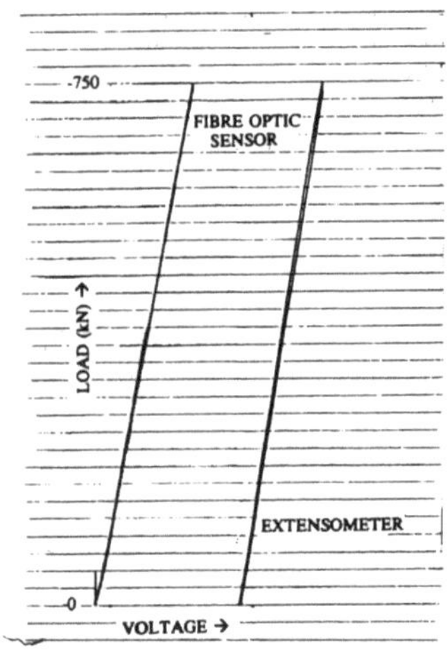

Figure 4. Comparison of the response of the pultruded rod sensor and the extensometer.

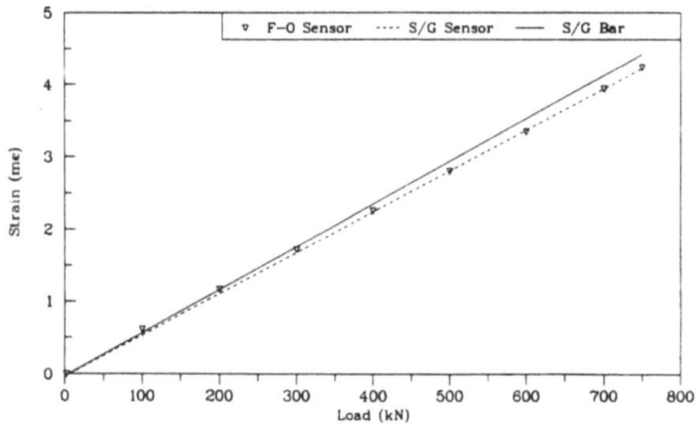

Figure 5. Comparison of the strain experienced by the pultruded rod sensor and the strain gauges on the steel bar and the sensor surface.

CONCLUSIONS

These initial experiments were undertaken to assess the potential of an RF subcarrier encoded system incorporating a pultruded rod type fibre optic sensor for use as a strain sensor in ground anchor applications. Some of the principal conclusions which can be derived from the preceding work are

- A suitably compensated RF modulated fibre optic strain sensing system has the potential of micrometre length resolution.
- A 2mm glass/epoxy pultruded sheathing forms a rugged, protective layer over the sensing fibre.
- This pultruded fibre rod sensor gave a linear strain response when attached to a high strength steel tendon which was loaded by a tensile tester.
- There was complete transfer of strain from the pultruded coating to the internal sensing fibre.
- For optimum strain transfer from the ground anchor to the sensing element alternative bonding methods may have to be employed.

ACKNOWLEDGEMENTS

This work was conducted under the auspices of the OSMOS project, a CEC funded BRITE-EURAM programme.

REFERENCES

[1] C.D. Butter and G.B. Hocker, *Fiber optics strain gauge*, Appl. Opt., 17, p.2867-2869, (1978).
[2] C.D. Burnside, *Electromagnetic Distance Measurement*, Crosby Lockwood, London, (1971).

EXPERIMENTS WITH FIBEROPTIC SMART SKIN SENSORS APPLIED TO MARINE VEHICLES

Leif Bjerkan, Kjetil Johannessen, and Helge Storøy

SINTEF DELAB, N-7034 Trondheim Norway

INTRODUCTION

Advanced composite materials are currently finding an increased use in marine vehicle construction. The introduction of new materials reduces hull weight, and is especially attractive for fast vehicles. By the introduction of new materials a need arises to design a cost effective construction which can withstand expected environmental stresses under operation with an optimum use of material for reinforcement. Furthermore, it is desired to monitor the long term integrity of the new materials when subjected to sea loads under operation. Data collected on these composite materials are still rather limited.

In order to address these issues model tests as well as real tests on full scale vehicles have been carried out where sensors are monitoring the stresses on the vehicle. The well known fiberoptic polarimetric sensor based on high birefringent fibers[1-3] has been evaluated for these purposes. This paper will present some results with this fiberoptic sensor.

EXPERIMENTS ON SCALED MODELS

A scaled elastic model of a 110 m long monohull marine vehicle was constructed using the composite material Divinycell and glass fiber/polyester resin[4]. The length of the model was 5.104 m, the width at waterline midships was 0.849 m and the hull thickness of the composite material was 30 mm. The weight in water was 420 kg. This model scaled as 1:21.55, and was constructed with correct stiffness in bending, shearing and torsion related to the actual vehicle. A 28 cm long polarimetric sensor fiber (York Hi-Bi 1300) was positioned midships away from the neutral axis. The sensor fiber was epoxied in a groove which was carved out in the outer side of the hull. A 1300 nm semiconductor laser was used as the light source. An all-fiber experimental setup was employed using polarizing single mode fibers as lead-in and lead-out. A conventional strain gauge was inserted at the same position on the opposite side of the hull for comparison. The model was subject to computer controlled sea wave motions of regular and impulse waves of different speeds and wave amplitudes in a 260 m long, 10.5 m wide, and 5-10 m deep towing tank. During the tests the model was mounted in a free-to-surge rig in front of the towing carriage. In all experiments the scaled model was moved with a constant speed of 2.77 m/s (corresponding to 25 knots for the actual vehicle) against the water waves. The response of the fiberoptic sensor was recorded and compared with the conventional strain gauge and theoretical predictions.

Figure 1 shows examples of the fiber and the strain gauge responses respectively for a

Applications of Photonic Technology
Edited by G.A. Lampropoulos *et al.*, Plenum Press, New York, 1995

regular wave motion of 10 cm amplitude and a period of 1.6 seconds. The fiber response is given in microstrain while the strain gauge response yields the midship vertical bending moment of the structure after a proper calibration. Unfortunately the same calibration of the sensor fiber was unreliable due to an unstable laser so only a relative comparison between those is possible. Figure 2 shows a similar comparison for an impulse wave of 20 cm amplitude. From these examples and others with different parameters (wave amplitude and period) the time series from both the fiber and the strain gauge agree well qualitatively. For regular wave motions a maximum induced strain variation of about 450 μstrain was found for a wave period of 1.6 m/s with a 10 cm wave amplitude which corresponds to a wavelength close to the length of the model. For higher wave speeds the induced strains decreased rather rapidly. The same maximum was observed with the conventional strain gauge, but the decrease in induced strains with increasing wave speeds was slower than measured with the fiber sensor. From theoretical calculations the fiber measurements agree better than those of the conventional strain gauge. However, for lower wave speeds some deviations between the two sensor responses were observed. The measured signals might to some extent be influenced by local stress concentrations which are sensed differently due to the different sensor lengths. For the impulse waves the two sensor responses correlate well for all measurements. Based on these results on the scaled models one can predict the expected stresses on the corresponding full scale marine vehicles. In this case the measured vertical bending moments should be multiplied by the scale factor to the fourth power.

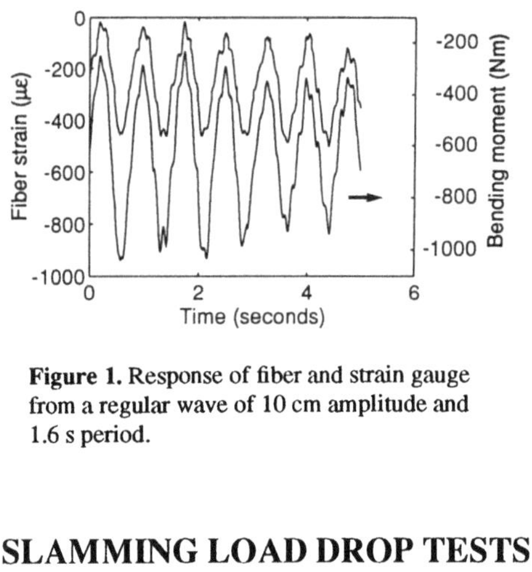

Figure 1. Response of fiber and strain gauge from a regular wave of 10 cm amplitude and 1.6 s period.

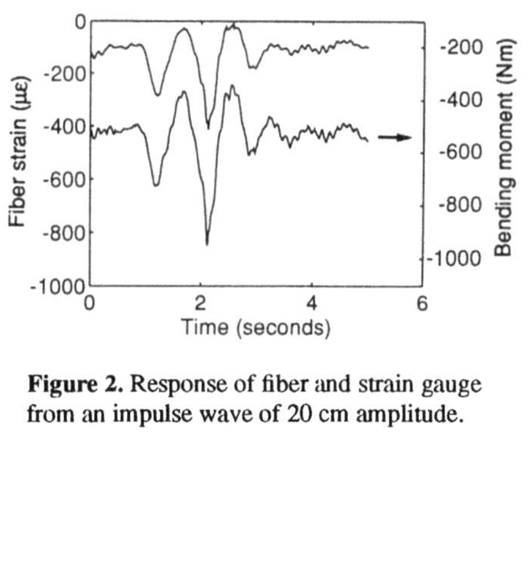

Figure 2. Response of fiber and strain gauge from an impulse wave of 20 cm amplitude.

SLAMMING LOAD DROP TESTS

An important consideration in the structural design of fast sea vehicles like catamarans is the ability of the water-exposed areas to withstand slamming loads. Individual slamming events can be simulated by means of a drop test where a model is released from a given height (and inclination) and can fall freely onto a water surface. In such tests the slamming impacts are defined by the model, the drop height and the speed and amplitude of the water wave, and the relationship between slamming loads and the induced responses can be determined. A series of slamming drop tests were performed in order to verify a numerical calculation model for slamming. One is primarily interested in the response in the short time frame at the slamming impact. Therefore sensors with fast responses are needed. A fiberoptic polarimetric sensor was adapted for this purpose. In these experiments a stable Nd:YAG laser operating at 1319 nm was used since the semiconductor laser used in the previous experiments was not sufficiently stable. For the tests we used a large stainless steel structure weighing 385 kg, and at the bottom an aluminium plate (50·10 cm) with a multiple T-profile facing inwards was attached. A 30 cm long polarimetric fiber sensor was epoxied in a carved groove in one of the T sections. Polarizing single-mode fibers were used as lead-in and lead-out from the control table to the drop rig. Conventional strain gauges were positioned on an adjacent T section.

The drop rig was dropped from different heights on a moving wave crest and the response of the sensors were recorded in a time frame of about 30 ms following the impact. The induced strains and the corresponding bending moments were calculated after proper calibration. Figure 3 shows a typical example of the fiber and strain gauge responses respectively. The results were strongly dependent on how precisely the drop rig hit the peak of the wave, and a maximum bending moment of 1000 Nm was found from the fiber measurement for a 72.5 cm drop height (distance from the bottom of the drop rig to the wave crest). For all measurements the fiber sensor gave about 60% higher values than the strain gauge positioned at the centre of the beam. This may be explained by the fact that the 30 cm fiber senses a different distortion pattern than the strain gauge which is close to a point sensor.

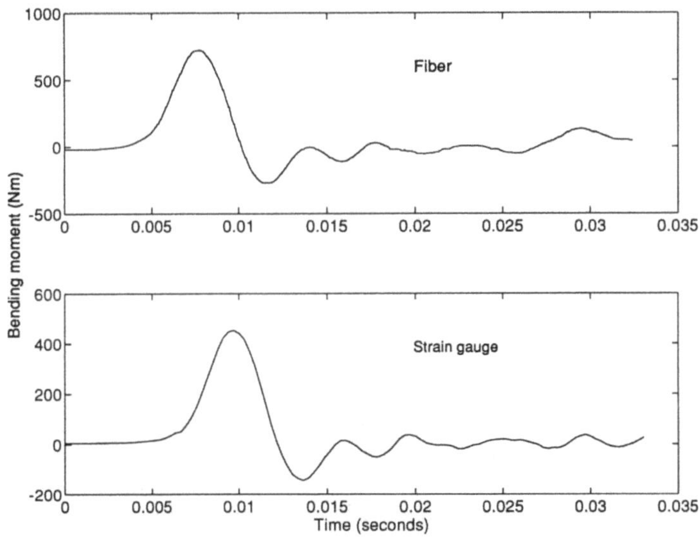

Figure 3. Response of fiber and strain gauge from a drop test. The drop rig hit a wave of 22.5 cm amplitude from a height of 62.5 cm.

FULL SCALE TESTS

Finally some experiments were carried out on an air cushion catamaran operating off the norwegian coast. Two fiberoptic sensors were attached between the two hulls in a tunnel between the two engine rooms. The fibers were prestrained and glued in the whole length to the fiberglass laminates with epoxy. In these experiments we used 3 m sensor lengths (York Hi-Bi fibers) and a 780 nm semiconductor laser in order to determine strain in the fiber-glass laminates of the air-cushion craft. The sensors operated for two days of on-site tests and provided useful information on the deformations of the ship in an unfriendly environment. In particular deformation between the two hulls was observed and related to global ship movements and vibrations due to the air cushion. Figure 4 shows an example of a 5 minute time sequence and the corresponding power spectrum. The frequency components around 1 Hz correspond to air cushion vibrations while the low frequency components correlate well with the measured roll of the craft. In general it was observed that the global strains were rather small, and less than 100 $\mu\varepsilon$ for about 1m sea waves.

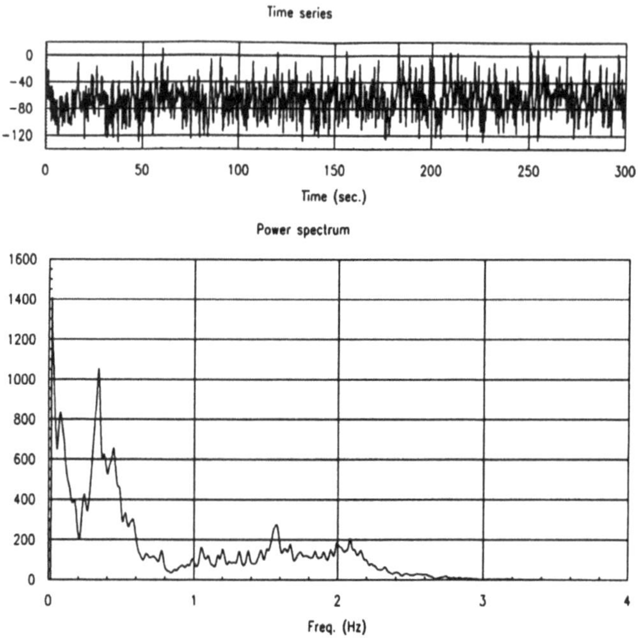

Figure 4. Measurements of strains in an air cushion catamaran operating in the sea with swell from the side.

CONCLUSIONS

The conclusion so far is that fiberoptic sensors are well suited for marine vehicle characterization. Based on the results on scaled models one can predict the expected stresses on the corresponding full scale marine vehicles by appropriate scaling. Such data provides useful input for an improved construction of marine vehicles particularly with the introduction of new composite materials. Furthermore fiberoptic sensors have proven capable for characterizing the induced material stresses and global deformation of marine vehicles during operation in the open sea.

Acknowledgements

This work was supported by the Norwegian Scientific Council. Furthermore, the authors acknowledge support from E. Jullumstrø and J. V. Aarsnes at SINTEF Division of Marine Vehicles for giving us access to their experiments in the water towing tank.

REFERENCES

1. M. P. Varnham, A. J. Barlow, D. N. Payne and K. Okamoto, "Polarimetric strain gauges using high birefringence fibre", Electron. Lett. 699:700 (1983).
2. S. C. Rashleigh, "Polarimetric Sensors: Exploiting the axial stress in high birefringence fibers", IEE Proc. of 1st Int. Conf. on Optical Fibre Sensors (London UK) 210:213 (1983).
3. W. D. Hogg, R. D. Turner, and R. M. Measures, "Polarimetric fibre optic structural strain sensor characterization", Proc. Fiber Optic Smart Structures and Skins, SPIE vol. 1170, 542:550 (1989).
4. E. Jullumstrø and J. V. Aarsnes, "Hull loads and response, hydroelasticity effects on fast monohulls", Presented at NEVA-93, Int. Conf. on Design and Operation of Ships and Offshore Structures, St. Petersburg, Russia (1993).

SMART SKINS - A STEP TOWARD A PRACTICAL FIBRE-OPTIC SENSOR

Claude Belleville[1], André Morin[1], Serge Caron[1] and Merv Edgecombe[2]

[1]Institut national d'optique
Sainte-Foy (Québec)
Canada, G1P 4N8

[2]Metricomp Systems
Calgary (Alberta)
Canada, T3E 5M6

INTRODUCTION

Intelligent structures are those which highly integrate sensors to provide information on the state of the structure[1]. The complete sensing of the structure may require strain, pressure, thermal, chemical or electrical sensors, while a mechanically intelligent structure is sensed mainly by strain sensors. In recent years, many efforts have been devoted to develop a fibre-optic strain sensor with characteristics suitable for intelligent structures. We have developed a white-light interferometric multimode fibre-optic sensor with great potential for use in smart structures. This sensor is very versatile since the gauges can be configured so as to be sensitive to different parameters such as strain, temperature, pressure, humidity, etc[2]. In the specific case of strain, our sensor is perfectly linear, can be thermally self-compensated, is not sensitive to transverse strain, and provides precise, absolute and stable measurements over long periods of time, at a cost competitive with conventional techniques. 54 Fabry-Perot strain gauges, configured as 18 strain rosettes, have been embedded inside a propeller blade of a Polar Class icebreaker. The large number of gauges serves to monitor the distribution of load induced by the impact of ice on the propeller. In this paper, the basic principle of operation of our sensor is described, and some preliminary results obtained with the smart blade are presented.

PRINCIPLE OF OPERATION

The optical strain sensor is based on an extrinsic Fabry-Perot cavity as shown in Figure 1. The Fabry-Perot gauges are fabricated by inserting a multimode optical fibre into both ends of a quartz microcapillary. The tips of the optical fibres, constituting the Fabry-Perot cavity, are adjusted to a given cavity length. The fibres are then fused directly to the microcapillary, the sensitivity being determined by the distance between the welding spots, hereafter referred to as gauge length. When bonded to a structure, the elongation of the gauge within the gauge length region is completely converted into cavity-length elongation. Thus, the cavity length is directly proportional to the strain.

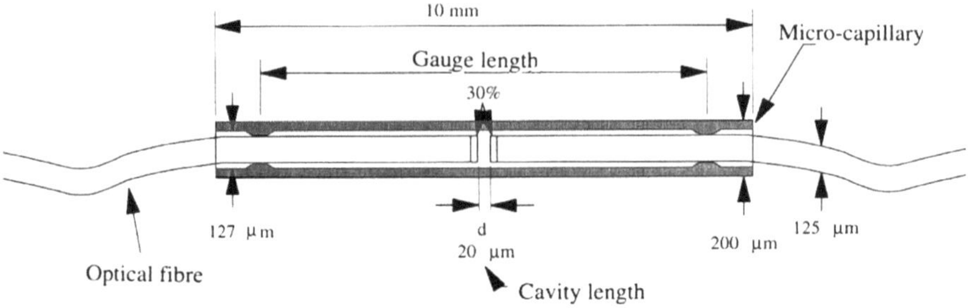

Figure 1. Construction of the multimode extrinsic Fabry-Perot strain gauge.

Relative variations of a Fabry-Perot cavity length have already been measured by fringe counting[3,4]. These schemes are not suitable because fringe counting is not sensitive, and sub-fringe counting is very complex, while they are not reliable for absolute measurements. Our sensor has the capability to measure the absolute cavity length of the Fabry-Perot with an accuracy of one hundredth (1/100) of a fringe. This characteristic is achieved by making use of an optical cross-correlator constituted of a Fizeau interferometer. The configuration of the sensor is illustrated in Figure 2. White light is directed toward the Fabry-Perot sensor with the help of a 2 x 2 coupler. The signal coded by the Fabry-Perot is then reflected back through the Fizeau cross-correlator which exhibits a cavity length varying as a function of the position on the wedge. The light reflected back by a Fabry-Perot strain gauge with a cavity length of d μm is transmitted minimally by the Fizeau interferometer exactly at the position where the distance between the optical flats of the Fizeau is also equals to d μm. If the Fabry-Perot spacing varies in response to a mechanical deformation, the position of the minimum intensity transmitted through the Fizeau will shift correspondingly. The cross-correlation is obtained by illuminating the whole width of the Fizeau interferometer with the returned light signal, and detecting the transmitted light with a silicon linear photodiode array. Finally, the strain is obtained by measuring the position of the minimum on the linear photodiode array.

The dynamic response of such a fibre-optic strain sensor epoxied to a cantilever is illustrated in Figure 3. The upper curve represents the strain as measured with a conventional foil gauge, while the bottom trace illustrates the signal obtained from the optical gauge. The reading rate of both sensors was 1 kHz. The cantilever moves freely on the left side of the curve, while a hammer hits the cantilever near the right-middle of the curve. The impact of the hammer produces some high-order vibration modes. These curves clearly show the excellent performance of the optical sensor. The small differences in the amplitude of the higher-order modes are related to different localizations of the sensors on the cantilever.

BASIC CHARACTERISTICS

This sensor is a good candidate for smart skins and structures since the most important characteristics are fully achieved. As described in a previous paper[5], the Fabry-Perot strain gauge can be configured to self-compensate for thermal expansion. Figure 4 shows results obtained with a thermally self-compensated gauge bonded to a stainless steel plate. The 1 με/°C residual thermal drift is caused by a mismatch between the thermal expansion of the metallic fibre and of the stainless steel plate. The extrinsic Fabry-Perot strain gauge is the only fibre-optic strain gauge not sensitive to tranverse strain because the air gap cavity exhibits no photo-elastic effect[6]. The strain measurement is absolute and perfectly linear because the Fizeau cross-correlator provides an absolute measurement of the Fabry-Perot cavity length. The stability of measurements relies only on the stability of the Fizeau interferometer, which can be made very stable with a shaped dielectric thin layer. Furthermore, the sensor is not sensitive to light intensity fluctuations, as the amplitude of the signal would drop without any shifting of the cross-correlated signal. The cavity length is measured with an accuracy of 0,003 μm, which corresponds to an accuracy of 0,025% of full scale. Moreover, the associated electronics which must be incorporated to extract the signal can be made competitive with conventional techniques both in terms of cost and size. This point is important because it otherwise tends to prejudice the use of fibre optics for measurement in smart structures[1]. The actual system can be enhanced to provide spatial multiplexing of a large number of gauges by using a 2-dimensional silicon detector array combined with a 2-dimensional Fizeau cross-correlator. It is worth noting that the technology required to achieve this goal, namely a custom 2-dimensional silicon detector array, is already mature. This configuration would result in a highly integrated system at a cost which is competitive with conventional techniques.

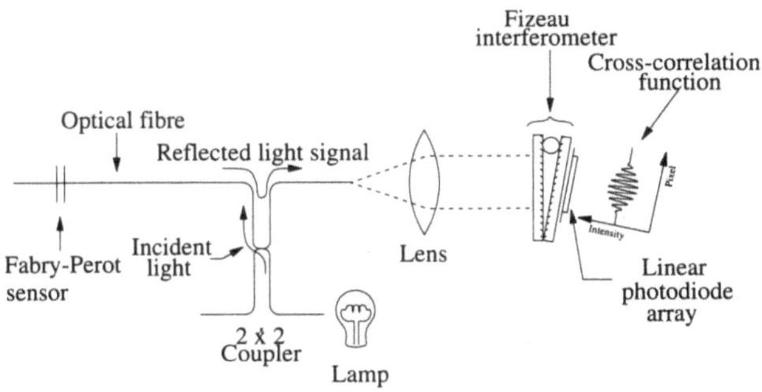

Figure 2. Optical fibre strain sensor configuration combined to the Fizeau interferometer.

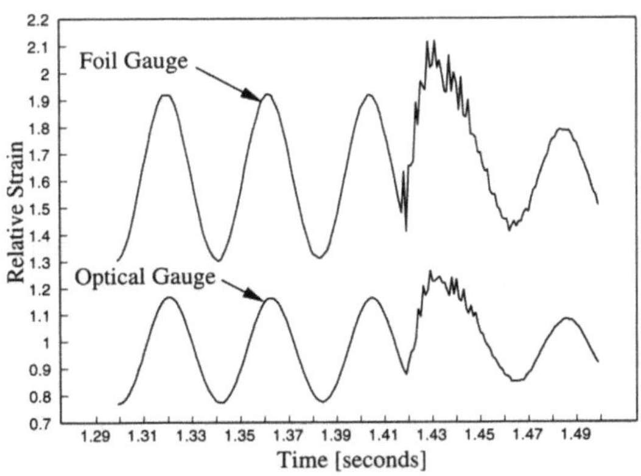

Figure 3. Dynamic response of an optical strain sensor (bottom curve) and a resistance foil gauge.

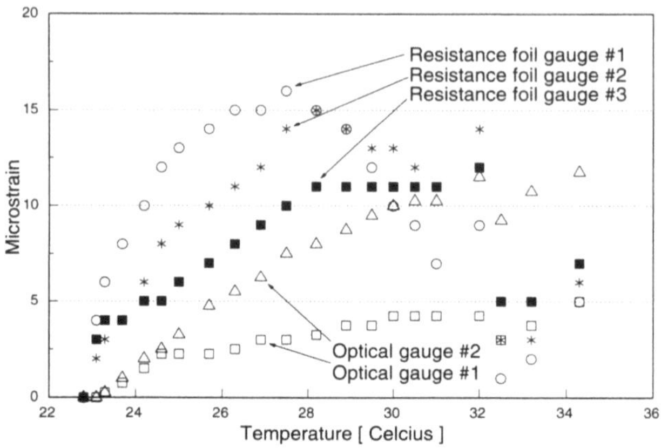

Figure 4. Strain measurement as a function of temperature. Measurements as read by two fibre-optic compensated strain gauges and three compensated resistance foil gauges.

AN EXAMPLE OF SMART SKIN - THE SMART BLADE

We have embedded 54 fibre-optic strain gauges to determine the ice loads that act on the propellers as a function of the ship, propeller and environmental parameters that cause such loads. It was concluded that the use of our sensor for this application would offer numerous advantages over conventional foil strain gauges, especially since the fibre link between the gauges and the electronics has to be routed through the blade, propeller hub, and the 20-metre long propeller shaft. These advantages include the following:

- The small dimension of the fibres can be accomodated within a small cable in comparison with the electrical wire gauge required for proper foil gauges signal transmission over 20 metres. This has permitted the total number of sensors in the blade to be maximised within the limit imposed upon the optical cable diameter routed through the propeller mechanism.

- The fibre-optic sensors and conductors are both strong and durable, and almost chemically inert in the presence of corrosive electrolytes such as sea water, which can attack and progressively destroy the foil strain gauges and lead wires. The fibre-optic sensors are expected to exhibit much longer useful life in such environments than conventional strain gauges.

- The fibre-optic strain gauges also offer exceptional immunity to all forms of electromagnetic and radio frequency interference noise, an important consideration in the electrically noisy shipboard environment.

- The fibre-optic sensors are more tolerant to high temperatures than conventional electrical-resistance strain gauges. This is of particular benefit since the gauges had to be protected by 6 mm thick welded cover plates. The distance between the gauges and the weld beads was as small as 6 mm.

The basic installation methodology to embed the 54 fibre-optic sensors in the blade involved the machining of three radial channels on both faces of the blade. These channels communicate with the central trunion bore via a drilled hole approximately 5 mm in diameter. In each channel, three fibre-optic strain gauge rosettes were installed, for a total of 18 rosettes. The fibre-optic leads from the sensors were led along the machined channels, through the drilled holes to access the trunion bore. Then, a multi-conductor cable brings the fibres through the 20-metre long propeller shaft, and exits through the shaft inside the ship.

All the reading units including the optical cross-correlator, associated electronics and data acquisition computer will be shaft-mounted. The data will be acquired from the 54 fibre-optic gauges at a rate of 1 kHz and will be telemetered off-shaft to a remote computer for processing. It is interesting to note that such a system will provide copious amount of information in real time. The 54 gauges sampled at a rate of 1 kHz will thus generate a 100 kbytes of data per second. A schematic of the complete system is illustrated in Figure 5.

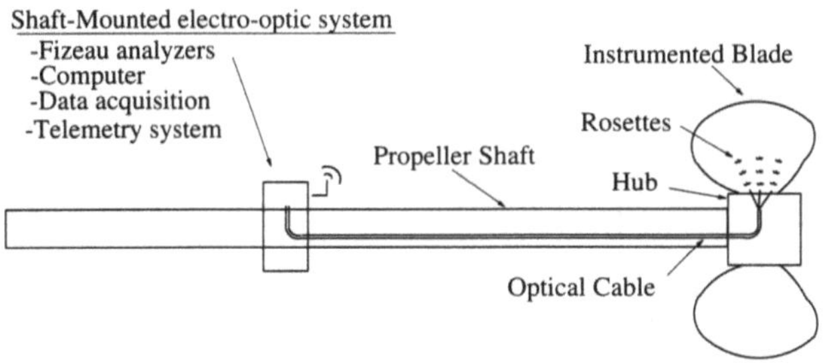

Figure 5. Schematic of the smart blade system.

SOME PRELIMINARY RESULTS

The blade has been calibrated by applying loads at seventy different locations on the surface of the blade. The purpose of this calibration is to make possible the conversion of strain measurements into blade load distribution. Results obtained for a load of 13.5 tons applied near the trailing edge of the blade are illustrated in Figure 6. The arrows indicate the major strain orientation and magnitude as read by the rosettes bonded on one side of the blade.

CONCLUSION

The development of this white-light multimode fibre-optic strain sensor constitutes an important step toward the development of fibre-optic-based smart skins and structures. We have demonstrated the applicability of our sensor in a hostile environment, and some of its technical benefits over conventional techniques. Further developments will prove its economical benefits by reducing the cost per channel and increasing the density by spatial multiplexing.

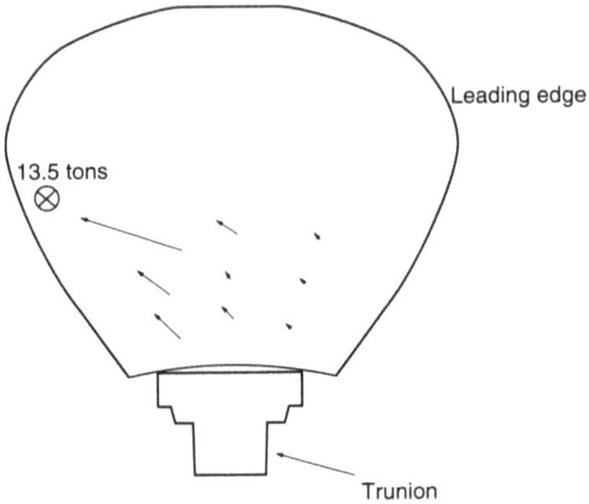

Figure 6. Major strains computed from 9 rosettes strain measurements with an applied load of 13.5 tons. The rosette close to the applied load show a strain magnitude of 335 $\mu\varepsilon$.

ACKNOWLEDGEMENTS

The authors would like to express their appreciations to the Centre de développement des transports du Canada for their financial supports.

REFERENCES

1. E.F. Crawley, Intelligent Structures - A Technology Overview and Assessment, AGARD Conference proceeding 531, (1993).
2. US patent #5,202,939.
3. K.A. Murphy, M.F. Gunther, A.M. Vengsarkar, and R.O. Claus, Quadrature Phase-Shifted, Extrinsic Fabry-Perot Optical Fiber Sensors, Opt. Lett., p. 273, (1991).
4. J.L. Lesko, G.P. Carman, B.R. Fogg, W.V. Miller III, A.M. Vengsarkar, K.L. Reifsnider, and R.O. Claus, Embedded Fabry-Perot Fiber Optic Strain Sensors in the Macromodel Composites, Opt. Eng., p. 13, (1992).
5. C. Belleville and G, Duplain, White-Light Interferometric Multimode Fiber-Optic Strain Sensor, Opt. Lett., p. 78, (1993).
6. J.S. Sirkis, Unified Approach to Phase-Strain-Temperature Models for Smart Structure Interferometric Optical Fiber Sensors: Part 2, applications, Opt. Eng., p. 762, (1993).

A DISTRIBUTED FIBRE OPTIC SENSING SYSTEM FOR HUMIDITY MEASUREMENT

A. Kharaz and B.E. Jones

The Brunel Centre for Manufacturing Metrology
Brunel University
Uxbridge, Middlesex UB8 3PH
United Kingdom

ABSTRACT

A novel multi-point distributed humidity sensing system has been constructed based on the principle of using the absorption spectrum of a colorimetric reagent (cobalt chloride) immobilised on the surface of the core of a multimode optical fibre by employing a thin gelatin film. Two-wavelength detection is employed to provide intensity referencing. Light attenuation at the reference wavelength is only 0.2 dB. The basic design, construction and testing of an experimental sensor in the humidity region 20 to 80 % RH and temperature range 25 to 50°C is described. Resolution can be better than 2% RH and time constant 1 second. Repeatability is at least as good as the resolution of the measurement. A two-wavelength optical time domain reflectometer (OTDR) has been developed and a four-sensor network has been built. The resolution of measurement on the network can be better than 4 % RH, but because of noise in the OTDR the resolution value is dependant on averaging time in the instrument.

INTRODUCTION

The measurement of relative humidity (RH) is rapidly increasing in importance in physical, chemical and biological processes as industry attempts to improve quality and production rates by better control of RH. Typical examples occur in the food, textile, chemicals, electronics and pharmaceutical industries. Fibre optic humidity sensors would not only have the advantages of freedom from electromagnetic interference, intrinsic safety and remote monitoring but offer the prospect of distributed sensing using the fibre itself as the sensor, and so it is not surprising that several optical fibre humidity sensing methods have been reported. These methods are mainly based on one of the following phenomena: colorimetric reagents that change colour with relative humidity[1]; change in physical size with humidity[2]; change of refractive index with humidity[3]; change of reflectivity of thin metal films with humidity[4]; change of florescence intensity with humidity[5].

BASIC SCHEME

The sensor described here[6] is based on the colorimetric interaction of cobalt chloride with water molecules. The anhydrous salt is blue and turns pink as water of hydration converts the salt to hexahydrate[1]. In the sensor cobalt chloride is immobilised on the surface of the core of the optical fibre using gelatin film with a slightly higher refractive index than that of the fibre core. Because of this light in the optical fibre will travel through the film.

As the external humidity level alters so does the moisture content of the film and the light absorption in the sensitive wavelength region. Thus attenuation experienced by the wavelength signal in this region will give an indication of humidity level, while the second wavelength signal is used to eliminate the common mode variations, notably temperature variation. Because the optical fibre is used to make the measurement this novel humidity sensor is an intrinsic optical fibre device.

A single optical fibre sensor was constructed using 600μm core HCS (hard clad silica) optical fibre. At the sensing point along the fibre the cladding was removed over 50mm and replaced by the cobalt chloride-gelatin thin film, as shown in figure 1. Since the refractive index of the film is greater than that of the fibre core, practically all light launched through the fibre will transverse through the film thus interacting with cobalt chloride molecules and giving good sensitivity to changes in atmospheric moisture.

A distributed sensing network can be constructed for a series of measurands that effect the attenuations of the optical fibre at various points along its length[7]. Thus a series of humidity sensors as previously described can be sited at various points along the optical fibre. An optical time domain reflectrometer (OTDR) technique can be used to establish the optical attenuation values of the multiple sensors to establish humidities at the various points. In this method a pulsed laser diode is coupled to the optical fibre via a directional coupler which also allows the return backscattered light (a function of localised attenuations) to be collected by a photo-diode detector.

An OTDR utilising two pulsed laser diodes operating at 670nm and 850nm, respectively, has been developed. For demonstration purposes a four-sensor network has been constructed using a 200μm core HCS optical fibre; the length of optical fibre from the OTDR to the first sensor is 20m and the distance between sensors is 20m.

EXPERIMENTS AND RESULTS

White light was launched into the sensing fibre and the intensities of spectra of the collected light were measured using a spectrum analyzer. The sensor (using 600μm core) was placed in an environmental chamber, and two spectral responses were obtained at two different RH% values, while keeping the temperature constant at 36°C (figure 2). It can be seen that a change of relative humidity effects the spectral absorption of the sensor in the wavelength region 600-740nm. Hence a second wavelength outside this region (at, say, 850nm) can be used to provide an intensity reference to enable rejection of changes in common mode light intensity variations (as for example caused by temperature variations). The light attenuation of the sensor at a wavelength of 850nm is 0.2dB.

With the sensor placed in the environmental chamber, the relative humidity was varied again with the temperature now kept constant at 25°C. At each humidity value the intensities of the collected light at the two wavelengths 670nm and 850nm were noted. The relative intensity (RI) plotted in the graph of figure 3 was obtained by dividing the intensity of light at 670nm by that at 850nm.

This graph can be considered to have three different operating regions, 0-40,40-70 and 70-100 %RH. The average change in the RI per 1% RH change for these three regions are 0.001, 0.007 and 0.002 RI units per 1% RH, respectively. The resolution of relative intensity measurement was about 1% relative intensity, determined by the resolution of the spectrum analyzer. When the experiment was repeated for a range of temperatures 25-50 °C no discernible change in these results was noted.

Repeatability tests have been undertaken over a 3-month period during which time the sensor has been exposed to cycles of humidity change in the operating range 20 to 80 %RH; repeatability has always been within the resolution limit. The time constant of the sensor, defined as the time taken to reach 90% of the final value, was found to be approximately 1 second. This measurement was achieved by applying two moisture streams (70% RH and dry 20% RH) alternatively at the sensor. The resolution of the sensor is a function of the $CoCl_2$ concentration within the film and was found to be better than 2% at RH greater than 40% RH, and better than 4% below 40% RH.

The sensors of the network were place in the humidity chamber and have been exposed

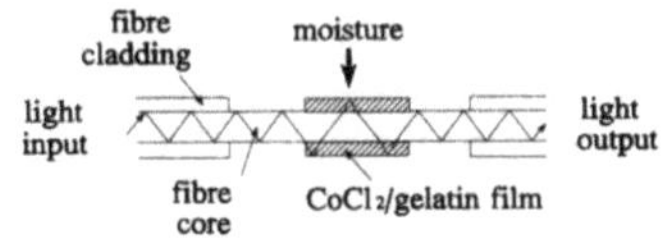

Figure 1. Basic sensor construction.

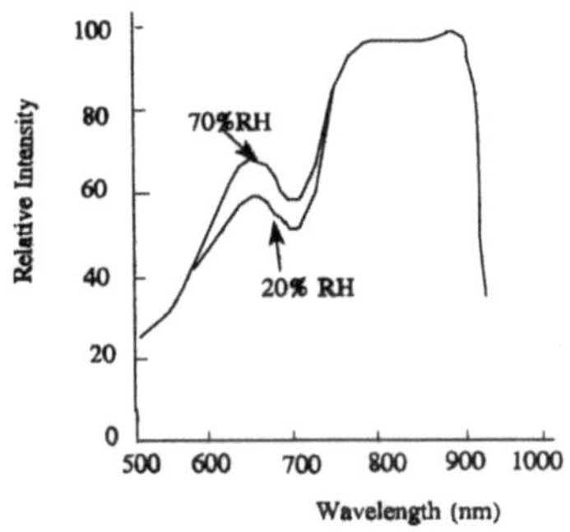

Figure 2. Spectral response of sensor at two RH % values.

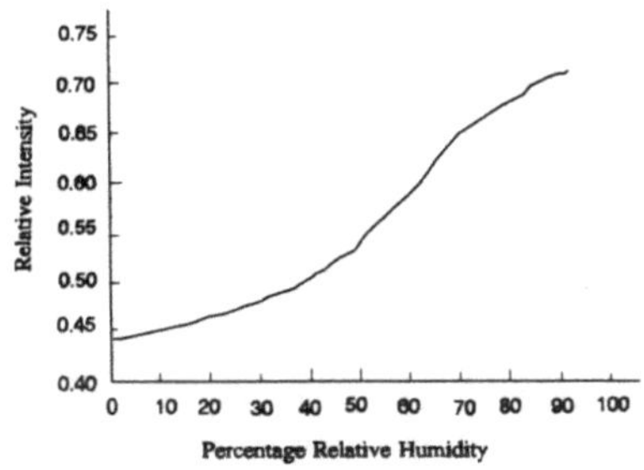

Figure 3. Sensor Characteristics.

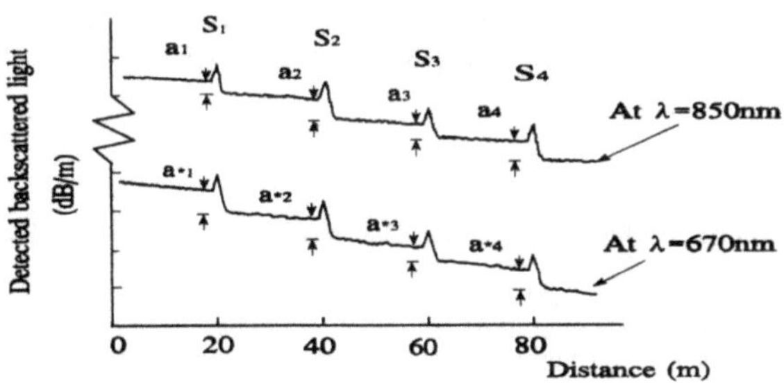

Figure 4. Typical OTDR trace, $a_1..a_4$ and $a^*_1..a^*_4$ are optical attenuations of sensors $S_1..S_4$ at wavelength 850nm and 670nm, respectively.

to cycles of humidity in the operating range 20 to 80 %RH. At each humidity level the attenuation caused by the sensors was measured at 670nm then at 850nm using the OTDR. A typical OTDR trace is shown in figure 4. From these two values a relative attenuation (RA) in dB has been calculated and tabulated in Table 1; the standard deviation had a value of 0.3 dB, caused by noise, when averaging over 1 second, but when averaging was taken over 2 minutes the standard deviation was reduced to 0.03 dB. The peaks of reflected light at each sensor are due to Fresnel reflections.

Table 1. Characteristics of the four sensor network

RH%	RA of S_1	RA of S_2	RA of S_3	RA of S_4
20	3.2	2.8	2.3	2.4
30	3.1	2.7	2.2	2.3
40	2.8	2.6	2.1	2.2
50	2.5	2.4	2.0	2.1
60	2.3	2.1	1.9	1.8
70	1.8	1.6	1.5	1.6
80	1.4	1.3	1.6	1.5

CONCLUSIONS

A novel multi-point distributed humidity sensing system has been developed. It is expected that more than 10 sensors can be sited on the single multimode optical fibre which might be for example 1km in length, the sensors could be placed within several meters of one another, if necessary. There are a number of applications where installation of a distributed humidity sensing system would be far easier than using multiple electronic humidity sensors. Further work is required to establish the long-term life of the sensor and to engineer an arrangement to minimise any effects due to atmospheric dirt and pollution, to improve the signal-to-noise ratio and to increase the operational temperature range.

ACKNOWLEDGEMENTS

The authors wish to acknowledge support for this work received from the UK Science and Engineering Research Council, the UK National Physical Laboratory, and Professor R.C. Spooncer.

REFERENCES

1. A.P. Russell, and K.S.Fletcher, Optical sensor for the determination of moisture, *Anal. Chim. Acta.* 170:209 (1985).
2. R.C. Spooncer, F.A. Al-Ramadhan and B.E. Jones, A humidity sensor using a wavelength-dependent holographic filter with fibre optic links, *Intern. J. of Optoelectronics* 7:449 (1992).
3. F. Mitschke, Fibre-optic sensor for humidity , *Optics Letters* 14: 967 (1989).
4. A.D. Stuart and P.E. Grazer, A fibre-optic relative humidity sensor, *Intern. J. of Optoelectronics,* 3:177 (1988).
5. C. Zhu, F.V. Bright, W.A. Wyatt and G.M. Hieftje, A new fluorescence sensor for quantification of atmospheric humidity, *J. Electrchem. Soc.*, 136:567 (1989).
6. A. Kharaz and B.E. Jones, A intrinsic optical fibre sensor,*Intern. Conf. on Moisture and Humidity Measurements (Sira Ltd, Chislehurst)* 1:1.1 (1993).
7. M.K. Banoski and S.M. Jensen, Fibre waveguide: a novel technique for investigating attenuation characteristic *Applied Optics*, 15:2112 (1976).

FRINGE BEATING EFFECTS IN A WHITE-LIGHT INTERFEROMETER

Y. N. Ning, K. T. V. Grattan, and A. W. Palmer

Department of Electrical, Electronic and Information Engineering
City University, London EC1V OHB, U.K.

ABSTRACT

The profile of the interference fringe packet obtained in a white light interferometer (WLI) may be dramatically altered by the so-called "fringe beating effect", as a result, the peak value of the central fringe will be equal to or even less than that of the other fringes and in consequence the WLI scheme will give an erroneous result for the determination of the central fringe position. This effect is analyzed theoretically using a computer simulation model, and the results from both the computer simulation and experiment are discussed and compared.

INTRODUCTION

A typical WLI system comprises a sensing interferometer, which converts the measurand into a variation in the optical path difference (OPD) and also a reference interferometer which measures this OPD by means of determining the position of the central fringe in the zero-order interference region[1,2] of the fringe pattern produced. This occurs when the values of the OPDs of both interferometers are the same and the fringe with the maximum intensity peak value may be obtained[3,4]. In order to increase the measurement accuracy in the determination of the measuand itself, it is important to be able to identify accurately the central fringe from the whole fringe packet and determine its corresponding position, in the so-called "fringe-order identification" problem[5,6].

When a conventional white-light (Michelson) interferometer is operated under ideal conditions i.e., the light beam is well collimated, and the two mirrors are normal to each other, if the frequency and amplitude of vibration of the mirror are carefully adjusted such that the amplitude of the vibration is slightly larger than that of the value of source coherence length, the whole fringe packet in the zero order interference region can be observed. Its profile shows a symmetric, smooth, distributed pattern which may be described by either a Lorentzian or a Gaussian distribution function[4,7].

If a WLI system is operated under non-ideal conditions, i.e., where the light beam is not perfectly collimated due to the quality of the collimation lens or its alignment, and the mirrors are slightly displaced at an angle, more typical of many industrial measurement situations, then the interference fringe pattern will exhibit some degree of distortion of its profile, and as a result, the accuracy of identifying the position of the central fringe will be directly affected. Since it is highly possible for such a non-ideal WLI system to be found in many practical applications, with or without the knowledge of the operator, its performance needs to be carefully examined.

In this work, it will be shown that under the certain conditions, the profile of the interference fringe packet may be altered by the so-called "fringe beating effect". In the worst cases, the peak value of the central fringe will be equal to or even less than that of the other fringes and in consequence the WLI measurement scheme in which the fringe of maximum intensity is identified with the central fringe cannot be operated accurately.

FRINGE BEATING EFFECT - EXPERIMENTAL RESULTS

It has been shown that if a single mode laser diode is driven with a current which is slightly less than the threshold value for lasing of the diode, its output will exhibit a multi-longitudinal mode and it can

Applications of Photonic Technology
Edited by G.A. Lampropoulos *et al.*, Plenum Press, New York, 1995

be used as a multimode laser diode[5]. In this work, a single mode laser diode (Sharp LT022MCO) was operated in the vicinity of its threshold current as a multimode light source. Light from the laser diode was collimated into a parallel beam which was divided into two by a beam splitter, and then were normally reflected by two mirrors in the two interference arms respectively, and finally recombined and detected by a photodiode. Under ideal operating conditions, a fringe profile pattern with a symmetric, smooth Gaussian distribution can be observed. However, if the intensity across the spot of the collimated light beam is non-uniform and the diameter of the spot is relatively large compared with the detection area of the photodiode, the two mirrors in the interferometer is not perpendicular to each other, and the plane of the vibrating mirror is not always normal to the light beam, the fringe beating effect may be observed in the experimental arrangement. Figures 1a, 1b, and 1c show the recorded interference fringe patterns obtained, revealing that the fringe profiles were modulated by the presence of the fringe beating effect, corresponding respectively to the different situations discussed above. It can be seen that under certain conditions the modulated profile obtained makes it more easily to identify the central fringe (e.g. fig. 1a); however, for other cases, the central fringe cannot be identified (e.g. fig. 1b) and an error in determining the central fringe could easily occur (e.g. fig. 1c). Clearly, this fringe beating effect not only can severely affect the operation of the WLI, but also give faulty information about the position and nature of the central fringe in the whole fringe pattern.

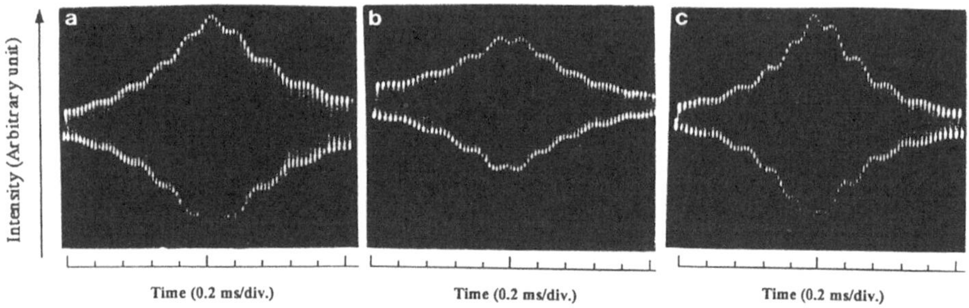

Figure 1. Photodiode outputs when the interferometer is adjusted at three different sets of the non-ideal operation conditions (Scale: x: 0.2 ms/div; y: arbitrary units).

THEORETICAL CONSIDERATIONS AND THE COMPUTER SIMULATION

In order to analyze theoretically a Michelson interferometer which is operated under the non-ideal conditions discussed, it is assumed that the light beam from the collimator can be represented by two separated light beams which are incoherent with each other, as shown in fig. 2. Thus each of these light beams generates a set of interference fringes at the output of the interferometer, and the superposition of the fringes is detected by the photodiode. If the two mirrors are not normal to each other, there is an additional OPD introduced between these two light beams. If only the zero-order interference region in the output of the interferometer is considered, the intensity detected by the photodiode can be written as

$$I_{output}=I_1\exp[-(2\,|\Delta L_1|/L_c)]\cos(2\pi\Delta L_1/\lambda)+I_2\exp[-(2|\Delta L_2|/L_c)]\cos(2\pi\Delta L_2/\lambda) \quad (1)$$

where I_1 and I_2 are the amplitudes of the zero-order fringes generated by the two separated light beams respectively, and λ and L_c are the central wavelength and the coherence length of the laser diode respectively. ΔL_1 and ΔL_2 are values of the OPDs induced in the two light beams respectively, and it was assumed that the relative position of the two mirrors is determined by the variations of the OPDs, given by

$$\Delta L_2 = k\Delta L_1+\Delta L_0 \quad (2)$$

where ΔL_0 is the OPD induced in the second light beam when ΔL_1 is zero, and k is a constant which determines the relative variation in the OPD between ΔL_1 and ΔL_2.

Based upon this simple theoretical model, by altering the values of the relative intensities of I_1 and I_2, and the values of k and ΔL_0, by which the relative mirror position may be defined, the characteristics of the fringe beating effect can be analyzed and simulated with a simple computer program. Under ideal operating conditions, the two mirrors are normal to each other, i.e., k=1 and ΔL_0 =0, and hence no fringe beating effect is observed. If the two mirrors are not perfectly normal to each other, but the variations of

ΔL_1 and ΔL_2 are the same, i.e., k=1 and $\Delta L_0 \neq 0$, it can be shown that the fringe beating effect still cannot be obtained. By contrast, for non-ideal conditions, where the variations of ΔL_1 and ΔL_2 are not the same, i.e., $k \neq 1$, the fringe beating effect can be observed, and the shape of the modulated fringe profile may be determined by the values of the relative intensities between I_1 and I_2, and the values of k and ΔL_0. Figures 3a, 3b and 3c show a series of simulated intensity profiles of the fringe in the zero interference region for the cases where the coherence length, L_c, is 30 mm, I_1 is 0.88, I_2 is 0.12, and the value of ΔL_1 varies from -30 to 30 mm. Figure 3a shows the modulated fringe profile when k=1.15 and ΔL_0 =0. This is the condition that occurs when the OPD is zero (the balance position), and the two mirrors are parallel to each other, whilst when the vibrating mirror moves away from the balance position, the OPDs induced in the two

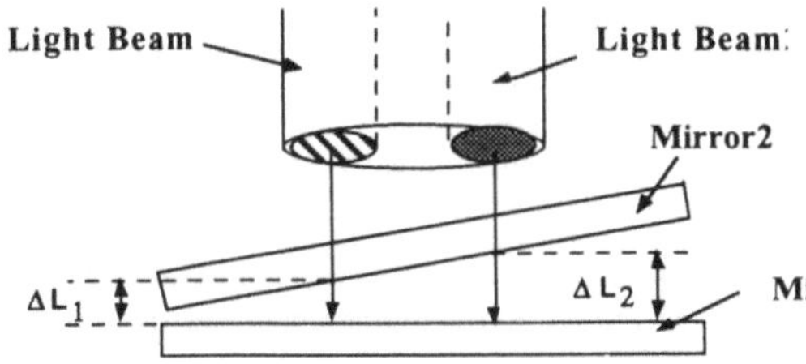

Figure 2. Schematic of the theoretical model used to analyze the fringe beating effect.

separated beams are different. Comparing this with figure 1a, it can be seen that if this type of fringe profile is obtained, the identification of the central fringe from the whole fringe packet becomes easier. However, for the case of k=1.15, and ΔL_0 =390 nm (half of the central wavelength), the intensity value of the central fringe becomes less than that of the other adjacent fringes (as shown in figure 3b) and hence the central fringe cannot be identified simply from the maximum intensity value of the output profile. For a more general case where k=1.15 and ΔL_0 =200 nm, the simulated pattern is shown in figure 3c. It can be seen that the fringe with the maximum intensity value has moved away from the position where ΔL_1 is zero. In this case, an erroneous result may be obtained if the maximum intensity value of the fringe is chosen to represent the central fringe in a measurement situation. Since the profile of the fringe packet is asymmetrical, the error caused in this case cannot be corrected directly by calculating the centre of gravity of the fringe profile.

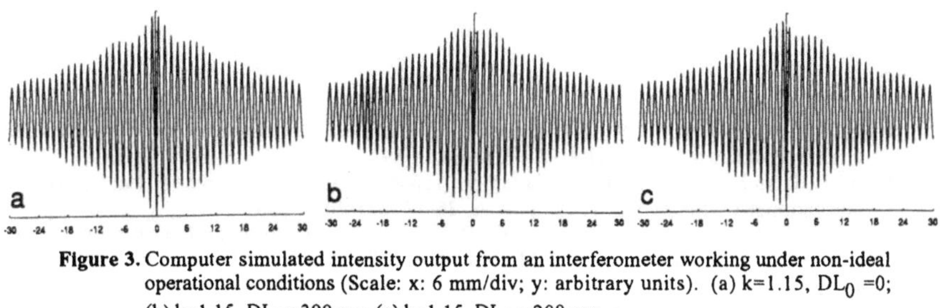

Figure 3. Computer simulated intensity output from an interferometer working under non-ideal operational conditions (Scale: x: 6 mm/div; y: arbitrary units). (a) k=1.15, DL_0 =0; (b) k=1.15, DL_0 =390 nm; (c) k=1.15, DL_0 =200 nm.

COMPARISON AND DISCUSSION

Comparing figures 1 with 3, it can be seen that the patterns of the fringe profile obtained from both experiment and computer simulation are in a good agreement with each other, showing that the peak value of the central fringe in the fringe beating patterns does vary considerably, corresponding to the relative position of the mirrors. This implies that when an interferometer is illuminated with a not-well-collimated laser beam with its intensity distribution across the output laser spot being non-uniform and the mirrors in the interferometer not being well aligned, the fringe beating effect will occur and the interpretation of the measuement from the WLI system in a real measurement situation can result in error. Based upon the theoretical model and the resulting computer simulation, the parameters of the arrangement which cause the fringe beating effect can be identified and thus a corresponding adjustment can be made, to ensure the WLI system operates in the optimal way.

It should be pointed out that one of the important assumptions in a typical theoretical model of the fringe beating effect is that the intensity across the spot of the collimated light beam is non-uniform and hence the light beam can be represented by two separated incoherent beams. In the experiment carried out in this work, it was found that the spot of the collimated light beam was actually divided into several separated spots where each spot has associated with it its own optical path. This experimental result partially supports the assumption used in the theoretical model. When the laser diode was dismounted from the collimator, only a single fundamental transverse mode laser light output was observed. Thus, this indicates that it was the use of the collimator itself that caused the separated transverse spots and care should be taken to avoid such effects in practical measurement situations.

From the results of the experiment and the associated computer simulation, the conditions under which the fringe beating effect is most likely to occur can be identified. The different regions identified in the spot of the collimated light beam were incoherent with each other. This may be due to the fact that the coherence length of the laser diode is very short (30 mm) and hence a slight OPD introduced between the different regions of the collimated light spot will make them become incoherent. Further, the displacements of different positions of the beam across the plane of the vibrating mirror are slightly different, hence $\Delta L_2 = k\Delta L_1$ and $k \neq 1$. This may happen when the mirror is not mounted exactly at the centre of the vibrating loudspeaker or its movement is not uniform, resulting in the displacements at corresponding positions of the loudspeaker not being the same. The third factor is that the detection area of the photodiode is relative small compared with that of the light spot. This is usually so because the fringe beating effect can only be detected over the area where the two separated light spots overlap with each other and this area is only a fraction of the whole spot size. If the detection area is relative large, the depth of the modulation of the beating profile will be reduced due to the intensity averaging being obtained over the whole spot falling on the detector. Avoidance of these adverse conditions and careful alignment can thus improve considerably the achievement of good fringe identification in WLI-based measurement systems.

In summary, the situation has been investigated both theoretically and experimentally for a WLI system operated under the non-ideal conditions. The results show that a fringe beating effect can occur, and in consequence, the position of the central fringe cannot simply be identified by comparing the magnitudes of fringes inside the profile of the fringe packet, as the WLI scheme could give an erroneous result for the determination of the central fringe position. As a result, the operation of the WLI in a real measurement situation can be dramatically affected. The possible parameters which could lead to the fringe beating effect have also been analyzed and discussed, with a view to their avoidance in practical systems.

ACKNOWLEDGEMENTS

This study is a part of a project supported by the UK Science and Engineering Research Council. The authors are grateful to Dr. K. Weir and Mr. D. N. Wang for their helpful discussions.

REFERENCES

1. Y. N. Ning, B. T. Meggitt, K. T. V. Grattan, and A. W. Palmer, "Characteristics of laser diodes for interferometric use", Appl. Opt. **28**, 3656-60, 1989.
2. A. Koch and R. Ulrich, "Fiber-optic displacement sensor with 0.02 mm resolution by white-light interferometry", Sensors and Actuators A, **25-27**, 201-207, 1991.
3. Y.N. Ning, K.T.V. Grattan, A.W. Palmer and B.T. Meggitt "Coherence length modulation of a multimode laser diode in a dual Michelson interferometer configuration", Appl. Opt. **31**, 1322-7, 1992.
4. Y.N. Ning, K.T.V. Grattan and A.W. Palmer, "Fibre-optic interferometric systems using low coherence light sources", Sensors and Actuators A, **30**, 181-192, 1992.
5. S. Chen, K. T. V. Grattan, B. T. Meggitt and A. W. Palmer, " Instantaneous fringe-order identification using dual broadband source with widely spaced wavelengths", Electron. Lett. **29**, 334-335, 1993.
6. Y. J. Rao, Y. N. Ning, and D. A. Jackson, "Synthesized source for white-light sensing system", Opt. Lett. **18**, 462-464, 1993.
7. A. S. Gerges, F. Farahi, T. P. Newson, J. D. C. Jones and D. A. Jackson, "Fibre-optic interferometric sensors using a low coherence source: dynamic range enhancement", International J. of Optoelectronics, **3**, 311-322, 1988.

ACTIVE WAVELENGTH DEMODULATION FOR BRAGG GRATING STRAIN SENSORS

T. Coroy and R. M. Measures

University of Toronto
Institute for Aerospace Studies
4925 Downsview, Ontario
Canada M3H 5T6

ABSTRACT

An active wavelength demodulation system based on an acousto-optic tunable filter has been demonstrated, suitable for use with intracore Bragg grating fiber optic sensors. A wavelength measurement precision of 0.02 nm was achieved for this initial system. Potential for further system improvement is discussed.

INTRODUCTION

Of the various optical fiber based sensors that have been developed to date, the intracore Bragg grating strain sensor has been identified as one of the most promising[1,2,3]. In the usual case where the Bragg grating is used for its reflective rather than its transmissive properties, the strain imposed on the Bragg grating is determined by measuring the wavelength of the narrow linewidth of light that the grating reflects. It is evident that any practical system using these sensors will require a means of measuring the wavelengths of their narrow linewidth back-reflected optical signals.

One such wavelength demodulation system (WDS) was demonstrated by Melle *et al.*, using a ratiometric technique with a passive wavelength filter[3]. The simplicity of this passive approach makes it attractive for a number of applications, but the passive filter imposes a fundamental trade-off between wavelength measurement range and resolution. This limitation can be overcome by using an active, tunable filter approach, wherein a high resolution may be maintained over a wide tuning range.

Such an active wavelength demodulation system (AWDS) can be built using an acousto-optic tunable filter (AOTF). In 1993, M. G. Xu *et al.* proposed an AWDS which used a frequency shift keying approach to track the wavelength of a Bragg grating[4]. However, in their article they presented results for open loop operation only; thus they did not in fact demonstrate an *active* WDS, although they did propose one and present experimental results to illustrate how it might work. At about the same time, J. Dunphy *et al.* reported demonstration of a tracking approach AOTF based AWDS[5], but no details on their system or their results were reported.

In this paper we disclose our initial results for a closed-loop tracking AOTF based AWDS, and discuss possibilities for further system improvement.

EXPERIMENT

The experimental setup for this work is shown in Fig. 1. Note that although the primary intended use for this AWDS is to demodulate signals from Bragg grating fiber optic sensors, it may also be used to measure the wavelength from another narrow linewidth source such as a tunable laser. Here a Santec TSL-300 tunable laser was used to provide an optical signal, whose wavelength, power, and spectral qualities were monitored by a Hewlett

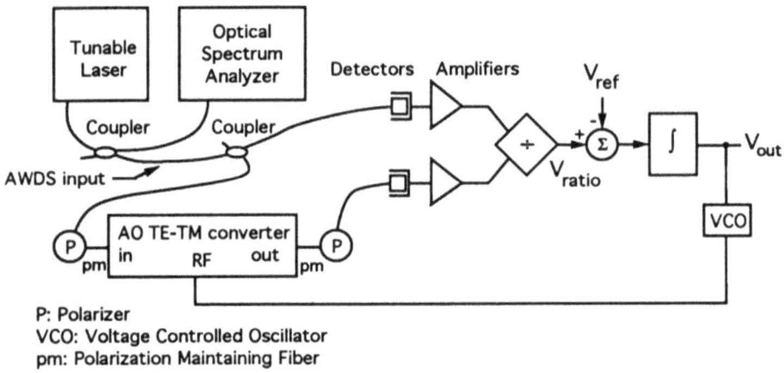

Figure 1. Experimental setup for demonstration of an AOTF based AWDS.

Packard 70951A optical spectrum analyzer. The AOTF was made up of an acousto-optic TE-TM converter made by New Focus, and two crossed polarizers (one made by Oz Optics and one by Wave Optics). A Hewlett Packard 8656B high frequency signal generator was used as a limited range voltage controlled oscillator (VCO).

The state diagram of Fig. 2 illustrates the operation of the AOTF based AWDS. A ratiometric signal, V_{ratio}, is obtained from the filtered and unfiltered detector signals to provide intensity self-referencing. This signal is continuously compared with a reference,

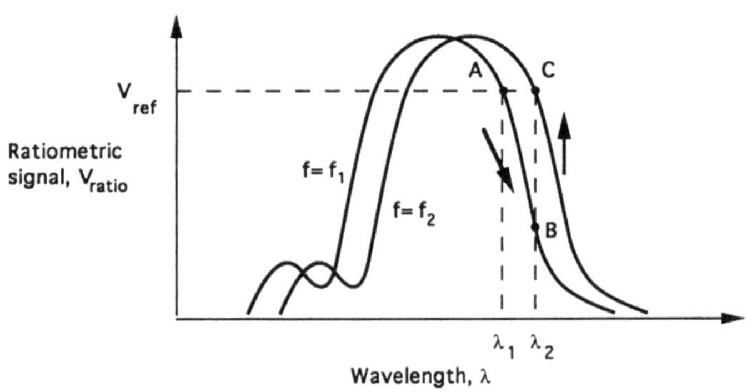

Figure 2. System state diagram illustrating wavelength tracking by an AOTF based AWDS. The curves shown represent the transmission functions of the AOTF for input RF signals of frequencies f_1 and $f_2 < f_1$.

V_{ref}. For an input wavelength of λ_1, when V_{ratio} equals V_{ref} the system is in equilibrium (state point A), and the system provides an RF signal of frequency f_1 to the AOTF. Now consider a change from input wavelength λ_1 to λ_2. V_{ratio} will drop below V_{ref} (state point B). The system responds by increasing the RF signal frequency input to the AOTF, shifting the spectral response until equilibrium is re-established (state point C). The magnitude of the bias applied to the VCO is taken as the output, from which the input wavelength can be determined.

RESULTS

The wavelength and power responses of the AWDS are shown in Figs. 3 and 4. The response of the AWDS was linear with wavelength and independent of input power within the error of the measurements taken. The measurement precision was 0.02 nm, equivalent to about 17 microstrain at around 1550 nm[6]. The error associated with the output of the AWDS was to a large part due to two factors. First, the AOTF used was extremely lossy, due to high loss in the acousto-optic TE-TM converter. The combined peak transmission of the tunable bandpass filter (including the TE-TM converter and the two polarizers) was only about -20 dB, whereas with New Focus' standard commercial TE-TM converter -10 dB or slightly better should be achievable. This high loss in the AOTF resulted in much lower light levels at one of the detectors, and thus translated into lower signal to noise ratios.

The other dominant factor contributing to the error in the AWDS output was that the electronics used were fairly elementary, and susceptible to 60 Hz pickup noise. We estimate that through the use of a better TE-TM converter and with minor improvements to the electronics, the resolution of the system could be significantly improved.

It should be noted that the experimental system, as it is shown in Fig. 1, is polarization sensitive. For the measurements reported, the input light was made polarization stable, and so polarization sensitivity was not a problem. In order to make this system polarization insensitive, at the very least a polarizer should be placed at the input to the AWDS. However, this does not protect against signal fading, which will arise if the input

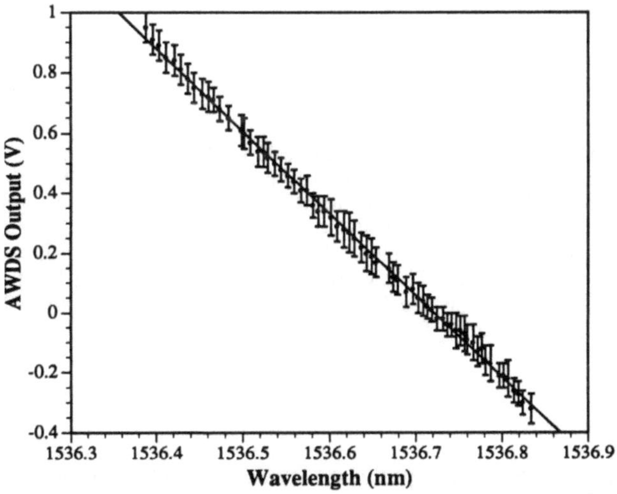

Figure 3. Wavelength response for the AOTF based AWDS. Data was taken over the wavelength range that the Santec tunable laser could be continuously tuned without mode-hopping or dual peak effects. Linear fit yields $f(V) = -2.738\ \lambda(nm) + 4207.532$, $R^2 = .9986$.

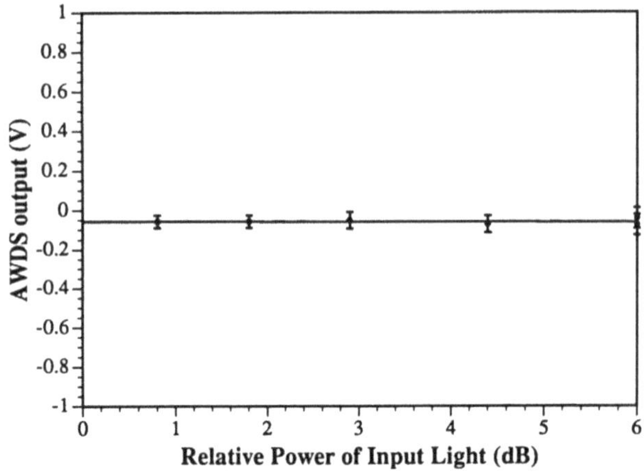

Figure 4. Power response for the AOTF based AWDS, for 1536.812 ± .003 nm light.

polarization is orthogonal to that transmitted by the polarizer. The best solution is to use a polarization insensitive AOTF[7] in place of the TE-TM converter and polarizers used in our experiments.

Finally, we note one key advantage of the ratiometric approach over the frequency shift keying approach of Xu *et al.* [4]. Consider that the frequency shift keying approach suggested in Ref. 4 makes no amplitude distinctions in attaining a locked-on state. Rather, it locks on whenever the dip (or peak if the Bragg is used in reflective mode) of the input Bragg grating spectrum overlaps a peak in the transmission spectrum of the AOTF. However, most AOTFs have secondary peaks, something like side lobes, besides their primary peaks. These secondary peaks are generally kept much lower in amplitude (roughly 9 dB to 15 dB) by the manufacturers of the AOTFs, but are nonetheless there. Unfortunately, the frequency shift keying approach of Ref. 4 has no provision for differentiating between primary and secondary peaks, and so may end up locking on using the wrong peak of the AOTF transmission spectrum. The ratiometric approach described in this work, on the other hand, does not suffer from this problem.

ACKNOWLEDGMENTS

We wish to offer warm thanks to Santec USA Corp. for the use of their TSL-300 tunable laser; to New Focus, Inc. for loaning us the acousto-optic TE-TM converter and one of the polarizers used in these experiments; to MPB Technologies, Inc. for providing us with a second polarizer; and to Peter Smith of the University of Toronto for the use of his HP 8656B high frequency signal generator. We also thank Kathy Li-Dessau for helpful discussions regarding the AOTF. This work was partially supported by the Ontario Laser and Lightwave Research Center, the Institute for Space and Terrestrial Science and the National Sciences and Engineering Research Council of Canada.

REFERENCES

1. R. M. Measures, "Smart composite structures with embedded sensors," *Composites Engineering*, Vol. 2, pp. 597-618 (1992).
2. R. M. Measures, S. Melle and K. Liu, "Wavelength demodulated Bragg grating fiber optic sensing systems for addressing smart structure critical issues," *Smart Mater. Struct.*, Vol. 1, pp. 36-44 (1992).

3. S. M. Melle, K. Liu and R M Measures, "Strain sensing using a fiber optic Bragg grating," Proc. SPIE, Vol. 1588, pp. 255-263 (1991).
4. M. G. Xu, H. Geiger, J. L. Archambault, L. Reekie and J. P. Dakin, "Novel interrogating system for fiber Bragg grating sensors using an acousto-optic tunable filter," *Electron. Lett.*, Vol. 29, pp. 1510-11 (1993).
5. J. Dunphy, G. Ball, F. D'Amato, P. Ferraro, S. Inserra, A. Vannucci, and M. Varasi, "Instrumentation development in support of fiber grating sensor arrays," Proc. SPIE, Vol. 2071, pp. 2-11 (1993).
6. The conversion factor used is 1.20 pm/με, from S.M. Melle, A.T. Alavie, S. Karr, T. Coroy, K. Liu and R.M. Measures, "A Bragg grating-tuned fibre laser strain sensor system," *IEEE Photonics Technol. Lett.*, Vol. 5, pp. 263-266 (1993).
7. For example, see A. d'Alessandro, D. A. Smith and J. E. Baran, "Polarisation-independent low-power integrated acousto-optic tunable filter/switch using APE/Ti polarisation splitters on lithium niobate," *Electron. Lett.*, Vol. 29, pp. 1767-9 (1993).

ANALYSIS OF MISALIGNMENT SENSITIVITY OF A HI-BI FIBER-BASED INTERFEROMETRIC DISPLACEMENT SENSOR

U. Minoni, G. Scotti, F. Docchio

Dipartimento di Elettronica per l'Automazione, Facoltà di Ingegneria
Università degli Studi di Brescia, Via Branze 38, I-25123 Brescia, Italy

INTRODUCTION

The use of birefringent (Hi-Bi) optical fibers makes it possible to implement interferometric displacement sensors which are miniaturized, flexible, and accurate over an extended measuring range. With respect to ordinary interferometric systems based on free propagation of light, these sensors provide a higher geometrical flexibility, which is particularly appealing with reconfigurable measuring machines. As recently presented[1], it is possible to keep the laser source and the photodetectors remote from the measuring head using a HiBi optical fiber. Theoretically, the approach presented has proven to be insensitive to fiber bending and to fiber temperature changes. In fact, HiBi fibers provide adequate control of the polarization state of the beams, if the input polarization state is correctly adjusted with respect to the orientation of the fiber polarization axes. However, residual angular misalignment of the fiber ends due to the complex procedure needed to identify the fiber optical axis[2,3], and a non perfect polarization separation of the laser beam may result in a sensitivity of the measuring system to fiber perturbations. The sensitivity to such misalignments must therefore be taken into account for high precision measurements: in this work an analysis of the residual sensitivity to fiber thermal and mechanical stresses due to angular misalignment fiber end is presented.

SENSOR DESCRIPTION

Figure 1 shows the optical layout of the system. A linearly polarized laser beam (with the polarization axis parallel to the drawing plane) is amplitude-divided by beam splitter BS1: one of the components is used to feed the sensor head. GRIN lens L1 is used to launch this beam into the HiBi optical fiber mounted, at the input end, with its fast axis parallel to the beam polarization. The Hi-Bi fiber is used (i) to deliver the polarized laser beam to the sensor head, and (ii) to deliver the interference signals back to the photodetectors. The measuring head is based on a Michelson interferometer. The output fiber end is oriented with its fast axis parallel to the drawing plane. The beam exiting the fiber is collimated by GRIN lens L2 and then amplitude-divided by beam splitter BS2. One of the resulting beams is directed to the signal arm and the other to the reference arm. A $\lambda/8$ retarding plate is inserted in the signal arm, to provide a returning beam circularly polarized; a λ/4 retarding plate is inserted in the reference arm, to provide a returning beam linearly polarized with the polarization plane at 45° with respect to the drawing plane. The two beams recombine at beam splitter BS2. Two interference signals (one from the parallel and one from the perpendicular polarization plane) out of phase by 90° are produced. The fiber, as the return path, keeps these two signals separated. Finally, the polarizing beam splitter PBS is used to direct the two signals to the photodetectors (PD1, PD2).

THEORY OF OPERATION

From the analytical description based on Jones matrix formalism[1], the interferometer is described by the following matrix equation:

$$\mathbf{E}_f \propto \mathbf{F}\begin{bmatrix} 1+e^{i2\Phi} & 1+e^{i(2\Phi+\pi/2)} \\ -\left(1+e^{i(2\Phi-\pi/2)}\right) & 1-e^{i2\Phi} \end{bmatrix}\mathbf{F}\mathbf{E}_{in} \tag{1}$$

where $\mathbf{E}_{in}=\begin{bmatrix}\mathbf{E}_x\\ \mathbf{E}_y\end{bmatrix}=\begin{bmatrix}1\\0\end{bmatrix}$ describes the electric field of the laser beam at the input end of the fiber, $\mathbf{F}$ is the matrix describing the fiber in both forward and backward direction, and $\mathbf{E}_f$ is the electric field vector in front of the polarizing beam-splitter of Figure 1. $\mathbf{F}$ has the form:

$$\mathbf{F}=\begin{bmatrix} \cos\alpha & \sin\alpha \\ -\sin\alpha & \cos\alpha \end{bmatrix}\begin{bmatrix} 1 & 0 \\ 0 & e^{i\delta} \end{bmatrix}\begin{bmatrix} \cos\beta & -\sin\beta \\ \sin\beta & \cos\beta \end{bmatrix} \tag{2}$$

where α is the fast axis orientation of the input fiber end with respect to the X axis (a left-hand XYZ coordinate system is assumed), β is the fast axis orientation of the exit fiber end with respect to the X axis, and $\delta=\delta_{fast}-\delta_{slow}$ is the phase difference between the E_X and E_y components of the beam after traveling through the fiber. Obviously, this phase difference is a function of the fiber stress induced by fiber bending and by temperature

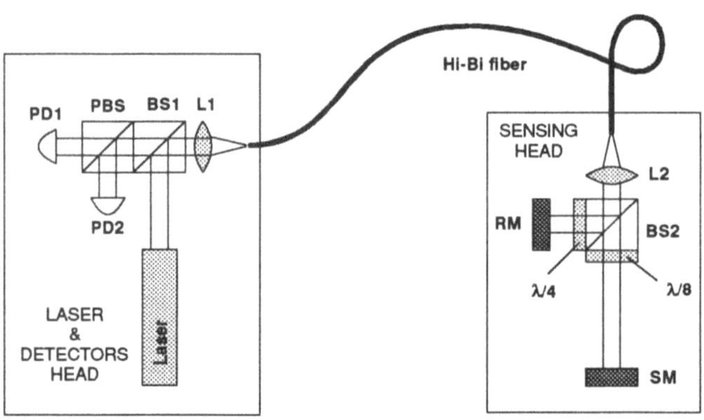

Figure 1. Optical layout of the displacement measuring system. Laser: HeNe linearly polarized laser; SM: signal arm mirror; RM: reference arm mirror; λ/8 and λ/4 retarding plates; BS1 and BS2: non-polarizing beam splitters; PBS: polarizing beam splitter; PD1 and PD2: photodetectors; L1 and L2: GRIN lenses.

changes. The signals at the photodiodes obtained by polarization separation are as follows:

$$V_{PD1} \propto E_{fx} E^*_{fx} = (I_1(\alpha,\beta) + a_1(\alpha,\beta)\cos(2\Phi) - a_2(\alpha,\beta)\sin(2\Phi) + e_1(\alpha,\beta,\delta,\Phi)) \tag{3}$$

$$V_{PD2} \propto E_{fy} E^*_{fy} = (I_2(\alpha,\beta) + a_3(\alpha,\beta)\cos(2\Phi) + a_4(\alpha,\beta)\sin(2\Phi) + e_2(\alpha,\beta,\delta,\Phi)) \tag{4}$$

where:

$$I_1(\alpha,\beta) = \cos^2(\alpha)\left[\cos^4\left(\beta+\frac{\pi}{4}\right)+\sin^4\left(\beta+\frac{\pi}{4}\right)+\cos^2\left(2\beta+\frac{\pi}{4}\right)\right]+\frac{1}{2}\sin^2(\alpha)\cos^2(2\beta)+\sin^2(\alpha)\sin^2\left(2\beta+\frac{\pi}{4}\right) \tag{5}$$

$$I_2(\alpha,\beta) = \sin^2(\alpha)\left[\cos^4\left(\beta+\frac{\pi}{4}\right)+\sin^4\left(\beta+\frac{\pi}{4}\right)+\cos^2\left(2\beta+\frac{\pi}{4}\right)\right]+\frac{1}{2}\cos^2(\alpha)\cos^2(2\beta)+\cos^2(\alpha)\sin^2\left(2\beta+\frac{\pi}{4}\right) \tag{6}$$

$$a_1(\alpha,\beta) = 2\cos^2(\alpha)\cos^2\left(\beta+\frac{\pi}{4}\right)\cos\left(2\beta+\frac{\pi}{4}\right)-\sin^2(\alpha)\cos(2\beta)\sin\left(2\beta+\frac{\pi}{4}\right) \tag{7}$$

$$a_2(\alpha,\beta) = 2\cos^2(\alpha)\sin^2\left(\beta+\frac{\pi}{4}\right)\cos\left(2\beta+\frac{\pi}{4}\right)-\sin^2(\alpha)\cos(2\beta)\sin\left(2\beta+\frac{\pi}{4}\right) \tag{8}$$

$$a_3(\alpha,\beta) = \cos^2(\alpha)\cos(2\beta)\sin\left(2\beta+\frac{\pi}{4}\right)+2\sin^2(\alpha)\sin^2\left(\beta+\frac{\pi}{4}\right)\cos\left(2\beta+\frac{\pi}{4}\right) \tag{9}$$

$$a_4(\alpha,\beta) = \cos^2(\alpha)\cos(2\beta)\sin\left(2\beta+\frac{\pi}{4}\right)+2\sin^2(\alpha)\cos^2\left(\beta+\frac{\pi}{4}\right)\cos\left(2\beta+\frac{\pi}{4}\right) \tag{10}$$

$$\begin{aligned}
e_1(\alpha,\beta,\delta,\Phi) = &-\frac{1}{2}\sin(2\alpha)\left[\frac{1}{2}\sin(4\beta)+\cos(4\beta)\right]\cos(\delta)+\frac{1}{2}\sin(2\alpha)\cos(2\beta)\sin(\delta)+ \\
&+\sin(2\alpha)\left[\frac{1}{2}\cos(2\beta)\cos\left(2\beta+\frac{\pi}{4}\right)-\sin\left(2\beta+\frac{\pi}{4}\right)\cos^2\left(\beta+\frac{\pi}{4}\right)\right]\cos(\delta)\cos(2\Phi)+ \\
&-\frac{1}{2}\sin(2\alpha)\left[\cos(2\beta)\cos\left(2\beta+\frac{\pi}{4}\right)+\sin\left(2\beta+\frac{\pi}{4}\right)\cos^2\left(\beta+\frac{\pi}{4}\right)\right]\sin(\delta)\sin(2\Phi)+ \\
&+\frac{1}{2}\sin(2\alpha)\left[\cos(2\beta)\cos\left(2\beta+\frac{\pi}{4}\right)+2\sin\left(2\beta+\frac{\pi}{4}\right)\sin^2\left(\beta+\frac{\pi}{4}\right)\right]\cos(\delta)\sin(2\Phi)+ \\
&+\frac{1}{2}\sin(2\alpha)\left[\cos(2\beta)\cos\left(2\beta+\frac{\pi}{4}\right)-2\sin\left(2\beta+\frac{\pi}{4}\right)\sin^2\left(\beta+\frac{\pi}{4}\right)\right]\sin(\delta)\cos(2\Phi)
\end{aligned} \tag{11}$$

$$\begin{aligned}
e_2(\alpha,\beta,\delta,\Phi) = &\frac{1}{2}\sin(2\alpha)\left[\frac{1}{2}\sin(4\beta)+\cos(4\beta)\right]\cos(\delta)-\frac{1}{2}\sin(2\alpha)\cos(2\beta)\sin(\delta)+ \\
&-\frac{1}{2}\sin(2\alpha)\left[\frac{1}{2}\cos(2\beta)\cos\left(2\beta+\frac{\pi}{4}\right)+2\sin\left(2\beta+\frac{\pi}{4}\right)\sin^2\left(\beta+\frac{\pi}{4}\right)\right]\cos(\delta)\cos(2\Phi)+ \\
&+\sin(2\alpha)\left[-\frac{1}{2}\cos(2\beta)\cos\left(2\beta+\frac{\pi}{4}\right)+\sin\left(2\beta+\frac{\pi}{4}\right)\sin^2\left(\beta+\frac{\pi}{4}\right)\right]\sin(\delta)\sin(2\Phi)+ \\
&+\sin(2\alpha)\left[-\frac{1}{2}\cos(2\beta)\cos\left(2\beta+\frac{\pi}{4}\right)+\sin\left(2\beta+\frac{\pi}{4}\right)\cos^2\left(\beta+\frac{\pi}{4}\right)\right]\cos(\delta)\sin(2\Phi)+ \\
&+\sin(2\alpha)\left[\frac{1}{2}\cos(2\beta)\cos\left(2\beta+\frac{\pi}{4}\right)+\sin\left(2\beta+\frac{\pi}{4}\right)\cos^2\left(\beta+\frac{\pi}{4}\right)\right]\sin(\delta)\cos(2\Phi).
\end{aligned} \tag{12}$$

If the fiber ends are correctly aligned and the component E_y is equal to zero, one has $I_1 = I_2 = 1$, $a_1 = a_2 = a_3 = a_4 = \sqrt{2}/2$ and $e_1 = e_2 = 0$. In this case the displacement obtained from these signals is not affected by perturbations induced into the fiber such as those due to mechanical or thermal stress. From Eq. (11) and Eq. (12) it can be seen that the alignment of the output end is less critical than that of the input one. In fact, if α=0, e_1 and e_2 are equal to zero for any β and δ.

As an example, Figure 2 shows the traces of signals from Eq. (3) and Eq. (4) in the case of α=3°. Traces in solid line correspond to the signals in the absence of perturbations induced into the fiber, while traces in dotted line show the signals in the presence of a perturbation corresponding to a δ=90° phase difference between the fast and the slow fiber optical axis. As expected, both amplitude and phase of the interferometric signals change with respect to the ideal case. As far as the measurement accuracy is concerned, the main parameter to be considered is the phase displacement between the signals from the two photodiodes.

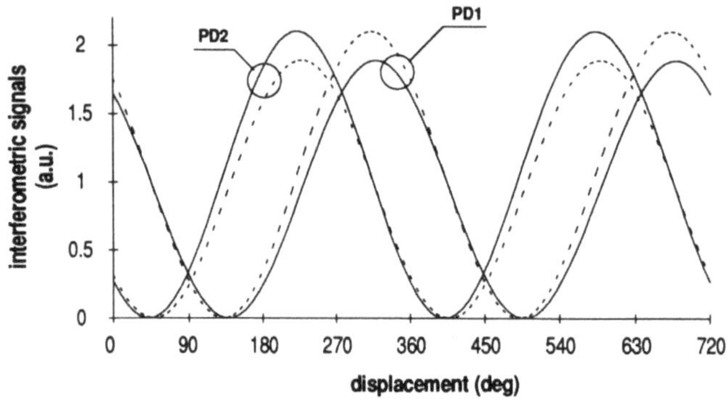

Figure 2. Simulation of the interferometric signals in presence of an input end misalignment of 3° drawn as a function of the interferometer displacement. Solid lines correspond to 0° in-fiber phase disturbance, dotted lines to 90° in-fiber phase disturbance.

EXPERIMENTAL APPARATUS AND RESULTS

Measurement of the phase-difference between signal from PD1 and from PD2 of Figure 1 has been carried out with the setup shown in Figure 3. As shown in the figure, the reference mirror of the interferometer has been mounted onto a piezoelectric transducer (PZT) driven by a sinusoidal signal at 480 Hz. In order to avoid any optical feedback of the modulated beam into the laser source, the interferometer has been mounted with a small tilt with respect to the exit face of the fiber. The interference signals have been detected by photodiodes PD1 and PD2 mounted on the sensor side of the system. This choice allowed us to perfectly isolate the effect of fiber induced stresses from any other sources of signal distortion such as optical feedback into the laser cavity or unwanted back reflections from fiber ends and/or collimating optics. The amplitude of the reference mirror movement is of about 3.5 μm allowing a sufficient number of fringes to be acquired for each vibration period.

Signals from PD1 and PD2 enter, respectively, the A and B channel of a digital phase-meter (SR620). To improve the phase measurement accuracy, the phase-meter has been

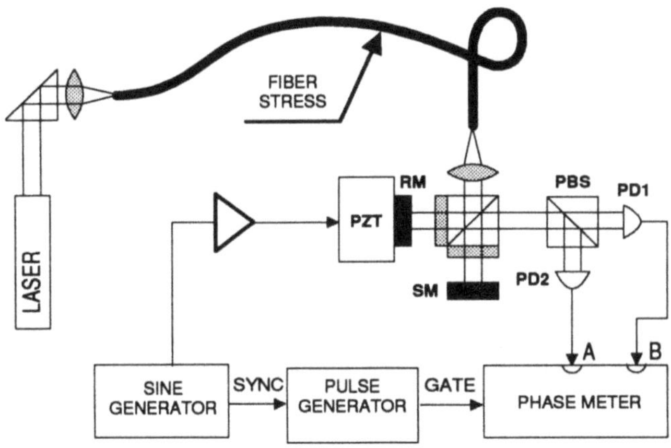

Figure 3. Experimental apparatus for the measurement of the phase difference between the photodiode signals.

operated in gated mode with an input gate signal generated by means of a pulse generator triggered by the sync signal from the sine generator. Actually the meter measure the phase inside a time window were the interferometric signals approximate very well a sine waveform. Figure 4 shows an example of the mentioned signals as acquired by a digital oscilloscope.

Figure 3 shows the phase difference between the two interferometric signals as a function of the phase difference δ induced into the fiber by slowly heating the fiber. Here α is equal to 3°. The deviation, induced by misalignment, from the ideal phase difference of 90°, is evident. A set of measurement for different values of α has shown that the best

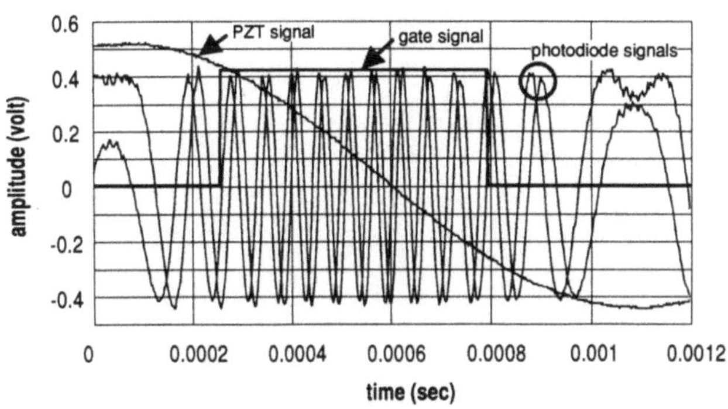

Figure 4. Measurement of the phase difference between the interferometric signals.

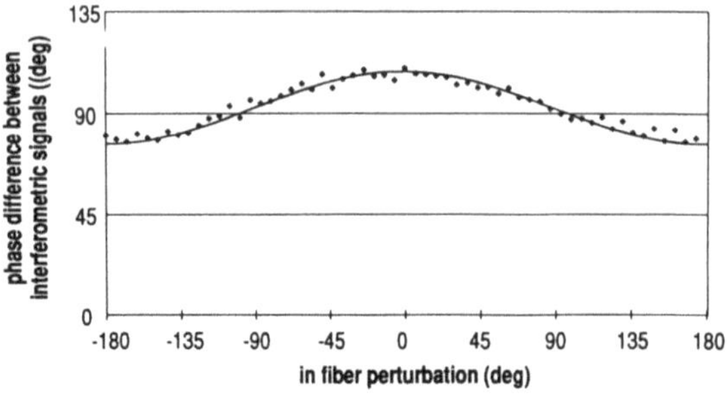

Figure 5. Effect of the input end misalignment on the phase difference between the photodiode signals as function of δ in the case of α=3°. Solid line theoretical data, diamonds experimental data.

system behavior, as expected is obtained when α is equal to zero, that is, when the input end of the fiber is perfectly aligned with the polarization axis of the laser beam feeding the system.

CONCLUSIONS

In this work a theoretical analysis and an experimental setup suitable to measure the influence of the input and output fiber end misalignments over the interferometric signals have been presented. Measurements carried out by using a prototype unit of the flexible, Hi-Bi fiber based interferometer has evidenced the importance of the correct fiber input end alignment in order to keep the phase difference of the interferometric signals near to the desired value of 90°.

REFERENCES

1. U. Minoni, Fiber-stress insensitive interferometric setup for displacement measurements, Opt. Lett. 17, 1382-1384 (1992).
2. A. Ebberg, R. Noe, Novel high precision alignment technique for polarization maintaining fibers using a frequency modulated tunable laser, Electron. Lett. 26, 2009-2011 (1990).
3. S.L.A. Carrara, Birefringent fiber splice alignment, Proc. SPIE 1267, 24-28 (1990).

ANALYSIS AND DESIGN OF INTERFEROMETRIC SENSOR CIRCUITS

Otto Schwelb

Department of Electrical and Computer Engineering
Concordia University
Montréal, Québec, H3G 1M8
Canada

INTRODUCTION

In recent years a large variety of optical circuits have been examined and proposed for sensing, filtering, channel separation, etc. Practically all of these circuits consist of the same basic building blocks, namely 2×2 or 3×3 couplers, mirrors and transmission lines. Until now each configuration has been treated as a separate entity. As the number of possible combinations is rather high, a plethora of articles appeared discussing the design equations and the performance characteristics of these systems.[1-5]

Using the scattering matrix and the scattering transfer matrix representation of couplers, transmission lines and reflecting elements, we present a generalized method to analyze interferometric sensor architectures. This method is particularly suitable for numerical simulation, but it also lends itself to analytical study on a symbolic algebra manipulation platform.

From the point of view of this treatment it is irrelevant whether bulk wave, surface wave (integrated) or optical fibre technology is used to fabricate the system. The focus is on such performance characteristics as, for example, the dependence of output and circulating intensities on wavelength, phase delay, coupling coefficient or splice loss. Furthermore, we provide the condition for resonance in ring type architechtures and in circuits with gain.

CIRCUIT COMPONENTS AND CONFIGURATIONS

Circuits using 2×2 couplers are described by 4×4 matrices while those based on 3×3 couplers are described by 6×6 matrices. When a configuration supports wave propagation in one direction only, the arithmetic simplifies considerably. E.g. for a 3×3 coupler described by a 6×6 matrix, only the upper left 3×3 submatrix is needed. Since the architectures encountered consist of cascaded $2N$ ports we make use of the scattering transfer matrices $\boldsymbol{A}$ and $\boldsymbol{T}$. For a four-port device these are defined by

$$[a_1 \quad b_1 \quad a_2 \quad b_2]^T = \boldsymbol{A}[b_3 \quad a_3 \quad b_4 \quad a_4]^T \text{ and } [a_1 \quad a_2 \quad b_1 \quad b_2]^T = \boldsymbol{T}[b_3 \quad b_4 \quad a_3 \quad a_4]^T \quad (1)$$

respectively, where a_i and b_i are incident and reflected waves at port i and the superscript indicates transposition. Matrix descriptions of couplers, waveguides and reflecting surfaces, as well as conversions between representations, have been documented in the literature.[6] The transfer matrix of concatenated four-ports or six-ports obtains by simply multiplying the transfer matrices of the components in the chain, in the order they are connected.

Feedback to the $2N$ port can be either internal or external. We speak of internal feedback when a reflecting element terminates port i:

$$b_i = r_i a_i \quad (2)$$

where r_i is the reflection coefficient. External feedback connects two ports, say i and j:

$$a_j = t_{ij} b_i \quad (3)$$

where t_{ij} is the transmission coefficient from port i to port j. In general, both r and t are complex. Feedback can be applied to a four-port in six different ways to result in a two-port network. Three of these, using external feedback, are shown in Fig. 1. For the configuration in the middle, for example, the resultant transfer matrix is

$$\boldsymbol{A}_r = \boldsymbol{A}_B + \boldsymbol{A}_A(\boldsymbol{A}_2^{-1} - \boldsymbol{A}_C)^{-1}\boldsymbol{A}_D \quad (4)$$

where $\boldsymbol{A}_A$, $\boldsymbol{A}_B$ $\boldsymbol{A}_C$ and $\boldsymbol{A}_D$ are the four 2×2 submatrices of $\boldsymbol{A}_4$. For the same circuit

$$\left|\frac{b_1}{a_1}\right|^2 = \frac{\left|(\boldsymbol{A}_r)_{21}\right|^2}{\left|(\boldsymbol{A}_r)_{11}\right|^2}, \quad \left|\frac{b_4}{a_1}\right|^2 = \frac{1}{\left|(\boldsymbol{A}_r)_{11}\right|^2} \quad \text{and} \quad \left|\frac{b_4}{a_1}\right|^2 = |t|^2 \frac{\left|\left[(\boldsymbol{A}_2^{-1} - \boldsymbol{A}_C)^{-1}\boldsymbol{A}_D\right]_{11}\right|^2}{\left|(\boldsymbol{A}_r)_{11}\right|^2} \quad (5)$$

give the relative reflected intensity, the relative output intensity and the relative circulating intensity, respectively. Simple expressions describe the conditions for a null in the reflected power and that in the output power. Similar formulas apply to the remaining feedback configurations using 2×2 couplers.

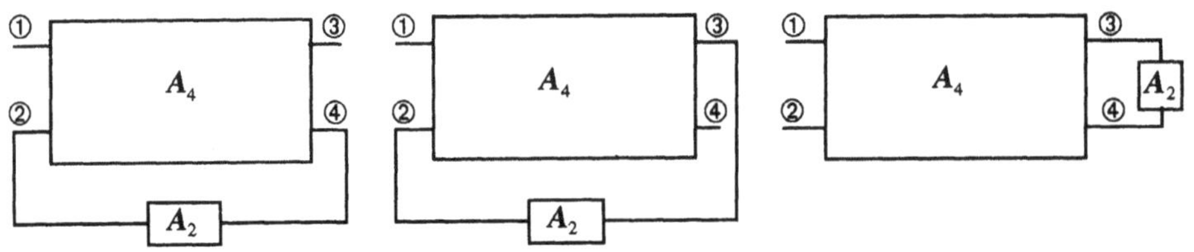

Figure 1. Three alternatives of external feedback applied to a four-port, represented by $\boldsymbol{A}_4$.

Note, that the four-ports shown in Fig. 1 can be composite structures themselves, such as a Mach-Zehnder interferometer, or a systolic array, etc.

Sensor architectures based on the 3×3 coupler are richer in variety, not only as a result of more ports beeing available, but also because a 3×3 coupler can be either of the so-called in-line geometry, where coupling between the outer waveguides is negligible, or of the equilateral triangle geometry, where the coupling between the waveguides is equal. Two double external feedback loop configurations are illustrated in Fig. 2. The corresponding transfer coefficients are found to be, respectively,

$$\frac{b_4}{a_1} = m_{11} + \boldsymbol{M}_{i1}\left[\boldsymbol{M}_f^{-1} - \boldsymbol{M}_{ii}\right]^{-1}\boldsymbol{M}_{i1} \qquad i = 1,2 \tag{6}$$

and

$$\frac{b_5}{a_1} = m_{21} + \boldsymbol{M}_{2i}\left[\boldsymbol{M}_f^{-1} - \boldsymbol{M}_{ji}\right]^{-1}\boldsymbol{M}_{j1} \qquad i = 2,3 \quad j = 1,3 \tag{7}$$

where $\boldsymbol{M}$, essentially the inverse of $\boldsymbol{A}$, is a 3×3 matrix relating the output wave amplitudes to the input wave amplitudes: $\begin{bmatrix} b_4 & b_5 & b_6 \end{bmatrix}^T = \boldsymbol{M}\begin{bmatrix} a_1 & a_2 & a_3 \end{bmatrix}^T$, $\boldsymbol{M}_f$ is a diagonal 2×2 matrix containing the transmission coefficients of the feedback lines, and $\boldsymbol{M}_{ji}$ is a submatrix of $\boldsymbol{M}$. The configurations shown in Fig. 2 do not cause waves to propagate in the reverse direction. As a result a 3×3 $\boldsymbol{M}$ matrix fully describes the coupler. When, however, reflecting components terminate some of the ports, or when external feedback lines are applied so as to cause counterpropagating power flow, the full 6×6 matrix description of the coupler is required. Inasmuch as reflection at a waveguide splice is always present, a pure forward wave model can only be considered as a first approximation.

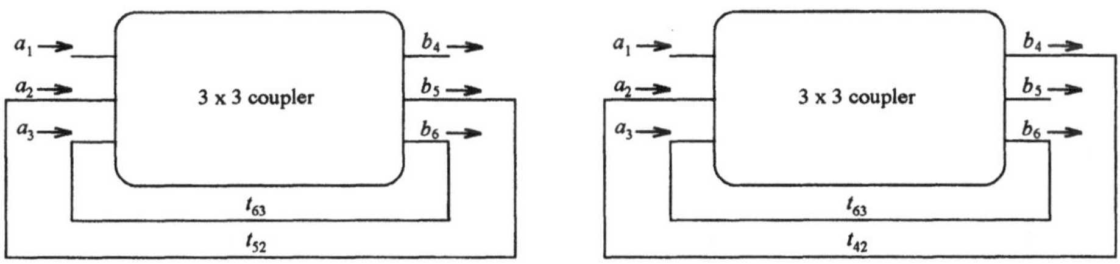

Figure 2. Two circuits using a 3×3 coupler with double feedback.

RESONANCE

Resonance occurs in a closed loop when the round trip phase delay is an integral multiple of 2π and when the rond trip gain is unity. Unless there is gain in the circuit the second condition cannot be fully satisfied and one observes resonance peaks with finite finesse, characteristic of the losses in the resonant ring. In interferometric circuits involving 2×2 couplers resonance can be achieved by terminating all four ports with mirrors and/or feedback lines. One example is shown in Fig. 3 for which the resonance condition can be expressed as

$$\det\begin{bmatrix} -r_1 & tA_{12} & A_{11} & A_{13} + r_4 A_{14} \\ -1 & tA_{22} & A_{21} & A_{23} + r_4 A_{24} \\ 0 & tA_{32} & A_{31} - t & A_{33} + r_4 A_{34} \\ 0 & tA_{42} - 1 & A_{41} & A_{43} + r_4 A_{44} \end{bmatrix} = 0 \tag{8}$$

In some systems degenerate two-wave mixing is used to amplify the primary signal. Here, instead of a net loss, one or more of the feedback lines exhibit gain and a threshold condition applies when the system gain balances out the losses.

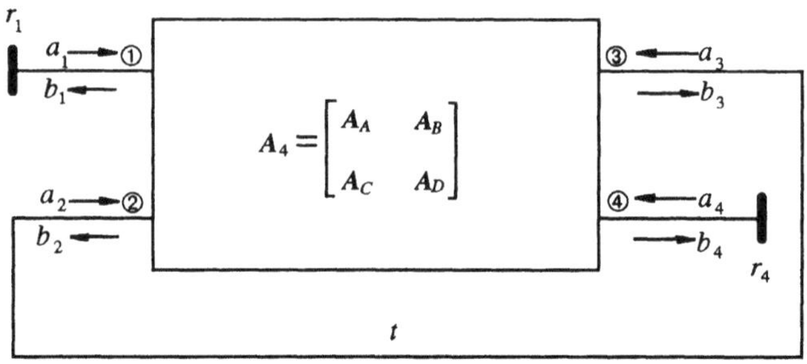

Figure 3. Resonant circuit obtained by terminating the ports with a feedback line and mirrors.

Most interferometric sensors include a resonant ring. These systems exhibit the well known comb filter characteristics with a finesse determined by the free spectral range and the bandwidth of the resonance curve. Several methods can be used to enhance finesse, such as injecting gain into the feedback loop, decreasing the coupling coefficient in a 3×3 in-line coupler, or making use of the so-called vernier effect. In the latter case the phase delay in the two external feedback lines of a 3×3 coupler with double feedback is controlled according to $N\varphi_1 = (N+1)\varphi_2$, where N is an integer.[7]

Software implementation of our approach requires a linear algebra package and graphics interface, both beeing easily available for the PC from a number of commercial suppliers.

Acknowledgement

This work was supported by the Natural Sciences and Engineering Research Council of Canada and by the Fonds pour la Formation de Chercheurs et l'Aide a la Recherche.

REFERENCES

1. L.F. Stokes, M. Chodorow, and H.J. Shaw, All-single-mode fiber resonator, *Opt. Lett.* 7:288 (1982).
2. F. Zhang and J.W.Y. Lit, Direct-coupling single-mode fiber ring resonator, *J. Opt. Soc. Am. A*, **5**:1347 (1988).
3. Y.H. Ja, A double-coupler optical fibre ring-loop resonator with degenerate two-wave mixing, *Optics Commun.*, **81**:113 (1991).
4. G Abd-El-Hamid and P.A. Davies, Fibre-optic double ring resonator, *Electronics Lett.*, **25**:224 (1989).
5. Y.H. Ja, A symmetric S-shaped double-coupler optical-fibre loop, *J. Modern Optics*, **37**:1297 (1990).
6. F. Sanchez, Matrix algebra for all-fiber optical resonators, *IEEE J. Lightwave Technol.*, **9**:838 (1991).
7. Y.H. Ja, A vernier S-shaped fiber double-loop resonator with double couplers and degenerate two-wave mixing, *IEEE J. Lightwave Technol.*, **11**:258 (1993).

ANGLE MEASUREMENT BY GEOMETRIC POLARISATION ROTATION IN A FIBRE LOOP REFLECTOR

D.A. Chapman, D.A. Gorham and G. Dickinson

Faculty of Technology
The Open University
Walton Hall
Milton Keynes, MK7 6AA, UK

SUMMARY

A combination of bends and twists can rotate the state of polarisation of light in standard single mode optical fibre [1]. It has previously been proposed to use the effect to measure rotation angle [2], but there were practical difficulties due in part to the need for a linearly polarised optical input. By incorporating the rotator in a fibre loop reflector [3], an elegant angle sensor has now been designed which is independent of input polarisation state and which delivers the output as optical intensity. The paper describes the theory of operation and an experiment to verify the sensor design.

GEOMETRICAL POLARISATION ROTATION

Geometric polarisation rotation has been reported elsewhere [1], but it is a subtle effect which is not widely appreciated. Because it is central to the operation of the proposed sensor, a simple description is included here.

Using the weakly guided approximation, the polarisation of light propagating in single mode fibre can be resolved into two orthogonal plane wave eigenstates. In standard single mode fibre, these two states are nominally degenerate, due to the circular symmetry of the fibre cross-section. One of the consequences of the degeneracy is that, to a first approximation (neglecting stress in the fibre), the polarisation of light does not follow axial twists. On the other hand, the state of polarisation relative to the frame of reference of the fibre is constant around a bend. It will be shown that combinations of bends and twists in the fibre path geometry can be used to effect a net polarisation rotation.

For the discussion which follows, it is necessary to identify a co-ordinate system for the polarisation states in the fibre. Because of the circular fibre cross section there are no natural axes, but we can imagine some means of identification being imposed on the outside of the fibre cable. In the figures accompanying this paper the cable is shown with a square cross-

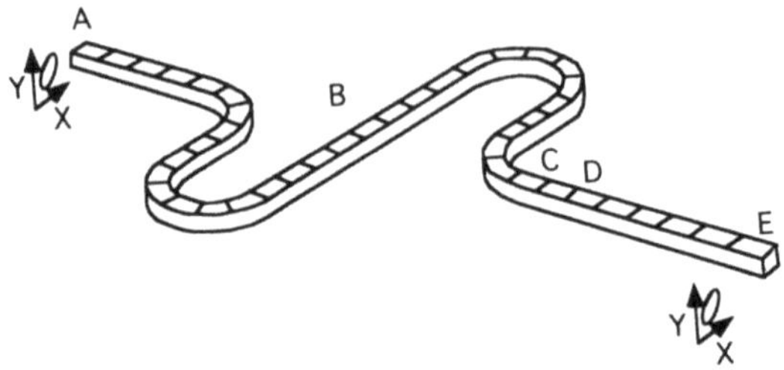

Figure 1. No net rotation of polarisation

section and with one side shaded. The axes for describing the polarisation are then taken as perpendicular to the cable sides, marked X and Y.

Consider first the case of a fibre with no twists, shown in Figure 1. Assuming that the bends are of sufficiently large radius that the bending birefringence may be ignored, the polarisation of light will be unchanged by propagation through the structure. This is shown on the figure by an arbitrary elliptical polarisation maintaining the same ellipticity and azimuthal orientation (referenced to X and Y).

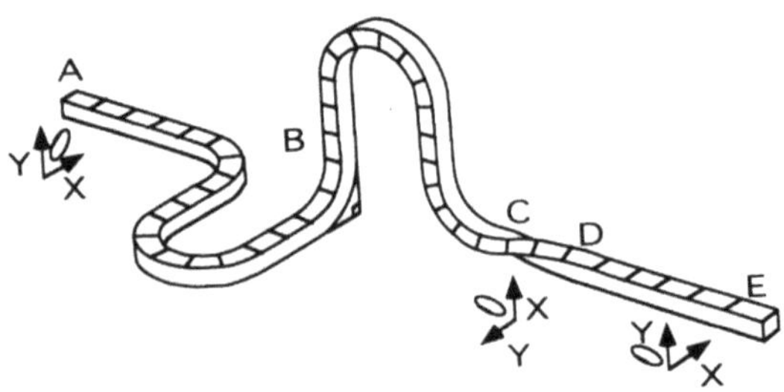

Figure 2. Fibre rotation generates polarisation rotation

Consider now Figure 2, where the second 'lobe' of the bends has been rotated through 90°, causing a bend in the fibre at B and a twist at D. Prior to the twist, at point C, the polarisation remains unchanged relative to the X and Y axes, although both the orientation of the polarisation and the axes have been rotated between A and C. The twist at D returns the axes to their original orientation, but (as explained above) the polarisation does not follow the twist. Comparing the output at E with the input at A, it can be seen that the polarisation has maintained the same ellipticity, but has been rotated through 90°.

Figure 2 shows the lobe bent through 90°, causing a rotation of 90°. More generally, bending the lobe through θ rotates the polarisation through θ.

It is emphasised that this discussion has assumed that all bends are of sufficiently large radius to allow bending birefringence to be ignored: the effects are caused only by the fibre path geometry.

FIBRE LOOP REFLECTOR

The theory of a fibre loop reflector is contained in Mortimore [3]. In the absence of any birefringence, the loop acts as a mirror. If polarisation rotation is introduced within a loop by the Faraday effect the transmitted/reflected power ratio can be altered. This will not be the case with the geometric polarisation rotation described above. The difference comes about because the Faraday effect is a non-reciprocal rotation, whereas the geometric rotation is reciprocal. However, if the loop contains a linearly birefringence element in addition to the reciprocal rotation, then the combination is, in effect, non reciprocal. (By non-reciprocal, we mean here that the effect is different for light traversing the loop clockwise and anticlockwise. In one direction the light encounters the birefringence prior to rotation, in the other direction it encounters the birefringence after rotation.)

Analysing the polarisation as in [3], it can be shown that, for a linear birefringence of ϕ (with eigenaxes parallel and perpendicular to the plane of the loop) and polarisation rotation through θ, the reflected and transmitted powers are as follows:

$$P_t = 0.5(1\text{-}\cos\phi)\sin^2\theta\ P_i,\quad P_r = (1 - 0.5(1\text{-}\cos\phi)\sin^2\theta\,)\ P_i$$

Where P_i, P_t and P_r are the incident, transmitted and reflected light powers respectively.

With the lobe lying flat ($\theta = 0$) there is no polarisation rotation and, regardless of ϕ, the loop acts as a mirror. As θ is varied up to 90°, a proportion, up to a maximum of $0.5(1\text{-}\cos\phi)$, is transmitted. With $\phi = \pi$ (the polarisation controller acting as a half-wave plate), the maximum modulation depth (as a function of θ) is obtained, giving:

$$P_t = \sin^2\theta\ P_i,\quad P_r = \cos^2\theta\ P_i$$

EXPERIMENT

An experiment to verify the model was conducted with the configuration of Figure 3. The light source was a 1300nm multimode laser diode and the loop of single mode fibre was constructed by splicing two arms from a 3dB coupler. The transmitted light (P_t) was measured at the output of the loop and the reflected light (P_r) was measured at a 3dB coupler inserted between the laser and the loop. Unconnected fibre ends were plunged into tetrahydrofuran to reduce fresnel reflections. A manual all-fibre polarisation controller was used to introduce the fixed linear birefringence of π radians. The experiment consisted of pivoting a lobe, contained within the loop, from 0° to 90° from the plane of the loop, and measuring the reflected and transmitted power.

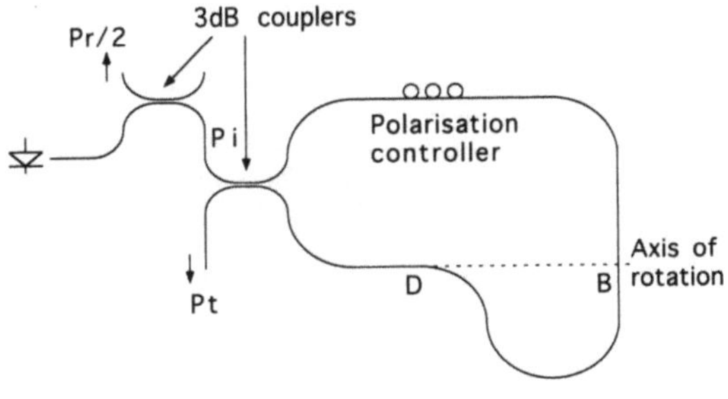

Figure 3. Experimental configuration

COMPARISON OF THE EXPERIMENT WITH THE MODEL

The experimental results are compared with theory in Figure 4. The transmitted and reflected powers have been normalised to the total output power ($P_r + P_t$) to remove the effects of systematic errors in the absolute power measurements (due to unknown splice losses and thermal drift in the laser output power). The horizontal error bars correspond to the approximately +/- 3° accuracy in the angle measurement. The remaining slight discrepancy between experiment and theory at $\theta = 0^o$ is thought to be due to reflection at the splice contained within the loop.

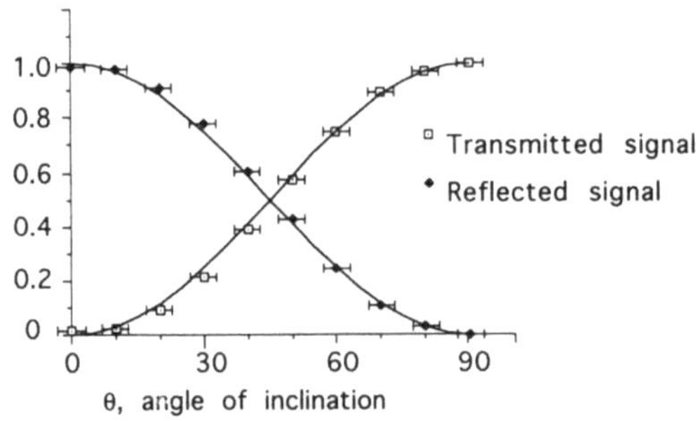

Figure 4. Results compared with theory

CONCLUSIONS

The polarisation of light in a single mode fibre can be rotated simply by the geometry of the fibre path. If this reciprocal polarisation rotation is used in conjunction with linear birefringence, the combined effect is non-reciprocal and can be converted into amplitude modulation of the transmitted and reflected signals.

The effect is independent of wavelength and input polarisation state. It is also independent of radius of curvature (although for a small radius of curvature, the separate effect of bending birefringence needs to be considered). Potential applications include the measurement of angles [2] and reflection modulation in passive optical networks [4].

REFERENCES

1. J.N. Ross, The rotation of the polarisation in low birefringence monomode optical fibers due to geometric effects *Optical & Quantum Electronics* **16** pp 455-461 (1984)
2. K. Itoh, T. Saitoh and Y. Ohtsuka, Optical rotation sensing by geometric effect of fiber-loop twisting *J. Lightwave Technol.* **5**(7) (1987) pp 916–919
3. D.B.Mortimore, Fiber Loop Reflectors *J. Lightwave Technol.* **6**(7) pp 1217-1224 (1988)
4. D.A. Chapman and D.W. Faulkner, Use of reflected light for low data rate upstream signalling in a single mode passive optical network, *IEE Proc. Pt J* **137**(2) pp108-114 (1990)

THE APPLICATION OF SUPER-RESOLUTION ADAPTIVE ALGORITHMS TO FRINGE ORDER ESTIMATION IN ALL-OPTICAL-FIBRE INTERFEROMETRIC SENSORS

M S Rizk, D Romare, K T V Grattan and A W Palmer

Dept. of Electrical, Electronic & Information Engineering
City University, Northampton Square
London EC1V 0HB UK

INTRODUCTION

Optical-fibre white-light interferometry (WLI) has recently been attracting significant attention in the research area of optical-fibre sensors[1]. In this application it is necessary to identify the centre fringe from an interference fringe pattern which is the output of an optical fibre interferometer incorporating a broadband optical source. As a result, the unambiguous range of the output signal is no longer limited to within half a fringe, and an absolute phase measurement over a large operating range can be achieved. The output fringe pattern (Figure 1) of such a system takes the form of a Gaussian amplitude-modulated cosinusoidal pattern, but the centre fringe and adjacent may not be easily distinguishable from the rest of the pattern. This problem is further exacerbated when the signal is buried in noise. Because of the difficulty of pinpointing the centre fringe, long-range absolute phase detection with a resolution in the subfringe region may not be possible. However, modern adaptive digital filtering techniques coupled with parametric estimation, which have been successfully applied in such diverse fields as communications, radar, sonar, seismology, and biomedical engineering, can be utilized to enhance the spatial fringe pattern in the presence of strong additive Gaussian noise even when such a pattern is buried in noise. This implies that it is possible to multiplex intrinsic sensors distributed along a fibre without stringent constraints on power requirements.

There is no unique solution to the adaptive filtering problem. Rather, we have a "kit of tools" represented by a variety of recursive algorithms, each of which offers desirable features of its own. The challenge we faced was to understand the capabilities and limitations of various adaptive filtering algorithms, when applied to the modelling and enhancement of the noise contaminated fringe pattern with Gaussian statistics, and to use this understanding in the selection of the appropriate algorithm for the central fringe identification.

In this paper we study the convergence and tracking performance of the recursive

Least Mean Square (LMS) adaptive algorithm in the presence of weak and strong noises, when the power of the signal is actually varying slowly with time.

The promotion of the LMS against other recursive Least Squares (LS) adaptation methods stems from the fact that it can easily be implemented in both software and hardware. Moreover, it provides a practical frame of reference for assessing any further improvement that may be attained through the use of more sophisticated adaptive filtering techniques such as the Kalman filtering algorithm.

The study includes both computer simulations and real-life "measured" data.

THE OPTIMUM FRINGE ORDER ESTIMATION PROCEDURE

On the left-hand side of Figure 2, within the dashed lines, is a mathematical model of the fringe pattern "pure" signal we wish to estimate. It covers the generation of the pure signal, $s[n]$, using an autoregressive (AR) model, and its noisy version $y[n]$, and is based on Candy's model[2]. $v[n]$ represents zero mean additive white noise, which describes the hidden laws that are supposed to govern or constrain the generation of the contaminated signal $y[n]$ (the only quantity available to us). On the right-hand side of Figure 2 is a linear processor, taking the form of a finite-impulse-response (FIR) digital filter which provides an output signal $\hat{y}[n]$ with spectral properties similar to those of the actual signal $y[n]$, and is an optimum estimate of the "pure" signal $s[n]$. The FIR filter provides an inverse for the AR model, and its modelling error is tracked by the stochastic-gradient least mean square (LMS) adaptation algorithm[3].

The inherent dynamic nature of the WLI fringe pattern resumes Gaussian statistics when the measurement environment is stationary. Any environmental changes or slow external dynamics, such as those resulting from temperature drifts in one of the constituent arms of the interferometric system, will inevitably lead to further perturbations of the fringe pattern.

Clearly any adaptive filtering operation must be capable of responding fast to both the inherent and external dynamics of such interferometric systems. This capability can only be accomplished by adopting optimization strategies peculiar to the nature of such dynamics. In so doing not only we have had to trade off between the speed of adaptation and the asymptotic accuracy of the adaptive filter, but also we have had to consider hardware implementation and tried to avoid any unnecessary increase in the computational cost of the optimized LMS filter.

Modifications to the Adaptation Algorithm

In order to improve performance and simplify implementation two modifications have been considered; (1) Nonlinear Correlation Multipliers and (2) Normalization of step size. The first leads to simplification of the hardware without compromising the convergence of the algorithm. The second takes into account the dependence of speed of convergence on the signal power and yields a continuous optimization strategy.

RESULTS BASED ON COMPUTER SIMULATIONS

The central-fringe identification technique described above was tried using a large number of various fringe patterns possessing different characteristics. The WLI system employed electronic scanning and the fringe patterns were imaged on a CCD array of 1024 pixels[4]. Figure 3 I shows the learning curves averaged over 300 trials for different SNR's, with the central fringe fixed at pixel 512. Figure 3 II shows representative contaminated fringe spatial signals before filtering. It has been demonstrated that, over 300 trials, the LMS succeeds in identifying the central fringe between two adjacent

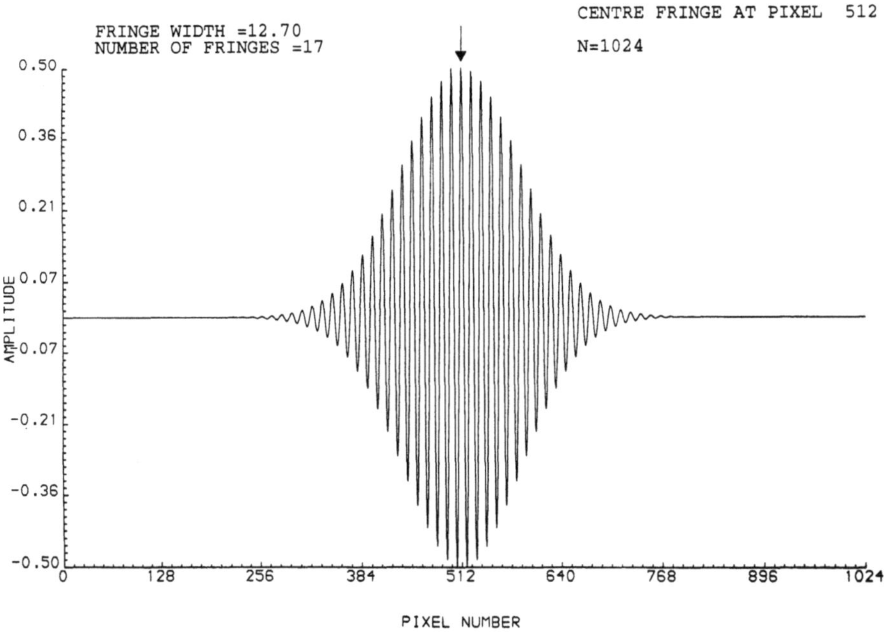

Figure 1. The ideal Gaussian amplitude-modulated fringe pattern.

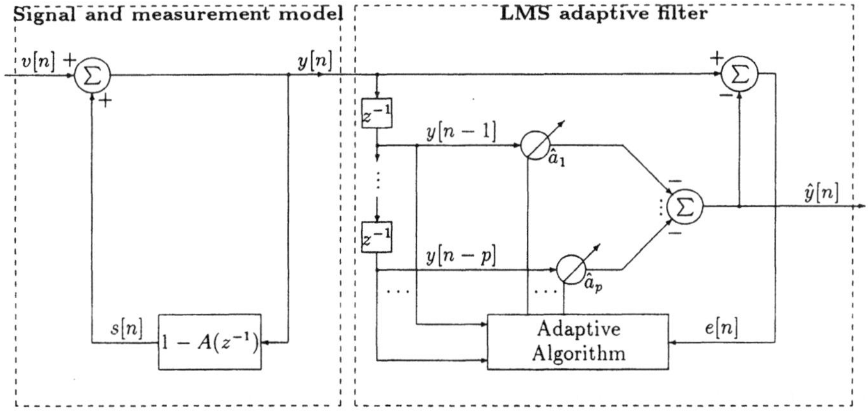

Figure 2. An assumed model for the generation of a noisy signal $y[n]$ and its estimation $\hat{y}[n]$ by a LMS filter.

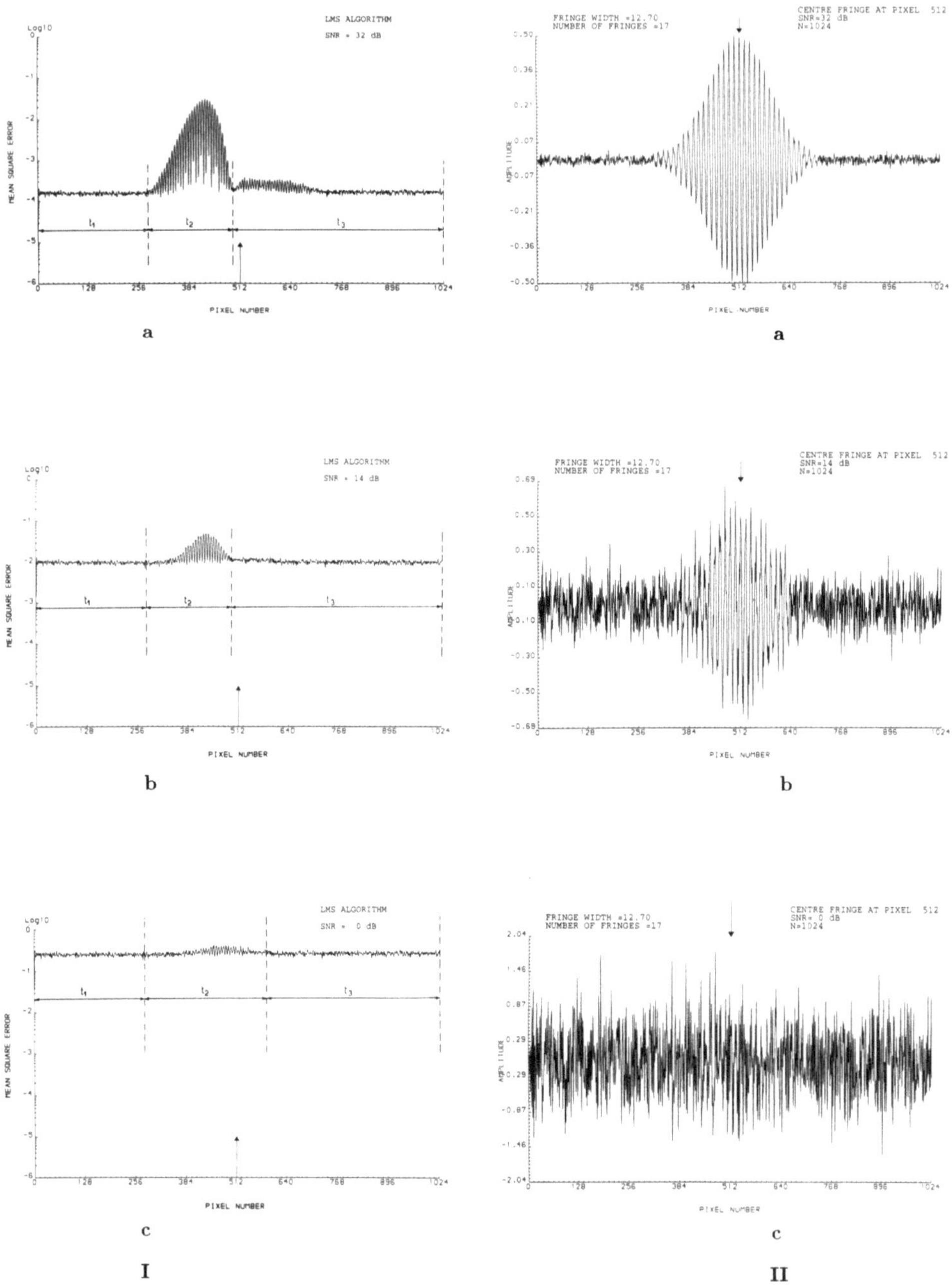

Figure 3. **[I]** The learning curves for the stochastic LMS-FIR filter with 25 taps, for different SNR's; t_1, t_2, t_3 correspond to the measured noise-only region, the data-triggered convergence region, and the tracking region, respectively. **[II]** A typical contaminated fringe spatial signal before filtering. The arrowheads point at the position of the central fringe after filtering.

pixels in the CCD, corresponding to a width of approximately one-thirteenth of a full cycle of a fringe in the spatial fringe pattern, for different SNR's, down to 0 decibels. Moreover, this accuracy is achievable when the central fringe is moved towards either end of the CCD frame, thus demonstrating the superiority over the centroid method[4].

CONCLUSIONS

A simple and accurate subfringe order estimation adaptive filtering technique based on the stochastic LMS and AR modelling criteria has been demonstrated using both computer simulations and real-life data. Further work involving more sophisticated techniques is now being carried out.

REFERENCES

1. R. Ulrich, Interferometric and polarimetric sensors using white-light interferometry, in: "Springer Proceedings in Physics", Vol. 44 *Optical Fiber Sensors*, pp 62-3 (1989).
2. J.V.Candy. "Signal Processing: The Model-Based Approach," ©McGraw-Hill, Inc. (1986).
3. B. Widrow et al., Adaptive noise cancelling: principles and applications, *IEEE Proc.*, Vol. 63, No. 12, pp 1692-1716 (1975).
4. S. Chen, A.W. Palmer, K.T.V. Grattan, and B.T. Meggitt, Digital signal-processing techniques for electronically scanned optical-fiber white-light interferometry, *Appl. Opt.*, Vol. 31, No. 28, pp 6003-6010 (1992).

DEPENDENCE OF GAIN SPECTRUM AND THRESHOLD IN SBS ON TEMPERATURE DISTRIBUTION IN FIBERS

Yoh Imai and Noriaki Shimada

Department of Computer Science and Electronics
Kyushu Institute of Technology
Iizuka, Fukuoka 820, Japan

INTRODUCTION

Stimulated Brillouin scattering(SBS) is the most notable nonlinear phenomenon occurred in optical fibers, because the threshold for SBS is the lowest compared with the other nonlinear phenomena.The frequency shift of SBS depends on external stimuli such as temperature and strain[1,2] as well as fiber parameters. This dependency has been investigated and applied to the distributed fiber sensing.[3,4] In these approaches, it is assumed that the temperature distributes uniformly along the fiber or along the sensing length at least. Attention has not been paid to non-uniformity in temperature distribution along fibers except for a few works.[5] This paper presents the dependency of SBS threshold and gain bandwidth on the temperature distribution in fibers.

EXPERIMENTS

Figure 1 shows the experimental setup for investigating the dependency of SBS on the temperature distribution in fibers. An LD pumped Nd:YAG ring laser of 1.319μm wavelength with 25mW output power was used as a pumping light source. The line width of the

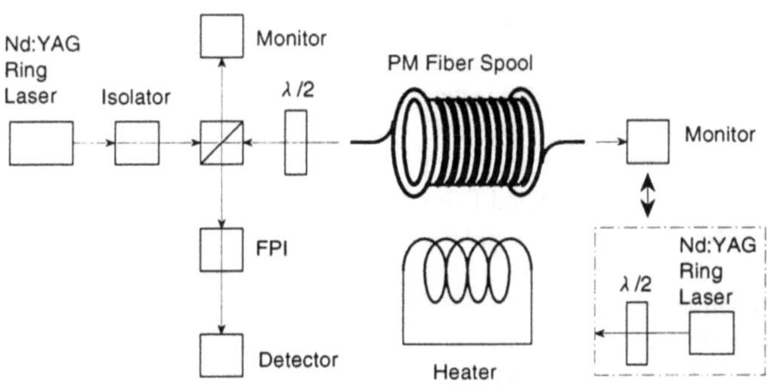

Figure 1. Experimental setup.

Applications of Photonic Technology
Edited by G.A. Lampropoulos *et al.*, Plenum Press, New York, 1995

laser source was 20kHz. Linearly polarized light from the laser source was launched into a polarization maintaining fiber of 1.6km length by adjusting the polarization direction to the birefringence axis of the fiber. The fiber spool was heated by a heater from its one side in order to apply periodically a graded temperature distribution along the fiber. The period of the periodic temperature distribution is 1.1m which is the length of one turn of the spool. The periodic temperature distribution is shown in Fig.2 in which the half period is depicted. A maximum temperature difference $\Delta T(=T_A-T_F)$ of the temperature distribution is used as a parameter to evaluate the SBS characteristics. The backward SBS was detected by passing through a Fabry-Perot interferometer. The polarization maintaining fiber used was of PANDA type and is characterized by 1.17μm cut-off wavelength, 7.5μm core diameter, 4.2×10^{-4} mode birefringence, and 0.51dB/km transmission loss at 1.3μm wavelength.

In the present setup, the SBS frequency was shifted from the pumping light frequency by an amount of 13.0GHz in the fiber without heating the fiber spool(with an uniform temperature distribution).

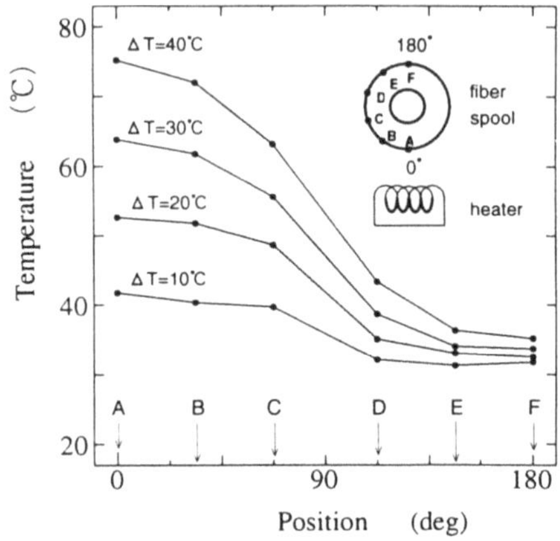

Figure 2. Temperature distribution applied to fiber.

RESULTS AND DISCUSSIONS

Figure 3 shows the dependence of the SBS intensity and the SBS threshold on the temperature difference. The temperature difference was changed from 0°C to 40°C where the highest temperature of the fiber was changed from 23°C to 63°C at point A and the lowest temperature was unchanged at point F. The SBS intensity decreases as ΔT increases. SBS disappears at ΔT=18°C. The SBS threshold changed from 9.1mW at ΔT=0°C to 13.4mW at

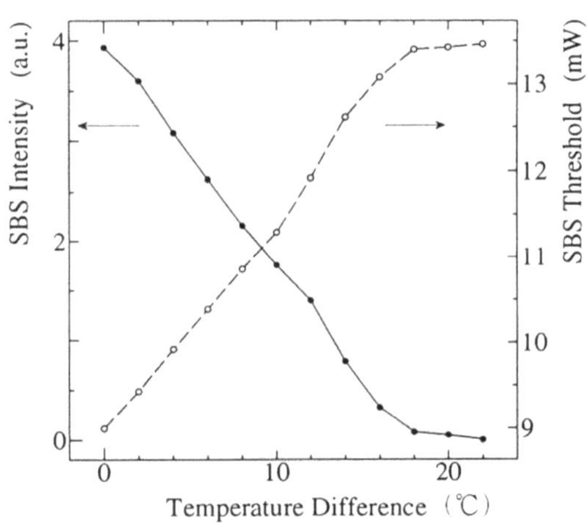

Figure 3. SBS threshold and SBS intensity as a function of temperature difference in fiber.

ΔT=18°C. In a preliminary experiments, it was confirmed that the SBS intensity was stable within 6% fluctuation by changing the temperature applied uniformly to the fiber in a range of 23°C ~ 60°C.

A typical Brillouin gain spectrum is shown in Fig.4 in which the temperature difference is ΔT=0°C in (a), 20°C in (b), and 40°C in (c). The solid line represents the theoretical curve obtained by integrating the Lorentzian spectrum over the temperature distribution along the fiber shown in Fig.2. In this measurements, a probe light shifted in frequency by an amount of the Brillouin frequency shift is launched into the fiber from the opposite side, as shown in an illustration inserted in Fig.1, so that the Brillouin amplified light is detected. The frequency of the probe light was scanned in order to obtain the gain bandwidth. The Brillouin gain bandwidth as a function of ΔT is shown in Fig.5. The Brillouin gain bandwidth was 62MHz at ΔT=0°C, then increases in proportion to the temperature difference with a ratio of 2.0MHz/°C, and became 103MHz at ΔT=20°C.

The SBS characteristics mentioned above are explained in relation to the frequency shift of SBS dependent on the temperature. In the present fiber, the coefficient of the change in SBS frequency shift to the uniformly applied fiber temperature has been measured to be 2.4MHz/°C, which is close to the ratio of the Brillouin gain bandwidth and the temperature difference. When the temperature distribution is applied to the fiber(ΔT>0°C), the SBS frequency shift is different at point by point in the fiber. Then, the SBS is not generated constructively in the fiber. This is why the SBS intensity decreases and the SBS threshold increases as the temperature difference increases. The SBS gain bandwidth also increases on the basis of this fact. The extension of the SBS frequency shift along the fiber with the temperature distribution at ΔT=20°C is estimated as 48MHz by using the coefficient of 2.4MHz/°C. This extension leads to the extended gain bandwidth of

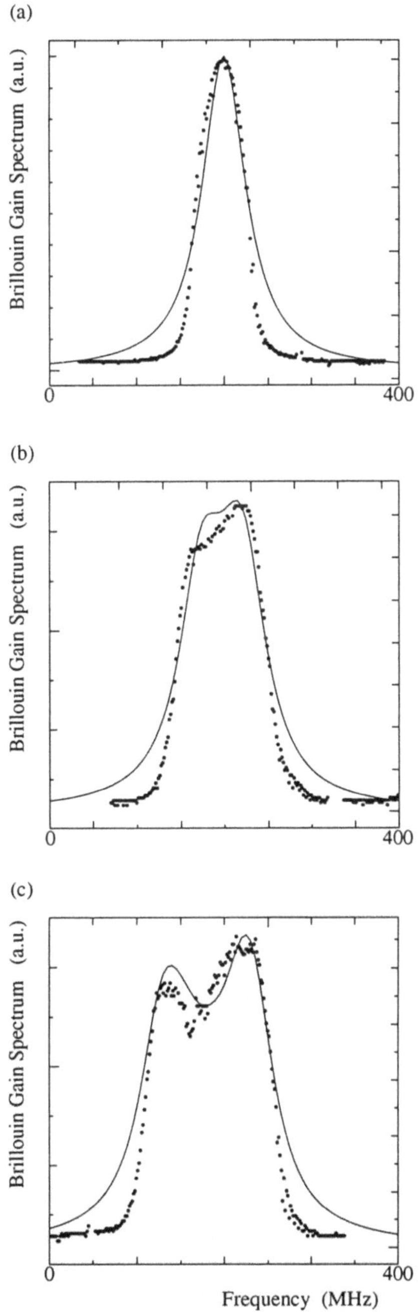

Figure 4. Brillouin gain spectrum.

110MHz at $\Delta T=20°C$ temperature difference, because the Brillouin gain bandwidth with the uniform temperature($\Delta T=0°C$) was 62MHz. This estimation agrees well with the measured SBS gain bandwidth of 103MHz at $\Delta T=20°C$.

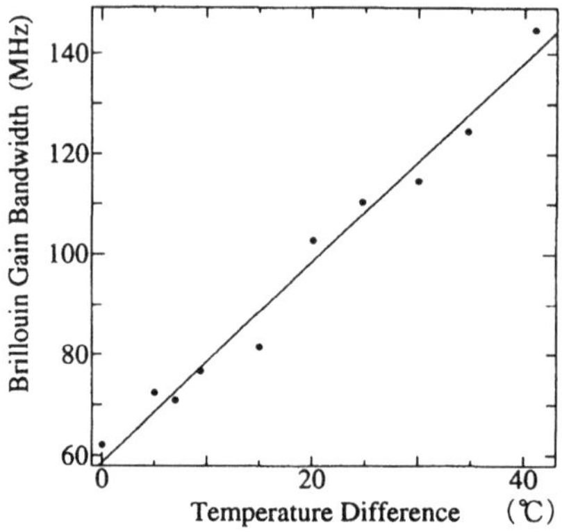

Figure 5. Brillouin gain bandwidth as a function of temperature difference in fiber.

CONCLUSIONS

The SBS intensity decreases and the SBS threshold increases as the temperature difference applied to the fiber increases. SBS disappears at 18°C temperature difference. The Brillouin gain bandwidth increases in proportion to the temperature difference with a ratio of 2.0MHz/°C. The present results can be applied to the suppression of SBS as well as to the sensing of temperature non-uniformity in the optical fiber communication system.

ACKNOWLEDGEMENTS

The present work was supported in part by the Izumi Science and Technology Foundation.

REFERENCES

1.T.Horiguchi, T.Kurashima, and M.Tateda, Tensile strain dependence of Brillouin frequency shift in silica optical fibers, IEEE Photon.Tech.Lett. 1:107(1989).
2.D.Culverhouse, F.Farahi, C.N.Pannell, D.A.Jackson, Potential of stimulated Brillouin scattering as sensing mechanism for distributed temperature sensor, Electron.Lett. 25:913(1989).
3.M.Tateda, T.Horiguchi, and T.Kurashima, and K.Ishihara, First measurement of strain distribution along field-installed optical fiber using Brillouin spectroscopy, J.Lightwave Tech. 8:1269(1990).
4.T.Kurashima, T.Horiguchi, and M.Tateda, Distributed-temperature sensing using stimulated Brillouin scattering in optical silica fibers, Opt.Lett. 15:1038(1990).
5.Y.Imai and N.Shimada, Dependence of stimulated Brillouin scattering on temperature distribution in polarization-maintaining fibers, IEEE Photon.Tech.Lett. 5:1335(1993).

MULTI-WAVELENGTH COMBINATION SOURCE
— A NOVEL TECHNIQUE FOR WHITE LIGHT INTERFEROMETRY

D. N. Wang, Y. N. Ning, A. W. Palmer, K. T. V. Grattan and K. Weir

Department of Electrical, Electronic and Information Engineering,
City University, Northampton Square, London EC1V 0HB

INTRODUCTION

White light interferometry (WLI) has been proved as a successful technique to improve the unambiguous range of a conventional interferometric system and to make absolute measurements. In WLI operation, the central fringe position in the output fringe pattern has crucial importance as a knowledge of such a position provides a reliable basic reference.

The multimode laser diodes frequently used as sources in such systems have a relatively long coherence length, whose use may reduce the accuracy in determining the central fringe position, but by the employment of a two wavelength combination source, the situation can be significantly improved[1,2], especially when the two sources have an "optimum" wavelength combination[3]. Further improvement can be achieved by using a three wavelength combination source[4] in fringe identification.

In this paper, a detailed comparison of different wavelength combination sources has been made to demonstrate clearly the improvement attainable in the central fringe identification, and with it the resulting benefits of sensor performance.

THEORETICAL ANALYSIS

The proposed system uses a series (two or three) laser diodes as sources in a WLI system. Assuming that the three laser diodes are chosen to have different wavelengths λ_1, λ_2 and λ_3, but the same coherence length, L_c, the normalized output ac signal intensity from an interferometer illuminated by a three wavelength combination sources, may be expressed as

$$I_{ac}(x) = (1/3)\exp[-(2x/L_c)^2][\cos(2\pi x/\lambda_1) + \cos(2\pi x/\lambda_2) + \cos(2\pi x/\lambda_3)] \quad (1)$$

where x is the optical path difference (OPD) introduced by the interferometer. From this equation, the characteristics of the interference fringe patterns can be modelled by means of several computer simulations and the optimum wavelength combinations can also be obtained by a simple analysis. In order to compare the results quantitatively, the minimum signal-to-noise ratio required by the system to identify the central fringe position, SNR_{min} (dB), may be used, and is defined by

$$SNR_{min}\,(dB) = -20\lg(\Delta I_{0n}) \quad (2)$$

where ΔI_{0n} is the normalized intensity difference between the central fringe and the largest side fringe.

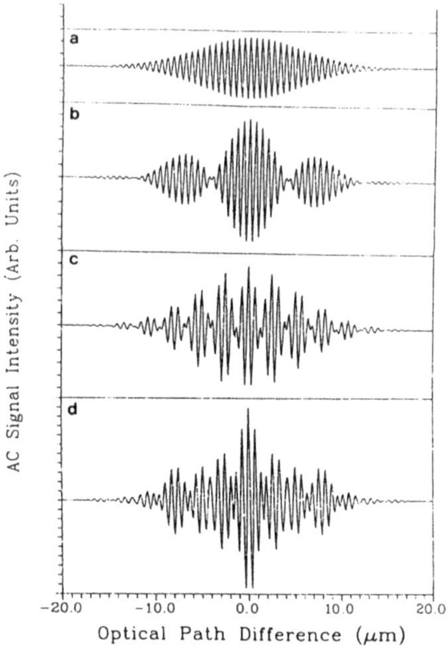

Fig.1 Computer Simulated Interference Fringe Patterns

(a) Single Wavelength: $\lambda = 635nm$, $L_c = 16\mu m$;

(b) Two Wavelengths: $\lambda_1 = 635nm$, $\lambda_2 = 688nm$, $L_c = 16\mu m$;

(c) Two Wavelengths: $\lambda_1 = 635nm$, $\lambda_2 = 830nm$, $L_c = 16\mu m$;

(d) Three Wavelengths: $\lambda_1 = 635nm$, $\lambda_2 = 688nm$, $\lambda_3 = 830nm$, $L_c = 16\mu m$

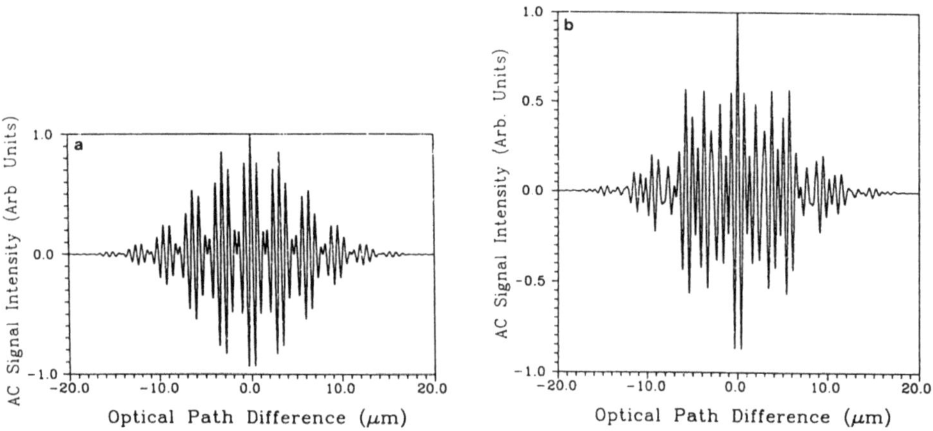

Fig.2 Computer Simulated Interference Fringe Patterns of Optimized Multiwavelength Combination Source

(a) $\lambda_1 = 635nm$, $\lambda_2 = 797nm$, $L_c = 16\mu m$

(b) $\lambda_1 = 635nm$, $\lambda_2 = 724nm$, $\lambda_3 = 959nm$, $L_c = 16\mu m$

A series of computer simulations has been carried out to generate the interference fringe patterns for different wavelength combination sources and the results are shown in Fig.1.

In these simulations, the light sources adopted are typical of commercially available laser diodes. Fig.1(a) reveals that, even for an ideal single laser diode, i.e. a system free of intensity noise and using a light source with a relatively short coherence length, the central fringe identification is still not straightforward.

By the use of two wavelength combination sources, the central fringe identification has become easier, particularly as is seen in Fig.1(c), where the wavelength difference between the two laser diodes has a relatively large value.

A further significant improvement can be made by using these three laser diodes together to generate a three wavelength combination source as shown in Fig.1(d), the output of which demonstrates an increasing central fringe dominance.

If, supposing the laser diodes of any wavelength can be chosen without restriction, then the optimum wavelength combination can be selected to increase the system performance.

The interference fringe patterns for the optimized two and three wavelength combination source are shown in Fig.2(a) and Fig.2(b) respectively, showing a significant improvement in the central fringe identification process and making the central fringe detection extremely straightforward.

In addition, it can be seen from Fig.2.(b) that only one fringe, the central fringe, is now absolutely dominant, thus showing a real signal resolution improvement.

EXPERIMENTAL RESULTS

The experimental arrangement used for investigation is depicted in Fig.3, representing a simple Michelson interferometer arrangement. The three laser diodes were all operated below their threshold in order to ensure the multimode nature of the output light beam and that the fringe patterns can be observed more easily on the oscilloscope.

A set of experiments was carried out by utilizing different combinations of three commonly available laser diodes, similar in characteristics to those used in the computer simulations. The fringe patterns obtained, as shown in Fig.4, are in close agreement with those in Fig.1.

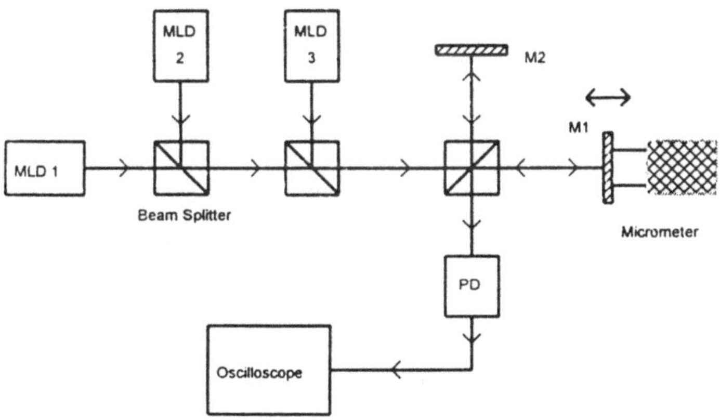

MLD : Multimode Laser Diode; M : Mirror; PD : Photodetector

Fig.3 Experimental Arrangement

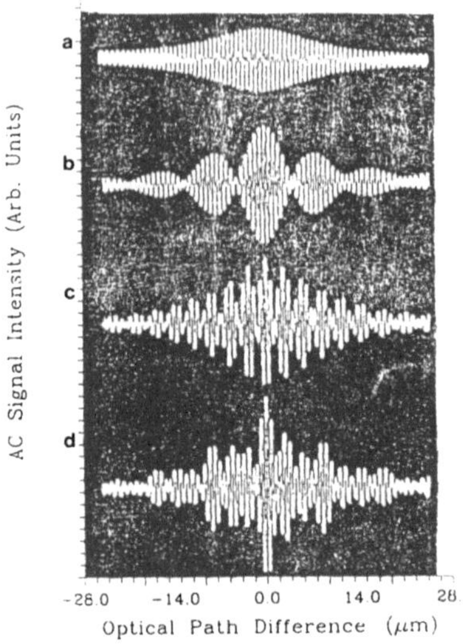

Fig. 4 Experimental Fringe Patterns (a) Single Wavelength: $\lambda = 635nm$, $L_c = 16\mu m$; (b) Two Wavelengths: $\lambda_1 = 635nm$, $\lambda_2 = 688nm$, $L_{c1} = 16\mu m$, $L_{c2} = 15\mu m$; (c) Two Wavelengths: $\lambda_1 = 635nm$, $\lambda_2 = 830nm$, $L_{c1} = 16\mu m$, $L_{c2} = 15\mu m$; (d) Three Wavelengths: $\lambda_1 = 635nm$, $\lambda_2 = 688nm$, $\lambda_3 = 830nm$, $L_{c1} = 16\mu m$, $L_{c2} = 15\mu m$, $L_{c3} = 15\mu m$

DISCUSSION

A summary and cross comparison of main theoretically and experimentally obtained SNR_{min} values is shown in Table 1.

As indicated, both from theory and experiment, a clear and progressive improvement has been made in central fringe identification by using multi-wavelength combination sources in a WLI system. With the optimum wavelength combination, the SNR_{min} can be greatly reduced, especially in the case of a three wavelength combination source, where only one fringe, the central fringe, becomes extremely dominant.

The obtained results shows that the use of a multi-wavelength combination source as an instant fringe detection technique is simple and effective and can be used effectively in a WLI sensor system. Moreover, because of the highly increased system output power and signal resolution, a number potential applications of this technique in coherence multiplexed interferometric sensor systems can also be expected.

Table 1 COMPARISON OF THEORETICAL AND EXPERIMENTAL RESULTS

Light Source λ(nm)	L_c(μm)	NR_{min} (dB) (Theory)	L_c(μm)	NR_{min} (dB) Experiment)
635	16	50.1	16	48.3±1.0
635 + 688	16, 16	34.4	16, 15	31.4±1.0
635 + 830	16, 16	21.2	16, 15	23.8±1.0
35+688 +830	16, 16, 16	18.5	16, 15, 15	18.1±1.0
635 + 797	16	*18.0		
35+724+959	16	*12.9		

* With Optimum Wavelength Combination

REFERENCES

1. S. Chen, K. T. V. Grattan, B. T. Meggitt and A. W. Palmer, "Instantaneous Fringe-Order Identification Using Dual Broad-Band Sources with Widely Spaced Wavelengths", Electronics Letters, Vol.29, No.4, pp.334-335, February 18, 1993

2. Y. J. Rao, Y. N. Ning and D. A. Jackson, "Synthesized Source for White-Light Sensing Systems", Optics Letters, Vol.18, No.6, pp.462-464, March 15, 1993
3. D. N. Wang, Y. N. Ning, K. T. V. Grattan, A. W. Palmer and K. Weir, "Characteristics of Synthesized Light Sources for White Light Interferometric Systems", Optics Letters, Vol.18, No.22, pp.1884-1886, November 15, 1993
4. D. N. Wang, Y. N. Ning, K. T. V. Grattan, A. W. Palmer and K. Weir, "Three-Wavelength Combination Source for White-Light interferometry", IEEE Photonics Technology Letters, Vol.5, No.11, pp.1350-1352, November 1993

AN OPTICAL FIBRE LINEAR POSITION SENSOR WITH CENTIMETRE SCAN UTILISING WHITE LIGHT INTERFEROMETRY AND TWO LIGHT SOURCES OF WIDELY SEPARATED WAVELENGTHS

K Weir, A W Palmer and K T V Grattan

Measurement and Instrumentation Centre
Department of Electrical, Electronic & Information Engineering
City University, Northampton Square
London EC1V 0HB UK

INTRODUCTION

The accurate measurement of displacement or position is an important procedure in many applications. Many sensors are based on inductive coupling (eg LVDT), however there is interest in using optical fibres for such measurements due to the reduced weight and electromagnetic immunity. Here an optical fibre based system using highly birefringent fibre is described with the fibre acting as the active sensing element. Using the technique of white light interferometry (WLI)[1] high resolution measurement may be made over a large range (centimetres).

BACKGROUND

Highly birefringent fibre is an important component in many optical fibre sensor systems. Its ability to deliver a defined polarisation state gives it many uses in interferometric sensor systems (where fixing the polarisation state avoids the problem of signal fading) and in possible coherent communications systems. It may also form the sensor element in optical fibre sensor systems as the birefringence is sensitive to many physical parameters (eg temperature and strain)[2].

In this system the highly birefringent fibre is used to provide the two arms of an interferometer. Linearly polarised light is launched into one of the eigenmodes of the fibre. Pressure is then applied to the fibre by squeezing it. The direction of the applied force is chosen to be at 45° to the eigenaxes of the fibre. This results in a small amount of light being coupled from the launch mode into the (originally) empty mode. A second coupling point is then applied to the fibre a distance L_s from the first, coupling of light between the two modes again occurs. At the output of the fibre the light propagating in the originally empty mode can be preferentially selected using a appropriately orientated polarising prism. The light transmitted by this polarising prism consists of two components which have an optical path difference (OPD) due to the different eigenmodes of travel within the fibre between the coupling points. This OPD is given by:

$$\mathrm{OPD} = \beta L_s \qquad (1)$$

where β is the group birefringence of the fibre and L_s is the separation between the coupling points.

If a laser light source is used interference fringes will be observed as the separation between the coupling points is changed. The technique may then be used to measure changes in the separation of the coupling points (using fringe counting or sub fringe techniques) however such a technique does not give an absolute measurement.

An suitable approach to overcome this and provide an absolute measurement is the technique of white light interferometry. In this case a broad band light source with a short coherence length is used. If the OPD introduced by the highly birefringent fibre is greater than the coherence length, no interference is observed between the coupled components at the output of the fibre. If however the light is introduced to a second interferometer (termed the 'recovery' interferometer) scanning of the path difference in this interferometer will result in interference being observed when the path difference in the recovery interferometer matches that introduced in the fibre. Maximum visibility is observed when the path difference, L_M, in the recovery interferometer matches exactly the OPD introduced in the fibre. That is

$$L_M = \beta L_s \qquad (2)$$

The requirement that the OPD within the fibre is greater than the coherence length, L_c, of the light source specifies a minimum separation between the coupling points L_{min} define by

$$L_{min} = \frac{L_c}{\beta} \qquad (3)$$

One problem for the accurate application of this technique is identifying the situation of maximum visibility in the interference fringes. The intensity difference between adjacent fringes in the interference pattern of a low coherence light source is small, and in the presence of noise in the system intensity considerations alone are not sufficient to identify the central fringe. One technique to improve this situation, which is applied to the system here, is to use two low coherence light sources of different wavelengths[3]. Each wavelength produces its own interference pattern which then add to give a beat pattern in which the fringe corresponding to zero path difference is accentuated.

EXPERIMENTAL ARRANGEMENT

The experimental arrangement is shown in figure 1. The light source used were laser diodes operating below threshold (to give a broad band output) with nominal wavelengths of 670nm and 780nm. The outputs from the light sources were combined and polarised such as to form the input to one of the eigenmodes of a 2m length of York HiBi fibre. The fibre was designed for use at a wavelength of 633nm so the launch efficiency was poor and attenuation high, however, for the short lengths of fibre used this was not a critical problem. The output from the fibre was collimated and passed through a polarising prism to select the light from the originally empty eigenmode. Finally the light entered a Michelson recovery interferometer. The mirrors in the Michelson interferometer were attached to piezo-electric transducers with a rage of 100μm. These PZTs incorporated a capacitive displacement sensor. The output from this sensor was used in a control loop to overcome hysteresis and

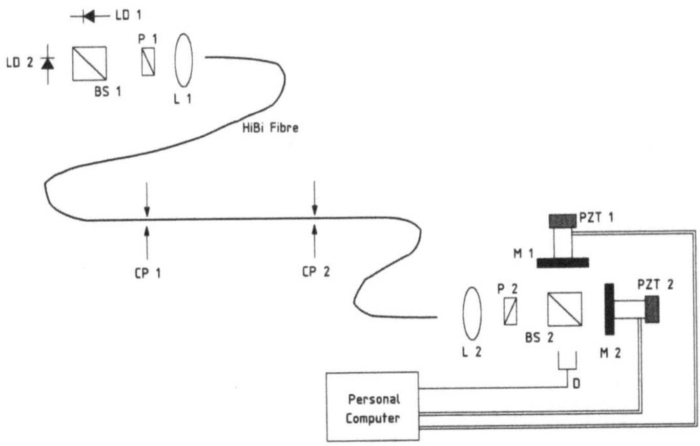

Figure 1. The experimental apparatus, LD laser diodes, BS beam splitters, P polarising prisms, L lenses, CP coupling points, M mirrors, PZT piezo-electric transducers (with displacement sensors) and D detector.

could be accessed directly from computer to give a direct measurement of the movement of the mirror. An ultimate resolution of 3nm was achievable using this displacement sensor. The intensity at the output of the Michelson interferometer was detected using a PIN photodiode and associated amplifier. The output form the detector circuit was again read by the computer via a 12 bit analogue to digital converter. Thus the system was under complete computer control. The path difference of the Michelson interferometer could be scanned and the mirror movement accurately recorded from the capacitive displacement sensor, and the corresponding intensity output of the interferometer also recorded, all automatically.

The coupling points, applied to the fibre, were in the form a narrow 'tooth' of 0.5mm width with a mass of 400g applied.

In operation, the Michelson interferometer was adjusted to zero path difference and the mirror adjusted to take the interferometer output out of the interference region (a mirror movement of typically 100μm). The second mirror was then scanned through its complete range of 100μm in steps of 10nm, and the intensity recorded. The position of maximum intensity was recorded as corresponding to the position of central fringe. No additional signal processing was required. A typical interference scan is shown in figure 2.

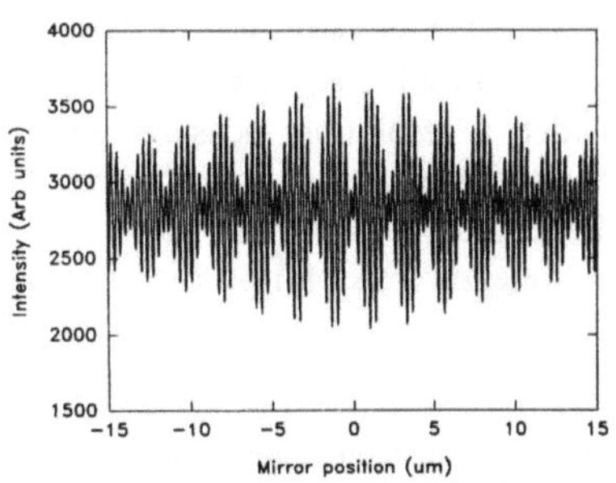

Figure 2. Typical interference pattern obtained.

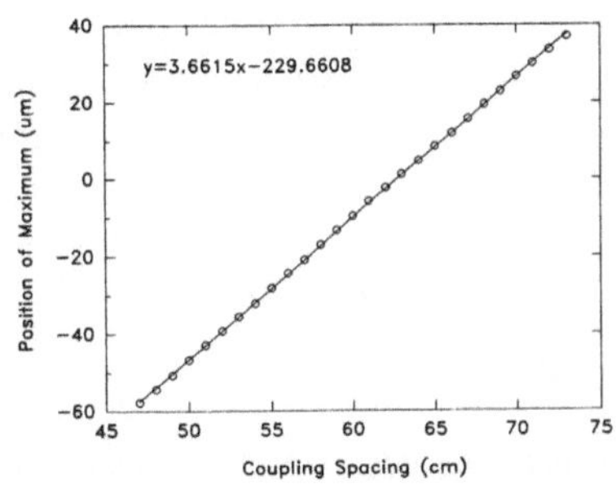

Figure 3. Calibration graph of path difference against coupling point separation.

The process was repeated for a range of separation between the coupling points from 47cm to 73 cm. A plot of mirror position in the Michelson corresponding to maximum intensity against the spacing between the coupling points is shown in figure 3. The mirror position is given as the value read from the capacitive displacement sensor by the computer.

From the calibration graph of figure 3, the separation corresponding to any position of maximum intensity may be calculated. Over the 100μm range of the PZT, a resolution of 10nm was easily achieved in reading mirror position via the capacitive displacement sensor. This translates to a range of 25cm with a potential resolution of 30μm. Experimentally, with a coupling point width of 0.5mm, a repeatable resolution of 0.2mm was achieved.

PRACTICAL CONSIDERATIONS

A number of practical points must be considered for the application of this technique:-

- The fibre must be fixed to ensure the applied force is always close to 45° to the eigenaxes of the fibre for efficient coupling. Perfect alignment need not necessarily be required as information is encoded in phase, not amplitude.
- To take full advantage of the potential resolution the point of contact with the fibre must be minimised. However, obviously it can not be too sharp so as to potentially damage the fibre. The contact must also slide easily over the fibre. Test have shown that fibres with suitable coatings will survive in excess of 1,000 sliding movements
- The use of two widely spaced wavelengths within a fibre means that the effects of dispersion may become important. In this study, with the short lengths of fibre used, no such effects were observed.

A further potential problem is thermal effects as the birefringence of the fibre is temperature dependent. The effect can however be reduced. By using elliptical core fibre (eg from Andrew Corporation) the thermal sensitivity is reduced by approximately a factor of 4. Alternatively a third coupling point may be introduced into the system. This coupling point is fixed and gives rise to another set of interference fringes. As the coupling point is fixed, any movement in the position of these fringes is due to environmental effects. This movement can then be used to determine the thermal sensitivity and used to correct the measurement of the position of the moving coupling point. This would necessarily reduce the range of the system (possibly by as much as a factor of 2) as room would need to be left within the scan range for the fringes from the fixed coupling point, reducing the free range for the fringes from the moving coupling point.

CONCLUSIONS

A simple white light interferometric system for the measurement of position has been described. The use of two broad band light sources provided an easy means of identifying the maximum fringe visibility. The use of a high resolution PZT control system to perform the linear scan in the recovery interferometer provided a resolution measurement. The potential resolution of the system was the order of 30μm (corresponding to a scan resolution of 10nm), however, experimentally, a resolution of the order 0.2mm was achieved. The range was 25cm. This represents a very useful range of measurement for position sensing.

REFERENCES

1. R Ulrich, 'Interferometric and polarimetric sensors using white-light interferometry', Springer Proceedings in Physics, Vol. 44 *Optical Fiber Sensors*, pp 62-3 (1989).
2. J Noda, K Okamoto & Y Sasaki, 'Polarization maintaining fibres and their applications', J. Lightwave Technol., **LT-4**, pp 1071-1089 (1986).
3. D N Wang, Y N Ning, K T V Grattan, A W Palmer, & K Weir, 'The optimized wavelength combination of two broadband light sources for white light interferometry' J. Lightwave Technol. in press (1993).

SIMULTANEOUS MEASUREMENT OF STRAIN AND ACOUSTIC EMISSION USING A PASSIVE HOMODYNE DEMODULATED FIBER-OPTIC SENSOR

Kexing Liu

Canadian Marconi Company
600 Doctor Frederik Philips Boulevard
Ville Saint-Laurent, Québec, H4M 2S9, CANADA

INTRODUCTION

Simultaneous measurement of strain and acoustic emission is often required in fiber optic structural sensing. This is of particular importance in aircraft structural integrity monitoring where weight saving, minimum perturbation, and safety are essential.[1] Over the past few years a number of different sensor configurations (with their respective advantages and disadvantages) have been proposed.[2] Most of the techniques require either a narrow linewidth laser that is continuously tunable over a wavelength range of about 20nm without mode-hops or a relatively high power broad-band lightsource that has low intensity noise. This paper proposes a technique that uses the standard fiber Michelson interferometer with a 3x3 directional coupler to obtain quadrature outputs and to process the strain and acoustic emission signals in parallel.

The principles of fiber optic strain sensing and acoustic emission sensing have been discussed very extensively in other publications.[3,4,5] Typically, strain measurement range is ±0.5% which corresponds to approximately ±800rad. of optical phase of the interferometer (assuming 10mm sensor, 850nm light source). The bandwidth is about 10kHz. On the other hand, the acoustic emission signals usually induce optical phase change on the order of 0.01 to 0.1 rad.; and the spectrum of these signals falls into a range of 50kHz to 2MHz. The sensor system must be able to cover the two regime in order to perform simultaneous measurement.

SYSTEM CONFIGURATION

The optical configuration of the system consists of a single-mode semiconductor laser, a balanced fiber Michelson interferometer as the sensor, and a monolithic 3x3 fiber directional coupler. The strain and acoustic emission signals from both quadratic outputs of the interferometer are separated by high/low bandpass filters. The quadratic fading in the (high frequency) acoustic emission signals can be corrected with respect to the low frequency phase information by a simple algorithm. The analysis of the measurement error

Applications of Photonic Technology
Edited by G.A. Lampropoulos *et al.*, Plenum Press, New York, 1995

due to non-quadrature offset in the fiber coupler is presented in the next section. The strain measurement can also be derived from the low frequency phase information by calculation or via a look-up table. Figure 1 illustrates the configuration of the experimental system.

The Michelson fiber interferometer has a short unbalanced optical path and therefore the optical phase noise is reduced. Ordinary single mode fiber (York SM800) is used in the system. Light from a pig-tailed laser diode (General Optics GO-ANA, 857nm) is coupled into the two interferometric arms through a 3x3 fiber directional coupler (Sifam 33S82C). The third arm of the coupler is index matched to remove Fresnel reflection. The two fibers are paired close together to improve common mode rejection. The localized sensing region is formed by the misaligned part between the mirrored ends of the two fibers (approximately 10mm).

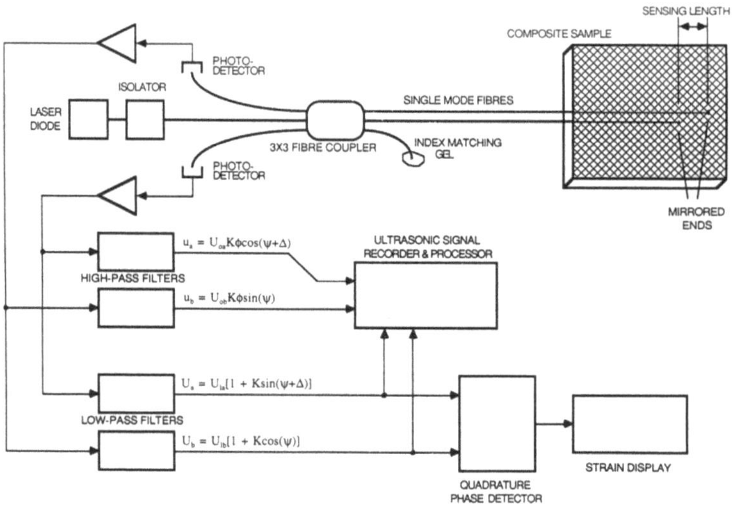

Fig.1. Schematics of the system configuration for the simultaneous measurement of strain and acoustic emission.

The two reflected beams are recombined and subject to interference in the 3x3 coupler. The relative phase of the optical interference signals, γ, distributed through the three output ports of the coupler is a function of power splitting ratio i.e. the coupling coefficients. If the splitting ratio is exactly 1:1:1, a relative phase of 120° exists between any two of the three output ports.[6,7] However, for a commercial 3x3 coupler, the variation in the splitting ratio allows us to choose the two outputs with the phase difference closest to satisfy the quadrature condition.

ALGORITHM AND ERROR ANALYSIS

In an ideal case of zero quadrature error, the optical intensities of the interference from the two pre-selected output arms of the coupler received by two photodiodes (Hamamatsu S1190-03) are:

$$I_a = I_{oa}[1 + K\sin(\phi+\psi)], \quad (1)$$

and

$$I_b = I_{ob}[1 + K\cos(\phi+\psi)], \quad (2)$$

respectively. Where I_{oa} and I_{ob} are the nominal output optical intensities, K is the fringe visibility, ϕ is the ultrasonic wave-induced optical phase change, and ψ is the strain induced (slowly varying) optical phase difference. In reality, the power splitting ratio and the relative phase retardation of a commercial 3x3 coupler are fixed and can be determined by a simple measurement. The quadrature error, Δ, is defined by:

$$\Delta = \gamma - 90^{\circ}. \tag{3}$$

Expanding equations (1) and (2) and taking the quadrature error into consideration, we arrived at:

$$I_a = I_{oa}[1 + K\sin(\phi)\cos(\psi+\Delta) + K\cos(\phi)\sin(\psi+\Delta)], \tag{4}$$

and $$I_b = I_{ob}[1 + K\cos(\phi)\cos(\psi) - K\sin(\phi)\sin(\psi)]. \tag{5}$$

Since the ultrasonic wave-induced optical phase change, ϕ, is a smaller amplitude (<0.1rad) and higher frequency signal (100kHz - 2MHz) compared with the strain and vibration induced optical phase change in the interferometer, it can be assumed that $\sin\phi \approx \phi$ and $\cos\phi \approx 1$. Passing through high pass filters (3db cut-off frequency at 70kHz), the two output voltage signals, u_a and u_b, corresponding to I_a and I_b become:

$$u_a = U_{oa}K\phi\cos(\psi+\Delta), \tag{6}$$

and $$u_b = U_{ob}K\phi\sin(\psi), \tag{7}$$

where U_{oa} and U_{ob} are the nominal voltages including the gains of the high-pass amplifiers. Similarly, voltage signals, U_a and U_b, after the low-pass filters (3dB cut-off frequency at 20kHz), can be written as:

$$U_a = U_{la}[1 + K\sin(\psi+\Delta)], \tag{8}$$

and $$U_b = U_{lb}[1 + K\cos(\psi)], \tag{9}$$

where U_{la} and U_{lb} are the nominal voltages including the gains of the low-pass amplifiers. Through quadrature phase detection[4] the low frequency phase change (e.g. strain induced phase signal) can be determined. The error Δ is a constant value within a large temperature range and can be compensated by calculation.

For the ultrasonic induced phase change, ϕ, a simple calculation is used to recover the digitised signal. The outputs from the low-pass filter, i.e. equations (8) and (9) are used to obtain $\sin(\psi+\Delta)$ and $\cos(\psi)$ as described by the following equations.

$$K\sin(\psi+\Delta) = [(U_a / U_{1a}) - 1] \tag{10}$$

$$K\cos(\psi) = [(U_b / U_{1b}) - 1] \tag{11}$$

If the absolute value of $\cos(\psi)$ is equal to or greater than 0.707, ϕ can be recovered from equation (6); and if $\cos(\psi)$ is smaller than 0.707, ϕ can then be recovered from equation (7). This simple algorithm is proclaimed by the following equations:

$$\phi = \begin{cases} u_a / [U_{oa}K\cos(\psi)] & |\cos(\psi)| \geq 0.707 \\ u_b / [U_{ob}K\sin(\psi+\Delta)] & |\cos(\psi)| < 0.707 \end{cases} \tag{12}$$

The quadrature error Δ can be compensated in the digital processing of the signal. If uncompensated, such as in a hardware approach to implement the above operation, the ultrasonic signal amplitude error, $\delta\phi$, due to this non-quadrature condition with quadrature error $\pm\Delta$ is then estimated as:

$$\delta\phi \leq |\sqrt{2}\,[u_b / (U_{ob}K)][1-1/\{\sin(\Delta)+\cos(\Delta)\}]|. \tag{13}$$

DISCUSSIONS

Compared to active homodyne,[2] the passive homodyne approach requires no active modulation components, and the two fibers can be paired together without any separation, thereby reducing the ambient effect. However, like all interferometric sensors based on ordinary single-mode fibers, this system suffers from fringe visibility variations induced by random fluctuations in the state of polarization (SOP) of the two interfering beams. The visibility change due to input orientation of the SOP was minimized by preadjustment and maintained by securing the lead fibers against any tight bends, twists or vibrations. Nevertheless, polarization maintaining fibers and components are recommended for future sensors. Other errors in the system such as light intensity (I_0) variation and digitizing errors also limit the performance of this system. A monolithic 4x4 coupler can also be used in lieu of the 3x3 coupler, therefore providing quadrature outputs with smaller error. The penalty is that the optical intensity level is then reduced by nearly 3dB, which is critical to a system with low signal-to-noise ratio margin.

This technique was applied to an experiment where the fiber-optic sensor was embedded in the composite specimen which was subject to a three-point bend loading.[8] Due to the relatively high noise level of the laser source only high level acoustic emission signals were recorded. The measured quiescent noise level of the system is approximately $1.0\mu V/\sqrt{Hz}$ which translates into about 10^{-4}rad./$\sqrt{Hz}$ optical phase noise. A better light source such as a diode-pumped single frequency solid state laser would be preferable in future experiments.

ACKNOWLEDGEMENT

This work was performed at the University of Toronto Institute for Aerospace Studies (UTIAS) with the support of Ontario Laser and Lightwave Centre (OLLRC) and the Natural Sciences and Engineering Research Council (NSERC). Encouragement and advice from Prof. R. M. Measures is appreciated.

REFERENCES

1. R.M. Measures and K. Liu, Interferometric fiber optic sensor demodulation for smart structures, *IEEE Circuits and Devices*, 8:4 (1989).
2. K. Liu and R.M. Measures, Signal demodulation techniques for localised interferometric strain sensors, *J. of Intelligent Material Structures and Systems*, 15:1255 (1990).
3. K. Liu, S.M. Ferguson, and R.M. Measures, Fiber-optic interferometric sensor for the detection of acoustic emission within composite materials, *Optics Letters*, 15:1255 (1990).
4. T. Valis, E. Tapanes, K. Liu, and R.M. Measures, Passive quadrature demodulated localized michelson fiber optic strain sensor embedded in composite materials, *J. Lightwave Technology*, 9:535 (1991).
5. S.M. Ferguson, The detection of damage induced acoustic emission in advanced composite materials using optical fiber sensors, Master's thesis, *UTIAS* (1990).
6. S.K. Sheem, Optical fiber interferometers with [3x3] directional couplers: analysis, *J. Appl. Phys.*, 52:3865 (1981).
7. K.P. Koo, A.B. Tventen, and A. Dandridge, Passive stabilisation scheme for interferometers using 3x3 fiber directional couplers, *Appl. Phys. Lett.*, 41:616 (1982).
8. ASTM, Standard test methods for flexural properties of unreinforced and reinforced plastics and electrical insulating materials (metric), ASTM Designation: D 790M-84, *Annual Book of ASTM Standards*, 8.01 (1984).

SEMICONDUCTOR LASER OVERVIEW

Richard Murison

EG&G Optoelectronics Canada
22001 Dumberry Road
Vaudreuil, Quebec J7V 8P7
CANADA

INTRODUCTION

A very brief overview is given of the large and wide-ranging field of semiconductor laser diodes, in the context of the development of the field from its earliest days. The key issues are described, and the interested reader is invited to consult the references cited for more thorough treatments. Following the first theoretical discussions, and practical demonstrations, of lasing in semiconductors in the early 1960s, it took over ten years before lasers could be fabricated with remotely useful levels of performance. By the early 1980s, however, materials and technology had progressed at a phenomenal pace to the point where semiconductor lasers were employed at the heart of the Compact Disc players which came to dominate consumer audio, optical fiber communications links which now route virtually all of the world's telephone trunk lines, and laser printers which have displaced mechanical printers from all but the lowest cost end of the market. Semiconductor lasers in the 1990s are expanding into new materials systems, exploiting mechanical strain as a design feature, and being implemented in many new device topologies. Despite these phenomenal successes, semiconductor lasers in particular, and photonics technology in general, still remain to be widely exploited.

SEMICONDUCTOR LASER BASICS

The possibility of exploiting stimulated emission in semiconductors, arising from transitions between directly coupled conduction and valence bands, was first proposed by Basov[1], and discussed in considerable detail by Bernard and Duraffourg[2], as early as 1961. They showed that strict population inversion was not necessary for net gain due to the continuous distribution of energy states either side of the forbidden band. It would only be necessary to separate the quasi Fermi levels by more than the photon energy. Since the separation of the quasi Fermi levels is governed by the density of states in each band, net gain could best be achieved by reducing the density of states in each band as much as possible. However, high values of gain could only arise due to transitions between levels with a high density of states, that is to say, levels somewhat above the band edge. Practical devices would therefore require the quasi Fermi levels to be separated by more than just the band-gap energy. The generation of sufficiently high gain for lasing (the threshold gain), arising from transitions between two appropriate energy levels, would therefore require the lower energy states in each band to be filled first, and implies therefore the presence of a high excess carrier density.

The generation of high excess carrier densities is most readily achieved at a forward biassed p-n junction. In order to minimize the bias current necessary to achieve a particular carrier density, the total volume in which this carrier density is generated must be minimized. This is done by sandwiching the p-n junction between two layers of a higher band-gap material, thereby confining the injected carriers to the lower band-gap junction or "active" layer. The total volume of this layer, and therefore the bias current needed to pump it to the required excess carrier density (the threshold current), can be minimized by reducing the thickness of this layer as much as possible. This configuration is commonly referred to as a double heterostructure, or DH.

The higher band-gap layers of the DH fulfil a second function. Since they have a lower refractive index than the active layer, they form an optical waveguide, whose core is the active layer. Any optical lasing mode propagating along the junction plane will be guided by this optical waveguide. Consequently, the optical gain experienced by the mode is not simply the gain inside the active layer. It is reduced by a factor describing the degree to which the guided mode spreads out into the wide band-gap layers where there is no gain. If the active layer is reduced in thickness below a certain value, typically of the order of 0.2μm, this interaction causes the threshold gain, and consequently the threshold current, to rise rapidly. There is therefore a significant tradeoff between the optical and electronic properties of a DH laser[3]. Several complex multi-layer heterostructures have been proposed which aim to decouple those design features which are responsible for the optical performance from those which control the electronic properties[4].

QUANTUM WELLS AND STRAIN

If the optical and electronic properties can be completely decoupled, then we are free to consider reducing the active layer thickness to arbitrarily thin layers. The equations which govern the behaviour of electrons or holes in a potential well are completely analogous to the confinement of light by a slab waveguide. Just as narrow core waveguides are found to confine light to certain discrete modes which propagate in the plane of the waveguide, an electron in a narrow potential well is confined to certain discrete (quantized) energy states, and is restricted to motion only in the plane of the well. In order for these "quantum confinement" effects to be significant, the active layer thickness would need to be reduced to the order of 100Å, about 30 or so atomic layers. Such thin layers are now referred to in common parlance as quantum wells. The confinement of the electron's motion to two dimensions has great significance for the available density of states. In bulk material, where the electron is free to move about in three dimensions, the density of states increases as the square root of its energy. In a quantum well, for each confined state, the density of states is constant, independent of the energy arising from its in-plane motion. Provided this constant density of states is high enough, and it turns out that it is, we can obtain sufficient gain from transitions between the lowest available energy states in the well, with lower values of threshold current than would be required for a bulk active layer[5]. The quantized states in a quantum well were first observed in 1974 [6], and quantum well lasers were postulated immediately after, but due to difficulties of producing good abrupt interfaces in materials of sufficiently high quality, it took until 1982 [7] until lasers could be fabricated which demonstrated properties significantly superior to those of bulk active layer devices. At the present time, quantum wells have become the preferred design approach for all new semiconductor lasers[8].

In the early days of ternary and quaternary compound semiconductors, great pains were taken to ensure a precise lattice match between each layer and the substrate. Repeated experiments showed a strong correlation between mismatched layers, which exhibited crystalline defects or dislocations, and poor reliability. These were often related to the enhanced formation and propagation of "dark line defects", regions of degraded material. Until late in the 1980s, the concepts of lattice mismatch, and the dislocations which arose through relaxation of the attendant strain, were seldom distinguished. It had, however, been shown in 1974 [9] that strains induced through lattice mismatch could in fact be accommodated elastically, provided that the mismatched layer did not exceed some misfit-dependent critical thickness. This critical thickness turned out to be more or less equivalent to the thickness of a quantum well. Due in large part to the widely held conviction that strained layers, whether accommodated elastically or not, would in any case relax into a mess of dislocations under the stress of laser operation, the properties of lasers fabricated with strained quantum wells was not investigated until 1984 [10]. By 1989 it was becoming clear that, surprisingly enough, reliability was not being compromised[11]. Although the primary motivation behind the initial experiments with strained layers was to gain access to previously inaccessible wavelengths, it happens that the strain also imparts significant modifications to band structure of the material[12,13]. Taken in total, these developments sparked a massive surge in interest in the properties of strained layer quantum well lasers. The presence of strain has enabled precise tailoring of the band structure within the quantum well, permitting highly desirable configurations which would not previously have been achievable. This "band-gap engineering" has now become *de rigeur* in all areas of laser diode research.

MATERIALS

For the efficient generation of photons in a semiconductor, a direct band-gap transition must be employed. In silicon, the most widely used semiconductor, the band structure is indirect, and the generation of photons is extremely inefficient. The development of direct band-gap compound semiconductors such as GaAs has proven to hold the keys to rapid advancement. In the early days of GaAs it was clear that it was the semiconductor of the future. Being inherently superior to silicon in so many critical respects, it appeared clear that GaAs would supersede silicon completely. Today, in the field of microelectronics, it remains the semiconductor of the future, but a future that looks increasingly further off as silicon continues to hold court. The picture is different in optoelectronics, where GaAs and many other compound semiconductors are firmly established. These materials have permitted the fabrication of semiconductor lasers and LEDs at almost any wavelength from 500nm to 100µm, and the range continues to widen.

In the beginning there was Gallium Arsenide. This was one of the candidate compound semiconductors proposed by Bernard and Duraffourg[2]. The first demonstrations of lasing action in a semiconductor used a p-n junction in a piece of GaAs deep inside a cryostat. This was accomplished independently, and virtually simultaneously, in 1962 in four separate laboratories[14,15,16,17]. From these early experiments, progress has been rapid. The key to this success has been the development of heterojunctions[18,19], where regions of a single crystal can be prepared with electronic and optical properties tailored precisely to the device's needs. For example, layers of higher band-gap $Al_xGa_{1-x}As$ either side of a layer of GaAs provide two functions. First, the higher band-gap traps injected carriers in the lower band-gap GaAs layer where they are needed. Second, the lower refractive index (considered to be a consequence of the higher band-gap, but not always necessarily so) confines the generated photons to the GaAs layer. By 1970, double-heterostructure lasers constructed in this manner had been demonstrated, operating continuously at room temperature[20,21]. $Al_xGa_{1-x}As$ alloys having any AlAs mole fraction x can be grown directly on GaAs because the lattice constant of AlAs is virtually identical to that of GaAs. Using this materials system, semiconductor lasers can be fabricated at any wavelength from about 780nm up to 905nm.

The initial motivation for the development of semiconductor lasers was in the field of fiber-optic communications. It soon became clear that significantly lower losses could be found in optical fibres around the 1.3µm and 1.55µm wavelength windows. At these wavelengths a different materials system would be required, and $In_xGa_{1-x}As_{1-y}P_y$ was identified as the obvious choice. These alloys could be grown on InP substrates, but unlike the situation with AlGaAs, the lattice constant of this range of alloys varies considerably according to the mole fractions x and y. Only those alloy compositions with $x \approx 0.53(1-y)$ were found to be lattice matched to the InP substrate, which made the growth of high quality dislocation-free heterojunctions a more demanding problem. Continuous operation at room temperature was first reported at 1.3µm wavelength in 1977[22], and at 1.55µm in 1979[23], but such was the rate of progress that by the early 1980s production devices were being installed in commercial telephone trunk lines[24]. $In_xGa_{1-x}As_{1-y}P_y$ lattice matched to GaAs is now being grown to eliminate Aluminium from the materials system[25]. Aluminium, which oxidizes rapidly, particularly at the exposed surface, has been identified as a culprit in many power limitation and reliability problems.

Although the $In_xGa_{1-x}As_{1-y}P_y$ materials system goes out to over 3µm, wavelengths longer than about 1.6µm cannot be reached with materials lattice matched to InP. Other materials systems have been characterized which emit at wavelengths out to 34µm, including $In_xGa_{1-x}As_{1-y}Sb_y$, $Al_xGa_{1-x}As_{1-y}Sb_y$, $Pb_{1-x}Sn_xTe$, $PbS_{1-x}Se_x$, $Cd_xHg_{1-x}Te$ and $Cd_xPb_{1-x}S$, and the solid solution $Bi_{1-x}Sb_x$ has emitted at wavelengths as long as 100µm. While some success has been achieved with these materials, the devices in general require to be cooled severely, which has limited their practical application, and has largely confined them to the periphery of the mainstream. Space restrictions prevent a fuller discussion in the context of this article.

Visible semiconductor lasers have been pursued almost from the beginning with the ZnS_xSe_{1-x} family of materials, more recently expanded to include $Cd_yZn_{1-y}Se$ and $Zn_{1-y}Mg_yS_xSe_{1-x}$, which emit at blue-green wavelengths. Only recently, however, has lasing action been demonstrated, and serious reliability concerns are being voiced. These II-VI materials have now been joined by the III-V Nitrides $(Al_yGa_{1-y})_xIn_{1-x}N$, and it will be interesting to see which materials system ultimately proves to be superior. At the red end of the visible spectrum, $(Al_yGa_{1-y})_{1-x}In_xP$ is producing commercial lasers, typically at 630nm and 670nm wavelengths, with performance almost on a par with that obtained in the near infrared with $Al_xGa_{1-x}As$.

SEMICONDUCTOR LASER STRUCTURES

Nature has been kind to us. In the zincblende crystal structure shared by most of the compound semiconductors used for laser diodes, if the laser structure is grown on the easiest crystal surface to prepare, the 100 surface, the orthogonally perpendicular 010 surfaces can be easily exposed by cleaving. These surfaces can then be employed as the end face mirrors required to establish a traditional Fabry-Perot laser cavity. All that remains to form a laser is to drive an electric current through the region between the mirrors. As has been stated above, a certain threshold current is required to generate the threshold gain needed to turn the laser on. Above this threshold current, the gain is clamped by the lasing action, and the additional current is converted, via stimulated emission, into photons which serve to amplify the lasing mode. With properly prepared materials, this can often be accomplished with close to 100% quantum efficiency.

A given epitaxial laser structure is characterized by a typical threshold current density, rather than a typical threshold current. In order to minimize the absolute threshold current, it is necessary to minimize the total area which the current has to pump, by minimizing the cavity length, and the width which is being pumped. Changing the cavity length changes the threshold gain, due to optical losses through the cavity end mirrors (the emitted beam); this effectively determines the optimum cavity length for any given epitaxial laser structure. The width which is pumped is most easily controlled by opening a stripe in a layer of insulating oxide, or some similar material, deposited on the surface of the wafer, usually directly below the contact metallization. This is the so-called stripe laser[26], which to this day remains the simplest means of fabricating a semiconductor laser diode.

In many applications it is necessary for the laser output to be in the form of a single spatial mode, usually the fundamental spatial mode. As described above, the epitaxial structure forms a waveguide perpendicular to the junction plane. This is usually selected to support only the fundamental mode. In the transverse direction within the junction plane (the direction of propagation is conventionally referred to as the longitudinal direction), with a conventional planar epitaxial structure, there is nominally no waveguide. However, under laser operation, the gain under the stripe is represented by a change in the imaginary part of the refractive index, which effectively serves to confine the light to the region under the stripe. Such devices are commonly referred to as "gain guided" lasers. These devices, however, have proven to be unsatisfactory for single spatial mode applications due to high threshold currents, and other undesirable properties. For this reason, single spatial mode lasers are normally constructed with a real refractive index waveguide in the transverse direction. Space limitations prevent a full discussion of the many various approaches to this problem[27,28,29], but among the most common and successful types are Ridge Waveguides (RWG)[30] and the many varieties of Buried Heterostructure (BH)[31].

Operation in the fundamental spatial mode is often referred to as "single mode", even though in actual fact such a device often operates in a plurality of longitudinal modes. Each mode corresponds to a separate line in the output spectrum. Even though such a laser under any specific drive condition may operate only in a single longitudinal mode, or single spectral line, the particular mode may hop from line to line in response to changes in temperature or drive current, or due to aging. If this behaviour is to be overcome, some mechanism must be incorporated in the laser design to select a specific spectral line, and preferably to suppress the fundamental Fabry-Perot modes of the cavity. The laser chip could be operated as the gain element in an external cavity, with anti-reflection coatings applied to its facets to suppress the F-P modes, and with a diffraction grating used as an external mirror to select the desired spectral line. While excellent results can be obtained in this way, the implementation is mechanically challenging, optically inefficient, and usually quite bulky. An alternative approach is to incorporate the wavelength selection mechanism into the laser diode itself, by incorporating a periodic variation in the refractive index, as seen by the guided wave, along the direction of propagation. Devices employing these periodic features fall into two camps, the Distributed Bragg Reflector (DBR)[32] where the periodic variations function as wavelength selective cavity mirrors, and are only implemented at either end of the cavity, and the Distributed Feedback (DFB)[33] laser where the periodic variations exist along the entire cavity length.

Throughout the 1980s many variations on the theme of a conventional laser diode have been explored. Surface emitting lasers[34,35], phase locked arrays[36], and superluminescent diodes[37] have been proposed and reported, without seeming to solve many real problems. The 1990s, however, have been witness to some advanced and creative thinking about different ways in which a laser cavity could be constructed. Of potentially great importance is the MOPA[38] (Master Oscillator - Power Amplifier) structure demonstrated by SDL Inc., in which the output of a small-signal DBR laser (alternatively, a DFB could have been employed) is integrated with a tapered travelling wave amplifier. This three-terminal single chip device combines single spatial mode, single spectral line output, with high output power levels previously obtainable only with multi-moded broad area lasers. The drawback to this device is significant curvature of the phase front at the output facet, but a number of solutions suggest themselves. Vertical Cavity Surface Emitting Lasers (VCSELs) are now being reported at various wavelengths by many laboratories[39,40,41]. These structures are extremely complicated, and involve epitaxially grown bragg reflectors for the cavity mirrors. VCSEL devices can in principle be designed with negligibly low threshold currents, and with an easily collimated output beam. In addition, the prospect for implementing these devices in individually addressable two-dimensional arrays is attractive for many applications. The integration of an echelle grating as a wavelength selectable element has been proposed by various workers. Perhaps the most interesting implementation of this concept, the so-called "*MAGIC laser*"[42], is an N+2 terminal device, comprising N individually addressable lasers each having its own single wavelength, internally coupled to a single output port. Early demonstrations by Bellcore have exhibited fifteen separately addressable wavelengths close to 1520nm, each separated by only 2nm. A similar structure has been implemented by the National Research Council of Canada[43] to demodulate eight wavelengths, each being passed to a separate detector. Also from SDL Inc., is a curious ring laser structure formed by multiple bragg reflectors[44], which has the interesting property of single spatial mode single spectral line operation, almost regardless of the size of the cavity, and which, when coupled to a surface-emitting output grating, exhibits a near diffraction-limited output beam. This device itself may be overly complex to find a commercial application, but illustrates some interesting and potentially useful properties of bragg mirrors, and provides some useful insights into the operation of broad area lasers.

These developments, and many others, are rapidly expanding the applications horizon for semiconductor lasers. It remains to be seen, however, whether sufficient new photonics-based applications, utilizing the specific properties of these remarkable devices, can be realized.

REFERENCES

1. **N. G. Basov** *et al*, Sov. Phys. JETP, **13**, p1320 (1961)
2. **M. G. A. Bernard and G. Duraffourg**, Phys. Status Solidi, **1**, p699 (1961)
3. **W. W. Anderson**, IEEE,JQE, **1**, 6, p228, (1965)
4. **G. H. B. Thompson** *et al*, J. Appl. Phys., **47**, 4, p1501 (1976)
5. **S. W. Corzine** *et al*, Chapter 1 in Ref. 8
6. **R. Dingle** *et al*, Phys. Rev. Lett., **33**, p827 (1974)
7. **W. T. Tsang**, Appl. Phys. Lett., **40**, p217 (1982)
8. "Quantum Well Lasers", **Peter S. Zory jr.** (Ed.), Academic Press (San Diego CA), (1993)
9. **J. W. Matthews and A. E. Blakeslee**, J. Crystal Growth, **27**, p118 (1974)
10. **W. D. Laidig** *et al*, Appl. Phys. Lett., **44**, p653, (1984)
11 **S. E. Fischer** *et al*, Appl. Phys. Lett., **54**, p1861, (1989)
12. **A. R. Adams**, Electron. Lett., **22**, p249, (1986)
13. **E. Yablonovitch and E. O. Kane**, IEEE-JLT **4**, p504, (1986)
14. **R. N. Hall** *et al*, Phys. Rev. Lett., **9**, p366, (1962)
15. **M. I. Nathan** *et al*, Appl. Phys. Lett., **1**, p62, (1962)
16. **T. M. Quist** *et al*, Appl. Phys. Lett., **1**, p91, (1962)
17. **N. Holonyak jr. and S. F. Bevacqua**, Appl. Phys. Lett., **1**, p82, (1962)
18. **H. Kroemer**, Proc. IEEE, **51**, p1782, (1963)
19. **Z. I. Alferov and R. F. Kazarinov**, Author's certificate 1032155/26-25, USSR, (1963)
20. **I. Hayashi** *et al*, Appl. Phys. Lett., **17**, p109, (1970)
21. **Z. I Alferov** *et al*, Fiz. Tekh. Polupov, **4**, p1826, (1970)
22. **K. Oe** *et al*, Jpn. J. Appl. Phys., **16**, p1273 (1977)
23. **G. D. Henshall and P. D. Greene**, Electron. Lett., **15**, p621, (1979)
24. **P. K. Runge and P. R. Trischitta**, IEEE-JLT **2**, p744, (1984)
25. **D. Z. Garbuzov** *et al*, IEEE-JQE **27**, 6, p1531, (1991)
26. **J. E. Ripper** *et al*, Appl. Phys. Lett., **18**, p155, (1971)
27. "Semiconductor lasers and heterojunction LEDs", **H. Kressel and J. K. Butler**, Academic Press (New York, NY), (1977)
28. "Heterostructure Lasers", Parts A & B, **H. C. Casey jr., and M. B. Panish**, Academic Press (New York, NY), (1978)
29. "Physics of semiconductor laser devices", **G. H. B. Thompson**, John Wiley & Sons (Chichester, UK), (1980
30. **I. P. Kaminow** *et al*, Electron. Lett., **15**, p763, (1980)
31. **T. Tasukada**, J. Appl. Phys., **45**, p4899, (1974)
32. **W. Tsang and S. Wang**, Appl. Phys. Lett., **28**, p596, (1976)
33. **H. Kogelnik and C. V. Shank**, Appl. Phys. Lett., **18**, p152, (1971)
34. **G. A. Evans** *et al*, IEEE-JQE **25**, p1525, (1989)
35. **M. Jansen** *et al*, Proc. SPIE, Vol **1418**, p32, (1991)
36. **D. Botez and D. E. Ackley**, IEEE Circ. & Devices Mag., **2**, p8, (1986)
37. **N. K. Dutta and P. P. Deimel**, IEEE-JQE **19**, p496, (1983)
38. **D. F. Welch** *et al*, Proc. LEOS '93, San Jose CA, paper SCL1.2, (1993)
39. **D. B. Young** *et al*, IEEE-PTL **5**, p129, (1993)
40. **K. Uomi** *et al*, IEEE-PTL **6**, p317, (1994)
41. **R. P. Schneider** *et al*, IEEE-PTL **6**, p313, (1994)
42. **J. B. D. Soole** *at al*, Appl. Phys. Lett., **61**, p2750, (1992)
43. **M. Fallahi** *et al*, IEEE-PTL **5**, p794, (1993)
44. **K. M. Dzurko** *et al*, Appl. Phys. Lett., **61**, p2389, (1992)

IMPLEMENTATION OF INTEGRATED OPTIC SYSTEMS WITH ACTIVE WAVEGUIDES, PASSIVE WAVEGUIDES, AND OPTICAL FIBERS

C.L. Balestra

McDonnell Douglas Aerospace - West
P.O. Box 516, St. Louis, MO 63166-0516

ABSTRACT

We currently stand at the threshold of a world society served and linked together by light. Fiber optic telecommunications is already a reality in most of the industrialized sector. Due to the high volume telephone communications market, optical fiber and associated optical sources and detectors are now off-the-shelf commodities. But telephone communications, now largely made up of voice and data transfer, is merely the embryo. There is a myriad of other practical, commercial applications for optical devices and systems. The scope of these applications will be determined by the economics, as well as ergonomics, of their implementation.

This report provides an overview of the implementation of guided-wave integrated optics systems in terms of constituent components. The emphasis is on semiconductor waveguide devices, which endow these systems with reconfigurability and thus make them active. We present our view of the performance limits and current technology challenges confronting the practical implementation of active waveguide devices and subsystems. We discuss applications of the active waveguide device concept to higher levels of network integration for space, avionics, and sensor applications. The roles of passive waveguides cast on a single substrate and optical fibers--the ultimate link for practically all guided wave optical systems--are also discussed.

INTRODUCTION

Future space and avionics systems will require high-speed reconfigurable networks in order to achieve system-level fault tolerance through redundancy, while simultaneously meeting objectives for allowable cost, size, and power. Commercially oriented telecommunications, distributed computing, and local area networks have similar performance and cost objectives that also can be satisfied by active waveguide devices. Network reconfigurability serves to reduce the overall component count and integration complexity of control, communication, or computing systems by allowing functional areas to share resources such as processors and memory. This feature also promotes system-level fault tolerance by permitting the network to route around failed components to their functioning counterparts. Although optical fibers, or other passive waveguides, that provide the high bandwidth point-to-point interconnectivity for such a network are available, switches that preserve that high bandwidth are not yet commercially available. Electronic interconnect switches can meet neither the bandwidth nor the configuration times required for future high data bandwidth systems. They reduce the large optical level bandwidth to that of electronics, add to system complexity, and introduce additional sources of signal error from electrical-to-optical and optical-to-electrical conversions.

Applications of Photonic Technology
Edited by G.A. Lampropoulos *et al.*, Plenum Press, New York, 1995

Directional coupler switches can be utilized for moderate levels of interconnectivity but are limited by inherent lossiness. Discrete device optical amplifiers can compensate for this loss, but at additional increase to system complexity.

In our laboratory we have demonstrated a monolithic 4 x 4 optical crossbar switch that can permit zero insertion loss.[1,2,3] The device is merely an example of a wide variety of reconfigurable optical interconnect devices that can be designed and laid out by conventional integrated circuit production techniques. Our preliminary devices have been based on well established AlGaAs optoelectronic device technology and operate at 850 nm and 930 nm. Commercial optical fiber telecommunication is established around the 1.3 μm and 1.55 μm wavelengths for optimum utilization of silica-core fiber. Since the same physical principles apply to semiconductor active waveguide devices operating at any optical wavelength, there is no fundamental obstacle to implementing optical crossbar switches, or similar devices, that can serve as switches for commercial fiber optic systems. Indeed, other researchers in Sweden[4,5], the United Kingdom[6], and Japan[7,8] have also shown that similar semiconductor active waveguide switches, with internal gain, can be fabricated on longer wavelength InGaAsP/InP material system.

SEMICONDUCTOR ACTIVE WAVEGUIDE CONCEPT

The semiconductor active waveguide optical interconnect device is based upon the integration of waveguides/optical amplifiers with interconnect structures such as turning mirrors, Y-branches, and T-branches on a single semiconductor chip in order to route a single mode optical signal from one or more waveguide-to-fiber ports to another set of one or more ports. The telephone crossbar switch is a made-to-order application for this concept.

The principle of operation of the active waveguide structure is shown in Figure 1. This example epitaxial InGaAsP/InP structure consists of a stepped index separate confinement heterostructure (SISCH) containing one or more quantum wells (QW). The InGaAsP SISCH layer, sandwiched between layers of InP, acts as a slab waveguide by virtue of its larger index of refraction. The concomitant lower bandgap in the SISCH layer, and still lower in the quantum well region, serves to drive both holes and electrons flowing vertically from the p- and n-doped InP layers above and below the SISCH layer, into the quantum well region. This results in an excess carrier population that can be optically stimulated to recombine there and provide optical gain over a characteristic wavelength range of about 60 nm--or about 7.5 THz for a center wavelength of 1.55 μm.[9] The relative gains of TE and TM guided waves in the structure are determined by the amount and sign of strain in the quantum wells, which is controlled by the chemical composition of the InGaAsP in the SISCH.[9,10,11] A variety of structures, utilizing these confinement

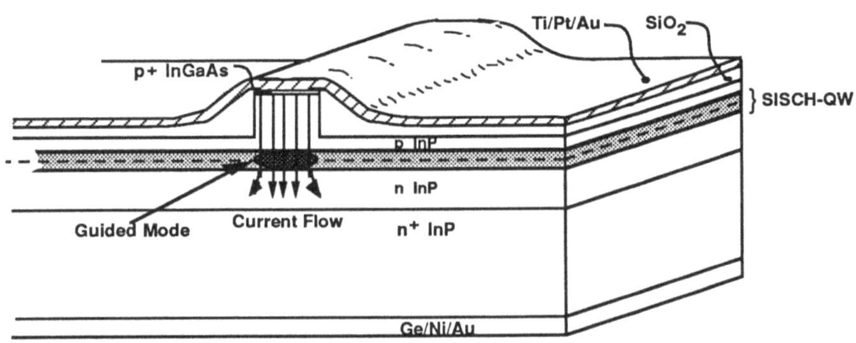

Figure 1. The semiconductor pn junction active waveguide structure serves as the basis for forming a monolithically integrated network of lossless waveguide/optical, optical amplifier, laser, modulator, detector, receiver, and transmitter devices.

principles, have beenfabricated to form slab waveguide optical amplifiers. By etching away a portion of the p-layer at the top, forming a ridge, the refractive index on either side in the underlying SISCH region is effectively reduced and lateral confinement is achieved. Since this index reduction is slight, only the lowest order mode of this waveguide structure is confined, resulting in single mode operation. Maintaining the guided mode well below the ridge minimizes perturbing effects of any residual roughness of the etched rib.

The active waveguide structure can be made to perform a variety of optical signal processing functions through variations in its configuration and operating conditions. One can vary the gain, or loss, of this structure by varying the current flowing from a contact at the top of the rib through the mode confinement region to the underlying substrate. With no forward current, or with a back bias, the quantum well absorbs light in its characteristic wavelength range and the device serves as an optical attenuator. In our laboratory we have observed 30 dB of extinction for zero bias and 40 dB for -4 V back bias from an 800 μm length of active waveguide. Other researchers have achieved extinction ratios of greater than 45 dB for similar active waveguide devices based upon InGaAsP/InP.[6,7] At the transparency current level the guide functions as a lossless light conduit. At higher currents the structure becomes an optical amplifier with small signal gains of as much as 80 cm^{-1} (35 dB/mm) for 100 Å GaAs quantum well devices.[2] Gains of in excess of 41 cm^{-1} (18 dB/mm) have been achieved for InGaAsP SISCH-QW devices operating at 1.55 μm.[9] Through control of gain or attenuation by the level of injection current, a signal can be impressed upon a guided wave through the contact at the top of the ridge. Several signals can be introduced, independently, through a series of separate contacts along the top of the ridge. Similarly, an electrical signal can be extracted from the ridge contact by detecting photocurrent generated by the guided optical signal. With reflecting facets at the ends of the ridge, the structure becomes an optical resonator (laser) and thus a signal source. Thus, the semiconductor active waveguide structure, which is amenable to high volume batch fabrication processing, is a building block for a variety of optical signal processing circuits.

The fabrication technology and operational characteristics of semiconductor waveguides, optical amplifiers, and related link structures are reaching maturity. Redirection of the on-chip guided signal can be accomplished by means of the total-internal-reflection (TIR) turning mirror depicted in Figure 2(a). The high refractive index of III-V semiconductors (~3.5 for InGaAsP) provides for a wide range of TIR-consistent redirection angles. Guided optical signals can be split or combined by means of the

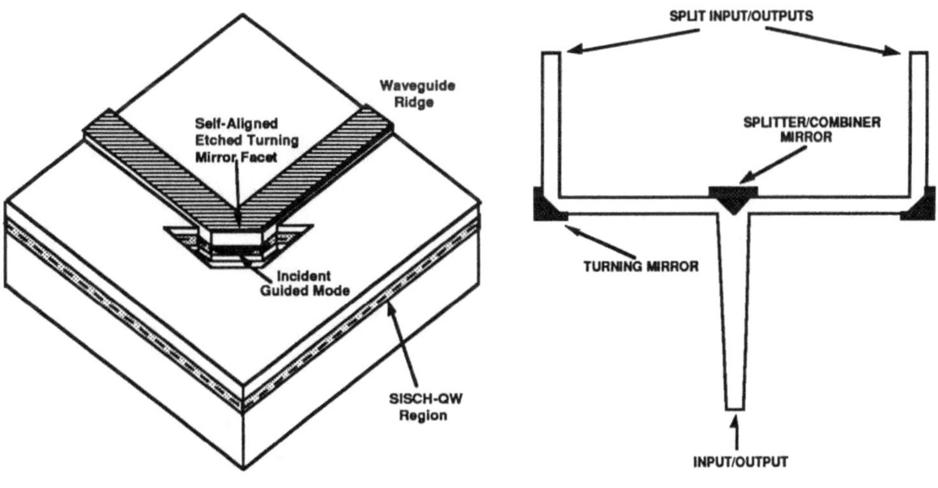

(a) Self-aligned etched-facet tuning mirror linking two active waveguides on a common substrate.

(b) Double faceted T-branch splitter/ combiner.

Figure 2. The turning mirror and T-branch structures permit active waveguides to be linked together into a wide range of interconnect networks that are lossless, purely optical, and monolithic.

double-faceted T-branch structure depicted in Figure 2(b). An incoming signal at the bottom of the figure is split and redirected into the two transverse guides by the two opposing 45° angle facets that join the three guides. A basic S-parameter analysis, accounting for the reciprocal nature of the associated devices and material properties of the SISCH-QW and ridge waveguide structure, shows that a 3 dB loss is incurred for a signal propagating through the T-branch in either a splitting or combining direction. In the combining direction, half of the guided signal couples into radiation modes and is absorbed in the surrounding unbiased confinement layer. By combining the basic building blocks--active waveguides, turning mirrors, and T-branches--a variety of interconnecting optical network structures can be realized at a single chip or multiple chip module level. With the support of passive polymer[12,13] or inorganic dielectric[14] waveguide technologies that have been demonstrated by others, large scale networks of active waveguide chips can be linked together with passive waveguides patterned on single substrates.

Figure 3 depicts a low level integration version of two monolithic active waveguide networks. The viability of these configurations was demonstrated in 1992 with the implementation of a monolithic 4 x 4 optical crossbar switch.[1,2] The optical tree splitter shown at the left accepts an optical input from the top and replicates it by a factor of 2^n by successively splitting it and amplifying it through n stages. Signal paths can be selected through appropriate configuration of the electro-absorption switch regions of the linking waveguides. As shown further on in this report, a substantial number (~20) of stages can be cascaded with a tolerable signal-to-noise (S/N) penalty. Figure 3(b) shows a waveguide network configured into an optical crossbar. The splitter/combiner arrangements at both the top and the bottom of the device structure link together through electro-absorption switches to form a switchable network that provides all conceivable interconnect patterns between the top and bottom ports. The device also utilizes in-plane crossing of optical signals--a physical impossibility for electrical signals. It permits bidirectional communication to occur between ports A and D, while a second, independent communication, occurs between ports B and C. By hierarchically extending the splitter/combiner structure in both the upward and downward directions, one can achieve higher levels of interconnectivity. Port-to-port links can be made in either single point-to-point or multiple link broadcast mode without blocking each other. This matrix vector multiplier crossbar configuration is strictly nonblocking, is bidirectional, and can exhibit 0 dB fiber-to-fiber loss by virtue of distributed internal gain.

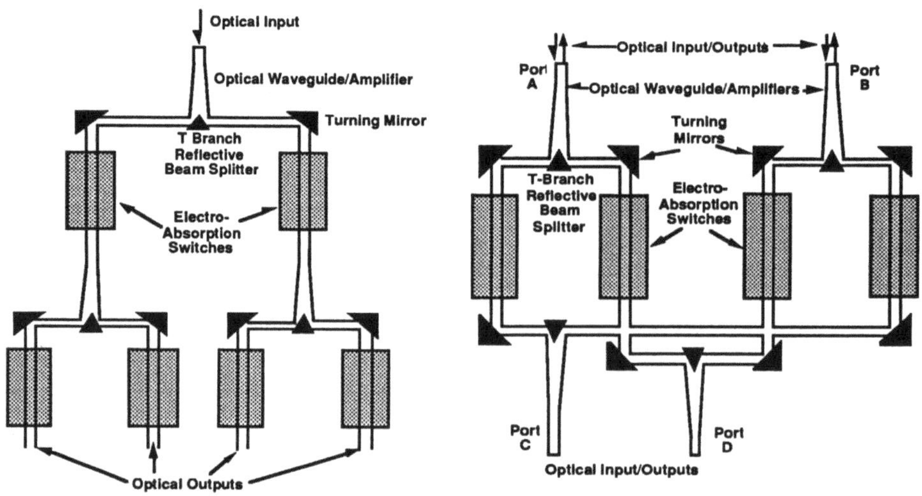

(a) Optical Tree Splitter (b) Optical Crossbar

Figure 3. Reconfigurable active waveguide devices can split and redirect ultrahigh bandwidth optical signals along purely optical paths, while preserving signal strength through optical amplification.

A variety of other active waveguide and waveguide interconnect structures also have been developed and demonstrated by other researches. AT&T Bell Laboratories has demonstrated an active waveguide balanced mixer incorporating gratings and zero-gap directional couplers.[15] Hitachi Central Research Laboratory has demonstrated a 4 x 4 optical switch with integrated single-stage optical amplifiers and single-slip-structure switches.[7,8] British Telecom Laboratories has demonstrated a monolithic InGaAsP/InP laser amplifier gate switch matrix with TIR turning mirrors and spread mode power splitters.[6] Ericsson Telecom Fiber Optics Research Center has demonstrated a monolithically integrated 4 x 4 InGaAsP/InP gate switch array using Y-branches.[4,5] Semiconductor active waveguide technology has passed through its infancy and is poised to transition from the laboratory and into the market place.

OPTICAL CROSSBAR SWITCH

The layout of our first optical crossbar chip,[1,2] along with a device microphotograph, is shown in Figure 4. The ridge waveguides of this device are 5 μm wide and the absorption modulators are 800 μm long. The shortest and longest paths through the chip are 2.5 and 4.4 mm. The chip has AR coated facets (R~0.1%) and is biased in three sections: long side, absorption switches, and short side. There are 16 separate absorption modulators to provide for 2^{16} conceivably possible cross-chip interconnect configurations. The short side and long side sections of the chip are operated at current densities of 1350 and 710 A/cm^2, respectively. Each absorption modulator is switched between zero current, when off, and 1900 A/cm^2, when on. These switches are simply sections of waveguide that are forward biased in the "on" state and unbiased or reverse biased in the "off" state. This device operates as a non-blocking switch by simultaneously routing any input to any output.

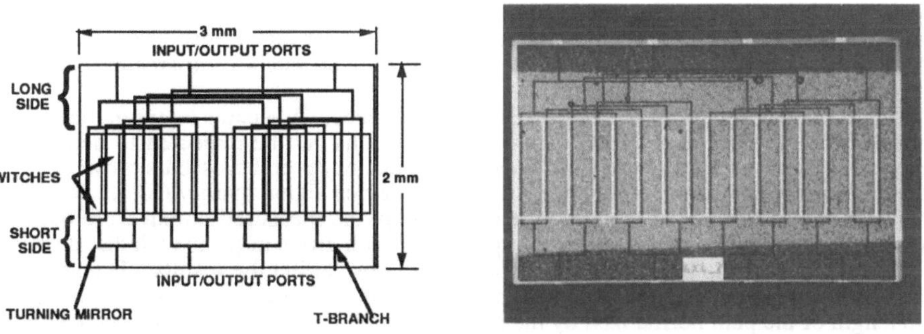

(a) Schematic active waveguide layout. (b) Photograph of crossbar chip.

Figure 4. The 4 x 4 active waveguide optical crossbar chip makes a connection from each input to each output using an active waveguide network with etched-facet T-branches and turning mirrors. Forward biasing a switch makes the connection between one input and one output, while zero biasing or back biasing the switch breaks the connection.

Devices were tested with TE polarized light, at a nominal wavelength of 850 nm. The crossbar switching function was demonstrated by using a Ti-sapphire laser as a CW source to inject about 2 mW into each port. The appropriate switches were then powered and the signal appeared at the designated port on the opposite side of the chip. The spectrum was measured for each output, and with an optical bandwidth of 0.1 nm (41 GHz) the signal was greater than 20 dB above the the spontaneous emission background level when the switch was turned on and was not discernable when the switch was turned off. A typical

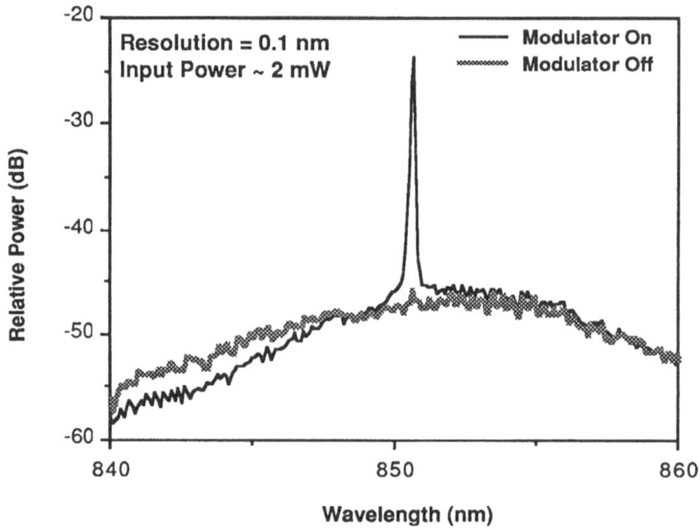

Figure 5. A typical channel of the 4 x 4 optical crossbar chip yielded an output signal that was greater than 20 dB above the spontaneous emission background. Measurements with a lock-in amplifier revealed an extinction ratio of 30 to 40 dB between the on and off states.

optical spectrum of the crossbar output is shown in Figure 5. An extinction ratio of 30 dB with the switch off was measured using a lock-in amplifier. This ratio increased to 40 dB when a -4 V was applied to the absorption switch. By measuring the outputs of the other ports we determined that port-to-port crosstalk is not greater than these figures. Previously performed beam propagation model simulations indicates that a scattering loss of less than 0.1% occurs at a 90° waveguide intersection.[16] Thus for either absorption switch leak-through or waveguide intersection scatter, the resulting crosstalk is no greater than -30 dB at zero V absorption modulator bias.

As can be observed from Figure 4, the sixteen conceivable paths between the opposite sides of the chip are of unequal path lengths and thus can be expected to demonstrate unequal gain, or transfer ratio. This situation, typical of several crossbars presently under development, can result in asymmetry in overall device signal quality (S/N). In order to evaluate the uniformity of the structure on the chip, the measured output from each port was normalized by the maximum output power for each group of ports with the same input port. In this way, the input coupling efficiency did not affect the results. The results for each path were then averaged over the data for six chips. This average was then divided by the length of the path normalized by the maximum path length. The result was 0.99 with a standard deviation of 0.19, indicating that the output power is proportional to length of the path, which is characteristic for a gain-saturated amplifier. These numbers also indicate that the quality of the T-branches and turning mirrors is uniform from device to device.

This crossbar was packaged with polarization-maintaining (PM) (single-mode) fiber optic pigtails and commercially available connectors on four short side and two long side ports. The two remaining ports were pigtailed with multimode fibers and connectors in order to minimize coupling loss. The best device fiber-to-fiber insertion loss observed for the crossbar was 4 dB, which was for a channel that had a single mode fiber on one side and a multimode fiber on the other. The coupling efficiently was measured at 7 dB from the chip to the single mode fiber and at 3 dB for the multimode fiber. Internal loss in the crossbar is due to splitting at the T-branches and to scattering at the etched facets. The net gain coefficient was determined from the current densities for various sections of the chip. Using all of the aforementioned information, we estimate turning mirror and T-branch losses of 3 dB and 6 dB, respectively, which is far from the best active waveguide devices we have produced.

We transmitted video signals through the crossbar to demonstrate its ability to route digital signals and to perform broadcasting. A digital FM video format, with a 28 MHz square wave modulated by a National Television System Committee (NTSC) analog video signal, was used in these experiments. The following modes of operation were demonstrated: 1) switching of one input channel, sequentially, to each of three outputs, 2) broadcasting of one input to two outputs, 3) simultaneous transmission from two independent video sources to two outputs, and 4) simultaneous bidirectional transmission of two signals in forward-forward, forward-backward, and backward-backward modes.

Pseudo-random non-return-to-zero (NRZ) data with a 2^{15}-1 pattern length was routed through the crossbar at 33, 100 137 Mbps. We set up a direct detection receiver for this measurement using a PIN photodiode and a transimpedance amplifier, but we made no attempt to optimize its performance. No optical filtering of the spontaneous emission from the crossbar output was done. The bit error rate (BER) curve for the 44 Mbps case is shown in Figure 6. The crossbar introduces a performance penalty of approximately 4 dB. BER's as low as 10^{-10} were also observed at 137 Mbps. BER data was taken at the limits of our measuring equipment, but, as indicated below, we are certain that the crossbar is capable of routing data streams at considerably higher rates.

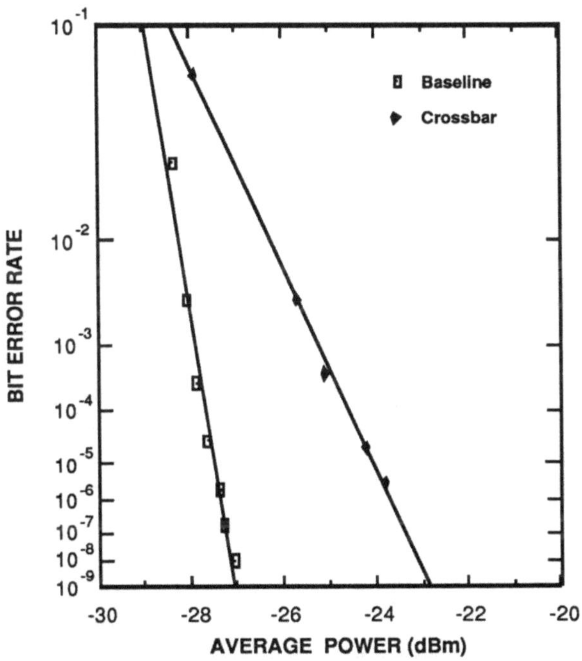

Figure 6. The bit-error-rate (BER) data for the crossbar at 44 Mbps with no optical filtering of the spontaneous emission shows about a 4 dB degradation in performance at a BER of 10^{-9}.

MULTICHIP IMPLEMENTATION

Active waveguide chips and electronic chips can be organized onto a multi-chip module (MCM). Silicon, which currently is available in quality wafers as large as eight inches in diameter, is an ideal substrate for a photonic/electronic multichip module. Chips mounted on this substrate can be optically interconnected by low pressure chemical vapor deposition (LPCVD) Si_3N_4 waveguides. The required deposition, photolithography, and etching processes can be well controlled and are well established. The electrical and optical properties of these materials are well known and understood, and they are readily available for theoretical simulations and device modelling of candidate MCM device configurations.

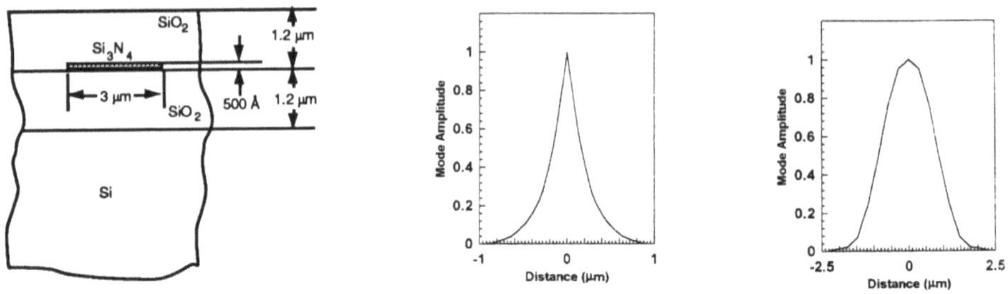

(a) Embedded Si_3N_4 waveguide (b) Vertical Confinement Lateral Confinement

Figure 7. Simulation of optical confinement of of lowest order transverse mode in a 500 Å x 3 µm Si_3N_4 passive waveguide embedded in SiO_2.

Figure 7(a) shows a sample single mode waveguide structure formed by a Si_3N_4 layer bounded by SiO_2 cladding layers. Figures 7(b) and (c) shows the waveguide modes for this structure in the transverse and lateral directions determined by an in-house numerical eigenmode solver. Because MCM techniques are based upon standard silicon IC design and fabrication technologies, micron-level tolerances can be achieved for interconnects within the MCM, and thus single mode waveguides are feasible for interconnection of MCM chips.

Passive Si_3N_4 optical waveguides can be formed on or embedded in a SiO_2 layer on top of a silicon substrate through utilization of standard semiconductor device fabrication processes. The resulting structure is planer between the chip mount positions, allowing metal traces to be fabricated on the same substrate, using standard semiconductor device fabrication processes. It is possible to incorporate electrical functions into the silicon substrate because of the flexibility and predictability of these well established processes. For example, this scheme permits the fabrication of a discrete photodetector at the termination of a waveguide. An alternate way to define the passive waveguide is to selectively phosphorous dope a photomask defined pattern into thermal oxide to slightly increase its index of refraction.[14] Also, titanium doped silica waveguides on silicon substrates are now commercially available.

A network of passive Si_3N_4 or doped SiO_2 waveguides can readily be defined onto an oxidized silicon wafer substrate. They can incorporate turning mirrors, Y-branches, evanescent couplers. By using the very fine control of etches and layer thicknesses available in semiconductor processing (on the order of hundreds of Å), one can make an

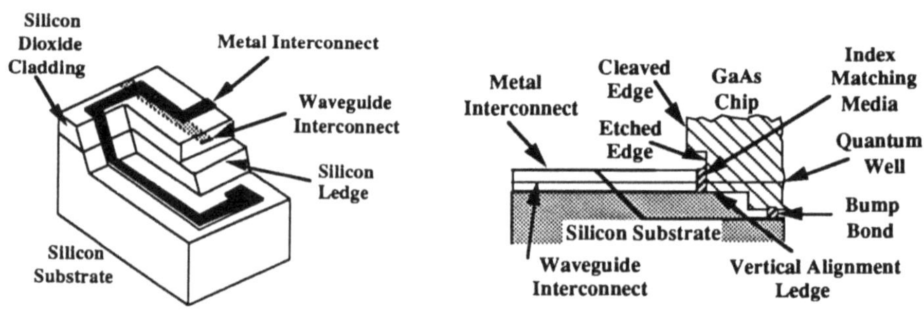

(a) Device recess structure on a silicon MCM substrate.

(b) Waveguide chip alignment to silicon MCM.

Figure 8. The silicon MCM concept uses well established and controlled device fabrication techniques to achieve automatic alignment between chip and MCM waveguide.

optical waveguide chip self aligning in the vertical direction, as illustrated in Figure 8. Optically active devices can be mounted with the active side down to take advantage of a precisely controlled distance between the emitting region and the chip surface. Theelectrical connections to the device side can be made using silver epoxy or solder bumps to bridge the gap between the bottom of the trench and the electrical contact on the chip.

Polymer materials are available in which waveguides can be fabricated, and which are compatible with PC board processes.[12] The waveguide interconnect patterns in these materials are defined using standard photolithographic techniques. On a given PC board, a polymer waveguide network can be fabricated and linked together, together with the semiconductor active waveguide devices to provide complete reconfigurability among devices on a single PC board, as shown in Figure 9.

SYSTEM-LEVEL IMPLEMENTATION

As external coupling losses are brought under control, considerably higher levels of reconfigurable optical interconnectivity can be achieved. Shown in Figure 9 is a schematic of a strictly non-blocking array of crossbar switches configured to form a 32 x 32 Clos network.[17] The number of optical external link-to-chip connections (320) presents a somewhat formidable fiber optic alignment and bonding task. However, this chip-to-chip optical interconnect network can be laid out on a single printed circuit (PC) board or silicon substrate with polymer or dielectric waveguides linking the chips. Rectangular cross section structures that match well to the ridge waveguide mode can be laid out, patterned, and formed with the same IC precision that goes into the fabrication of the active waveguide chips.

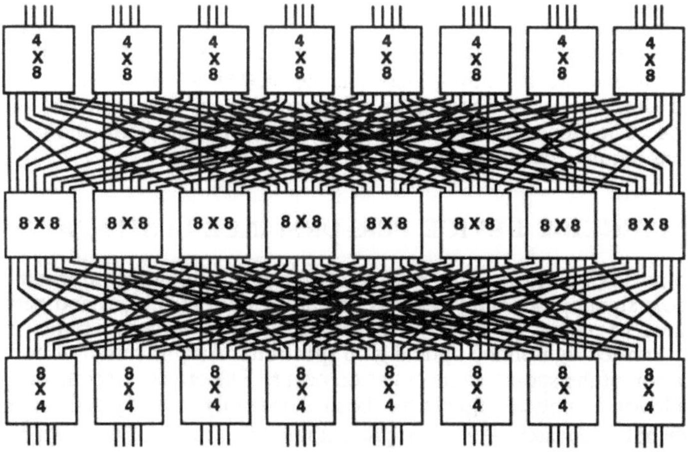

Figure 9. A strictly non-blocking Clos switching network can be implemented with a multi-chip optical crossbar switch architecture. The "rat's nest" chip-to-chip interconnection network can be implemented in polymer or inorganic dielectric waveguide by lithographic techniques amenable to batch processing.

Figure 10 shows a scalable implementation of the crossbar switch utilizing an array of passive dielectric or polymer waveguides to establish interconnectivity. To date, this concept appears as the optimal scheme for large scale and customizable interconnectivity for

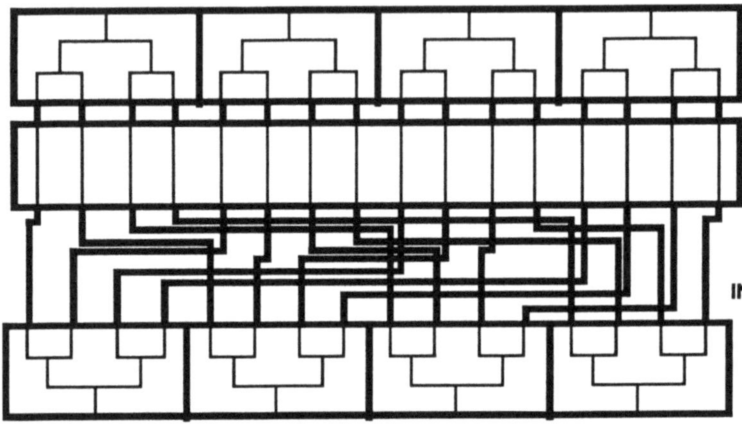

Figure 10. The distributed gain matrix vector multiplier architecture is scalable, provides uniform length of active waveguide paths, and utilizes a minimum of gain medium.

active waveguide networking. Inter-waveguide minimum distances are limited by the lateral penetration, and thus unwanted coupling, of the evanescent portions the of guided modes. However, this distance (~25 μm) is considerably smaller than minimum fiber separation distances encountered in external coupling. Thus the absorption switches and divider/combiner chips can be made fairly small in size. Their layout is simple and scalable; all active waveguide paths have the same length and therefore same gain and noise characteristics; the chips make optimum usage of semiconductor real estate; they are amenable to production of large numbers of the same device on a single wafer.

A space division switch analysis performed by Kalman, Kazovsky, and Goodman indicated that the DGMVM switch is a close second to the Benes switch for providing the maximum usable number of switches for a given level of optical amplifier saturation power.[17,18] It is also worth noting, also that the DGMVM architecture is strictly non-blocking and thus does not require any interruption of interconnect paths during reconfiguration. The Benes architecture is rearrangable nonblocking and thus for certain reconfigurations requires path interruptions.

ACTIVE WAVEGUIDE DEVICE PERFORMANCE LIMITATIONS

Signal-to-noise (S/N) ratio is the key parameter determining the information handling capacity of most signal processing systems, including optical devices and networks. The key contributor of noise to the optical amplification process is the unavoidable spontaneously emitted light in the optical gain spectrum. Amplification of both signal and the guided portion of the spontaneous emission occur. The effect of spontaneous emission on S/N degradation of optical signals has been thoroughly investigated for AlGaAs and InGaAsP fiber optic repeaters.[19,20] We present here the results of our own model, based upon the work of the previous authors, to simulate the effect of signal splitting on signal fidelity for a semiconductor active waveguide network.[21]

The waveguide sections, turning mirrors, and T-branch combiners/splitters can be modeled as lumped parameter loss or gain elements, with appropriate loss or gain and spontaneous emission noise contribution being assigned to each element. For a path of optical amplifiers and loss elements, as depicted in Figure 11, one can describe the composite effect upon overall optical output and the corresponding noise level in terms of Equations (1) and (2). The overall S/N ratio of the detected signal, as indicated in Equation (3), is a function of the shot noise from the signal itself (Nout), the beat noise from the interaction between the signal and the spontaneous emission ($N_{sig\text{-}spon}$ and $N_{spon\text{-}spon}$), and the equivalent thermal noise of the detection circuit. We have developed a multiple-

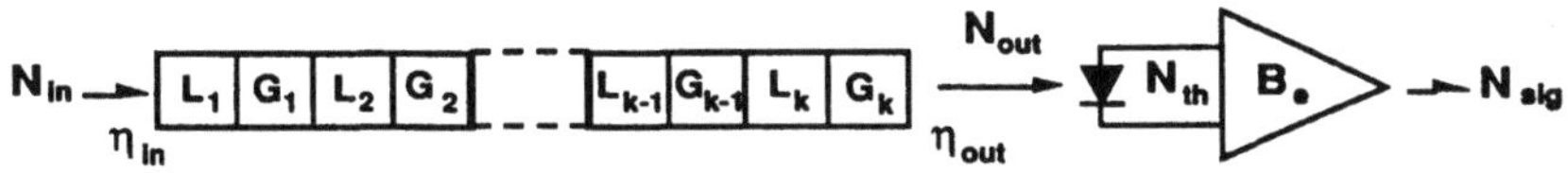

Figure 11. A series of optical amplifiers joined by lossy links, such as turning mirrors and T-branches, can be modeled as a series of linkages, wherein the gains and losses in both signal and spontaneous emission are compounded as the signal propagates from source to destination.

$$N_{out} = \left[\eta_{in} N_{in} \prod_{l=1}^{k} (L_l G_l) + B_o \sum_{l=1}^{k} \left\{ (G_l - 1)\, N_{sp,l} \prod_{j=l+1}^{k} (L_j G_j) \right\} \right] \eta_{out} \tag{1}$$

$$\sigma_{OA}^2 = N_{out} + \left[\begin{array}{c} 2\, \eta_{in} N_{in} \sum_{l=1}^{k} \left\{ \prod_{j=1}^{l} L_j\, G_j\, (G_l - 1)\, N \prod_{l=l+1}^{k} (L_l G_l)^2 \right\} \\ + B_o \left\{ \sum_{l}^{k} (G_l - 1)\, N_{sp,l} \prod_{j=l+1}^{k} L_j G_j \right\}^2 \end{array} \right] \eta_{out}^2$$

$$= N_{out} + N_{sig\text{-}spon} + N_{spon\text{-}spon} \tag{2}$$

$$\frac{\text{Signal}}{\text{Noise}} \equiv \frac{S}{N} = \frac{N_{sig}^2}{2\, B_e \left[N_{out} + N_{sig\text{-}spon} + N_{spon\text{-}spon} \right] + N_{th}} \tag{3}$$

N_{in}	-	Optical input
N_{out}	-	Total optical output = Signal + ASE
$N_{sig\text{-}spon}$	-	Signal-spontaneous emission beat noise
$N_{spon\text{-}spon}$	-	Spontaneous-spontaneous emission beat noise
η	-	Coupling loss at input (in) and output (out)
L_k, G_k	-	Loss (L) and gain (G) factors
N_{th}	-	Equivalent thermal noise of detection circuit
B_o	-	Bandwidth of optical amplifier
B_e	-	Electrical bandwidth of detection circuit
N_{sig}	-	Electrical signal output

amplifier model, based upon these equations for a cascade network of active waveguide elements linked with T-branch splitter/combiners and turning mirrors. The T-branch has an inherent coupling loss of about 3 dB, along with an additional scattering loss of 1 dB (practical lower limit). The total internal reflection (TIR) etched facet turning mirror can be fabricated with a scattering loss of about 0.5 dB. Any path through the crossbar consists of a cascade of T-branches and turning mirrors linked in an alternating series by optical amplifiers. The results of an InGaAsP device simulation are shown in Figure 12. The basic elemental link is depicted in the upper right hand corner. Gains G_1 and G_2 are selected for a net link insertion loss of 0 dB. The inversion ratio value of N_{sp} equal to 3 chosen for this particular simulation is large for quantum well amplifiers and is more typical of double heterostructure devices. Saitoh and Mukai have shown that N_{sp}'s of less than 2 can be achieved by InGaAsP traveling wave amplifiers.[19] Since our model does not account for the variation of Nsp along the active waveguide with signal loss and amplification, we use the value of 3 to add a degree of conservatism to the simulation. It is readily apparent from the data that a -5 dBm input signal can be split and amplified back to its original level more than twenty times before dropping to the detector/amplifier S/N floor for a 10^{-15} BER at 2.5 GHz. Note that this simulation assumes an optical bandwidth of 10 nm (1.25 THz), readily achievable with a commercially available interference filter. Considerable improvement beyond these results can be realized by tightening B_o to within a factor of two of B_e--e.g. through the use of a Fabry Perot or grating filtering process.

The second fundamental performance limit to a chain of splitter-optical-amplifier links is gain saturation by the overall optical signal level (processed input signal plus amplified spontaneous emission (ASE)). Output saturation powers of in excess of 10 dBm have been observed, and are thus achievable, from polarization insensitive strained well InGaAsP devices measuring 2.5 µm x 625 µm and operating at 1.55 µm.[9] Figure 13 shows a simulation of the ASE level, distributed over the 60 nm bandwidth of the optical amplifier, accompanying the signal along the amplifier/splitter cascade. The zigzag characteristic of the plot is due to the signal splitting that occurs at each T-branch. Coupling loss limits the ASE to well below amplifier saturation levels for our design. The simulation data in Figures 12 and 13 show that there is sufficient performance margin to permit operation of a monolithic 1024 x 1024 optical crossbar, which would involve twenty splitter/combiners in the waveguide path from one side of the chip to the other. A tighter constraint on the optical bandwidth could push this figure to a greater level.

As can be discerned from Equations (1) and (3), keeping the net optical gain of each stage close to unity minimizes the ASE level of the signal and the associated beat noise contributions. The purpose of the optical amplifier in a reconfigurable interconnect network is restoration of signal level, not high gain amplification. Thus, if scattering losses, which are determined by the fabrication process, can be controlled to the minimum levels already demonstrated in the laboratory, optical interconnect networks providing ultrahigh transmission bandwidth and overall system functional versatility and fault tolerance can be realized with monolithic chips amenable to high volume production processes.

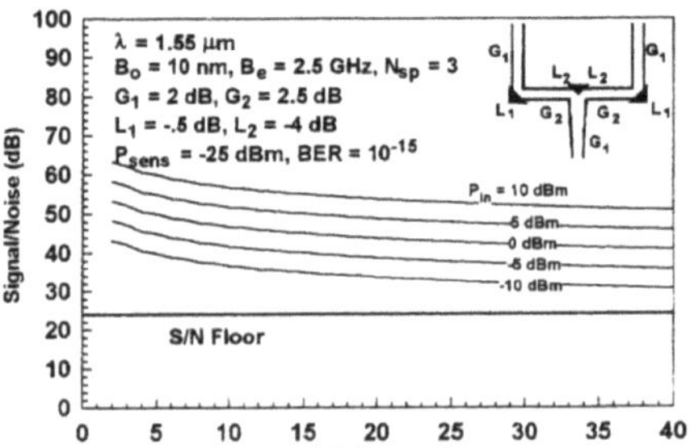

Figure 12. Signal degradation due to ASE for a series of unity gain splits (as depicted in the upper right). A signal can undergo a substantial number of splits or combines (2^{20}~10^6), and maintain a bit error rate (BER) of less than 10^{-15}.

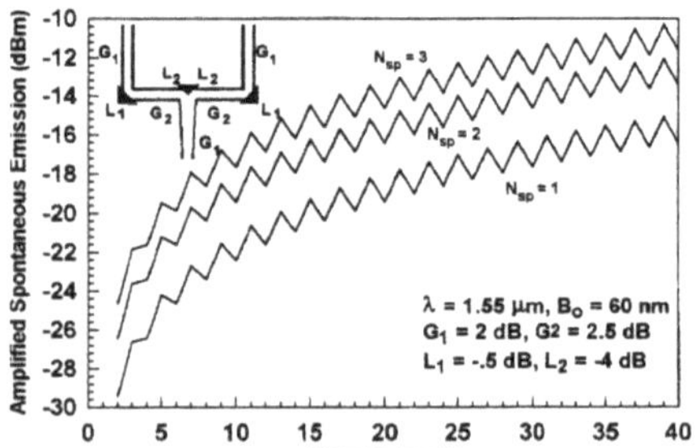

Figure 13. ASE levels in a monolithic optical waveguide network can be kept well below amplifier saturation levels if scattering loss and compensatory gain can be controlled to a low level.

APPLICATIONS

Optical intersatellite links are gaining favor due to their small terminal size and weight coupled with the attainability of very high data rates. With lighter terminals projected for the near future, laser communications is in good position for a number of upcoming crosslink applications. Laser crosslink technology is still maturing while RF technology, the present standard, has matured to where significant advances are fewer and less frequent. Most significant, laser output power is rapidly increasing, which makes for easier link closure with small optics. With the advent of high power semiconductor traveling wave amplifiers,[22] it is conceivable that CW optical outputs of several watts will become available in the near future. Such source power levels makes 10,000 km crosslink distances possible. We anticipate that in the future systems will use RF uplinks, downlinks will be split between RF and laser communications, while crosslink technology will be primarily optical.[23] Integrated photonics offers optical intersatellite links the potential for increased speed and the ability to accomplish a number of functions optically which are very challenging electronically. The ultrahigh bandwidth functional versatility and fault tolerance afforded by active waveguide devices considerably enhance satellite communications terminal performance and reliability. The pn-junction nature of active waveguide devices lends itself readily to signal modulation and demodulation.

Trends in military avionics are towards enhanced system performance, increased use and integration of sensors for added capability, and smaller integrated packaging, while at the same time improving reliability, maintainability, and overall reduced life-cycle cost. The utilization of photonic devices and fiber optic technology for avionics will provide: (1) new or improved sensors with high reliability based on a multitude of optical effects; (2) significantly higher bandwidth for data communication and distribution; (3) immunity to electromagnetic radiation (EMI/EMP); (4) high resistance to nuclear radiation; (5) electrical isolation; (6) no sparks or fire hazards; (7) corrosion resistance; (8) size and weight savings. Currently avionic fiber optic applications are multimode point-to-point sensor and control systems. For hard-fibered point-to-point links operating at less then a few hundreds of Mbps over distances of less than a 100 meters, multi-mode fibers are sufficient. As bandwidth requirements increase to multi-gigabit levels and reconfigurability of interconnect patterns is required, single mode fibers and active waveguide switches become more desirable for avionic applications. More sophisticated photonic components are required to interconnect and network multiple subsystems with a reconfigurable bus architecture. Because of lack of current demand for military fiber optic systems, most current photonic components are not designed to meet military packaging requirements--particularly with regard to mechanical shock and temperature range. The emphasis of the Pave Pace initial assessment is on photonic components that will enable the integration of advanced optical sensors and subsystems into optical architectures.[24]

Optical interconnections between digital avionic circuits are currently under development at the system level, the board level, and even the chip level. The optical interconnect as an interface between electronic circuits, has several advantages over wire connections, including wide bandwidth, reduced crosstalk, and immunity to electrical interference and ground loops. A major cause of avionics failures is connector failure (solder joint, edge-connector, etc.). The use of optical fiber reduces the number of connections (and hence failure rate) and could prove to be more reliable, being substantially free from problems of corrosion and thermal cycling failures. These characteristics can have a significant impact on system design and hardware implementation, and they permit more reasonable interconnect and control configurations than are practicable with electronic approaches. The most dramatic advantage of optical communication will be its capacity to easily handle high density interconnects by multiplexing numerous high speed channels onto a digital channel for chip-to-chip, board-to-board, and board-to-system applications.

Monolithic semiconductor active waveguide networks can provide a cost effective means of linking fiber optic sensors into a total system for structure monitor and control. The wide variety of fiber optic sensors in use and under development offers an *in situ* means of monitoring strain, temperature, or vibration in aircraft, buildings, bridges, and dams. The electromechanical bandwidths associated with the signals monitored are considerably smaller than inherent fiber capabilities. Figure 14 schematically depicts a laboratory demonstration[25] of the integration of a group of fiber Fabry Perot vibration sensors time shared through an active waveguide 1 x 10 splitter. A substantial cost saving, through reduction of the amount of diagnostic equipment required, can be achieved by multiplexing sensors through active waveguide chips. In addition, sensing, phase adjustments, and amplitude adjustments can be incorporated into the chips.

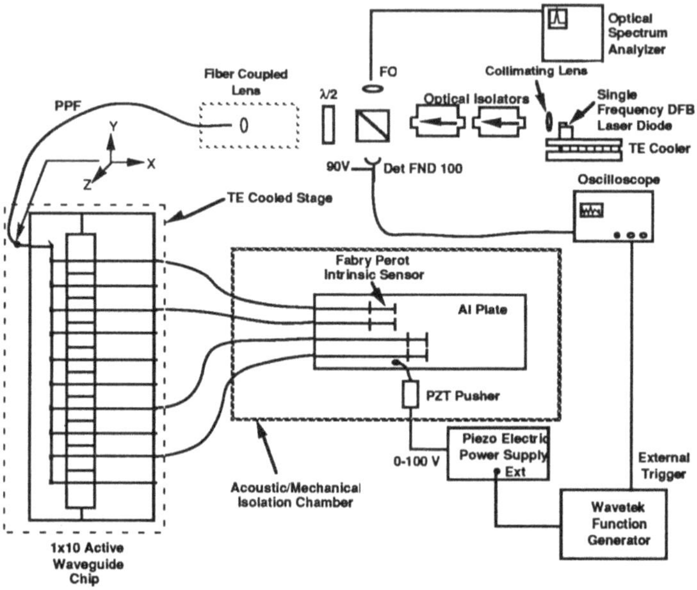

Figure 14. As shown by this laboratory demonstration layout, a single active waveguide chip can integrate a number of fiber optic sensors from separate points into a single integrated structural integrity monitoring system.

FUTURE TECHNOLOGY

A few researchers have demonstrated the viability of integrated active waveguide technology at a small scale integration level. With the incorporation of materials and device

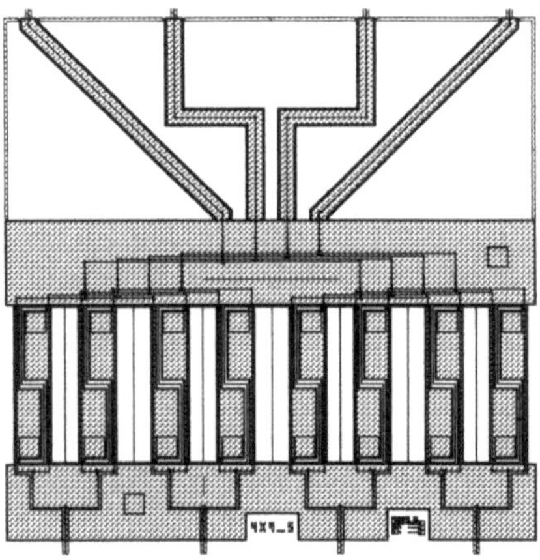

(a) Equi-path optical crossbar layout.

(b) Equi-path crossbar mounted in hybrid package.

Figure 15. The advanced design 4 x 4 optical crossbar chip, measuring 4 mm x 4 mm, makes use of process and test experience gained form the earlier proof-of-concept version.

technology currently under development, we foresee a progression certainly to a medium scale level, and possibly beyond. In our own laboratory we already have designed and fabricated an advanced version of our 4 x 4 optical crossbar. This second generation device has port-to-port paths all of equal length, providing uniformity of gain and S/N characteristics Shown in Figure 15 are a layout and a photograph of the new crossbar, which is currently undergoing characterization. We also redesigned the layout of the plated metal interconnect layer, shown as a stippled pattern, to provide negligible ohmic loss while not covering the entire chip in order to optimize yield. Turning mirrors and T-branches have been expanded in width to intercept more of the evanescent portion of guided waves. Refinements in our optical fiber alignment and bonding capability, since this chip's design, will now permit us to move the input/output waveguide elements closer together, which will permit a 50% reduction in total chip real estate for the same 4 x 4 function.

Our concept for the optical crossbar switch, and related devices, is based upon the application of established wafer growth and optoelectronic device fabrication techniques to the semiconductor active optical waveguide device structure. Through the combination of the design and fabrication disciplines of silicon integrated circuit technology with currently emerging integrated optoelectronic technology, high-volume low-cost integrated active waveguide devices can be made readily available. High quality custom epitaxial starting material is now commercially available. In particular, the metal organic chemical vapor deposition (MOCVD) growth process is applicable to high-volume low cost manufacturing techniques, while at the same time it provides a very high degree of thickness uniformity and interfacial abruptness for yield/performance optimization. Wafer fabrication equipment tailored to III-V materials also are now available. At present, the primary obstacle to integrated optoelectronic device fabrication is the lack of high volume automated fiber optic packaging. Through our own related efforts, we have demonstrated that over twenty single mode fibers can be aligned and bonded into a single hybrid package.

We have previously demonstrated an in-house fabrication capability for AlGaAs optical crossbar devices. Others throughout the world have demonstrated capability for fabrication InP-based 1.3 μm and 1.55 μm devices. We have demonstrated etched facets with an RMS surface roughness of the order of 100 Å--less than 1/20 of a wavelength in the guide.[26] As demonstrated by Lincoln Laboratory, the InGaAsP material family is amenable to surface mass transport processing, which produces atomically smooth surfaces.[27] Thus even better mirror quality is achievable for InGaAsP devices. Figure 16

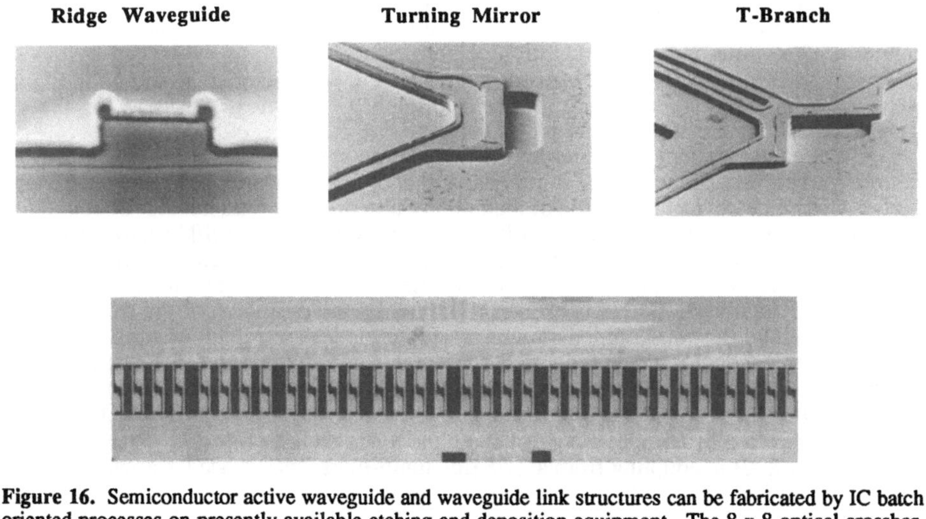

Figure 16. Semiconductor active waveguide and waveguide link structures can be fabricated by IC batch oriented processes on presently available etching and deposition equipment. The 8 x 8 optical crossbar, below, is a good example of what can be configured from these basic optoelectronic building blocks.

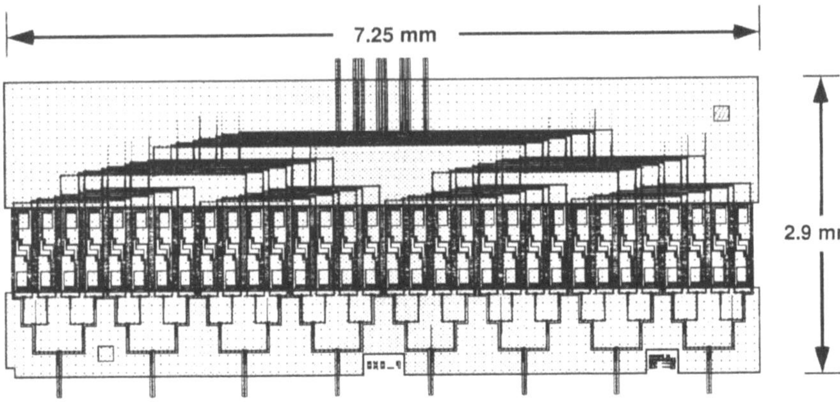

Figure 17. An updated design for an 8 x 8 optical crossbar chip has been laid out with conventional IC CAD equipment and in accordance with a documented set of layout design rules that can be verified by IC CAD methods.

shows SEM photographs of active waveguide structure and an active waveguide network configured to form an 8 x 8 optical crossbar switch from a 100 Å GaAs quantum well in an AlGaAs/GaAs GRINSHCH structure. The SEM photos show well-defined waveguide structure can be fabricated. An updated design, shown in Figure 17, making use of reduced input/output waveguide-to-waveguide spacing has an area of 21 mm^2 and if implemented in 1.55 μm InGaAsP could exhibit 0 dB of insertion loss with an estimated power dissipation of about 5 W.

Electrical power density and dissipation currently are a serious limitations to the level of integration of active waveguide devices. Since signal power levels are at the mW level, there is substantial room for improvement here. Data on high power laser diodes would tend to indicate that 50% is a practical upper level to efficiency. Waveguide gain or attenuation is essential only for compensation of input/output coupling losses, splitter/combiner loss, and for scattering loss. Others have shown that native waveguide absorption can be reduced by vertically displacing the active region to either above or below the guide and confining it to only those portions of the structure where gain is required.[15] Thus this section of the semiconductor waveguide network can be made passive, not requiring any transparency level current. In our own laboratory we have demonstrated that if the quantum well is displaced outside of the normal depletion region of the structure's pn junction, absorption induced generation of hole-electron pairs, as evidenced by device current, is extremely low.[28] As seen from the device IV characteristics, shown in Figure 18, such a junction/quantum-well profile can amplify, be optically passive, or absorb, depending upon externally controlled bias voltage.

The greatest amount of signal loss in fiber optic systems occurs in device-to-fiber coupling. This fact, alone, makes it imperative that optical networks be implemented with active waveguide devices, by virtue of their ability to reduce the total number of couplings. Nevertheless, this loss poses a serious limitation to overall system S/N performance. A number of schemes have been suggested for the reduction of fiber-to-device loss. Refractive index mismatch is largely correctable through conventional optical coating techniques. Coupling efficient is primarily limited by the mismatch of the shapes of the modes guided by the two coupled structures. Fiber modes are round, or square, in cross section. Semiconductor waveguide modes are rectangular with high (~5:1) aspect ratios. Workers at AT&T Bell Laboratories have shown that an asymmetric microlenses can be formed at the end of an optical fiber through micromachining with a pulsed CO_2 laser, as the fiber is directed by computer control about the focussed laser beam.[29] They observed a loss of only 2 dB in coupling from a 1.48 μm multi-transverse mode laser into the fiber.

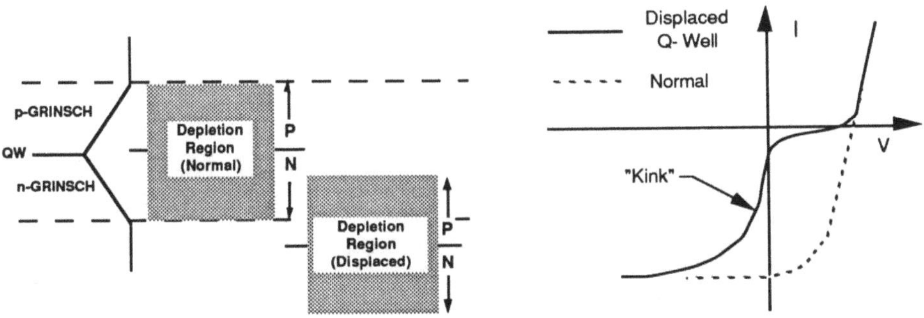

(a) Epitaxial GRINSCH-QW structure illuminated

(b) Device IV characteristics of displaced junction device.

Figure 18. Displacement of the pn junction depletion region from the quantum well considerably reduces its optical absorption and results in a corresponding reduction in photocurrent.

Yamada and Hanafusa of NTT have demonstrated that a mode shape converter could be introduced at the end of PANDA fiber by means of a controlled thermal diffusion process.[30] They observed coupling losses as low as 0.7 dB. AT&T has also described how a stepped reduction in thickness of the layer of material above the waveguide core close to the edge of the chip, along with a low index "outrigger" region below the core, can be used to vertically spread the mode at the device edge.[31] In our own laboratory we have fabricated and tested active waveguide devices with low index outriggers above and below the active core. We reduced beam far field divergence to 23° and achieved a device-to-fiber coupling loss of 2 dB. By virtue of the outriggers, vertical mode confinement may be less restrictive and critical angle of reflection may be reduced. If this is the case, one might expect such a structure to guide a smaller percentage of spontaneously emitted light, and thus it should exhibit a lower level of ASE.

SUMMARY AND CONCLUSIONS

We have developed design and fabrication capabilities for implementing semiconductor active waveguide devices. The associated processes are amenable to high volume production. We have verified our technology through the demonstration of a 4 x 4 optical crossbar chip. We have further verified its potential through the fabrication of over a hundred different designs of active waveguide devices including an 8 x 8 crossbar. Our semiconductor device technology appears to be compatible with higher levels of integration including passive waveguide networking and optical fiber pigtailing. Design and fabrication were executed in accordance with a standardized set of rules and procedures. We developed a S/N model of our active waveguide device technology. We have shown through simulation that integration levels of up to a 1024 x 1024 crossbar switch appear to be feasible.

Other research laboratories in the US, Europe, and Asia also have been involved for about the past five years in the development of active waveguide networks on semiconductor substrates. Passive waveguide fabrication and optical fiber mode matching techniques have also been investigated. Many of the components and techniques demonstrated are amenable to large volume production.

There is a broad range of applications for active waveguide devices in telecommunications, distributed computing, sensor integration, avionics, control systems, and space systems. Their niche is in no-loss/no-gain optical interconnect devices[21] and in special signal processing, such as signal mixing and detection.[15] There is substantial potential for improved performance and reliability. Not surprisingly, the extent and range of market penetration of active waveguide devices will be limited primarily by cost per unit function delivered *vis a` vis* the need for the level of performance it provides.

REFERENCES

1. Probst, D.K., Perrymore, L.G., Johnson, B.C., Blackwell, R.J., Priest, J.A., and Balestra, C.L., Demonstration of an integrated, active 4 x 4 photonic crossbar, *IEEE Photonics Technol. Lett.* 4:1139 (1992).
2. Balestra, C.L., Blackwell, R.J., B.C. Johnson, B.C., and Perrymore, L.G., Monolithic semiconductor active waveguide optical crossbar switch, *SPIE Meeting on Photonics for Space* 1953:18 (1993).
3. Rice, R.R., Zediker, M.S., Balestra, C.L., Priest, J.A., Monolithic active waveguide optical crossbar switch, *US Patent No. 5,283,844* (1994).
4. Gustavsson, M., Lagerstrom, B. Thylen, L., Janson,M., Lundgren, L. Morner, A.C., Rask, M., Stoltz, B., Monolithically integrated 4 x 4 InGaAsP/InP laser amplifier gate switch arrays, *Electron Lett.*, 28:2223 (1992).
5. Gustavsson, M., Janson, M., Lundgren, L., Digital transmission experiment with monolithic 4 x 4 InGaAsP/InP laser amplifier gate switch array, *Electron. Lett.* 29:1083 (1993).
6. Burton, J.D., Fiddyment, P.J., Robertson, M.J., and Sully, P., Monolithic InGaAsP-InP laser amplifier gate switch matrix, *IEEE J. Quantum Electron.* 29:2023 (1993).
7. Kirihara, T., Ogawa, M., Inoue, H., Kodera, H., Lossless and low-crosstalk characteristics in an InP-based 4 x 4 optical switch with integrated single-stage optical amplifiers, *IEEE Photon. Technol. Lett.* 6:218 (1994).
8. Kirihara, T., Ogawa, M., Tsuji, Inoue, H., High-speed signal-transmission performance in a lossless 4 x 4 optical switch for photonic switching, *OFC '94 Technical Digest* 4:55 (1994).
9. Newkirk, M.A., Miller, B.I., Koren, U., Young, M.G., Chien, M., Jopson, R.M., and Burrus, C.A., 1.5 μm multiquantum-well semiconductor optical amplifier with tensile and compressively strained wells for polarization-independent gain, *IEEE Photon. Technol. Lett* 4:406 (1993).
10. Dubovitsky, S., Mathus, A., Steier, W.H., Dapkus, P.D, Gain saturation properties of a polarization insensitive semiconductor amplifier implemented with tensile and compressive strain quantum wells, *IEEE Photon. Technol. Lett.* 6:176 (1994).
11. Olsson, N.A., Kazarinov, R.F., Nordland, W.A., Henry, C.H., Oberg, M.G., White, H.G., Garbinski, P.A., and Savage, A., Polarization-independent optical amplifier with buried facets, *Electron. Lett.* 25:1048 (1989).
12. Booth, B.L., Optical interconnection polymers, in: "Polymers for lightwave and Integrated Optics," L. Hornak, ed., Marcel Dekker, New York (1992).
13. Krchnavek, R.R., Lalk, G.R., Denton, R., Photo-polymerized acrylic waveguides for optical interconnects, *Mat. Res. Soc. Symp. Proc.* 228:95 (1992).
14. Henry, C.H., Blonder, G.E., Kazarinov, R.F., Glass waveguides on silicon for hybrid optical packaging, *J. Lightwave Technol.* 7:1530 (1989).
15. Koch, T.L., Koren, U., Semiconductor photonic integrated circuits, *IEEE J. Quantum Electron.*, 27:641 (1991).
16. Herrick, R.J., Beam propagation modeling of intersecting rib waveguides, McDonnell Douglas Corporation, Unpublished (1988).
17. Spanke, R.A., Architectures for guided-wave optical space switching systems, *IEEE Commun. Mag.* 25:42 (1987).
18. Kalman, R.F., Kazovxky, L.G., Goodman, J.W., Space division switches based on semiconductor optical amplifiers, *IEEE Photonics Technol. Lett.* 4:1048 (1992).
19. Saitoh, T. and Mukai, T., 1.5 μm GaInAsP traveling-wave semiconductor laser amplifier, *IEEE J. Quantum Electron.* QE-23:1010 (1987).
20. Stevens, P.J. and Mukai, T., Predicted performance of quantum-well GaAs(GaAl)As optical amplifiers, *IEEE J. Quantum Electron.* QE-26:1920 (1990).
21. Balestra, C.L., Atkinson, T.L. , Johnson, B.C., and Perrymore, L.G., Semiconductor active waveguide reconfigurable optical interconnect devices and networks, *SPIE Meeting on Photonics for Space Environments* 1953:19 (1993).

22. Walpole, J.N., Kintzer, E.S., Chinn, S.R., Wang, C.A., and Missaggia, L.J., High-power strained-layer InGaAs/AlGaAs tapered traveling wave amplifier, *Appl. Phys. Lett.* 61:740 (1992).
23. Crofts, D., Balestra, C., Dreisewerd, D., Lambert, S., Rice, R., Simmons, S., Current capabilities and photonics trends in optical intersatellite links, *AIAA - 15th Satellite Communications Conference* (1993).
24. Advanced systems avionics (ASA) technology and applications assessment (Pave Pace), *Final Report, Contract No. F33615-89-C-1110, Rep. MDC 91B0643*, (1992).
25. Haake, J.M., Fairles, R.T., Development of a fiberoptic interferometer demodulation/multiplexing system using an active 1 x 10 photonic polarization preserving fiber coupler, *SPIE Conference on Smart Structures* 1918:54 (1993).
26. Herrick, R.W., Sabo, L.G., Levy, J.L., Novel method for measuring and analyzing surface roughness on semiconductor laser etched facets, *J. Va. Sci Technol.* B9:2778 (1991).
27. Liau, Z.L. and Zeiger, H.J., Surface-energy-induced mass-transport phenomenon in annealing of etched compound semiconductor structures: theoretical modeling and experimental confirmation, *J. Appl. Phys.*, 67:2434 (1990).
28. No, K.H., Blackwell, R.J., Herrick, R.W., Levy, J.L., Monolithic integration of an amplifier and a phase modulator fabricated in a GRINSCH-SQW structure by placing the junction below the quantum well, *IEEE Photon. Technol. Lett.* 5:990 (1993).
29. Presby, H.M., Giles, C.R., Asymmetric fiber microlenses for efficient coupling to elliptical laser beams, *IEEE Photon. Technol. Lett.* 5:184 (1993).
30. Yamada, H., Hanafusa, H., Mode shape convertor produced by the thermal diffusion of different dopants, *IEEE Photon. Technol. Lett.* 6:531 (1994).
31. Koch, T.L., Koren, U., Eisenstein, G., Young, M.G., Oron, M., Giles, C.R., Miller, B.I., Tapered waveguide InGaAs/InGaAsP multiple-quantum-well lasers, *IEEE Photon. Technol. Lett.* 2:88 (1990).

TECHNIQUES FOR MONOLITHICALLY FABRICATING PHOTONIC INTEGRATED CIRCUITS

Emil S. Koteles

Institute for Microstructural Sciences
National Research Council of Canada
Ottawa, Ontario K1A 0R6
Canada

INTRODUCTION

The impact, on information processing and distribution, of a practical technique for integrating semiconductor active and passive photonic devices cannot be overstated. It will be as significant as was the development of the electronic integrated circuit on the modern, technological world. Despite considerable efforts by a number of groups worldwide over the years, this utopian vision has yet to be realized, highlighting the magnitude of the technological challenges involved. However, recent progress in several areas is very encouraging.

State-of-the-art **discrete** semiconductor components, whether high speed or high power laser diodes, fast and/or sensitive detectors, low power, fast, modulators, polarization insensitive optical amplifiers, etc., have attained remarkable performance levels in recent years. The chief obstacles to their widespread utilization in real life systems performing more than simple emission and detection functions are rooted in cost and reliability issues. In particular, implementing somewhat more sophisticated optoelectronic functions (such as optical switching, wavelength routing, optical add and drop nodes, wavelength division multiplexing, etc.) by packaging optimized **discrete** components has generally proven to be very difficult and costly with little possibility of improvement. This is a consequence of the so-called packaging problem, the requirement that these devices be aligned and maintained under operating conditions to within a faction of a micron - no mean task.

A possible solution to this dilemma, realized early on by analogy with the development of semiconductor electronic circuits, is through the **monolithic integration** of several optoelectronic devices with different functionalities on a single semiconductor wafer. It's easy to visualize that this has the potential of solving both cost and reliability issues simultaneously. The packaging problem is essentially eliminated since, once the **Photonic Integrated Circuit** (PIC) is completed, the only active alignment needed is with fibers which bring the external optical signal to the PIC and/or receive it from the PIC. All internal alignments are optimized during the design and fabrication stages and remain fixed thereafter. Reliability is also enhanced automatically. Monolithic integration also has the **potential** of cost reduction provided that the integration scheme is simple and reproducible. Simplicity implies a small number of growth and processing steps (preferably uncomplicated, straightforward, and, if possible, standard). The corollary of reproducibility in processing is high yield . None of these requirements have been easy to achieve and lie at the root of the slow progress made toward practical PICs.

CRITERIA FOR PRACTICAL PHOTONIC INTEGRATED CIRCUITS

It's appears evident that to achieve a practical PIC some tradeoffs between device performance and integratability will be required. For example, it's been demonstrated by several groups (e.g., reference 1) that a reversed biased, standard laser structure can perform as an efficient p-i-n detector and thus can be easily and monolithically, integrated to act as a rear facet monitor of a laser diode. However, since the device is not optimized to perform as a detector, there is some reduction in response speed.

In principle, and it has been demonstrated in practice in a few laboratories, it is possible to integrate different **optimized** devices using elaborate growth, etch, regrowth, etc. sequences. An example of such a PICs will be presented later. However, as growth and processing complexity increases, the yield, manufacturability, and hence, practicality rapidly decreases. Practicality mandates low manufacturing cost and high reliability (low operating cost) even at the expense of somewhat lowered performance. With this in mind, we can establish the most important requirement for monolithically integrating optoelectronic devices of differing functionality in order to achieve practical waveguided PICs. (We will concern ourselves here only with waveguided PICs, in which the signal carrying optical beam is confined within a waveguide on the surface of the wafer. All signal processing takes place in this plane. This is believed to be the most likely form for telecommunications PICs as waveguide optical modes are the most compatible with optical fibers and on chip optical signal processing most easily implemented.)

The primary requirement for waveguided PICs is bandgap compatibility among the various optoelectronic devices, as illustrated schematically in Fig. 1 for a "generic" PIC. In such PICs, the signal carrying light beam is directed by passive, transparent, optical waveguides to various active optoelectronic devices in the plane of the wafer. Optimum performance demands that the bandgap energies of the various components relate to each other in a specific fashion. For example, in Fig 1, it is essential that the passive waveguide receiving the light from the laser diode be as transparent as possible to the laser wavelength (i.e., have a larger bandgap energy than that of the laser structure) so that a minimum amount of optical power is lost during transit. On the other hand, the external electro-

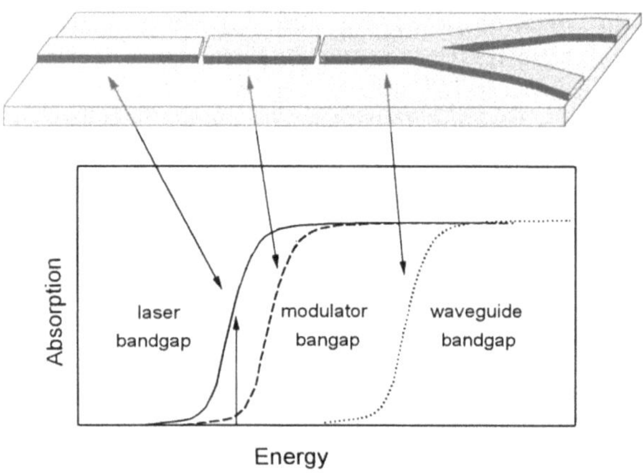

Fig. 1. Schematic diagram of a "generic" PIC (photonic integrated circuit) containing a laser diode whose output is directed into a transparent waveguide 3 dB splitter through an external electroabsorptive modulator. The optimum operation of this PIC requires that the bandgap energies of these three components possess the specific relationship illustrated in the figure.

absorptive modulator (a quantum confined Stark effect (QCSE) modulator - essentially a reverse biased laser structure) should have a slightly larger bandgap than the laser for optimum operation (low loss and good contrast). This very simple PIC (essentially a distributed feedback (DFB) laser diode integrated with an QCSE modulator) has been fabricated using a variety of techniques and provides a useful test case. The advantage of even such a simple PIC as this lies in the increased operational bandwidth, reduced chirp, and possibly reduced demands on the drive circuits that external modulation provides.

The supreme goal of monolithic photonic integration is to achieve the correct bandgap alignment of the various components in a practical (i.e., cheap and reliable) fashion. To this end, a number of fabrication techniques of varying complexity have been developed over the years. Each has its advantages and disadvantages. None is a panacea. These can be loosely grouped into four broad catagories and are discussed briefly below with representative PICs chosen somewhat arbitrarily from the literature. The discussion will be focused on fabrication issues; other important considerations (such as electrical, optical, and thermal crosstalk, heat dissipation, etc.) are beyond the scope of this article.

FABRICATION TECHNIQUES FOR PICs

Selective Area Regrowth

Conceptually, this is the simplest technique, and it has the potential of providing the best overall performance since the device structures being integrated are optimized for each individual function. In simplest terms, devices of differing functionality are fabricated in a series of growths, one for each function. In between growths, processing is used to prepare the original substrate for regrowth in selected areas. For example, in Fig. 1 the laser diode structure could be grown first, followed by an etch step to prepare a trench for the modulator (which is grown next inside the trench), followed by etching another trench for the passive waveguide (grown last).

Conceptually simple, in practice, this is, by far, the most challenging technique to implement since it involves several difficult steps. Even simple epitaxial growth on a virgin substrate, under the best of circumstances, demands high skill and much experience for satisfactory, reproducible results. Growth on a processed substrate containing etched and also masked regions adds considerable complexity to the process. A particular problem occurs where the regrowth layer abuts the original active device. For example, the interface between the passive waveguide and active modulator structure in Fig. 1 is susceptible to irregularities since growth must proceed on several crystallographic planes simultaneously and smoothly. Any irregularities present may disrupt the coupling between the two sections and increase optical power loss. Overgrowth on a patterned substrate, in which there is a single growth of a double (one active, one passive) waveguide structure on an etched, stepped virgin substrate has similar difficulties.

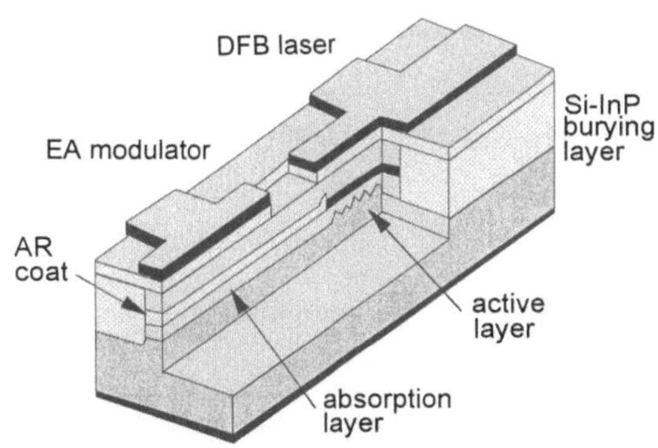

Fig. 2. Schematic diagram of a DFB laser monolithically integrated with an external electroabsorption modulator using selective area regrowth.[2]

In spite of these complications, several groups have produced excellent results through, it seems at times, heroic efforts. (There is one instance of an optoelectronic integrated circuit which involved 25 mask levels, seven epitaxial growths, and 170 processing steps!) Fig. 2 illustrates schematically the test case PIC (a DFB laser monolithically integrated with an external electroabsorptive modulator) fabricated with selective area growth.[2] The performance of this particular PIC is impressive. However the complexity of the process, and consequent (probable) low yield in a commercial environment, makes it unlikely that PICs fabricated using this technique will soon be seen in large numbers in the marketplace. However, this result and others in the literature (including tunable three section MQW-DBR lasers, MQW balanced heterodyne receivers, wavelength division multiplexed (WDM) sources, bidirectional WDMs, etc.) illustrate the capability of the technology today and prospects for the future.

Selective Area Growth

By initiating epitaxial growth on a virgin substrate through a dielectric mask, it is possible to control the growth rate and/or the composition of epilayers in a spatially selective fashion. Thus, for example, the width of the well layer of a quantum well (QW) can be increased in a waveguide region relative to other open regions on the wafer by adding a variable width dielectric mask on either side of the guide. The process is illustrated schematically in Fig. 3.[3] During MOCVD (metalorganic chemical vapour deposition) growth, under a precisely controlled set of growth parameters, source material impinging on the dielectric mask does not adhere since growth nucleation is inhibited. Rather the material moves over to the unmasked regions before contributing to the epitaxial growth. Thus, in Fig. 3, the area between the non-growth dielectric surfaces receives more flux than normal and grows at a more rapid rate. (An inverse process, in which the width of the narrow, unmasked, growth area is varied, achieves similar results.) In fact, by tailoring the width of the dielectric mask, it is possible to taper the central region so that there is a smooth progression from one layer thickness, at the front of Fig. 3, to the other at the back. If QWs are present, the resulting increase in QW width decreases the effective bandgap energy in the QW relative to that of the beginning of the waveguide. Thus, any light emitted by a laser fabricated on the rear portion of the guide could pass through the front portion unabsorbed if the bandgap energy difference is large enough. By judicious masking prior to growth, it is possible to grow device structures for a PIC as in Fig. 1 with the required bandgap energies. Fig. 4 illustrates the test case PIC fabricated using selective area growth.[4] This PIC exhibited a wide frequency bandwidth (14 GHz) at a small

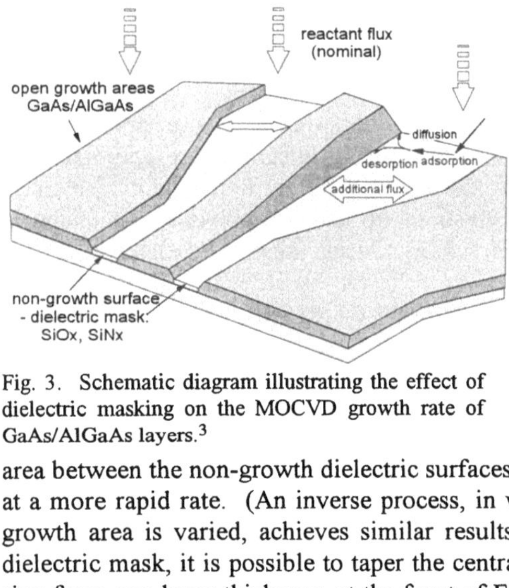

Fig. 3. Schematic diagram illustrating the effect of dielectric masking on the MOCVD growth rate of GaAs/AlGaAs layers.[3]

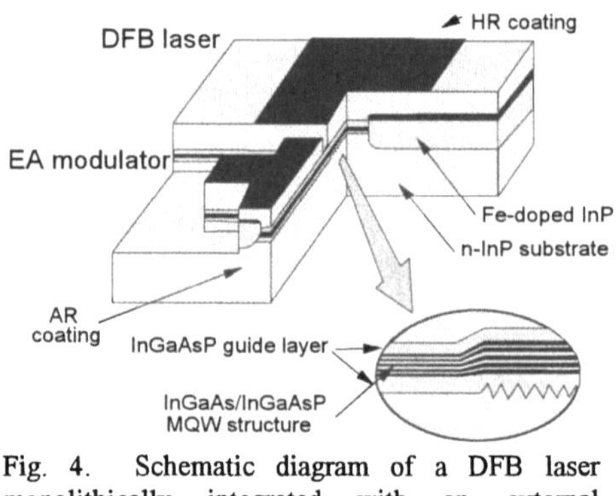

Fig. 4. Schematic diagram of a DFB laser monolithically integrated with an external electroabsorption modulator using selective area growth.[4]

modulation voltage ($1V_{pp}$), high modulation efficiency (13 dB/2V), and low chirp.

The advantage of this technique over the one discussed previously is that, in principle, only one growth is required. The disadvantage is that this growth is relatively sensitive to the growth parameters and may be difficult to control as a result. Also, since there is only one growth, it is not possible to optimize individual device design. However, it's been demonstrated that the structures for most active devices, such as lasers, modulators, and detectors are quite similar. The major difference lies in the required bandgap. Generally, passive structures, such as waveguides, can be based on any waveguiding layers, provided the bandgap is large enough to be transparent to the light.

This technique is relatively new and initial results have been very promising. Provided that the proper growth conditions are not too stringent and/or can be controlled so that reproducibly is not a problem, it could develop into a very powerful tool for monolithically fabricating PICs.

Selective Area QW Intermixing

It has been demonstrated by many groups that it is possible to change the effective bandgap energy of a QW by modifying its profile from abrupt (normally the as-grown shape) to gradual (or error-function shape) in a spatially selective manner by using a variety of techniques for enhancing QW intermixing (i.e., the interdiffusion of well and barrier material driven by thermal effects).[5]

In the simplest picture, the idea is to produce some type of crystal damage in the structure in a spacially selective manner. This has been accomplished by a variety of techniques including impurity diffusion, ion implantation (both low and high energy), and vacancy generation in GaAs at elevated temperatures through then use of a SiO_2 cap. Masking provides spatial selectivity. When the wafer is subsequently annealed (usually by rapid thermal annealing), the presence of the defects, vacancies, interstitials, etc. enhances the thermally driven intermixing taking place at the interface between the well material and the barrier layers thereby "rounding" the QW. The annealing process also repairs much of the initial damage thereby returning the wafer to a state near the virgin condition. In areas where there is no damage, "rounding" is significantly smaller provided critical temperatures are not exceeded. In general, the "rounded" QW has a larger bandgap energy than the as-grown structure as illustrated in Fig. 5. Under optimum conditions in some material systems, this bandgap increase can be as much as several hundred meV and can be precisely controlled by adjusting the damage in each section of the PIC. This technique has recently been employed to fabricate the test PIC discussed above.[6] Several other PICs, such as extended cavity lasers, laser arrays with each laser having a unique wavelength, transparent waveguides, etc.) have also been successfully fabricated using this technique.

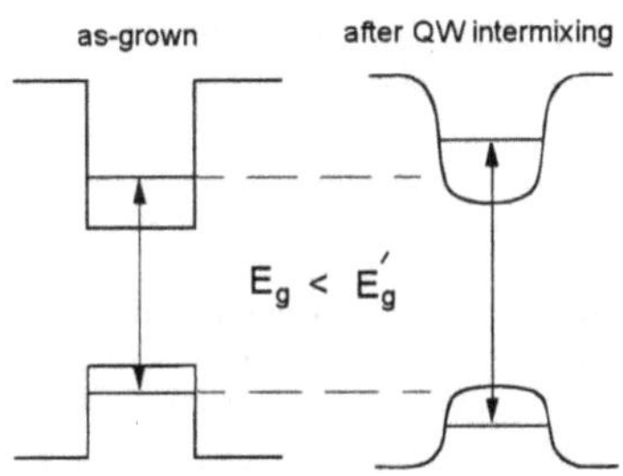

Fig. 5. Schematic diagram illustrating the effect of QW shape modification on QW bandgap energy.

Bandgap shifting using QW intermixing is, perhaps, the simplest of all the techniques investigated so far for accomplishing monolithic integration of photonic devices. It has the considerable advantage that the growth is very simple. The basic structure is grown in one standard growth run and then modified post-growth in a conceptually simple manner. However, as with all techniques except the first, simplicity comes at the price of the inability to optimize individual device performance. But, as discussed in the previous section, a single device structure can perform a number of different functions, such as light emission, detection, and modulation adequately if not superbly. Thus, provided the bandgap energies can be fine tuned sufficiently so that the operations of the various devices are optimized, satisfactory performance from the PIC should be achievable.

A potential difficulty lies in the presence of some residual damage which remains in the structure after incomplete thermal treatment. Provided that care is taken to minimize this damage, passive structures such as transparent waveguides should prove little problem. However, although preliminary results on active devices are encouraging, further study is required to ensure that no long term detrimental effects remain.

The simplicity of this technique makes it a very attractive solution to the monolithic integration problem, especially where it is used solely to produce transparent waveguides.

Other Techniques

A number of other schemes for the monolithic integration of PICs have been attempted over the years, none with significant success. For example, the operating wavelength of p-i-n waveguide detectors can be tuned to some extent by applying an external bias. Also, waveguides made from active material can be driven into transparency through the application of large currents. While simple, these approaches have serious problems and, it appears to me, little future.

CONCLUSION

Progress in developing techniques for the monolithic integration of optoelectronic circuits has been especially rapid recently. However, since even after several decades of intensive development, heterostructural epitaxy still may be said to be more of an art than a science, it's probable that those techniques which place minimum demands on sophisticated growth and processing will come to dominate, - at least initially and at least for relatively simple PICs containing moderate performing optoelectronic devices.

REFERENCES

1 D. Moss, D. Landheer, D. Conn, D. Halliday, S. Charbonneau, G. Aers, R. Barber, and F. Chatenoud, High-speed photodetection in a reverse biased GaAs/AlGaAs GRINSCH SQW laser structure, *IEEE Photon. Tech. Lett.* 4:609 (1992).

2 H. Soda, M. Furutsu, K. Sato, N. Okazaki, Y. Yamazaki, H. Nishimoto, and H. Ishikawa, High power and high speed semi-insulating BH structure monolithic electroabsorption modulator/DFB laser light source, *Electron. Lett.* 26:9 (1990).

3 E. Colas, A. Shahar, B.D. Soole, W.J. Tomlinson, J.R. Hayes, C. Caneau, and R. Bhat, Lateral and longitudinal patterning of semiconductor structures by crystal growth on nonplanar and dielectric-masked GaAs substrates: application to thickness-modulated waveguide structures, *J. Crystal Growth* 107:226 (1991).

4 M. Aoki, H. Sano, M. Suzuki, M. Takahashi, K. Uomi, and A. Takai, Novel structure MQW electroabsorption-modulator/DFB-laser integrated device fabricated by selective area MOCVD growth, Electr. Lett. 27:2138 (1991).

5 e.g., see E.S. Koteles, B. Elman, P. Melman, J.Y. Chi, and C.A. Armiento, Quantum well shape modification using vacancy generation and rapid thermal annealing, Opt. Quant. Electron. 23:S779 (1991).

6 A. Ramdane, Ph. Krauz, E.V.K. Rao, A. Hamoudi, A. Ougazzaden, D. Robein, A. Gloukhian, J. Landreau, M. Carré, A. Mircéa, Monolithic integration of strained layer multi-quantum well distributed feedback laser and external modulator by selective disordering, CLEO '94 Summaries of Papers Presented at the Conference on Lasers and Electro-Optics, Volume 8, 1994 Technical Digest Series, Optical Society of America, Washington, DC, page 239 (1994).

A NEAR TRAVELING WAVE SEMICONDUCTOR LASER AMPLIFIER FOR SIMULTANEOUS AMPLIFICATION, DETECTION AND DISCRIMINATION

M. Le Ligné, M. J. Chawki, T. Valkova

FRANCE TELECOM/CNET RIO/ARO
BP 40 22301 Lannion, France

INTRODUCTION

The detection properties of the semiconductor laser amplifier have been examined by several authors, e. g. Gustavsson et al.[1] have analysed a Near Travelling Wave Optical Amplifier (NTWOA) for simultaneous amplification and detection. Shoa and Koch[2] have theoretically studied the static and dynamical characteristics of semiconductor narrow-band tunable resonant optical amplifiers as active filters and receivers.

In this letter, a NTWOA has been theoretically and experimentally studied as a discriminator/photodetector in a lightwave system.

THEORETICAL MODELLING OF A DETECTION SYSTEM USING A NTWOA

The theoretical description of the NTWOA as a detector is based on Gustavsson's study[1]. A very simple rate equation model has been used for evaluating the total carrier density N in the active region of volume V.

$$\frac{\partial N}{\partial t} = \frac{J}{eV} - \Gamma g_m S - U(N) \quad (1)$$

where Γ is the optical confinement factor, J the current, and g_m the material gain per unit length. The term U(N) includes different recombination processes inside the semiconductor material.

The injected optical signal is included in one of the longitudinal modes according to $S=S_{sp}+S_{sig}$, where S_{sp} is the averaged spontaneous photon flux density[1]. To arrive at a description of the NTWOA photodetector in terms of the junction voltage, we use the approximation given by Joyce and Dixon[3] for the quasi-Fermi levels for electrons and holes, denoted by E_{F_c} and E_{F_v}, respectively.

$$eV(N) = E_{F_c}(N) - E_{F_v}(N) \quad (2)$$

The time dependent simulation of the structure is obtained by solving the equation set of (1) which takes into account the different epitaxial layers. Numerical transient resolution uses the Runge-Kutta method and gives the voltage V(N) from equation (2) versus time for an ASK and a FSK transmission system.

1. Modelling of a NTWOA as a photodetector : ASK receiver

In the first case, the input signal $S_{sig}(t)$ from (1) takes into account that the input optical power, P_{in}, is modulated by a binary sequence with 100% modulation index (ASK system). The wavelength of the signal, λ_{in}, is assumed constant. The amplifier gain spectrum of the NTWOA/photodetector is calculated without any ripples.

2. Modelling of a NTWOA as a discriminator/photodetector : FSK receiver

In the second case, the input signal, $S_{sig}(t)$ takes into account that the input wavelength λ_{in}, is modulated with displacement $\Delta\lambda_d$ (FSK system). The input power, P_{in}, of the NTWOA is assumed constant. Figure 1 shows the gain spectrum of the NTWOA for different values of I/I_{th}(bias current / threshold current).

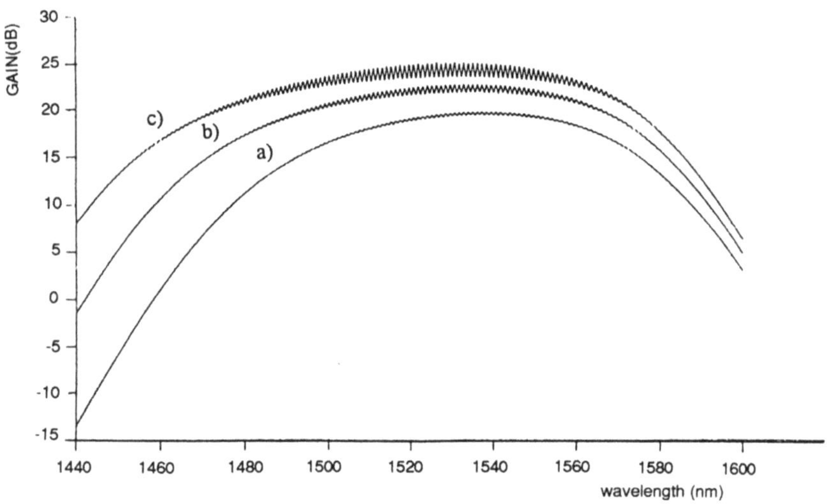

Figure 1. Gain of the NTWOA versus wavelength with $R_1=R_2=5.10^{-4}$, P_{in}=-10 dBm and for different values of bias current : a) $I=2I_{th}$; b) $I=2.5I_{th}$; c) $I=3I_{th}$.

We note that there are many ripples due to facet reflectivities ($R_1=R_2=5.10^{-4}$). The amplitude of the ripples increases with the injection current, and this results in the slope of the ripple becoming more rigid. The discrimination function is obtained between a maximum and a minimum on one of these ripples.

The output signal V(N) is therefore digitally processed to simulate the reception filter whose frequency cut-off is adjusted to digital bit rates. After the FFT filtering process, the eye diagram is simulated for a 2^5-1 bit configurations. The relation between Bit-Error-Rate (BER), eye diagram and noise allows us to evaluate the power penalty. The power penalty is defined as the receiver sensivity degradation in decibels at a fixed BER due to the eye diagram, as compared with the sensivity without intersymbol interference.

THEORETICAL RESULTS

Curves A and B of Figure 2 present the power penalty at BER=10^{-9} versus bit rate for the NTWOA used as a photodetector and a discriminator /photodetector , respectively (P_{in}=-10 dBm).

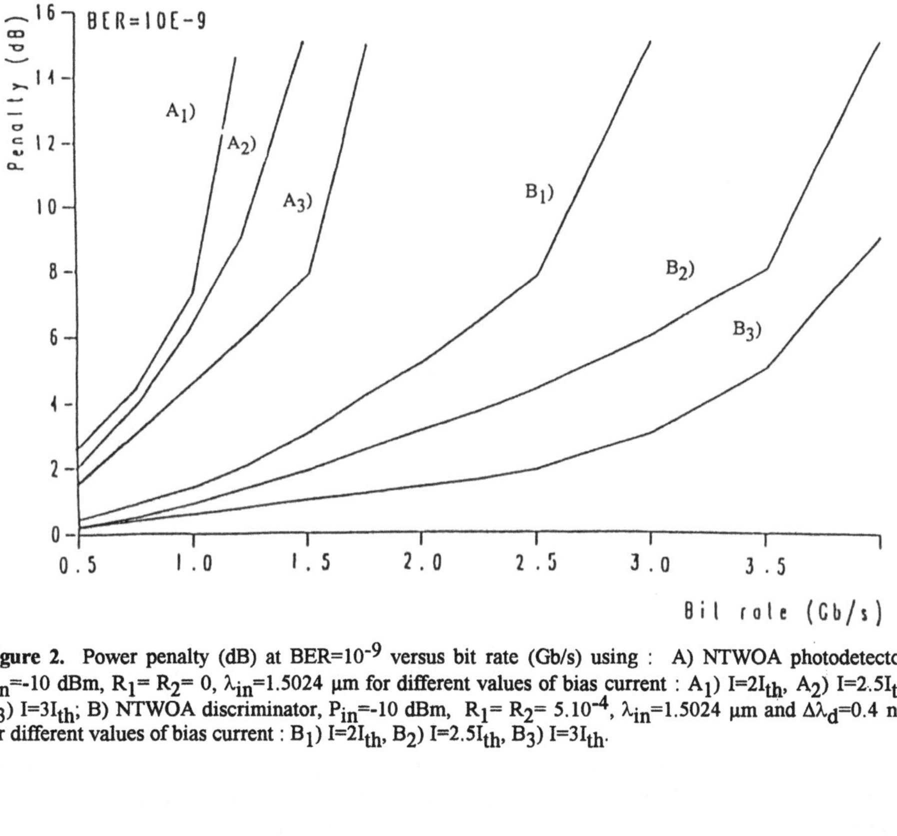

Figure 2. Power penalty (dB) at BER=10^{-9} versus bit rate (Gb/s) using : A) NTWOA photodetector, P_{in}=-10 dBm, R_1= R_2= 0, λ_{in}=1.5024 µm for different values of bias current : A_1) I=$2I_{th}$, A_2) I=$2.5I_{th}$, A_3) I=$3I_{th}$; B) NTWOA discriminator, P_{in}=-10 dBm, R_1= R_2= 5.10^{-4}, λ_{in}=1.5024 µm and $\Delta\lambda_d$=0.4 nm for different values of bias current : B_1) I=$2I_{th}$, B_2) I=$2.5I_{th}$, B_3) I=$3I_{th}$.

We note that the system performance improves with increasing injection current. For a power penalty of 5 dB, I=$3I_{th}$, the bit rate is 1.1 Gb/s for a NTWOA photodetector (A_3) against 3.4 Gb/s for a NTWOA discriminator (B_3). It can be seen that the system performance of the NTWOA discriminator/photodetector is better than that of the NTWOA photodetector for different injection currents.

EXPERIMENTAL RESULTS

In the first case of our experiment, the AM response of the FP laser amplifier receiver is obtained using the FP laser with antireflection coated facets with 10^{-5} reflectivity (considered as a TWOA).

Figure 3 shows the AM response of the FP laser amplifier at I=$2I_{th}$ for different values of input power (P_{in}=-10 dBm, P_{in}=-20 dBm). From Figure 3 the AM response of the laser receiver falls by 3 dB at about 450 MHz corresponding to a 800 Mb/s NRZ system.

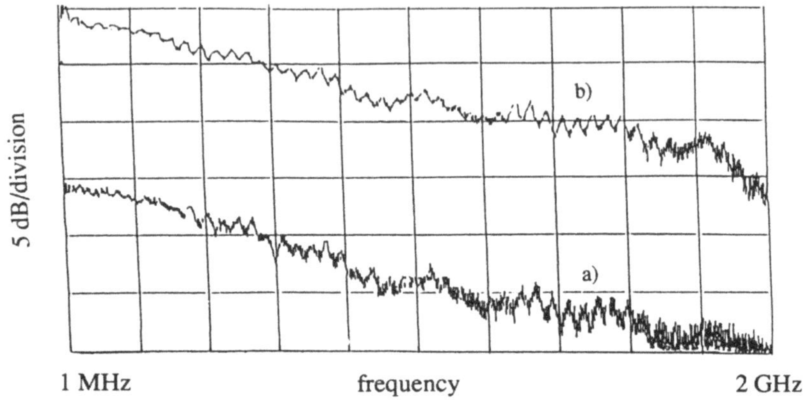

Figure 3. AM responses of the FP laser amplifier with $I=2I_{th}$ for different values of input power : a) P_{in}=-20 dBm; b) P_{in}=-10 dBm.

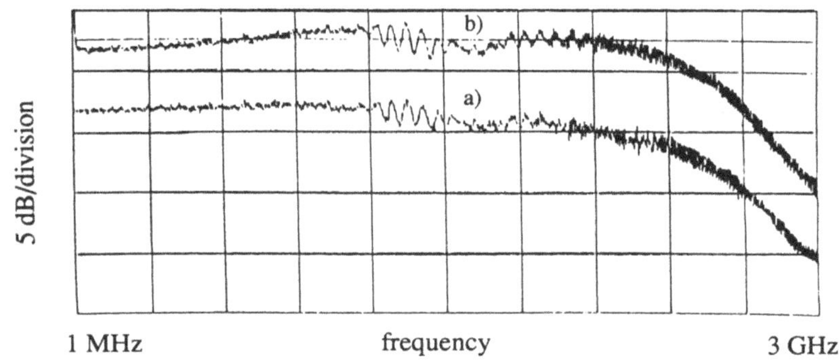

Figure 4. FM responses of the FP laser amplifier with P_{in}=-10 dBm for different values of bias current : a) $I=2I_{th}$; b) $I=2.5\ I_{th}$.

In the second case, the experiment is carried out to measure the FM response of the FP laser with two antireflection coated facets with 10^{-3} reflectivity (NTWOA). The FM response is measured with the FP laser used as a FM discriminator for two different bias current : $I=2I_{th}$, $I=2.5I_{th}$ at P_{in} =-10 dBm (Figure 4). We note that the FM response is relatively flat up to 2 GHz and presents a cut off frequency of about 2.6 GHz, corresponding to 3.7 Gb/s NRZ system. The theoretical study is in agreement with the experimental work.

REFERENCES

1. M. Gustavsson, A. Karlsson and L. Thylén, "Traveling Wave Semiconductor Laser Amplifier Detectors", IEEE, J. of Lightwave Technology, LT-8:610 (1990).
2. F. S. Choa and L. Koch, "Static and Dynamical Characteristics of Narrow Band Tunable Resonant Amplifiers As Active Filters and Receivers", IEEE, J. of Lightwave Technology, LT-9:73 (1991).
3. W. B. Joyce and R. W. Dixon, "Analytic approximations for the Fermi energy of an ideal Fermi gas", Applied Physics Letters, 31:354 (1977).

SMALL SIGNAL ANALYSIS OF FREQUENCY CHIRP IN MULTIPLE QUANTUM WELL LASERS

Marek S. Wartak* and Toshihiko Makino**

*Department of Physics and Computing, Wilfrid Laurier University
Waterloo, Ontario, Canada N2L 3C5

**Advanced Technology Lab., Bell-Northern Research Ltd.
P.O. Box 3511, Station C
Ottawa, Canada K1Y 4H7

INTRODUCTION

In modern optical communication systems, one of the primary parameters that determines the system performance is the frequency chirping [1]. It can eventually limit the system performance. Therefore it is essential to understand how the frequency chirp depends on laser design and how it can be optimized. The step in this direction was taken by Shen and Agrawal[2], who analyzed the effect of the current pulse on the dependence of frequency chirp. Furthermore, for high modulation bandwidth of quantum well lasers, carrier capture and escape time into and out of quantum well must also be taken into account [3], [4].

MODEL

In this paper we have performed for the first time the analysis of frequency chirp in multiple quantum well (MQW) lasers with the quantum capture and quantum escape effects included. First, we have formulated general model of transport in MQW region. In that model (model A) we have introduced 3 dimensional (3D) electron's density above the wells which we have assumed to be nonuniform above barrier region as well as quantum well region. For comparison, we have also formulated a simpler model (model B) where we have assumed that 3D electrons are distributed uniformly over MQW. 3D electrons enjoy coupling to 2 dimensional (2D) electrons which occupy QW via the quantum capture and quantum escape processes.

Chirp is introduced as

$$\delta\nu(t) = \frac{1}{(2\pi)^{3/2}} \int_{-\infty}^{+\infty} (-i\omega)\phi e^{-i\omega t} d\omega \tag{1}$$

Applications of Photonic Technology
Edited by G.A. Lampropoulos *et al.*, Plenum Press, New York, 1995

The phase of electric field is described by

$$-i\omega\phi = \frac{1}{2}\alpha G_N \sum_{k=1}^{M} n_k^{(2)} \tag{2}$$

We will now formulate two models:

Model A (density of electrons above the wells is nonuniform):

3-dimensional electrons above wells (M is the number of wells)

$$-i\omega n_k^{(3)} = -\Gamma_{cap} n_k^{(3)} + \Gamma_{esc} n_k^{(2)} - \Gamma_{sp} n_k^{(3)} - \Gamma_d \{n_k^{(3)} - n_{k-1}^{(3)}\} - \Gamma_d \{n_k^{(3)} - n_{k+1}^{(3)}\} \tag{3}$$

2-dimensional electrons:

$$-i\omega n_k^{(2)} = \Gamma_{cap} n_k^{(3)} - \Gamma_{esc} n_k^{(2)} - \Gamma_n n_k^{(2)} - \Gamma_t \{2n_k^{(2)} - n_{k-1}^{(2)} - n_{k+1}^{(2)}\} - \frac{v_g S_0 G_N}{1+\varepsilon S_0} n_k^{(2)} - \frac{v_g G_0}{1+2\varepsilon S_0} s \tag{4}$$

Model B (density of electrons above the wells is uniform):

electrons above wells

$$-i\omega n_Q = \frac{j_{into}}{e} - M\Gamma_{cap} n_Q + \Gamma_{esc} \sum_{k=1}^{M} n_k^{(2)} - \Gamma_{sp} n_Q \tag{5}$$

2-dimensional electrons

$$-i\omega n_k^{(2)} = \Gamma_{cap} n_Q - \Gamma_{esc} n_k^{(2)} - \Gamma_n n_k^{(2)} - \Gamma_t \{2n_k^{(2)} - n_{k-1}^{(2)} - n_{k+1}^{(2)}\} - \frac{v_g S_0 G_N}{1+\varepsilon S_0} n_k^{(2)} - \frac{v_g G_0}{1+2\varepsilon S_0} s \tag{6}$$

Dynamical equation for photons is the same in both models:

$$-i\omega s = \frac{\Gamma v_g S_0}{1+\varepsilon S_0} G_N \sum_{k=1}^{M} n_k^{(2)} + \frac{\Gamma_P}{1+\varepsilon S_0} s - \Gamma_P s \tag{7}$$

The dynamics in the separate confinement region is described by classical diffusion equation characterized by diffusion constant and spontaneous lifetime.

RESULTS

The analysis has been performed for sinusoidal modulation, also logarithmic optical gain expression has been used as $G = \Gamma_f v_g g_m(N)$ with $g_m(N) = g_0 \ln N/N_0$. Logarithmic gain coefficient $g_0 = 3000 cm^{-1}$ [5]. We have found that both models give very similar results, thus in practice simpler uniform model B can be used.

In Fig.1 we have plotted frequency chirp vs modulation frequency for different number of quantum wells. Separate confinement width was chosen to be $L_{SCH} = 1000\overset{o}{A}$ and parameter $R = \tau_{cap}/\tau_{esc} = 0.1$. There is unnoticable difference between both models. As the number of wells increases chirp decreases for frequencies up to 10 GHz. The reverse is taking place after relaxation-oscillation peak, i.e. chirp increases as the number of wells increases.

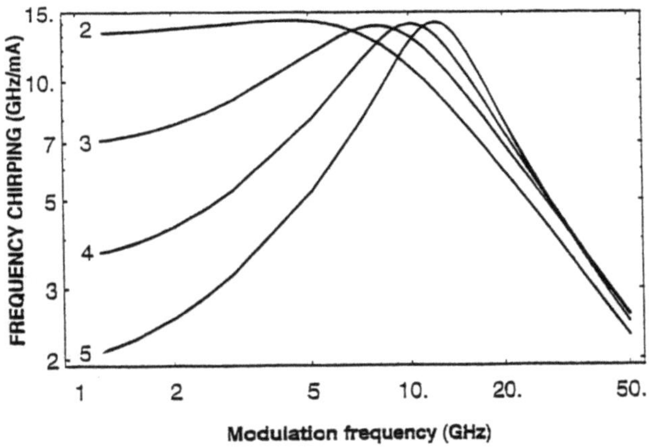

Fig. 1. Frequency chirp vs. modulation frequency for 2, 3, 4, and 5 wells with transport effects.

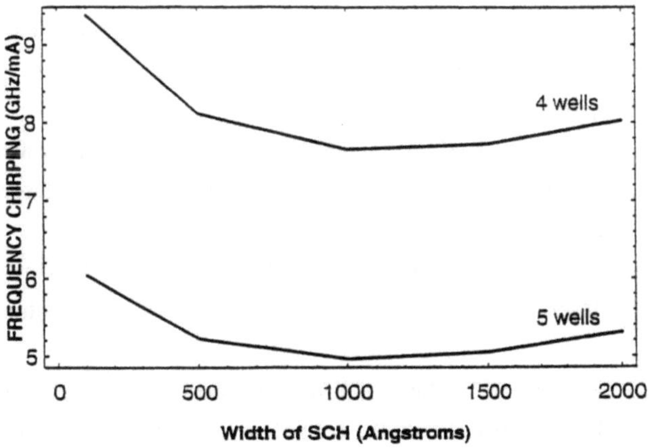

Fig.2. Frequency chirp as a function of SCH width for frequency 4.8 GHz.

In Fig.2 we have demonstrated the effect of separate confinement region where we have plotted frequency chirp as a function of SCH width for frequency f=4.8 GHz. At small values of SCH width the value of chirp is quite large indicating the very bad optical confinement, it then decreases up to a certain value and increases again. The increase reflects the fact that for large values of SCH width, carriers need more time to move across SCH region and this fact degrades chirp.

CONCLUSIONS

For the first time we have performed the analysis of carrier capture and escape as well as the effect of SCH width on frequency chirp in multiple quantum well semiconductor lasers. It has been also found that there is an optimum value of separate confinement width at which frequency chirp becomes minimum.

ACKNOWLEDGMENTS

MSW acknowledges partial support from the Natural Sciences and Engineering Research Council of Canada operating grant.

REFERENCES

1. G.P. Agrawal and N.K. Dutta,"Long-wavelength semiconductor lasers", Van Nostrand, 1986.
2. T.M. Shen and G.P. Agrawal,"Pulse-Shape Effects on Frequency Chirping in Single-Frequency Semiconductor Lasers Under Current Modulation", IEEE J. Lightwave Technology,v.4, 497 (1986).
3. S.C. Kan, D. Vassilovski, Ta C. Wu and K.Y. Lau,"On the effect of carrier diffusion and quantum capture in high speed modulation of quantum well lasers", Appl. Phys. Lett.,v.61, 752 (1992).
4. R. Nagarajan, M. Ishikawa, T. Fukushima, R.S. Geels and J.E. Bowers,"High Speed Quantum-Well lasers and Carrier Transport Effects", IEEE J. Quantum Electron.,v.28, 1990 (1992).
5. J. Hong, W.-P. Huang, and T. Makino,"Effect of Linear Gain Saturation on Small-Signal Dynamics od MQW DFB Lasers", IEEE Photon. Technol. Lett. **5**, 1373 (1993).

BROAD BANDWIDTH MICROWAVE MEASUREMENTS OF MATERIALS USING PHOTOCONDUCTIVELY-PULSED ANTENNAS

W. M. Robertson[1] and G. Arjavalingam[2]

[1]Institute for Information Technology, National Research Council Canada
M-50 Montreal Road, Ottawa, Ontario K1A 0R6
[2]IBM Research Division, Thomas J. Watson Research Center, P. O. Box 218
Yorktown Heights, New York 10598

INTRODUCTION

Accurate measurement of the microwave dielectric properties of materials is an important issue because of the continued advances in high-speed electronic and optoelectronic devices. In packaging conventional electronics, materials such as polymers or ceramics are used as dielectric interlayer spacers. The complex microwave dielectric properties of these surrounding media determine the delay, attenuation, and dispersion of fast electrical signals between devices. Because at present the fastest electrical devices have risetimes of less than 10 picoseconds,[1] it is important to be able to measure the properties of dielectric packaging materials to frequencies greater than 100 GHz. Similarly a knowledge of the dielectric properties of polymers and crystalline materials are required in optoelectronics because these materials are being used in high-speed electro-optic modulators and switches. The optical response of these materials is generally very fast so that the ultimate performance of the electro-optic device is determined by the switching time which depends on the capacitative time constant and the microwave group velocity.[2]

EXPERIMENTAL TECHNIQUE

The technique of coherent microwave transient spectroscopy (COMITS) is one example of a class of recently developed methods for measurements in the GHz and THz frequency regimes based on the radiation and detection of electromagnetic transients by photoconductively-pulsed antennas.[3-4] The experimental configuration for COMITS measurements is shown in Figure 1. The transmitting and receiving elements consist of coplanar strip feed lines terminated by exponentially-flared antennas. The feedlines are 5 μm wide and they are separated by a 10 μm gap. The antenna section is flared to an opening of about 3 mm. The antennas are fabricated in 0.5 μm thick Al on silicon-on-sapphire wafers using standard photolithographic techniques. After fabrication the silicon epilayer is ion implanted to reduce the carrier lifetime of photogenerated carriers to less than 1 ps.

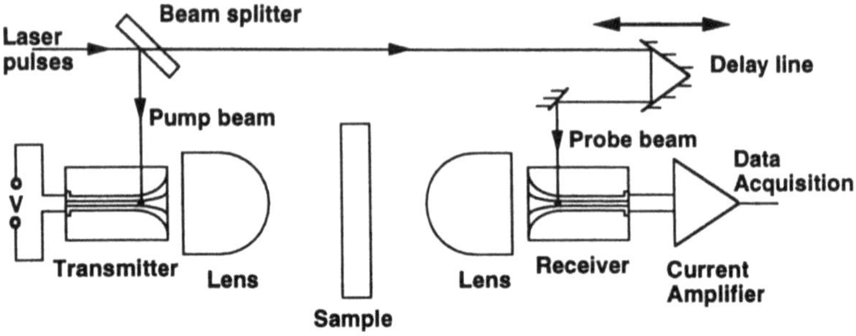

Figure 1. Experimental configuration for COMITS transmission experiments

Short pulse optical radiation, typically from a modelocked Ti:Sapphire laser, is configured in a standard pump-probe configuration as shown in Fig. 1. The pump pulse, focused so as to illuminate the small gap between the feedlines of the dc-biased transmitter, photoconductively generates a short current transient on the coplanar strip transmission line. The transient propagates down the feedline, spreading because of dispersion to about 7 ps, and is radiated into free space in the exponentially-flared antenna section. Hyperhemispherical fused silica lenses are used to collimate the radiation from the transmitter and to focus the signal onto the receiver. The signal voltage arriving at the receiver is determined as a function of the variable delay between the pump and probe by photoconductive sampling. The recorded signal is a time representation of the transient voltage pulse propagated between the transmitter and receiver. A time-domain waveform is shown in Figure 2. The inset to the figure contains the amplitude spectrum of the time-domain waveform, calculated by a numerical Fourier transform, demonstrating that the transient contains frequency components over the range 10-150 GHz. Because the measured signal at the receiver is proportional to the instantaneous electric field of the transient, phase information is preserved in the Fourier transform.

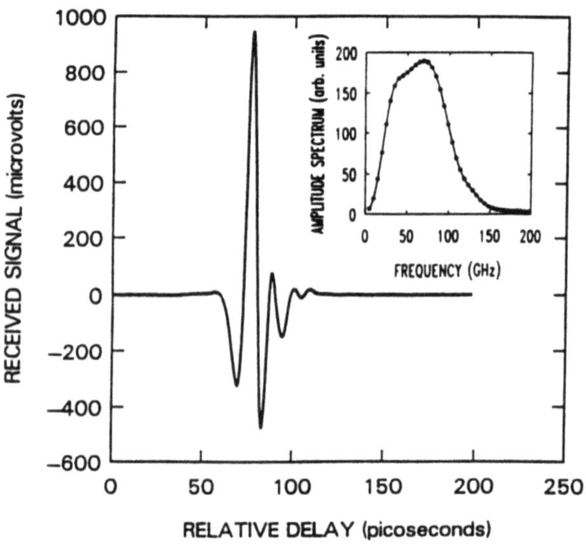

Figure 2. COMITS time-domain waveform. The inset shows the corresponding amplitude spectrum calculated by a numerical Fourier transform.

To measure the dielectric properties of a material, two samples of different, accurately measured, thicknesses are inserted in turn between the transmitter and receiver. Time-domain waveforms are recorded. The complex Fourier transforms of the time-domain data are divided to yield the transmission function (the amplitude and phase of the transmitted radiation at each frequency) due to a piece of material whose thickness is the difference of the thicknesses of the two samples. The use of two samples eliminates the need to correct for reflection losses at the sample surfaces. From the net phase delay of the transmission function the real part of the microwave refractive index can be calculated at each frequency; similarly, the reduction in amplitude provides a value for the loss coefficient. Although the coplanar strip antennas sacrifice some of the bandwidth available from the short optical pulses they compensate for this by emitting radiation that is quasi-directional requiring only a single microwave lens to obtain well-collimated radiation. Furthermore, the planar antenna structures emit radiation that is polarized with its E-field in the plane of the antenna. Thus, in a single experiment the dielectric constant and loss of isotropic or anisotropic materials can be measured over a frequency range from 10-150 GHz.

RESULTS

An example of the use of the COMITS technique for measuring the dielectric properties of a bulk material is shown in Figure 3 which plots the refractive index and E-field loss for the ordinary and extra-ordinary axes of $LiNbO_3$. These data were obtained by Fourier analyzing time-domain waveforms obtained with different thickness single-crystal $LiNbO_3$ samples in the COMITS transmission set up. For the experiments the crystals were oriented either with the optic axes parallel or perpendicular to the plane of the antenna (which defines the direction of the E-field of the radiated transient) to measure the extra-ordinary (n_e = 5.11) and ordinary (n_o = 6.64) refractive indices respectively. A more complete description of the COMITS measurements on $LiNbO_3$ are given in reference [5].

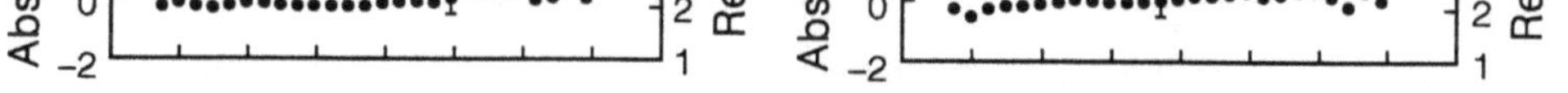

Figure 3. Microwave refractive index and absorption for $LiNbO_3$.

In general, the resolution of the measurements increases for thicker samples because the larger time-delay and amplitude reduction can be more accurately measured. However, it is also possible to perform measurements on thin films by using a precision delay line[6] to measure the small time delays produced by transmission through free-standing thin films. The analysis of the thin film data is more complicated than measurements on bulk samples because of the effect of multiple internal reflections within the film. In the bulk sample case these multiple reflections are well separated in time and they can be eliminated by time-windowing. For very thin films these multiple reflections occur within a time smaller than the width of the transient pulse itself. To extract the dielectric constant values from this data the transmission function (amplitude and phase) are fit with a theoretical model that accounts for the multiple internal reflections. The data plotted in Figure 4 show the experimentally determined transmission function for a 190 μm thick film of the polymer poly (p-phenylene bi phenyltetracarbox-amide) [BPDA-PDA]. The solid and dashed lines are fits to the amplitude and phase data respectively using a model that accounts for the multiple internal reflections. The measured in-plane dielectric constant for this material is $\varepsilon_r = 3.6$, $\varepsilon_i = 0.0$.

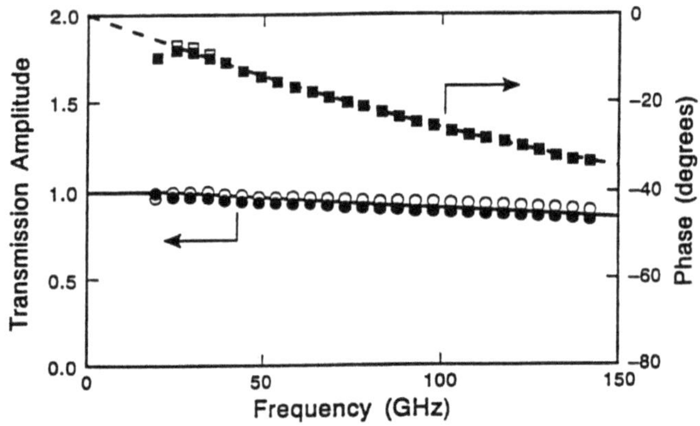

Figure 4. Amplitude and phase data for transmission through 190 μm film of BPDA-PDA

CONCLUSION

This paper has briefly reviewed the capabilities of the COMITS technique for the broadband microwave measurement of the dielectric properties of materials. The technique has also proven useful for a variety of other microwave applications including reflection spectroscopy of semiconductors, scattering experiments, microwave grating characterization, and the measurement of the photon dispersion relation in regular dielectric arrays.

REFERENCES

1. G. L. Patton et. al., 75-GHz f_T SiGe-base heterojunction bipolar transistors, *IEEE Electron Device Lett.* 11:171 (1990).
2. D. S. Chemla and J. Zyss, "Nonlinear Optical Properties of Organic Molecules and Crystals", Academic Press, Orlando (1987).
3. Y. Pastol, G. Arjavalingam, J.-M. Halbout, G. V. Kopcsay, Coherent broadband microwave spectroscopy using picosecond optoelectronic antennas, *Appl. Phys. Lett.* 54:307 (1989).
4. G. Arjavalingam, Y. Pastol, J.-M. Halbout, G. V. Kopcsay, Broadband microwave measurements with transient radiation from optoelectronically pulsed antennas, *IEEE Trans.* MTT-38:615 (1990).
5. W. M. Robertson, G. Arjavalingam, and G. V. Kopcsay, Broadband microwave dielectric properties of lithium niobate, *Electron. Lett.* 27:175 (1991).
6. D. C. Edelstein, R. B. Romney, and M. Scheuermann, Rapid programmable 300 ps optical delay scanner and signal averaging system for ultrafast measurements, *Rev. Sci. Instrum.* 62:579 (1991).

SOLITON SWITCHING IN SEMICONDUCTOR DIRECTIONAL COUPLERS

Falk Lederer, Wolfram Biehlig, Ralf Muschall, and Lutz Leine

Friedrich-Schiller-Universität Jena
Physikalisch-Astronomisch-Technikwissenschaftliche Fakultät
Max-Wien-Platz 1
D-07743 Jena, Germany

INTRODUCTION

There is an urgent need in all-optical signal processing to implement fast switching devices with high switching contrast. Nonlinear directional couplers (NLDC) have attracted a great deal of interest as basic all-optical elements because the output may be routed between the two channels as a function of the input power launched into one channel. The stationary response of that device has been discussed in detail for numerous situations (for a summary, see e.g. the review paper[1] and the references therein). That approach holds also in the nonstationary regime provided that the pulses are sufficiently long and/or the waveguides are short. In that case the dispersion of the group velocity (GVD) may be neglected. The time in the reference frame of the pulse enters the coupled mode equations merely as a parameter. This has the consequence that the coupling behavior (efficiency and coupling length) depends on the instantaneous intensity of the pulse at a certain time leading to the detrimental pulse break up[2] and thus to an incomplete switching between both channels. This phenomenon limits considerably the opportunity of cascading different switching elements.

The formation of stable, robust solitons that maintain their shape in the course of propagation even in the presence of small perturbations has been shown to appear in fused-silica optical fibers in the region of anomalous GVD. Due to that robustness solitons switch completely in NLDC without the detrimental pulse breakup[3,4]. Unfortunately, fibre directional couplers require a considerable device length due to their extremely weak Kerr nonlinearity. In that respect it is interesting to study whether soliton switching can also be achieved in planar or rib waveguide configurations.

Currently, semiconductors are considered as promising materials for all-optical applications because they may exhibit ultrafast nonlinearities being much stronger than that encountered in silica optical fibers. Recently, ultrafast switching in planar and rib waveguide couplers has been reported in semiconductors (AlGaAs) operated below half the band gap[5]. In that frequency domain the nonlinearity responds quasi-instantaneously and is

Kerr-like. Moreover, for a typical AlGaAs system the half gap frequency meets extremely well the communication wavelength of $\lambda = 1.55$ μm. Unfortunately, pulse break up cannot be avoided because the group velocity dispersion of semiconductors at that wavelength is normal. Conventional waveguides cannot compensate for that large normal GVD. But, recently, we have shown that one can form a directional coupler by two Anti-Resonant-Reflecting-Optical-Waveguides (ARROWs)[6,7] that may be used to couple modes over remote distances due to the peculiar dispersion properties.

The aim of this paper is to design an ARROW coupler structure that may be used for soliton switching. To our best knowledge, this is the first paper where soliton switching in planar waveguides has been shown.

GROUP VELOCITY DISPERSION IN SEMICONDUCTOR WAVEGUIDES

If β denotes the propagation constant of a guided wave and ω its frequency the GVD is defined as

$$D = \left.\frac{\partial^2 \beta}{\partial \omega^2}\right|_{\omega_0} \tag{1}$$

where ω_0 is the mean frequency of the pulse and the GVD is measured as ps^2/km. Standard optical fibres exhibit a GVD of $D \approx -20\ ps^2/km$. In order to provoke soliton switching different length scales have to coincide. Roughly speaking, the dispersion length (L_d), the nonlinear length (L_{nl}) and the half-beat length of the coupler (L_h) have to be equal, at least approximately. The respective lengths are defined as:

$$L_d = \frac{T_o^2}{|D|}, \quad L_{nl} = \frac{1}{\gamma P_0}, \quad L_h = \frac{\pi}{\Delta\beta} \tag{2}$$

where T_0 is the pulse length, P_0 the pulse peak power, $\Delta\beta$ the difference of the propagation constants of the guided supermodes and γ the effective nonlinear coefficient. If one limits the device length (L_h) to 1 cm and uses pulses with 100 fs the dispersion is required to be $D = -1000\ ps^2/km$ exceeding considerably those found in silica fibres. Moreover, the GVD in AlGaAs at $\lambda = 1.55$ μm is in the same order of magnitude, but unfortunately normal, as can be seen in Fig.1.

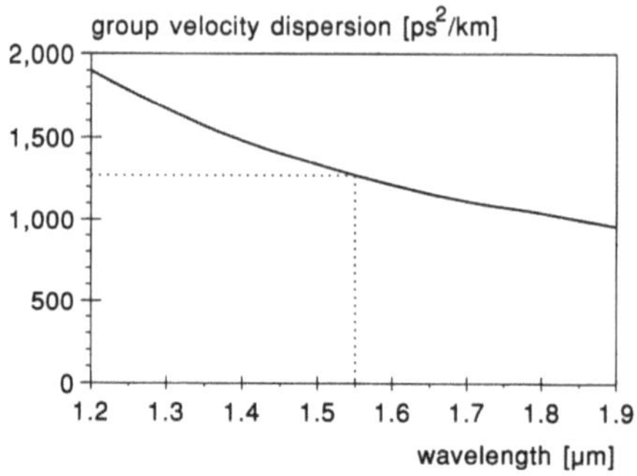

Figure 1. Group velocity dispersion of bulk $Al_xGa_{1-x}As$ (x=0.2)

But, an ARROW[8] that consists of a low-index core surrounded by a few alternating high-and low-index reflectors exhibits a particular dispersion behavior[6] being of main concern here. In the domain where optimum guiding is provided the dispersion curves are extremely flat whereas they exhibit a fairly strong curvature if the guide gets detuned. The key idea consists now to exploit that peculiarity in order to identify domains with large anomalous GVD. The price one has to pay is that one must add some more reflector films because one has left the region of optimum guidance. In Fig.2 the optimized configuration together with the GVD of both the symmetric and the antisymmetric mode is depicted. It is evident, that the ARROW has the potential to overturn the "wrong" material dispersion. Note, that both GVDs are slightly different and that the third-order dispersion is zero for λ=1.55 µm being important for fs-pulses.

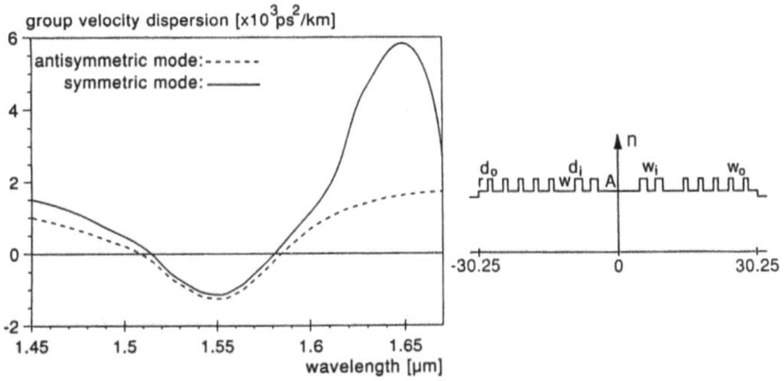

Figure 2. ARROW configuration and GVD (material and waveguide) for the symmetric as well as antisymmetric modes in an AlGaAs - coupler at λ = 1.55 µm

THEORETICAL MODEL

We restrict ourselves to TE-polarized fields and may then write the y-component of the electric field as

$$E(x,z,t) = A_s(z,t)\overline{E}_s(x)\exp\left[i(\beta_s z - \omega_0 t)\right] + A_a(z,t)\overline{E}_a(x)\exp\left[i(\beta_a z - \omega_0 t)\right] \quad (3)$$

where A_j, $\overline{E}_j$ and β_j are the slowly-varying amplitudes, the field profiles and the propagation constants of the symmetric (s) as well as antisymmetric (a) mode, respectively. In the framework of a coupled supermode approach, but taking into account GVD, we end up with a system of modified, coupled Nonlinear Schrödinger Equations

$$\left(\frac{\partial}{\partial z} + \frac{1}{v_{gj}}\frac{\partial}{\partial t} + i\frac{D_j}{2}\frac{\partial^2}{\partial t^2}\right)A_j = iC_{1j}\left|A_j\right|^2 A_j + 2iC_{2j}\left|A_{3-j}\right|^2 A_j + iC_{3j}A_{3-j}^2 A_j^* \exp\left[(-1)^j 2i\Delta\beta z\right] \quad (4)$$

where the C_{ij} are the nonlinear coupling coefficients and j=1(s),2(a). By an appropriate normalization (4) may be normalized (soliton units) and solved with a standard BPM code. The resulting equations are similar to that describing the pulse evolution in birefringent fibres. Note that all nonlinear terms are in the same order of magnitude and that the GVDs are different. The evolution of the pulses is shown in Fig.3 for an input power in channel 1 below and above the critical power. The parameters of our configuration are

$L_h = 13.74$ mm, $L_{ds} = 14.46$ mm , $L_{da} = 13.09$ mm , $T_0 = 0.1279$ ps. The critical power for switching was estimated to be: $P_c \approx 31$ W/μm.

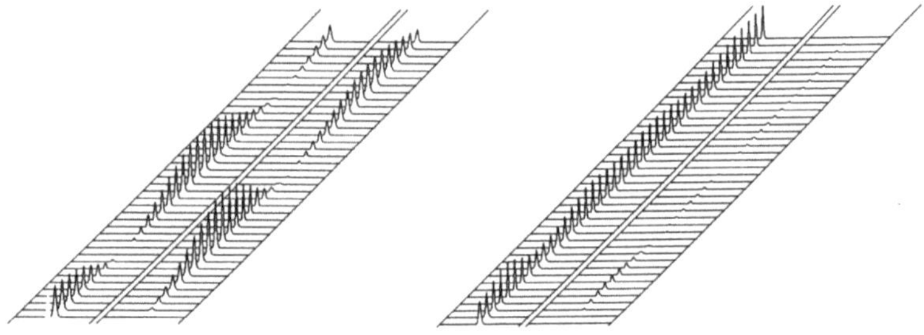

Figure 3. Evolution of the pulses in an ARROW NLDC for different input peak powers over $z=4L_h$

It is evident from Fig.3 that we may achieve a similar switching behaviour as familiar from fiber nonlinear directional couplers. The most important result is that the detrimental pulse break up can be avoided. We may expect that both the half-beat length, the critical switching power and the device thickness can be decreased if we are using rib ARROW couplers[9]. The only difference with respect to the modeling consists in the change of the overlap integrals as well as in the GVD coefficients.

In conclusion, we have shown that soliton switching in nonlinear semiconductor ARROW directional couplers can be achieved. A proper design of the ARROW coupler provides the anomalous GVD being a prerequisite for soliton propagation.

REFERENCES

1. I. Romagnoli, S. Trillo, and S. Wabnitz, Soliton switching in nonlinear couplers, *Opt. Quant. Electr.* 24:1237 (1992).
2. S. Trillo, S. Wabnitz, R.H. Siolen, G. Assanio, C.T. Seaton, and G.I. Stegeman, Experimantal observation of polarization instability in a birefrigent optical fiber, *Appl. Phys. Lett.* 49:1224 (1986).
3. K.J. Blow, N.J. Doran, and B.K. Nayar, Experimental demonstration of optical soliton switching in an all-fiber nonlinear Sagnac interferometer, *Opt. Lett.* 14:754 (1989).
4. S. Trillo, S. Wabnitz, E.M. Wright, and G.I. Stegeman, Soliton switching in fiber nonlinear directional couplers, *Opt. Lett.* 13:672 (1988).
5. J.S. Aitchison, A. Villeneuve, and G.I. Stegeman, All-optical switching in a nonlinear GaAlAs X-junction, *Opt. Lett.*18:1153 (1993).
6. M. Mann, U. Trutschel, Ch. Wächter, L. Leine, and F. Lederer, Directional coupler based on antiresonant reflecting optical waveguides, *Opt. Lett.* 16:805 (1991).
7. U. Trutschel, M. Mann, F. Lederer, Ch. Wächter, and A.D. Boardman, Nonlinear remote switching in nonlinear waveguides, *Appl. Phys. Lett.* 59:1940 (1991).
8. M.A. Duguay, Y.A. Kokubun, T.L. Koch, and L. Pfeiffer, Antiresonant reflecting optical waveguides in SiO_2-Si multilayer structures, *Appl. Phys. Lett.* 19:13 (1986).
9. J. Gehler, A. Bräuer, and W. Karthe, Remote coupling over 93 μm using ARROW waveguide in strip configuration, *Electron. Lett.* 30:218 (1994).

DESIGN, FABRICATION AND CHARACTERISATION OF HIGH SPEED GaAs MSMs FOR OEIC APPLICATIONS

François Gouin, Christian Beaulieu and Julian Noad

Communications Research Centre, 3701 Carling Ave
Ottawa, Ont, Canada, K2H 8S2

INTRODUCTION

GaAs optoelectronic integrated circuits (OEICs) are rapidly finding use in a variety of communications and signal processing applications such as high speed, broadband switching (up to 10 Gbits/s and higher), high speed interconnects for multi-processor LANs and space-based, optically controlled phased array radars and antennas for next generation personal communications. One of the key elements in any of these circuits is the photodetector, which serves as the front end receiver, converting optical input signals from the fibre optic network, into electronic signals which can be processed by subsequent electronic circuitry. Planar Metal-Semiconductor-Metal (MSM) photodetectors are particularly attractive as the receiving element in these circuits. They offer a combination of high speed, wide bandwidth and low power consumption and can be easily integrated with conventional GaAs MESFET circuit technology[1]. In this paper the design, fabrication and characterisation of GaAs MSMs and MSM-based OEICs is discussed.

MSM CHARACTERISTICS

Frequency Response of MSM on Various Materials

These MSMs have been fabricated both directly on undoped semi-insulating GaAs substrates as well as on high quality, thick (1-2 μm) undoped epitaxial layers ($N_d < 1x10^{14}$ cm^{-3}). Results from finite element modelling of these MSM structures indicate that their bandwidth is very sensitive to the density of deep traps within the semiconductor[2]. To verify these simulations, MSMs were fabricated in exactly the same manner on a number of semi-insulating substrates from different manufacturers. The devices consist of an interdigitated metal-contact pattern defined over an area 100x100 μm^2. To investigate the effects of electrode spacing, devices with gaps ranging from 2 to 20 μm were defined. In all cases, the electrode finger width was 3 μm. To avoid some of the problems associated with determining accurate analogue bandwidths using pulse response/Fourier transform techniques, true ac frequency response measurements were carried out using a dc laser diode/high frequency E-O modulator combination to accurately measure the frequency response of the detectors over the 0.1-15 GHz range.

Applications of Photonic Technology
Edited by G.A. Lampropoulos *et al.*, Plenum Press, New York, 1995

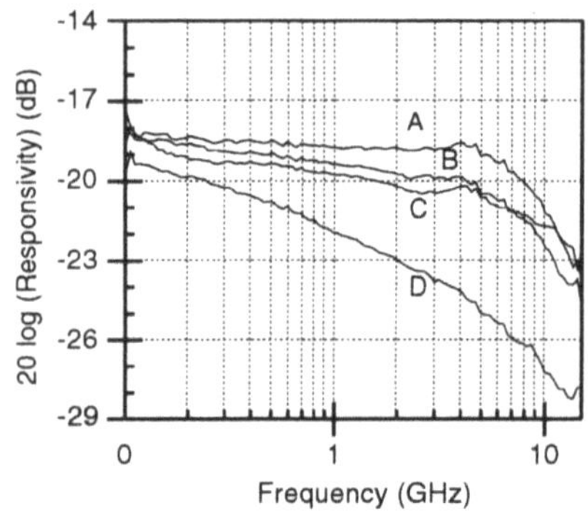

Figure 1. Frequency response of 2 μm gap MSM photodetectors fabricated on a GaAs epitaxial layer (A) and on three semi-insulating substrates (B, C and D). The bias is 5 V.

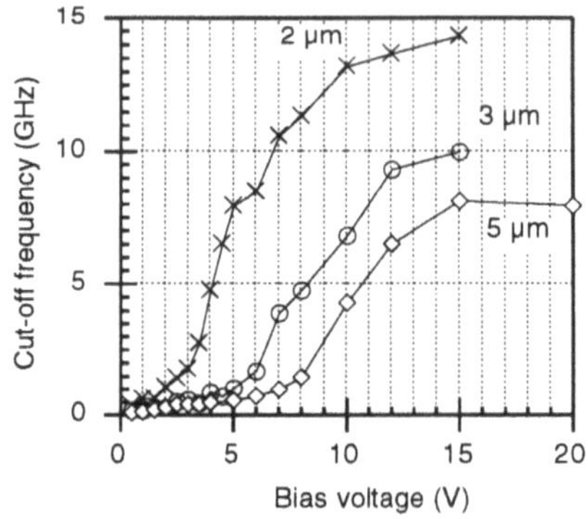

Figure 2. Bandwidth of MSM photodetector sample B as a function of bias for three values of finger spacing.

It has been observed that the 3-dB bandwidths can vary widely, from as low as 500 MHz to over 15 GHz, suggesting that the trap levels within these materials can vary markedly from manufacturer to manufacturer and even batch to batch. Fig. 1 shows the frequency response of MSMs with 2 μm electrode spacing fabricated with Ti-Au contacts on a variety of high-resistivity substrates from four different suppliers. On the vertical axis, a value of -20 dB corresponds to a responsivity of 0.1 A/W. As expected, the best performance was obtained from nominally, trap-free epitaxial layers, with these devices exhibiting 3-dB bandwidths in the neighbourhood of 12 GHz. The responsivity and bandwidth of detectors with electrode spacings ranging from 2 to 5 μm exhibit interesting variations. The bandwidth of a typical device from wafer B is shown in Fig. 2 as a function of bias for three electrode spacings. As expected the bandwidth of these detectors is a function of both the bias and the electrode spacing. This is consistent with other results that suggest that the transit time of the photogenerated carriers dominates their high frequency performance.

The responsivity of sample B at 100 MHz is shown in Fig. 3 as a function of bias for three finger spacings. Initially, it rises rapidly then saturates softly, i.e. continues to rise but

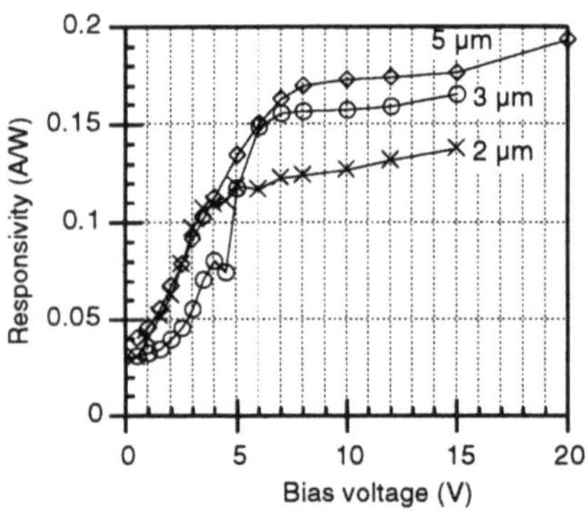

Figure 3. Sample B responsivity at 100 MHz as a function of bias for three finger spacings.

Table I. Saturated responsivity of sample B at 100 MHz for different finger spacings.

gap	V_{sat}	E_{sat}	Resp.	Normalized Resp.
μm	V	kV/cm	A/W	
2	5.5	27	0.124	0.31
3	6.5	22	0.157	0.31
5	7.5	15	0.173	0.28

with a much shallower slope. Some of the parameters at saturation are shown in Table I. The normalized responsivity is the responsivity divided by the fraction of the exposed area. The variation in saturated responsivity as a function of spacing is linear and is simply due to the larger exposed semiconductor area in those devices with wider gaps and hence fewer fingers resulting in reduced metal coverage. An average value of the field is calculated although, as shown in Ref. 2, the field is not expected to be uniform. In some cases, the slope of the 0.1 GHz responsivity vs bias starts to increase again at larger biases. This is the result of a second, low-frequency gain mechanism observed in semi-insulating GaAs[1]. This rise in 0.1 GHz responsivity accounts for the apparent decrease in bandwidth of the 5 μm MSM at larger bias.

MSM equivalent circuit

The design of microwave integrated circuits requires accurate device models which can be used for reliable circuit simulation. MSMs are often modelled by the circuit represented in Fig. 4. The photogeneration process is represented by a current source, I_p, in parallel with a capacitor, C_p, representing the capacitance between the contacts and a resistor, R_p,

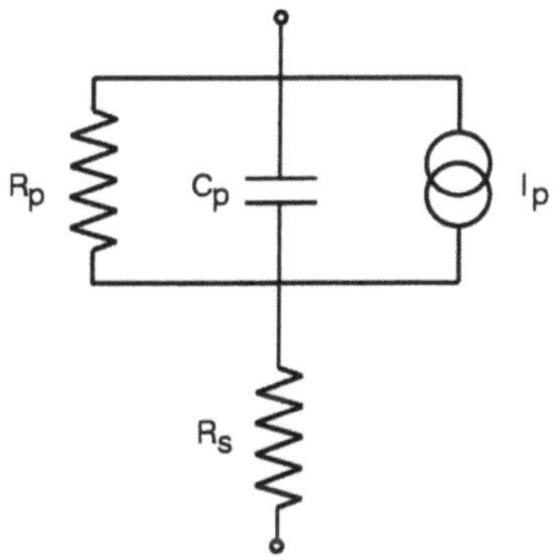

Figure 4. MSM photodetector equivalent circuit.

Table II. MSM equivalent circuit values for a 5 V bias.

gap μm	C_p fF	R_p KΩ	R_s Ω	illum. μW
2	188	936	1.96	0
2	185	101	1.92	42
3	125	988	2.36	0
3	127	103	2.10	42

representing the internal source resistance of the device. A resistor in series, R_s, is also added to take into account the resistance of the contacts. To determine the values of these parameters, the electrical reflection coefficient S_{11} was measured as a function of frequency using the network analyzer and the values of C_p, R_p and R_s were extracted using standard RF parameter extraction software. The equivalent circuit parameters obtained from typical photodetectors with 2 and 3 μm gaps are shown in Table II. As expected, the capacitance of the MSMs decrease with increasing electrode spacing from approximately 185 fF for the 2 μm device to less than 130 fF for the detector with 3 μm gaps. The capacitance remains unaffected by the dc illumination condition once the bias exceeds a fraction of a volt but the internal resistance can be reduced substantially. As a result, when designing a circuit, it may be important to consider the average light intensity the detector will see during the operation.

PERFORMANCE OF MSM-BASED OEICS

To investigate the high-speed performance of MSM-based receivers, a number of OEIC circuits were designed at CRC and fabricated at Bell Northern Research under the auspices of the Canadian MSM Working Group (BNR, CRC, DEC, TRLabs). Included in these circuits are MSMs followed by multi-stage amplification as well as a number of 1x4, 1x8 and 3x3 multi-detector arrays with on-board amplification and bias switching. As a compromise between speed and responsivity, 100x100 μm^2 detectors with 1 μm finger width and 3 μm spacing were selected. Contrary to the individual MSMs which were fabricated using wet processing, the MSMs which were integrated with an amplifier were fabricated using a standard MESFET ion-implantation process which included dry etching. It has been reported[3] that the MSM photodetector frequency response is affected by the surface processing.

Nine different circuit configurations were implemented, all based on the transimpedance design: there were three 2-stage common source amplifiers and three 2-stage cascode amplifiers. Four-stage amplifiers were also included, two of the common source type and one of the cascode type. In Fig. 5, the frequency response of an MSM is shown with and without amplification. Gains of 38.5 dB have been achieved. The bandwidth is limited by that of the MSM. Some circuits were optimized to operate with a single MSM at the input while others were designed to be more tolerant of the larger capacitance arising from many MSMs in parallel

The 3x3 array is also of great interest for switching applications. Each row has a common output connected to an amplifier. Fig. 6 shows the frequency response of a particular MSM under illumination and various biasing conditions. The isolation of the switch is the ratio of the signal detected when enabled to that when disabled. The isolation is 35 dB when the MSM is disabled by grounding the bias line. The isolation reaches 60 dB if the device is left floating. The apparent decrease in isolation at higher frequency is caused by the reduced signal level which results in a worsening of the signal-to-noise ratio of the measuring system.

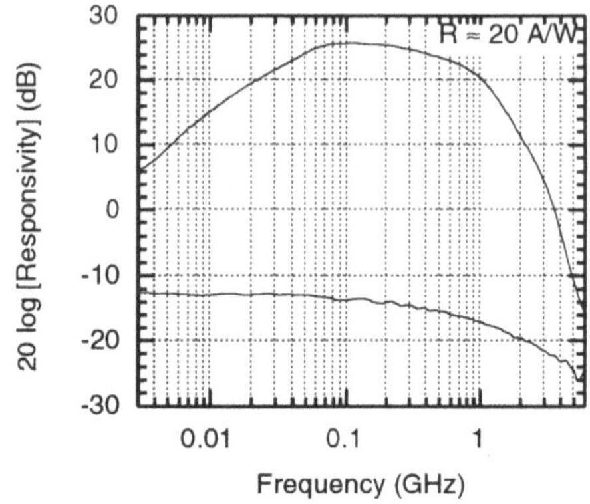

Figure 5. Frequency response of a MSM combined with a capacitance-coupled cascode amplifier.

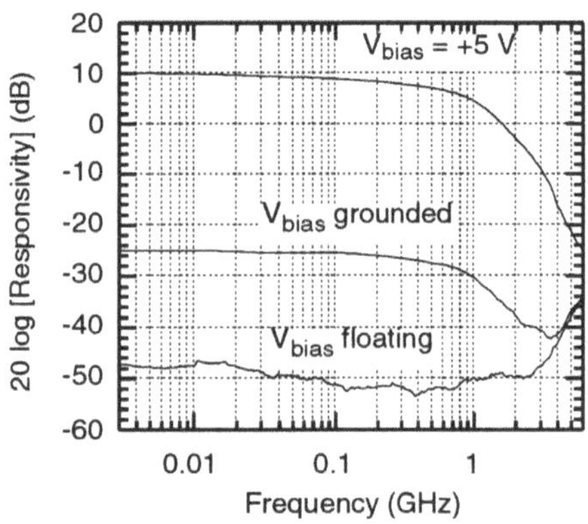

Figure 6. Frequency response and isolation of one of the elements in the 3x3 optical receiver array.

ACKNOWLEDGMENTS

We would like to thank the members of the MSM Working Group which made the fabrication of the OEIC possible. Thanks also go to CRC colleagues K. Verver for the MSM parameter extraction and M. Veilleux for circuit design and simulation.

REFERENCES

1. D.L. Rogers, "Monolithic Integration of a 3-GHz Detector/Preamplifier Using a Refractory-Gate Ion-Implanted MESFET Process", *IEEE Electron. Dev. Lett.*, **EDL-7**:600 (1986).
2. C.M. Hurd, W.R. McKinnon, "Modeling the Behavior of Photogenerated Charge in Semi-insulating GaAs", *J. Appl. Phys.* **75**:596 (1994).
3. J.H. Burroughes, "H-MESFET Compatible GaAs/AlGaAs MSM Photodetector", *IEEE Photon. Techn. Lett.* **3**:660 (1991).

A UNIQUE CAPABILITY OF QUANTUM WELL INFRARED PHOTODETECTORS: MULTICOLOR AND MULTIBAND RESPONSE

H. C. Liu, J. Li, M. Buchanan, Z. R. Wasilewski

Institute for Microstructural Sciences
National Research Council
Ottawa, Ontario K1A 0R6, Canada

P. H. Wilson, J. G. Simmons

Centre for Electrophotonic Materials and Devices
McMaster University
Hamilton, Ontario L8S 4M1, Canada

INTRODUCTION

Quantum well infrared photodetectors (QWIPs)[1] have recently attracted a great deal of attention. Although the absolute performance[2] of single element detectors is somewhat inferior than the standard HgCdTe technology for the long wavelength infrared (IR), advantages of QWIP approach based on GaAs and Si multiple quantum wells (MQWs) are obvious, including, e.g., mature materials and processing, and extremely uniform and low defect densities suitable for large array fabrication. Moreover we believe that with the advances in our understanding of the physics involved, the QWIP's performance related, e.g, the dark current and operating temperature will be improved. Another distinct feature of the QWIP approach is the capability of producing multicolor and multiband detectors.

Several approaches have been explored in realizing a QWIP with multicolor or multiband response. Köck et al. [3] made a direct electrical contact to the intermediate conducting layer separating two one-color QWIPs by etching a double mesa. This leads to a separately addressable two-color QWIP with multiple-electrical terminals (as opposed to a two terminal device)[3]. The advantage of this approach is that the different one-color QWIPs have no cross talks. The drawback of this approach is the difficulty of fabricating a many color version. The approach used by Gravé et al. [4, 5] used a tuning of the electric field distribution within a QWIP by the applied voltage. The voltage division is accomplished by high-low field domain formation[6, 7]. This approach is very attractive because the device is structurally simple and can provide many response colors, but a quantitative model of the phenomenon of high-low field domain formation is highly non-trivial[8]. Other approaches using special shapes of quantum wells[9, 10, 11] or special band structure effects[12, 13] have also been investigated.

In this paper, we briefly present our multicolor voltage tunable QWIP[14, 15] using a three-color embodiment. Our multicolor QWIP is realized simply by stacking the usual one-color GaAs/AlGaAs QWIPs having different detection wavelengths. Separating the stacks is a heavily doped conducting layer which does not need to be thick for making an electrical contact. The basic principle of operation can be used to produce detectors with many response colors.

A THREE COLOR QWIP MADE BY STACKING ONE COLOR DETECTORS

Our three-color QWIP was grown by molecular beam epitaxy. The quantum wells and barriers within each one-color are identical. The GaAs well widths for the QWIP stacks are 5.5, 6.1, and 6.6 nm, respectively. The $Al_xGa_{1-x}As$ barriers are all 46.8 nm with alloy factions of 0.26, 0.22, and 0.19, respectively. Each QWIP consists of a multiple quantum well structure with 32 repeats. The separation between the one-color QWIPs is a 93.4 nm thick GaAs layer doped with Si to 1.5×10^{18} cm^{-3}. We measured a detector with an active area of 120×120 μm^2 in the usual 45-degree facet geometry[1]. Detailed results on this three-color QWIP have been reported in Refs. [14, 15].

To show the voltage tunability, raw spectral response curves at selected voltages are shown with an arbitrary scale in Fig. 1. Three clearly resolved spectral peaks at about 7, 8.5, and 9.8 μm are displayed in the raw spectra given in Fig. 1. These response curves show several interesting features. First, the shortest wavelength detector is turned on at zero bias which is due to a built-in field caused by dopant segregation which we now can correct for[16]. The most interesting behavior is the relative strength of different spectral peaks as a function of voltage: at low voltages (zero to -1 V) the 7-μm response dominates, in the range from -4 to -8 V the only spectral peak visible on the linear scale is the 8.5-μm one, and at high voltages (from -9 V on) the response at 9.8 μm starts to dominate.

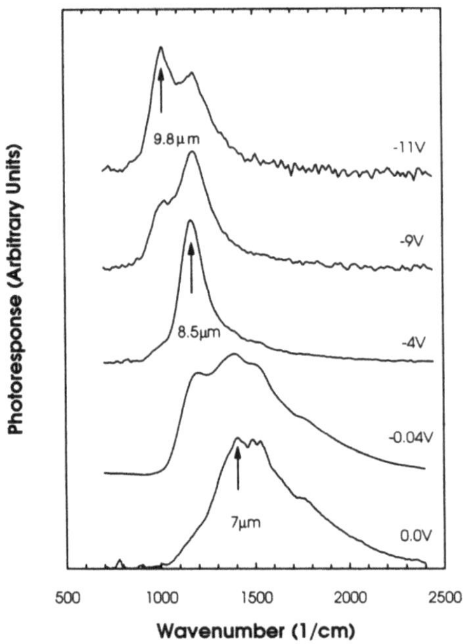

Figure 1. Raw spectral response curves at 82 K. The bias polarity is defined by choosing the bottom mesa contact as the ground.

To understand this behavior, we use the equivalent circuit shown in Fig. 2 where each one-color QWIP is modeled by a photocurrent source $i_{p,i}$ and a parallel source resistance r_i ($i = 1,\ 2,$ and 3) which is the differential resistance of the corresponding one-color QWIP. The measured photoresponse is the voltage developed across the load resistor due to the IR illumination:

$$v_{meas} = R_s \frac{i_{p1}r_1 + i_{p2}r_2 + i_{p3}r_3}{R_s + r_1 + r_2 + r_3}, \tag{1}$$

where R_s is the resistance of the series load resistor. From the above expression, the measured response is not only related to the intrinsic photocurrents (i_{p1}, i_{p2}, and i_{p3}) but also weighted by the differential resistances (r_1, r_2, and r_3). If in a given voltage range one of the differential resistances dominates, the photoresponse will be dominated by this one-color QWIP. More specifically, if r_1 is much larger than both r_2 and r_3 at low voltages, r_2 is much larger than both r_3 and r_1 at intermediate voltages, and r_3 is much larger than both r_1 and r_3 at high voltages, one obtains the behavior shown in Fig. 1. The above interpretation has been tested experimentally by examining two individual one-color QWIPs with similar device parameters as two of the three MQW stacks[15].

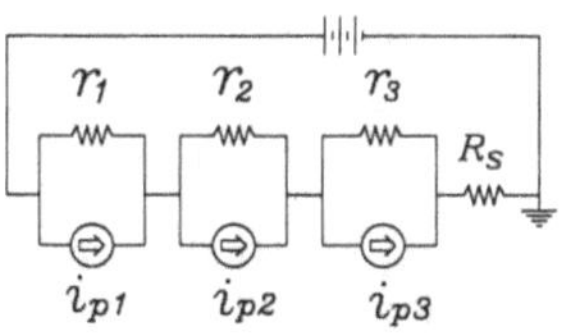

Figure 2. Equivalent circuit model. The three one-color detectors are modeled by three photocurrent sources in parallel with three resistances which are the differential resistances of the corresponding one-color detector. The series load resistor is used for measurement.

CONCLUSIONS

We have briefly presented our multicolor quantum well IR photodetector for the long wavelength infrared region. We believe that this capability is unique to the quantum well approach to IR detection. Further work is needed and is underway both in optimizing the detector performance and in investigating advanced practical applications.

Acknowledgements — We thank P. Chow Chong, P. Marshall, J. Stapledon, and J. Thompson for device fabrication. This work was supported in part by DND DREV.

References

[1] B. F. Levine, J. Appl. Phys. **74**, R1 (1993).

[2] H. C. Liu, Appl. Phys. Lett. **61**, 2703 (1992).

[3] A. Köck, E. Gornik, G. Abstreiter, G. Böhm, M. Walther, and G. Weimann, Appl. Phys. Lett. **60**, 2011 (1992).

[4] I. Gravé, A. Shakouri, N. Kruze, and A. Yariv, Appl. Phys. Lett. **60**, 2362 (1992).

[5] K. L. Tsai, K. H. Chang, C. P. Lee, K. F. Huang, J. S. Tsang, and H. R. Chen, Appl. Phys. Lett. **62**, 3504 (1993).

[6] L. Esaki and L. L. Chang, Phys. Rev. Lett. **33**, 495 (1974).

[7] K. K. Choi, B. F. Levine, R. J. Malik, J. Walker, and C. G. Bethea, Phys. Rev. B **35**, 4172 (1987).

[8] H. C. Liu, J. Li, M. Buchanan, Z. R. Wasilewski, and J. G. Simmons, Phys. Rev. B **48**, 1951 (1993).

[9] K. K. Choi, B. F. Levine, C. G. Bethea, J. Walker, and R. J. Malik, Phys. Rev. B **39**, 8029 (1989).

[10] E. Martinet, E. Rosencher, F. Luc, Ph. Bois, E. Costard, and S. Delaitre, Appl. Phys. Lett. **61**, 246 (1992).

[11] K. Kheng, M. Ramsteiner, H. Schneider, J. D. Ralston, F. Fuchs, and P. Koidl, Appl. Phys. Lett. **61**, 666 (1992).

[12] Y. H. Wang, S. S. Li, and P. Ho, Appl. Phys. Lett. **62**, 621 (1993).

[13] H. C. Liu, P. H. Wilson, M. Lamm, A. G. Steele, Z. R. Wasilewski, J. Li, M. Buchanan, and J. G. Simmons, Appl. Phys. Lett. **64**, 475 (1994).

[14] H. C. Liu, J. Li, J. R. Thompson, Z. R. Wasilewski, M. Buchanan, and J. G. Simmons, IEEE Elect. Dev. Lett. **14**, 566 (1993).

[15] H. C. Liu, J. Li, Z. R. Wasilewski, M. Buchanan, P. H. Wilson, M. Lamm, and J. G. Simmons, *A three-color voltage tunable quantum well intersubband photodetector for long wavelength infrared*, in **Quantum Well Intersubband Transition Physics and Devices**, Kluwer, Dordrecht, Netherlands, 1994, edited by H. C. Liu, B. F. Levine, and J. F. Andersson.

[16] H. C. Liu, Z. R. Wasilewski, M. Buchanan, and H. Chu, Appl. Phys. Lett. **63**, 761 (1993).

SOFRADIR MCT IRFPAS :

TECHNOLOGY AND PRODUCTS

Jean-Pierre Chatard

Sofradir
43-47, rue Camille Pelletan
92290 Chatenay Malabry - France

INTRODUCTION

Infrared applications require a large range of specific types of infrared detectors including linear arrays or staring arrays sensitive in different wavelength regions, mainly 3 to 5 μm and 8 to 12 μm.

The second generation infrared focal plane array (IRFPA) incorporates a large number of pixels thanks to the marriage of photodiodes arrays and silicon readout circuits which include signal processing like gain control, offset control or (time delay and integration - TDI). With this new approach, the system architecture is simplified at the same time the performance is improved.

France developed such a technology based on photovoltaic (PV) detectors built on mercury cadmium telluride (MCT).

The overall IRFPA technology involves MCT epilayer metallurgical growth, infrared photovoltaïc detector technology and an indium bump interconnection technique to hybridize the MCT chip with the signal processor chip[1,2]

These types of IRFPA operate at low temperatures therefore cryogenic techniques must be mastered using different types of coolers like Joule Thomson, closed cycle Stirling or thermoelectric coolers[2].

All these technologies are now used in production at Sofradir where detector dewar assemblies (DDA) are produced in quantities each month[3].

The following descriptions consider the main products which are already in production or ready for production and products which are at the development level.

IN PRODUCTION OR READY FOR PRODUCTION

Linear arrays

High performance arrays for seeker applications and for FLIR for surveillance, identification or targeting[4,5,6] commonly use long linear arrays with or without TDI such as 240 x 1, 288 x 1, 288 x 4, 480 x 4, 576 x 4, 960 x 4 and 1152 x 1 sensitive either in the 3 to 5 μm or in the 8 to 12 μm spectral band.

Applications of Photonic Technology
Edited by G.A. Lampropoulos *et al.*, Plenum Press, New York, 1995

At the present time, the 288/240 x 4 LWIRFPA using the TDI concept is one of the most mature second generation detectors based on the number of DDAs operating in FLIRs and on the fact that it has been in production at Sofradir for almost one year[3].

It is a hybrid IRCCD (chart 1) consisting of a photovoltaic diode array associated with a CCD processor which for this demonstration operates at 80K with a field of view (FOV) of 0.25 sr, an integration time of 20 μs and a spectral band of 8 to 10.5 μm.

Chart 1
288 x 4 LWIRFPA

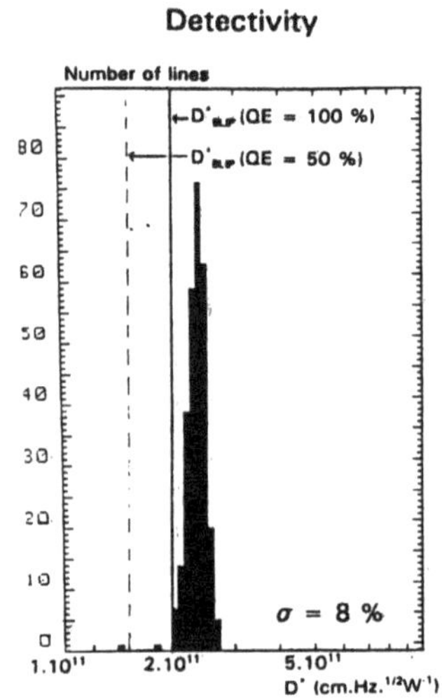

Chart 1 - 288 x 4 LWIRFPA

Channel to channel detector response in front of a 20°C uniform background temperature shows the good homogeneity of this IRFPA as the offset dispersion is less than 250 mV which represents a fixed pattern noise lower than 5 % and a 100 % connection yield. The responsivity homogeneity is also very good as the standard deviation is lower than 3 %. RMS peak detectivity (2.2 $10^{11} W^{-1} cmHz^{1/2}$) is higher than the background limited detectivity thanks to the TDI function.

By just changing the MCT material the same array is manufactured sensitive in the 3 to 5 μm spectral band. Operating at 80K, with a FOV of 0.5 sr and an integration time of 20 μs this array gives a good uniformity of response in front of a 20°C blackbody and a very homogeneous responsivity as the channel to channel response standard deviation is better than 3%. RMS peak detectivity (6 10^{11} w^{-1} $Hz^{1/2}$ cm) is very close to BLIP detectivity.

Increasing the size of the array is possible by staggering several MCT subarrays on the same silicon circuit. These subarrays are perfectly aligned thanks to the indium bump technique which is used.

For example, the 480 x 4 LWIRCCD array[6] (chart 2) shows that under the same conditions, exactly the same results are obtained as the ones obtained with the 288 x 4. The responsivity is very homogeneous all along the 14 cm length of the array, the standard deviation is better than 3 %. The detectivity itself is once more higher than the BLIP detectivity. In addition no difference can be detected between the two subarrays.

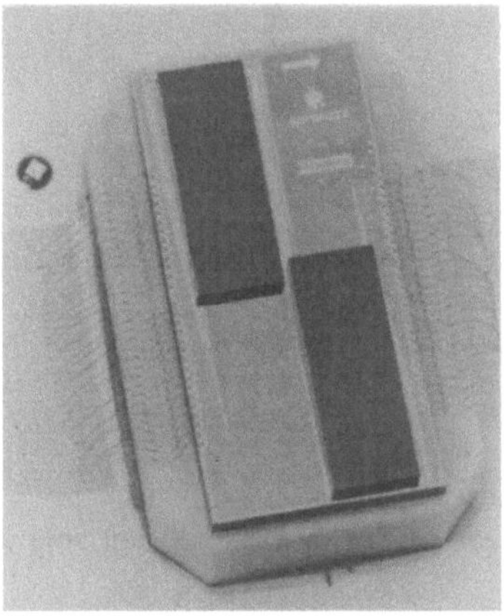

Chart 2
480 x 4 LWIRFPA

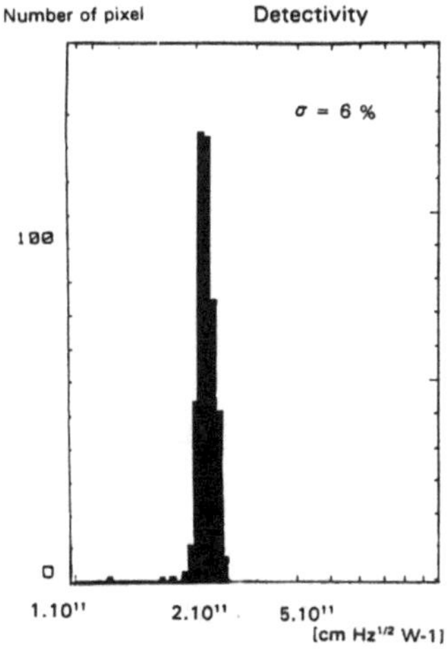

Chart 2 - 480 x 4 LWIRFPA

Staring arrays

Applications using staring arrays in France are mainly orientated to seeker head applications[7]. Of course this concept could be applied to FLIR applications and some effort is in progress at the French research center (LIR).

To date IRCMOS and IRCCD 64 x 64 LWIRFPA have been produced presenting equivalent levels of performance. For 80K operating temperature conditions, a FOV of 0.6 sr and a frame rate of 500 Hz, the 20°C responsivity homogeneity is very good as 100% of the pixels are within 0.7V for a total swing of 5V. The result also show also a 100% pixel connection yield.

Pixel to pixel 500K responsivity is also very homogeneous as all the figures are within ± 25% (σ = 3 %).

Average peak detectivity in such conditions is 5 10^{10} w^{-1} cm $Hz^{1/2}$ and the number of defective pixels is lower than 0.5%.

MWIRFPA 128 x 128 have also been produced (chart 3). For 80K operating temperature conditions, a field of view of 0.3 sr and a frame rate of 250 Hz, the number of defective pixels is limited to 0.1%, (defective pixels are defined as figures outside of ± 20%). The standard deviation of the channel responses is about 3%. RMS peak detectivity (chart 1) outlines a high detector performance level (2.3 10^{11} w^{-1} $Hz^{1/2}$ cm).

Chart 3
128 x 128 MWIRFPA

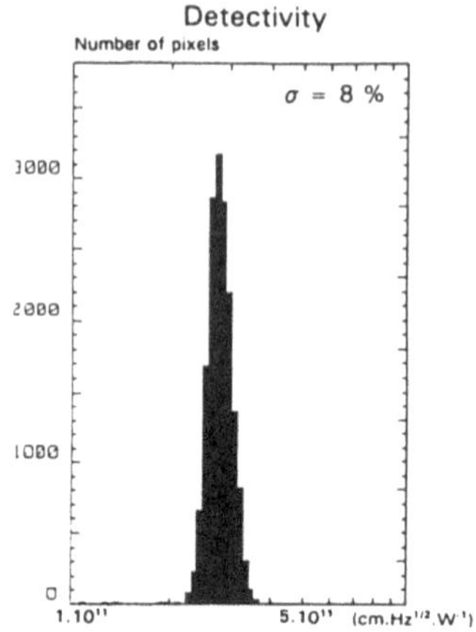

Chart 3 - 128 x 128 MWIRFPA

UNDER DEVELOPMENT

More advanced DDA are under development which include IRFPA, dewar and from time to time coolers in cooperation with cooler manufacturers.

Largely, as now at Sofradir large IRFPA are already developed, new development activity is focused on complete detector dewar cooler assemblies in order to adapt complete systems to the needs of applications.

Flat dewar cooler assembly (8)

Nowadays, most Joule Thomson coolers have a long cylindrical structure to allow infrared detectors to be cooled to a cryogenic temperature of 77K. Consequently, the associated dewar has to be adapted to the shape of the cooler by means of a long stem at the top of which the detector is cooled. This slim structure makes the dewar very sensitive to vibrations caused by environmental conditions. Such a cold stem can be considered as an embedded beam with a mass at the top. The resonant frequency is known to be a decreasing function of the mass for this geometry. To increase the resonant frequency above 2000 Hz, for example, means shortening the stem or thickening its walls, which is not recommended thermally. A solution frequently used is to adapt some stem stiffeners which complicate the dewar by increasing the number of its parts and of course, its cost. A new generation of cooler, based on a flat structure, can be a solution to the problem of mechanical vibrations.

Such coolers reach cryogenic temperatures in the range of 90-100K and could be valuable for lower cost applications.

Their flat structures should prevent them from being prone to vibrations but moreover, as they are part of the dewar, they should ensure good mechanical behavior of the whole detector dewar assembly. The particular shape of the cooler should then yield reduced dimensions and compactness to the assembly.

The diameter of such a complete dewar cooler assembly is around 2.5 cm and the height is 2 cm.

The actual result is really as planned as no pixel displacement is evident can be focused up to 2000 Hz and no resonant frequency appears.

The cooldown time obtained is also very interesting as it is less than two seconds.

<u>200K operating conditions (9)</u>

Second generation IRFPA costs follow the microelectronic production rules. Therefore by entering into the production phase the cost of the FPA itself starts to be negligible in comparison to the dewar or the cooling system which still remains expensive. The solution to decrease the cost is to eliminate or to simplify these two last components.

As midwave MCT can operate at temperatures as high as 200K with good performance such array cooldown with a thermal electric cooler has been considered.

A complete detector cooler dewar assembly has been developed showing good electro-optical performance. This approach can be considered for arrays like the 288 x 4 and the 128 x 128 but also certainly for even larger array like the 256 x 256 which is under development.

As an example the 288 x 4 MWIR component gives for a cutoff wavelength of 4 μm an average detectivity better than $4.4 10^{11} W^{-1} Hz^{1/2}$ cm.

CONCLUSION

The new infrared world is based without any doubt on second generation focal plane array as the system demonstrations using this new generation have already shown their superiority.

As a matter of fact using a Sofradir 288 x 4 instead of using a classical common module detector allows one to improve the probability of recognition by at least a factor three [10].

As such detectors are now available in quantities, most new programs or retrofit programs will use the second generation infrared focal plane array.

REFERENCES

1. Sofradir IRFPA technology
   J.P. Chatard - Sofradir
   Spie Proceedings - San Diego July 1992

2. The microwave engineering handbook
   Edited by Chapman and Hall (Paragraph 18 page 385)

3. MCT TDI arrays producibility
   D. Zenatti and P. Radisson
   SPIE Proceedings - Orlando April 1994

4. Design of a high performance scanning in sensors
   NDRE
   SPIE Proceedings - San Diego July 1992

5. Non linearities and non uniformities corrections for the IRIS family of IRCCD thermal images
   Sagem
   SPIE Proceedings - San Diego July 1992

6. Development and demonstration of a Sofradir 480 x 4 infrared charge
   coupled device (IRCCD) focal plane array (FPA) for the Losat second generation FLIR
   Texas Instruments
   SPIE Proceedings - San Diego July 1992

7. Use 2F 2D CMT array for seeker applications
Thomson - DEM
SPIE Proceedings - San Diego July 1992

8. Flat structure cooled detector assembly
N. Reeb - Sofradir
SPIE Proceedings - Orlando April 1994

9. IRCCD cooled with thermoelectric cooler
L.P. Angebault - Sofradir
SPIE Proceedings - Orlando April 1994

10. Focal plane arrays come of age in FLIR devices
Walter E. Chase - Laser Focus World - March 1992

AN IR-VISIBLE CONVERTER FOR SPATIALLY AND TEMPORALLY RESOLVED IR-IMAGE DETECTION

H. Willebrand[1, 2], Y. Astrov[3], L. Portsel[3], S. Teperick[3], T. Gauselmann[2] and H.-G. Purwins[2]

[1] Dept. for Electr. Eng., University of Colorado, Boulder 80309-0425, USA
[2] Institute of Applied Physics, University of Münster, 48149 Münster, Germany
[3] A. F. Joffe Physical Technical Institute, St. Petersburg, Russia

Abstract

Electron-optical devices are broadly used nowadays in experimental physics, technology research and technical applications for high-speed recording. Fast processes can be studied with these devices in framing and streak modes[1], thus giving vast information on the spatio-temporal evolution of optical fields in a broad spectral range. Especially, in the visible, ultraviolet (X-ray) and near infrared (IR) spectral ranges, these methods are efficient. But unfortunately, many of them lose sensitivity for wavelengths of incoming light larger than $\lambda >$ 1.3 µm. In this contribution we present a technique that will allow spatio-temporal resolved high-speed recording even at wavelengths beyond 1.3 µm. We will also demonstrate, that this technique can be used for the spatially and temporally resolved measurement of temperature distributions on the surface of heated bodies.

Setup and Principle of Operation

The detection technique we use is based on the principle of IR-visible conversion. A schematic drawing of the experimental setup is shown in figure 1. For the IR-visible conversion process the converter uses a self-sustained (Townsend-) gas-discharge which is initiated in the thin discharge gap (3) when a constant or pulse voltage of sufficiently high amplitude is applied to the electrodes. The cathode is an electrically homogeneous semiconductor layer (2) with photoconducting properties in the infrared spectral range. On one side the cathode is provided with an ohmic, infrared transparent boron contact (1). As cathode material we used silicon doped with zinc or Indium and dopant concentrations in the order of 10^{16}. The light emission from the discharge can be observed through a transparent anode (5) that is covered with a thin Indium-Tin-Oxide (ITO) layer on the surface facing the discharge gap. The discharge gap thickness is determined by the thickness of a dielectric foil (4), which is in the range (50-100) µm. The typical diameter of the photosensitive area is (20-25) mm. In the operation mode the discharge gap is filled with an inert gas at pressures about 100 hPa. The semiconductor cathode is cooled with liquid nitrogen to ensure a high resistivity of the semiconductor layer, which is an essential requirement for the proper function of the converter. A more detailed description of the setup is given in reference 2.

Applications of Photonic Technology
Edited by G.A. Lampropoulos *et al.*, Plenum Press, New York, 1995

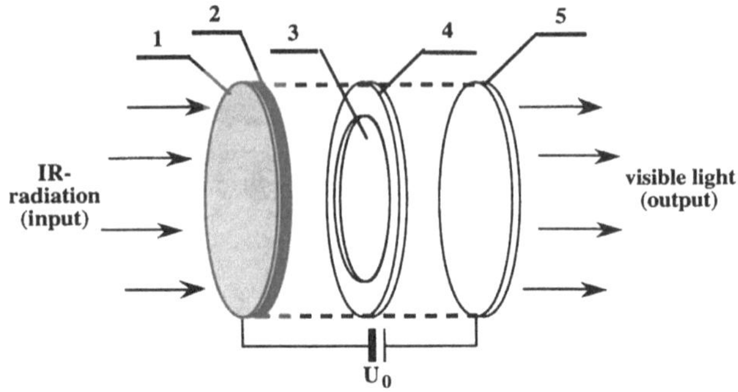

Figure 1. Schematic drawing of the experimental setup used for IR-visible conversion process. The numbers will be explained in the text.

Without (or with homogeneous) external illumination a spatially homogeneous visible light emission from the discharge gap can be observed by applying a voltage of sufficiently high amplitude. For our typical experimental conditions the voltage in the order of a few hundred volts up to one kilovolts. The current density in the converter cell can be increased by enlightening the photoconducting semiconductor through a transparent boron contact in the spectral range of its sensitivity. By projecting a spatially inhomogeneous light intensity distribution (e.g. an IR-image) it is possible to generate a resistivity distribution in the cathode layer, that is a true representation of the projected image. Due to the fact that the semiconducting layer serves as a spatially distributed "load resistor" for the gas discharge layer, the light emission from the gas subsequently follows the resistivity distribution. In this way we get the image in the visible spectra, that is true to an original image in the IR spectral range. In[3-6] a more detailed description of the physical processes involved in this compound semiconductor gas discharge structure can be found.

Experimental Results

For the demonstration of the system capabilities in technical applications, we used two rather simple experimental setups. In the first experimental setup a heated soldering iron tip was projected on the semiconductor cathode, and the resulting visible image was captured by a CCD-camera that was connected to an image processing system or by simply taking a photo. The typical result of such a measurement can be seen in the figure 2a. The shape of the tip can be clearly seen. The bright circle around the tip marks the boundary of the detector area. The average soldering tip surface temperature, which was measured with a Ni-Cr-Ni probe, was about 450 K. The intensity variations in the image display the spatially inhomogeneous temperature distribution on the surface of the soldering tip. The "hot spot" in the middle of the tip holder shows the location of the heating element.

Preliminary experimental investigations show that the spatial resolution of this technique is approximately 5 - 10 pairlines/mm, depending on the system parameters like gas pressure or applied voltage. For a quantitative measurement the soldering tip was replaced by a black body source. The device can be sensitive in a the broad IR spectral range, because the conversion process results from the internal photoeffect mechanism in photoconductive semiconductor materials. In figure 2b a typical temperature - light emission characteristic is shown. This characteristic was measured by projecting the radiation emitted from a black body source (aperture 22.2 mm) on the semiconductor surface. For the projection we used a BaF_2 lens, the diameter and focal length of which were 50 mm and 230 mm, respectively. The aspect ratio was chosen to be 1 : 1. The value for the light emission was found from the video signal amplitude. The doted line shows an exponential fit of the measurement data.

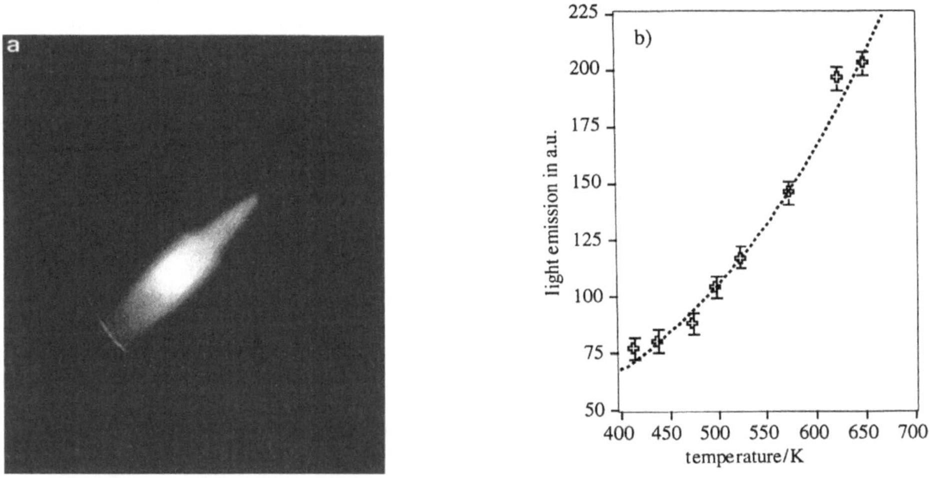

Figure 2. Photo of a soldering iron tip (a) observed with the IR-visible converter. The average soldering iron surface temperature was about 450 K. (b) shows the change of light emission from the gas discharge gap as a function of the black body temperature. The doted line represents an exponential fit of the measurements data.

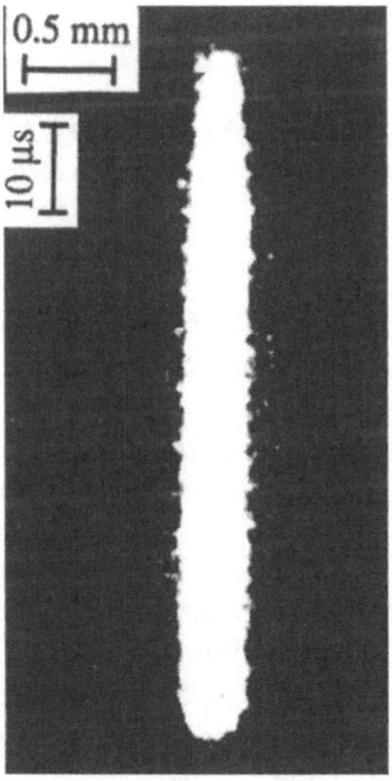

Figure 3. Streak image of the IR-converter response on the excitation with a laser diode pulse

A second experimental setup was used to estimate the temporal resolution of the IR-visible converter. For this purpose the light beam of an IR-laser diode ($\lambda = 1.3\ \mu m$) was projected on the surface of the semiconductor cathode via a silica lens. The visible light emission from the discharge gap, resulting from the light pulse excitation of the cathode by the laser diode, was measured by a streak camera system. The streak camera was equipped with a conventional S 20 photocathode. A typical experimental result is shown in figure 3. The total length of the laser pulse was about 80 μs and the excitation starts in the upper part of the image. These measurements show, that the temporal resolution of the system can be in the order of 1 μs or even higher. Similar to the spatial resolution, the rise time of the IR-converter depends on the system parameters like gas pressure or applied voltage. A detailed discussion of the speed possibilities can be found in reference. 4.

Conclusions

The IR-visible conversion technique presented in this contribution may be useful for two different kinds of applications:

1. Direct spatially and temporally resolved observation of temperature distributions of heated bodies. In combination with a standard CCD-camera an electronic readout system for reel time (video frequency) applications be achieved easily.
2. High speed two dimensional IR-image detection at frame rates in the order of 100 kHz or even higher in combination with an high speed readout system (CCD- or electro-optical cameras).

Due to the fact that the converter structure is a highly nonlinear system, it can display different kinds of spatio-temporal instabilities. The finding of a parameter constellation for the optimal function of the device will be one of the most important task in the future development of this conversion technique.

Acknowledgments

The authors would like to thank the BMFT for the financial support of this work (project number 13N6183).

One of us (H.W.) especially likes to thank the DAAD for giving the financial support for his staying at the University of Boulder in Colorado/USA.

Literature

1 Y. Tsuchiya, *IEEE J. Quantum Electron.* **QE-20** (1984) 1516

2 Yu. A. Astrov, H. Willebrand, L. Portsel, S. P. Teperick, H.-G. Purwins, *J. Appl. Physics* **74** (1993) 2159

3 Yu. A. Astrov, L. M. Portsel, *Study of of the stability of the homogeneous discharge in a resistive electrode - gas discharge gap structure*, deposited in VINITI N 1494 (1980) (in Russian)

4 Yu. Astrov, L. M. Portsel, S. P. Teperick, H. Willebrand, H.-G. Purwins, *Speed Pro perties of a Semiconductor-Discharge Gap IR Image Converter Studied with a Streak Camera System*, J. Appl. Phys. **74**, 2159 (1993)

5 H. Willebrand, Yu. Astrov, L. M. Portsel, S. P. Teperick, T. Gauselmann, H.-G. Purwins, *Exitation of wave patterns in an infrared-visisible converter*, J. of Physics D: Appl. Phys. **27**, 1354 (1994)

NANO-COMPOSITES: NEW MATERIALS FOR PHOTONICS?

Frank Marlow

Institute of Applied Chemistry
Rudower Chaussee 5, 12489 Berlin, Germany

ABSTRACT

Molecular sieves are inorganic nano-porous crystals with a structure-specific inner pore system. The ordering of guest molecules by the channels of molecular sieves has attracted considerable interest because of the optical effects resulting from the alignment of the molecules. It is found that para-nitroaniline (pNA) molecules in the channel of the molecular sieve AlPO-5 can form solid-like dipole chains, in which the orientation of the length axis of the incorporated pNA molecules differs from the channel axis by less than 9°.

The occurrence of nonlinear optical effects of second order (second harmonic generation) does furthermore prove that the dipolar molecules have a preferred direction in all channels additionally to the alignment of the molecules. Therefore, modified molecular sieve crystals are very promising for all $\chi^{(2)}$-applications. The advantages are the high nonlinearities of organic guest molecules and the tunability of the material by varying both host and guest in this nm-scale composite.

This material could be used for micro-optical devices in waveguide structures.

1. INTRODUCTION

The interconnected pores and cavities of zeolite-like molecular sieves form a highly regular micropore system. The access to the pore system is controlled by channels with diameters between 3 and 13 Å. More than two hundred different synthetic and natural molecular sieve structures are known. Due to their chemical composition, mainly as alumophosphates or alumosilicates, zeolite-like molecular sieves show no resonances in the visible region. Therefore, one can use them as an inert matrix to incorporate optically interesting guests, e.g. molecules, clusters or polymers, into their pore system. The resulting guest-host structure is determined by the pore geometry. Consequently, these composites is highly ordered. Ordered systems have promising properties for optical applications, especially in nonlinear optics.

In 1988 second harmonic generation (SHG) of molecular sieves loaded with organic molecules was found[1] indicating that adsorbed molecules do not only form an ordered array, but in addition there is a breaking of inversion symmetry in this system. Otherwise the SHG would be forbidden. Consequently, molecular sieve based guest-host complexes exhibit all the properties one needs to design new nonlinear optical materials: Tunable electronic resonances (by the guest molecules) and an ordered and directed arrangement of these chromophores (variable by the molecular sieve).

Although there is a variety of nonlinear optical materials, non of them allows optically integrated circuits comparable to electronical ones[2]. So, there is a need to find new materials for optical information processing. Nanometer-scaled molecular sieve based composites are a good

Applications of Photonic Technology
Edited by G.A. Lampropoulos *et al.*, Plenum Press, New York, 1995

candidate. In this paper the properties of these materials are discussed and possible applications are proposed.

2. THE ALIGNMENT OF MOLECULES

In the last few years a variety of techniques[3] ranging from the inclusion of dye molecules during the molecular sieve crystallization to post-synthesis modifications like intracrystalline polymerization or the building up of ship-in-the-bottle like complexes in zeolite cavities have been reported. In our case, we have introduced the dye molecules into the molecular sieve pores by an adsorption process from the gas phase or from a solution. The evacuated molecular sieve crystals were brought into contact with pNA and dimethylaminobenzonitrile (DMABN) (6h, 100°C). The molecular sieves AlPO-5[4] with pores of 7.3 Å, ZSM-5[5] with 5.5 Å and VPI-5 with 13 Å have been used.

The simplest test for the alignment of the molecules is polarization microscopy[6]. The pNA loaded crystals act as polarizers. pNA has only one transition in the visible region which is polarized parallel to the long axis of the molecule. Since the molecules are aligned parallel to the channel axis the loaded molecular sieve crystal only absorbs the light polarized parallel to the channel direction. This property of anisotropic absorption can be regarded more quantitatively with microfocus absorption spectroscopy. Here the alignment and the electronic levels of more complicated molecules can be analysed.

Infrared spectroscopy[7] also gives alignment information. In addition to that, intermolecular interactions and the deformation of the molecule can be analysed.

Raman spectroscopy on the micro-crystallites is very well suited for the analysis of the adsorbate structure. The pronounced polarization dependences are caused by the alignment of the transition moments of the electronic states which are generally well known. On the other hand, the position of the Raman bands sensitively reflects the intermolecular interactions. The molecule pNA can form structures with or without intermolecular hydrogen-bonds within the channels. The alignment of the molecules ranges from 9° to 20° (mean deviation from the channel axis)[8].

Fluorescence spectroscopy[9] is suited for the investigation of the dynamics of the excited states of the molecules. The photochemical behavior is modified by the formation of molecular chains in the channels.

3. NONLINEAR OPTICAL EFFECTS

Among the NLO effects SHG[1,10,11] and hole burning are studied most extensively. The investigation of electro-optical effects an optical rectification is in progress. Since these effects are related to SHG by the $\chi^{(2)}$-tensor, they must exist. The SHG is regarded as a demonstrating example with a more general meaning ($\chi^{(2)}$-material). Therefore, the SHG measurements are described in more detail here.

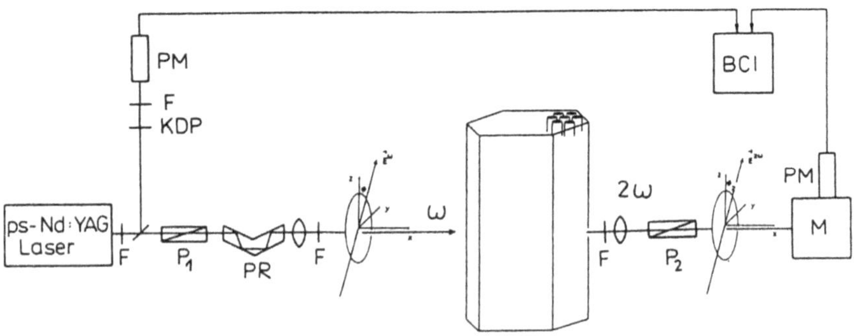

Figure 1. Experimental setup: A dye loaded AlPO-5 single crystal of about 0.1 mm length in the laser beam.

In our investigations, an individual crystal loaded with pNA or DMABN was positioned in the beam of a ps-Nd:YAG laser (see Fig. 1). As expected, the SHG of such crystals shows a pronounced polarization dependence. The relation of the different signal components, however, depends on the details of the electronic levels of the guest molecule as well as on the geometric arrangement of the molecules.

Perfectly Aligned 2-Level Systems. This case is realized for pNA in AlPO-5 and VPI-5, loaded from the gas phase. The produced second harmonic light is fully polarized parallel to the channel direction (z-direction). When the polarization plane of the incident laser light is varied, the second harmonic intensity passes a maximum when the crystal is excited parallel to the same direction. The measured values are well fitted by $\cos^4\varphi$ where φ is the angle between the channel direction in the molecular sieve and the polarization plane. The occurrence of SHG proves that the pNA is not incorporated centro-symmetrically. That means that different pNA dipole chains in the molecular sieve must have the same preferred direction. Whereas the formation of dipole chains is energetically favoured, the same preferred direction of the chains in neighboring zeolite pores seems to be the result of the pNA uptake mechanism. Obviously, there must be a strong preference of the pNA molecules to enter the molecular sieve with either their nitro or amino group in front. As a result, the molecules form two domains of different dipole direction (e.g. an upper and a lower half of the 100 μm long crystals). This macroscopic structure was verified by pyroelectric studies[12].

Aligned 3-Level Systems. This case is realized for DMABN in AlPO-5. For this system two components of the SHG have been found. One component of the frequency-doubled radiation is polarized parallel to the channel direction (z-axis, φ =0°) of the molecular sieve and exhibits the maximum intensity when the excitation beam has the same polarization. The second component, however, is polarized perpendicular to the z-axis (φ =90°) and shows the maximum efficiency for mixed excitation (φ =45°). Obviously, different components of the hyperpolarizability tensor contribute to the resulting SHG. The reason for the different tensor elements are the two close-lying lowest excited states in DMABN having transition dipole matrix elements perpendicular to each other. In contrast to perfectly aligned 2-level systems which are able to mix waves of the same polarization only, for the DMABN based materials a mixing of waves of different polarizations is observed. This is a promising property for phase-matching as a prerequisite for a technical application.

Zigzag Arrangement of 2-Level Systems. This case is realized for pNA in Silicalite I and for pNA in AlPO-5, loaded from a solution. In the molecular sieve Silicalite I the positions of the adsorbed molecules are fully determined by the crystal structure. The channel crossings were found to be the favoured adsorption sites. At these places the alignment of the molecules is disturbed by the interaction with the host causing a zigzag arrangement.

In AlPO-5 the channels are very smooth and, therefore, there are likely to be no favoured adsorption places. However, one can influence the adsorbate structure by the density of adsorbed molecules, which is determined in the loading procedure. If the molecules are entering the crystal from a dye solution, higher concentration of pNA in the channels can be reached and the molecule will arrange in a zigzag manner. Although the molecular hyperpolarizability has only one component, the emsemble has different tensor components.

4. APPLICATION IN WAVEGUIDE STRUCTURES

In this section promising applications of optically nonlinear molecular sieve crystals are discussed. None of them has been realized yet, mainly due to the inhomogeneities in the optical properties. However, the strong nonlinearities of the nano-composites give rise to some hope that they can lead to purpose-dependent micro-optical devices. The nonlinear susceptibilities of organic materials are often much larger than the values for inorganic ones. The nano-composites use this advantage and, therefore, susceptibility values up to 10^{-8} cm/V (this is $200\times\chi_{KDP}$) can result. This would remarkably shorten the interaction length in integrated optical devices.

Incorporation in Waveguide Structures. A disadvantage of the molecular sieve crystals is the difficulty to manufacture extended substrates serving as support for the waveguides. Realistically one can only expect molecular sieve crystals of maximum 0.2 mm. Only in exceptional cases larger crystals can be synthesized. To make arrangements of NLO devices by nano-composite crystals, two ways are thinkable. One consists in the preparation of ordered arrays of crystals[13] embedded into a supporting material, e.g. into a polymer film. This array has to be used as the substrate for the waveguide arrangement, where all NLO devices have to be on the appropriately oriented crystals. The NLO material is then in the cladding of the waveguide.

The second way to use the crystals is the fabrication of a device arrangement on an appropriate non-NLO material. The place for the NLO material has to be empty. Into these small holes one puts the nano-composite crystals afterwards. This requires a perfect control of the crystal growths. Attempts towards this direction are in progress.

Electro-Optical Switching and Frequency Conversation. Phase modulators, modulators on the basis of interferometers and switchable couplers are well known elements for controlling the propagation of light with electrical signals. Using nano-composites one can make these elements smaller.

On the basis of nano-composites also mixers for different waves (sum frequency generation) can be built. To this aim the phase-matching of the mixing process has to be realized. The tunability of the material enables this.

Optical Amplification. Photonic devices will have losses. If one would like to realize photonic computation, many operations have to be carried out. Therefore, one must amplify the light from time to time. Optical micro-amplifiers which can be integrated on a photonic chip would be ideal for such a purpose. The material for such a device could be a composite of a rare-earth containing zeolite crystal and organic guests in the molecular sieve pores.

The availability of optical micro-amplifiers could also influence the concepts for the nonlinear optical devices (modulators or frequency converters). Now one seeks to operate outside of resonances to avoid absorption losses. However, near the electronic states the NLO effects are strongly enhanced (some orders of magnitude larger than the values given above). To compensate the resulting absorption losses optical amplification could be used simultaneously or afterwards.

ACKNOWLEDGEMENT

The author thanks J. Caro for his contributions to this work. The work was supported by the Deutsche Akademie der Naturforscher Leopoldina and by the AiF via the DECHEMA (grand 315D).

REFERENCES

1. S.D. Cox, T.E. Gier, G.D. Stucky, J. Bierlein, *J. Am. Chem. Soc.* 110: 2986 (1988).
2. R. Dorn, D. Baums, P. Kersten and R. Regener, *Adv. Mater.* 4: 464 (1992).
3. For a review see: G.A. Ozin, *Adv. Mat.* 4: 613 (1992).
4. G. Finger, J. Richter-Mendau, M. Bülow and J. Kornatowski, *Zeolites* 11: 443 (1991).
5. J. Kornatowski, *Zeolites*, 8: 77 (1988).
6. F. Marlow and J. Caro, *Zeolites* 12: 433 (1992).
7. F. Schüth, *J. Phys. Chem.* 96: 7493 (1992).
8. F. Marlow, W. Hill, J. Caro, G. Finger, *J. Raman Spec.* 24: 603 (1993).
9. K. Hoffmann, F. Marlow, J. Caro, S. Dähne, *Chem. Phys. Lett.*, subm.
10. L. Werner, J. Caro, G. Finger, J. Kornatowski, *Zeolites* 12: 658 (1992).
11. F. Marlow, J. Caro, L. Werner, J. Kornatowski, S. Dähne, *J. Phys. Chem.* 97: 11286 (1993).
12. F. Marlow, M. Wübbenhorst and J. Caro, *J. Phys. Chem.*, subm.
13. J. Caro, G. Finger, J. Kornatowski, J. Richter-Mendau, L. Werner, B. Zibrowius, *Adv. Mater.* 4: 273 (1992).

PERFORMANCE OF GERMANIUM AND III-V AVALANCHE PHOTODIODES IN THE DETECTION OF SINGLE-PHOTONS IN THE NEAR-INFRARED

A. Lacaita, F. Zappa, and S. Cova

Dipartimento di Elettronica e Informazione
Politecnico di Milano
p. Leonardo da Vinci 32, 20133 Milano (Italy)

INTRODUCTION

Silicon Avalanche Photodiodes (APD's), biased above the breakdown voltage to work in Geiger-mode, are excellent single-photon detectors in the 0.5-1.1μm wavelength range. They find applications in photon correlation spectroscopy, velocimetry, Optical Time-Domain Reflectometry, Laser Ranging. However, the need of fast and more sensitive photodetectors working beyond 1μm are increasingly required in many fields of science and technology, not to mention all the applications related to optical communications, with careful testing of optical components and networks at 1.3μm and 1.55μm. Cooled S1 photocathode and silicon Single Photon Avalanche Diodes (silicon SPAD's) detect single photons up to 1.3μm, but with a residual quantum efficiency of 10^{-6}. Therefore, we have investigated the single-photon counting performance of germanium and III-V APD's.

In this paper we show how commercially available germanium and InGaAs/InP APD's must be properly operated in Geiger-mode, in order to get the best performance. We demonstrate that selected germanium APD's, cooled at 77K, can achieve 10%-quantum efficiency and 85ps-Full Width at Half Maximum (FWHM) time resolution in the detection of the arrival time of single photons at 1.3μm, with a *NEP* of $8 \cdot 10^{-16}$W/Hz$^{1/2}$. We show that InGaAs/InP APD's, at 150K, attain a 3%-quantum efficiency, a 1ns FWHM-time resolution and a *NEP* of $8 \cdot 10^{-15}$W/Hz$^{1/2}$ at 1.3μm. Since the tested devices were purposely designed for room temperature operation in the usual analog multiplication regime and, therefore, they were not optimized for single photon detection in cooled environment, we are very confident that III-V APD's could achieve single-photon response even faster than germanium APD's.

In Geiger-mode regime, the APD works biased above the breakdown voltage, V_b. As soon as a carrier reaches the high field region of the junction, it can trigger a diverging avalanche process. A suitable external circuit senses the onset of the current and quenches the avalanche, by lowering the bias below V_b. If the first carrier is photogenerated, the current leading edge marks the photon absorption time, otherwise it is a spurious event, a so-called

Applications of Photonic Technology
Edited by G.A. Lampropoulos *et al.*, Plenum Press, New York, 1995

dark-count. After a well controlled hold-off time, the bias is restored and the APD is able to detect another photon. The intensity of the incident optical power is measured by counting the triggering rate of the APD. This technique can also measure fast waveforms of repetitive optical pulses, just collecting the time delay between the emission of the optical pulse and the avalanche ignition. The histogram of repetitive measurements gives the shape of the optical waveform. With ultra short laser pulses, the Full Width at Half Maximum (FWHM) of the histogram is defined as the time resolution of the experimental set-up.

GEIGER-MODE OPERATION OF GERMANIUM AND III-V APD's

We have studied the performance of germanium APD's from different suppliers: the Fujitsu FPD13R31 with a 30μm-diameter (Fig.1), the Judson J16A with a 100μm-diameter and a few developmental devices from Siemens with a 50μm-diameter, all with $V_b \approx 32$V at room temperature. We tested some InGaAs/InP APD's from EG&G[1] (Fig.2) with 50μm-diameter, and 30μm-Fujitsu FPD15R31 ($V_b \approx 60$V at room temperature). For III-V APD's, the photons in the 1.3-1.6μm wavelength range are only absorbed within the InGaAs layer, thus holes can pile-up at the InGaAs/InP-heterointerface and may recombine before jumping across the step by thermoionic emission, and triggering the avalanche in the InP layer. Among the EG&G samples, we selected those devices with the highest breakdown voltage (about 100V at room temperature) and the lowest voltage at which the junction depletion region reaches the heterointerface just beneath the active area (about 32V). In this way, at the operating conditions, the electric field at the heterointerface is maximized, thus lowering the residual potential barrier for holes, and improving the speed of the detector response.

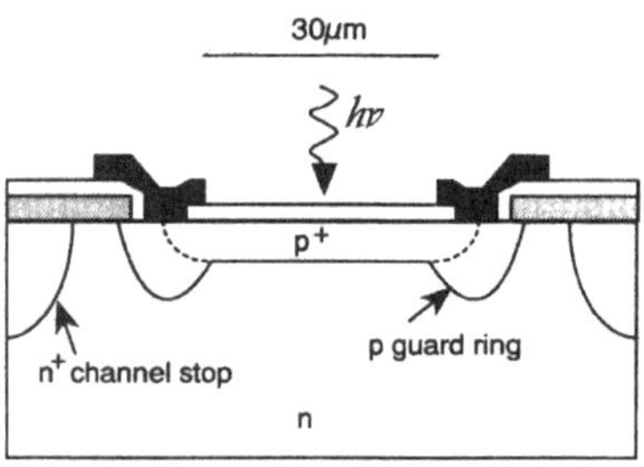

Fig.1 Cross-section of a Fujitsu germanium APD.

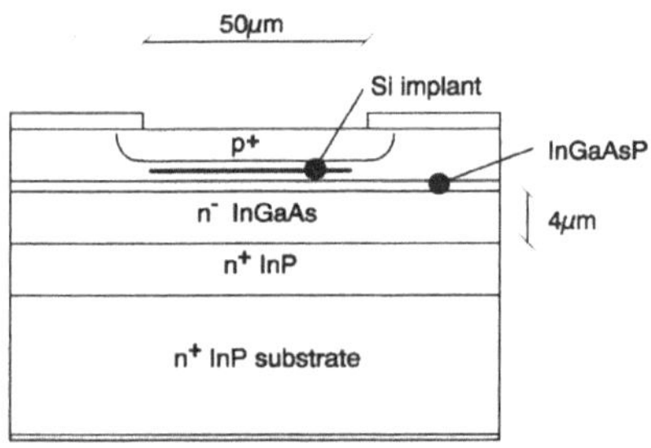

Fig.2 Cross-section of an EG&G InGaAs/InP APD.

The current pulses of APD's operated in Geiger-mode ought to be rectangular since, as the avalanche is triggered, the current rises in a few nanoseconds to $I=(V_0-V_b)/R_s$, where V_0 is the bias above V_b and R_s is the diode resistance. Instead, in some devices the avalanche current first reaches a peak and then decays to a steady-state value. This behaviour is explained since, as the current flows through the junction, V_b increases due to trapping of majority carriers in the junction depletion layer.[2] For the Judson germanium APD's, we estimated a density of majority carrier traps of $8.4 \cdot 10^{13}$cm^{-3}, while Fujitsu germanium APD's did not suffer this effect. We selected the APD's with the lowest trapping effect, and we used an hold-off time sufficiently long to recover the standard V_b after each avalanche pulse.

Dark-counts are essentially due to: i) trapping levels; ii) thermal processes; iii) band-to-band tunnelling. The first effect is caused by a carrier, trapped in a previous avalanche pulse, which is released when the bias is again above V_b, thus triggering the avalanche again (afterpulse). This phenomena depends on the density of trapping centers located in the high field region of the junction.[2] In both germanium and InGaAs/InP devices operated at 77K, thermal generation is negligible and the sensitivity is impaired by trapping and tunnelling processes. The larger is the charge flowing through the device, the higher is the number of carriers trapped and the more recurrent are the afterpulses. Moreover, the electric field can lower the potential well of the deep levels. In conclusion, afterpulsing can be minimized: i) by lowering the excess bias above V_b (i.e. the avalanche current), but at the expense of worse quantum efficiency and timing performance; ii) by shortening the duration of each avalanche pulse or, better, by adopting an active quenching circuit, which quenches the avalanche in less than 20ns;[3] iii) by using a longer hold-off time; iv) by quenching the avalanche just below V_b.

PERFORMANCE OF GERMANIUM APD's

Fig.3 shows how the dark-counting rate of a germanium Fujitsu APD, at 0.6V above V_b and at 77K, can be reduced from 1MHz down to 400kHz, just by changing the bias level during the hold-off time from 99% V_b (200mV below V_b) to 50% V_b. On the basis of the measured dark-counting rate and absolute detection efficiency,[3] we estimated an optimum *NEP* of $8\cdot10^{-16}$W/Hz$^{1/2}$, at an excess bias of 0.2V, with a 10%-detection efficiency. By lowering the excess bias, the *NEP* increases due to a decrease of the detection efficiency, while at higher excess biases it rises due to the growing dark-counting rate. If trapping effects were completely avoided, the *NEP* would be limited only by tunnelling to $4\cdot10^{-16}$W/Hz$^{1/2}$.

Fig.4 shows the dependence of the measured time resolution of the different germanium samples on the avalanche current. At 3V of excess bias (i.e. 7.5mA) the Fujitsu reaches the best 85-ps FWHM time resolution so far reported. With this detector, we have reached a 5-cm spatial resolution in the localization of fused-splice losses, along an optical fiber path, by Optical Time-Domain Reflectometry at 1.3μm-wavelength. This is only one of the many applications waiting for these promising germanium APD's, operated in Geiger-mode.

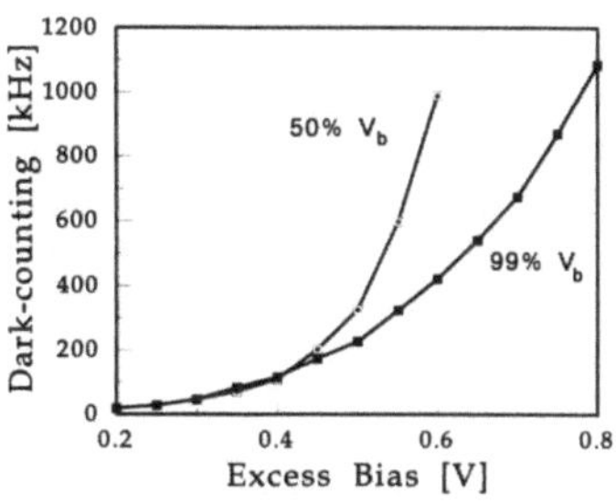

Fig.3 Dark-counting rate of a Fujitsu germanium APD, biased 0.2V above V_b and cooled at 77K.

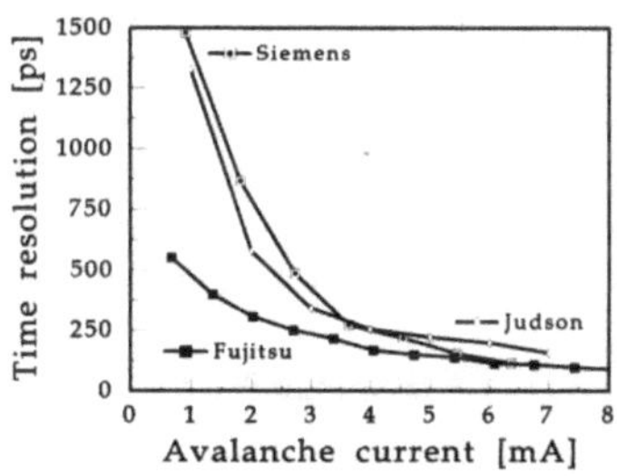

Fig.4 Dependence of the time resolution of the tested germanium samples on the avalanche current.

PERFORMANCE OF InGaAs/InP APD's

As the temperature decreases, V_b lowers, namely from 100V at 300K to 40V at 50K, for the EG&G APD's. Thus, for temperature lower than 100K, both the electric field at the InGaAs/InP heterointerface and the thermoionic activation process are so weak that most of the photogenerated holes do not sweep over the potential barrier and they recombine.[4] This effect causes the η roll-off at low temperature, which can be seen in Fig.5. The same Figure reports the temperature dependence of the measured *NEP*. The minimum of $1 \cdot 10^{-14} W/Hz^{1/2}$ at $1.3\mu m$ is reached at 150K. Similar values can be obtained at $1.55\mu m$.

In both InGaAs/InP devices, the afterpulsing effect was so high that we decided to gate the APD's at a repetition frequency of 1kHz. Even shorter hold-off times could have been adopted, but at the expenses of a much higher dark-counting rate. Therefore, during the fabrication processing, more efforts are required to reduce the trap density in these materials.

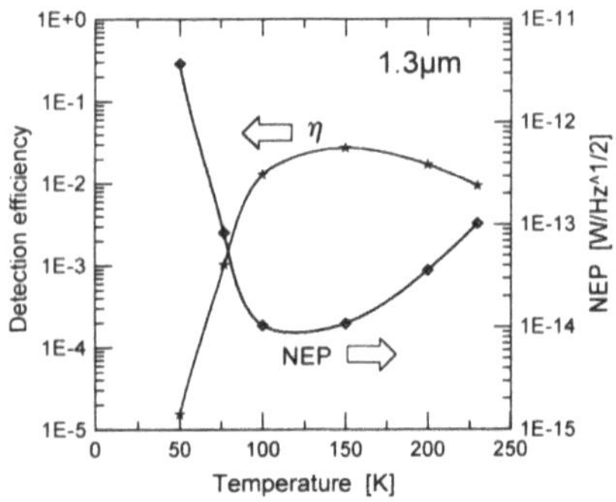

Fig.5 Detection efficiency and *NEP* for the tested EG&G InGaAs/InP APD's.

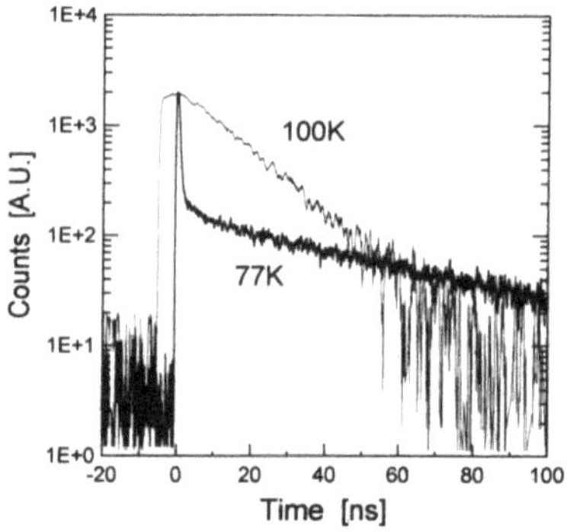

Fig.6 Time response of EG&G InGaAs/InP APD's at 1.3μm.

Fig.6 shows the time responses of an EG&G APD biased 6V above V_b. Above 150K the time resolution is 1ns FWHM (not shown in Fig.6). Instead, as the temperature decreases, the pulse response gets broader. It shows a long tail due to the increasing difficulty that photogenerated holes experience in crossing over the potential barrier at the heterointerface. Below 80K, a new feature appears for the EG&G APD's: a sharp peak due to residual absorption in the quaternary InGaAsP layer. This is a spurious effect and the detection efficiency is very low. The same timing performance is expected at 1.55μm.

Since the tested devices were not purposely designed for Geiger-mode operation, we believe that by designing InGaAs/InP APD's with a smoother compositional grading, and by reducing the trap density of both germanium and III-V APD's, they can provide better performance, working even closer to room temperature. In this way, the InGaAs/InP APD's could reach the same time precision of the silicon APD's, down to few tens of picoseconds.

ACKNOWLEDGEMENTS

The work was supported by CNR (The Italian Research Council), and by The Italian Ministry of University and Scientific Research. We wish to acknowledge R.J. McIntyre for many useful discussions and suggestions and P. Webb for having provided the EG&G InGaAs/InP photodiodes.

REFERENCES

1. P. P. Webb and W. R. Clark, *SPIE Solid State Laser III Conf.*, SPIE Vol.1627: 32 (1992).
2. A. Lacaita, P. A. Francese, F. Zappa, and S. Cova, "Single Photon Detection beyond 1μm: Performance of commercially available Germanium Photodiodes," in press in Applied Optics (1994).
3. G. Ripamonti and A. Lacaita, *SPIE Distributed and Multiplexed Fiber Optic Sensors II*, SPIE Vol.1797, 38 (1992).
4. F. Zappa, A. Lacaita, S. Cova, and P. Webb, "Nanosecond Single-Photon Timing with InGaAs/InP Photodiodes," in press in Optics Letters (1994).

LIMITS TO THE TIMING PERFORMANCE OF SINGLE PHOTON AVALANCHE DIODES

A. Lacaita, S. Longhi, A. Spinelli

Dipartimento di Elettronica e Informazione
Politecnico di Milano
piazza L. da Vinci 32 - 20133 Milano, Italy

INTRODUCTION

Semiconductors p-n junctions with uniform breakdown voltage are gaining acceptance as single-photon detectors in the visible and near infrared wavelength range in many applications, such as measurements of fast optical signals, laser ranging, Optical Time-Domain Reflectometry and in all cases where fast and/or weak optical signals must be detected[1,2]. These devices are operated reverse biased above their breakdown voltage, V_b. As soon as a first carrier is generated in the junction depletion layer, a diverging avalanche discharge is triggered and sensed by a fast discriminator. If the carrier is photogenerated, the current leading edge marks very precisely the photon arrival time. A suitable arrangement then quenches the avalanche by lowering the diode bias below V_b, and after a hold-off time the operating bias is restored, thus making possible the detection of another photon.

The silicon devices nowadays employed can be classified as: i) shallow junctions Avalanche PhotoDiodes (APD), with depleted region of about 1μm, and ii) commercial reach-through structure APD with about 30μm depleted region. In terms of performance, the shallow junction detectors make possible the measurement of the photon arrival time with a precision ranging from 20ps to 50ps Full Width at Half Maximum (FWHM)[2,3,4]. However due to their thin depletion layer the quantum efficiency is less than 10% at 850nm. Instead, with fully depleted reach-through APD's the quantum efficiency can be close to 80% in the 500-800nm range, but the best timing performance is about 150ps FWHM[5,6].

Recently, experimental and theoretical studies have shown that quantum efficiency and timing performance are conflicting requirements. However, it is still not well defined what is the actual achievable trade-off, its dependence on the device geometry and in general what are the design criteria for attaining the best timing performance. This situation is mainly due to the numerical difficulties met in the simulation of all the physical effects involved in the operations of these detectors. In fact a detailed description of the impact ionization, of its

Applications of Photonic Technology
Edited by G.A. Lampropoulos *et al.*, Plenum Press, New York, 1995

spreading from the point where the first photon impinges to the whole junction area and, finally, of the detector current waveform sensed by the timing discriminator, would at least require the coupled solution of the three-dimensional Poisson equation and of the time dependent semiconductor continuity equations with both the avalanche and the optical generation terms. This approach in not numerically feasible. Therefore, aim of our work was to develop simplified models of these transient phenomena in order to achieve a reasonable compromise between quantitative accuracy and numerical complexity. In this paper we report the model of the avalanche dynamics in shallow junctions APD's. The models for reach-through APD's has been reported in details elsewhere[9].

AVALANCHE SPREADING SIMULATION

It has been experimentally demonstrated that in shallow junction devices, the leading edge of the avalanche current depends on the spreading of the avalanche from the point where it is triggered to the whole junction area. More precisely, the closer is the avalanche seed point to one edge of the sensitive area the slower is the current rise[7]. The dependence leads to a jitter in the time at which the current leading edge crosses the threshold of the timing discriminator, thus impairing the time resolution of the detector and making this performance dependent on the shape and size of the sensitive area[8]. It follows that the proper design of these devices cannot be performed without a computer model of the effect which should make possible to predict the actual shape of the avalanche current waveform and therefore the timing performance.

To this purpose we have first studied in details the transient avalanche dynamics in a simplified device geometry (Fig.1a). The active area has a rectangular shape, with the y-side much longer than the z-side, and junction depletion region extending in the x direction. The dopant profiles are the profiles of a shallow junction APD with V_b=16V. The transient simulation of the carrier transport has been performed with a two-dimensional numerical simulator coupling the continuity equations for both electrons and holes to the Poisson equation. In order to numerically handle the problem some simplifying assumptions were made. The boundaries of the space charge region were assumed equipotential thus neglecting the slight change of the depletion layer width during the transient. A lumped resistance models the ohmic path to the external contact. The time dependent continuity equations were further simplified by taking into account that, due to the high electric field in the x-direction, the carriers move at saturated velocity and the x-component of the diffusion current can be correspondingly neglected. Instead, in the y-direction, the electric field is weaker and the carrier transport is dominated by diffusion. Then, the resulting electron transport equation in the depleted region becomes:

$$\frac{\partial n}{\partial t} = -v_n \frac{\partial n}{\partial x} + D_n \frac{\partial^2 n}{\partial y^2} + \frac{n}{\tau_n} + \frac{p}{\tau_p}, \qquad (1)$$

where τ_n and τ_p are the carrrier generation rate due to impact ionization, and D_n is the average transverse diffusion coefficient for hot electrons. The simulation starts with the generation of an electron hole pair at (x,y)=(0,0). Fig. 1 shows the hole distribution 200ps after avalanche triggering.

The result highlights some peculiar features of the carrier transient. For $0<y<8\mu m$ the hole density has already reached a steady state value. For $y>11\mu m$ there are regions still quiescent where the free carrier density is zero and the junction electric field is above the breakdown value. Around y=10μm there is the diffusing wavefront, and the hole density rises over the stationary value due to a boundary space charge effect. The electron density profile shows a similar behavior.

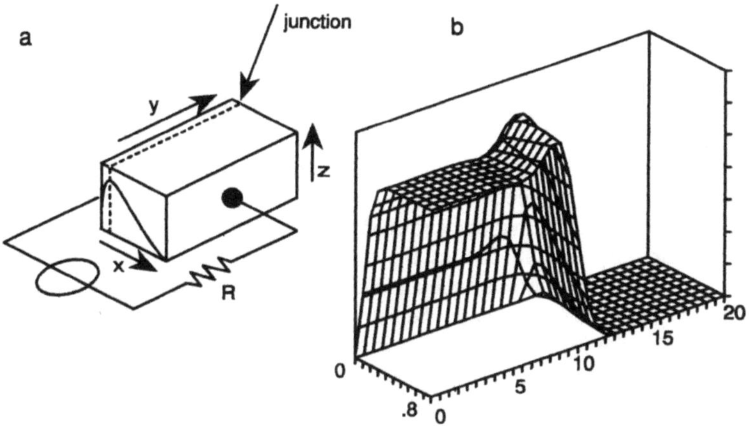

Figure 1. a): Simplified device geometry, b): Holes density 200ps after triggering at (x,y)=(0,0). Due to the symmetry only the region y>0 is shown.

It can be questioned whether the number of carriers in the wavefront is high enough to neglect the statistics which may arise from carriers which diffuse out of the wavefront but fail to impact ionize. This effect could cause statistical fluctuations in the spreading velocity of the wavefront and an additional noise on the leading edge of the avalanche current. To investigate this point, we developed a Monte Carlo version of the simulation program, where the diffusion walk and the avalanche multiplication were treated statistically. The spreading dynamics obtained with the Monte Carlo program was in agreement with the results obtained from the deterministic simulation for excess bias higher than 1V; therefore we conclude that the statistics of the diffusion process can be neglected at the operating bias of conventional shallow junction devices.

SIMPLIFIED MODEL

The next step towards an efficient simulation of realistic devices was to develop a simplified computer model. The basic idea was to describes the depletion layer of a shallow junction detector as an ensemble of elementary diodes connected in parallel. Let us say Δ_y and Δ_z the size of the elementary diodes in the y and z directions, respectively (see inset in Fig. 2). The current flowing through each elementary diode may be written as:

$$\frac{dI_{i,j}}{dt} = \frac{I_{i,j}}{\tau_{i,j}} + D\frac{I_{i+1,j} + I_{i-1,j} - 2I_{i,j}}{\Delta_y^2} + D\frac{I_{i,j+1} + I_{i,j-1} - 2I_{i,j}}{\Delta_z^2}, \qquad (2)$$

where $I_{i,j}$ is the avalanche current and $\tau_{i,j}$ the multiplication time constant of the element in position i,j. With D we indicate the average transverse diffusion coefficient of carriers. It can be shown that the multiplication time constant can be approximately written as[9,10]: $\tau_{i,j}=\tau_0/(K-1) = \tau_0 F_b/(F_m-F_b)$, where K is the well known ionization integral, F_m and F_b are the maximum and breakdown electric field at the junction of the element and τ_0 is the only fitting parameter. The effects of space charge are all included in the parameter $\tau_{i,j}$, which depends in turn on the avalanche current flowing only in the element i,j, and not on adjacents elements.

In order to verify if this model could effectively replace the detailed simulation, we have compared the time dependence of the avalanche current of the device in Fig.1 with those obtained from the two dimensional simulator discussed in the previous section. Since in the simplified model adjacents diodes are directly coupled only by the diffusion term in Eq.(2), it is expected that the simplified model should fail in describe space charge effect related to the electrostatic coupling of adjacent elements. In fact, it does not account for the peculiar peaking of the carrier density just before the shifting wavefront. However this discrepancy is negligible for our purposes. In fact as the avalanche current increases the charge in the transition region between the avalanching and the non-avalanching areas becomes a few percent of the overall free charge in the depletion layer. In practice it does not affect the time at which the avalanche leading edge crosses the discriminator threshold. Furthermore, the value of τ_0, which could be in principle a fitting parameter of the simplified model, was always set in the simulations equal to the value obtained from theoretical calculations[10,11]. Therefore, in practice the simplified model had no adjustable parameters and it has been always proved in agreement with the detailed simulation of the current transient.

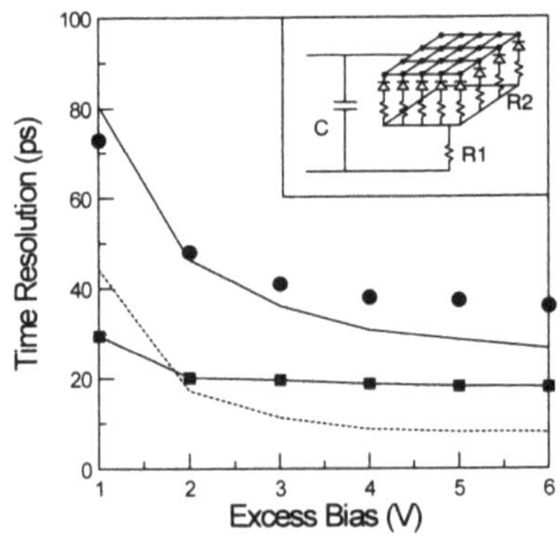

Figure 2. Contributions to the time resolution of a 8μm diameter shallow junction SPAD. The figure shows the experimental values (●), the electronics and laser contributions (■), the computed contributions arising from the spreading dynamics (—) and the contributed values arising from the multiplication noise (---). A schematics of the simplified model is also shown in the inset.

On the basis of these results, we have adopted the simplified model to compute the time resolution of realistic devices with circular geometries, already tested in our laboratories[8]. Note that the model can take into account single resistors in series at each element (R_2) as well as resistor paths common to all diodes (R_1), and a stray capacitance due to the guard ring and the external bondings, thus describing better the real detector geometry. Fig.2 shows the results obtained for a circular device with 8μm diameter, Vb=16V, R_1=300Ω, R_2=1400Ω, C=4.5pF, τ_0=0.3ps[12]. The time resolution was computed at different bias by sorting the point of photon absorption over the area with uniform probability. The simulation results are in good agreement with the experimental values, thus confirming that the spreading dynamics sets the main limit to the achievable timing performance of this type of detectors. However, also the timing electronics gives a significant contribution which should be further reduced. The same figure also shows that the time jitter arising from the statistics of the first ionizing events[13] is negligible in present devices at the usual operating bias.

CONCLUSIONS

We have developed a computer model for the avalanche spreading dynamics in shallow junction single photon avalanche diodes. This model has been validated by comparison with both detailed simulation of the carrier transient in simplified device structures and the measurements on fast devices. The model may become a computer tool for the design of new device. More precisely, it can be useful to quantitatively assess the dependence of the time resolution on the device geometry and to find a trade off between the timing performance and the quantum efficiency for this class of detectors.

REFERENCES

1. T. A. Louis, G. Ripamonti and A. Lacaita, "Photoluminescence lifetime microscope spectrometer based on time-correlated single-photon counting with an avalanche diode detector", *Rev. Sci. Instr.* 61:11 (1990).
2. S. Cova, A. Lacaita, M. Ghioni, G. Ripamonti and T. Louis, "20ps timing resolution with single photon avalanche diodes", *Rev. Sci. Instr.* 60:1104 (1989).
3. A. Lacaita, S. Cova, M. Ghioni, F. Zappa, "Single-photon avalanche diodes with ultrafast pulse response free from slow tails", *IEEE Electron Device Lett.* 14:360 (1993).
4. S. Cova, G. Ripamonti and A. Lacaita, "Avalanche semiconductors detector for single optical photons with a time resolution of 60ps", *Nucl. Instrum. Methods* A253:482 (1987).
5. M. Ghioni and G. Ripamonti, "Improving the performance of commercially available Geiger-mode avalanche photodiodes", *Rev. Sci. Instr.* 62:1 (1991).
6. Li-Qiang Li and Lloyd M. Davis, "Single photon avalanche diode for single molecule detection", *Rev. Sci. Instr.* 64:1524 (1993).
7. A. Lacaita, M. Mastrapasqua, M. Ghioni and S. Vanoli, "Observation of avalanche propagation by multiplication assisted diffusion in p-n junctions", *Appl. Phys. Lett.* 57:489 (1990).
8. A. Lacaita and M. Mastrapasqua, "Strong dependence of the resolution on detector diameter in single photon avalanche diodes", *Electron. Lett.* 26:2053 (1990).
9. A. Lacaita, S. Cova, A. Spinelli and F. Zappa, "Photon assisted avalanche spreading in reach-through photodiodes", *Appl. Phys. Lett.* 62:606 (1993).
10. C. A. Lee, R. L. Batdorf, W. Wiegmann and G. Kaminsky, "Time dependence of avalanche processes in Silicon", *J. Appl. Phys.* 38:2787 (1967).
11. R. Kuvås and C. A. Lee, "Quasistatic approximation for semiconductor avalanches", *J. Appl. Phys.* 41:1743 (1970).
12. A. Lacaita, M. Ghioni, S. Cova, "Double epitaxy improves single-photon avalanche diodes performance", *Electron. Lett.* 25:841 (1989).
13. A. Lacaita, M. Ghioni, F. Zappa, G. Ripamonti and S. Cova, "Recent advances in the detection of optical photons with silicon photodiodes", *Nucl. Instrum. Methods* A326:290 (1993).

AVALANCHE PHOTODIODES (APD) AS PROPORTIONAL DEVICES

R.S. Gilmore, A.R. Duell, T.J. Llewellyn,
R.J. Tapper, S. Nash and K. Xyloparkiotis

H.H. Wills Physics Lab
Bristol University
England

ABSTRACT

APDs may be used in proportional detectors for particles. Lowering the temperature gives both a reduction in thermal ionization and an increase in gain. Selective control of the ratio of electron to hole ionization would allow the construction of diodes with larger proportional gains.

INTRODUCTION

Avalanche photodiodes (APDs) are just diodes operated with so high an electric field in the reverse biased depletion region that electrons and holes produce further ionization and so give charge gain. They share the high sensitivity of other semiconductor detectors with the addition of charge gain, which will improve the size of the signal relative to any noise produced in the readout electronics.

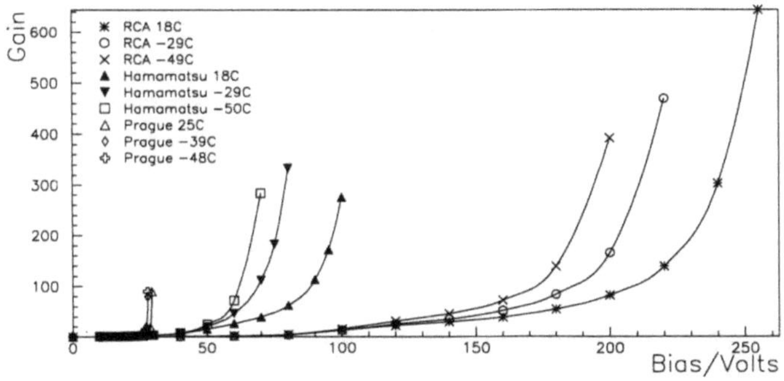

Figure 1. APD gain as a function of bias voltage
for three different diode types, at various temperatures as indicated.

Applications of Photonic Technology
Edited by G.A. Lampropoulos *et al.*, Plenum Press, New York, 1995

Our recent investigations in Bristol have concentrated on the use of APDs in the proportional mode. We have investigated 3 types of APD: an EG&G (RCA) type C30921E, a Hamamatsu type S2382 and one made by the Czech Technical University in Prague. Figure 1 shows that for each type of diode there is a voltage, the breakdown voltage, above which the diode will conduct continuously after it detects a signal.

In most earlier applications APDs were operated in the geiger mode, where the diode is biased slightly above the breakdown voltage. This can give a very large signal initiated by a single photon but must have an external quenching circuit. This geiger mode gives a saturated signal and information about the amount of primary ionization is lost.

If the diode is biased below the breakdown voltage then the avalanches will be limited and the output charge will be proportional to the number of such avalanches and hence to the initial ionization, apart from the inevitable fluctuations in gain.

An APD working in proportional mode may be used to detect light from an external scintillator (Lorenz, 1993) operating as a compact and robust substitute for a photomultiplier which would be insensitive to magnetic fields. It might alternatively be used as a semi-conductor detector for fast charged particle passing through it, with enough gain to enhance the signal and remove the need for very low noise electronics.

The proportional gain in available APDs is rather low. As can be seen from figure 1 the RCA diodes give a maximum useable gain of a few hundred and the Hamamatsu diodes about 100, while the Prague diodes give little proportional gain. Such gains are much less than from a photomultiplier, but still provide a useful increase in signal relative to any noise from electronics. Because silicon and other semiconductors have a low ionization energy in there is a serious background at room temperature which is produced by thermal ionization of the material.

THERMAL EFFECTS

In semiconductor devices the amount of thermal ionization may be reduced by lowering the temperature. In an APD close to breakdown the dark current *rises* dramatically as the temperature is decreased. This is because the *gain* increases as the temperature goes down. The mean free path for the electrons (or holes) increases at reduced temperature, so the electrons gain more energy from the field and ionize more readily at a given field strength. This increase in gain more than compensates for the decrease in the thermal generation of primary electron-hole pairs as the temperature is reduced. At lower temperatures the thermal noise does decrease relative to the signal, but both currents will increase. A change in temperature of 1 ^{0}C will give a change of more than 1 % in signal size for a silicon APD operated with a gain of about 100.

The variation of the gain in an APD is unpleasantly large, especially close to breakdown, both as a function of bias voltage and of temperature. It would be useful to have an APD which could be operated with an adequate gain while still biased well below the breakdown voltage. This requires an APD which will not break down until it is biased to a voltage significantly higher than needed to give the required gain.

FEEDBACK AND BREAKDOWN IN APDS

In a silicon diode the ionization coefficient for electrons α_e is appreciably greater than the coefficient for holes α_h, so that avalanches are dominantly due to ionization by electrons. If this were all that could happen the gain would rise exponentially with the bias voltage without any limit. The limit to the gain attained in practice is caused by some feedback process which means that the generation of one avalanche causes another avalanche to start and so on.

This breakdown condition is very similar to that encountered in the gas-filled multigap avalanche chambers developed as particle detectors (Charpak 1978). Here the feedback process is provided by positive ions which return to the electrodes at the start of the avalanche region and release electrons from the metal surface. In multigap chambers it is possible to inhibit the onset of breakdown (Gilmore, 1981) by dividing the gain between two gaps separated by wire mesh electrodes through which some of the charge can pass. The gain in either gap will be too low to produce breakdown. If the transparency of the electrodes for passing positive ions is reduced then the average final charge can be greater before a sufficient fraction is fed back to start a new avalanche.

In a semiconductor both electrons *and* holes may cause ionization, so the feedback process is intrinsic in the material. In silicon the ionization is dominantly by electrons and $\alpha_e \gg \alpha_h$, but at higher field values holes will ionize also. Breakdown results if on average *one* of the holes produced in an avalanche should generate a further electron-hole pair close to the cathode end of the high field region. The gain is very unstable near breakdown. Figure 2 gives a distribution of pulse areas for a Hamamatsu APD operating just below breakdown and shows the tail of abnormally large pulses caused by extra avalanches from ionization by holes.

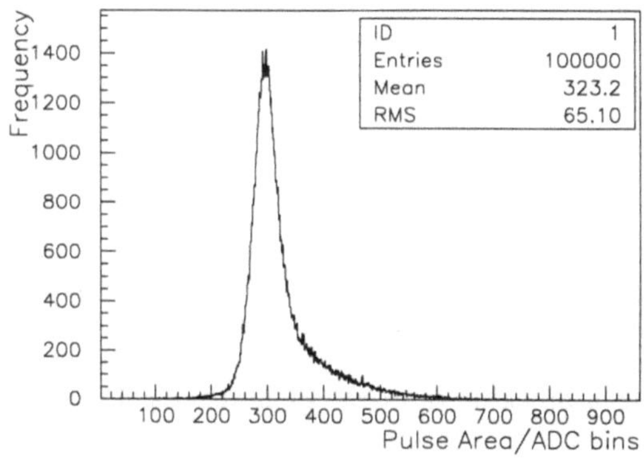

Figure 2. Pulse size distribution near breakdown.

The standard condition for breakdown in a depletion layer of width W in a semiconductor such as silicon where the ionization is mostly by electrons is (Sze, 1981, p 99)

$$\int_0^W \alpha_e \exp\left(-\int_x^W (\alpha_e - \alpha_h)dx'\right) dx = 1$$

In the case where $\alpha_e = \alpha_h$ this reduces to a limit where the mean number of ionizations per electron is one. This would correspond to a gain of just 2, but when both electrons and holes cause equal ionization it results in breakdown. An APD giving a high gain before break down should have a ratio α_e/α_h as far as possible from unity. In silicon the ratio α_e/α_h is greater at lower fields (Sze, 1981, p 47) so a high gain diode should have a wide depletion layer with a relatively low field. The rate of ionization per unit length will be lower than at higher field, but in a wide enough layer the same gain could be achieved and if α_h is comparatively low then breakdown will be inhibited. Ideally the material should also have a greater capture cross-section for holes than for electrons, so that holes may be captured before they cause ionization.

The different diodes we have investigated show striking variations in their performance, as summarised by figure 1. The EG&G diodes give a maximum gain in excess of 200 for a bias around 230 volts. The Hamamatsu diodes give gains of around 100, but at a much lower bias, around 100 volts. The diodes built by the Technical University in Prague have a very low breakdown bias, about 30 volts, but give proportional gains of little more than 5.

These results are consistent with the arguments outlined above. The Hamamatsu APD is a comparatively normal diode with the high field in a depletion layer at a junction. The EG&G (ex RCA) diode has a *reach-through* structure where the high field is in a lightly doped region which would tend to give a relatively wide gap with a reduced field. This needs a higher bias voltage to produce the same gain, but the maximum gain which can be obtained before breakdown is the greatest among the three types. The Prague diode contains germanium doped silicon. In silicon the ionization coefficient for electrons is greater than that for holes, whereas in germanium the reverse is true. In a mixture the coefficients may be comparable and breakdown consequently sets in at a very low value of gain. This is advantageous for an APD operated in geiger mode, since it may be operated from a low voltage supply, but is unhelpful for a proportional device.

We are developing a computer simulation of the avalanche process in APDs, for comparison with measured values for the gain and pulse variation. We plan also to measure the linearity and absolute quantum efficiency of the detectors.

CONCLUSION

APDs show promise as particle detectors because they have high quantum efficiency and specific ionization combined with intrinsic gain and the ability to operate in a magnetic field. To obtain reasonably stable operation as proportional detectors the diodes should be biased well away from the breakdown voltage and so it is important to develop diodes which break down at the highest possible gain.

REFERENCES

Charpak, G. and Sauli, F., 1978, The multistep avalanche chamber: a new high-rate high-accuracy gaseous detector, *Phys. Lett.* **78B**;523.

Gilmore, R.S., Lavender, W.M., Leith, D.W.G.S, and Williams, S.H., 1981, High gain multigap avalanche chambers for cherenkov ring imaging *IEEE Trans. Nucl. Sci.* **NS28**;435-438.

Lorenz, E., Natkaniez, S., Renker, D., Schwarz, B., 1993, Fast readout of plastic and crystal scintillators by avalanche photodiodes, *Max Planck Institute report* MPI-PhE/93-23.

Sze, S.M., 1981, "Physics of Semiconductor Devices" Publisher, John Wiley.

DETECTION EFFICIENCY AND DARK PULSE RATE OF ROCKWELL (SSPM) SINGLE PHOTON COUNTERS [1]

P.H. Eberhard,* P.G. Kwiat,* M.D. Petroff,†
M.G. Stapelbroek† and H.H. Hogue†

*Physics Group, Lawrence Berkeley Laboratory
Berkeley, California 94720
†Rockwell International Corporation Science Center
3370 Miraloma Ave., Anaheim, CA 92803

ABSTRACT

This paper describes direct photon counting data on several pixels of the Rockwell SSPM detector array, about which absolute single photon detection efficiencies and time responses were reported previously by Kwiat et al. (ref 1,2). The new data shows a substantially larger efficiency, 85%, and suggests that damage in the optical fiber employed in the earlier data was responsible for the lower number, 70%. When the results of this paper are analyzed using reasonable assumptions, one concludes that several improvements of the SSPM detector are still possible to bring the photon counting efficiency close to or larger than 95%.

INTRODUCTION

A series of measurements of the absolute efficiency and time response of two different types of single-photon detectors was performed at Berkeley in August 1992 (ref. 1,2). The detectors were chosen because of their reported high quantum efficiencies and speed of response. One of these consisted of an array of solid-state photo-multipliers (SSPMs) from Rockwell (ref. 3) and the the other a pair of Geiger mode avalanche photodiodes from EG&G (ref. 4). The SSPMs and their later versions, referred to by Rockwell as Visible light photon counters (VLPCs) (ref. 5,6) utilize the same impurity band avalanche effect in silicon at liquid helium temperature. These types of detectors were known to have a quantum efficiency of better than 60% and response time shorter than 10 ns. The Berkeley absolute measurement provided a minimum detection efficiency value of 70.9% at a wavelength of 633 nm. However, subsequent tests on the SSPMs suggested a possible degradation of transmission in the input fibers. The measurements reported here used the same SSPM chip, but equipped with improved fibers, less subject (though not quite immune) to degradation of transmission. These measurements allow a re-evaluation of the achievable SSPM or VLPC detection efficiencies. The subject of high photon detection efficiency is important for quantum cryptography and loophole-free tests of Bell's inequalities.

[1] This work was supported by the Director, Office of Energy Research, Office of High Energy and Nuclear Physics, Division of High Energy Physics of the U.S. Department of Energy under Contract DE-AC03-76SF00098

MEASUREMENTS

Three SSPM detectors on an 8-pixel chip were available for testing. Each of the three pixels was connected to a different optical fiber to bring light from an optical arrangement at room temperature to the device operating at a temperature of 6 K. Each of them came with its output connected to a separate charge integrator at room temperature. The three devices were studied in various conditions. In almost all the tests, the wavelength was selected to be 660 nm by a filter of 10 nm FWHM. Data on efficiencies and background count rate were recorded. Typical signal pulse heights were about .7 mV at the output of the charge integrator.

In the first test setup, no additional optical feature was added to the assembly as it came from Rockwell International Corporation. Efficiencies were measured and found to fall in a range from 70 to 76%, depending on the device, on the discriminator level, on device bias voltage, and on the temperature. A curve of efficiency versus discriminator level is shown on Fig. 1. The horizontal scale on Fig. 1 indicates the pulse height at the output of the charge integrator that triggers the discriminator.

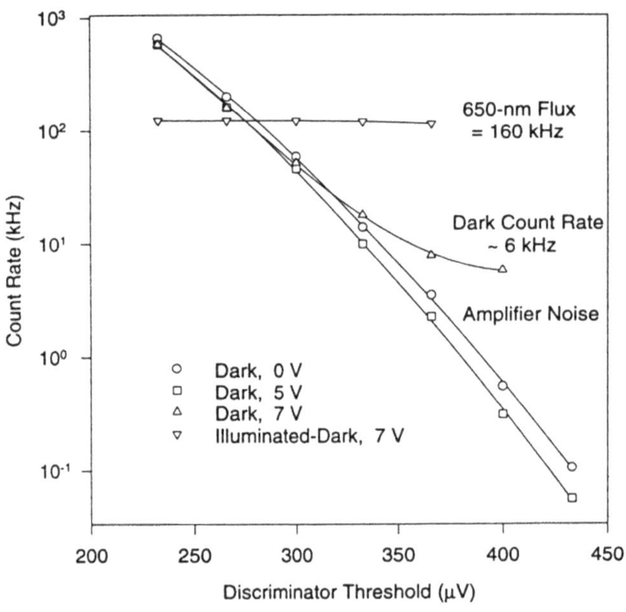

Figure 1."Measured count rate vs. discriminator threshold for different applied biases, with and without illumination"

The rate of background (all light blocked at the top of the fiber) was measured as a function of discriminator threshold for different bias voltages. The result is also shown on

Fig. 1, in units of the charge integrator output. At bias voltages of 0 and −5 V, the spectrum is rapidly decreasing as a function of pulse height. This background has a gaussian shape of 90 microvolt standard deviation. Its shape and magnitude fit what is expected from an amplifier with input noise of 2 nV/SQRT(Hz) and the circuits time constants. At 0 and at −5 V, the curves are slightly different, as expected because the device capacitance depends on the bias voltage. At −7 V, the noise spectrum results from the superposition of the noise seen at 0 and −5V, and of another noise far less dependant on discriminator level. The latter phenomenon may be due to electrons liberated in the device drift region by thermal agitation and to infra red radiation emitted at room temperature, transmitted down to the device via the fiber. The dependance of the latter background on discriminator level is expected to resemble the dependance of the efficiency, shown on that same Fig. 1. At the discriminator level where most of the efficiency measurements were taken (.37 mV), the total amount of background is in the range of 6000 to 8000 counts/sec, with a significant contribution of the noise component which we attribute to the amplifier.

Tests were also performed using a spherical mirror with the sphere center located at the top of the fiber. The purpose was to catch photons reflected at various places in the apparatus and to reflect them back toward the device (giving them other chances to be counted). The efficiency was found to be increased to values in the range of 80 to 86%, depending on the values set for the operating parameters. Sometimes, for two of the devices, consecutive measurements were taken with and without mirror, not changing any other experimental condition, just measuring the effect of the mirror. The mirror enhancement factor was systematically found to be 1.168 + or − .01 for one device and 1.160 for the other.

Using the known index of refraction of the different materials on the path of the photons, one can compute the Fresnel reflections at all the interfaces. We can then estimate what happens to photons in the absence and in the presence of the spherical mirror. The mirror enhancement factor depends on the absorption in the fiber, which can therefore be determined. According to our measurements and this analysis, the internal transmission factor (i.e. not counting reflections) is found to be 91 + or − 3% in the fiber associated with one device, 90 + or − 1.6% in the other. After the tests, the devices were dismounted and the absorption at 660 nm was measured directly. The internal transmission factor was 94.6% in the first fiber and 91% in the other, quite consistent with the values determined from the mirror enhancement factors. The transmission factor of virgin fibers was also measured and found consistent with the manufacture specifications, which correspond to a transmission of 99% in a fiber of the length used in our assembly, 19.0 cm.

From the above analysis, we can conclude that the fibers have aged or have been damaged since they were manufactured and the amount of light absorbed has increased from 1% to about 8%. The observed small degradation is ascribed to handling in assembly and to temperature cycling. The intrinsic efficiency of the silicon device with an optimum AR coating would be 96 + or − 3%, (taking into account a possible normalization error of our calibration apparatus estimated at 2%).

DISCUSSION

We believe our results represent a significant breakthrough in visible light photon counting detectors. With the actual efficiency obtained in our test, 85%, with a background of 7 kHz as we measured it, and a timing accuracy of 10 ns (as determined in the test of last year), it should be possible to perform an EPR loophole free experiment if, in addition, the loss of one photon could be less than 3% and the visibility be 95% or better. However the margin of safety would be small. Therefore it is most important to look at possible improvements. Such improvements of the detector seem possible. Therefore EPR loophole free experiments are definitely in the realm of possibilities.

There are areas where improvements are possible

1. The plastic fiber could be replaced by a glass one if one can be found that absorbs infrared efficiently but does not age.
2. Each interface could be coated with a proper AR coating, to avoid the need of a spherical mirror. Timing accuracy would be improved too.
3. A cold amplifier could be used as a preamp to reduce amplifier noise.
4. A more sophisticated coincidence circuit (relying on zero crossing for instance) would improve timing accuracy.

The authors are indebted to T. Weber for his help in getting the measurements done and to Prof. R. Chiao for lending them equipment for the test.

REFERENCES

1. P.G. Kwiat, A.M. Steinberg, R.Y. Chiao, P.H. Eberhard, and M.D. Petroff, "High-efficiency single-photon detectors", *Phys Rev A,* **48**, R867 (1993).
2. P.G. Kwiat, A.M. Steinberg, R.Y. Chiao, P.H. Eberhard, and M.D. Petroff, "Absolute efficiency and time-response measurement of single-photon detectors", *Applied Optics*, **38**, 1844 (1994).
3. M.D. Petroff, M.G. Stapelbroek and W.A. Kleinhans, "Detection of individual 0.4–28 micrometer wavelength photons via impurity-impact ionization in a solid state photomultiplier," *Appl. Phys. Lett.* **51**, 406–408 (1987).
4. A.W. Lightstone, A.D. MacGregor, D.E. MacSween, R.J. McInty, C. Trottier, and P.P. Webb, "Photon counting modules using RCA silicon avalanche photodiodes," *Electron. Eng.* **61**, 37–47 (1989).
5. G.B. Turner, M.G. Stapelbroek, M.D. Petroff, E.W. Atkins, and H.H. Hogue, "Visible Light Photon Counters for Scintillating Fiber Applications: I. "Characteristics and Performance", Proceedings of the Workshop on Scintillating Fiber Detectors, Notre Dame Univ. October 24–28, 1993, Editor R. Ruchti, World Scientific Publishing Co Pte Ltd. 1994.
6. M.G. Stapelbroek and M.D. Petroff, "Visible Light Photon Counters for Scintillating Fiber Applications: II. Principles of Operation", Proceedings of the Workshop on Scintillating Fiber Detectors, Notre Dame Univ.,October 24–28, 1993, Editor R. Ruchti, World Scientific Publishing Co Pte Ltd. 1994.

ABSOLUTE RADIOMETRY USING CORRELATED PHOTONS

A. Migdall[1], R. Datla[1], A. Sergienko[2], and Y. H. Shih[2]

[1]Radiometric Physics Division, 221/B208
National Institute of Standards and Technology
Gaithersburg, MD 20899, U.S.A.

[2]Department of Physics
University of Maryland, Baltimore County
Baltimore, MD 21228, U.S.A.

INTRODUCTION

Correlated photons have shown promise in several areas of radiometric measurements[1]. In particular, several groups have demonstrated that correlated pairs of photons can be used to determine the absolute quantum efficiency of photon counting detectors without reference to any externally calibrated detector standard[2-6]. The ultimate goal of our work is to determine the maximum accuracy that can be achieved with this method, as well as to develop additional radiometric applications of correlated light. To carry out a test of the method, we have set up a system at the National Institute of Standards and Technology employing this method to map out the spatial variations of the absolute quantum efficiency of a photomultiplier (PMT) photocathode. To our knowledge, this is the first time that this method has been used to provide a detailed spatial response map consisting of direct absolute measurements at each spatial position, rather than a relative response map which is tied to an absolute scale at only one or a few points. Toward the goal of determining the ultimate accuracy of the method, the measured efficiency was compared with an estimate using the specifications supplied with the detector. Further comparisons are planned employing an independent calibration to verify the method to better accuracy. In addition, the effects of the pulse height distribution of the PMT, the discriminator threshold level setpoint and the PMT supply voltage on the detection efficiency were investigated.

This method is based on the process of parametric down conversion, which produces optical photons in pairs within a nonlinear crystal. Within the crystal, photons from a pump laser beam, in effect, "decay" into pairs of photons under the restrictions of energy and momentum conservation. These restrictions, while not quite rigorous (when transfer of energy and momentum to the crystal is neglected), are useful to visualize the process. Since the photons are created in pairs, the detection of one photon indicates the existence of the other photon. Because of the near energy and momentum conservation requirements, the direction and energy of the detected photon can be used to predict not only the existence, but also the direction and energy of the other photon of the pair.

EXPERIMENT

To produce correlated pairs of photons, a linearly polarized Ar^+ laser, power stabilized with 30 mW of output at 351 nm, was used to pump a KDP crystal (see Fig. 1). The input and output ends of the crystal were antireflection coated for 351 nm and 702 nm respectively. The crystal was cut so that its optic axis was inclined at 52° to the pump beam direction. The output angles of the correlated photons can be obtained using the energy and momentum constraints, the index of refraction data for the crystal and the orientation of the pump beam with respect to the crystal optic axis. This particular configuration yields correlated pairs of photons covering the visible region of the spectrum emitted within a fairly narrow range of angles centered around ~4° to the pump beam direction. For our measurements, we selected degenerate pairs of photons, that is where the wavelengths of the photons in a correlated pair are both 702 nm. In addition to the energy and momentum requirements already

Applications of Photonic Technology
Edited by G.A. Lampropoulos *et al.*, Plenum Press, New York, 1995

described, the polarization of the pump beam must be oriented parallel to the plane defined by the optic axis and the laser direction. This polarization orientation is required by the parametric downconversion process defined as Type 1, which uses pump radiation polarized along the fast axis of the crystal to produce downconverted light polarized along the slow axis[1]. A 90° rotation of the pump polarization turns off the production of downconverted light.

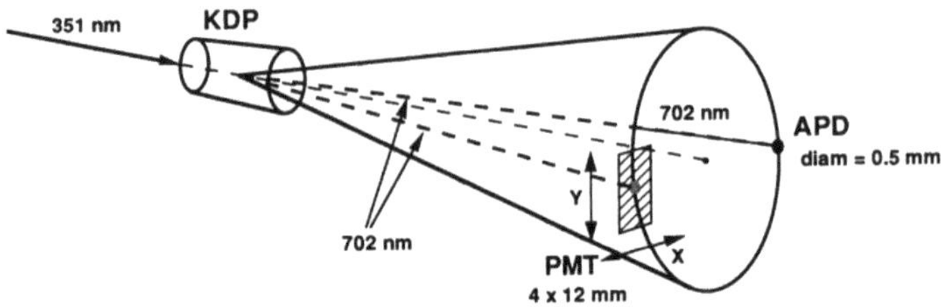

Figure 1. Scheme for x-y mapping of detector efficiency. The gray spot on the PMT photocathode indicates the region containing photons correlated to those detected by the APD. Note that the lens used to image this spot onto the PMT is not shown.

A dry ice cooled Si avalanche photodiode (APD) was used to detect the presence of the down converted light. To detect individual photons, the APD was operated in the Geiger mode[7] with a bias voltage of 221.9 V (this is about 40 V above breakdown). The output pulses, without any electronic amplification, were very smooth pulses about 1.5 V in amplitude and less than 3 ns in width. These pulses were sent directly to the start input of a time-to-amplitude converter(TAC). The TAC "valid starts" output was sent to a counter to record the APD counts. No lens was used in front of the APD, so the detector diameter of 0.5 mm and 1.85 m distance from the crystal defined the 0.27 mrad collection angle. A 702 nm interference filter with a 2 nm bandpass was used to limit the light on the APD. The APD was mounted on a manually operated X-Y stage to allow the APD position to be centered on the peak of the 702 nm downconverted light.

The PMT (Burle[8] model C31034A)*, whose spatial response was to be measured, was placed to intercept the portion of the downconverted light correlated to the light detected by the APD. A lens positioned in front of the PMT was used to focus all of the light correlated with the APD to a small spot, ~0.25 mm at the PMT photocathode. A 702 nm interference filter with a 10 nm bandpass was used to limit the light on the PMT. The PMT was cooled thermoelectrically to ~ -30 C. The PMT was mounted on a computer controlled X-Y positioning stage.

The PMT socket was built following recommendations for producing fast pulses with minimal ringing[9]. The pulses were amplified using a 200x fast pulse amplifier. The signal was then sent to an amplifier/discriminator with a manually set threshold level. The amplifier/discriminator provided about 40 ns of delay between the input and output pulses. The discriminator output was then sent to the TAC stop input and to a counter to record the PMT counts.

The TAC output was sent to a multichannel analyzer (MCA) producing a histogram of delay times between start and stop times. Correlated photons were seen as a coincidence peak in this histogram at a delay time corresponding to the total time difference between the two channels due to optical and electronic path differences. The counts in a region containing the coincidence peak and the counts in a background region away from the peak were recorded by computer. Both the MCA and the counters

---

* Certain trade names and company products are mentioned in the text or identified in an illustration in order to adequately specify the experimental procedure and equipment used. In no case does such identification imply recommendation or endorsement by the National Institute of Standards and Technology, nor does it imply that the products are necessarily the best available for the purpose.

recording the individual APD and PMT counts were gated simultaneously under computer control so that all gate times were identical.

The APD and PMT counts due to uncorrelated sources such as dark counts and scattered light were determined by rotating the pump polarization 90° with a half wave plate, effectively turning off the downconverted light, so any residual signal was due to uncorrelated light. These uncorrelated counts were subtracted off the individual counts of the detectors with the proper pump polarization to obtain the true correlated counts. The ratio of the coincidence counts to the APD counts after zero subtraction is the absolute quantum efficiency of the entire PMT optical system including all optics such as windows, the lens and the filter. The absolute quantum efficiency of the PMT was spatially determined as its position was scanned in x and y in 0.25 mm steps. At each x,y point a 10 s gate time was used.

It is important that the bandpass of the PMT filter is wider than the APD filter bandpass, so that all of the photons correlated to the APD light pass though the PMT filter. With this configuration, the quantum efficiency measurement is insensitive to slight shifts of the center wavelength of the PMT filter.

Measurements of both the PMT count rates and the coincidence count rates were also made as a function of PMT discriminator level. For these measurements, the PMT was placed so that the correlated light spot was positioned near the center of the PMT photocathode. The measurements were then made as the discriminator level was varied manually.

RESULTS

The spatial map of the absolute efficiency of the PMT and optics system is shown in Fig. 2. The observed rectangular shape and size of the responsive region is what was expected for this particular photocathode. The sharpness of the edges in both the horizontal and vertical directions indicates that the spot size at the photocathode which determines the spatial resolution of the measurements is less than the scan step size of 0.25 mm. The PMT to lens distance was adjusted to produce the sharpest edges. The overall responsivity of the PMT/optical system, of about 5.4% with a 1σ spread of 0.5%, is consistent with the estimated value of 6.6%. This estimation includes the PMT responsivity at 702 nm from the manufacturer's data sheet (13%), the peak filter transmittance (65%) and the losses in the optical path. (6 surfaces with 4% reflectance). Since the uncertainty of the data from the manufacturers is unknown, an independent measurement of the PMT/optical system response will be needed to improve the accuracy of this comparison.

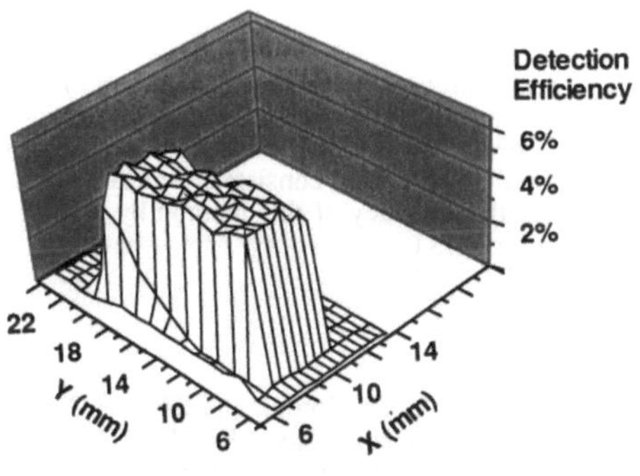

Figure 2. Absolute responsivity map of the PMT/optical system. The efficiency on the plateau region varies from about 4.9% to 5.9%.

The dependence of the efficiency on discriminator threshold level and PMT supply voltage was tested in several ways. First, the PMT pulse height distribution was measured to verify that the PMT was operating properly and that a photoelectron peak could be seen at the supply voltage used. Once this was established, an attempt was made to set the discriminator threshold at a point where the PMT response variation versus threshold level was as low a possible. This was done by looking at both the single counts out of the PMT and the absolute efficiency determined from the correlated counts. As shown in Fig. 3, both of these tests yielded an observed variation in the response of ~4%/decade. This data, along with how precisely the threshold level can be set, will provide a limit to the ultimate accuracy with which a PMT efficiency can be determined. For example, if the threshold could be defined with an accuracy of 5%, the limit on the efficiency accuracy would be 0.2%.

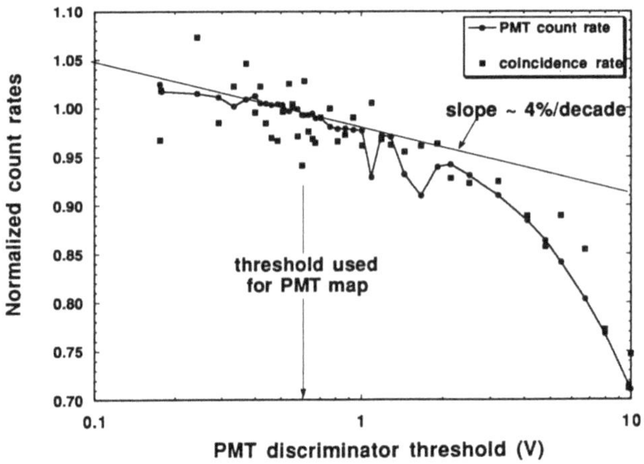

Figure 3. The relative coincidence count rates and total count rates of the PMT versus discriminator threshold setting. The straight line indicates the relative variation with threshold, at the threshold setting used for the x-y efficiency map, is ~4 %/decade for both data sets.

CONCLUSIONS

The results presented here demonstrate that the spatial variation of the absolute response of a photon counting system can be measured using correlated light. The ultimate accuracy of the method is yet to be determined, although the results so far indicate internal consistency in terms of response variation with discriminator threshold and external consistency of the overall absolute response. Further work will be undertaken to quantify the accuracy of this method by independently calibrating the PMT system against the radiometric base at NIST.

REFERENCES

1. D. N. Klysho, "Photons And Nonlinear Optics, Gordon and Breach", New York (1988).
2. A. A. Malygin, A. N. Penin, and A. V. Sergienko, *Sov. Phys. JETP Lett.* **33**:477 (1981).
3. J. G. Rarity, K. D. Ridley, and P. Tapster, *App. Opts.* **26**:4616 (1987).
4. Y. H. Shih and C. O. Alley, *Phys. Rev. Lett.*, **61**:2921 (1988).
5. A. V. Sergienko and A. N. Penin, *App. Opts.* **30**:3528 (1991).
6. P. G. Kwiat, A. M. Steinberg, R. Y. Chiao, P. H. Eberhard and M. D. Petroff, "High-efficiency single-photon detectors", *Phys. Rev. A*, 48:R867 (1993).
7. W. Budde, "Optical Radiation Measurements", Academic Press, Inc., New York (1983).
8. Burle Industries, Inc., Lancaster, PA 17601
9. R. E. McHose, "Fast photomultiplier tube techniques", Burle Application Note TP-133.

FAST PHOTOCONDUCTIVE PHOTODETECTORS EMPLOYING IRON-DIFFUSION INTO EPITAXIAL GaAs

Jun Ohsawa, Nobuhiko Hashimoto,
Motohisa Nakamura, and Masatoshi Migitaka

Department of Information and Control
Toyota Technological Institute
2-12 Hisakata, Tempaku, Nagoya, 468 Japan

INTRODUCTION

Iron in gallium arsenide forms a deep acceptor level at 0.46 ev above the valence band.[1] This energy level can be used to overcompensate n-GaAs layers resulting in high-resistivity p-type material.[2,3] Although the level was identified as a hole trap,[4] it is expected in p-type materials to function as a recombination center rather than a trap. Hence fast response without a long tail can be realized. The photodetectors on the iron-diffused layers exhibit short decays limited by the carrier lifetime rather than the carrier transit time: Voltage pulses with full widths at half-maximum (FWHMs) as short as 72 ps were observed in spite of the large spacing of 15 μm between the fingers of interdigitated electrodes. An analysis revealed that the deconvolved FWHM is about 50 ps and the transit distance of electrons is less than 5 μm.[5] Then an improvement in responsivity can be expected with a finer electrode pattern fitted to the transit distance. This approach is important in increasing the "reduced" sensitivity of lifetime-limited devices. Here we report on the characteristics of the devices with a finger spacing/width (L/W) combination of 7μm/3μm in comparison with those of 15μm/5μm.

EXPERIMENT

An undoped GaAs layer of 2.5 μm thick was grown on an LEC-grown semi-insulating substrate by liquid phase epitaxy(LPE). The layer was n-type with the electron concentration of 5×10^{15} cm^{-3}. Iron diffusion was performed at 800 °C for 10 min from a spin-on-glass source. Iron diffused through the LPE layer to render it highly resistive p-type material (π-type). SIMS measurement revealed that the iron concentration was about 7×10^{16} cm^{-3}. AuZn ohmic electrodes were used to avoid the complication of current transport in the so-called MSM structure. The detection area is 100 μm $\times$ 200 μm for both

Applications of Photonic Technology
Edited by G.A. Lampropoulos *et al.*, Plenum Press, New York, 1995

electrode patterns, so that the number of fingers is doubled in 7μm/3μm combination. No anti-reflection coatings were applied on the detector surface. A chip mounted on a printed board was directly attached to an SMA connector, which was connected to a digital oscilloscope (HP54124T) through a bias tee (Picosecond Pulse Lab. 5550B). Optical pulses from a diode laser emitting at 783 nm were focused on the detector. The FWHM of the pulse was 39 ps, and the repetition rate was 10 MHz. Photoresponse was evaluated by observing the voltage pulse in the 50 Ω line.

RESULTS AND DISCUSSION

Figure 1 compares the photoresponses of detectors with and without iron diffusion at the same bias of 5V. The device on the undoped n-type layer exhibits a long fall time over 10 ns, which decreases as the bias is increased, suggesting that the response speed is dominated by the carrier(hole) transit time. Meanwhile the device on the iron-diffused layer shows a sharp decay with an FWHM less than 100 ps. The FWHM showed very weak dependence on the bias voltage, indicating that the carrier lifetime limits the response speed. The iron level is, as mentioned before, identified as a hole trap. In p-type materials, however, the level has already been filled with a hole, and the hole population at the level would not change after illumination. This means that no excess hole is trapped, and hence no subsequent photocurrent. On the other hand, the level depth of 0.46 eV is not at the midgap, but is still deep enough to be a recombination center for a carrier concentration as low as 10^{12} cm^{-3}. Thus we see the iron level in p-type materials should act as a carrier recombination center but not as a carrier trap.

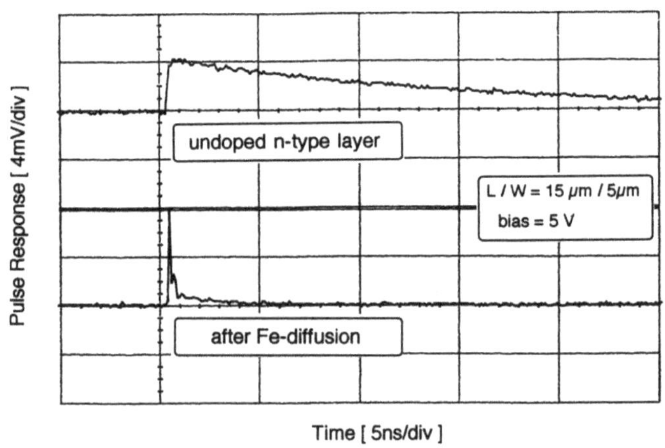

Figure 1. Comparison of the pulse responses of photodetectors on epitaxial GaAs with(bottom) and without(top) iron diffusion. Spacing and width of the electrode fingers are respectively 15μm and 5μm.

Figure 2 shows the dependence of responsivity on the average field intensity between the electrode fingers, whose spacing(L) is 15 or 20 μm. In a lifetime-limited device, the collection efficiency of photogenerated carriers mainly limits the responsivity. This means that the responsivity is proportional to the carrier transit distance during its lifetime, and hence to the electric field until the drift velocity saturates. The result of our experiment was that the responsivities varied almost in parallel with the electron transit distance in its lifetime, which was calculated by use of an expression for electron velocity.[6] The figure also indicates that the transit distance in the higher bias range amounts only to several micrometers even when we assume a longer lifetime of 50 ps. Then the finger spacing of 15 μm or more is too wide to collect the whole photogenerated carriers.

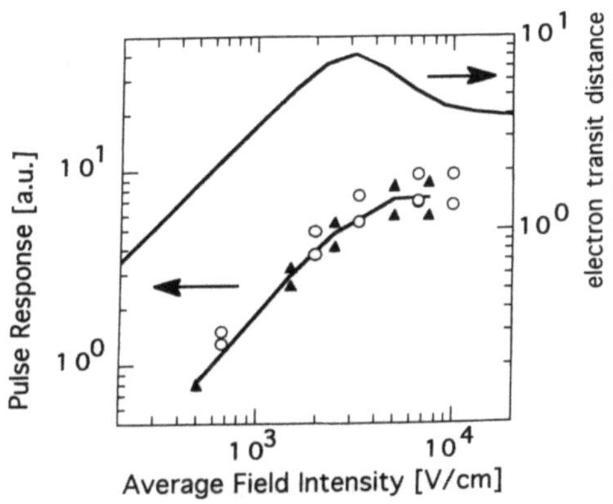

Figure 2 Pulse response vs. average electric field intensity as compared with a calculated electron transit distance in 50 ps.

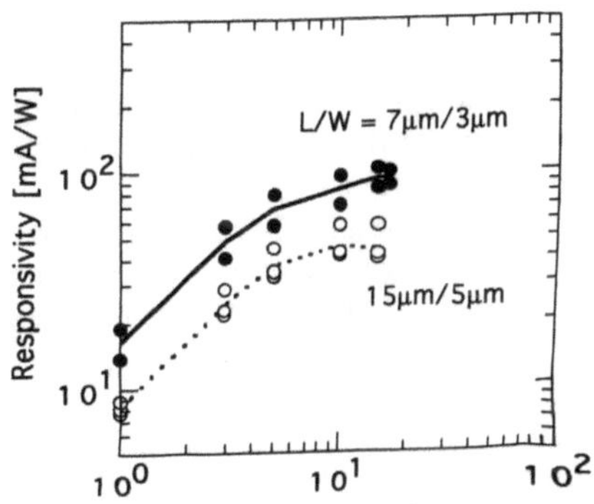

Figure 3 Responsivity vs. bias voltage showing the increased responsivity by use of a finer electrode pattern.

Figure 3 compares the responsivities of the devices with 7μm/3μm and 15μm/5μm combinations. With 7μm/3μm, it is doubled for all the voltage range measured. Since the exposed detecting area is about the same and the number of fingers is doubled, twofold increase is expected provided the electron drift velocity does not change. This is true in the higher bias range where the electron velocity saturates. At the lower bias, however, the simple drift model suggests it be proportional to the bias. In this case the increase should be fourfold. This discrepancy is most probably due to the non-uniform field distribution arising from a space charge region developed before the anode.[6] In fact the current-voltage characteristics become nonlinear in the bias range over 1 V.

The increase in responsivity was confirmed by low frequency measurements of spectral response, which result is shown in Fig. 4. Again twofold increase was obtained with the combination of 7μm/3μm. This is reasonable since the electron velocity and hence the transit time are saturated at the bias of 5 V for both configurations.

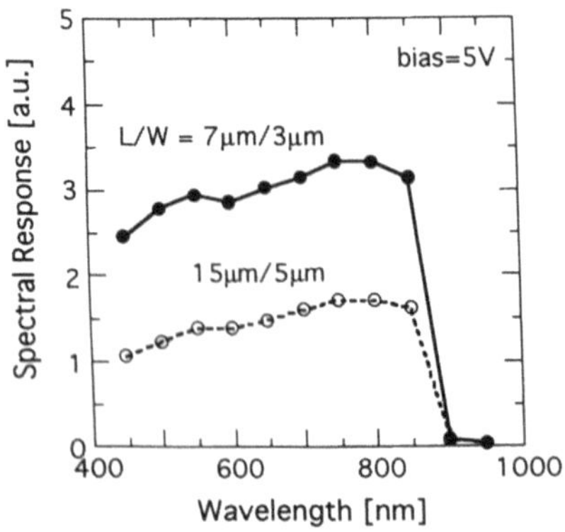

Figure 4 Spectral responses showing the increase by use of a finer electrode pattern.

CONCLUSIONS

In conclusion, we have demonstrated that overcompensation of epitaxial GaAs by the deep acceptors of iron can be used for the fabrication of fast photoconductive detectors, and that fitting the electrode spacing to the carrier transit distance is effective in enhancing the responsivity of lifetime-limited devices.

At present the device suffers from relatively large dark current of more than 1μA. This is because ohmic electrodes were used to examine the proper material characteristics of iron-diffused layer. Reduction of the dark current will be realized by adopting Schottky electrodes of appropriate metal instead of ohmic ones.

REFERENCES

1. R-S. Tang, S.B. Saban, and J.S. Blakemore, Melt-grown p-type GaAs with iron doping, *J. Appl. Phys.* 73:7416(1993).
2. J. Ohsawa, H. Kakinoki, H. Ikeda and M. Migitaka, Diffusion of iron into GaAs from a spin-on source, *J. Electrochem. Soc.* 137:2608(1990).
3. J. Ohsawa, P. Z. Lee, H. H. Wieder and M. Migitaka, Observation of a hole trap in liquid-phase epitaxial GaAs by intentional doping and space-charge-limited current analysis, *Jpn. J. Appl. Phys.* 31:L877(1991).
4. D.V. Lang and R.A. Logan, A study of deep levels in GaAs by capacitance spectroscopy, *J. Electron. Mater.* 4:1053(1975).
5. J. Ohsawa, N. Hashimoto and M. Migitaka, Carrier-lifetime-limited fast photoconductive photodetectors on iron-diffused GaAs, *Jpn. J. Appl. Phys.* 32:L1066(1993).
6. C.J. Wei, H-J. Klein, and H. Beneking, A study on the photoresponses in GaAs n-channel optical detectors, *IEEE Trans. Electron Devices* ED-29:1431(1982).

ULTRASENSITIVE SUB-NANOSECOND TIME-GATED DETECTION USING A SINGLE PHOTON AVALANCHE DIODE

Lloyd M. Davis[1] and Li-Qiang Li

University of Tennessee Space Institute
Center for Laser Applications
Tullahoma, TN 37388-8897
[1]e-mail: ldavis@utsi.edu

INTRODUCTION

Ultrasensitive detection and spectroscopy experiments require single photon detectors with sub-nanosecond timing jitter. Since the first experiments demonstrating the detection of single chromophore dye molecules in solution,[1] a significant improvement in the signal rate from each molecule has been achieved by using a high quantum efficiency single photon avalanche diode (SPAD). The EG&G SPCM-200 SPAD with custom-modified passively-quenched bias circuit has been characterized for this purpose[2] and has been found to provide a quantum efficiency of 55% at 532 *nm* and a single photon timing jitter of 168 *ps* when the input light is focused to a spot of diameter < 20 μm.

For single molecule detection in solution, the increased quantum efficiency of the SPAD yields about an order of magnitude increase in the photon count rate. A higher rate of fluorescence detection allows the use of a shorter processing time for each molecule and thus provides an important advantage for analytical applications, in which a large number of molecules are to be individually detected and counted.[3] Also, the high quantum efficiency can be used to obtain a larger number of fluorescence photons from each molecule, allowing more precise spectroscopic measurements to be made on each individual.[4]

SINGLE MOLECULE DETECTION IN SOLUTION

The experimental configuration used for detecting single molecules in solution is shown in Fig. 1. A modelocked laser producing picosecond pulses at 76 *MHz* is focused into the sample flow cell. There are many thousands of laser pulses during the transit time of each molecule through the beam. Since the fluorescence lifetime is shorter than the time between laser pulses, each passing molecule is cycled many times and generates a burst of fluorescence photons.

Applications of Photonic Technology
Edited by G.A. Lampropoulos *et al.*, Plenum Press, New York, 1995

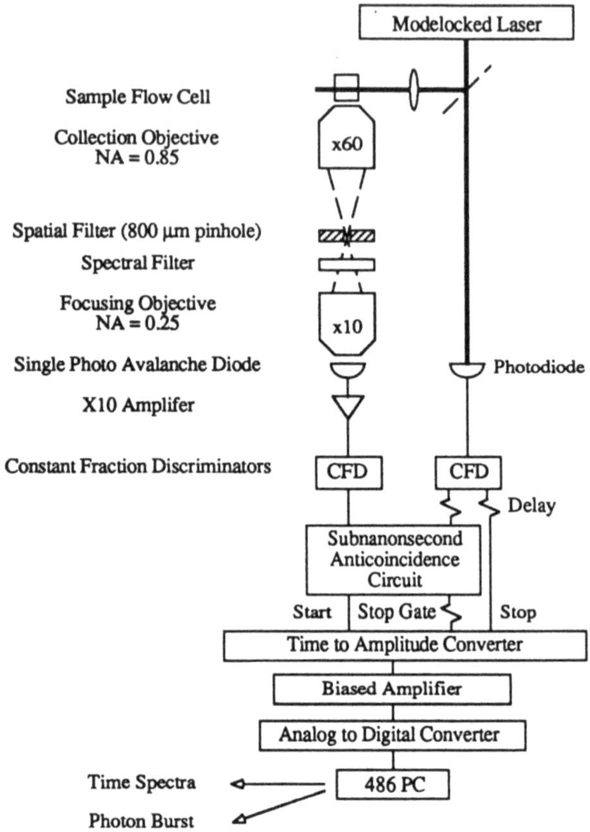

Figure 1. Optics and electronics for single molecule detection.

The laser beam is imaged through a 800 μm pinhole using a high numerical aperture microscope objective. The intersection of the field of view of the pinhole and the laser beam defines a probe region of volume ≈ 1.8 *pl*, inside which there are ≈ 6×10^{13} solvent molecules. The large number of solvent molecules gives rise not only to Rayleigh scatter, which is rejected using a spectral filter, but also to Raman scatter, which spectrally overlaps the weak single molecule fluorescence. Time-gated detection is used to discriminate promptly scattered Raman photons from delayed fluorescence photons.

The time delay between each laser excitation pulse, as measured by a photodiode, and each subsequently emitted photon, as detected by the SPAD, is measured using a time-to-amplitude converter (TAC). The output from the TAC is amplified, digitized and histogramed on a personal computer to generate the time spectrum. Fig. 2 shows the time spectra obtained from the solvent alone and from an ultradilute solution, together with the time window used for single molecule detection.

The computer allows the time window to be set using software and it also enables the stream of photons which occur inside the window to be processed by a digital filter and graphed in real time. Figure 3 shows photon bursts that have been processed by a Gaussian weighted sliding sum filter. Each burst indicates the transit of a single molecule of sulforhodamine 101. The transit time is only 0.63 *ms*, about 50 times shorter than that used in the original experiments.[1]

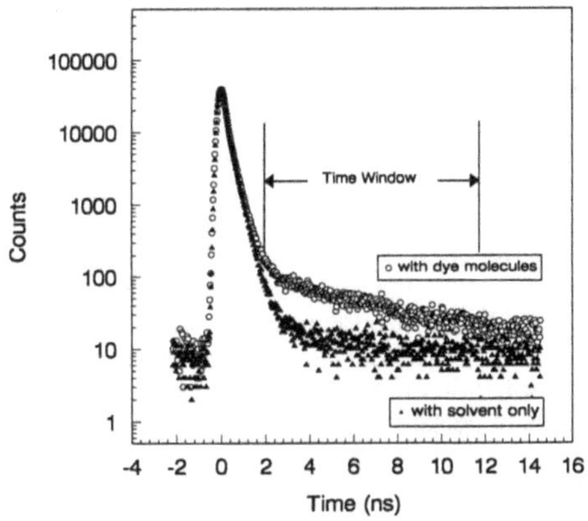

Figure 2. Time-gated detection.

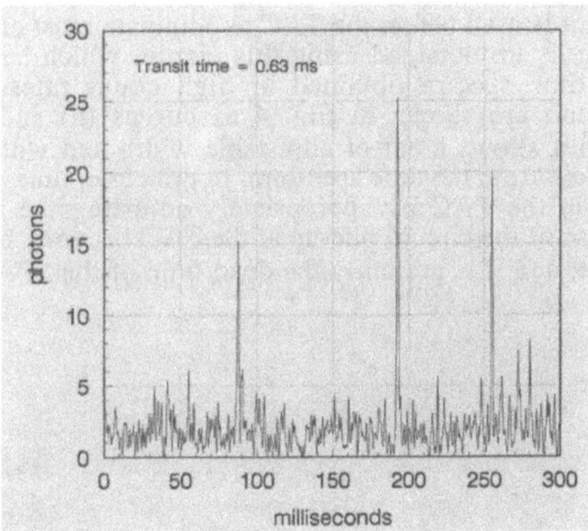

Figure 3. Photon bursts from single molecules of sulforhodamine 101.

DETERIORATION OF TIMING

Ultrasensitive detection and spectroscopy experiments can utilize the SPAD detector only if the probe region is small enough to be imaged onto the ≈ 100 *μm* central active region of the SPAD. If a high numerical aperture lens or objective is used to collect as much light as possible from the probe region, then another high numerical aperture lens must be used to focus this light onto the SPAD. In the single molecule detection experiment of Fig. 1, the diameter of the probe region, ≈ 13 *μm*, is defined by the × 60 collection objective and the 800 *μm* pinhole of the spatial filter. The light which passes through the pinhole is then imaged using a × 10 microscope objective to a disk of diameter ≈ 80 *μm* at the center of the active region of the SPAD. As noted in Ref. 2, the timing of the SPAD deteriorates as the area of illumination increases, even if all of the light falls within the active region. Thus, as seen in Fig. 2, the width of the prompt has increased from its optimal value of 168 *ps* full width at half maximum to ≈ 250 *ps*.

Internal mounting of the SPAD closer to the window of its TO-8 package would permit the use of a smaller working distance higher magnification objective, which would give tighter focusing of the collected light onto the active area and hence faster time response. This would permit the time window to be set earlier for greater fluorescence throughput, it would permit the detection of dye molecules with shorter fluorescence lifetimes and it would enable the use of larger probe regions in certain experiments.

DEAD TIME OF ELECTRONICS

In addition to increasing the rate of fluorescence photons, the high quantum efficiency of the SPAD gives an equally large increase in the rate of Raman photons. Although these are subsequently eliminated using time-gated detection, the high rate of Raman counts gives rise to significant pile-up at the TAC, which has a dead time of $D \approx 5\ \mu s$. For a laser power of ≈ 14 *mW*, the rate of Raman counts is $R \approx 3 \times 10^5\ s$-1, so that the loss of counts due to pile-up, given by $1 - e^{-RD}$, is ≈ 75%.

To avoid this problem, as shown in Fig. 1, a locally constructed subnanosecond anticoincidence circuit is used before the TAC to eliminate most of the Raman counts. The results shown in Fig. 3 are obtained using this circuit, which has a dead time of a few nanoseconds. The time spectra obtained at high count rates without and with the anticoincidence circuit are shown in Fig. 4 as curves (a) and (b) respectively. The anticoincidence circuit allows a cut of adjustable width and with a precision of tens of picoseconds to be taken from the time spectrum. In principle, time-gated detection could be accomplished without the TAC by appropriately adjusting the anticoincidence circuit. However, with the use of the circuit, pile-up at the TAC becomes negligible, even during a burst of fluorescence, and it is presently the dead time of the SPAD that limits the signal throughput.

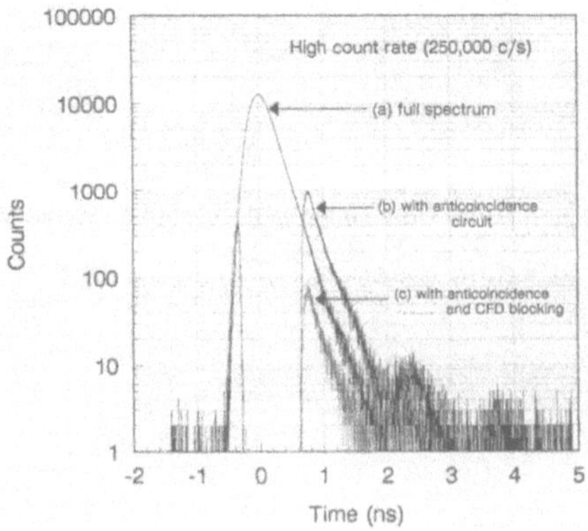

Figure 4. Time spectra obtained with anticoincidence circuit.

DEAD TIME OF SPAD

The passive quenching circuit of the SPAD causes the bias to quickly drop after each photon is detected so as to quench the avalanche current. The SPAD will not detect subsequent photons until after the bias has sufficiently recharged.

The dead time of the SPAD is measured using a constant light flux which gives a mean count rate of $\approx 2 \times 10^{-5}$ s^{-1} by recording and overlaying successive sweeps of the SPAD output voltage using a boxcar integrator. As shown in Fig. 5, the SPAD is completely unresponsive for $\approx$ 200 *ns* and has reduced response as the bias voltage recovers with a time constant of $\approx$ 700 *ns*. If the bias voltage is reduced, the SPAD is found to give lower quantum efficiency and output pulses of smaller voltage. Furthermore, the prompt shifts slightly to later times. Thus, if a photon is detected soon after a prior photon, it will give an output pulse that is smaller and somewhat delayed.

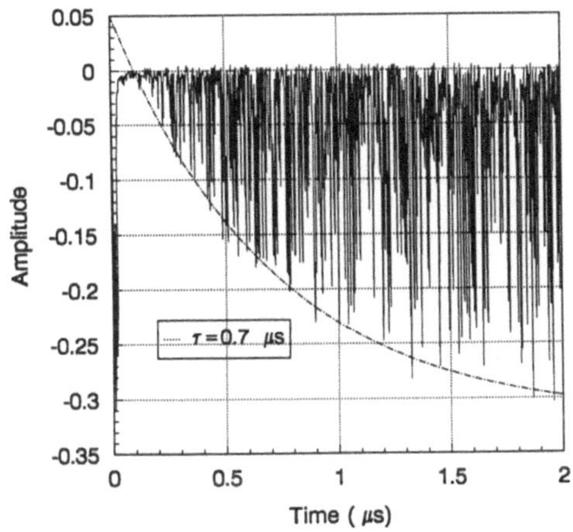

Figure 5. SPAD dead time.

From curve (b) in Fig. 4, it may be noted that when the anticoincidence circuit is used, the tail of the prompt increases. This effect is due to photons which are detected during the time when the bias has not fully recovered. As the count rate is lowered, the tail of the prompt decreases because the probability of detecting a photon during this time becomes less. Even at high count rates, photons which are detected soon after prior photons can be ignored simply by increasing the blocking width of the constant fraction discriminator that is used to condition the SPAD pulses. If the blocking width is adjusted to 700 *ns*, the tail of the prompt drops to that shown as curve (c) in Fig. 4. This permits the window for the time-gated detection to be set at earlier times, giving greater throughput of fluorescence. However, fluorescence photons that occur during the dead time imposed by the blocking width are then lost.

FUTURE IMPROVEMENTS

The laser power used to obtain Fig. 3 was limited by the blocking width dead time rather than the saturation of the absorption of the single molecule. For a dead time of 700 *ns*, almost 20% of the counts are lost when the rate is $\approx 3 \times 10^5$ s^{-1}. A Monte Carlo simulation of the single molecule detection experiment[5] predicts that if the dead time of the SPAD could be reduced to < 50 *ns*, a further order of magnitude increase in the processing rate of single molecules could be obtained by increasing the laser power. Ongoing research on an actively quenched bias circuit for the EG&G SPCM-200 SPAD indicates that this should be realizable in the near future.

REFERENCES

1. E. B. Shera, N. K. Seitzinger, L. M. Davis, R. A. Keller, and S. A. Soper, "Detection of single fluorescent molecules," *Chem. Phys. Lett.* 174:553 (1990).
2. L. Q. Li and L. M. Davis, "Single photon avalanche diode for single molecule detection," *Rev. Sci. Instrum.* 64:1524 (1993).
3. L. Q. Li and L. M. Davis, "Single molecule detection for analytical applications," *Opt. Soc. Am. Tech. Dig.* 16:1 (1993).
4. S. A. Soper, L. M. Davis, and E. B. Shera, "Detection and identification of single molecules in solution," *J. Opt. Soc. Am. B* 9:1761 (1992).
5. L. M. Davis and L. Q. Li, "Monte Carlo model of a single molecule counting experiment," *Opt. Soc. Am. Tech Dig.* 5:206 (1994).

PHOTON COUNTING USING A LARGE AREA AVALANCHE PHOTO-DIODE COOLED TO 100 K

Nathan G. Woodard, Eric G. Hufstedler, and Gregory P. Lafyatis

Department of Physics
The Ohio State University
Columbus, OH 43210

INTRODUCTION

We have recently succeeded in counting single photon signals using 5 mm diameter avalanche photodiodes (APDs). While photon counting with APDs is not new [1], our work represents the first success with *large area* devices. For light from the ultraviolet to the near infrared the expected photon detection efficiency is outstanding; this combined with the large area of the device make it a very attractive detector for many applications. As we describe below, in our particular application, our system outperforms the best commercially available photomultiplier tubes by well over an order of magnitude.

We reported initial results in Woodard, Hufstedler, and Lafyatis, [2] which we refer to as "I." In the present paper we begin by outlining the application behind our interest in photon-counting with large area APDs. We then go on to discuss important technical aspects of our APD work. Next, we present results. We have begun to try to interpret our results, theoretically, and note some difficulties in these efforts. Most recently we have experimentally addressed the possibility of "afterpulsing" in devices cooled to LN_2 temperatures and find very encouraging results for most applications. We conclude with a summary.

The goal of our application is to use the "photon burst" method to detect individual potassium atoms in an atomic beam. The basic idea is to use a CW laser to repeatedly excite a potassium atom in the atom beam as it passes in front of a detector and then detect as many of the fluorescence photons as possible. If our light collection and detection system is sufficiently efficient, we will be able to tell when an atom passes because *several* photons, a "burst," will be detected in the short, $\sim 20\mu s$ time the atom is actually in front of the detector. A CW laser excited potassium atom can scatter about 200 fluorescence photons, of which we can collect 20%. The best PMT available (multi-alkali photocathode) has a quantum efficiency of 2% at 770 nm.[1] This means, on the average, 0.8 photons will be detected by a PMT as an atom passes: *not* a burst.

[1]The GaAs PMTs can have up to 14% QE, but they have a very narrow acceptance angle for light and actually prove inferior detectors for this application. See I.

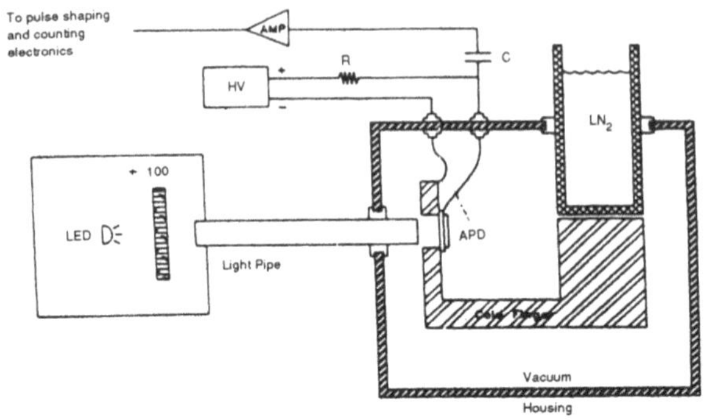

Figure 1. The experimental apparatus.

We get a photon detection efficiency of about 50% at 770 nm with our APD detector. For this detector, we can expect to see bursts of 20 photons from an average atom as it passes: the large area photon counting avalanche photodiode makes this application possible.

EXPERIMENTAL

We begin with some definitions. Note: we are defining terms a little differently than was done in I, but more in harmony with standard usage. First, we operate our APDs in "sub-Geiger" mode (called "gain mode" in I). That is, the avalanche created by an initial photoelectron is *not* self-sustaining, but rather produces a finite number of electrons at the device's anode. We are ultimately interested in *photon detection efficiency*, "P_d," the probability our *system* detects a photon.

$$P_d = \eta P_p. \quad (1)$$

Here η is the (device) quantum efficiency — the probability that a photon creates an electron-hole pair and P_p is the probability that the pair is detected.

Fig. 1 is a diagram of our experimental apparatus. An APD is mounted in vacuum to a liquid nitrogen cooled "cold finger." "100 K" refers to the temperature we measure in the APD mounting block when the reservoir is filled with LN_2. A red LED is used as a signal source. Its light is introduced into the vacuum and onto the device in a controlled fashion using a glass "light pipe." The data in this paper were taken with about 8 kHz of photons incident on the device.

Photon-generated pulses from the APD anode are capacitively coupled into a charge-sensitive preamp followed by pulse shaping and counting electronics. We typically work with symmetric pulses having 1 μs rise and fall times. The charge noise of our amplifier system (referred to the input) is about 300 electrons rms. To measure gains and dark currents, we insert an electrometer in the ground (negative) leg of the power supply circuit. Gain measurements, for example, are made by comparing electrometer currents from the device when it is operated at high ($\sim$ 2000 V) and low (100 V) bias voltages. We determine the absolute photon flux on the device by operating it as a conventional photodiode (i. e. 100 V bias) and using the manufacturer's quantum efficiency calibration. Using a pulse height analyzer, we find the distribution of avalanche sizes created by individual photon events. We calibrate our pulse height system by injecting test pulses of known magnitude at the input of the charge sensitive preamp. Good shield-

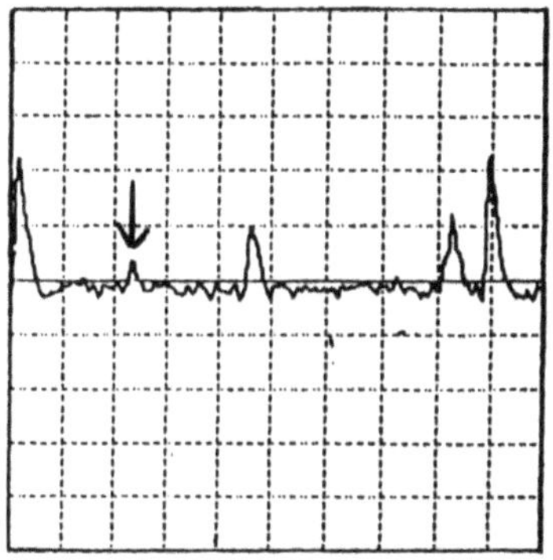

Figure 2. An oscilloscope trace showing single photon pulses. The horizontal scale is 10μs per box. The pulse indicated by the arrow could be either a small signal pulse or a large noise pulse.

ing practices are essential in this work. Our apparatus is above a ground plane and critical areas are carefully shielded. We do *not* work in a shielded room.

We use 5 mm diameter APDs obtained from Advanced Photonix. They were specially selected for "high gain." These are bare chips: the last time we tried them, we found packaged devices to be susceptible to catastrophic thermal shocks. At room temperature, these devices are too noisy to be used for photon counting. Cooling the devices with LN_2 virtually eliminates the detector noise except for a small number of amplified pulses. In going from room temperature to 100 K, we find:

- Small changes in η. At 650 nm, η is down (at 100 K) 18% from the room temperature value. At 770 nm, η *increases* by about 5% upon being cooled.
- Roughly the same usable maximum gain before breakdown,
- A substantial reduction in the dark current: at a gain of 4000, the dark current drops from 3 μA to 1 pA.

RESULTS AND DISCUSSION

Fig. 2 is an oscilloscope trace showing single photon pulses. The figure also suggests our major experimental difficulty: many pulses are small and in the "hair" of the amplifier noise. Is the pulse indicated by the arrow a real photon generated pulse or a peak in the noise? Neither we nor our electronics can say for sure. So far as we can tell, the contribution of the APD to the noise (at LN_2 temperatures) are dark counts (amplified events) and the expected increase in the charge preamp noise due to the 15 pF of device capacitance. For this system, the photon detection efficiency is a strong function of where we set the discriminator level of our pulse counting electronics. The trade-off is that we can get a higher detection efficiency if we use a lower discriminator setting but we will pass more false counts from the amplifier noise. In our "photon-burst" application, high quantum efficiency is at a premium and false counts are not much of a problem, so long as there are not many in any 20 μs interval. Thus we would choose a low discriminator setting. A setting of 1200 electrons, for example,

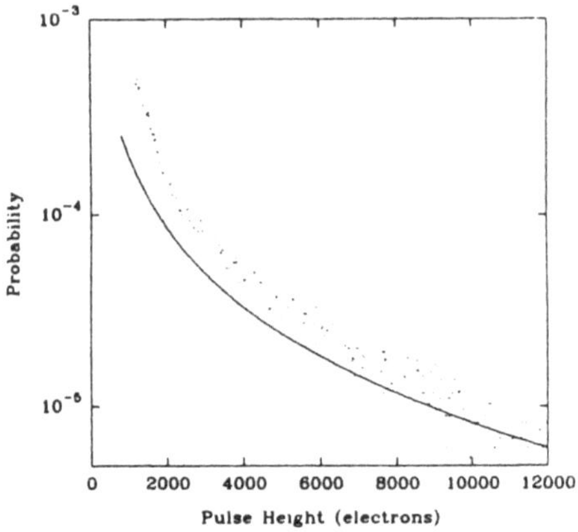

Figure 3. The probability that a photon generated electron hole pair will result in a given pulse height. The dots correspond to data collected with a measured gain of 4000. The solid line shows McIntyre's prediction assuming $k = 0.002$ with a gain of 4000.

gives us a photon detection efficiency of 50% (see below for caveats) and about 5 kHz of false counts. This represents an improvement of a factor of *twenty-five* over what can be done with the best commercially available PMT technology. For applications in which false counts are a problem, one could operate with a discriminator setting of 2000 electrons. Here we see a photon detection efficiency of 25% — still much better than present PMTs — with 300 Hz false counts.

The potentials and limitations of this way of doing things appear fairly well defined at this point. Assuming we stick with 1 μs pulse shaping time constants, by mounting the charge sensitive preamp's input FET to the cold finger and getting state of the art performance from our electronics, it should be possible to reduce our noise by about a factor of four. This would get most of our photon pulses out of the amplifier hair for the present detector. What are the prospects for larger area detectors? The principal engineering problem is that larger detectors necessarily have larger capacitances ($\sim$ detector area). The larger capacitances will lead to more noise at the charge preamp's input; in the regime of interest, the rms charge noise goes a little bit less than linearly with the device capacitance. Practically speaking, low noise, high efficiency devices up to about one cm in diameter seem ultimately realizable; much larger detectors could be constructed from several 1 cm sized channels in parallel.

We have begun trying to interpret, theoretically, our results. We compare to the McIntyre theory, which has in the past provided a very useful framework for understanding APD behavior.[3] The McIntyre theory predicts the pulse height distribution of single photon generated avalanches in terms of two parameters: M, the average gain of the device, and k, the ratio of hole to electron ionization probabilities. We have measured this pulse height distribution. As noted above, we measure the average gain of our device. We know of no relevant data for k at LN_2 temperatures. But if we use a typical room temperature value, $k = 0.002$, we may use the theory to make predictions. Fig. 3 is a comparison between theory (line) and our measurements (dots). Though theory and experiment show the same qualitative behavior, that is, there are more small pulses than big ones, the general (lack of) agreement between the two is not very satisfying. By picking other values of k, we can essentially make the theory match the data above 2000 electrons. We are unable to reconcile our measurements around 1500 electrons with any simple version of the McIntyre theory: the discrepancy is at the

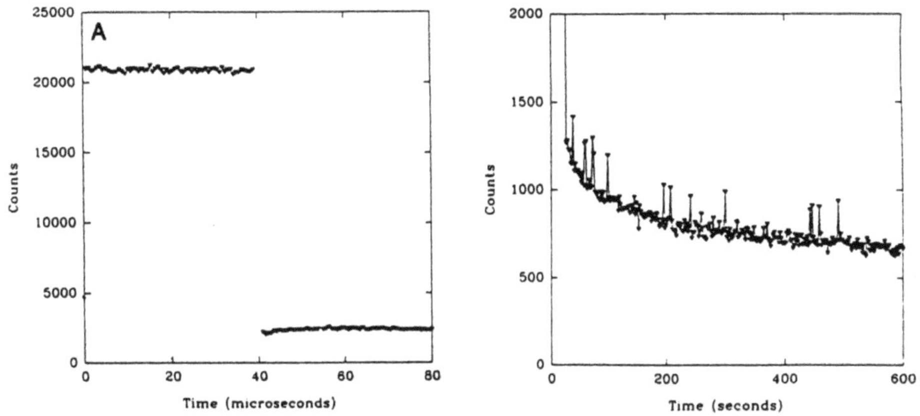

Figure 4. A- Multichannel scalar data showing no measurable after-pulsing with 640ns bins. (The light is turned off abruptly at 40µs.) B- Long time scale after-pulsing data. The light is turned off at 25 seconds. The multichannel scalar is set for two second bins. The last bin with light on (not shown) contained 71,000 counts.

"factor of two level." The failure to have a satisfactory theoretical interpretation of our measurements remains an "outstanding problem." On the experimental side, we note the discrepancy occurs for pulses in the noise hair and there may be some subtle issues. We (electronically) make corrections for dead time, baseline restoration, and consider pulse pile-up. As we discuss shortly, "afterpulsing" does not explain the discrepancy. Our present direction is to reduce the amplifier noise and then revisit the issue.

Most recently, we have experimentally investigated the possibility of "afterpulsing" in the cooled APD. The idea is that during a photon generated event, a carrier may become "trapped" for some period of time and released later to generate a second avalanche. For many applications afterpulsing would represent a severe device drawback. For example, in our application we are most sensitive to afterpulsing occuring one to twenty microseconds after the initial avalanche — two pulses within twenty microseconds look to us like a little "burst." On the other hand, the electronics integrate any charge created within a microsecond of the initial avalanche, so we are insensitive to "prompt" afterpulsing. Afterpulsing on a time scale much longer than twenty microseconds would appear to us as a relatively unimportant increase in dark counts.

We have studied afterpulsing by observing the arrival rate of pulses following the rapid turnoff of our LED light source. Experimentally, we feed photon pulses from the APD into a multichannel scalar which is set to cyclically scan through its channels. During the first several channels, the LED shines about 150kHz of photons on the APD and is then rapidly shut off (100 ns). Excess dark counts in succeeding channels are identified as "afterpulses." Fig. 4a shows representative data taken to probe 2 to 100 μs time scales. The sharp turnoff seen in the APD signal indicates that to the limit of our ability to measure, there is *no* afterpulsing at these time scales. The data of Fig. 4a indicates that fewer than 1% of all photon-generated avalanches produce an afterpulse within 100 μs. We have performed this experiment on several different time scales and generally find null results. The exception is for extremely long time scales. In Fig. 4b we make the same measurement, but now observe for several minutes after the light was turned off. We see that after the LED was turned off (at 25 seconds) there is a small residual signal that decays with a characteristic time of a couple of minutes. This long time scale afterpulsing is the *only* type we have observed. Quantitatively, about 0.008 of all photon-generated avalanches result in an afterpulse that is released on a time scale of minutes.

CONCLUSION

We have demonstrated photon counting using large area APDs cooled to LN_2 temperatures. For many applications, these are superior detectors to their PMT counterparts. We find good photon detection efficiency: competitive with PMTs at all wavelengths — truly superior in the near infrared. There exists a "factor of two" discrepancy between our measurements and the present theoretical understanding of APD operation. We looked for afterpulsing on timescales from microseconds to hours and generally found none, with the exception that about 1% of the photon events produced an afterpulse on a "minutes" timescale. For ours (and we expect most) aplications, this is fairly harmless. Future improvements in devices (specifically — larger gains) and the use of lower noise electronics should enhance these already very good detectors.

ACKNOWLEDGEMENTS

We acknowledge the invaluable assistance of Chuck Rush with our electronics. This research was supported by the MISCON program of DOE. G. P. L. acknowledges the support of an NSF Presidential Young Investigator Award. E. G. H. was supported by an NSF REU grant.

References

[1] J. Conradi, IEEE Trans Electron Devices **ED-19**, p713 (1972). R. J. McIntyre, Measurement **3**, p146 (1985).

[2] N. G. Woodard, E. G. Hufstedler, and G. P. Lafyatis, Appl. Phys. Lett. **64**, p1177 (1994).

[3] R. J. McIntyre, IEEE Trans Electron Devices **ED-19**, p703 (1972).

ULTRA-STABLE NARROWBAND FILTERS FOR TELECOMMUNICATION APPLICATIONS

Allan J. Waldorf,[1] Brian T. Sullivan,[1] Llynne M. Plante[1] and David Cushing[2]

[1]National Research Council of Canada
Institute for Microstructural Sciences
Montreal Road, Ottawa, Ontario, Canada K1A 0R6
[2]JDS FITEL Inc., 570 Heston Drive,
Nepean, Ontario, Canada K2G 5W8

INTRODUCTION

Wavelength division multiplexing (WDM) techniques are commonly used in fiber optic telecommunications where the information transmitted down the fiber is carried on different channels at specific wavelengths. As the number of channels increase, and the wavelength separation between the channels decrease, it becomes more challenging to isolate one wavelength channel from another.

The next generation of WDMs will require temperature stable wavelength separations of 0.3 nm to 4 nm for high speed multi-channel telecommunication systems. Filters can play an important role in WDM transmission systems if they are stable over time and stable with respect to changes in the temperature and humidity. Hence, the challenge for thin film manufacturers is to fabricate narrowband filters with a narrow square bandpass, a low bandpass loss and a high temperature stability.

Filters fabricated with conventional resistive heating or e-beam techniques can have, with the proper materials, excellent stability over time (less than 0.1 nm wavelength shift over 3 to 10 years). However, these filters usually have a relatively high fractional wavelength shift/°C, $\alpha = 1/\lambda \cdot d\lambda/dT$, which makes them unsuitable for the next generation of WDM filters. More energetic thin film deposition processes such as sputtering, ion-assisted evaporation and ion-plating result in denser thin films that have near bulk-like properties. Filters deposited with these energetic processes have a smaller fractional wavelength shift/°C, and, hence, are more stable with respect to temperature.[1]

Narrowband single and double cavity filters, with 3 dB bandwidths ($\Delta\lambda$) ranging from 0.5 to 10.0 nm, have been fabricated for the wavelength region of 1550 nm using reactive ion plating. The single cavity filters are essentially Fabry-Perot structures with a half-wave spacer (cavity) layer sandwiched between two high reflectance quarterwave stacks.[2] The desired bandwidth and rejection region for a single cavity filter was obtained by adjusting the reflectance of the two stacks and the thickness of the spacer layer. The double cavity filter, corresponding to two Fabry-Perot structures coupled together by a quarterwave layer, provides a steeper transition from the bandpass to the rejection region. The next section describes the deposition of these narrowband filters and the setup used to characterize the temperature stability of these filters. The following section describes the results and this is followed by the conclusions.

EXPERIMENTAL

A Balzers BAP 800 reactive ion plating system at the National Research Council of Canada was used for the deposition of the reactive ion-plated narrowband filters.[3] Briefly, this system contains a plasma source and two e-beam guns. After an e-beam source has been heated up, argon and oxygen are introduced into the chamber and a low voltage (~60V), high current (~60A) plasma arc is ignited between the plasma source and the e-beam source (anode). Along with the argon and oxygen, a fraction of the evaporant flux is ionized. The positive ions are then accelerated towards the negatively self-biased substrate and the bombardment of the growing thin film by these energetic ions and neutrals results in a film density similar to that of the bulk material.

The narrowband filters were deposited on 150 × 150 × 2 mm glass substrates. During deposition, the substrates were rotated to improve the thickness uniformity. In addition, the substrate-to-source distance was adjusted until a thickness uniformity better than 0.1% was achieved over a 75 mm diameter area. Most-sensitive-wavelength optical monitoring was chosen to accurately deposit the required layer thicknesses.

Two high index materials, Ta_2O_5 and Nb_2O_5, were studied to determine which material in combination with SiO_2 provides the best performance. The narrowband filters should ideally have a minimum overall stress, low absorption and low scattering losses along with a high refractive index ratio to minimize the number of layers required.

To determine the maximum peak transmittance that could be achieved with the above materials, two single cavity filters were designed for a bandwidth of 0.5 nm at a peak wavelength of ~1540 nm. The first filter was fabricated using Ta_2O_5 and SiO_2 while Nb_2O_5 and SiO_2 were used for the second filter. Next, double cavity filters with 1 and 10 nm bandwidths were constructed using Ta_2O_5 and SiO_2 for a peak wavelength of 1553 nm. After deposition, the filters were then measured either with an Anritsu MS9701A or an Ando optical spectrum analyzer. The single cavity filters were then annealed in ambient air at 400 °C for up to twelve hours and then remeasured.

To measure the thermal stability of the single cavity narrowband filters the following setup was used : light from a tungsten source was focused onto a 9 mm core diameter single mode fiber. The light exiting this fiber was collimated with a JDS fiber optic beam expander, transmitted through the filter under test, and was focused onto another fiber connected to an Ando optical spectrum analyzer. The collimator and filter were placed in an oven and the temperature was varied from 25 to 85 °C. The wavelength shift in the filters was measured and the fractional wavelength shift/°C, accurate to ± 10%, was calculated.

The experimental setup, material deposition conditions and the optical monitoring strategy will be discussed in more detail in another paper.

Table 1. As-deposited and annealed single cavity filters.

Materials	Layers	λ_{Peak} [nm]	T_{Peak}[%]	$\Delta\lambda$ [nm]	State
Ta_2O_5 / SiO_2	35	1540.4	80±5	0.55	as-deposited
		1545.2	80±5	0.64	annealed
Nb_2O_5 / SiO_2	31	1547.7	80±5	0.54	as-deposited
		1554.6	92±5	0.57	annealed

Table 2. As-deposited double cavity filters.

Material	Layers	λ_{Peak} [nm]		$\Delta\lambda$ [nm]		T_{peak}[%]	
		Desired	Measured	Desired	Measured	Desired	Measured
Ta_2O_5 / SiO_2	40	1578	1580	10.0	10.0	100	>95
Ta_2O_5 / SiO_2	63	1558	1558	1.0	1.0	100	70

RESULTS AND DISCUSSION

A summary of the transmittance measurements, made with respect to uncoated substrates, of the as-deposited and annealed single cavity narrowband filters are listed in Table 1. For both filters, annealing resulted in a permanent peak wavelength shift of ~5 nm and an increase in the bandwidth. The annealed Nb_2O_5 filter transmittance had a 0.4 dB insertion loss, while the passband transmittance of the as-deposited and annealed Ta_2O_5 filter had a 1.0 dB insertion loss.

The as-deposited 1 and 10 nm double cavity Ta_2O_5/SiO_2 filters are shown in Fig. 1 and a summary is listed in Table 2. The 10 nm filter included an AR coating and had an insertion loss less than 0.2 dB. The first attempt at fabricating the 1 nm filter was successful with a passband transmittance of 70% being achieved at the desired wavelength and with the desired bandpass. It should be possible to increase the passband transmittance even further if the layer thickness errors can be reduced with improved optical monitoring strategies.

Table 3 lists the fractional wavelength shift/°C for single cavity filters deposited by e-beam, resistive heated sources and reactive ion-plating. The large negative wavelength shift (41.0 nm) observed for the e-beam filter, deposited on a BAK 760 deposition system, arises from the water vapour content of the porous films in this filter being affected by humidity changes in the oven as the temperature is increased. By comparison, the ZnS/ThF_4 filter with a ZnS spacer, deposited at JDS Fitel using resistive heated sources, shifted by only 1.7 ± 0.02 nm over the same temperature range. However, the best performance was achieved by the reactive ion-plated filters which only exhibited a 0.16 to 0.19 ± 0.02 nm wavelength shift, an order of magnitude improvement over the filters deposited by resistive heated sources. The value of $\alpha \sim 1.9 \pm 0.2 \times 10^{-6}$ for the reactive ion-plated filters is nearly a factor of two better than a previously reported value of $\alpha \sim 4 \times 10^{-6}$ for ion-assisted e-beam deposited narrowband filters. Annealing the reactive ion-plated filters did not change the temperature stability of these filters significantly.

In a temperature/humidity exposure test, a 10 nm Ta_2O_5/SiO_2 double cavity filter was subjected to 75 °C and 85% relative humidity for a period of 1 year. No significant differences in the spectral characteristics of this filter were found after this period of time.

Table 3. Fractional wavelength shift/°C for filters deposited by different processes.

Process	$\Delta\lambda$ [nm]	Materials	α [°C^{-1}]	
			as-deposited	annealed
e-beam	50.0	ZrO_2/SiO_2-SiO_2 spacer	-4.3×10^{-4}	—
resistive heated	10.0	ZnS/ThF_4-ZnS spacer	$+1.74\times10^{-5}$	—
ion-plating	0.6	Ta_2O_5/SiO_2-SiO_2 spacer	$+1.83\times10^{-6}$	$+1.85\times10^{-6}$
ion-plating	0.6	Nb_2O_5/SiO_2-SiO_2 spacer	$+1.98\times10^{-6}$	$+1.73\times10^{-6}$

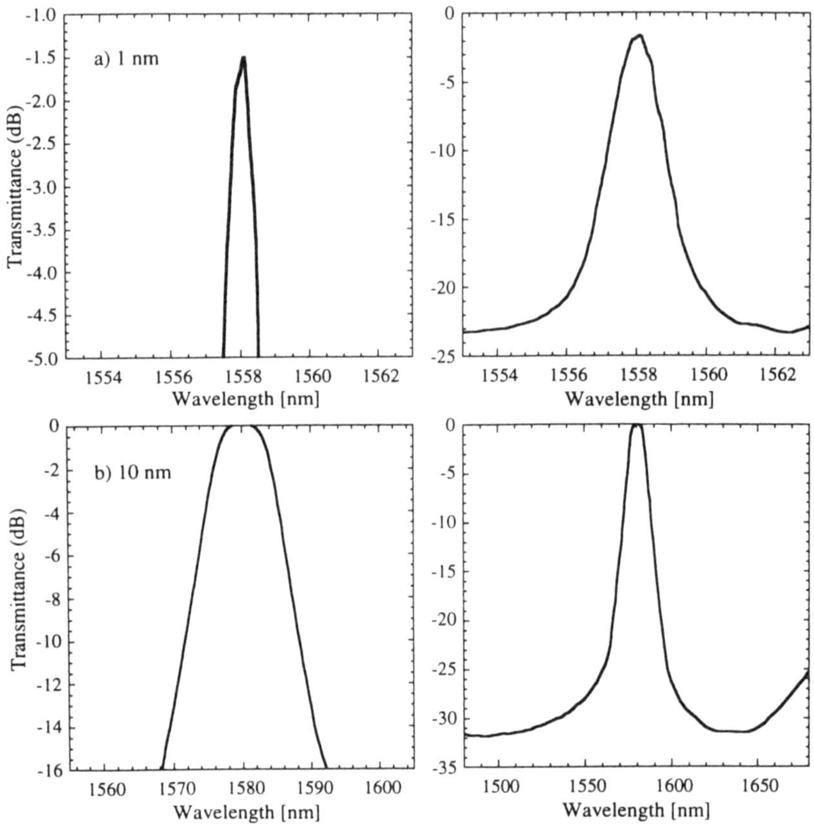

Figure 1. Measured transmittance of a a) 1 nm and b) 10 nm Ta_2O_5/SiO_2 double cavity filter.

CONCLUSION

Narrowband single cavity filters, with bandwidths of 0.6 nm, and double cavity filters, with bandwidths of 1 and 10 nm, have been fabricated using reactive ion-plating for the wavelength region of 1500 to 1600 nm. The transmittance measurements of these filters and their small wavelength shift (≤ 0.2 nm from 25 to 85 °C) indicate that the single and double cavity narrowband filters deposited by reactive ion-plating are suitable for the next generation of WDMs.

ACKNOWLEDGEMENT

We would like to thank Dr. P. Myslinski for performing some of the transmittance measurements on the narrowband filters.

REFERENCES

1. J. R. Potter and J. C. Simons, Stability of IAD refractory oxide narrowband interference filters, *SPIE* 1952:186 (1993).
2. H. A. Macleod, "Thin Film Optical Filters," 2nd ed., Macmillan, New York (1986).
3. A. J. Waldorf, J. A. Dobrowolski, B. T. Sullivan, and L. M. Plante, Optical coatings deposited by reactive ion plating, *Appl. Opt.* 32: 5583 (1993).

A LIGHT CONTROLLED OPTICAL MODULATOR IN SILICON TECHNOLOGY

Z. Xiao and O. Engström

Department of Solid State Electronics
Chalmers University of Technology
S-412 96, Göteborg, Sweden

ABSTRACT

A light controlled optical modulator is demonstrated which is based on the micromachining and thermal bonding of silicon. It consists of two silicon wafers, one n-type and one p-type bonded to each other so that a pn junction is formed. Using micromachining technique, an optical cavity is prepared with a wall depth of 1 μm bordered on one side by a membrane of about 8.5 μm thickness. Using two light sources with different wavelengths, the intensity of transmitted light from one light source, can be controlled by non-penetrating light from the second light source. The device structure is analyzed by simulation.

INTRODUCTION

A number of different physical effects and materials have been used in recent years to obtain solid state devices for the modulation of light. They include electric field induced variation of refractive index in lithium niobate [1], Faraday rotation in magnetooptical materials [2], polarization effects in liquid crystals [3, 4] and more recently quantum confined Stark effects in III-V related materials [5]. Silicon, known as an optically passive material, has a possibility to play a role on this stage by utilizing micromachining. Modulators prepared as deformable gratings or as interferometers have been presented in the literature [6 - 8]. In the present paper we demonstrate a novel silicon device for optical control of the modulation of light. The device structure is formed as an interferometer with a cavity positioned at the metallurgical interface of a thermally bonded p-n junction.

DEVICE FABRICATION

The modulator device is prepared from a 1673-2065 ohm-cm, n-type silicon wafer, *A*, with (100) orientation and a 0.015-0.025 ohm-cm, p-type wafer, *B*, with (100) orientation joined together by thermal bonding as shown in Fig. 1. To prepare the cavity step, we used an earlier described multi-oxidation technique [7]. In this procedure, wafer *B* was furnished with a thermal oxide. Openings in the oxide were made by lithography and etching. The structure thus obtained was oxidized again followed by oxide stripping over the whole wafer surface. As the oxide growth is faster in the openings of the original oxide layer than on the oxidized areas of the surface, steps with well controlled depths were obtained in the openings after oxide stripping. The cavities were formed by thermally bonding the n-type silicon of wafer *A* to the p-type silicon of wafer *B*. The cavity depth was about 1μm measured by a Tencor Long Scan Profiler. In wafer *A*, precise etching was performed to get a thin silicon membrane. The preferential etch used for this step was a mix of ethylenediamine, pyrocathecol, pyrazine and water, known as "EDP" etch. Alignment of the two wafers to each other was performed using infrared light when bonded at room temperature. The thermal annealing was made in nitrogen atmosphere at 900°C for 30 minutes. No contact doping was made. The device cavity was hermetically closed from the surrounding with lateral membrane dimensions of 1.0 x 1.0 mm^2.

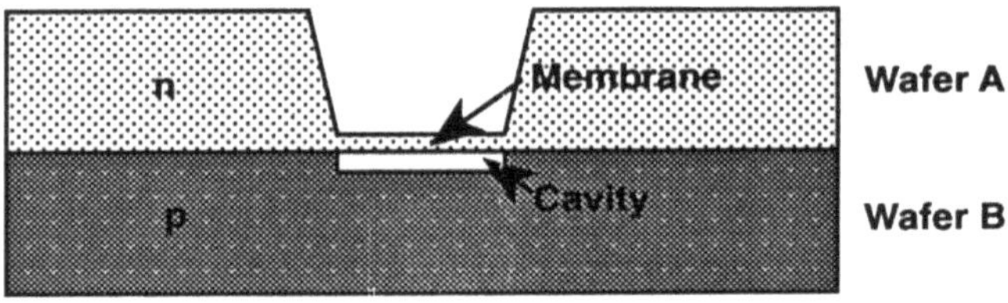

Figure 1. Cross-sectional view of the device structure

THEORY

When the pn-junction is back biased, an electrostatic force will bend the membrane until equilibrium is obtained with the elastic restoration force of the silicon crystal. The electrostatic force, F, is expressed by

$$F(x,y) = -\frac{\partial U(x,y)}{\partial z} = \frac{V^2}{2}\frac{\partial C(x,y;z)}{\partial z} \qquad (1)$$

where U is the energy per unit area stored by the capacitor, V is the applied voltage, z is the coordinate perpendicular to the (x,y) plane defined by the membrane surface [9] and C is the capacitance per unit area as measured between the membrane and the bottom of the cavity.

Due to the bending of the membrane, the depth, D, of the cavity changes by an amount δ(x,y) at different points (x,y). Therefore, the capacitor governed by the cavity has a varying capacitance per unit area at different points (x,y) given by

$$C(x,y;z) = \frac{\varepsilon}{D-\delta(x,y)} \qquad (2)$$

where ε is the dielectric constant of air.

The total potential energy of the system is the sum of the electrostatic energy and the elastic energy. The latter quantity was calculated by approximating the shape of the bent membrane with a spherical geometry and using the Rayleigh-Ritz method to obtain the deflection by minimizing the total energy [9]. The interference properties of the cavity were simulated by considering the cavity as a Fizeau interferometer [10]. The intensity of a Fizeau fringe at a point P is given by

$$I = \left|\sum TR^n \exp(-i\delta_n)\right|^2 = T^2\{(\sum R^n \cos\delta_n)^2 + (\sum R^n \sin\delta_n)^2\} \qquad (3)$$

where T and R denote the transmittivity and reflectivity, respectively, of the membrane (n-type silicon) and substrate (p-type silicon). Fig. 2 shows the simulated results of transmitted intensity distribution across the cavity area at different voltages. For zero voltage the transmitted intensity is uniform over the whole area but is reduced to about $I_{out}/I_{in} = 0.56$. When increasing the applied voltage to 3 volts, the membrane is bent and the transmitted intensity decreases as seen in Fig 2a. A further decrease is seen for 6 volts bias (Fig. 2b) and for higher voltages a peak starts to occur in the central part of the membrane area thus increasing the transmitted intensity again. For this specific device therefore, in order to obtain a single valued modulation function, a suitable voltage range is 0 - 6 volts.

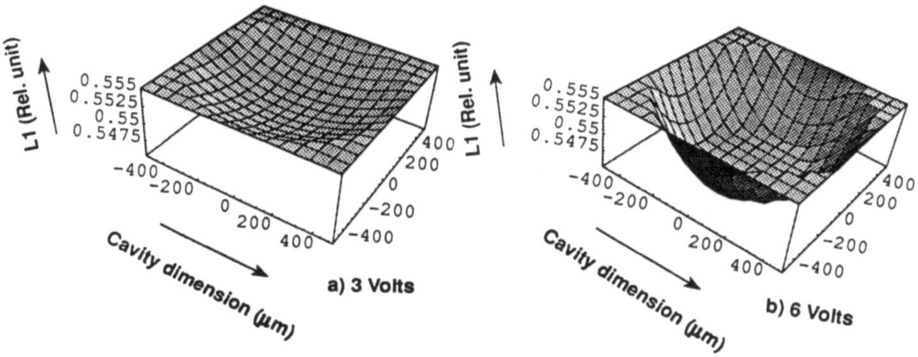

Figure 2. Simulated intensity distribution across the cavity area for (a) 3volts and (b) 6 volts applied voltage

EXPERIMENTS AND RESULTS

A light intensity L1 from a light emitting diode with a peak wavelength of 1.27 µm was fed to the device structure through an optical fiber with a diameter of 1000 µm. A certain portion of the light intensity is subject to multiple reflection and interference inside the cavity. For the detection of the modulated light, either the reflected or the transmitted light beam can be used. The light reflected from the cavity was detected by using a split fiber carrying the light to a photodiode, connected to a current amplifier. For the transmitted light, an optical fiber was attached to the opposite side of the device and the light intensity transmitted by the fiber was detected in the same way as for the reflected light. The p- and n-sides of the device were connected to a DC voltage source.

Fig. 3 shows the measured output signal (solid curve) as a function of applied voltage between the two parts *A* and *B* of Fig. 1. The dashed curve in Fig. 3 was obtained by integrating simulated intensity distributions as shown in Fig. 2 across the membrane area to obtain the totally transmitted photon flux.

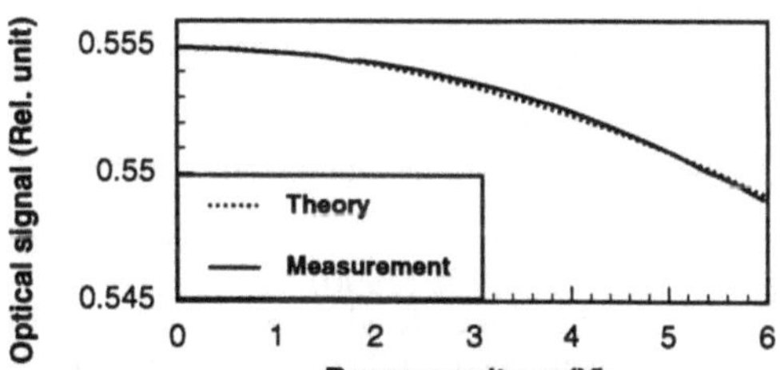

Figure 3. Measured output signal as a function of applied voltage together with simulated results

In a second test the modulator device was connected in series with a resistor and a voltage was applied to backbias the pn-junction as shown in Fig. 4. Besides the light with the wavelength 1.27 µm and intensity L1, passing through the optical fiber as described above, a second light source with the intensity L2 was added. This light, with a wavelength of 0.6 µm such that it was absorbed close to the silicon surface, created a photocurrent in the pn-junction thus changing the resistance of the device. This changes the portion of the totally applied voltage falling across the pn-junction which in turn will change the bending of the membrane and the interference properties inside the cavity. Fig. 5 demonstrates

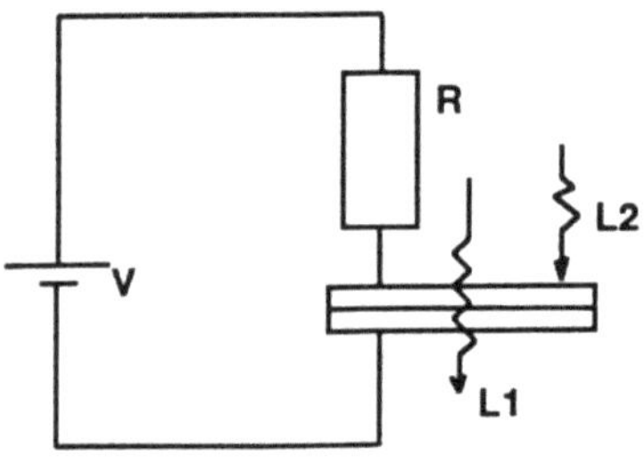

Figure 4. Experimental configuration for the modulation by a second light source

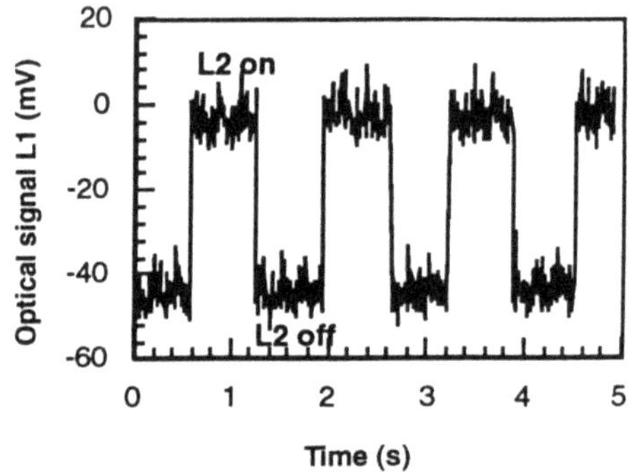

Figure 5. Experimental result from the measurement of transmitted light intensity, L1, modulated by a second light intensity, L2

how the device can be used to optically control the reflected intensity, L1, when the second light source with intensity L2 is switched on and off.

DISCUSSION

The principal function of a novel light controlled optical modulator has been demonstrated. The device is prepared by micromachining of silicon and constitutes an interferometer where an artificial cavity is obtained by thermal bonding technique. As the structure is based on silicon technology, it has the potential of serving as a picture element of a matrix together with electronic components integrated on the same device for optical parallel data processing.

REFERENCES

1. N.A.F. Jaeger, and L.Young, "Voltage-induced optical wave guide modulator in lithium niobate", IEEE J. Quantum Electron., QE-23:1673 (1989)
2. J.A.Davis, J.Ganlieli and G.W.Bach, "Optical transmission and contrast ratio studies of the magnetooptical spatial light modulator", Appl. Opt. 27:5194 (1988)
3. T.D.Hudson, R.K.Worcester and D.A.Gregory, "Performance characteristics of an optically addressed ferroelectric liquid-crystal spatial light modulator", Appl. Opt. 30:2867 (1991)
4. S.-T. Wu, "Nematic liquid crystal modulator with response time less than 100 μs at room temperature", Appl.Phys.Lett. 57:986 (1990)
5. H.K.Kang, J.B.D.Soole,, H.P.LeBlanc, R.Bhat, M.A.Koza and I.H.White, "Efficient InGaAsP/InP multiple quantum well waveguide optical phase modulator", Appl.Phys.Lett. 57: 2285 (1990)
6. O.Solgaard, F.S.A.Sandejas and D.M.Bloom, "Deformable grating optical modulator" Opt. Lett. 17:688 (1992)
7. Z. Xiao S. Norrman and O. Engström, "A micro pn junction structure sensor for measuring voltage on high potentials", Sensors and Actuators A 41:432 (1994)
8. Z. Xiao, S. Norrman and O. Engström, "A novel micro sensor structure sensor for voltage on high potentials or low pressure", Proceeding of "The 7th International Conference on Solid State Sensors and Actuators (Transducers'93)", Yokohama, Japan, (1993)
9. Hiroyuki Fujita and Toshiaki Ikoma "Numerical determination of the electromechanical field for a micro servosystem" Sensors and Actuators, A21:215 (1990)
10. K. Kinosita, "Numerical evaluation of the intensity curve of multiple-beam Fizeau fringe", J. Phys. Soc. Japan 8:219 (1953)

ASYMMETRIC SILICON INFRARED WAVEGUIDE SWITCHES BASED ON TOTAL INTERNAL REFLECTION

Liu Enke, Liu Yuliang, Li Guozheng, Zhang Shengliang

Department of Electronic Engineering
Xi'an Jiaotong University
Xi'an, Shaanxi, 710049, P. R. China

Zhou Fan, Cheng Meiqiao, Li Bingchen, Ge Huang

National Research Centre for Optoelectronic Technology
P. O. Box 912, Beijing, 100083, P. R. China

INTRODUCTION

Semiconductor optical switches utilizing carrier-induced refractive index change have tremendous potential for the application of optical processing because of their small size, single-mode operation, polarization-independence and wavelength-insensitivity. These switches are generally based on total internal reflection, and have a requirement of precise control of reflection interface[1,2]. The requirement causes a difficulty in developing silicon switches of the same kind. Because silicon is an indirect-gap semiconductor, the injected carriers in it diffuse a long distance due to their long lifetime. They can not be restricted within a specific region in the normal structure like what has been taken in InGaAsP or AlGaAs. Although some authors have reported that attempt to make use of the free-carrier effect in silicon for optical switching, the developed switches are all based on mode-interference other than internal reflection[3,4]. These switches are devoid of the advantages mentioned above. From the viewpoint of practical application, a kind of silicon switch with these advantages is more essential.

This paper proposes a novel silicon switch based on free-carrier effect as well as total internal reflection. We fully considered the optical tunnelling effect caused by the surface optical declining wave and leaky wave, and the reflection-region-position effect which results from the Goos-Haenchen shift. A four-port asymmetric silicon optical waveguide switch was designed and fabricated successfully. And its optical characteristics was studied.

DEVICE STRUCTURE

The schematic diagram of the switch is depicted in Fig. 1. The two rib waveguides with a width of W and composed of a n-type silicon guiding layer grown on a n^+ substrate intersect at an angle of θ. The intersection region of the waveguide is widened to a unified width of 2W. An abrut p^+n junction is formed below the top surface of the waveguide at the intersection region. Being different from the normal structure of the reflection type switch, the collection electrode in this switch is placed beside the p^+n junction and on the surface of the guiding layer other than the substrate. This forms the transverse injection structure. When carriers are injected from the p^+n junction into the waveguide layer, they diffuse and drift towards the collector transversely, the refractive index in this region decreases and a reflection region is formed. Because the reflection region is not located in the centre of the intersection region, the device operates as an asymmetric 2×2 switch. In other words, only the incident light from port 1 is reflected by the reflection region.

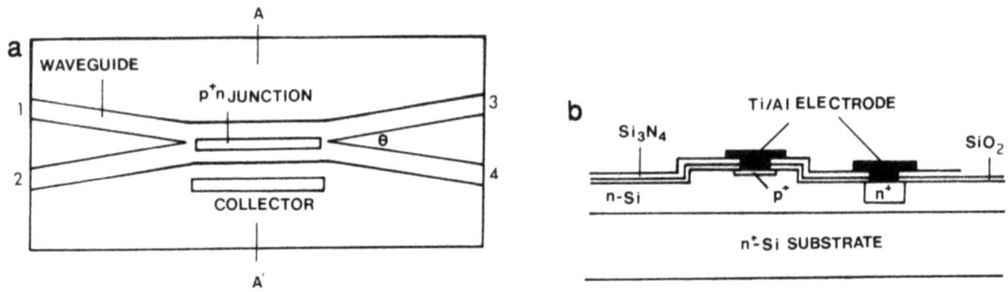

Figure 1. Schematic diagram of (a) asymmetric silicon waveguide switch and (b) cross-section of intersection region

Being compared with the normal structure, this asymmetric transverse injection structure has two main advantages. The first, an electric field whose direction is from the p^+ region towards the collector is induced. The field causes the holes to drift towards the collector and the diffusion to the reverse direction is restrained. This leads to the carrier distribution at the interface of the reflection region being improved. The second, the holes cover a longer distance in this structure than in normal one before they run off the n type layer. This means that the hole utilization ratio increases, or the reflection region can be widened without increasing the injection current. Therefore, the optical tunnelling effect due to the limited width of the reflection region can be restrained. As a consequent result, the injection intensity can be reduced.

DEVICE FABRICATION

The starting material for the switch was 3. 6 μm epitaxial silicon with a P concentration of $10^{16}/cm^3$ grown on a heavily doped silicon substrate with a Sb concentration of $3\times 10^{18}/cm^3$. The rib waveguides were fabricated by dry-etching technique. To ensure a single-mode operation, the waveguides had a ridge height of 1. 8 μm and a ridge width of 6 μm. The cover of the waveguide was SiO_2/Si_3N_4, as shown in Fig. 1(b), which was grown by employing the plasma enhanced chemical vapor deposition method. The intersection angle was 4°. B^+ and As^+ ion implantation were performed through the openings of the SiO_2/Si_3N_4 mask to form the p^+n junction and collector respectively. The junction area was 4 × 170 μm². For a maximum carrier utilization, the collector was placed to be 12 μm from the p^+n junction. This value was equal to the diffusion length of the holes. The p^+ and n^+ ohmic contact were obtained by sputtering a Ti/Al layer. In order to improve the response speed, the electron irradiation was performed with electron energy of 14 MeV. Then the wafer was diced to an overall length of 7 mm, but the real operation length of the switch was less than 200 μm. To couple light into the switch, the chip was end polished. After that, it was mounted on a Cu heat sink, so that no temperature rise was observed during the experiment.

DEVICE MEASUREMENT

The device performance was evaluated at the wavelength of 1.3 μm. The input light was end coupled directly to the waveguide port 1 from a laser diode's single-mode pigtail fiber. The magnified near-field radiation pattern was imaged on an infrared vidicon and displayed on a video monitor. Fig. 2 shows the output intensity profiles at 0 and 85 mA injection currents. At zero injection current, the switch was in a cross state. Fig. 2(a) shows the output light distribution which was imaged from waveguide post 4. The extinction ratio was 18.6 dB, and the insertion loss was 4.8 dB which did not include Fresnel loss. At 85 mA injection current, the device reached a maximum optical switching. Fig. 2(b) shows the output light distribution which was imaged from wavegeuide port 3. The crosstalk was less than -11.4 dB and the swtching time was measured to be about 100 ns.

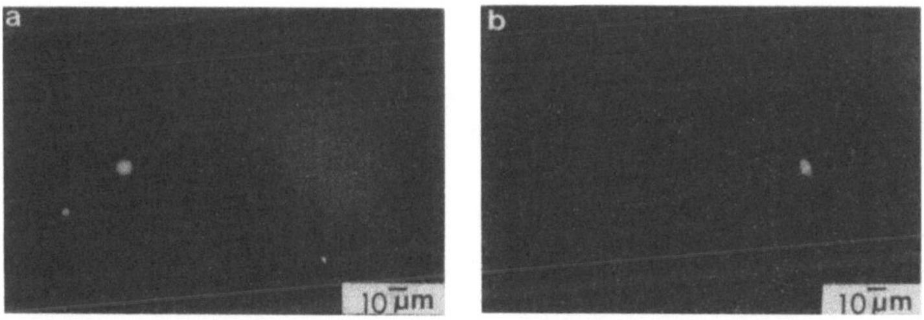

Figure 2. Output light distribution at (a) O mA, (b) 85 mA of injection current

The polarization-dependence was investigated by changing the polarization direction of the input light. But it was not observed. The wavelength band-width was not measured in this experiment. However, we expect that the device can operate from 1.3 μm to at least 1.6 μm, but the injection current has some changes. The reason for this is that the refractive index change induced by injected carriers is not very sensitive to wavelength within this range.

In addition to above result, we observed that the excellent optical switching can also be realized if input light was coupled into waveguide port 2 but a much larger injection current about 570 mA was needed.

CONCLUSION

The silicon asymmetric 2×2 waveguide switches based on total internal reflection have been proposed and fabricated successfully, which utilize transverse injection structure. The crosstalk was measured to be less than -11.4 dB at 85 mA injection current for a wavelength of 1.3 μm. The switching time is 100 ns. This is the best result obtained on silicon so far. The switches are very useful for monolithic integration with other silicon devices because of their small aize, single-mode operation, wavelength-insensitive and absence of polarization dependence.

ACKNOWLEDGMENTS

The authors are indebted to Zhao Changwei, Wang Xin and Shen Hong for their excellent technical work. We also thank Gao Yong for the help of the measurement. This work was supported by EST plan of China.

REFERENCES

1. K. Ishida, H. Nakamura, H. Matsumura, T. Kadoi, and H. Inoue, InGaAsP/InP optical switches using carrier induced refractive index change, *Appl. Phys. Lett.*, 50 : 141(1987).
2. O. Mikami, H. Nakamure, Waveguided optical switch in InGaAsP/InP using free-carrier plasma dispersion, *Electron. Lett.*, 20 : 229(1984).
3. J. P. Lorenzo, R. A. Soref, 1.3 μm electro-optic silicon switch, *Appl. Phys. Lett.*, 51 : 6(1987).
4. Liu Enke, Liu Yuliang, Zhang Shengliang, and Li Guozheng, 1.52 μm silicon 2×2 electrooptic switch, *Photonics'* 92 : "The Second International Workshop on Photonic Networks, Components and Applications", J. Chrostowski, L. Terry, Editors, OCRI, Ottawa, Ontario Canada, 3.10.1 (1992).

ANALYSIS AND FABRICATION OF 45° SELF-ALIGNED INTEGRATED MIRRORS ON DIELECTRIC OPTICAL WAVEGUIDES

L. Joannes, B. Grimbert, J. Harari, J.P. Vilcot, W. Tchana
and D. Decoster

Institut d'Electronique et de Microélectronique du Nord
UMR - CNRS 9929 - Dép^t Hyperfréquences et Semiconducteurs
Domaine Universitaire Scientifique de Villeneuve d'Ascq
Avenue Poincare, BP69
F-59652 Villeneuve d'Ascq Cedex

INTRODUCTION

Integrated corner mirrors are of great interest in integrated optics since they can reduce the distance in which changes of propagation direction of light is done. A corner mirror consists of an etched vertical facet producing total internal reflection and abrupt change of the propagation direction.

The first corner mirror was reported by Buchmann et al.[1]. To overcome alignments problems, a self-aligned process was proposed by Albrecht et al.[2] and successfully used on GaAsInP/InP optical waveguide by Gini et al.. Losses as low as 0.3dB was achieved for a 45°-corner mirror (change of direction of propagation of 45°)[3]. Technological characteristics such as facet verticality, smoothing surface and undercutting are known to be the main contribution to losses of the corner mirror. Nevertheless, no evaluation of the intrinsic losses of the self-aligned technics (i.e. design of self-aligned shape) has been reported. We propose an evaluation of such losses through a comparison of three self-aligned structures on dielectric optical waveguides but semiconductor optical waveguide can also be considered[4]. From the results obtained, we have fabricated low-loss 45° self-aligned integrated mirrors (SAIM) on a dielectric optical waveguide.

SELF-ALIGNED INTEGRATED MIRRORS ANALYSIS

From a technological point of view, SAIM technique consists of using the same mask to define both waveguides and mirror facet. A second mask defines areas for the mirror facet deep etching, but its position is not critical [2].

Self-aligned technique takes advantage of low lateral diffraction of light at the end of a waveguide to widen rib near the corner and define mirror facet and ribs with the same mask. Nevertheless, diffraction contribute to SAIM intrinsic losses. Different shapes have already been experimentally used; nevertheless, no systematic comparison of self-aligned structures has been carried out yet.

Using a two dimensional Finite-Difference Beam Propagation Method (FD-BPM)[5] - combined with effective index method-, we compare three different self-aligned shapes described in Figure 1. For all simulations, mirror is fixed at the optimum position d which have been previously calculated (-0.2μm for quasi-TM mode and -0.08μm quasi-TE mode) and computational window size is 40μm, the number of mesh points is 2500, the number of step over 100μm is 500. FD-BPM accuracy in the analysis of diffraction has been found better then 0.02dB.

Fundamental eigenmode is injected at z=0 with phase correction for tilted propagation. At the end of the propagation (z=100μm), final field is compared to expected field with overlap integral to obtain reflection losses.

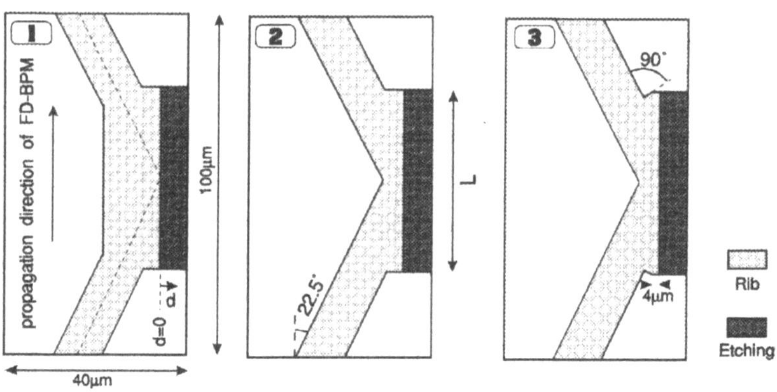

Figure 1. Description of the three studied SAIM shapes (top view).

The two monomode polyimide waveguides considered are shown in the Figure 2. We consider a wavelength of 1.3μm. Rib height of the narrow waveguide has been increase to have a good confinement of light under the ridge.

Figure 3 and 4 shows theoretical reflection losses of a SAIM on a the 6μm-wide and the 3μm-wide rib waveguides, for the three considered shapes, as a function of mirror width L. We observe a rapid increasing in reflection losses with the reduction of mirror width due to the fact that all the propagated light is not reflected by narrow mirrors. For wider mirrors, diffraction makes reflection losses growing up slowly with increasing L for the 6μm-wide waveguide but drastically increasing with L for the 3μm-wide rib waveguides.

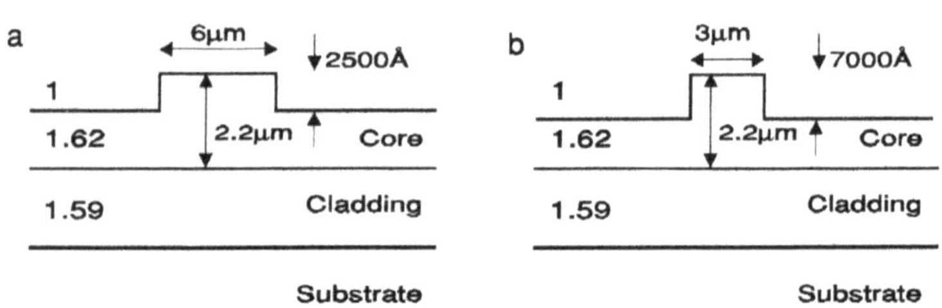

Figure 2. Monomode polyimide waveguides of 6μm (a) and 3μm (b) rib width.

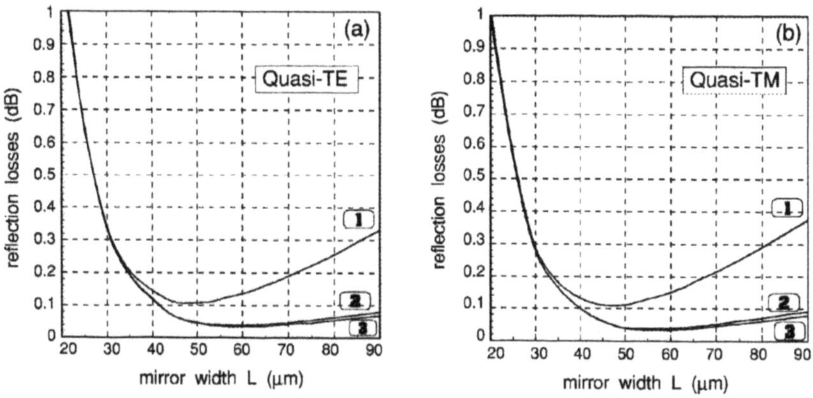

Figure 3. Reflection losses in SAIM on 6μm-wide polyimide rib waveguide as a function of mirrror width for the three shapes.

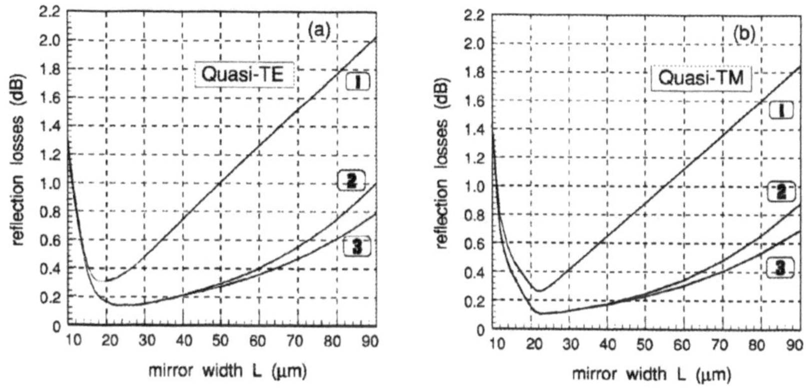

Figure 4. Reflection losses in SAIM on 3μm-wide polyimide rib waveguide as a function of mirror width for the three shapes.

As expected, diffraction effects are more important in the first shape : reflection losses increase more rapidly for this shape and the optimum mirror width is smaller for this shape then for the other ones. Shapes two and three are quasi equivalent; and a very small difference is observed between quasi-TE and quasi-TM polarisation modes.

Through this analysis we have show that a suitable choice of SAIM structure can reduce reflection loss bellow 0.05dB in both quasi-TE and quasi-TM polarization modes. We also show that for narrow waveguides (3μm-wide rib) and for a non-optimized SAIM structure, loss can theoretically reach 0.3dB per reflection.

INTEGRATED MIRROR FABRICATION

From these results, we have designed masks to fabricate 45° SAIM on dielectric optical waveguide with an optimized reactive ion etching process. Since at the time of the fabrication we had only one polyimide available, we have chosen to fabricate the optical waveguide with a core in silicon nitride sandwiched between two polyimide layers. The fabrication of that optical waveguide is totally compatible with microelectronic technology. The structure of the optical waveguide is described in Figure 5 and the self-aligned structure is close to the second one considered in the calculation (see Figure 1).

A calculation of reflection losses in 45° SAIM in that waveguide has shown that minimal reflection losses are also around 0.05dB.

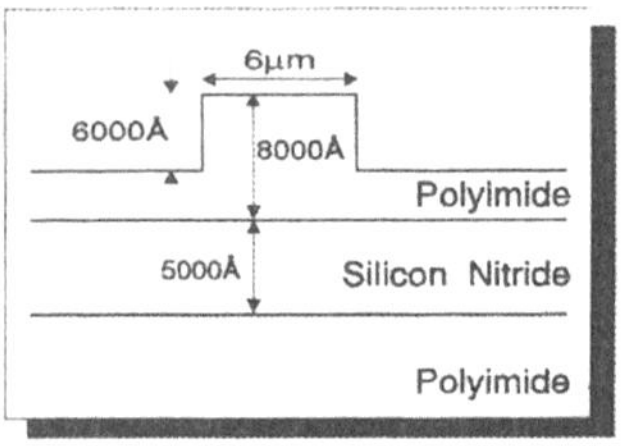

Figure 5. Fabricated dielectric optical waveguide

Figure 6. Photography of two successive reflexions through 50μm-wide mirrors.

Reflection losses have been evaluated from a comparison of transmitted light through waveguides including 2, 4, 6 mirrors and a straight reference waveguide. We have measured 0.6dB and 0.75dB for quasi-TE and quasi-TM modes respectively and for λ=830nm. Figure 6 shows two successive reflexions through 50μm-wide mirrors.

Losses in the fabricated device are mainly attributed to scattering on reflection facet roughness and could be reduced with an electron beam masking.

CONCLUSION

Intrinsic reflection losses of 45° Self-Aligned Integrated Mirrors were studied using Finite Difference Beam Propagation Method. We have compared the reflection losses expected for three self-aligned shapes and optical waveguides of different kinds. It is shown that the contribution of self-aligned technique to losses can reach 0.3dB for the simplest structure and for a narrow waveguide, and decreases to 0.05dB for other structures and a wider rib waveguide. These results are promising for the fabrication of future integrated mirrors with losses below 0.1dB, which is needed to fabricate complex photonic circuits with several tens of mirrors.

We have also designed masks and fabricated 45° SAIM on a dielectric optical waveguide compatible with microelectronic technology. Losses of 0.6dB and 0.75dB for quasi-TE and quasi-TM modes respectively have been measured at λ=830nm. That experimental result is among the best reported so far.

REFERENCES

1. P. Buchmann. and H. Kaufmann, GaAs single mode rib waveguide with reactive ion-etched totally reflecting corner mirrors, *Journ. Light. Tech.*, 3:785 (1985).
2. P. Albrecht, W. Döldissen, U. Niggebrügge, H.P. Nolthing, and H. Schmid, Self-aligned mirrors for optical integration on InGaAsP/InP, *Proc. ECOC*, 242 (1987).
3. E. Gini, G. Guekos and H. Melchior, Low loss corner mirrors with 45° deflection angle for integrated optics, *Elect. Lett.*, 28:499 (1992).
4. L. Joannes, J. Harari, J.P. Vilcot and D. Decoster, Analysis of intrinsic losses of 45° self-aligned integrated mirror using finite-difference beam propagation method, *Proc. SPIE* 2212 (1994) to be published.
5. Y. Chung and N. Dagli., Analysis of integrated optical corner reflectors using a finite difference beam propagation method , *IEEE Photon. Tech. Lett.*, 3:150 (1991).

POTENTIAL APPLICATIONS OF IMPLANTED FUSED SILICA AS A PHOTOSENSITIVE MATERIAL FOR PLANAR LIGHTWAVE CIRCUITS

Marc Verhaegen[1], John L. Brebner[1], and Jacques Albert[2]

[1]Groupe de recherche en physique et technologies des couches minces
Département de Physique, Université de Montréal
Montréal, Québec, Canada H3C 3J7
[2]Communications Research Center, P.O. Box 11490, Station H
Ottawa, Ontario, Canada K2H 8S2

Fused silica is used extensively as a material for optical devices and waveguides. Current optical fibers for communication applications are made of doped silica and planar optical lightwave circuits of similar composition are being developed. With the deployment of communication systems using these components, the need is growing for sophisticated optical signal processing devices. In recent years, a procedure has been developed to fabricate such devices by a photoimprinting technique based on the discovery that refractive index of certain doped silica glasses can be modified permanently by exposure to ultraviolet light.[1]

Several conditions must be met for the silica glass to be photosensitive. However, it is still not clear why certain glasses or processing techniques lead to changes in sensitivity of several orders of magnitude.[2-3] Our goal is to study the changes in the glass properties associated with exposure to ultraviolet light in order to better understand how efficient refractive index modifications are generated.

In our experiments, we use pure synthetic bulk fused silica implanted with germanium and silicon ions at energies between 3 and 5 MeV using the Universté de Montréal 5 MeV Tandem accelerator. The doses of implanted ions were varied from 10^{12} to 10^{14} ions/cm^2. The substrate is initially transparent to light across the spectral range from 160 to 1300 nm but implantation induces strong absorption bands near 163, 212 and 245 nm. In addition, implantation increases the refractive index of the silica in a 3 to 5 micron thick layer near the implanted surface, leading to the formation of an optical waveguide. Exposure of the implanted samples to intense ultraviolet light at a wavelength close to an absorption band bleaches the absorption and changes the refractive index permanently.[4] Refractive index changes of the order of $2X10^{-3}$ have been observed in preliminary experiments,[5] as well as evidence of dichroic absorption changes when using polarized bleaching light.[6]

In our recent work, high dose germanium and silicon implantation were achieved at 5 MeV. Absorption spectra were measured between 190 and 400 nm using a CARY-5 spectrophotometer. Absorption bands induced by high dose silicon implantations (up to $3X10^{17}$ ions/cm^2) are presented in Fig. 1. One can note that the absorption coefficient drastically increases across dose from $5X10^{16}$ ions/cm^2 to $3X10^{17}$ ions/cm^2. This important increase is believed to be due to one or several absorption bands in the vacuum

Applications of Photonic Technology
Edited by G.A. Lampropoulos *et al.*, Plenum Press, New York, 1995

ultraviolet region. It is also interesting to note that higher doses (> $1X10^{17}$ ions/cm^2) led to a yellow colouring at the surface of the silica substrates. Carbon contamination during implantation was ruled out by XPS measurements. Further studies will be necessary to determine the correlation between this colouring and the absorption coefficient increase.

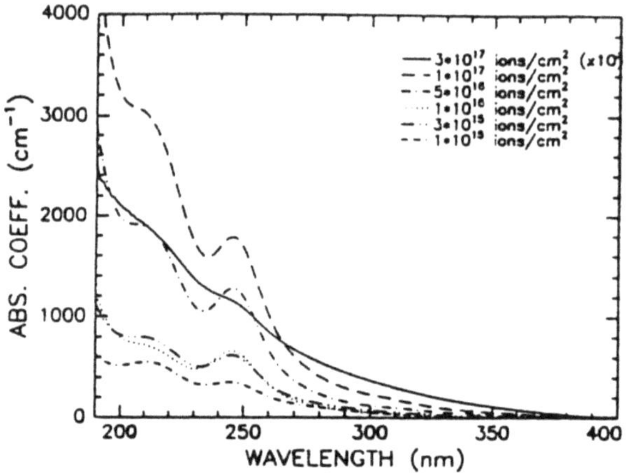

Fig. 1. Absorption coefficient as a function of wavelength for silica implanted with different silicon (5 MeV) doses.

Volume and index change induced by implantation of silicon ions at 5 MeV were studied as a function of the dose. The compaction resulting from ion implantation occurs only along the direction of incidence of the implanted ion and gives rise to a change in sample thickness. Thickness changes were measured with a DEKTAK 3030 ST profilometer and are presented in Fig. 2 for different doses.

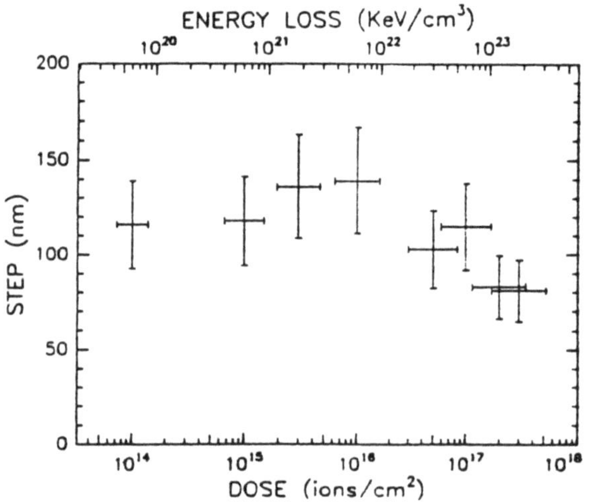

Fig. 2. Thickness change in silicon (5 MeV) implanted silica versus doses and total energy loss.

The results showed an increase in the thickness change with the dose. A maximum is reached near $1X10^{16}$ ions/cm^2 which corresponds to a total energy loss of $6X10^{23}$ KeV/cm^3. For higher doses, the thickness change tended to decrease.Index change measurements were performed using a Abbe refractometer and are shown in fig. 3. One can see that the index change has similar characteristics as the thickness change.

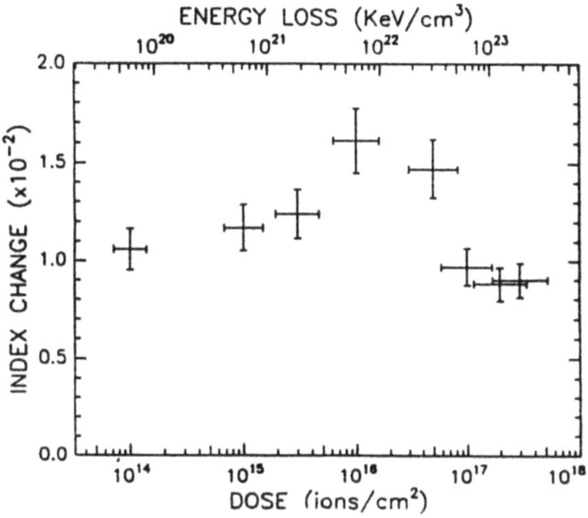

Fig. 3. Index change in silicon (5 MeV) implanted silica versus doses and total energy loss.

Finally, channel waveguides have been fabricated by germanium implantation at 5 MeV through a thin gold film deposited on the silica substrate and patterned with 4 micron-wide openings using a conventional photolithography. Optical guiding was achieved by coupling He-Ne laser light in the channel waveguide. The measured propagation losses are about 2 dB/cm which is fairly high. Preliminary mode profile was measured using a CCD camera. The mode profile for a silica substrate implanted with $1X10^{14}$ ions/cm^2 germanium atoms at 5 MeV is presented in fig. 4.

In summary, our experimental results show that absorption bands induced by high dose silicon and germanium implantation increase with dose. It has also been seen that the volume change and the index change behaved similarly as a function of the dose. Furthermore, priliminary mode profile measurements evidenced optical guiding in germanium implanted channel waveguide. Taking all these properties into account, the waveguides obtained from high dose high energy implantation suggests they are strongly photosensitive.

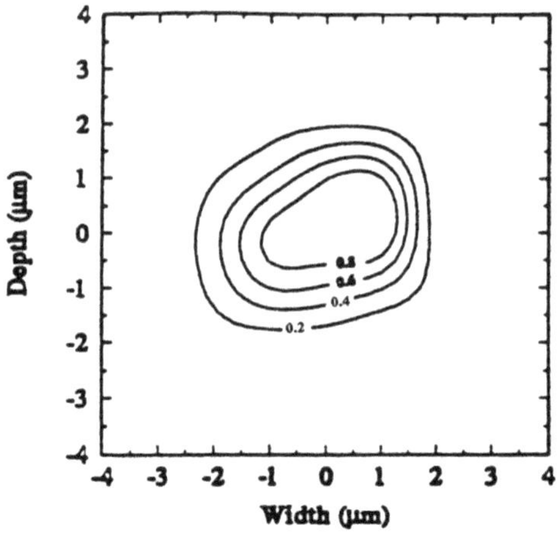

Fig. 4. Mode profile of germanium (5 MeV / 1X10^{14} ions/cm^2) implanted channel waveguide in silica. Near field intensity pattern at 632.8 nm.

References

1. K. O. Hill, Y. Fujii, D. C. Johnson, and B. S. Kawasaki, *Appl. Phys. Lett.* **32** (1978) 647

2. F. Bilodeau, B.Malo, J. Albert, D. C. Johnson, K. O. Hill, Y. Hibino, M. Abe, and M.Kawachi, *Opt. Lett.* **18** (1993) 953

3. R. M. Atkins, P. J. Lemaire, T. Erdogan, and V. Mizrahi, *Electron. Lett.* **29** (1993) 1234

4. J. Albert, K. O. Hill, B.Malo, D. C. Johnson, J. L. Brebner, Y. B. Trudeau, and G. Kajrys, *Appl. Phys. Lett.* **60** (1992) 148

5. J. Albert, B.Malo, D. C. Johnson, K. O. Hill,J. L. Brebner, and R. Leonelli, *Opt. Lett.* **17** (1992) 1652

6. J. Albert, B.Malo, D. C. Johnson, F.Bilodeau, K. O. Hill, J. L. Brebner, and G. Kajrys, *Opt. Lett.* **18** (1993) 1126

OPTICAL DELAY LINES IN HIGH-SILICA (SiO_2/Si) WAVEGUIDES

S. Paquet[1], F. Chenard[1], Z. Jakubczyk[2], M Bélanger[3], M. Têtu[4], C. Bélisle[5]

1 N.O.I., 369 Franquet, Ste.-Foy, Québec, Canada, G1P 4N8
2 Corporation Optiwave, 360 Franquet, suite 10, Ste.-Foy, Québec Canada, G1P 4N9
3 Teleglobe Cable Systems, 1000 de la Gauchetière Ouest, Montréal Québec, Canada, H3B 4X5
4 C.O.P.L., Département de Génie Électrique, Pavillon Pouliot, Université Laval, Ste.-Foy, Québec, Canada, G1K 7P4
5 Communication Research Center, 3701 Carling ave., P.O. Box 11490 Station H, Ottawa, Ontario, Canada, K2H 8S2

INTRODUCTION

Recently, there has been a resurgence of interest in the field of optically-controlled phased-array antennas. In some cases, true time delays instead of phase shifters are needed to control the phase of the radiating elements to avoid significant losses in the direction of emission. This kind of phase adjustment cannot be made using conventional microwave technology because of high losses and large dimensions of the waveguides. True time delays have been realised optically using different techniques such as high dispersion fibre[1], and micro-mirrors[2]. However, these methods require prohibitive number of discrete components when used in antennas with large number of radiating elements. In this paper, a serial approach, similar to that of Sullivan and al[3], which provides multiple delays in a single integrated-optic device working at 1.3 μm, is explored.

DEVICE ARCHITECTURE

The architecture of the device is shown in Fig. 1. Thermooptic interferometric switches are used for selecting between a short reference path and a longer delayed path. Time-delay values are generated by the path difference between the reference and delayed path. Six switches are used, the last one being use as a combiner. Therefore, 32 different delay times can be selected with proper switching configurations. There are two main considerations in the design of this device: individual switch performance and small bend radius.

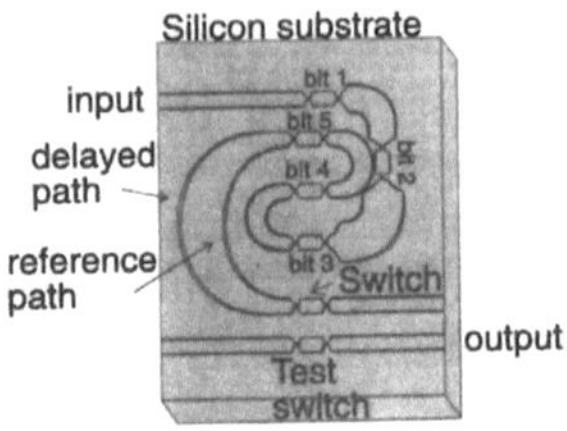

Figure 1. 5 bits integrated optic time delay device with thermooptic switches

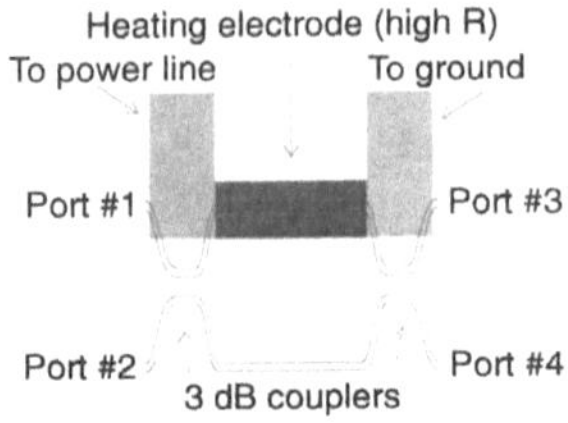

Figure 2. Layout of the thermooptic interferometric switch

SWITCH MODELLING, DESIGN AND TOLERANCE STUDY

The layout of the thermooptic interferometric switch is shown in Fig. 2. Two 3-dB couplers are separated by two straight parallel waveguides. A heating electrode is placed over one of these parallel waveguides. For convenience, the two left ports, labelled port 1 and 2, are designated as inputs, whereas ports 4 and 5 are the outputs. When a control voltage is applied to the electrodes, a phase shift is induced between the signals in the two arms. From coupled-mode theory, the following equations can be used to model the switch:

$$\begin{aligned} C &= \left|Ae^{j\phi}\left[e^{j\Psi}\cos^2(\kappa L)-\sin^2(\kappa L)\right]+Bj\left[-e^{j\Psi}\cos(\kappa L)\sin(\kappa L)-\cos(\kappa L)\sin(\kappa L)\right]\right|^2 \\ D &= \left|Aje^{j\phi}\left[-e^{j\Psi}\cos(\kappa L)\sin(\kappa L)-\cos(\kappa L)\sin(\kappa L)\right]+B\left[-e^{j\Psi}\sin^2(\kappa L)+\cos^2(\kappa L)\right]\right| \end{aligned} \quad (1)$$

where C and D are the wave intensity at ports 3 and 4 respectively and A and B are the wave amplitude at ports 1 and 2 respectively, ϕ is the phase difference between signals at port 1 and port 2 and ψ is the phase shift due to the heating electrode.

Given a height of 5 μm and a width of 4.5 μm for the waveguide, the coupling coefficient κ can be calculated from the index difference between the core, substrate and cladding. κ was found to be 257.86. From equ. 1, the length of the coupling region L is then 3.05 mm. For fabrication purpose, we have two correct this length to account for coupling in the s shape waveguides by using a software implemented beam propagation method. The actual length of the straight waveguides in the coupling region is 2.35 mm. We can now compute the light intensity curves of ports 3 and 4, shown in Fig. 3, when light is entering in port 1. The two switching states are labelled "bar state" (output maximum at port 3) and "cross state" (output maximum at port 4).

In order to estimate fabrication tolerances, it is important to investigate the switch efficiency when waveguide dimensions and shape are changing. Such changes in dimensions can be represented as variation in κ. Results of calculations of extinction ratio vs. κ are presented in Fig. 4. Extinction ratio deteriorates when $\kappa \neq 257.86$.

CURVED WAVEGUIDE THEORETICAL ANALYSIS

From a fabrication perspective, the objective is two make the radius of curvature as small as possible while keeping low radiation losses. Simulation has been conducted using an index transformation technique[4] that models a curved waveguide by a straight waveguide with an index profile that emulates the curvature. Losses were computed using a beam propagation method. Results are shown in Fig. 5. A minimum radius of 7 mm was chosen for the first prototype, giving losses of less than 0.07 dB for a 180° turn. The total insertion loss for the device, including fibre to waveguide coupling, were calculated to be 13 dB.

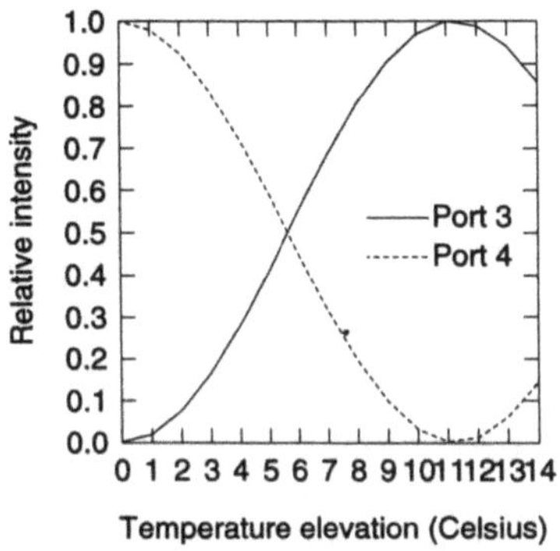

Figure 3. Light intensity curves at ports 3 and 4 for light entering at port 1

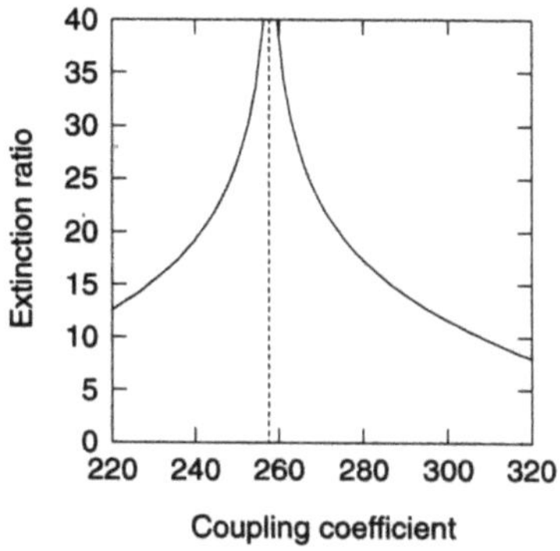

Figure 4. Variation of extinction ratio vs. coupling coefficient κ.

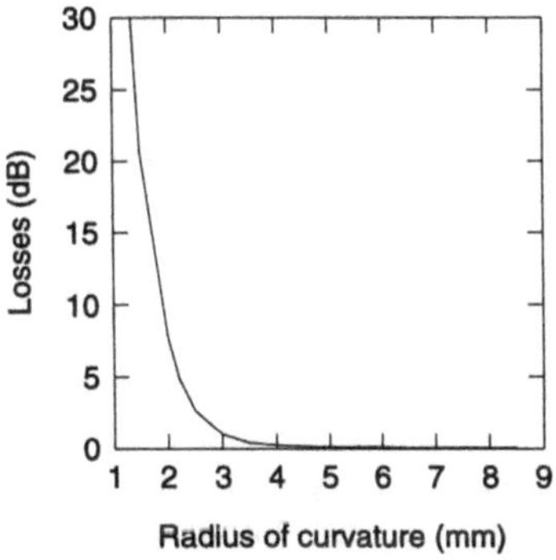

Figure 5. Losses in 180° curved waveguides with respect to radius of curvature

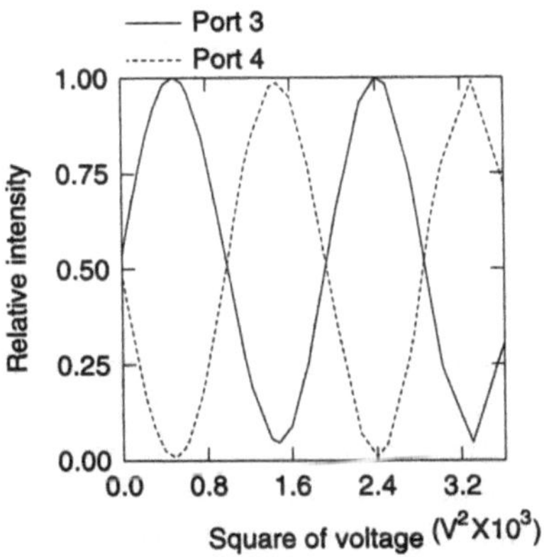

Figure 6. Measured intensity curves at ports 3 and 4 when light enters at port 1

DEVICE FABRICATION AND CHARACTERISATION

The device has been fabricated on a 4 inches silicon substrate. Photolithography and Chemical Assisted Ion Beam Etching were used to fabricate the waveguides. Gold and chrome electrode deposition was done by e-beam sputtering and the pattern was obtained using a lift-off technique.

Measured intensity curves are shown in Fig. 6. Table 1 gives characteristics of the switch. Best results of extinction ratio were obtain for TM polarisation; measured rise time of the switch is 1 ms; average power consumption is 500 mW. The measured waveguides dimensions are different from the ones prescribed. A new value of κ computed with measured waveguides dimensions was used in equ.1 to fit the experimental data to the theory. Results are shown in Fig 7. The actual value of κ used is 291.24 instead of the calculated 257.86. In addition, irregularities in the centre parallel waveguides induce a initial phase shift. This causes the shift of the switching curves and an augmentation of power consumption because the initial state is no longer at V = 0. Therefore, we have to improve the fabrication process in order to obtain a switch with higher extinction ratio (probably about 20 dB). For the present prototype, the extinction ratio of the switches did not permit measurement of the delays.

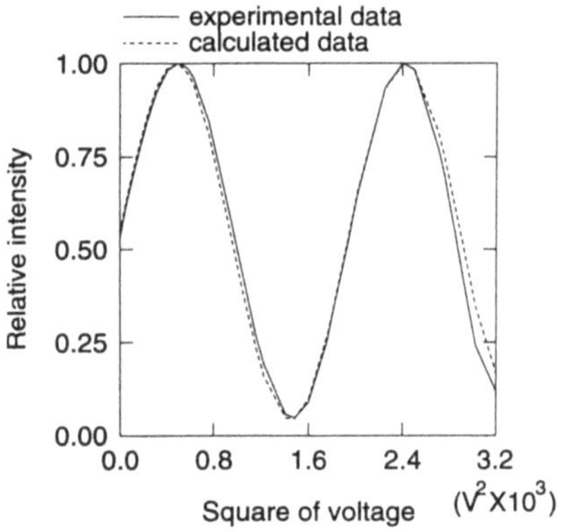

Figure 7. Curve fit of the intensity curve at output 3.

Table 1. Measured switch characteristics

Ext. ratio (bar state)	20 dB
Ext. ratio (cross state)	13 dB
Applied voltage (bar state)	22.5 V
Applied voltage (cross state)	38.5 V
Total switch length	17.3 mm

Total insertion losses of the device was measured to be 20 dB. The difference from the calculated value can be attributed to waveguide imperfections, resulting in higher propagation losses.

CONCLUSION

An integrated-optic, thermooptic device, providing multiple time delays for phased-array-antenna control, has been fabricated on SiO_2/Si and characterised. The measured extinction ratio of 13 dB can be improved to 20 dB by better control of key fabrication parameters which have been determined through modelling and simulation. While the compactness of the integrated-optic device is attractive, the power requirement in a thermooptic switch, 500 mW, is high for some applications. Similar devices using electrooptic materials are being considered.

Acknowledgements

Thanks are due to CRC for the financing of the project and N.O.I. for the use of their facilities.

REFERENCES

1. R. Soref., Optical dispersion technique for time-delay beam steering, *Applied Optics*, vol. 31, no. 35, pp. 7395-7397 (1992)
2. E. Toughlian and H. Zmuda, A photonic variable RF delay line for phased array antennas, *J. Lightwave Tech.*, vol. 8, no. 12, pp. 1824-1828 (1990)
3. C. T. Sullivan and al., Switched time delay elements based on AlGaAs/GaAs optical technology at 1.32 μm for optically controlled phased array antennas, *SPIE- Optoelectronic Signal Processing for Phased Array Antennas III*, vol. 1703, no. 36 (1992)
4. K. Petermann, Microbending loss in monomode fibres, *J. Lightwave Tech.*, vol. 12, no. 4, pp. 107-109 (1976)

AN EXPERIMENTAL APPARATUS FOR THE CHARACTERIZATION OF THICK-FILM OPTICAL WAVEGUIDES

F. Docchio, G. Sansoni, D. Marioli, A. Taroni, M. Perini and S. Corini

Dipartimento di Elettronica per l'Automazione, Facoltà di Ingegneria
Università degli Studi di Brescia, via Branze, 38, I-25123 Brescia, Italy

INTRODUCTION

Optical sensors are largely used in modern metrology and quality control in manufacturing. Generally these sensors are based on the use of (i) discrete optical components, (ii) optical fibres and (iii) integrated optical devices. Main limitations of these sensors are the cost of bulk optics and the difficulties in the manufacturing of integrated optics.

The use of thick film waveguides has recently become very interesting for the development of optical sensors. In fact, the manufacturing technique for their realization is simple and low cost, and the waveguides can be integrated over substrates which are suitable for sensor application. The realization of such optical thick film sensors requires the knowledge of the optical properties of thick film waveguides.

In the past, thick film waveguides have been investigated as a low cost approach to optical communications devices and their attenuation characteristics have been investigated.[1,2] In view of their use as sensors, we manufactured new thick film linear waveguides of different widths and equal thickness. A suitable measuring setup has been developed to test the waveguides, especially their attenuation and light polarization properties. In this work, the experimental apparatus for the waveguides characterization is presented. New experimental data about the attenuation and the polarization properties of the waveguides are reported.

WAVEGUIDE STRUCTURE

Figure 1 shows the structure of the developed waveguides. The waveguides have been produced by screen printing and firing of different dielectric layers on alumina substrates. EMCA Glass 92 has been used as the waveguide material while Hereaus IP041 has been used as the glazing material. These materials present refractive indices equal to 1.82 and 1.69 respectively, well-matched thermal expansion coefficients and firing temperatures of about 750°C and 1000°C respectively.

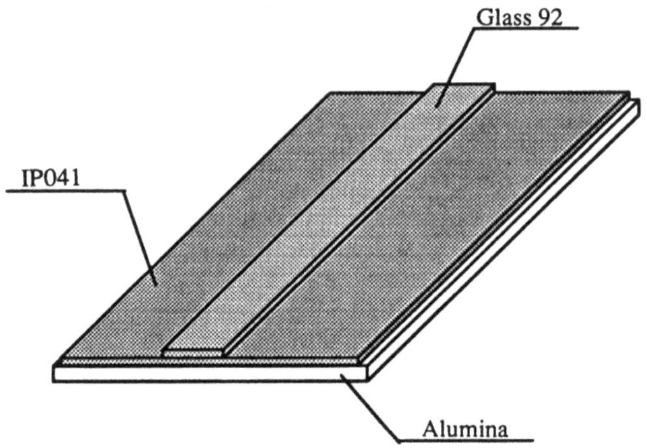

Figure 1. Basic structure of the developed thick film waveguides.

A number of waveguides of different widths have been manufactured, whose geometrical parameters are reported in Table 1.

Table 1. Geometrical parameters of the waveguides.

Guide number	Width (mm)	Length (mm)	Thickness (μm)
1	0.3	10.0	5.0
2	0.6	10.0	5.0
3	0.9	10.0	5.0
4	1.0	10.0	5.0
5	2.0	10.0	5.0

EXPERIMENTAL SETUP

Apparatus for the measurement of waveguide attenuation

Waveguide attenuation is measured by means of the Scattering Detection Method.[3] The intensity of the light scattered from the waveguide is proportional to the guided light. Thus, the detection of the light scattered from the waveguide along a direction perpendicular to the propagation direction, allows us to measure the guide attenuation.

In Figure 2, the developed experimental setup is shown.[4] The light is injected into the guide by using the End Coupling Method.[3] A 5 mW Helium-Neon laser is used as the light source. The beam is spatially filtered and focused into the waveguide by using a 40X microscope objective. The laser, the microscope objective and the guide are mounted on calibrated position actuators to maximize the efficiency of the coupling. The cross section of the illuminated waveguide is imaged onto a CCD, black and white 2-D video camera (Sony Mod. XC-77 CE) interfaced to a frame grabber (Imaging Technology Mod. AFG 100), plugged into a personal computer. The images acquired by the video camera are sampled on a pixel matrix 512 by 512 wide. The pixel gray level is represented by an 8 bit binary code. A monitor connected to the frame grabber is used to display the images.

The measurement of the waveguide attenuation is performed by means of suitable procedures. Firstly, the image of the waveguide under measurement is acquired in continuous mode, thus simplifying the adjustment of the system components and optimizing the alignment of the guide with respect to the laser beam. Secondly, an averaged acquisition of the guide illuminated by the laser is performed, and a grey level diagram of the scattered light intensity is plotted. Detection of the scattering at the entrance of the guide and verification of the presence of structural defects in the waveguide are thus possible. Thirdly, a selection of the area of the waveguide to be elaborated is on-line performed, thus eliminating from the measurement both the light scattered in correspondence of a defect and the light that

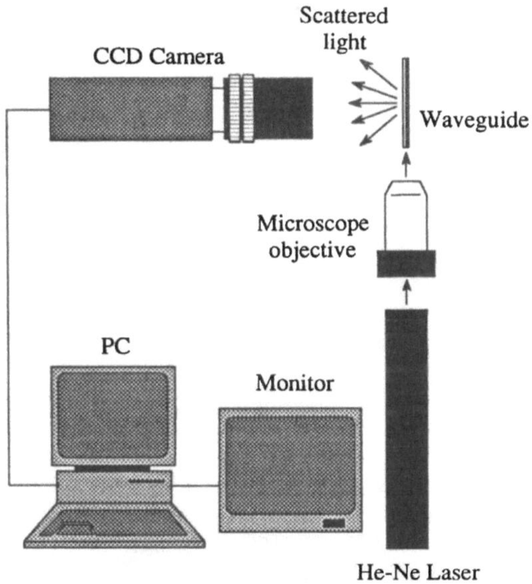

Figure 2. Optical layout of the experimental apparatus for the waveguide attenuation measurement.

saturates the video camera. Finally, the attenuation of the guide is evaluated by means of an automatic procedure: the mean value of the intensities of the pixels in each column of the matrix corresponding to the selected area is calculated. As the scattered light distribution well fits an exponential curve, a logarithmic regression is performed on these values. The slope of the regression line, in logarithmic scale, yields the attenuation of the waveguide.

Figure 3 shows the acquired image of the light scattered from one of the tested waveguides: the presence of a structural defect is fairly well observable. In this example, the scattered light at the input guide had previously been eliminated by a suitable selection of the guide area. A plot of the 3D-distribution of the scattered light along the transversal section of the guide is presented in Figure 4. Both Figure 3 and Figure 4 are referred to the guide no. 5 in Table 1.

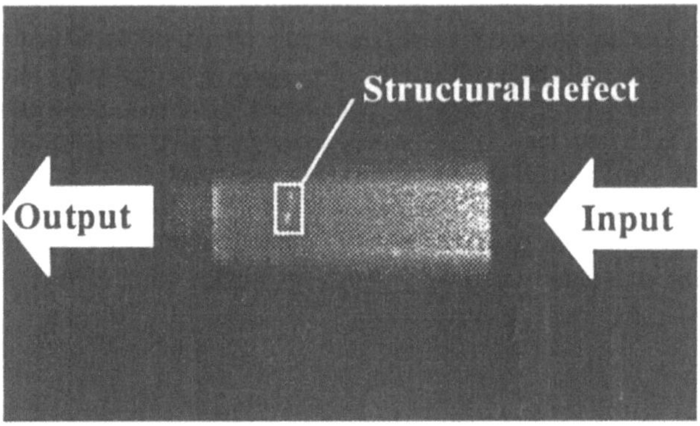

Figure 3. Image of the light scattered from the waveguide no. 5.

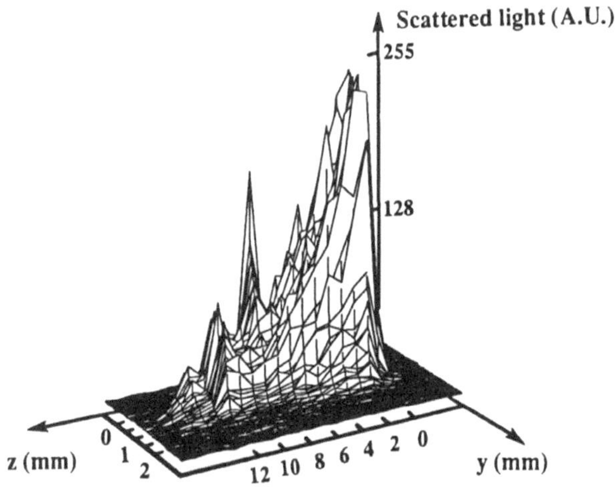

Figure 4. 3-D plot of the distribution of the light scattered from waveguide no. 5.

Apparatus for the detection of the polarizing properties of the waveguides

The experimental apparatus developed for investigating the polarizing properties of the waveguides is based on Malus law:

$$I^2=I_0{}^2\cdot\cos^2\theta \quad (1)$$

In Eq. (1), I is the output light intensity, I_0 is the input light intensity and θ represents the polarization direction. Thus, by injecting polarized light into the waveguides and by analyzing the polarization state of the light emerging from them, it is possible to detect the polarization properties of the waveguides.

The optical setup of the developed apparatus is shown in Figure 5. A 5 mW Helium-Neon polarized laser is used as the light source. It is mounted on an angular positioner, in order to inject the light into the waveguide at different, adjustable polarization directions. After the guide a rotating polarizer has been placed, and the intensity of the light passing through it is measured by an optical power meter.

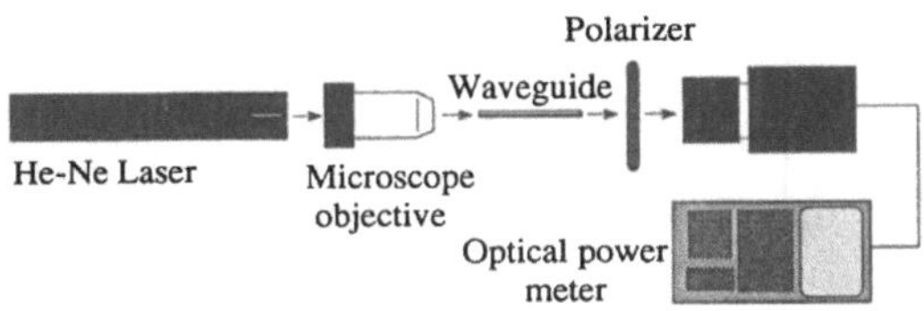

Figure 5. Experimental apparatus for the measurement of the polarizing properties of the waveguides.

EXPERIMENTAL RESULTS

The experimental setup for the attenuation measurement has been characterized as far to its time stability. Firstly, the system has been used to perform the attenuation measurement of the waveguide no. 5, according to the procedure above described. The resulting plot of the scattered light intensity (in logarithmic scale) with respect to the distance from the entrance of the guide is shown in Figure 6. The regression line is also shown, whose slope is assumed to be the attenuation parameter of the thick film waveguide.

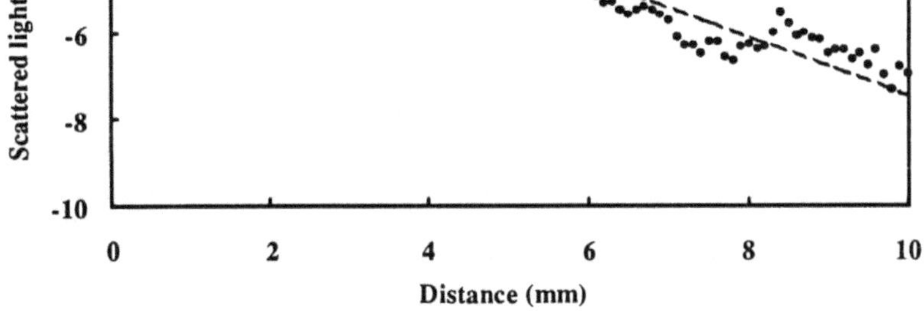

Figure 6. Logarithmic plot of the light scattered from the waveguide no. 5 as a function of the distance from the input. The dashed line represents the regression line.

Secondly, the same measurement has been repeated, without modifying anything of the system, for twenty times, with an interval of half an hour between two consecutive measurements. The result is presented in Figure 7: the flatness of the curve reveals a good stability (1%) of the measuring setup on a period of ten hours.

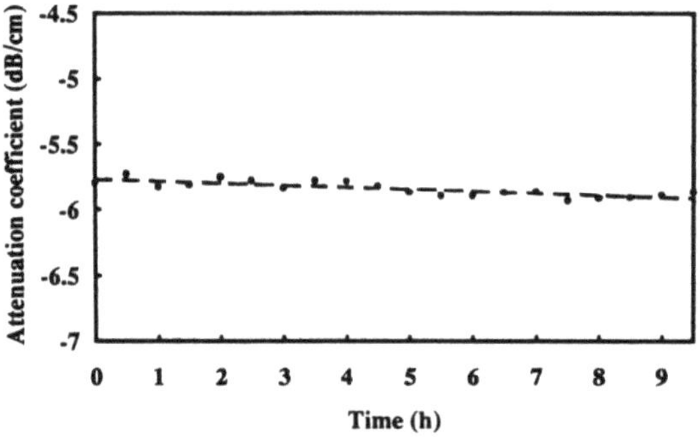

Figure 7. Stability of the experimental setup for the attenuation measurement. The dashed line represents the regression line.

The attenuation coefficient of all the waveguides has been evaluated. Table 2 shows the results: except for the narrowest guide, the attenuation is lower than the values reported in literature.[1] This is promising for the realization of thick film optical sensors.

Table 2. Attenuation values for thick film waveguides

Guide number	Attenuation (dB/cm)
1	9.1
2	8.7
3	6.5
4	3.8
5	6.8

The polarization characteristics of the waveguides have been tested: as an example, Figure 8 shows the results for the waveguide no. 1. The three curves represent the intensity of the light exiting from the guide as a function of the rotation of the polarizer after the guide, when the light injected into the guide is polarized respectively parallely (curve (a)), perpendicularly (curve (b)) and at 45° (curve (c)) with respect to the plane of propagation of the light into the guide.

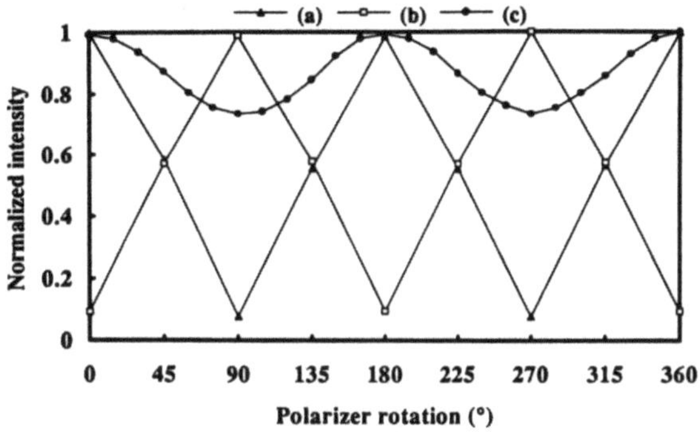

Figure 8. Polarization measurement for the waveguide no. 1

The polarization measurements have outlined a marked birefringent behaviour of the waveguides. In fact, only in the case of curves (a) and (b) the polarization direction of the light is maintained passing through the guide. This birefringence has been explained in terms of the stress induced in the waveguide materials during the manufacturing process. This aspect is also particularly relevant in view of the potential use of the waveguides as sensors based upon polarization changes.

CONCLUSIONS

In this work, the experimental apparatuses for the measurement of both the attenuation and the polarizing properties of thick film optical waveguides has been presented. The guides have been purposely manufactured in view of the possibility of their application for the realization of optical sensors.

A good stability of the guide attenuation measuring setup has been found, and attenuation coefficients about 6 dB/cm have been experimentally measured. The polarization properties of the guides have been investigated: the birefringent behaviour observed is likely to stimulate further investigation of new waveguides with different geometrical parameters and shapes.

REFERENCES

1. R.B. South and H. McDonough, Thick film optical waveguides, *in:* "Proc. of the 5th European Hybrid Microelectronics Conference," Stresa (1985).
2. L. Guerra. "Componenti a Film Spesso: - Caratterizzazione Ottica e Realizzazione di Guide," Dissertation Thesis, University of Modena (1987).
3. H. Nishihara, M. Haruna, and T. Suhara. "Optical Integrated Circuits," McGraw-Hill, New York (1985).
4. F. Docchio, D. Marioli, G.Sansoni, and A. Taroni, Measurement of attenuation losses and of light distribution in thick film optical waveguides, accepted for publication in *Sensors and Materials* (1994).

INVESTIGATING THE PERFORMANCE OF OFF-AXIS DIFFRACTIVE OPTICAL ELEMENTS (DOEs)

K.E. Hamilton and E.J. Fjarlie

The Royal Military College of Canada
The Department of Mechanical Engineering
Kingston, Ontario K7K 5L0, Canada

INTRODUCTION

The behavior of DOEs used in an off-axis configuration has been investigated. The phase coefficients that are needed to establish the profile in the aperture for either on-axis or off-axis elements were determined. Detailed phase function design was done for on-axis wavefronts(1,2). The option was to use the element with an incident off-axis wave or to design the phase coefficients specifically for off-axis operation with the detector on the mechanical axis of the element as shown in Figure 1.

DISCUSSION

A three-dimensional concept of an off-axis designed DOE is sketched in Figure 2. The required phase profile was approximated by an nth order general polynomial in each axis using a custom designed Mathcad model that combines the two orthogonal planes, as shown in Figure 3. The model also investigated the imaging effectiveness in the focal plane.

It can be seen from Figure 2 that the feature sizes steadily decrease as the optical path difference (OPD) limits become more extreme. The manufacturing technique becomes a fundamental limitation for maintaining the integrity of the phase profile.

On-axis designs with apertures of 8 mm and F/Nos ranging from F/4 to F/32 were designed, fabricated and tested; a 10° off-axis design with 2 mm aperture at F/16 was designed and tested. This particular DOE was fabricated using micro-lithographic focused ion beam technology. The off-axis design was established on a plat quartz substrate, the normal to the substrate defined the mechanical axis, the optical phase factors were arranged to give focusing along this mechanical axis for a 10° off optical axis incident energy case.

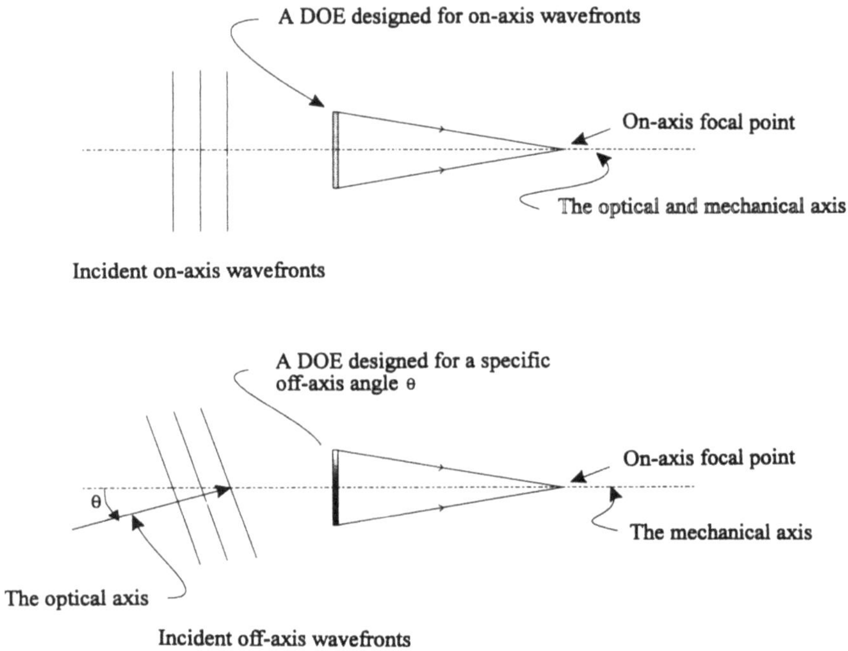

Figure 1. A Diagrammatic explanation of on and off the optical and mechanical axis's

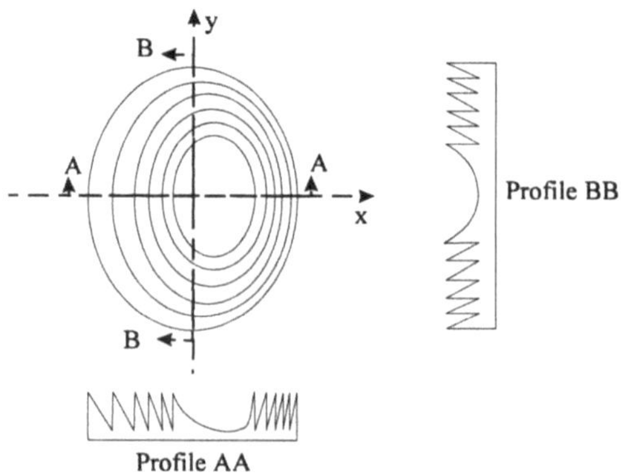

Figure 2. A Sketch showing the effect of an off axis condition on the DOE feature size

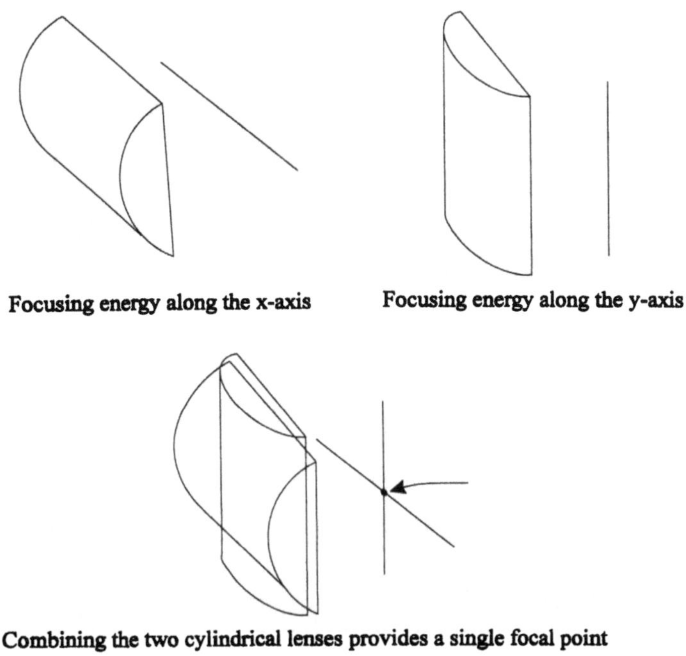

Figure 3. A pictorial explanation of how the math model determines the on or off-axis phase coefficients for first the y-axis of the DOE, then the x-axis, and then combining the two axis data

Blur circles were measured for up to 15° off the mechanical axis for each of the DOEs; thus the off-axis design was tested from +25° to -5°, and the on-axis designs from ± 15°. The DOEs were rotated with reference to the incident 1064 nm input. Fields-of-view (FOV) were arbitrarily established based on the off-axis blur circle $1/e^{-2}$ values increasing to three times the diffraction limited spot diameter.

The incident laser beam was moved off the mechanical axis, but kept parallel to the optical axis. This was found to be superior to using a beam expander to fill the aperture - wherein the wavefront had a slight curvature requiring a convolution operation. The scan in the image space proceeded in as small as 4 μm steps.

SUMMARY

Table 1 summarizes the results.

Table 1. A summary of the test results

How Made	Phase Profile Type	Aperture Size	F/Nos	On or Off axis design	Theoretical Diffraction Efficiency	Measured Diffraction Efficiency	FOV @ $\xi = 3$ (half angle)
		mm			%	%	degrees
#1	Contin-uous	8	8	On	100	88.2	6
			16			87.2	7
			32			86.4	13
#2	2 level	8	4	On	40.5	34.0	N/A
			8			N/A	7
			16			37.2	8
			32			39.1	10
#3	4 level	8	4	On	81	N/A	N/A
			8			N/A	5
			16			64.6	6
			32			N/A	10
#4	4 level	2	16	Off	81	6 - 8	9 - 15

Notes: - Manufacturing Technique #1 is direct laser write
- Manufacturing Technique #2 is micro-lithographic with reactive ion etch - two level
- Manufacturing Technique #3 is micro-lithographic with reactive ion etch - four level
- Manufacturing Technique #4 is micro-lithographic with focused ion beam etch - four level

ACKNOWLEDGMENT

Financial support for this work was provided by DND through DASP, CRAD and DRDA on project number 021W0.

REFERENCES

1. Codere, JRM and Fjarlie, EJ. *"Optimized IR Diffractive Optical Elements"*, ICAPT, Session W3, IEEE, Toronto (Jun 94)

2. Verreault, JJM and Fjarlie, EJ. "*Diffractive Optical Beam Steering/Scanning Device*", ICAPT, Session Th2, IEEE, Toronto (Jun 94)

DYNAMIC RANGE EXTENSION OF PHOTOSENSITIVE DEVICE ARRAY

Tomasz W. Wysokiński[1], Andrew H. Rawicz[1], and Stanisław Łętowski[2]

[1]Simon Fraser University
Dept. of Applied Sciences
School of Engineering Science
Burnaby, B.C. V5A 1S6 CANADA

[2]University of Gdańsk
Institute of Experimental Physics
Wita Stwosza 57
Gdańsk, PL 80-952 POLAND

ABSTRACT

A locally adapting filter/photosensitive detector array system is analyzed for use in image acquiring systems. Different transmittance characteristic filter models are investigated to extend the global light intensity operating range as well as to enhance detection in less illuminated regions of the image. This is confirmed by the results of experiments using glass and plastic Photochromic Adaptation Filters.

INTRODUCTION

Many modern vision systems work in conditions where the light intensity changes in the range of 14 decades - the intensity range we encounter everyday. However, only biological visual systems, such as the human eye, can effectively operate under such conditions [1]. Even so-called intelligent vision systems with built-in adaptive properties can effectively operate only in part (max. 7 decades) of this range[2]. Human photoreceptors are very sensitive to extremely weak sources of light. It is well established that they are sensitive enough to detect a few photons, and with more light coming in, the sensitivity of the visual system decreases due to a variety of adaptation mechanisms. Generally, we can divide those mechanisms into two types: global adaptation [3] and local adaptation [4]. The analysis of different adaptation mechanisms that are built into the human visual system stimulated the idea to modify the photosensitivity dynamic range of the array of the photodetectors by means of a locally adapting filter. A locally adapting filter is a type of filter that can change its light transparency preferentially to space intensity variation. On the other hand a globally adapting filter has space uniform transparency. Both kinds of filters can be passive (self-adjusting to different light conditions) or active (externally controllable).

Applications of Photonic Technology
Edited by G.A. Lampropoulos *et al.*, Plenum Press, New York, 1995

Photochromic materials are a special kind of passive photosensitive materials that can change their darkness in proportion to incident illumination. When embedded in glass or plastic they can adjust the global or local transmittance of an optical component. As the darkening energy is removed, the system reverts to its initial state. Global adaptation of photochromic materials, whose spectral transmittances depend on previous and current radiant exposures, have so far found their principal application in spectacle lenses. It has also been suggested that silver-halide based photochromic glasses may be used in some special technical applications [5],[6]. We postulate the use of these materials as a component in opto-electronic photosensitive device arrays for image parameter transformations.

A locally adapting filter, on the one hand, can be used in conditions where large differences in light intensity in the image occur, to extend the dynamic range of the filter/photodetector device. On the other hand, their application can be foreseen to change the characteristic of operation of the filter/photodetector in a given light intensity region. As a global filter, the Photochromic Adaptation Filter (PAF) can shift the operating range of the photodetector, mimicking an autonomous diaphragm.

MODELLING

We model a passive system using a PAF placed in the focal plane of a CCD or some other array of photosensitive devices. Applying a coating of a thin PAF film on top of a photosensitive array in the focal plane of the visual system will result in the negative, neutral gray image of the scene within the PAF. More intensely illuminated parts of the image will produce a proportional darkening of the photochromic agent within the film thus blocking a fraction of the light. This phenomenon can be described as equivalent to the local adaptation mechanism. With the average light intensity increase, the increased average absorbance of the film limits the light coming to photodetectors thus mimicking global adaptation. Thus, the PAF acts both as a locally and globally adapting filter. Depending on the location of the PAF in the optical system the two functions can either be separated or used jointly.

Given a photosensitive array, the light intensity after passing a photochromic filter is described by:

$$I_{i,j} = I^0_{i,j} * \beta_{i,j}(I^0_{i,j}) \tag{1}$$

Where $I^0_{i,j}$ is the light intensity on the surface of the filter and $\beta(I^0_{i,j})$ is the transmittance at a given location.

We assume that the filter has uniform thickness, and that the photochromic agents of one type are uniformly distributed within the thin film. To preserve uniqueness of the image, we seek transparency of the filter as a continuous, strictly decreasing function of light intensity. Normalization requires that the transmittance of the filter with no light entering is less then or equal to 1. Substituting these conditions in Eq. 1 we find one solution for transmittance as a class of function:

$$\beta_{i,j}(I^0_{i,j}) = \frac{const_1 - const_2 * I^0_{i,j}}{const_3 + const_4 * I^0_{i,j}} \tag{2}$$

The value of the constants dictate the form of the changes in the transmission. For example taking $const_{2,4} = 0$ corresponds to an ordinary gray filter, $const_2 = 0$ and $const_{1,3} = 1$ translates to very high suppression of the most illuminated part of the image.*

*A paper accepted for publication in "Microwave and Optical Technology Letters" will describe in more detail the results of this modelling.

For selected theoretical transmittance curves as a function of light intensity, we modelled the transformation of several real images. The results show that including a PAF in the focal plane locally changes the characteristics of operation of the filter/photodetector and thus results in the expansion of the global light intensity range. Some of the modelling results were confirmed by designed and performed a simple experiment.

EXPERIMENTATION

In order to investigate local changes in the transmittance of PAF due to radiant intensity from sample images, we designed and constructed a simple test stand. (Fig.1) This stand follows the general rules of the similarity conditions for our concept. The pictures of the image created on a translucent screen just behind the PAF is taken with a "COHU" RGB CCD camera.

As a filter, we used commercial glass and plastic photochromic materials used for

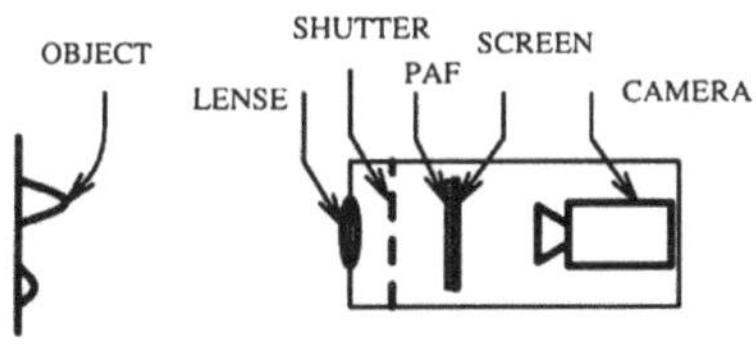

Figure 1. Schematic of the stand for the testing of PAF

ophthalmic lenses. To illustrate local adaptation, we have made a picture of two different light sources varying in light intensity over about 1 decade. The more intense mercury lamp (A) nearly saturates our camera, the other is a halogen lamp (B). The spectrum of these lamps is different, especially in the near UV region. As one can see in Fig.2a, lamp (A) is quite visible with our filter fully transparent while lamp (B) is not. With time, part of the filter illuminated by the high intensity lamp (A) darkens, due to the UV radiation emitted by this lamp. Fig.2b shows the image taken after a few seconds. Both lamps are barely visible as light from the lamp (A) is highly blocked. At that point we open the diaphragm of the camera wider and take one more picture. As one can see, we can now easily observe more detailed images of both lamps - Fig.2c. Comparing Fig.2a and 2.b, we conclude that local adaptation of the PAF is occurring. Using this phenomena we can say that the PAF-CCD system has changed its operating range, with no part of the array being saturated.

Figure 2. Image of the mercury lamp on the left, and halogen lamp on the right .

Fig.3a shows a supersaturated picture of a test chart used by ophthalmologists. We illuminated this chart with intense light with the UV part of the spectrum. Reflected light from the image caused the PAF to reduce locally the amount of light delivered to the camera, thus increasing the amount of detectable details in the scene. These results closely follow the results of simulation we made using an analytical form of the PAF transmittance. It is clear that application of local adaptation improves the quality of an initially saturated image.

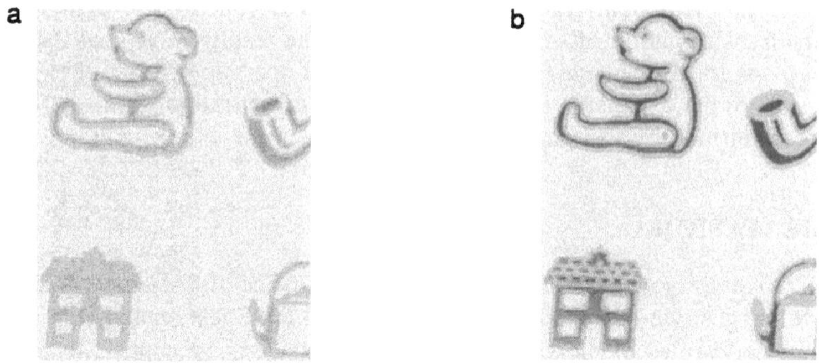

Figure 3. Image of the test chart

The results of these experiments show that it is feasible to construct a new type of camera using a passive local adaptation device (PAF filter) placed on the top of photosensitive device array.

CONCLUSIONS

The results from modelling and experimentation show that a filter/detector array system having nonlinear light vs output signal conversion can operate over a larger dynamic range of light intensities compared to the detector array alone. Controlling the character of this conversion allows the manufacture of different types of cameras for use in specific light environments. Such a system can also be used to enhance the less visible (dark) part of the image. This goal can be achieved with both passive and active types of filters, with the active type giving us more freedom to adjust for different light conditions.

At the present stage of development photochromic materials are still relatively slow, particularly for most real time visual processes. Another limiting factor is sensitivity mostly to highly energetic UV radiation. However, there is much research into developing new photochromic agents, sensitive to the visible part of light and respond much faster. The main advantages of using these are no additional energy consumption and automatic adjustment to different absolute illumination levels. These two features make thin film PAFs particularly attractive as potential components of the "Artificial Eye" project under investigation by OPCOM.

BIBLIOGRAPHY

[1] Alan Fein and Z. Ete Szuts. *Photoreceptors, their role in vision.* Cambridge [Cambridgeshire] ; New York : Cambridge University Press, 1982.

[2] J. Man. Implementing early visual processing in analog vlsi: light adaptation. In *SPIE Proc., Visual Information Processing: From Neurons to Chips*, volume 1473, 1991.

[3] K. N. Leybovic, editor. *Science Of Vision.* New York : Springler-Verlag, 1990.

[4] Donald I. MacLeod and Sheng He. Visible flicker from invisible patterns. *Nature*, 361:256–258, jan 1993.

[5] I. V. Tunimanova, F. V. Voronin, N. P. Chetvergov, M. M. Shevlyagina, and N. A. Papunoshvili. Photochromic glasses in systems for controlling high-intensity emitting objects. *Sov. J. Opt. Technol.*, 58(38–41):290–292, may 1991.

[6] Patent: 5055871. *Method and apparatus for enhancing illumination uniformity in wafer steppers using photochromic glass in the optical path.* 1991.

REMOTE COUPLING IN A RIB ARROW COUPLER

Barbara Holtz, Wolfram Biehlig, Christoph Wächter, Lutz Leine, and Falk Lederer

Friedrich-Schiller-Universität Jena
Physikalisch-Astronomisch-Technikwissenschaftliche Fakultät
Max-Wien-Platz 1
D-07743 Jena, Germany

INTRODUCTION

Almost a decade ago a novel guiding scheme which relies on resonance effects such as one finds in Fabry-Perot resonators could be identified[1]. Those so-called Anti-Resonant-Reflecting-Optical-Waveguides (ARROWs) did not attract a considerable attention among the waveguide community until recently. They exhibit several striking peculiarities as quasi-monomode behavior, particular dispersion relations that can be exploited to implement attractive devices as e.g. remote couplers[2] ,remote switches[3], polarizers[4], polarization splitters and extremely narrow-band wavelength filters.

Until recently, almost all ARROW devices were modeled and implemented in film configurations. But, very recently, remote coupling in rib ARROWs over 93 μm could be achieved experimentally[5].

Having in mind the various attractive devices that might rely on the ARROW concept it is of particular interest to develop powerful and accurate theoretical tools to identify the field profiles and propagation constants as well as to model the evolution of the fields in the course of propagation in those devices. Because the most interesting applications are based on the ARROW coupler the main interest is focused on its modelling.

Mathematically, due to the very nature of the ARROW modes that are leaky modes with complex propagation constants one has to use the supermode concept rather than the conventional coupled-mode theory in order to model a directional coupler. Because the half-beat length of a coupler is determined by the inverse difference of the real parts of the propagation constants of the fundamental symmetric and antisymmetric modes those propagation constants have to be calculated very precisely in using fully vectorial approaches. Similarly, the study of the evolution of the fields requires vectorial Beam Propagation methods (BPM).

The aim of this paper is to show that a reasonable accurate design of a rib ARROW coupler can be achieved in using vectorial Finite-Element-Methods (FEM) and that the coupling length predicted agrees excellently with that obtained by a vectorial BPM as well as the experiment.

MODELLING OF A RIB ARROW COUPLER

For a film configuration it is well-understood that for any spacing distance A between two ARROWs that meets approximately the antiresonance condition the beating of the two fundamental supermodes controls the output of an ARROW directional coupler. Those supermodes are equally excited if one launches e.g. an appropriate Gaussian beam into one channel. For a rib configuration the electric field may be written as

$$\mathbf{E}(x,y,z)=\mathbf{E}_s(x,y)e^{ik\beta_s z}+\mathbf{E}_a(x,y)e^{ik\beta_a z} \tag{1}$$

where the subscripts s,a label the symmetric and antisymmetric mode, respectively, and β_i are the normalized complex propagation constants, the half-beat length of the coupler is given by

$$L_h=\frac{\lambda}{2|\Re\beta_s-\Re\beta_a|} \tag{2}$$

Hence, the aim consists in determining the propagation constants and the vectorial field profiles. The accuracy needed to predict the half-beat length appropriately is in the order of 10^{-5}. Furthermore, for experimental convenience it is required to identify configurations that lead to a weak hybridization of the modes.

The rib ARROW configuration we are going to investigate is shown in Fig.1.

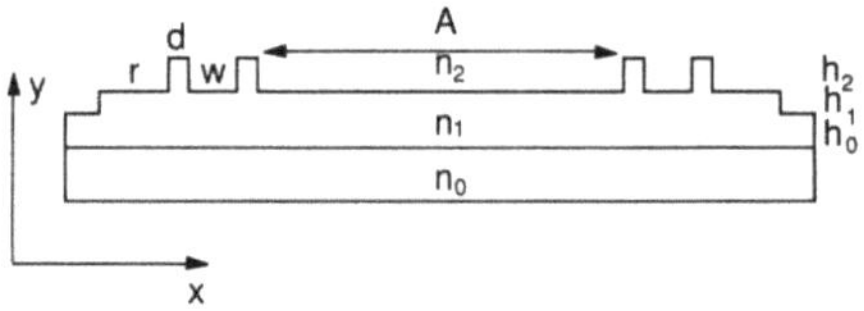

Figure 1. Rib ARROW configuration

The most popular method to model rib waveguides is the Effective-Index-Method (EIM). For most experimental situations this method provides sufficient accuracy although it represents a scalar approximation to a vectorial problem. Full vectorial Finite-Difference- or Finite-Element-Methods tackle directly the wave equation. They yield both the vectorial mode profiles and extremely accurate effective indices. The price one has to pay consists in large CPU times and extreme memory requirements. Recently, it has been shown[6] that for single rib ARROWs the difference in the effective indices provided by the EIM and the FEM differ by about 10^{-3}. Moreover, it turned out that the fields are quasi-scalar provided that the ARROW was appropriately designed. Hence, we may conclude that the EIM suffices if the design of single rib ARROWs is envisaged. But, as mentioned earlier, the most interesting ARROW devices are relying on coupler geometries. It is evident that the accuracy provided by the EIM is insufficient for that task. Consequently, we have implemented a FEM approach to determine the relevant quantities. To this end it is convenient to employ the wave equation for the magnetic field because all field compoments cross continously the interfaces. One gets

$$\nabla \times \left(\varepsilon^{-1}(x,y)\nabla \times \mathbf{H}\right) = k_0^2 \mathbf{H} \tag{3}$$

$$F = \int \left[(\nabla \times \mathbf{H})\varepsilon^{-1}(\nabla \times \mathbf{H})^* - k_0^2 \mathbf{H}\mathbf{H}^* + p^2 \nabla \mathbf{H} \nabla \mathbf{H}^* \right] d\Omega \tag{4}$$

The standard Rayleigh-Ritz procedure in its discretized representation (as the sum over all the finite elements forming the cross section under consideration) to minimize F leads to a high-dimensional generalized eigenvalue problem

$$\hat{\mathbf{A}}\mathbf{x} = \lambda \hat{\mathbf{B}}\mathbf{x} \tag{5}$$

with sparse, real-valued symmetric matrices . The 'penalty term' $\sim p^2$ is introduced to enforce the condition $\nabla \mathbf{H} = 0$ that is not fulfilled by the solutions of (3) automatically. It removes the 'spurious modes' from the guided mode spectrum thus allowing to use highly efficient iterative solvers. In order to increase the accuracy and to reduce the number of elements a mesh optimization was implemented basing on the minimization of the error introduced by the discontinuity of the first derivatives along the element boundaries. This procedure results in smooth field profiles even for a moderate number of nodes. A convergence of the effective indices with the number of nodes could be identified allowing for an extrapolation.

By applying the EIM the rib coupler geometry can be designed. The exact field profiles as well as effective indices are then calculated by the FEM. The corresponding two-dimensional field profiles ($H_x(x,y)$) for the configuration shown in Fig.1 are plotted in Fig.2 for the symmetric as well as antisymmetric quasi-TM modes (H_x - dominant). Similar to the results for the single ARROW almost no hybridization of the modes could be found provided that the configuration was quasi-symmetric.

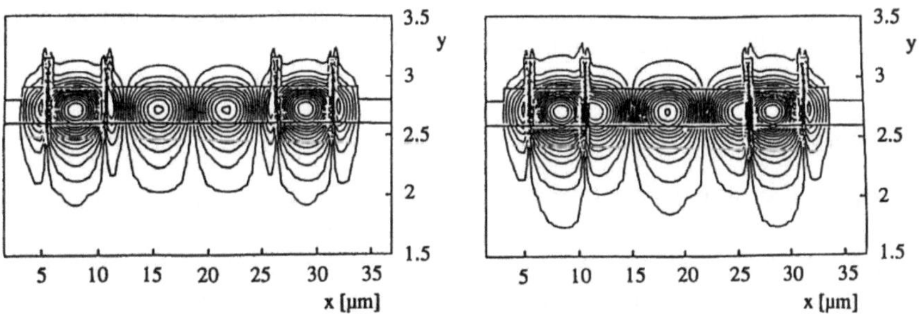

Figure 2. Quasi-TM-modes of a rib ARROW coupler

Additionally, we have investigated the evolution of the fields along the coupler if one channel was excited by a Gaussian beam. To this end we have used a full vectorial series expansion BPM. The coupling behaviour can clearly be recognized in Fig.3 (L_h=1.5 mm)

Eventually, we have compared the coupling lengths provided by the EIM, the FEM, the BPM and the experiment[5], respectively, for two different couplers. The couplers differ by the number of outermost reflectors used (coupler 1 - 2 reflectors, coupler 2 - 1 reflector) The results are listed in Table 1. It is evident that only fully vectorial methods yield reliable results for the design of rib ARROW coupler configurations.

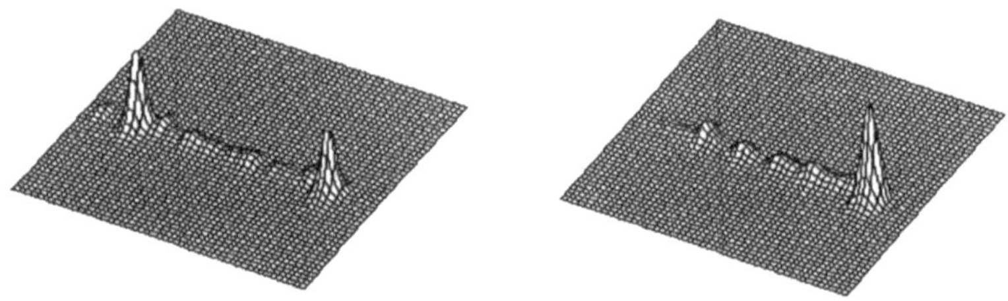

Figure 3. Full vectorial BPM simulation of the coupling process (intensity), z: propagation distance

Table 1. Half-beat length of rib ARROW couplers

	coupler 1	coupler 2 (Fig.1)
	L_h	L_h
EIM	9.09 mm	2.55 mm
FEM	3.84 mm	1.48 mm
BPM	-	1.50 mm
experiment[5]	4.0 mm	-

CONCLUSIONS

We have shown that the ARROW coupler concept can be extended towards rib configurations. Similar to planar configurations remote coupling could be proven to exist. It turned out that the exact prediction of the half-beat length requires fully vectorial methods. By using a vectorial BPM we have shown that the relevant supermodes can be excited by an appropriate Gaussian beam. It can be anticipated that various interesting two-dimensional ARROW devices, as e.g. polarizers, polarization splitters and filters, can be designed by using the theoretical methods presented.

REFERENCES

1. M.A. Duguay, Y. Kokubun, T.L. Koch ,and L. Pfeiffer, Antiresonant reflecting optical waveguides in SiO_2-Si multilayer structures, *Appl. Phys. Lett.* 19:13 (1986).
2. M. Mann, U. Trutschel, Ch. Wächter, L. Leine, and F. Lederer, Directional coupler based on antiresonant reflecting optical waveguides, *Opt. Lett.* 16:805 (1991).
3. U. Trutschel, M. Mann, F. Lederer, Ch. Wächter, and A.D. Boardman, Nonlinear remote switching in nonliner waveguides, *Appl. Phys. Lett.* 59:1940 (1991).
4. U. Trutschel, M. Cronin-Golomb, G. Fogarty, F. Lederer, and M. Abraham, Analysis of metal-clad antiresonant reflecting optical waveguides for polarizer applications, *IEEE Photonics Tech. Lett.* 5:336 (1993).
5. J. Gehler, A. Bräuer and W. Karthe, Remote coupling over 93 µm using ARROW waveguide in strip configuration, *Electr. Lett.* 30:218 (1994).
6. Ch. Wächter, W. Biehlig, A. Bräuer, J. Gehler, B. Holtz, F. Lederer, and L. Leine, Design and Fabrication of rib ARROW couplers, *Proc. SPIE* 2150-25 (1994), in press.

INVESTIGATING MICROMACHINED THERMAL EMITTERS FOR INFRARED APPLICATIONS

Larrie S. Carr and Marek J. Syrzycki

School of Engineering Science
Simon Fraser University
Burnaby, B.C. V5A 1S6

INTRODUCTION

The creation of sub-millimeter sized thermal spots and the generation of continuously changing two-dimensional thermal scenes using thermal spots are essential in calibrating complex infrared optical systems used in research, medicine and military applications. Hence, there is a need for fabricating arrays of miniature heater elements, featuring a large controllable range of infrared radiation and small crosstalk between array elements.

In the paper, we will investigate various CMOS micromachined thermal emitter structures operating in vacuum conditions. The results will also cover aspects of different clocking frequencies for the driving currents of thermal emitters. We will also discuss possible control schemes that assure expanded temperature range and high reliability of components for example applications.

MICROMACHINED THERMAL EMITTERS

Integrated thermal emitters may be fabricated using a variety of techniques. Typical methods include bulk silicon micromachining and the removal of sacrificial layers. The purpose of these fabrication steps is to thermally isolate the heating element from the bulk silicon substrate surrounding it. This isolation provides fixes the selectivity in the source of emission to single locations.

Bulk silicon micromachining techniques has certain advantages over sacrificial layer processes because bulk silicon micromachining may be performed on commercial CMOS processes by using a single post-process micromachining etch.[2] An opening to the silicon substrate is created by overlapping all the contact cuts of a standard CMOS process (substrate contact, polysilicon contacts, metal contacts and glass openings) in the layout. While such a mask combination violates standard CMOS layout design rules, it does not affect the circuitry on the wafer allowing control electronics to be integrated with the thermal emitter. An anisotropic etchant, such as EDP, creates a inverted pyramid pit whose surfaces are defined by the <111> planes of the silicon crystal lattice. With the proper layout, a suspended plate consisting of field oxide, polysilicon and aluminum can be

Applications of Photonic Technology
Edited by G.A. Lampropoulos *et al.*, Plenum Press, New York, 1995

created. Therefore, devices manufactured using this technique should be less expensive than full custom techniques due to the economies of scale.

Thermal emitters used for this research were manufactured using a commercially available digital 3μm CMOS process supplied by the Canadian Microelectronics Corporation (CMC). The emitter is formed using a polysilicon heater filament on a field oxide plate suspended over a micromachined cavity. Aluminum wires on the arms supporting the plate provide an electrical connection to the polysilicon heater. The cavity was formed using a hydrazine etch for about one hour. The suspended plate is 75x75μm supported by 9μm wide arms with an overall emitter size of 115x115μm.

Generally, the yield of the micromachined thermal emitters is high with most devices functional after etching. In some cases, the device will fail due cracks in the field oxide arms that suspending the plate above the cavity. If excessive cracking occurs, electrical connectivity to the heater or the entire plate may be lost during the etching post-processing. Fortunately, the cracking mechanism appears to be related to the layout of the thermal emitter and can be reduced with careful design.

THERMAL EMITTER CHARACTERISTICS

Thermal isolation provided by the cavity is sufficient for the polysilicon heater filament to maintain a temperature ranging from ambient to about 400°C above ambient. At these temperatures, the thermal emitter is capable of generating emissions of radiation in the sub-millimeter infrared wave band. The intensity of the emitted infrared radiation is close to a black body relationship, which has been experimentally verified, with a reported emissivity constant between 0.7 and 0.8 for the 2μm to 14μm spectral bands.[3]

The simple heat flow calculations for the ideally isolated polysilicon heater resistor show that its temperature is proportional to the resistive power dissipation, I^2R, resulting in the infrared radiation intensity varying approximately with $I^{0.5}$. In reality, thermal losses due to non-ideal thermal isolation between the heater element and the substrate, as well as changes in the heater resistance versus temperature make the above relationship more complex.

Figure 1 shows the a plot of the emitter's temperature as a function of the heating element's power dissipation which was measured using a HP Semiconductor Parameter Analyzer (SPA). The temperature of the heating element can be estimated using the temperature coefficient of resistance (TCR) of polysilicon.[4] For the 3μm digital process used in this research, the TCR of polysilicon is reported to be 0.175%/°C.[5] Therefore, for accurate regulation of the emitter's temperature, the control electronics must regulate the power dissipation instead of the current which would require a constant resistance for proper operation.

A number of emitter driver designs have been reported using a closed-loop control of either the voltage or current to provide a constant power.[6] However, by using the thermal time constant of the emitter as a low-pass filter, the power dissipation may be controlled using pulse-width (PWM) or higher-order sigma-delta (ΣΔ) modulation techniques. The thermal time constant of the test structures was measured to be in the order of a few milliseconds, depending on the structure of the emitter. Work on pulse control techniques are still underway with preliminary results displaying good control over the heater temperature.

To improve the thermal isolation of micromachined polysilicon heaters, operation in a vacuum has been recently proposed. The vacuum reduces thermal losses between the heater and the surrounding substrate and gases, allowing higher temperatures to be achieved for the same power dissipation when compared to the operation at atmospheric pressure. Figure 1 shows the emitter's temperature as a function of pressure for a constant power dissipation. A Corona Vacuum Coater sputtering machine was used as a vacuum

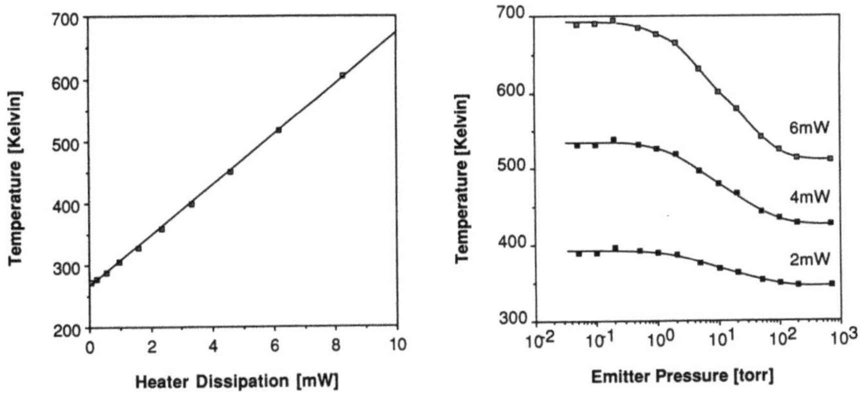

Figure 1. The left plot shows the thermal emitter temperature versus power dissipation in the heating element at room pressure. The right plot shows the effect of pressure on the thermal emitter temperature for a constant heater dissipation. Emitter temperature was determined using the TCR of polysilicon.

chamber while the emitter was controlled with the SPA. From the graph, we see the emitter's temperature increase as the pressure decreases for a constant power dissipation. The relative increase in temperature is reflective on the power dissipation as expected since conduction and convection mechanisms are driven by temperature differences: the higher the temperature, the greater the effect.

Thus, by using the emitter in a vacuum, a given heater temperature can be generated using a smaller heater power dissipation. This provides an opportunity to construct larger arrays of thermal emitters without exceeding the power dissipation limits of the chip. From the graph in Figure 1, we see that the temperature increase rolls off for pressures less than 1 torr. This roll-off effect is expected as a good vacuum is achieved at 10^{-1} to 10^{-2} torr. Therefore, thermal emitter systems using this vacuum effect to improve efficiency requires only a good quality roughing pump and not a large, expensive diffusion pump system.

On the other hand, the higher operation temperature can degrade the emitters' performance and induce reliability failures if special precautions are not taken. Experiences with the test structures show that aluminum structures located directly on polysilicon heater elements may diffuse into the polysilicon, causing a rapid decrease in heater resistance, at estimated temperatures as low as 400°C. Such a failure is essentially fatal as a significant amount of current must be used to generate the same power dissipation level. Other mechanisms include the polysilicon-aluminum contact or polysilicon failure as well as resistance drift of the polysilicon heater element. Work in this area is continuing as more devices with different structures are manufactured and stressed to their failure point.

THERMAL EMITTER ARRAYS

Additional design consideration must be taken when assembling these thermal emitters in arrays. The largest problem is that the amount of energy required power an entire array will usually exceed the chip's maximum power dissipation. For example, a 100x100 array of thermal emitter, each consuming 5mW of power, will require 50W of power. While the total chip dissipation may be reduced by pulsing different groups of emitters at different time, the emitter must be pulsed fast enough to provide a dynamic scene. Thus, extremely large arrays of emitters will still consume significant amounts of power.

Figure 2 shows a photomicrograph of an 8x8 arrays of thermal emitters. These arrays are used to test some of the design considerations of thermal emitter arrays such as possible

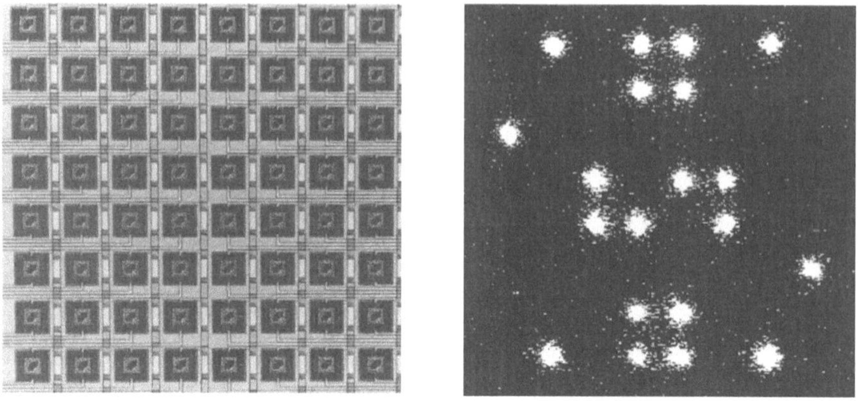

Figure 2. On the left is a photomicrograph of a thermal emitter array used for exploring thermal scene simulation. The right is an enhanced video capture, from a near infrared camera, of the emitter array operating in a vacuum.

control electronic configurations. Figure 2 also shows an enhanced video capture, from a Hammamatsu infrared camera, of an image generated using the thermal array. The temperature of the thermal emitters is about 400°C. While the infrared camera response rolls off around the 1.8µm spectral band, sufficient energy from the emitter is picked up by the camera to generate the video image.

In order to observe the thermal emitter array in operation in a vacuum, a special quartz cover for the chip package was manufactured. The quartz cover, which has a hose connection to a vacuum pump, replaces the package's original lid allowing a localized vacuum to be formed in the chip's cavity. Since the quartz material allows infrared to pass with little attenuation, the infrared camera can easily focus on the emitter array.

CONCLUSIONS

An evaluation of the characteristics of thermal emitters manufactured using a commercially available digital CMOS process was presented. The temperature changes due to the change in vacuum pressure were focused on. We also demonstrated an infrared image created by an 8x8 array operating in a vacuum captured with a near infrared camera system. The design of different thermal emitters and larger emitter arrays are continuing to research to use in the application of thermal scene simulation.

REFERENCES

1. M. Parameswaran, M. Syrzycki and R. Chung, "Integrated CMOS transducers for thermal scene generation", Microelectronics Journal, vol. 23, pp. 443-9, 1992.
2. M. Parameswaran, H.P. Baltes, L.J. Ristic, A.C. Dhahed and A.M. Robinson, "A new approach for the fabrication of micromachined structures", Sensors and Materials, vol. 5, pp. 113-21, 1993.
3. M. Gaitan, M. Parameswaran, R.B. Johnson and R. Chung, "Commercial CMOS foundary thermal display for dynamic thermal scene simulation", Proceedings of SPIE, vol. 1969, pp. 363-9, 1993.
4. H.A. Schafft and J.S. Suehle, "The measurement, use and interpretation of the temperature coefficient of resistance of metallizations," Solid-State Electronics, vol. 35, pp. 403-10, 1992.
5. P.P.S. Junja, "A micromachined wide-range vacuum sensor employing phase-sensitive detection", M.A.Sc. Thesis, Simon Fraser University, November, 1993.
6. D. Xie, M. Parameswaran and I. Filanovsky, "A simple CMOS power stabilizer for integrated sensor applications", Proceedings of 1993 Canadian Conference on Electrical and Computer Engineering, vol. 2, pp. 1178-1179, 1992.

OPTIMIZED IR DIFFRACTIVE OPTICAL ELEMENTS

J.R.M. Codère and E.J. Fjarlie

The Royal Military College of Canada
The Department of Mechanical Engineering
Kingston, Ontario, K7K 5L0, Canada

INTRODUCTION

Diffractive Optical Elements (DOEs) [1] were designed and fabricated to operate at 1.064 μm with apertures of 8 mm and F/Nos that varied from F/32 to F/4[2,3]. The designs were made using two-level and four-level lithography as well as the laser write technique.

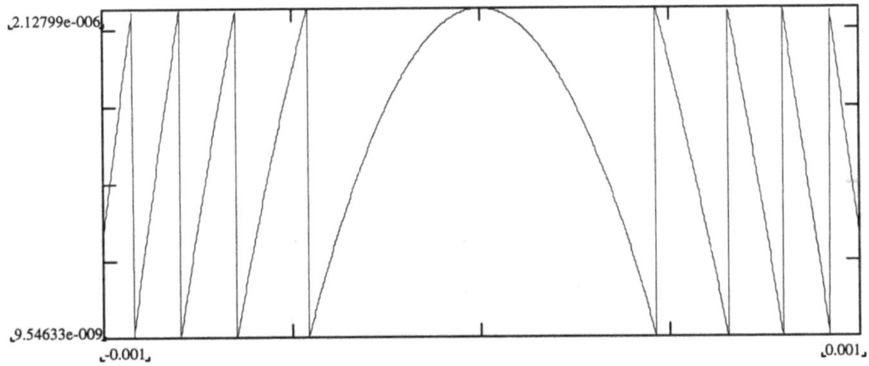

Figure 1. DOE Thickness profile.

Unlike Soret zone plates, DOEs are built by compensating for the alternate half-zones by adjusting the phase instead of the transmittance for the lateral geometry of the zones and for the optical path difference (OPD) from the object to the focal point of the device. Focusing action may be thougth of as the result of a chirped two-dimensional "blazed" grating optimized to achieve 100 % diffraction efficiency in the first diffraction order located at the focal point[4]. A typical thickness profile variation is shown in Figure 1. The vertical and horizontal scales are not to scale, there is a diameter-to-thickness ratio of 2000.

Applications of Photonic Technology
Edited by G.A. Lampropoulos *et al.*, Plenum Press, New York, 1995

DESIGN AND TEST

The on-axis optimized, circularly symmetrical DOE phase profile, was described for manufacturing purposes by using a Taylor series expansion of the OPD equation. The first six non-zero coefficients were used which came from a polynomial of order 12. DOEs optimized for different angles of incidence yielding non-symmetrical thickness profiles were also designed. Polynomial phase coefficients for these were determined numerically with a Fortran program.

The diffraction efficiency is influenced by the fabrication method. The continuous blazed grating shown in Figure 1 can only be approximated by a certain number of discrete levels from the lithography process. A more continuous approximation is achieved with the laser write process[5]

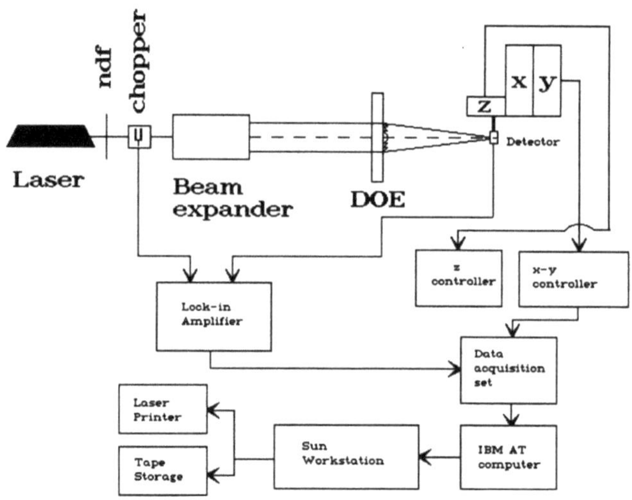

Figure 2 . Experimental Equipment Block Diagram.

Measurements were made following the arrangement shown in Figure 2. The radiance from the Nd:YAG laser was controlled via calibrated neutral density filters (ndf); the beam was expanded through ZnSe optics to give a beam that was about 5 cm in diameter to be incident on the DOE. A pinhole, 4 μm in diameter, was arranged directly in front of the Si APD and the detector-pinhole combination was moved around the image space using an X-Y-Z table programmed to move in 0.125 μm steps.

DOE blue circle scans for the on-axis elements are shown in Figures 3 and 4. Note that the size of the scaned regions in X and Y-axis for Figure 3 is 120 μm and for Figure 4 is 540 μm. Efficiency measured were close to the theoretical performance with 37 % for the two-level DOEs and 87 % for the laser write DOEs. The four-level DOEs had a mask alignment defect and produced efficiencies of 65 %. These compared to 41 %, 90 % and 81 % respectively for the theoritical values. Blur circle diameters averaged 13 % larger than the classic Airy disk.

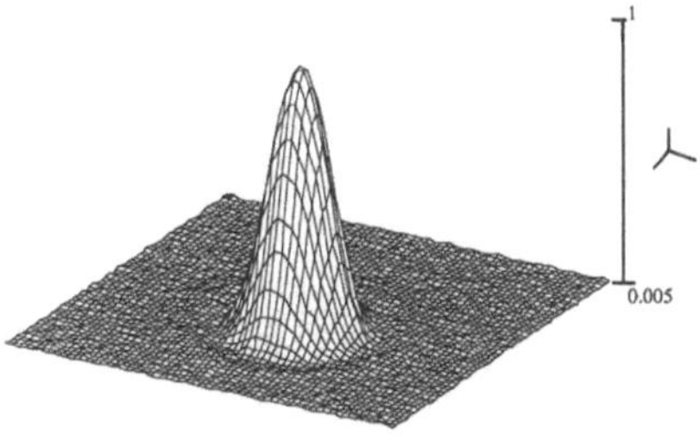

Figure 3. F/16 on-axis blur circle.

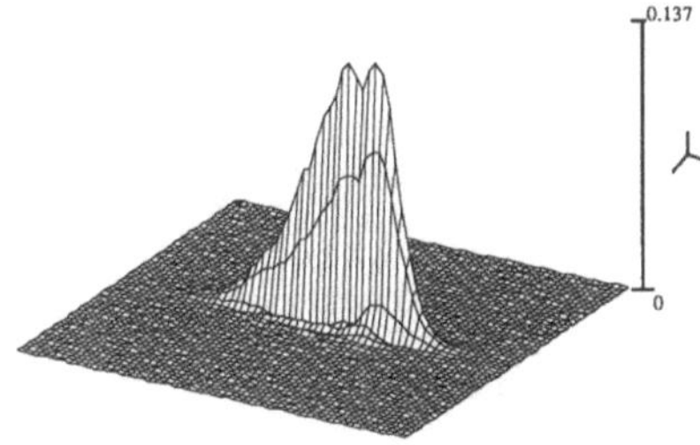

Figure 4. F/16, 10 ° incidence blur circle.

DOEs optimized to achieve diffraction-limited performance at 10 ° incidence angle were also designed, manufactured and tested. Efficiency measured for the four-level F/16 DOE manufactured using ion etching was 65 %, and the blur circle diameter was within 2 % of the diffraction limit value of 42 μm. For comparison with Figures 3 and 4, the blur circles measured with the off-axis optimized DOE are shown in Figures 5 and 6. The scanned regions in X and Y-axis for Figure 5 is 150 μm, and for Figure 6 is 120 μm.

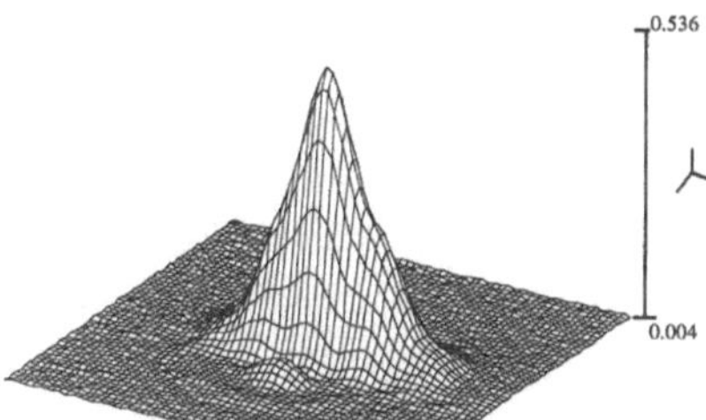

Figure 5. F/16 optimized DOE on-axis blur circle.

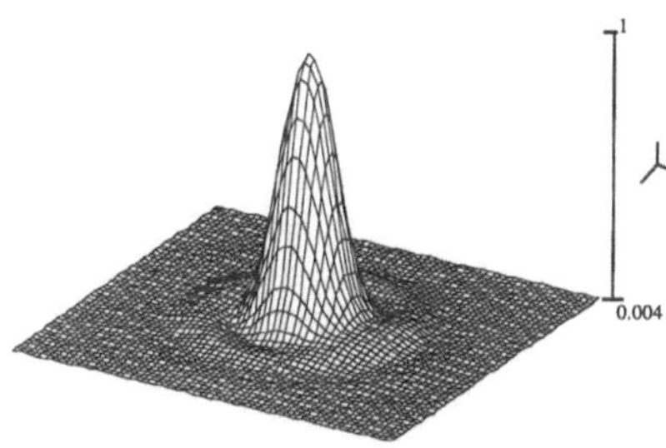

Figure 6. F/16 optimized DOE 10 ° blur circle.

Since the optimized DOEs were not circularly symmetrical, their performance is also asymmetric: the performance is diffraction-limited at 10° in only one azimuth. Note that focusing action is still present in other azimuth and the diffraction efficiency should still be around the same value.

The programs for controlling the X-Y-Z table were built on earlier work [2]. That work was done using a PC where the limitations on speed and memory soon caused the designs to be curtailed to cylindrical lens geometries[6]. The programs for the designs here were done using the SUN SPARC 10 computer system and Fortran language.

ACKNOWLEDGEMENT

Financial suport for this work was provided by DND through DASP and CRAD agencies via DRDA project number 021W0.

REFERENCES

1. G.J. Swanson, in *Binary Optics Technology : The theory and Design of Multi-level Diffractive Optical Elements*, MIT Lincoln Lab Tech Rpt **854**, MIT, (Aug 89).

2. J.J.M. Verreault and E.J. Fjarlie in *Diffractive Optical Element Beam Steering/Scanning Devices*, ICAPT, Session Th2, IEEE, Toronto (Jun 94).

3. K.E. Hamilton and E.J. Fjarlie in *Investigating the Performance of Off-Axis Diffractive Optical Elements (DOEs)*, ICAPT, Session Th5, IEEE, Toronto (Jun 94).

4. C.P. Daley and E.J. Fjarlie in *Binary Optical Elements for 10.6 μm wavelength*, Opcon **89**, Orlando, (Oct 89).

5.D. Falklis in Telcon, Falklis - RMC, RPS, Rochester - RMC, Kingston, (Dec 93).

6. N. Bell and E.J. Fjarlie in *Hybrid F/6 Binary Elements*, ICO XV, Garmish, (Aug 1990).

MEASUREMENT OF FEMTOSECOND TIME DELAY IN OPTICAL MATERIALS USING TYPE-II PARAMETRIC DOWN CONVERSION

A.V. Sergienko and Y.H. Shih

Department of Physics, University of Maryland Baltimore County
Baltimore, MD 21228, U.S.A.

Introduction

Correlated pairs of photons generated in Type-II nonlinear spontaneous parametric down conversion (SPDC) were used to develop a new approach for the measurement of time delay between two orthogonally polarized waves (linear or circular) in optical materials with femtosecond time resolution. This technique can be used to improve the accuracy of measurement of polarization properties of optical fibers, birefringent materials including nonlinear crystals, waveguides, Pockel's cells and Faraday rotators in the visible and IR spectrum.

This technique is based on the intrinsically quantum nature of the two-photon field (two-photon entangled quantum states) generated in the nonlinear process of SPDC. Two-particle entangled states have been known since the early days of quantum mechanics. Entangled states are states of two or more particles that can not be written as products of single particle states. These states play an important role in the study of the Einstein-Podolsky-Rosen (EPR) paradox[1] and in the test of Bell's inequalities.[2] Type-I spontaneous parametric down conversion (SPDC) was used intensively as a convenient source of two-photon entangled states.

In spontaneous parametric down conversion, a pump beam is incident on a birefringent crystal. The nonlinear optical effects lead to the spontaneous emission of a pair of entangled light quanta by means of the well known phase matching condition[3,4] $\omega_p = \omega_1 + \omega_2$, $\mathbf{k}_p = \mathbf{k}_1 + \mathbf{k}_2$. The down conversion is called Type-I or Type-II depending on whether the photons in the pair have parallel or orthogonal polarization. The light quanta of the pair that emerges from the nonlinear crystal may propagate in different directions or may propagate collinearly. The frequency and propagation directions are determined by the orientation of the nonlinear crystal according to the phase matching relations.

Several practical applications in optical photometry were developed based on quantum feature of Type-I SPDC for the measurement of quantum efficiency of photodetectors[5-7], of intensity of optical radiation[8], and of photon tunneling time.[9]

It was shown recently[10,11] that Type-II SPDC provides much richer physics due to the two photon entanglement both in space-time and in spin. Using Type-II SPDC we have developed a new practical approach to the measurement of time delays in optical materials due to the difference in group velocities of the two polarization states. This technique allows sub-femtosecond time resolution without the use of sophisticated femtosecond laser pulse sources.[12]

Theory

A theoretical description of this technique can be developed based on the quantum mechanical treatment of optical two-photon state. For collinear Type-II SPDC the two-photon part of the state that exits the down conversion crystal may be calculated from the standard theory of SPDC (first order perturbation theory) to be,[3,4]

Applications of Photonic Technology
Edited by G.A. Lampropoulos *et al.*, Plenum Press, New York, 1995

$$| \Psi > = \int d\omega_1 \; \delta(\omega_1 + \omega_2 - \omega_p) \; g(k_1 + k_2 - k_p) \; a_o^\dagger(\omega_1(k_1)) a_e^\dagger(\omega_2(k_2)) | \; 0 > \tag{1}$$

where ω and k represent the frequency and the wave number for signal (1), idler (2), and pump (p). The subscript indices o and e for the creation operators indicate the ordinary and extraordinary rays of the down conversion, traveling along the same direction as the pump, the z-direction. The coordinate axes x and y are chosen along the polarization direction of the o-ray and the e-ray, respectively. The use of small apertures make state (1) a good one dimensional approximation. The frequency phase matching condition is explicitly displayed by the delta function; the wave number phase matching condition is not of the form $\delta(k_1+k_2-k_p)$ because of the finite length of the crystal.[3,4] The function g determines the natural spectral width of the two-photon state. Taking the origin of the coordinates at the output surface of the down conversion crystal,

$$g(\Delta k) = [1 - \exp(-i\Delta k \cdot L)] \; / \; (i\Delta k \cdot L) \tag{2}$$

where L is the length of the crystal and $\Delta k \equiv k_1 + k_2 - k_p$.

Because of the finite bandwidth of the two-photon state we may expand k_1 and k_2 to the first order in ν. Considering that u_o (u_e) is the group velocity for the ordinary (extraordinary) ray, eq. (2) can be written as,

$$g(\nu) = [1 - \exp(-i\nu DL)] \; / \; i\nu D \tag{3}$$

where $D \equiv 1/u_o - 1/u_e$, we assume a negative crystal (BBO), i.e., $u_e > u_o$.

Consider the experimental set up of fig. 1. First suppose no quartz plates are placed between the BBO crystal and the beamsplitter. The average coincidence counting rate is given by,

$$R_c = (1/T) \iint_0^T dT_1 dT_2 \; |\Psi(t_1,t_2)|^2 \tag{4}$$

where $t_i \equiv T_i - \tau_i$, T_i is the detection time of the i-th detector. T is the duration time of the measurement. An effective two-photon wave function $\Psi(t_1,t_2)$:

$$\Psi(t_1,t_2) = < 0 \; | \; E_1^{(+)} \; E_2^{(+)} | \; \Psi >. \tag{5}$$

where $E_1^{(+)}$ and $E_2^{(+)}$ represent electric fields at the detector 1 and 2. Taking the analyzer angle to be 45° we can show the *rectangular* shape of effective two-photon wave function (*biphoton*):

$$\Psi(t_1,t_2) = \alpha_t \alpha_r \; v(t_1+t_2) \; [u(t_1-t_2) - u(-t_1+t_2)] \tag{6}$$

where

$$v(t) = v_0 \exp(-i\omega_p t/2) \tag{7}$$

$$u(t) = u_0 \int_{-\infty}^{\infty} d\nu [1-\exp(-\nu DL)]/(i\nu DL) \exp(-i\nu t) = \begin{cases} u_0 & DL > t > 0 \\ 0 & \text{Otherwise} \end{cases} \tag{8}$$

where v_o, u_o are constants (normalization). We have approximated the pump to be a plane wave (single mode) in the calculation. If the pump beam were taken to be a Gaussian with bandwidth σ_p, the constant v_o will be replaced by a Gaussian function $v_0 \exp(-\sigma_p^2 t^2/8)$.

Now consider the case when an additional birefringent material is placed between BBO and the beamsplitter. For example, we place a piece of birefringent material in such a way that the fast axis of the material coincides with the slow axis of the BBO, as is shown in fig. 1. In this case the effective wave function becomes (again the analyzers are set at 45^0),

$$\Psi(T_1,T_2) = \alpha_t \alpha_r \; v[(T_1+T_2] \; [u(T_1-T_2+\delta) - u(-T_1+T_2+\delta)] \tag{9}$$

where $\delta \equiv (l_e - l_o)/c$ is the optical delay in the crystal quartz. It is easy to see from eq. (9) that the two terms in the bracket now can overlap. When $\delta = DL/2$, they are completely overlapped and because of the minus sign in (6), they cancel each other.

The measured in experiment coincidence counting rate R_c would be:

$$R_c = R_{c0}[1 - \rho(\delta)] \tag{10}$$

where ρ is a Λ-shape function of δ (*self-convolution of the rectangular wave function*),

$$\rho = \begin{cases} 0 & -\infty < \delta < 0 \\ \kappa\,\delta & 0 < \delta \le DL/2 \\ 1 - \kappa(\delta\text{-}Dl/2) & DL/2 \le \delta < DL \\ 0 & DL < \delta < \infty \end{cases} \tag{11}$$

and $\kappa \equiv 2/DL$, resulting in a V-shaped R_c. It is easy to see that the width and the shape of the biphoton can be evaluated by the width and the shape of R_c.

If now any sample of optical material is placed in the optical path, the time delay $\delta \equiv (l_e - l_o)/c$ between the o-ray and the e-ray in the sample can be estimated by measuring the displacement of the vertex of the V-shape function, as is shown in fig. 2.

Experiment

A schematic of the experiment demonstrating the main physical principle of this technique is shown in fig.1.

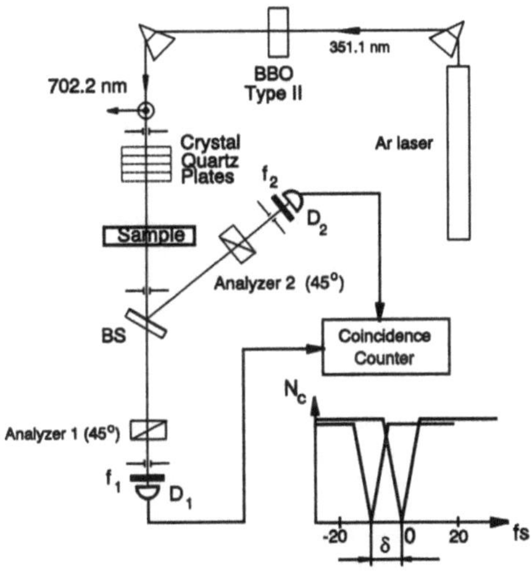

Figure 1. Schematic experimental setup.

A CW Argon ion laser line of 351.1 nm was used to pump an 8 mm x 8 mm x (0.56 ± 0.05) mm BBO (β-BaB_2O_4) nonlinear crystal. The BBO was cut at a Type-II phase matching angle to generate a pair of collinear orthogonally polarized signal and idler photons at 702.2 nm wavelength. They could be easily converted into two oppositely polarized circular waves using $\lambda/4$ waveplate. The down converted beams were directed at a near normal incident angle to a polarization independent beamsplitter which has 50% - 50% reflection and transmission coefficients. A single photon detector is placed in each transmission and reflection output port of the beamsplitter. The photon detectors are dry ice cooled avalanche photodiodes operated in photon counting Geiger mode.

A Glan-Thompson linear polarization analyzer, followed by a UV cutoff filter, is placed in front of each of the detectors. The polarization analyzers are oriented at 45^0 relative to the o-ray and e-ray polarization planes of the BBO crystal. The output pulses of the detectors are then sent to a coincidence circuit with a 3 nsec coincidence time window.

A set of crystal quartz plates is placed between the down conversion crystal and the beamsplitter for changing the optical delay δ between the signal and idler photons. Each of the quartz plates is (1 ± 0.1)mm in thickness, resulting in an optical delay $\delta \cong 30$ fsec between the o-ray and e-ray at wavelengths around 700 nm. Fig. 2 reports observed "V-shape" coincidence rate measurements as a function of the optical delay δ. The coincidence counts are direct measurements, with no "accidental" subtractions or any other theoretical corrections. Each of the data points corresponds to different numbers of quartz plates remaining in the path of the signal and idler beams.

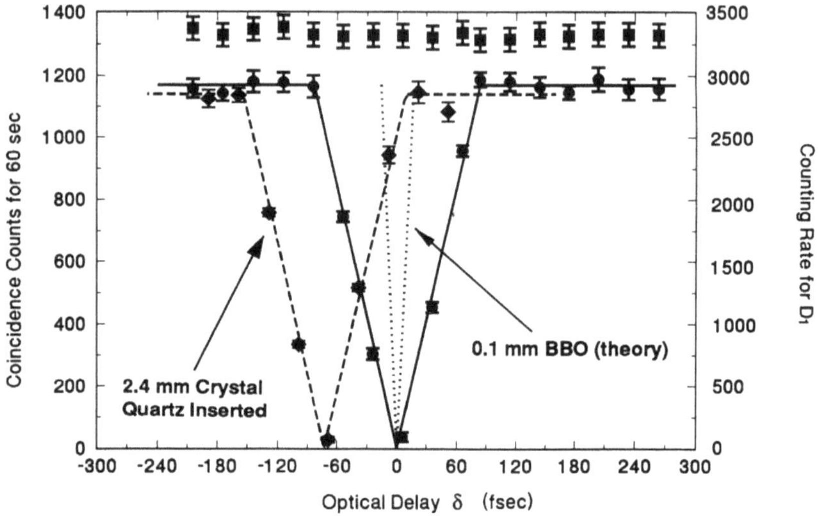

Figure 2. Measurement of femtosecond optical delay in crystal quartz.

Comparing the results indicated by solid and dashed curves, we find that the vertex has a displacement of (72 ± 3) fs, which corresponds to the propagation time delay in a 2.4 mm quartz testing sample. A similar result was observed for the circular polarized two-photon state created by the $\lambda/4$ waveplate. It is not difficult to achieve a sub-fs resolution (dotted curve) when a shorter crystal (smaller DL) is used.

Conclusion

The natural *rectangular* shape of the space-time structure of the two-photon effective wave function (*biphoton*) in Type-II parametric down conversion was verified by the observation of the V-shape correlation function. The optical delay between orthogonally polarized waves in birefringent material was measured with a 3 fs resolution. Sub-femtosecond resolution can be achieved by the use of a shorter nonlinear crystal. This approach could bring new perspectives to the problem of accurate control of polarization parameters of optical materials, especially polarization preserving fibers and nonlinear waveguides. Polarization parameters of birefringent and of optically active materials could be controlled as well. One can also independently calibrate devices for the measurement of femtosecond pulses such as nonlinear autocorrelators. Correlated photon pairs could be produced in a very broad spectrum (0.5 μm - 20 μm) by appropriate selection of nonlinear crystal and pump radiation.

Acknowledgements

We thank Prof. D.N.Klyshko for helpful discussions. This work is supported by the U.S. Office of Naval Research Grant No. N00014-91-J-1430.

References

1. A. Einstein, B. Podolsky and N. Rosen, *Phys. Rev.* **47**:777 (1935).
2. J.S. Bell, *Physics*, **1**:195 (1964).
3. D.N.Klyshko, "Photons and Nonlinear Optics", Gordon and Breach, New York (1988).
4. A. Yariv, "Quantum Electronics", Wiley, New York (1967).
5. A.V.Sergienko and A.N.Penin, *Applied Optics*, **30**:477 (1991)
6. J.G.Rarity, K.D.Ridley, and P.Tapster, *Applied Optics*, **26**:4616 (1987)
7. P.G.Kwiat, A.M.Steinberg, R.Y Chiao, P.H.Ebrhard, M.D.Petroff, *Phys.Rev.A*, **48**:R867 (1993).
8. D. N. Klyshko and A. N. Penin, *Sov.Phys.Usp.*, **152**:653 (1988).
9. A.M.Steinberg, P.G.Kwiat, and R.Y.Chiao, *Phys.Rev.Lett.*, **71**:708 (1993).
10. T.E.Kiess, Y.H.Shih, A.V.Sergienko, and C.O.Alley, *Phys.Rev.Lett.*, **71**:3893 (1993).
11. Y.H.Shih and A.V.Sergienko, *Phys.Lett. A*, **186**:29 (1994).
12. Picosecond and Femtosecond Spectroscopy, SPIE Proceedings, v. 1209 (1990).

ULTRA SHORT PULSE PROPAGATION ON NONUNIFORM PLANAR TRANSMISSION LINES

R. A. Olivares Véliz and J. R. Souza

Center for Telecommunication Studies
Pontifical Catholic University of Rio de Janeiro
Rua Marquês de São Vicente, 225
22453-900 Rio de Janeiro - RJ, Brazil

ABSTRACT

This paper presents a rigorous investigation of the propagation characteristics of short electric pulses on nonuniform planar transmission lines. For this purpose, frequency and time domain analyses are carried out by means of the Spectral Domain Approach and Fast Fourier Transform, respectively. This combined analysis shows if and how a pulse is distorted as it propagates down the nonuniform transmission line, and in consequence how the pulse distortion can be controlled through adjustments on the geometry of the transmission line. The results of this investigation find applications in high speed optoelectronics.

INTRODUCTION

The significant progress achieved by high speed optoelectronics made it possible to generate and manipulate pulses with pico- and sub-pico-second duration. Such progress gave rise to a continuos challenge in the design and fabrication of the microwave circuits (MICs, MMICs, and HMICs) used in high speed systems, due to the requirements of increasingly larger band-widths. Recently, an impedance transformer, using nonuniform planar transmission lines, was proposed to improve the efficiency in the pumping of semiconductor laser diodes by very short electric pulses[1], and also the temporal response of PIN photodiodes[2]. As the frequency contents of very short electric pulses can extend to several hundreds of GHz, the impedance transformer must have large band-width. A knowledge of the propagation characteristics of short electric pulses along nonuniform planar transmission lines is necessary for the proper design of such impedance transformers.

This paper presents a rigorous investigation of the propagation characteristics of very short electric pulses on nonuniform planar transmission lines by means of the Spectral

Domain Approach (SDA). Initially, a frequency domain analysis is carried out to determine the variation of the propagation constant along the length of the nonuniform transmission line, over a sufficiently wide frequency band. Next, a time domain analysis is performed, using the Fourier Transform, to simulate the propagation of the pulse. Although different planar transmission line configurations were considered, results are presented here for microstrip lines, and for the combined CPW/slotline, which has a typical CPW configuration on one side of the substrate, and that of a slotline on the other side.

FORMULATION

The type of transmission line considered here can be treated as a linear, time invariant system, whose temporal response, V(t,z), to an input signal V(t,z=0) can be written as[3,4]:

$$V(t, z) = F^{-1}\{F[V(t, z = 0)] \cdot H(\omega, z)\} \quad (1)$$

where F represents the Fourier Transform, and H(ω,z) is the transmission line transfer function. In the case of uniform transmission lines, this function is simply a phase lead, which equals the product of the (complex) propagation constant, γ(ω), by the propagating length z. However, in the case of nonuniform transmission lines, such function must incorporate the longitudinal variation of the line parameters. For example, in the case of a tapered transmission line, H(ω,z) can be written as[4]:

$$H(\omega, z) = \left\{1 - [\ln(Z_L/Z_o) \cdot F[\theta(\omega, z)]]^2/4\right\}^{0.5} \cdot \exp[-j\theta(\omega, z)] \quad (2)$$

where Z_o and Z_L represent the initial and final impedances, respectively, and F[θ(ω,z)] depends on the chosen impedance profile. In the case of an exponential profile[5],

$$F[\theta(\omega, z)] = \sin[\theta(\omega, z)]/[\theta(\omega, z)] \ , \quad [\theta(\omega, z)] = \int_0^z \gamma(\omega, u)du \quad (3)$$

Equation (1) describes the algorithm developed to simulate pulse propagation on nonuniform transmission lines. For the calculation of the propagation constant, which is a function of the frequency and of the position along the transmission line, γ(ω,z), the nonuniform transmission line is represented as a series of sections of uniform line of length Δz. For each section, the propagation constant is then calculated via the well known Spectral Domain Approach[6,7].

RESULTS

First, the SDA was used to determine the dispersion characteristics of tapered transmission lines with exponential impedance profile. Two tapered microstrip lines were considered, on a 0.635mm thick substrate, with ε_r=9.5. The strip widths vary from 0.22mm to 1.27mm, and from 0.22mm to 2.20mm, over a length z=25mm. Such strip width variations correspond to impedance transformation ratios of 2 and 3, respectively. Figure 1 shows, for the former, the variation of the normalized propagation constant with the position along the line and with frequency, as calculated by the SDA[7]. For this calculation, the taper was divided in 25 sections of 1.0mm. The dispersive nature of the structure, represented by the nonlinear variation of the normalized propagation constant with frequency, is clearly seen.

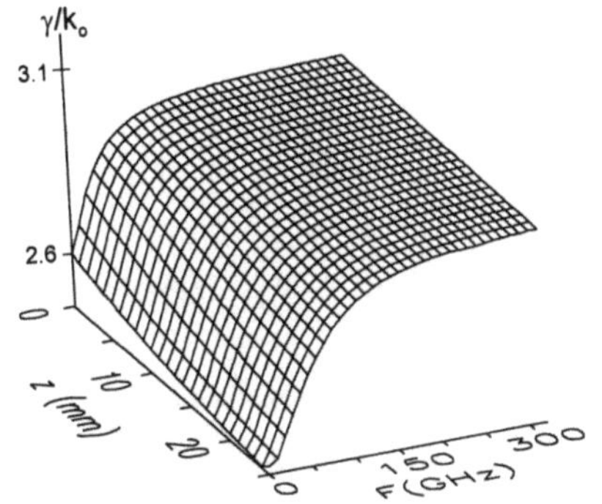

Figure 1. Variation of the normalized propagation constant (γ/k_o) with frequency (F) and position (z) along a tapered microstrip line.

Next, pulse propagation was simulated, as explained above. For both tapers, a 30ps FWHM Gaussian pulse is launched at z=0, and the dispersed pulses are shown in Figure 2. For comparison, the pulse obtained at the same length of uniform line ($Z_o/Z_L=1$) is also shown. It is seen that the main cause of distortion of the pulse is the transmission line dispersion. The variation of impedance along the taper does not change the shape of the pulse considerably, but introduces a time delay, and reduces the pulse amplitude, due to mismatching. These effects become more pronounced for higher ratios of impedance variation.

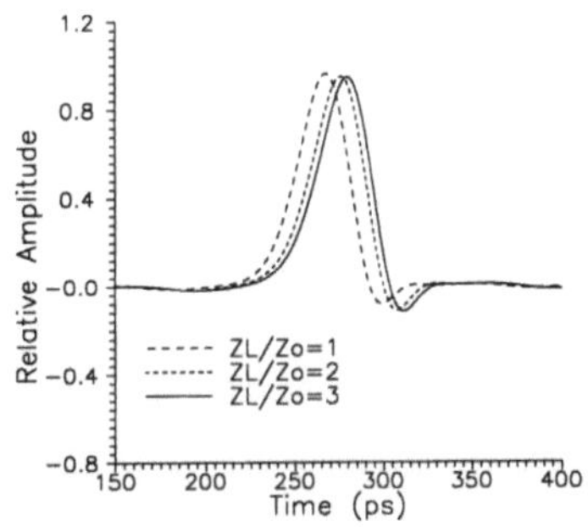

Figure 2. Dispersed pulses for exponential microstrip tapers.

A combined CPW/slotline 50Ω-to-7Ω impedance transformer[1,2], 25mm in length, was also considered, on a 0.635mm thick substrate, with $\varepsilon_r=38$. For the SDA calculation, the transformer was divided in 20 sections of 1.25mm. After this frequency domain characterization, a 50ps FWHM Gaussian pulse was launched at the low impedance side of this transmission line transformer (TLT). The dispersed pulse is shown in Figure 3. For comparison, two other pulses are also shown in Figure 3: the one obtained after the same lenght of uniform line, whose transversal dimensions are those of the low impedance side of the TLT, and one obtained with a dispersion free TLT, with the same lenght and impedance transformation ratio. It is again seen in Figure 3 that the main cause of pulse distortion is the transmission line dispersion, which introduces some ringing at the head and tail of the pulse, and so reduces its ampiltude, due to a redistribution of energy.

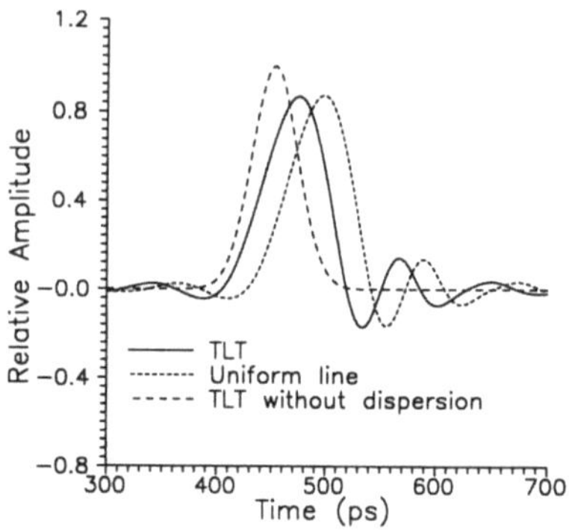

Figure 3. Dispersed pulses for exponential CPW-slotline taper.

CONCLUSION

Short electric pulse propagation on nonuniform (or tapered) transmission lines was rigorously simulated with the Spectral Domain Approach. It was observed that the main cause of pulse distortion was the transmission line dispersion, while the impedance variation along the taper introduced a small attenuation, due to mismatching, and a progressive time delay. The latter effects become more pronounced for faster impedance variation. The formulation developed here is quite general, and can be applied to different types of planar structures, including lossy ones.

ACKNOWLEDGEMENT

This work was supported by Telecomunicações Brasileiras S.A. under the contract PUC/TELEBRÁS 513/93.

REFERENCES

1· M. C. R. Carvalho, W. Margulis, and J. R. Souza, A new, small-sized transmission line impedance transformer, with applications in high-speed optoelectronics, *IEEE Microwave and Guided Wave Lett.* 2:428 (1992)
2· M. C. R. Carvalho, W. Margulis, and J. R. Souza, 2.8 Time reduction in the temporal response of PIN photodiodes with a transmission line transformer, *Proc. 1993 URSI Radio Science Meeting*, Michigan (1993)
3· T. Leung and C. A. Balanis, Pulse dispersion in open and shielded microstrips using the spectral-domain method, *IEEE Trans. Microwave Theory Tech.* 36:1223 (1988)
4· P. Pramanick and R. R. Mansour, Dispersion Characteristics of square pulse with finite rise time in single, tapered, and coupled microstrip lines, *IEEE Trans. Microwave Theory Tech.* 39:2117 (1991)
5· R. E. Collin, "Foundation for Microwave Engineering", McGraw-Hill, New York (1966)
6· T. Uwaro and T. Itoh, Spectral domain approach, in: "Numerical Techniques for Microwave and Millimeter-Wave Passive Structures", T. Itoh, ed., John Wiley & Sons, New York (1989)
7· J. R. Souza, Spectral domain analysis of printed transmission lines in multilayered substrate and superstrate configuration, *Proc. 8th Annual Rev. Progress in Applied Computational Electromagnetics*, Monterey (1992)

DIFFRACTIVE OPTICAL ELEMENT BEAM STEERING/SCANNING DEVICE

J.J.M. Verreault and E.J. Fjarlie

The Royal Military College of Canada
The Department of Mechanical Engineering
Kingston, Ontario, K7K 5L0, Canada

THEORY AND DESIGN

Discrete phase kinoform lenses, also known as DOEs[1], have been designed and fabricated on quartz substrates for operation at 1.064 micrometers (aperture diameters were 8 mm). Pairs of these lenses were arranged in a Keplerian geometry (Figure 1). The individual elements were measured separately and have already been reported[1,2]. An F/8 lens, manufactured by direct laser write, followed by a two-level or binary F/4 lens, manufactured by a photolithographic technique, were measured. Theoretical studies were also made of F/4-F/2 and F/2-F/1 combinations.

Agile beam steering (ABS) is achieved by laterally displacing the second (scan) lens relative to the first (input) lens. The movement is minimal compared to other techniques used for FOV scanning[3] (required displacements are of the order of millimeters). The result is a wavefront travelling at a non-zero angle to the optical axis. By going to a faster lens combination, larger FOVs are attainable; the kinoform lens may also be designed to minimize aberrations, notably third and higher order spherical. Such configurations may not be easily arranged for classical lenses.

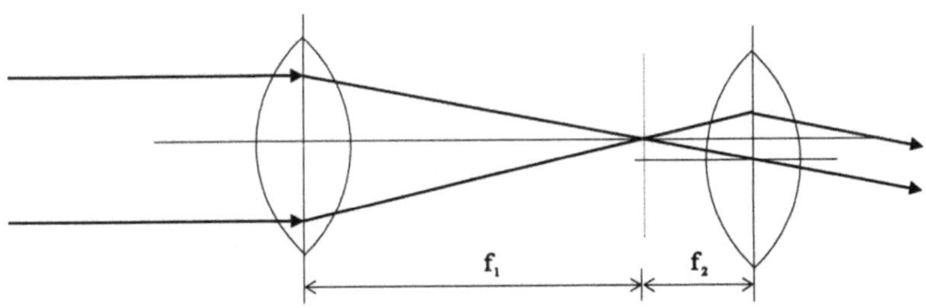

Figure 1. Lens pair beam steering concept using Keplerian geometry.

Applications of Photonic Technology
Edited by G.A. Lampropoulos *et al.*, Plenum Press, New York, 1995

One may use the ABS design to scan in two directions so that a detector may sequentially see the entire FOV. The ABS may be used conversely in an active system to rapidly and accurately direct a laser beam onto a given position. To achieve beam agility, the use of flexure-based optomechanical controls driven by piezo-electric actuators is envisioned. Typical FOVs are of the order of 10°.

EXPERIMENTAL RESULTS

Results for beam steering efficiency and wavefront quality for single-axis scanning are presented. Beam steering efficiency is defined as the ratio of the energy directed along the desired optical path to the total energy incident on the ABS kinoform pair. The optical beam source was a 150 mW Nd:YAG laser coupled to a 1.0 mm collimator.

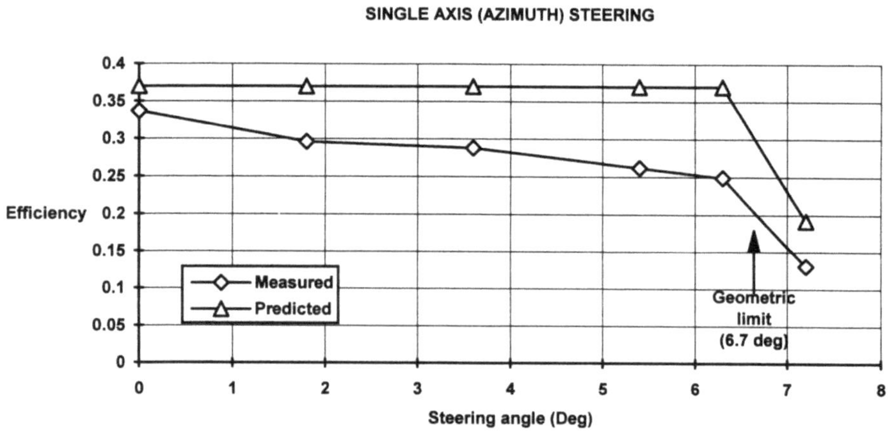

Figure 2. Single-axis beam steering efficiency for an F/8-F/4 ABS lens pair. The geometric limit defines the steering configuration at which vignetting starts to occur.

Predicted and measured first order efficiencies are shown in Figure 2. The sudden drop at the larger angle is due to vignetting; the drop at the lower steering angle is likely due to the m = 1 and m = -1 orders being within the FOV. As the scan proceeds, both orders move in opposite directions. Of course, only the m = 1 order is focussed on the detector. Average steering efficiency, over the 1.8 to 6.3 deg range, measured 0.26 compared to theoretical value of 0.37. Beam quality essentially remained the same as on-axis (see Figures 3 and 4); the drop in efficiency is possibly attributed to fabrication limitations, the feature tolerances becoming more noticeable towards the kinoform extremities, and edge effects.

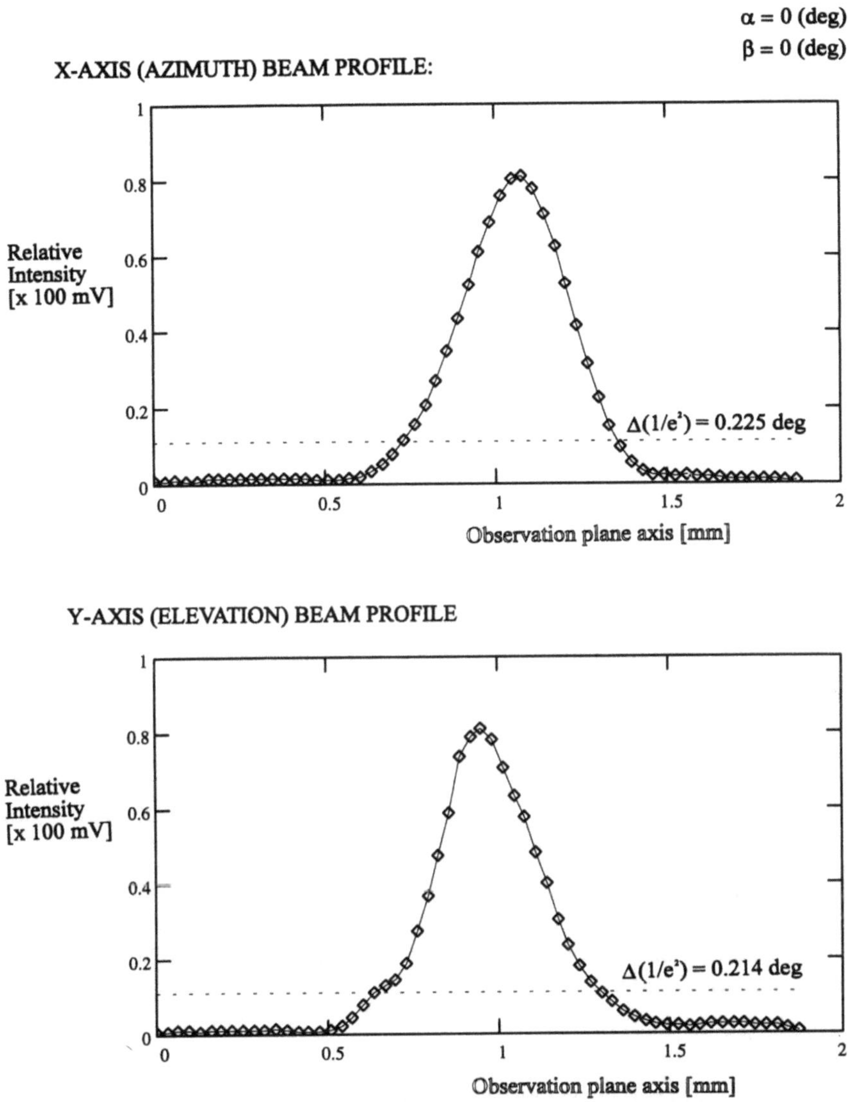

Figure 3. X-axis and Y-axis beam intensity profiles of output beam of prototype F/8-F/4 (pos-pos) lens pair for $\alpha = 0$ deg (azimuth) and $\beta = 0$ deg. Measured at observation plane 170 mm from scan lens.

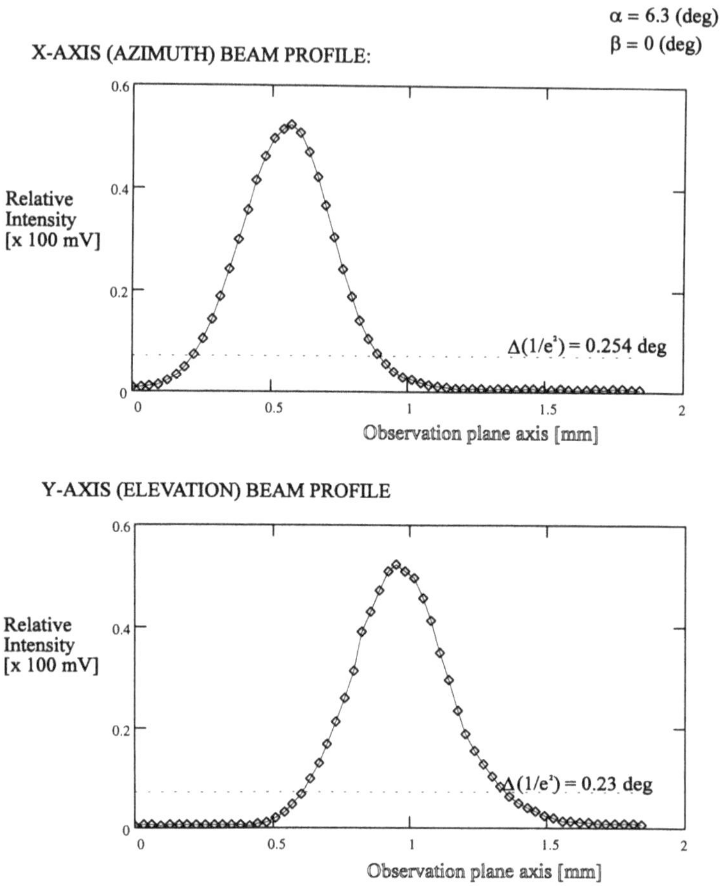

Figure 4. X-axis and Y-axis beam intensity profiles of output beam of prototype F/8-F/4 (pos-pos) lens pair for α = 6.3 deg (azimuth) and β = 0 deg. Measured at observation plane 170 mm from scan lens.

ACKNOWLEDGEMENT

Financial support for this work was provided by DND (via DASP) and CRAD (via DRDA) on project number 021WO.

REFERENCES

1. J.R.M. Codère and E.J. Fjarlie, "Optimized IR diffractive optical elements," ICAPT, Session W3, IEEE, Toronto (Jun 94).

2. K.E. Hamilton and E.J. Fjarlie, "Investigating the performance of off-axis diffractive optical elements (DOEs)," ICAPT, Session Th5, I.E.E.E., Toronto (Jun 94).

3. E.A. Watson, "Analysis of beam steering with decentered microlens arrays," *Opt. Eng.*, **32**: 2665-2670 (Nov 1993).

INTEGRATED LUMINESCENT-CONCENTRATOR PHOTODETECTOR

S.A. Evenson and A.H. Rawicz

School of Engineering Science, Simon Fraser University
Burnaby, British Columbia, Canada V5A 1S6

ABSTRACT

This work focuses on the development of a photosensitive device that utilizes the luminescent-concentrator (LC) approach to collect light energy. The paper presents a brief overview of the theory behind LCs, discusses computer simulation of an LC element, describes the design of a simple integrated LC photodetector (ILCP) on a Silicon substrate and presents the results of experimentation using this design.

INTRODUCTION

A luminescent-concentrator (LC) is a non-imaging optical device for collecting and concentrating light energy (see figure 1). It is essentially a planar optical matrix embedded with a luminescent dye. Photons incident on the LC are absorbed by this dye at the molecular or ionic level. A significant number of these *luminescent centers* will then emit new photons at longer wavelengths, a large fraction of which are trapped within the LC and guided to its edges by total internal reflection (TIR).

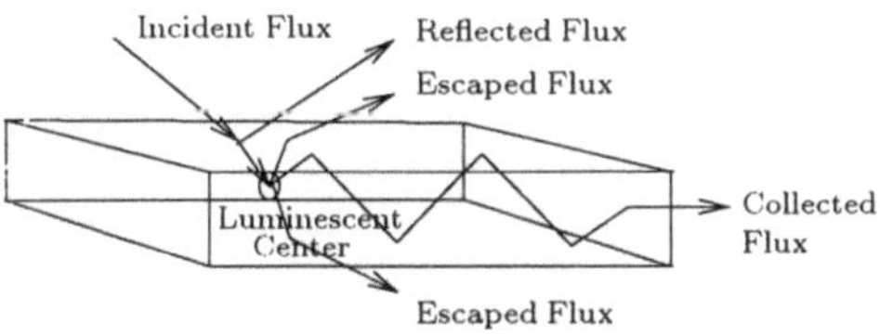

Figure 1. The Luminescent Concentrator

The application of LCs has been under investigation for several years.[1] Early research attempted to make use of LCs as radiance amplifiers for scintillation counters, in instrumentation for astronomical observations and in the collection of solar energy. More recently however, it was suggested that LCs might be used to construct optical devices for color or simple spectral analysis.[2] Such devices exhibit stable operation under either direct or diffuse light, improved sensitivity and quantum efficiency and the ability to provide wavelength filtering and shifting. Furthermore, through judicious choice of dye material, the flux output from an LC can be spectrally matched to a photosensitive element coupled to its output. In this way, it is possible to make more efficient use of photodiodes whose response is spectrally selective.

THEORY

The output intensity (Watts) of an LC is a function of incident angle and wavelength, and is given by the equation:

$$I_{out} = \Psi_{out} A_e = \Psi_{in} A_s \eta T Q_A Q_C \quad [W] \tag{1}$$

Applications of Photonic Technology
Edited by G.A. Lampropoulos *et al.*, Plenum Press, New York, 1995

where Ψ_{in} and Ψ_{out} are the incident and output flux densities (Watts per unit area), T is the surface transmission coefficient, Q_A is the probability of absorption by the dye, η is the quantum efficiency of the dye, Q_C is the probability of collection by internal reflections and A_s and A_e are the areas of the LC surface and edges respectively.

The output flux gain of an LC can be expressed as:

$$G_{out} = \frac{\Psi_{out}}{\Psi_{in}} = G_g \eta T Q_A Q_C \quad [-] \tag{2}$$

where

$$G_g = \frac{A_s}{A_e} \quad [-] \tag{3}$$

is the *geometric gain* of the LC. Owing to its planar structure, A_e is less then A_s. Therefore Ψ_{out} has a greater flux density then Ψ_{in} and the flux incident on the surface of an LC is available for collection in concentrated form at its edges.

Since the ability of an LC to concentrate flux is primarily a function of G_g, simply by adjusting the area of the LC face relative to that of its edges, its gain may be tailored for specific applications. Of course G_{out} is also a function of the LC's ability to absorb and collect flux. Therefore, the actual output flux gain of the LC will always be less then ideal. The composition of the flux leaving the edges of an LC is largely determined by the spectral characteristics of its luminescent dye.

SIMULATION

An LC element was simulated using a simple two dimensional ray-tracing algorithm. The objective of this simulation was to estimate the probability of flux propagating a distance L (from the point of incidence) through the LC as a function of incident angle θ_i and wavelength λ_i and to determine the spectral composition of output flux.

The LC structure modelled is shown in figure 2 (a). Here, d is the LC thickness, L is its length, n_1, n_2 and n_3 are the refractive indices of the upper cladding, core and lower cladding materials and C is the molar concentration of the fluorescent dye embedded in the core material. For the purposes of this simulation $d = 145\ \mu$m, $L = 1$ cm and $C = 1.177 \times 10^{-3}$ moles per liter was used. The upper cladding layer was modelled as air, the matrix material as polymethyl methacrylate (PMMA), the bottom cladding layer as SiO_2 and Rhodamine 6G was chosen as the fluorescent dye. Solvent effects on the absorption and emission spectra and reabsorption of fluorescence by the dye were neglected. Furthermore, the simulation is only considered valid for $\frac{d}{\lambda} \gg 1$.

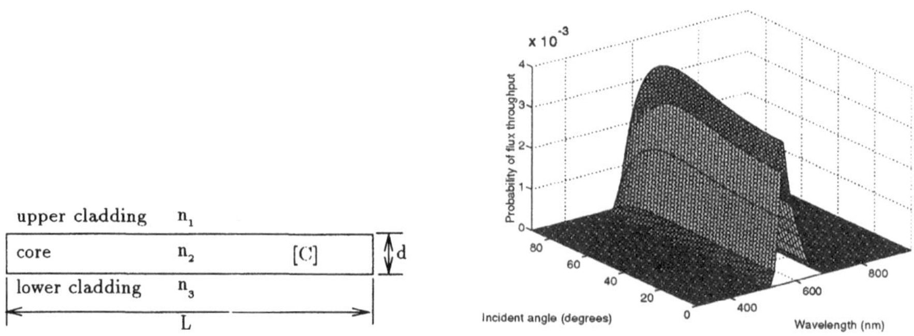

(a) LC Modelling Structure (b) Simulation Results

Figure 2. LC Simulation

The probability of flux of wavelength λ_i existing outside the LC was considered unity between 300 and 950 nm and zero elsewhere. The input flux was modelled at a single point incident on the top surface. θ_i was varied between 0° (normal) and 90° (tangential) to the LC surface.

The intensity reflection and transmission coefficients for unpolarized light were calculated at the surface of the LC from the Fresnel equations. These were used to determine the amount of flux entering the LC. The propagation angles θ_p of the transmitted flux were then calculated from Snell's law.

The intensity reflection coefficients were calculated at both dielectric surfaces inside the LC and for each θ_p and λ_i. These were then used to calculate the amount of incident flux remaining inside the LC after each reflection.

Once within the LC, the transmitted flux is partially absorbed by the luminescent dye with a probability given by the Beer-Lambert relation. The path-length between reflections in the LC was calculated and using the molar concentration C of Rhodamine 6G and its molar extinction coefficients (absorption characteristics),[3] the amount of flux absorbed following each pass was determined. The probability of absorption was considered independent of polarization and incident angle and absorption due to the LC matrix material (heat loss) was assumed negligible.

The dye was assumed to emit isotropically in the plane of the simulation and only flux emitted in the direction of the LC output was propagated. The spectrum of the emitted flux was calculated from the emission spectra of Rhodamine 6G in a methanol solution[3] multiplied by the quantum efficiency η of the dye. For the purposes of this simulation $\eta = 0.98$[4] was assumed. The intensity reflection coefficients were again calculated at both dielectric surfaces and for each θ_p of the luminescent flux. These were then used to calculate the amount of flux remaining after each reflection inside the LC.

Finally, the probability of flux propagating to the output edge of the LC was calculated by summing the probabilities of luminescent and nonluminescent flux reaching a distance L within the LC by internal reflections. The results (see figure 2 (b)) indicate that luminescent emission from the dye (peak at 560 nm) is the dominant component of the output flux. Since light absorbed by the dye is reemitted and collected by the photodiode, the responsivity of an LC photodetector will be a strong function of the absorption spectrum of the dye.

EXPERIMENTATION

Figure 3 shows the geometry of a simple integrated LC photodetector (ILCP) similar to what we fabricated in our clean room facilities at SFU. Several photodiodes were diffused into a Si wafer using standard planar IC technology. During this processing, a 1.9μm SiO_2 cladding layer was also formed by a combination of wet oxidation and RF sputtering. This oxide layer was subsequently etched from the surface of the photodiodes to facilitate collection of the guided light.

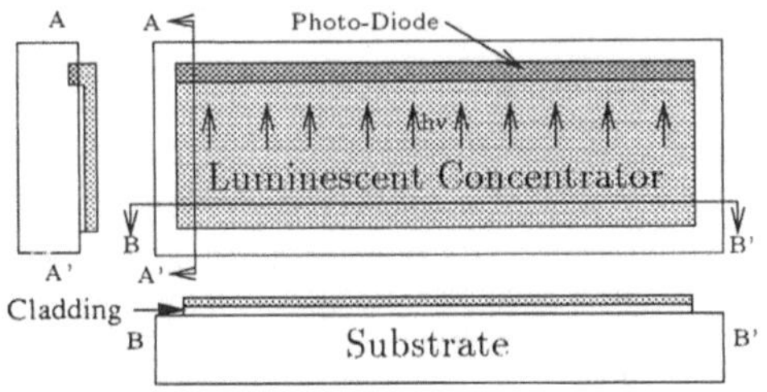

Figure 3. Simple ILCP

The LC core was then formed by spin coating a photosensitive aromatic acrylated epoxy[5] doped with Rhodamine 6G (concentration 1.177×10^{-3} moles per liter) at 1400 rpm for 90 seconds over the entire wafer surface. The resulting film was allowed to settle for several minutes. Then it was photo-cured by proximity photolithography for 5 minutes in an nitrogen environment using a 100W high pressure mercury arc lamp. After patterning, the film was developed by rinsing with acetone (to remove the unexposed epoxy) and annealed at 100°C for 5 hours. The final thickness of the LC core (after annealing) was 145μm. The thickness of the final film can be adjusted by diluting the uncured polymer with methanol and by varying the spin speed. Methanol can also be used as a vehicle to carry the dye into the polymer solution.

In order to test the ILCP, light from a 75W high pressure xenon arc lamp was focused onto the input of a grating monochromator (ISA HR640 with 1800 g/mm) and its output was used to illuminate the sample inside a light-tight enclosure. The photodiodes on the ILCP were accessed by wire bonding using a conductive epoxy (Acheson Electrodag 415) and attached to the input port of a high sensitivity transresistance amplifier (Newport 815 digital power meter). Finally, a computer data acquisition system was set up to control the monochromator and to sample the output of the amplifier.

The spectral response of the IC photodiodes was determined by positioning the bare IC devices 40 cm from the output slit of the monochromator. The photovoltaic output current from the diodes was then measured for incident wavelengths between 300 and 9500 nm. The power (W) incident normally ($\theta_i = 0°$) on the surface of the diodes was also measured using a commercial photodiode (Newport

818-SL). This was then used to determine the responsivity of the diodes (A/W) as a function of wavelength (see figure 4 (a)). A fourth order polynomial was fit to the data points using the method of least-squares.

A fully fabricated ILCP was then mounted in the same position in the monochromator behind a 1 cm^2 aperture so that no light was allowed to fall directly on the photodiodes. The distance from the edge of the aperture to the edge of each photodiode was 0.5 cm. The photovoltaic output current and incident power were again measured and used to determine the response characteristics of the ILCP. The data points were interpolated using a cubic spline technique and the results (see figure 4 (b)) indicate that the responsivity of the ILCP is a strong function of the absorption characteristics of the dye (peak at 530 nm) as expected.

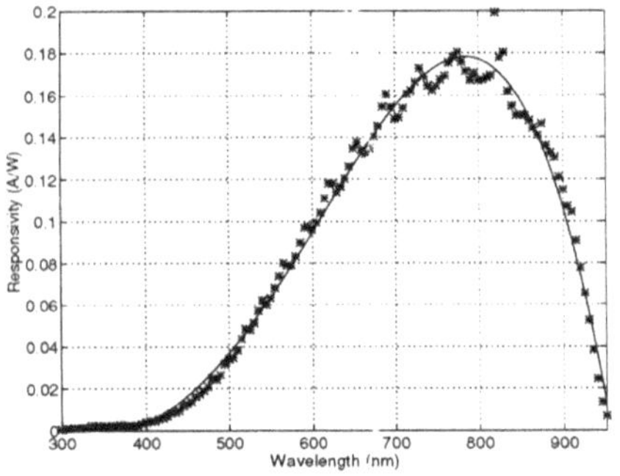

(a) Diode Responsivity

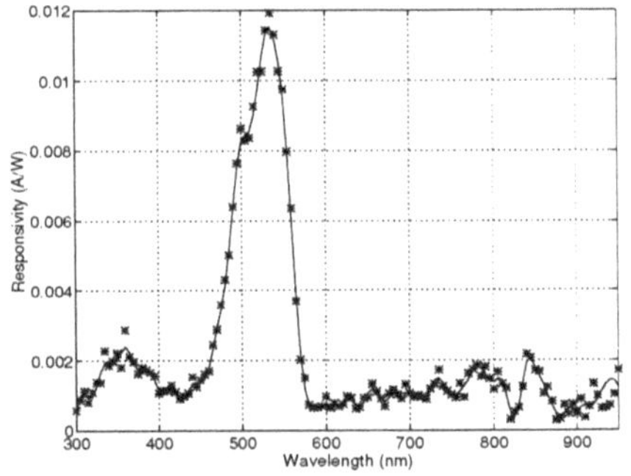

(b) ILCP Responsivity

Figure 4. Experimental Results

CONCLUSIONS

The possibility of fabricating integrated devices using LC elements for simple spectral or colorimetric analysis has been demonstrated. By tailoring device geometries and through appropriate selection of dye this approach can now be applied to specific applications.

Losses related to the fabrication process degrade the overall performance of the LC. These include the presence of scattering centers, reflective mismatches, geometrical trapping effects and incomplete TIR, inadequate absorption bandwidth, imperfect quantum efficiency and self-absorption of luminescence by the dye and by the matrix material. Further work to improve the processing techniques will reduce these limitations.

ACKNOWLEDGEMENTS

The authors would like to thank Dr. Eva Czyzewska, Dr. Marek Gnatowski, Dr. Romuald Lakowski, William Li, Dennis Michaelson, Dr. M. Parameswaran, Bill Woods and Tomasz Wysokinski for their continuing support.

REFERENCES

1. M.B. Levine, A.S. Cherkasov, and V.K. Baranow. Luminescent light concentrators. *Soviet Journal of Optical Technology*, 55(3):180–189, March 1988.

2. V.K. Baranow. Luminescent concentrators as spectral elements. *Soviet Journal of Optical Technology*, 58(8):472–474, August 1991

3. I.B. Berlman. *Handbook of Fluorescence Spectra of Aromatic Molecules*. Academic Press, New York, second edition, 1971.

4. A. Baczynski, T. Marszalek, H. Walerys, and B. Zietek. *Acta Physica Polonica A*, 44(6):805, 1973.

5. K. Nakagawa, T. Kowalewski, C.W. Phelps, D.L. Rode, and R.R. Krchnavek. Optical channel waveguides based on photo-polymerizable di/tri acrylates. Preprint: SPIE OE/LASE '94, Optical Interconnects II, Los Angeles, CA, January 1994.

INDEX

MIX
Papier aus verantwortungsvollen Quellen
Paper from responsible sources
FSC® C105338

If you have any concerns about our products,
you can contact us on
ProductSafety@springernature.com

In case Publisher is established outside the EU,
the EU authorized representative is:
Springer Nature Customer Service Center GmbH
Europaplatz 3, 69115 Heidelberg, Germany

Printed by Libri Plureos GmbH
in Hamburg, Germany